BIOCHEMISTRY
OF SMOOTH MUSCLE
CONTRACTION

BIOCHEMISTRY

OF SMOOTH

MUSCLE

CONTRACTION

EDITED BY

Michael Bárány

College of Medicine
Department of Biochemistry
University of Illinois at Chicago
Chicago, Illinois

1996

Academic Press

San Diego   New York   Boston   London   Sydney   Tokyo   Toronto

Academic Press, Inc.
A Division of Harcourt Brace & Company
525 B Street, Suite 1900, San Diego, California 92101-4495

United Kingdom Edition published by
Academic Press Limited
24-28 Oval Road, London NW1 7DX

Library of Congress Cataloging-in-Publication Data

Biochemistry of smooth muscle contraction / edited by Michael Bárány.
        p.   cm.
    Includes bibliographic references and index.
    ISBN 0-12-078160-3 (alk. paper)
    1. Smooth muscle--Physiology.   I. Bárány, Michael.
QP321.5.B54   1995
611'.0186--dc20   95-23967
                                                                    CIP

PRINTED IN THE UNITED STATES OF AMERICA
95   96   97   98   99   00   EB   9   8   7   6   5   4   3   2   1

This book is dedicated to
Professor Albert Szent-Györgyi,
1893–1986,
the founder of functional muscle biochemistry.

Contents

CONTRACTILE PROTEINS

CHAPTER 1

Myosin Structure and Function

ROBERT S. ADELSTEIN AND JAMES R. SELLERS

CHAPTER 2

Myosin Light Chains

KATE BÁRÁNY AND MICHAEL BÁRÁNY

C H A P T E R

3

Myosin Regulation and Assembly

KATHLEEN M. TRYBUS

C H A P T E R

4

Actin and the Structure of Smooth Muscle Thin Filaments

WILLIAM LEHMAN, PETER VIBERT, ROGER CRAIG,
AND MICHAEL BÁRÁNY

THIN FILAMENT AND CALCIUM-BINDING PROTEINS

C H A P T E R

5

Tropomyosin

LAWRENCE B. SMILLIE

C H A P T E R

6

Caldesmon

STEVEN B. MARSTON AND PIA A. J. HUBER

CHAPTER

7

Calponin

MARIO GIMONA AND J. VICTOR SMALL

CHAPTER

8

Calcium Binding Proteins

RAJAM S. MANI AND CYRIL M. KAY

ENZYMES OF PROTEIN PHOSPHORYLATION–DEPHOSPHORYLATION

CHAPTER

9

Myosin Light Chain Kinase

JAMES T. STULL, JOANNA K. KRUEGER, KRISTINE E. KAMM,
ZHONG-HUA GAO, GANG ZHI, AND ROANNA PADRE

MOTILE SYSTEMS

CHAPTER

14

In Vitro Motility Assays with Smooth Muscle Myosin

JAMES R. SELLERS

CHAPTER

15

Permeabilized Smooth Muscle

GABRIELLE PFITZER

CALCIUM MOVEMENTS

CHAPTER

16

Calcium Channels and Potassium Channels

HARM J. KNOT, JOSEPH E. BRAYDEN, AND MARK T. NELSON

CHAPTER

17

Molecular Biology and Expression of Smooth Muscle L-Type Calcium Channels

FRANZ HOFMANN AND NORBERT KLUGBAUER

CHAPTER

18

Electromechanical and Pharmacomechanical Coupling

CHRISTOPHER M. REMBOLD

CHAPTER

19

Calcium Pumps

LUC RAEYMAEKERS AND FRANK WUYTACK

SIGNAL TRANSDUCTION

20

The Nitric Oxide–Cyclic GMP Signaling System

THOMAS M. LINCOLN, TRUDY L. CORNWELL,
PADMINI KOMALAVILAS, LEE ANN MACMILLAN-CROW,
AND NANCY BOERTH

21

Inositol 1,4,5-Trisphosphate Production

MICHAEL BÁRÁNY AND KATE BÁRÁNY

CONTRACTION AND RELAXATION

CHAPTER

27

Calcium Sensitivity of Contraction

KRISTINE E. KAMM AND ROBERT W. GRANGE

CHAPTER

26

Regulation of Cross-bridge Cycling in Smooth Muscle

JOHN D. STRAUSS AND RICHARD A. MURPHY

CHAPTER

28

Pharmacological Regulation of Smooth Muscle by Ion Channels, Kinases, and Cyclic Nucleotides

PAUL J. SILVER AND DOUGLAS S. KRAFTE

ENERGETICS

CHAPTER

29

Energetics of Smooth Muscle Contraction

PER HELLSTRAND

CHAPTER

30

^{31}P Nuclear Magnetic Resonance Spectroscopy

PATRICK F. DILLON

Contributors

Numbers in parentheses indicate the pages on which the authors' contributions begin.

S. Thomas Abraham (143), Sigfried and Janet Weis Center for Research, Geisinger Clinic, Danville, Pennsylvania 17822

Leonard P. Adam (167), Boston Biomedical Research Institute, Boston, Massachusetts 02114

Robert S. Adelstein (3), National Health Lung and Blood Institute, National Institutes of Health, Bethesda, Maryland 20892

Kate Bárány (21, 269, 321), Department of Physiology and Biophysics, College of Medicine, University of Illinois at Chicago, Chicago, Illinois 60612

Michael Bárány (21, 47, 269, 321), Department of Biochemistry, College of Medicine, University of Illinois at Chicago, Chicago, Illinois 60612

Nancy J. Boerth (257), Department of Pathology, Division of Molecular and Cellular Pathology, University of Alabama at Birmingham, Birmingham, Alabama 35294

Joseph E. Brayden (203), Department of Pharmacology, Ion Channel Group, Medical Research Facility, College of Medicine, University of Vermont, Colchester, Vermont 05446

Trudy L. Cornwell (257), Department of Pathology, Division of Molecular and Cellular Pathology, University of Alabama at Birmingham, Birmingham, Alabama 35294

Roger W. Craig (47), Department of Cellular Biology, University of Massachusetts Medical School, Worcester, Massachusetts 01655

Joseph Di Salvo (283), Department of Medicine and Molecular Physiology, School of Medicine, University of Minnesota at Duluth, Duluth, Minnesota 55812

Patrick F. Dillon (393), Departments of Physiology and Radiology, Michigan State University, East Lansing, Michigan 48824

Ferenc Erdödi (131), Department of Medical Chemistry, University of Debrecen, School of Medicine, H-4026 Debrecen, Hungary

Antony Galione (295), Department of Pharmacology, Oxford University, Oxford OX1 3QT, United Kingdom

Zhong-Hua Gao (119), Department of Physiology, University of Texas Southwestern Medical Center, Dallas, Texas 75235

Mario Gimona (91), Cold Spring Harbor Laboratory, Cold Spring Harbor, New York 11724

Robert W. Grange (355), Department of Physiology, University of Texas Southwestern Medical Center, Dallas, Texas 75235

David J. Hartshorne (131), Muscle Biology Group, Department of Animal Sciences, University of Arizona, Tucson, Arizona 85721

Per Hellstrand (379), Department of Physiology and Biophysics, University of Lund, S-223 62 Lund, Sweden

Franz Hofmann (221), Institut für Pharmakologie und Toxikologie der Technischen Universität München, D 80802 München, Germany

Pia A. J. Huber (77), Department of Cardiac Medicine, National Heart and Lung Institute, London SW3 6LY, United Kingdom

Masaaki Ito (131), 1st Deparment Internal Medicine, Mie University School of Medicine, Tsu, Mie 514, Japan

Kristine E. Kamm (119, 355), Department of Physiology, University of Texas Southwestern Medical Center, Dallas, Texas 75235

Nihal Kaplan (283), Department of Medicine and Molecular Physiology, School of Medicine, University of Minnesota at Duluth, Duluth, Minnesota 55812

Cyril M. Kay (105), MRC Group in Protein Structure and Function, Department of Biochemistry, University of Alberta, Edmonton, Alberta, Canada T6G 2H7

Raouf A. Khalil (307), Cardiovascular Division, Department of Medicine, Harvard Medical School, Beth Israel Hospital, Boston, Massachusetts 02215

Norbert Klugbauer (221), Institut fur Pharmakologie und Toxikologie der Technischen Universität München, D-80802 München, Germany

Harm J. Knot (203), Department of Pharmacology, Ion Channel Group, Medical Research Facility, College of Medicine, University of Vermont, Colchester, Vermont 05446

Padmini Komalavilas (257), Department of Pathology, Division of Molecular and Cellular Pathology, University of Alabama at Birmingham, Birmingham, Alabama 35294

Douglas S. Krafte (367), Boehringer Ingelheim Pharmaceuticals, Ridgefield, Connecticut 06877

Joanna K. Krueger (119), Department of Physiology, University of Texas Southwestern Medical Center, Dallas, Texas 75235

William Lehman (47), Department of Physiology, Boston University School of Medicine, Boston, Massachusetts 02118

Thomas M. Lincoln (257), Department of Pathology, Division of Molecular and Cellular Pathology, University of Alabama at Birmingham, Birmingham, Alabama 35294

Lee Ann MacMillan-Crow (257), Department of Pathology, Division of Molecular and Cellular Pathology, University of Alabama at Birmingham, Birmingham, Alabama 35294

Rajam S. Mani (105), MRC Group in Protein Structure and Function, University of Alberta, Edmonton, Alberta, Canada T6G 2H7

Steven B. Marston (77), Department of Cardiac Medicine, National Heart and Lung Institute, London SW3 6LY, United Kingdom

Kathleen G. Morgan (307), Boston Biomedical Research Institute, and Cardiovascular Division, Department of Medicine, Harvard Medical School, Beth Israel Hospital, Boston, Massachusetts 02215

Richard A. Murphy (341), Department of Molecular Physiology and Biological Physics, University of

Virginia Health Science Center, Charlottesville, Virginia 22908

Mark T. Nelson (203), Department of Pharmacology, Ion Channel Group, Medical Research Facility, College of Medicine, University of Vermont, Colchester, Vermont 05466

Roanna Padre (119), Department of Physiology, University of Texas Southwestern Medical Center, Dallas, Texas 75235

Gabrielle Pfitzer (191), Institut für Physiologie, Medizinische Fakultät, Humboldt Universität, D-10115 Berlin, Germany

Luc Raeymaekers (241), Laboratorium Voor Fysiologie, K. U. Leuven, B 3000 Leuven, Belgium

Christopher M. Rembold (227), Department of Internal Medicine and Physiology, University of Virginia Health Science Center, Charlottesville, Virginia 22908

Charles M. Schworer (143), Sigfried and Janet Weis Center for Research, Geisinger Clinic, Danville, Pennsylvania 17822

James R. Sellers (3, 181), National Health Lung and Blood Institute, National Institutes of Health, Bethesda, Maryland 20892

Lori A. Semenchuk (283), Department of Medicine and Molecular Physiology, School of Medicine, University of Minnesota at Duluth, Duluth, Minnesota 55812

Jaswinder Sethi (295), Department of Pharmacology, Oxford University, Oxford OX1 3QT, United Kingdom

Paul J. Silver (367), Department of Vascular and Biochemical Pharmacology, Sterling Winthrop Incorporated, Collegeville, Pennsylvania 19426

Harold A. Singer (143, 155), Sigfried and Janet Weis Center for Research, Geisinger Clinic, Danville, Pennsylvania 17822

J. Victor Small (91), Department of Physics, Institute of Molecular Biology, Austrian Academy of Sciences, A-5020 Salzburg, Austria

Lawrence B. Smillie (63), MRC Group in Protein Structure and Function, Department of Biochemistry, University of Alberta, Edmonton, Alberta, Canada T6G 2H7

John D. Strauss (341), Department of Molecular Physiology and Biological Physics, University of Virginia, Health Science Center, Charlottesville, Virginia 22908

James T. Stull (119), Department of Physiology, University of Texas Southwestern Medical Center, Dallas, Texas 75235

Kathleen M. Trybus (37), Rosenstiel Basic Medical Sciences Research Center, Brandeis University, Waltham, Massachusetts 02254

Peter J. Vibert (47), Rosenstiel Basic Medical Sciences Research Center, Brandeis University, Waltham, Massachusetts 02254

Frank Wuytack (241), Laboratorium Voor Fysiologie, K. U. Leuven, B 3000 Leuven, Belgium

Gang Zhi (119), Department of Physiology, University of Texas Southwestern Medical Center, Dallas, Texas 75235

Introductory Note

The study of muscle, more than that of any other tissue, has been at the origin of some of the most basic discoveries in biology. This is largely due to its abundance and diversity of types, richness in enzyme content, readily observable changes in appearance and structure during contraction, and the ease with which one can measure the conversion of chemical energy into a mechanical response. It was by studying the contraction of ventricular muscle that Ringer obtained the first indication, more than 100 years ago, of a calcium requirement for this event. Subsequently, Hogkin, Loeb, Loewi, Heilbrunn, and others demonstrated the involvement of calcium in nerve excitability and other cellular and tissue processes. These studies eventually led to the evidence that calcium is an essential second messenger of signal transduction following its interaction with calmodulin and many other calcium-binding and calcium-modulating proteins. Many of the intricate mechanisms of calcium uptake and release, transport, and translocation have been elucidated today.

Muscle was also the tissue in which fundamental metabolic and regulatory events were uncovered. Ninety years ago, Fletcher and Hopkins showed that lactic acid was produced during contraction. In 1930, Meyerhof demonstrated that this compound was generated during anaerobic glycogenolysis, shown later to provide the energy required to maintain contraction. In the mid-fifties, muscle glycogen phosphorylase was shown to be regulated by phosphorylation–dephosphorylation brought about by the opposing action of a protein kinase and phosphatase. At that time, it was not known whether reversible protein phosphorylation represented a unique occurrence restricted to the control of that enzyme only or, at best, of carbohydrate metabolism. As it turned out, it represents one of the most prevalent mechanisms by which physiological processes can be regulated. Approx-

imately 30% of all proteins in the cell can exist in a phosphorylated form, and it is estimated that more than 3% of all genes encode protein kinases and phosphatases.

In the late seventies, protein tyrosine phosphorylation was discovered and shown to be intimately implicated in cell growth, proliferation, differentiation, and transformation. These processes were found to be initiated by tyrosine kinases of cellular or viral origin or linked to receptors for growth factors, mitogenic hormones, or cytokines. Signaling was also shown to involve adaptor proteins containing conserved structural motifs (such as the src homology SH2 and SH3 domains, the PH domain, etc.) displaying high affinity and specificity for sequences found in a great number of enzymes and cellular proteins. The dynamic interactions among these elements, acting in concert with an increasing number of second messengers (Ca^{2+}, cAMP, cGMP, IP_3, NO, etc.), trigger a cascade of downstream reactions on several interrelated pathways. The cross-talk that takes place within this intricate combinatorial system of interacting molecules ultimately allows the very fine tuning of a given physiological response.

This book, edited by one of the most prominent investigators in the field of muscle research, has been designed to offer a comprehensive and up-to-date perspective of smooth muscle contraction. Within its 8 sections, containing a total of 30 chapters, it covers all aspects of the fine structure of the contractile apparatus, signal transduction and regulation by calcium and protein phosphorylation, and the detailed mechanism and energetics of contraction. From the vantage point of an outsider, it has been exciting for me to see how extensively this field has developed. The application of new methodologies, including those of molecular biology, has led to remarkable progress in elucidating the structure and assembly of the contractile machin-ery, including the complex interactions that take place between contractile elements. This book, *Biochemistry of Smooth Muscle Contraction*, illustrates well how far these new technologies have taken us and where they might lead us tomorrow.

Edmond H. Fischer
University of Washington, Seattle

Foreword

Smooth muscle has long been studied by physiologists, but until the 1960s few biochemists were brave enough to use it as the starting material for their studies. Skeletal muscle, conveniently packaged in compact masses of homogeneous tissue, was a much more satisfactory source of the proteins and enzymes of the contracto-regulatory system. Smooth muscle, widely distributed in relatively small amounts in different organs and usually associated with large amounts of connective tissue, was a different matter. Although the structure of the smooth muscle cell is clearly different from that of striated muscle, it was assumed even in early studies that the contractile process, at least, was similar. Actin, tropomyosin, and myosin comparable in properties to the skeletal isoforms were present, but proteins corresponding to the troponin system of striated muscle could not be detected. Further, there were problems in obtaining good, consistent preparations of myosin, and when purified, these usually had very low ATPase activity compared to that of the skeletal isoform. Indeed because of these difficulties, it was considered good advice to tell a young biochemist, eager to start research on muscle, to keep off the smooth tissue. Now, all is changed, as this book bears striking witness.

Interest in the biochemical mechanisms of smooth muscle developed alongside the study of contraction in nonmuscle systems that expanded in the 1960s. It became apparent that the contracto-regulatory system of smooth muscle exhibited more basic similarities to nonmuscle systems than to those of striated tissue. A further stimulus was provided by reports (Perrie et al., 1972, 1973) of the phosphorylation of the light chain of striated muscle myosin, which was catalyzed by a specific myosin light chain kinase (Pires et al., 1974; Pires and Perry, 1977). A phosphatase involved in the dephosphorylation of the phosphorylated light chain

was also identified (Morgan et al., 1976). Although the kinase was widely distributed in striated muscles and changes in light chain phosphorylation accompanied contractile activity, particularly in fast skeletal muscle, its role appeared uncertain. Striated muscle actomyosin has high MgATPase activity in the absence of phosphorylation and model systems will contract in its absence. The kinase was shown to catalyze the phosphorylation of the light chain of smooth muscle myosin (Frearson et al., 1976) and its widespread nature was indicated by the report (Adelstein et al., 1973) of a similar enzyme involved in the phosphorylation of the 20-kDa light chain of platelet myosin.

A major step forward that revitalized the study of smooth muscle was the demonstration that the low activity of the MgATPase of smooth muscle actomyosin preparations was markedly increased when the regulatory light chain was phosphorylated (Sobieszek and Small, 1976; Gorecka et al., 1976; Chacko et al., 1977). This hinted at a new type of activation in smooth muscle, the importance of which in vivo was confirmed by the demonstration of an increase in the incorporation of ^{32}P in the P light chain on contraction of intact arterial smooth muscle (Barron et al., 1979).

Regulation by myosin phosphorylation, which is of particular importance for smooth muscle, is a feature of the less specialized actomyosin contractile systems. In striated muscle it would appear that light chain phosphorylation has been relegated to a modulatory role in the cross-bridge cycle, and response to stimulation has been accelerated by the evolution of the troponin system located in the I filament. The consequence is that the rapid binding of calcium to one molecule of troponin C renders seven actin molecules available to interact with as many heads of myosin molecules as they can accommodate. In smooth muscle, binding of calcium to calmodulin merely activates the enzyme, which in turn must activate each myosin head independently—a much slower process.

Knowledge of the special biochemistry of smooth muscle has been further expanded by the discovery in the tissue of two other proteins, namely caldesmon (Sobue et al., 1981) and calponin (Takahashi et al., 1986). Both proteins bind to actin and calmodulin and have the capacity, in vitro at least, to inhibit the Mg-ATPase of smooth muscle actomyosin in a calcium-regulated manner. This suggests that in smooth muscle also there is a mechanism for the regulation of modulation of contractile activity involving the thin filaments. It is clear that despite the enormous advances in knowledge there is much to be learned about the mechanism of regulation, and areas of controversy still exist. For this reason alone, a treatise such as Biochemistry of Smooth Muscle Contraction with detailed contributions by international experts is a very timely and important contribution to the literature.

The biochemistry of contraction and its regulation are dealt with at length in the book, and each of the many facets is the subject of a separate section contributed by a recognized international expert. As knowledge of the more specialized aspects of smooth muscle has progressed, there have been many important advances in the understanding of general cell biochemistry and pharmacology. These have occurred in aspects such as cell signaling, the widespread role of kinases in cell regulation, the use of model contractile systems, and the application of molecular biological methodology, to mention a few. Developments in these areas and their special significance for the biochemistry of smooth muscle are discussed in sections that together form the major part of this volume. The result is that the book provides a very comprehensive treatment of the subject. Its value is enhanced by details, in the sections where it is appropriate, of preparative methods and the specialized techniques available.

Michael Bárány is to be congratulated for undertaking the responsibility for compiling this treatise and for persuading so many outstanding workers in the field to contribute. He is extremely well qualified to do this, for in association with his wife Kate, he has been at the leading edge of muscle research for almost forty years. In particular, he is well known for his landmark paper correlating the ATPase of myosin with the speed of muscle shortening (Bárány, 1967) and his pioneering work in the application of phosphorus NMR to the study of intact living muscle (Burt et al., 1977). I have no doubt that this volume will become the international source for those who wish to update themselves or start research on smooth muscle.

S. V. Perry
University of Birmingham, UK

References

Adelstein, R. S., Conti, M. A., and Anderson, W. (1973). *Proc. Natl. Acad. Sci. USA* **70**, 3115–3119.

Bárány, M. (1967). *J. Gen. Physiol.* **50**, 197–218.

Barron, J. T., Bárány, M., and Bárány, K. (1979). *J. Biol. Chem.* **254**, 4954–4956.

Burt, C. T., Glonek, T., and Bárány, M. (1977). *Science* **195**, 145–149.

Chacko, S., Conti, M. A., and Adelstein, R. S. (1977). *Proc. Natl. Acad. Sci. USA* **74**, 129–133.

Frearson, N., Focant, B. W. W., and Perry, S. V. (1976). *FEBS Lett.* **63**, 27–32.

Gorecka, A., Askoy, M. O., and Hartshorne, D. J. (1976). *Biochem. Biophys. Res. Commun.* **71**, 325–331.

Morgan, M., Perry, S. V., and Ottoway, J. (1976). *Biochem. J.* **157**, 687–697.

Perrie, W. T., Smillie, L. B., and Perry, S. V. (1972). *Biochem. J.* **128**, 105–106.

Perrie, W. T., Smillie, L. B., and Perry, S. V. (1973). *Biochem. J.* **135**, 151–164.

Pires, E., Perry, S. V., and Thomas, M. A. W. (1974). *FEBS Lett.* **41**, 292–296.

Pires, E., and Perry, S. V. (1977). *Biochem. J.* **167**, 137–146.

Sobieszek, A., and Small, J. V. (1976). *J. Mol. Biol.* **102**, 75–92.

Sobue, K., Muramoto, Y., Fujita, M., and Kakiuchi, S. (1981). *Proc. Natl. Acad. Sci. USA* **78**, 5652–5655.

Takahashi, K., Hiwada, K., and Kokubu, T. (1986). *Biochem. Biophys. Res. Commun.* **141**, 20–26.

Preface

Within the past decade, considerable progress has been made in the biochemistry of smooth muscle contraction. The primary role of myosin light chain phosphorylation in the contractile event has been established. Analysis of Ca^{2+} indicators inside intact muscle has allowed quantification of the relationship between intracellular Ca^{2+} concentration and contraction or relaxation, and Ca^{2+} imaging techniques have shown the movement of Ca^{2+} in functioning muscle. Genetic engineering of smooth muscle proteins has delineated their active site regions and opened new avenues for studying the molecular mechanism of muscle contraction. The three-dimensional structures of actin and of the myosin head have been determined, and computer modeling has provided insights into how amino acid residues on the actin and myosin surfaces may change their position when muscle contracts. Highly sophisticated *in vitro* motility assays have made it possible to measure minute forces generated by the interactions of a few molecules of actin, myosin, and ATP. In fact, such assays are now routine, just as determination of actomyosin MgATPase activity was some time ago. New smooth muscle proteins, such as calponin and caltropin, have been discovered, and new enzymes, calcium/calmodulin-dependent protein kinase II and mitogen-activated protein kinase, are now studied by smooth muscle biochemists. It has been demonstrated that inositol trisphosphate, a second messenger, generates internal Ca^{2+} signals and thereby initiates smooth muscle contraction. Conversely, the nitric oxide-induced signal transduction pathway has been shown to be involved in smooth muscle relaxation through generation of cGMP, which is now the focus of the relaxation studies. These new results, along with the significantly enhanced understanding of previous results, justify publication of a comprehensive handbook that summarizes the current

knowledge of the biochemistry of smooth muscle contraction.

This book leads the reader from the properties of pure proteins, determined in the test tube, to the complex patterns of muscle energetics, followed in the intact tissue. The book is divided into eight parts. *Contractile Proteins* discusses the motor of the muscle, myosin, in three chapters: one dealing with myosin's structure and function, another with myosin's light chain subunits, and a third with myosin's assembly into filaments. The fourth chapter describes smooth muscle actin and the thin filaments that are composed mainly of actin. *Thin Filament and Calcium-Binding Proteins* consists of four chapters, dealing in turn with the structures and possible physiological functions of tropomyosin, caldesmon, calponin, caltropin, and other calcium-binding proteins. *Enzymes of Protein Phosphorylation–Dephosphorylation* first reviews myosin light chain kinase, including a proposal for the three-dimensional structure of the catalytic core, followed by a discussion of the opposing enzyme, myosin light chain phosphatase. The remaining three chapters are again concerned with protein kinases: calcium/calmodulin-dependent protein kinase II, protein kinase C, and the mitogen-activated protein kinase of smooth muscle. *Motile Systems* acquaints the reader with *in vitro* motility and the permeabilized smooth muscle, which has lost its soluble metabolites but maintains its capacity for receptor–effector coupling. The dynamics of *Calcium Movements* are covered next: Calcium and potassium channels and the expression of smooth muscle calcium channels, i.e., the molecular basis for Ca^{2+} channel diversity, are reviewed. The chapter on electro- and pharmacomechanical coupling presents the regulation of myoplasmic Ca^{2+} by changes in membrane potential, as well as regulation, which is independent of the potential. Furthermore, the structure and regulation of the calcium pumps are covered. The coupling of the receptors in the smooth muscle plasma membrane to intracellular second-messenger systems is described under *Signal Transduction*, with chapters covering: nitric oxide–cyclic GMP signaling systems, inositol 1,4,5-trisphosphate production, and protein tyrosine phosphorylation. A further chapter is devoted to a new second messenger, cyclic ADP-ribose, which may play a role in smooth muscle. A final chapter is on enzyme translocations during smooth muscle activation. *Contraction–Relaxation* covers protein phosphorylation, the most important intracellular control mechanism of smooth muscle contractility. This is discussed in chapters on protein phosphorylation during contraction and relaxation, regulation of cross-bridge cycling in smooth muscle, and calcium sensitivity of contraction. The chapter on pharmacological regulation of smooth muscle classifies special drugs that are useful in both smooth muscle physiology and human disease therapy. The final part of the book covers *Energetics* of smooth muscle contraction investigated either by chemical methods or by nuclear magnetic resonance techniques.

Major progress in the biochemistry of smooth muscle contraction was facilitated by key experimental methods, which are woven into the chapters of this book. Furthermore, an effort has been made both to present data in a way that illustrates how conclusions were drawn and to identify problems remaining to be solved. Correlations to cellular contractile motility and diseased states are pointed out whenever appropriate. The several thousand citations in the text should function as a useful reference base for years to come. It is our hope that research workers in various fields of biochemistry, physiology, pharmacology, biology, and medicine will find much here to inspire their future efforts.

I greatly appreciate the contributions of the authors in the preparation of the chapters. I am indebted to Dr. Kate Bárány for constant consultations during the editing process. I am grateful to Charlotte Brabants, Acquisitions Editor at Academic Press, for sponsoring and arranging rapid publication of this book. I also thank Anna M. Pravdik for dedicated assistance.

Michael Bárány
University of Illinois, Chicago

CONTRACTILE
PROTEINS

1

Myosin Structure and Function

ROBERT S. ADELSTEIN and JAMES R. SELLERS

Laboratory of Molecular Cardiology
National Heart, Lung, and Blood Institute
National Institutes of Health
Bethesda, Maryland

I. INTRODUCTION

Smooth muscle myosin is a bifunctional molecule composed of six polypeptide chains (Fig. 1). The two most outstanding biological properties of this molecule are (1) the ability to convert the chemical energy of MgATP into mechanical work, an enzymatic property that resides in the amino-terminal globular head domain, and (2) the ability to organize itself into polar filaments, a property that resides on the carboxyl-terminal, rodlike end of the molecule, and which will be discussed in Chapter 3 of this volume. The consequences of the interaction of smooth muscle myosin filaments with actin filaments in a variety of organ systems distributed throughout the body are manifold. They include processes determining life and death, such as the expulsion of the newborn fetus from the muscular uterus and the spasm of the diseased coronary artery that results in a fatal cardiac arrhythmia. In addition, smooth muscles play critical roles in more mundane physiological events such as digestion, evacuation of waste products, respiration, and sexual reproduction. Thus, an increased understanding of the structure and function of the smooth muscle myosin molecule should provide important information about the mechanisms underlying a number of different physiological processes. Reviews that have focused on smooth muscle myosin include those by Stull *et al.* (1991), Sellers (1991), Trybus (1991), Somlyo (1993), and Somlyo and Somlyo (1994). An extensive, but more dated, review by Hartshorne (1987) still provides important basic information.

In this chapter we will describe the general features of the molecule that is the molecular motor of vertebrate smooth muscle cells. As shown in Fig. 1, this myosin molecule is composed of two heavy chains of approximately 220 kDa each and two pairs of light chains that are located in the neck region, just carboxyl-terminal to the globular myosin head domain. The properties and structure of the myosin light chains are covered in Chapter 2 of this volume.

We will first concentrate on the standard methods for purifying vertebrate smooth muscle myosin, as well as methods for generating its proteolytic subfragments, heavy meromyosin (HMM) and subfragment-1 (S-1) (see Fig. 1). We then outline a procedure for phosphorylating the smooth muscle heavy chain (HC) using casein kinase II. Section II concludes with a description of how to raise isoform-specific antibodies to the HCs.

Section III discusses structural information revealed by an analysis of the primary structure of the smooth muscle HC. The derived amino acid sequences from cDNA cloning of two different smooth muscle HCs and a nonmuscle HC are compared to the HC sequence of chicken skeletal muscle. In this section, we speculate about a putative smooth muscle three-dimensional (3D) structure by relating it to the recently derived 3D structure for chicken skeletal S-1 (Rayment *et al.*, 1993a).

Section IV introduces the important role of alternative splicing in generating smooth muscle HC isoforms. As will be discussed, there are two different smooth muscle HC isoforms that can be separated by

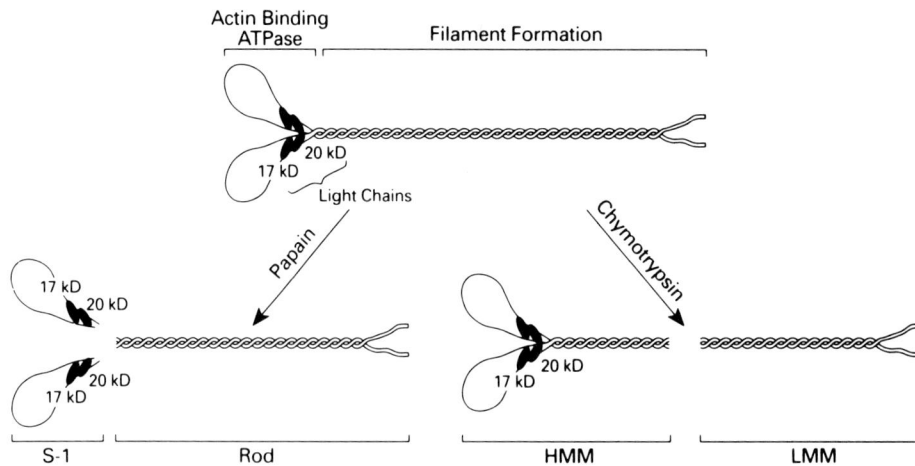

FIGURE 1   Diagrammatic representation of the smooth muscle myosin molecule and its proteolytic fragments. The globular head, the rodlike tail, and the nonhelical carboxyl terminus regions are not drawn to scale. The methods used for generating the indicated myosin fragments by proteolytic cleavage are outlined in the text.

SDS–PAGE, with MWs of 200,000 and 204,000. These, as well as another set of isoforms, are generated by alternative splicing of pre-mRNA. The dimeric nature of smooth muscle myosins will then be addressed, with reference to whether each smooth muscle myosin molecule is composed of homodimers of 204- and 200-kDa HCs, heterodimers, or a mixture of each.

In the following section, we will focus on the "heart" of this molecular motor molecule, its ability to hydrolyze ATP and convert this chemical step into the mechanical motion involved in propeling actin filaments. The mechanism underlying the regulation of the actin-activated MgATPase activity by phosphorylation of 20-kDa smooth muscle myosin light chain (LC20) will be introduced, though it will be expanded on in a number of chapters later in the book. The chapter ends with a short perspective on the future of smooth muscle HC research.

II. METHODS

A. Purification of Smooth Muscle Myosin and Myosin Fragments

The critical steps for preparation of smooth muscle myosin are: (1) Preparation of myofibrils, a step that is useful in the purification of avian gizzard myosin but is omitted when purifying myosin from tissues with less abundant amounts of smooth muscle, such as bovine aorta. (2) Extraction of myosin in the presence of ATP. (3) Fractionation of the extracted myosin with ammonium sulfate (40–60%) following addition of MgATP to dissociate actin and myosin. (4) Chromatography of the ammonium sulfate fraction on Sepharose 4B in 0.5 M NaCl (or KCl) to separate myosin from actin, myosin light chain kinase and phosphatase activities. These methods are outlined for smooth muscle myosin prepared from different sources such as turkey gizzards (Sellers et al., 1988) and turkey aortas (Kelley and Adelstein, 1990). To maximize separation of myosin from actin and contaminating myosin light chain kinase and phosphatase activities, the volume of the 40–60% ammonium sulfate fraction should not exceed 2% of the gel filtration column volume. To dissociate any residual actin from myosin, the ~10 mg/ml sample to be gel filtered is made 1–5 mM with respect to ATP and MgCl$_2$ in a pH 7.0 buffer in 0.5 M KCl. Typically, 100 mg of total protein is chromatographed on a 5×90-cm column of CL Sepharose 4B. By limiting the time that the crude myosin is exposed to MgATP in the presence of the contaminating myosin light chain kinase (MLCK) and by carrying out all procedures at 4°C or on ice in the presence of EGTA to chelate Ca^{2+} ions, the purified myosin is invariably found to have no phosphate on the LC20. Myosin that is properly column purified should be devoid of phosphatase activity and phosphorylated myosin will remain stably phosphorylated even in the absence of phosphatase inhibitors.

The fractions containing myosin, detected by K-EDTA-stimulated ATPase activity measured in 0.5 M KCl (or by analysis in SDS–PAGE), are pooled and dialyzed against a low ionic strength buffer (25 mM NaCl or KCl and 10 mM MgCl$_2$ at pH 7.0), which results in filament formation and precipitation. The sample is stored on ice or quick frozen in small aliquots and stored in liquid N$_2$ in 0.5 M NaCl, 10 mM MOPS (pH 7.0), 0.1 mM EGTA, and 5 mM dithiothreitol (DTT).

To obtain a soluble myosin fragment that, unlike intact myosin, is suitable for a variety of kinetic studies, HMM is prepared as follows. Starting with 50 ml of a 20 mg/ml preparation of crude turkey gizzard myosin (i.e., prior to gel filtration), the LC20 is phosphorylated by the addition of Ca^{2+} to 0.1 mM, calmodulin (10^{-7} M), and ATP (5 mM) using the activity of the endogenous MLCK. The sample is brought to 0.5 M in NaCl and 2 mM in EGTA. Chymotryptic digestion (final concentration 0.07 mg/ml) is carried out for 7 min at 25°C and terminated with 0.2 mM phenylmethylsulfonyl fluoride (PMSF) followed by dialysis against a low ionic strength buffer. LC20 is dephosphorylated during this dialysis by endogenous phosphatases. Undigested myosin as well as light meromyosin (LMM) are removed following centrifugation and the supernate is purified by gel filtration on Sephacryl S300 or Ultragel AcA34. It is important to perform the digestion in the absence of Ca^{2+} to avoid generation of a calcium-insensitive MLCK. Prepared as described, column-purified HMM should be unphosphorylated and essentially free of kinase and phosphatase activity. The purified HMM is concentrated by ammonium sulfate precipitation and stored following dialysis against 10 mM imidazole (pH 7.0), 0.1 mM EGTA, 3 mM NaN$_3$, and 1 mM DTT at 0°C for 2–3 weeks (Sellers *et al.*, 1982). It can also be quick frozen in small aliquots and stored in liquid N$_2$.

Two different enzymes have been used for the preparation of smooth muscle S-1. Ikebe and Hartshorne (1985) employed *Staphylococcus aureus* protease, which has the advantage of preserving both myosin light chains. Papain also produces an S-1, but LC20 is degraded to a form that can no longer be phosphorylated by MLCK. This fragment has been used in a number of kinetic studies (Sellers *et al.*, 1982; Greene *et al.*, 1983) and is produced by digesting crude turkey gizzard myosin filaments with papain for about 5 min. The digestion is terminated by addition of iodoacetate and 10 mM MgCl$_2$ is added to aid in the precipitation of myosin rod fragments, which are separated from the soluble S-1 by centrifugation. Interestingly, neither of the S-1 preparations is regulated by phosphorylation, which suggests that the two myosin heads are required for this property. The details for this preparation are given in Sellers *et al.* (1988).

B. Phosphorylation of Smooth Muscle Myosin Heavy Chains

All smooth muscle 204-kDa HCs examined to date, with the notable exception of the chicken gizzard 204-kDa HC, contain a consensus sequence for casein kinase II near its carboxyl-terminal end (see Fig. 6). This serine residue, which is also present in both vertebrate nonmuscle HC-A and HC-B (Murakami *et al.*, 1990), is located in the nonhelical tail part of the rod. *In vitro* phosphorylation can be carried out with ~50 nM casein kinase II at 25°C with 6.4 μM myosin in 50 mM Tris-HCl (pH 7.5), 40 mM NaCl, 7.5 mM MgCl$_2$, 0.2 mM EGTA, and 0.5 mM ATP ([γ-^{32}P]ATP, 192 Ci/mol) for 30 min (Kelley and Adelstein, 1990). Phosphorylation by casein kinase II is specific for the 204-kDa HC and 0.6 mol of phosphate can be incorporated into the single residue phosphorylated, provided the myosin is first dephosphorylated. Techniques similar to those described by Bárány and Bárány in Chapter 2, this volume, but substituting buffers that are appropriate for chromatography of HC peptides, were used to map the single tryptic peptide generated following phosphorylation of HC204.

C. Preparation of Antibodies to the 204- and 200-kDa Heavy Chains

Polyclonal antibodies that detect both the 204- and 200-kDa HCs were raised in rabbits using purified bovine aortic smooth muscle myosin, which is a 1:1 mixture of both isoforms as antigens. These antibodies have approximately equal affinity for HC204 and HC200 (Kelley and Adelstein, 1990). Immunoblot analysis of smooth muscle HMM and LMM using the 50% ammonium sulfate immunoglobulin (IgG) fraction revealed that practically all the epitopes were located in the LMM region (Christine A. Kelley, unpublished results, NHLBI).

To prepare antibodies that specifically recognized the 204- and 200-kDa HCs, synthetic peptides were generated that duplicated sequences near the carboxyl termini of the two heavy chains (see Fig. 6). The following HC204 peptides were synthesized: RRVIENADGSEEEVDAR and DADFNGTKSSE. The first of these peptides includes the sequence of the tryptic phosphopeptide generated following casein kinase II phosphorylation of bovine aortic smooth muscle myosin. It is identical to the rabbit uterine HC sequence. The second peptide was constructed based on this latter sequence (Nagai *et al.*, 1989). The sequence used to generate antibodies to HC200 was GPPPQETSQ. In each case, the peptides were conjugated to keyhole limpet hemocyanin with glutaraldehyde (Goldsmith *et al.*, 1987) and the product was used to immunize rabbits. The antibodies generated to synthetic peptides were purified using an affinity column of the relevant peptide coupled to Affi-Gel 15. They were eluted in 0.1 M glycine buffer at pH 2.5.

Purification of enzymatically active homodimers of

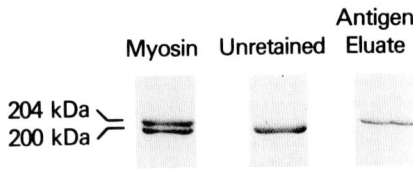

FIGURE 2   Separation of HC204 and HC200 by protein G-HPLC immunoaffinity chromatography. The three panels show Coomassie blue-stained SDS–5% polyacrylamide gels of bovine aorta myosin before (left panel) and following HPLC chromatography (right two panels). The antibodies used in this purification were raised to the amino acid sequence DADFNGTKSSE, which is at the carboxyl terminal of the HC204. The middle panel shows the purified HC200 dimers, which were not retained by the antibodies, and the panel on the right shows the HC204 dimers that were bound to the column and were eluted using the peptide antigen. The methods used are outlined in the text. Modified from Kelley et al. (1992).

HC204 and HC200 was carried out using affinity chromatography of myosin on a protein G-agarose high-performance liquid chromatography (HPLC) column containing immobilized HC204 antibodies (Kelley et al., 1992). The purified HC200 myosin was obtained in the unretained fraction (Fig. 2). The bound HC204 isoform was then eluted by addition of the peptide that was used to generate the antibody. The peptide is easily separated from the myosin by ultrafiltration through a 100-kDa cutoff membrane and can be reused for subsequent purifications.

Using similar techniques, antibodies were generated to the seven-amino-acid insert that is introduced by alternative splicing of pre-mRNA into the head region of a number of visceral smooth muscle myosins (Kelley et al., 1993) (see Fig. 7). Since these antibodies are raised to only seven amino acids, the titer of the serum obtained from different rabbits has not been consistently high.

III. ANALYSIS OF THE PRIMARY SEQUENCE OF SMOOTH MUSCLE MYOSIN HEAVY CHAIN

A. Sequence Comparison of Smooth, Nonmuscle, and Skeletal Muscle Heavy Chains

The overall domain organization of smooth muscle myosin is similar to that of skeletal muscle myosin when viewed by rotary shadowing in high salt in the electron microscope (Trybus et al., 1982). There are two slightly elongated heads connected to a long tail. Analysis of the primary sequence of the smooth and skeletal muscle and nonmuscle HCs confirms this gross structural similarity (Fig. 3). Each of the molecules has conserved motor domains at the amino terminus fol-

lowed by two light chain binding regions in the neck. This portion of the sequence accounts for S-1, which forms the globules seen in the electron microscope. Each molecule has a tail consisting of approximately 120 kDa of α-helical coiled-coil-forming sequence, ending in a short sequence that is not predictive of coiled-coil structure (Fig. 1).

The amino acid sequences of chicken smooth muscle HC, rabbit smooth muscle HC, chicken nonmuscle myosin-B HC, and chicken skeletal muscle HC are aligned in Fig. 3. We have chosen to begin the sequence of the smooth muscle and nonmuscle HCs with the amino acid following the initiating methionine. This is consistent with the chicken skeletal sequence, which, unlike the other sequences, was not derived from cDNA. This results in a single amino acid displacement when residue numbers are compared to those used in many of the references. Despite the low level of direct sequence identity in the rod regions, the sequences of all three types of myosin in this area show strong heptad helical repeats as described by McLachlan (1984). In the HC rod hydrophobic residues are preferentially located at positions a and d, which form the interchain contact regions required for dimerization of the two HCs (Yanagisawa et al., 1987).

The primary sequences of the head regions of smooth and skeletal muscle myosins are more conserved than their rod regions (Fig. 3), although the sequence of smooth muscle myosin is clearly more homologous to that of nonmuscle myosin. Eight areas of strong homology (shown boxed) are detected by visual inspection of the aligned sequences (Fig. 3). These areas show at least 85% sequence identity between smooth and skeletal muscle myosin and include residues thought to be involved in phosphate or nucleotide binding, such as GESGAKT and LEAFGNAKT (see Fig. 3, amino acid 176 and amino acid 230 for Ch Sm, respectively) (Warrick and Spudich, 1987).

Two regions located approximately 25 and 75 kDa from the amino termini of most types of myosin are susceptible to digestion by a variety of proteases (Fig. 7) (Bonet et al., 1987; Mornet et al., 1989). Digestion at these sites does not lead to dissociation of the polypeptide chains under nondenaturing conditions, but gives rise to polypeptides of approximately 25, 50, and 20 kDa from S-1 preparations, when electrophoresed on SDS polyacrylamide gels. These three regions were thought by some to be independently folded domains (Muhlrad, 1991), but the crystal structure of skeletal muscle S-1 demonstrated that this was not the case (Rayment et al., 1993a). Instead, all three segments of the myosin head are intertwined in space and the two proteolytically sensitive sites are located in separate surface loops.

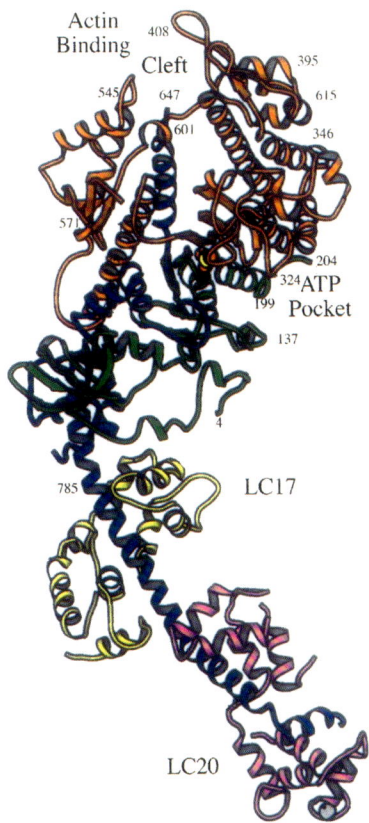

FIGURE 4 Three-dimensional structure of chicken skeletal muscle S-1. The position of areas referred to in the text are marked. The HC is color coded to correspond to the three tryptic fragments of skeletal muscle S-1. The amino-terminal 25-kDa region is colored in green, the 50-kDa central region is colored in red, and the 20-kDa carboxyl-terminal region of S-1 is colored in blue. The light chains are labeled according to the smooth muscle nomenclature. Modified from figure provided by Ivan Rayment (University of Wisconsin).

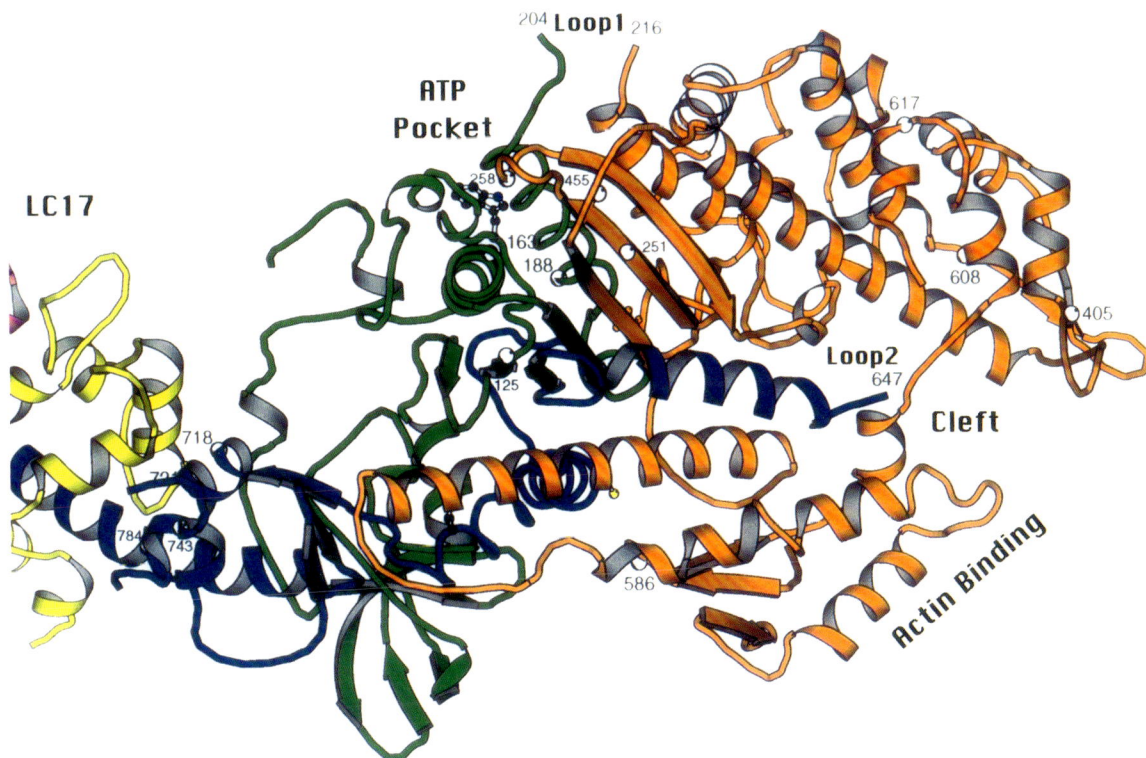

FIGURE 5 Three-dimensional structure of the head of chicken skeletal muscle S-1. A more detailed view of the motor domain of chicken skeletal muscle S-1. The positions of the missing loop 1 and loop 2 are indicated.

B. Comparison of Domains of the Heavy Chains to Three-Dimensional Structure of Chicken Skeletal Muscle Subfragment-1

At this point, it is useful to discuss the 3D structure of chicken skeletal muscle S-1, which will be used as a reference point for further discussion of the comparative structure of the head of the smooth muscle HC. A representation of the 3D crystal structure of skeletal muscle S-1 resolved to 2.8 Å is shown in Figs. 4 and 5. The heavy chains are color coded according to the respective tryptic fragment. There are several prominent topological features of the asymmetric molecule. The amino-terminal portion of the 25-kDa region forms a six-stranded, antiparallel β-sheet structure of unknown function. The 50-kDa tryptic fragment forms most of the structure at the tip of the myosin head and is divided into an upper and lower domain that is split by a prominent cleft. The 20-kDa tryptic fragment begins near the tip of the head and runs the length of the head through a series of helices. From residue 771 to 826, the 20-kDa fragment forms a long continuous helix that forms the binding sites for the two LCs. In Fig. 4, we label these as LC17 and LC20, corresponding to the molecular weights of the smooth muscle isoforms. This elongated structure gives rise to the marked asymmetry of the molecule.

The polypeptide chain at the 25/50- and 50/20-kDa junctions is not visualized in the structure derived from X-ray diffraction, which suggests that these junctions form flexible structures at the surface. Their position is marked in Fig. 5 as loop 1 and loop 2 (Spudich, 1994). Several potential actin binding regions on myosin have been implicated by cross-linking or other biochemical data or deduced from a computer docking model using the crystal structures of myosin S-1 and actin. These areas generally lie on both sides of the large cleft at the tip of the head and involve residues primarily from the 50-kDa region, but one of the sites is also formed from the sequence at the junction of the 50- and 20-kDa regions. The primary sequences thought to be involved in actin binding in skeletal muscle S-1 are in bold type in Fig. 3.

The nucleotide binding pocket is located on the opposite side of the head from the actin interface. It is formed mostly by seven β-sheets that are contributed by all three tryptic fragments of the myosin molecule. The loop at the junction between the 25- and 50-kDa fragments is located near the top of the pocket (see Fig. 5).

A final region deserves consideration. The sulfhydryl side chains of Cys-707 (SH1) and Cys-697 (SH2) are very reactive and have been covalently modified by a host of compounds (Wells and Yount, 1982).

They lie in what is thought to be a crucial area for mechanochemical transduction since these two cysteines can be cross-linked by a variety of reagents, and can be oxidized to a disulfide in the presence, but not the absence, of ATP. In the skeletal muscle S-1 crystal structure, which does not have ATP bound, they are found on an α-helix with their side chains pointing in opposite directions (Rayment et al., 1993a).

Given the high homology of the amino acid sequence in the head regions of smooth and skeletal muscle myosin, it is probable that the 3D structure of smooth muscle myosin is very similar to that of skeletal muscle myosin. In fact, when the regions identified as having the highest sequence homology between the two myosins (Fig. 3) are mapped onto the backbone of the 3D structure of the chicken skeletal muscle S-1, it is seen that they lie mostly in and around the nucleotide binding pocket and the region containing SH1 and SH2. Thus, one can use the 3D structure of skeletal muscle S-1 as a framework for discussion of the probable smooth muscle structure.

Sequence alignments reveal that the primary sequence of amino acids in the two loop regions varies dramatically among different myosins and may be a source of isoform diversity (Fig. 3) (Spudich, 1994). This is supported by the location of the two loops in the myosin head. Loop 1, connecting the 25- and 50-kDa segments, is located near the nucleotide binding site, whereas loop 2, connecting the 50- with the 20-kDa fragment, is located within an actin interface. The sequences in these two regions are overlined in bold in the sequence alignment shown in Fig. 3. The 50/20-kDa junction (loop 2) is of particular interest. The sequence in skeletal muscle myosin contains nine glycine, five lysine, and two glutamic acid residues. Smooth muscle and nonmuscle HCs have a longer loop region in that eight additional amino acids are present. The character of the sequence of these latter two myosins also differs from that of skeletal muscle HC in that there is no preponderance of glycine residues. Smooth muscle loop 2 does contain four lysines, one arginine, one glutamic acid, and one aspartic acid. This region is known to bind to actin and is postulated to constitute a nonstereospecific contact between actin and myosin in the weakly bound M·ATP and M·ADP·P_i states that will be discussed in more detail in section V.

The binding of skeletal muscle myosin to actin in the presence of ATP is extremely ionic strength dependent. Increasing the ionic strength from 23 to 150 mM decreases the affinity of skeletal muscle S-1·AMP·PNP for actin by a factor of 100, whereas the affinity of smooth muscle S-1·AMP·PNP for actin decreases only

```
        25 kD region ――――▶
Ch Sk   ASPDAEMAAF GEAAPYLRKS EKERIE..AQ NKPFDAKSSV FVVHPKESFV   48
Ch Sm   ....SQK.PL SDDEKFLF.V DKNFVNNPLA QADWSAKKLV WVPSEKHGFE   44
Rb Sm   ....AQKGQL SDDEKFLF.V DKNFINSPVA QADWVAKRLV WVPSEKQGFE   45
Ch NMB  ....AQRSGQ EDPERYLF.V DRAVIYNPAT QADWTAKKLV WIPSERHGFE   45


Ch Sk   KGTIQSKEGG KVTVK.TEGG ETLTVKEDQV FSMNPPKYDK IEDMAMMTHL   97
Ch Sm   AASIKEEKGD EVTVELQENG KKVTLSKDDI QKMNPPKFSK VEDMAELTCL   94
Rb Sm   AASIKEEKGD EVVVELVENG KKVTVGKDDI QKMNPPKFSK VEDMAELTCL   95
Ch NMB  AASIKEERGD EVLVELAENG KKALVNKDDI QKMNPPKFSK VEDMAELTCL   95


Ch Sk   HEPAVLYNLK ERYAAWMIYT YSGLFCVTVN PYKWLPVYNP EVVLAYRGKK  147
Ch Sm   NEASVLHNLR ERYFSGLIYT YSGLFCVVIN PYKQLPIYSE KIIDMYKGKK  144
Rb Sm   NEASVLHNLR ERYFSGLIYT YSGLFCVVVN PYKQLPIYSE KIVDMYKGKK  145
Ch NMB  NEASVLHNLR DRYYSGLIYT YSGLFCVVIN PYKNLPIYSE NIIEMYRGKK  145
                         Homology region 1


Ch Sk   RQEAPPHIFS ISDNAYQFML TDRENQSILI TGESGAGKTV NTKRVIQYFA  197
Ch Sm   RHEMPPHIYA IADTAYRSML QDREDQSILC TGESGAGKTE NTKKVIQYLA  194
Rb Sm   RHEMPPHIYA IADTAYRSML QDREDQSILC TGESGAGKTE NTKKVIQYLA  195
Ch NMB  RHEMPPHIYA ISESAYRCML QDREDQSILC TGESGAGKTE NTKKVIQYLA  195
                         Homology region 2
            Loop 1
                                    50 kD region ――――▶
Ch Sk   TIAASGEKKK EEQSGKMQ.. ..GTLEDQII SANPLLEAFG NAKTVRNDNS  243
Ch Sm   VVASSHKGKK DTSITQGPSF SYGELEKQLL QANPILEAFG NAKTVKNDNS  244
Rb Sm   VVASSHKGKK DTSIT..... ..GELEKQLL QANPILEAFG NAKTVKNDNS  238
Ch NMB  HVASSHKGRK DHNIP..... ..GELERQLL QANPILESFG NAKTVKNDNS  238
                                    Homology region 3


Ch Sk   SRFGKFIRIH FGATGKLASA DIETYLLEKS RVTFQLPAER SYHIFYQIMS  293
Ch Sm   SRFGKFIRIN FDVTGYIVGA NIETYLLEKS RAIRQAKDER TFHIFYYLIA  294
Rb Sm   SRFGKFIRIN FDVTGYIVGA NIETYLLEKS RAIRQAREER TFHIFYYLIA  288
Ch NMB  SRFGKFIRIN FDVTGYIVGA NIETYLLEKS RAVRQAKDER TFHIFYQLLA  288
                         Homology region 4


Ch Sk   NKKPELIDML LITTNPYDYH YVSQGEITVP SIDDQEELMA TDSAIDILGF  343
Ch Sm   GASEQMRNDL LLEGFN.NYT FLSNGHVPIP AQQDDEMFQE TLEAMTIMGF  343
Rb Sm   GAKEKMRNDL LLEGFN.NYT FLSNGFVPIP AAQDDEMFQE TVEAMSIMGF  337
Ch NMB  GAGEHLKSDL LLEGFN.NYR FLSNGYIPIP GQQDKDNFQE TMEAMHIMGF  337


Ch Sk   SADEKTAIYK LTGAVMHYGN LKFKQKQREE QAEPDGTEVA DKAAYLMGLN  393
Ch Sm   TEEEQTSILR VVSSVLQLGN IVFKKERNTD QASMPDNTAA QKVCHLMGIN  393
Rb Sm   SEEEQLSVLK VVSSVLQLGN IVFKKERNTD QASMPDNTAA QKVCHLMGIN  387
Ch NMB  SHDEILSMLK VVSSVLQFGN ISFKKERNTD QASMPENTVA QKLCHLLGMN  387
```

FIGURE 3   Alignment of the complete amino acid sequences of skeletal muscle, smooth muscle, and nonmuscle myosin HCs. Chicken fast skeletal muscle myosin (Ch Sk), GeneBank accession number P13538 (Maita *et al.*, 1991); chicken smooth muscle myosin (Ch Sm), accession number P10587 (Yanagisawa *et al.*, 1987); rabbit smooth muscle myosin (Rb Sm), accession number M77812 (Babij *et al.*, 1991); and chicken nonmuscle HC-B (Ch NMB), accession number M93676 (Takahashi *et al.*, 1992). The chicken smooth muscle myosin sequence has an insert of 7 amino acids starting at residue 209 (see text) and the sequence has been corrected according to Kelley *et al.* (1993). The rabbit smooth muscle sequence is from uterus and does not contain the inserted amino acids. (For the amino acid sequence of HC200 isoforms, see Fig. 6.) Regions of very strong homology among the four proteins are boxed. The regions corresponding to loop 1 or loop 2 are overlined and labeled. Regions in skeletal muscle myosin that have been postulated to interact with actin are shown in bold for that sequence only (Rayment *et al.*, 1993b). The start of the 25-, 50-, 20-kDa, S-2, and LMM regions are indicated, as are the LC17 and LC20 binding regions. A typical 28-residue repeat and the nonhelical tail region are underlined. The proline residue initiating the α-helical tail is indicated by an arrow and the two reactive cysteine residues are indicated by arrowheads.

```
Ch Sk   SAELLKALCY  PRVKVGNEFV  TKGQTVSQVH  NSVGALAKAV  YEKMFLWMVI  443
Ch Sm   VTDFTRSILT  PRIKVGRDVV  QKAQTKEQAD  FAIEALAKAK  FERLFRWILT  443
Rb Sm   VTDFTRSILT  PRIKVGRDVV  QKAQTKEQAD  FAVEALAKAT  YERLFRWILS  437
Ch NMB  VMEFTRAILT  PRIKVGRDYV  QKAQTKEQAD  FAVEALAKAT  YERLFRWLVH  437


Ch Sk   RINQQLD.TK  QPRQYFIGVL  DIAGFEIFDF  NSFEQLCINF  TNEKLQQFFN  492
Ch Sm   RVNKALDKTK  RQGASFLGIL  DIAGFEIFEI  NSFEQLCINY  TNEKLQQLFN  493
Rb Sm   RVNKALDKTH  RQGASFLGIL  DIAGFEIFEV  NSFEQLCINY  TNEKLQQLFN  487
Ch NMB  RINKALDRTK  RQGASFIGIL  DIAGFEIFEL  NSFEQLCINY  TNEKLQQLFN  487
                                            Homology region 5


Ch Sk   HHMFVLEQEE  YKKEGIEWEF  IDFGMDLAAC  IELIEKPM..  .GIFSILEEE  539
Ch Sm   HTMFILEQEE  YQREGIEWNF  IDFGLDLQPC  IELIERPTNP  PGVLALLDEE  543
Rb Sm   HTMFILEQEE  YQREGIEWNF  IDFGLDLQPC  IELIERPNNP  PGVLALLDEE  537
Ch NMB  HTMFILEQEE  YQREGIEWNF  IDFGLDLQPC  IDLIERPANP  PGVLALLDEE  537


Ch Sk   CMFPKATDTS  FKNKLYDQHL  GKSNNFQKPK  PAKGKAEAHF  SLVHYAGTVD  589
Ch Sm   CWFPKATDTS  FVEKLIQEQ.  GNHAKFQKSK  QLKDKTE..F  CILHYAGKVT  590
Rb Sm   CWFPKATDKS  FVEKLCTEQ.  GNHPKFQKPK  QLKDKTE..F  SIIHYAGKVD  584
Ch NMB  CWFPKATDKT  FVEKLVQEQ.  GTHSKFQKPR  QLKDKAD..F  CIIHYAGKVD  584
        Homology region 6
                                                      Loop 2


Ch Sk   YNISGWLEKN  KDPLNETVIG  LYQKSSVKTL  ALLFATYG..  ......GEAE  631
Ch Sm   YNASAWLTKN  MDPLNDNVTS  LLNQSSDKFV  ADLWKDVDRI  VGLDQMAKMT  640
Rb Sm   YNASAWLTKN  MDPLNDNVTS  LLNASSDKFV  ADLWKDVDRI  VGLDQMAKMT  634
Ch NMB  YKADEWLMKN  MDPLNDNVAT  LLHQSSDKFV  AELWKDVDRI  VGLDQVTGIT  634


        Loop 2
                              20 kD  region ——→
Ch Sk   GGGGKKGGKK  KGSSFQTVSA  LFRENLNKLM  ANLRSTHPHF  VRCIIPNETK  681
Ch Sm   ESSLPSASKT  KKGMFRTVGQ  LYKEQLTKLM  TTLRNTNPNF  VRCIIPNHEK  690
Rb Sm   ESSLPSASKT  KKGMFRTVGQ  LYKEQLGKLM  TTLRNTTPNF  VRCIIPNHEK  684
Ch NMB  ETAFGSAYKT  KKGMFRTVGQ  LYKESLTKLM  ATLRNTNPNF  VRCIIPNHEK  684
                                            Homology region 7


Ch Sk   TPGAMEHELV  LHQLRCNGVL  EGIRICRKGF  PSRVLYADFK  QRYRVLNASA  731
Ch Sm   RAGKLDAHLV  LEQLRCNGVL  EGIRICRQGF  PNRIVFQEFR  QRYEILAANA  740
Rb Sm   RSGKLDAFLV  LEQLRCNGVL  EGIRICRQGF  PNRIVFQEFR  QRYEILAANA  734
Ch NMB  RAGKLDPHLV  LDQLRCNGVL  EGIRICRQGF  PNRIVFQEFR  QRYEILTPNA  734
                  Homology region 8


Ch Sk   IPEGQFMDSK  KASEKLLGSI  DVDHTQYRFG  HTKVFFKAGL  LGLLEEMRDD  781
Ch Sm   IPKG.FMDGK  QACILMIKAL  ELDPNLYRIG  QSKIFFRTGV  LAHLEEERDL  789
Rb Sm   IPKG.FMDGK  QACILMIKAL  ELDPNLYRIG  QSKIFFRTGV  LAHLEEERDL  783
Ch NMB  IPKG.FMDGK  QACERMIRAL  ELDPNLYRIG  QSKIFFRAGV  LAHLEEERDL  783
```

FIGURE 3 (*Continued*)

by a factor of 3 over this range (Greene *et al.*, 1983). In skeletal muscle myosin, this ionic strength dependence was thought to reflect the ionic nature of the sequence around loop 2. However, the binding of smooth muscle and nonmuscle myosin to actin is not nearly as ionic strength dependent, yet each of these myosins also contains a large number of charged residues in loop 2. In addition, nonmuscle myosin binds considerably more strongly to actin at low ionic strength in the presence of ATP than does smooth muscle myosin, although it has a fairly similar sequence around loop 2 (Sellers *et al.*, 1988). This suggests that the determinants of the affinity of the weakly bound M·ATP and M·ADP·P$_i$ states are not solely based on the net charge of the loop regions.

The sequence of the postulated actin binding sites (Rayment *et al.*, 1993b) for skeletal muscle S-1 are shown in bold in Fig. 3. In general, the other actin binding sites that are postulated to be more stereospecific are considerably more conserved between

```
                    LC 17 Binding Region              LC20 Binding Region
Ch Sk   KLAEIITRTQ ARCRGFLMRV EYRRMVERRE SIFCIQYNVR SFMNVKHWPW   831
Ch Sm   KITDVIIAFQ AQCRGYLARK AFAKRQQQLT AMKVIQRNCA AYLKLRNWQW   839
Rb Sm   KITDVIMAFQ AMCRGYLARK AFAKRQQQLT AMKVIQRNCA AYLKLRNWQW   833
Ch NMB  KITDIIIFFQ AVCRGYLARK AFAKKQQQLS ALKILQRNCA AYLKLRHWQW   833

                ↓  S-2 region ──────▶
Ch Sk   MKLFFKIKPL LKSAESEKEM ANMKEEFEKT KEELAKSEAK RKELEEKMVV   881
Ch Sm   WRLFTKVKPL LQVTRQEEEM QAKDEELQRT KERQQKAEAE LKELEQKHTQ   889
Rb Sm   WRLFTKVKPL LQVTRQEEEM QAKEDELQKI KERQQKAESE LQELQQKHTQ   883
Ch NMB  WRVFTKVKPL LQVTRQEEEL QAKDEELMKV KEKQTKVEAE LEEMERKHQQ   883


Ch Sk   LLQEKNDLQL QVQAEADSLA DAEERCDQLI KTKIQLEAKI KEVTERAEDE   931
Ch Sm   LCEEKNLLQE KLQAETELYA EAEEMRVRLA AKKQELEEIL HEMEARIEEE   939
Rb Sm   LSEEKNLLQE QLQAETELYA EAEEMRVRLA AKKQELEEIL HEMEARLEEE   933
Ch NMB  LLEEKNILAE QLQAETELFA EAEEMRARLA AKKQELEEIL HDLESRVEEE   933


Ch Sk   EEINAELTAK KRKLEDECSE LKKDIDDLEL TLAKVEKEKH ATENKVKNLT   981
Ch Sm   EERSQQLQAE KKKMQQQMLD LEEQLEEEEA ARQKLQLEKV TADGKIKKME   989
Rb Sm   EDRGQQLQAE RKKMAQQMLD LEEQLEEEEA ARQKLQLEKV TAEAKIKKLE   933
Ch NMB  EERNQILQNE KKKMQGHIQD LEEQLDEEEG ARQKLQLEKV STEAKIKKME   933


Ch Sk   EEMAVLDETI AKLTKEKKAL QEAHQQTLDD LQVEEDKVNT LTKAKTKLEQ  1031
Ch Sm   DDILIMEDQN NKLTKERKLL EERVSDLTTN LAEEEEKAKN LTKLNKHES  1039
Rb Sm   DDILVMDDQN NKLSKERKLL EERISDLTTN LAEEEEKAKN LTKLNKHES  1033
Ch NMB  EEILLLEDQN SKFLKEKKLM EDRIAECTSQ LAEEEEKAKN LAKLKNKQEM  1033


Ch Sk   QVDDLEGSLE QEKKLRMDLE RAKRKLEGDL KLAHDSIMDL ENDKQQLDEK  1081
Ch Sm   MISELEVRLK KEEKSRQELE KIKRKLEGES SDLHEQIAEL QAQIAELKAQ  1089
Rb Sm   MISELEVRLK KEEKSRQELE KLKRKMDGEA SDLHEQIADL QAQIAELKMQ  1083
Ch NMB  MITDLEERLK KEEKTRQELE KAKRKLDGET TDLQDQIAEL QAQIEELKIQ  1083


Ch Sk   LKKKDFEISQ IQSKIEDEQA LGMQLQKKIK ELQARIEELE EEIEAERTSR  1131
Ch Sm   LAKKEEELQA ALARLEDETS QKNNALKKIR ELESHISDLQ EDLESEKAAR  1139
Rb Sm   LAKKEEELQA ALARLEDETS QKNNALKKIR ELEGHISDLQ EDLDSERAAR  1133
Ch NMB  LAKKEEELQA ALARGDEEAV QKNNALKVIR ELQAQIAELQ EDLESEKASR  1133
                            Typical 28 residue repeat
Ch Sk   AKAEKHRADL SRELEEISER LEEAGGATAA QIEMNKKREA EFQKMRRDLE  1181
Ch Sm   NKAEKQKRDL SEELEALKTE LEDTLDTTAT QQELRAKREQ EVTVLKRALE  1189
Rb Sm   NKAEKQKRDL GEELEALKTE LEDTLDTTAT QQELRAKREQ EVTVLKKALD  1183
Ch NMB  NKAEKQKRDL SEELEALKTE LEDTLDTTAA QQELRTKREQ EVAELKKAIE  1183


Ch Sk   EATLQHEATA AALRKKHADS TAELGEQIDN LQRVKQKLEK EKSELKMEID  1231
Ch Sm   EETRTHEAQV QEMRQKHTQA VEELTEQLEQ FKRAKANLDK TKQTLEKDNA  1239
Rb Sm   EETRSHEAQV QEMRQKHTQV VEELTEQLEQ FKRAKANLDK TKQTLEKENA  1233
Ch NMB  EETKNHEAQI QEIRQRHATA LEELSEQLEQ AKRFKANLEK NKQGLESDNK  1233
```

FIGURE 3   *(Continued)*

smooth and skeletal muscle myosin than is the site at loop 2 that has already been discussed.

The two heads of smooth muscle HMM can be cross-linked by 1-ethyl-3-(3-dimethylaminopropyl)carbodiimide·HCl (EDC), a zero-length cross-linker, if the HMM is complexed to actin under rigor conditions (Onishi *et al.*, 1989a, b). The amino acids that are cross-linked are Glu 168 on one head to Lys 65 on the adjacent head (Onishi *et al.*, 1990). In the skeletal muscle S-1 structure, Lys 63 (corresponding to Lys 65 of smooth muscle S-1) is located in the β-barrel protuberance at the amino terminus, discussed earlier, whereas Glu 171 (corresponding to Glu 168 of smooth S-1) is located 12 Å apart on the opposite side of the myosin head. In the modeled actin binding reconstruction, the two residues on different S-1 heads appear to be in close proximity (Rayment *et al.*, 1993b).

```
Ch Sk   DLASNMESVS KAKANLEKMC RTLEDQLSEI KTKEEQNQRM INDLNTQRAR 1281
Ch Sm   DLANEIRSLS QAKQDVEHKK KKLEVQLQDL QSKYSDGERV RTELNEKVHK 1289
Rb Sm   DLAGELRVLG QAKQEVEHKK KKLEVQLQEL QSKCSDGERA RAELNDKVHK 1283
Ch NMB  ELACEVKVLQ QVKAESEHKR KKLDAQVQEL TAKVTEGERL RVELAEKANK 1283
```

LMM region ———▶

```
Ch Sk   LQTETGEYSR QAEEKDALIS QLSRGKQGFT QQIEELKRHL EEEIKAKNAL 1331
Ch Sm   LQIEVENVTS LLNEAESKNI KLTKDVATLG SQLQDTQELL QEETRQKLNV 1339
Rb Sm   LQNEVESVTG MLSEAEGKAI KLAKEVASLG SQLQDTQELL QEETRQKLNV 1333
Ch NMB  LQNELDNVSS LLEEAEKKGI KFAKDAASLE SQLQDTQELL QEETRQKLNL 1333


Ch Sk   AHALQSARHD CDLLREQYEE EQEAKGELQR ALSKANSEVA QWRTKYETDA 1381
Ch Sm   TTKLRQLEDD KNSLQEQLDE EVEAKQNLER HISTLTIQLS DSKKKLQ.EF 1388
Rb Sm   STKLRQLEDE RNSLQEQLDE EMEAKQNLER HISTLNIQLS DSKKKLQ.DF 1382
Ch NMB  SSRIRQLEEE KNNLQEQQEE EEEARKNLEK QMLALQAQLA EAKKKVD.DD 1382


Ch Sk   IQRTEELEEA KKKLAQRLQD AEEHVEAVNA KCASLEKTKQ RLQNEVEDLM 1431
Ch Sm   TATVETMEEG KKKLQREIES LTQQFEEKAA SYDKLEKTKN RLQQELDDLV 1438
Rb Sm   ASTVESLEEG KKRFQKEIES LTQQYEEKAA AYDKLEKTKN RLQQELDDLV 1432
Ch NMB  LGTIEGLEEN KKKLLKDMES LSQRLEEKAM AYDKLEKTKN RLQQELDDLM 1432


Ch Sk   VDVERSNAAC AALDKKQKNF DKILAEWKQK YEETQTELEA SQKESRSLST 1481
Ch Sm   VDLDNQRQLV SNLEKKQKKF DQMLAEEKNI SSKYADERDR AEAEAREKET 1488
Rb Sm   VDLDNQRQLV SNLEKKQKKF DQLLAEEKNI SSKYADERDR AEAEAREKET 1482
Ch NMB  VDLDHQRQIV SNLEKKQKKF DQMLAEEKNI SARYAEERDR AEAEAREKET 1482


Ch Sm   KALSLARALE EALEAKEELE RTNKMLKAEM EDLVSSKDDV GKNVHELEKS 1531
Ch Sk   ELFKMKNAYE ESLDHLETLK RENKNLQQEI ADLTEQIAEG GKAVHELEKV 1538
Rb Sm   KALSLARALE EALEAKEELE RTNKMLKAEM EDLVSSKDDV GKNVHELEKS 1532
Ch NMB  KALSLARALE EALEAKEEFE RQNKQLRADM EDLMSSKDDV GKNVHELEKS 1532


Ch Sk   KKHVEQEKSE LQAALEEAEA SLEHEEGKIL RLQLELNQIK SEIDRKIAEK 1581
Ch Sm   KRTLEQQVEE MKTQLEELED ELQAAEDAKL RLEVNMQAMK SQFERDLQAR 1588
Rb Sm   KRALETQMEE MKTQLEELED ELQATEDAKL RLEVNMQALK VQFERDLQAR 1582
Ch NMB  KRTLEQQVEE MRTQLEELED ELQATEDAKL RLEVNMQAMK AQFERDLQAR 1582


Ch Sk   DEEIDQLKRN HLRIVESMQS TLDAEIRSRN EALRLKKKME GDLNEMEIQL 1631
Ch Sm   DEQNEEKRRQ LLKQLHEHET ELEDERKQRA LAAAAKKKLE VDVKDLESQV 1638
Rb Sm   DEQNEEKRRQ LQRQLHEYET ELEDERKQRA LAAAAKKKLE GDLKDLELQA 1632
Ch NMB  DEQNEEKKRM LVKQVRELEA ELEDERNERA LAVAAKKKME MDLKDLEGQI 1632


Ch Sk   SHANRMAAEA QKNLRNTQGT LKDTQIHLDD ALRTQEDLKE QVAMVERRAN 1681
Ch Sm   DSANKAREEA IKQLRKLQAQ MKDYQRDLDD ARAAREEIFA TARENEKKAK 1688
Rb Sm   DSAIKGREEA IKQLLKLQAQ MKDFQRELED ARASRDEIFA TAKENEKKAK 1682
Ch NMB  EAANKARDEA IKQLRNVQAQ MKDYQRKLEE ARASRDEIFA QSKESEKKLK 1682
```

FIGURE 3 (*Continued*)

IV. ALTERNATIVE SPLICING OF SMOOTH MUSCLE HEAVY CHAIN PRE-mRNA

A. Introduction

A single gene encodes the smooth muscle HC and in humans this gene is located on chromosome 16p13.13–13.12 (Deng *et al.*, 1993). The 5′ end of the gene, including the promoter element, has been par-tially characterized in rabbits (Babij *et al.*, 1991; Katoh *et al.*, 1994) and transcription was found to initiate from a single site. The smooth muscle HC gene differs from the nonmuscle HC-A gene (Kawamoto, 1994) in having a canonical TATAAA sequence 26 nucleotides upstream of the putative start site. Similar to the skeletal muscle genes, it has an unusual exon/intron organization at the 5′ end. The first eight contiguous exons are located within a region of at least 70 kb. The

```
Ch Sk   LLQAEVEELR  GALEQTERSR  KVAEQELLDA  TERVQLLHTQ  NTSLINTKKK  1731
Ch Sm   NLEAELIQLQ  EDLAAAERAR  KQADLEKEEM  AEELASANSG  RTSLQDEKRR  1738
Rb Sm   SLEADLMQLQ  EDLAAAERAR  KQADLEKEEL  AEELASSLSG  RNALQDEKRR  1732
Ch NMB  GLEAEILQLQ  EEFAASERAR  RHAEQERDEL  ADEIANSASG  KSALLDEKRR  1732


Ch Sk   LETDIVQIQS  EMEDTIQEAR  NAEEKAKKAI  TDAAMMAEEL  KKEQDTSAHL  1781
Ch Sm   LEARIAQLEE  ELDEEHSNIE  TMSDRMRKAV  QQAEQLNNEL  ATERATAQKN  1788
Rb Sm   LEARIAQLEE  ELEEEQGNME  AMSDRVRKAT  QQAEQLSNEL  ATERSTAQKN  1782
Ch NMB  LEARIAQLEE  ELEEEQSNME  LLNERFRKTT  LQVDTLNSEL  AGERSAAQKS  1782


Ch Sk   ERMKKNMDQT  VKDLQLRLDE  AEQLALKGGK  KQLQKLEARV  RELEGEVDAE  1831
Ch Sm   ENARQQLERQ  NKELRSKLQE  MEGAVKSKFK  STIAALEAKI  ASLEEQLEQE  1838
Rb Sm   ESARQQLERQ  NKELKSKLQE  MEGAVKSKFK  STIAALEAKI  AQLEEQVEQE  1832
Ch NMB  ENARQQLERQ  NKELKAKLQE  LEGSVKSKFK  ATISTLEAKI  AQLEEQLEQE  1832


Ch Sk   QKRSAEAVKG  VRKYERRVKE  LTYQCEEDRK  NILRLQDLVD  KLQMKVKSYK  1881
Ch Sm   AREKQAAAKT  LRQKDKKLKD  ALLQVEDERK  QAEQYKDQAE  KGNLRLKQLK  1888
Rb Sm   AREKQAAAKA  LKQRDKKLKE  MLLQVEDERK  MAEQYKEQAE  KGNAKVKQLK  1882
Ch NMB  AKERAAANKL  VRRTEKKLKE  VFMQVEDERR  HADQYKEQME  KANARMKQLK  1882


Ch Sk   RQAEEAEELS  NVNLSKFRKI  QHELEEAEER  ADIAESQVNK  LRVK......  1925
Ch Sm   RQLEEAEEES  QRINANRRKL  QRELDEATES  NDALGREVAA  LKSKLRRGNE  1938
Rb Sm   RQLEEAEEES  QRINANRRKL  QRELDEATES  NEAMGREVNA  LKSKLRRGNE  1932
Ch NMB  RQLEEAEEEA  TRANASRRKL  QRELDDATEA  NEGLSREVST  LKNRLRRG.G  1931


Ch Sk   ......SREI  HGKKIEEEE.  ..........  ..........  .....       1995
Ch Sm   PVSFAPPRRS  GGRRVIENAT  .DGGEEEIDG  RDGDFNGKAS  E....       1978
Rb Sm   T.SFVPTRRS  GGRRVIENA.  .DGSEEEVDA  RDADFNGTKS  SE...       1971
Ch NMB  PITFSSSRSG  RRQLHIEGAS  LELSDDDAES  KGSDVNEAQP  TPAE.       1975
```
Nonhelical tail region

FIGURE 3   (*Continued*)

exon/intron boundaries are conserved between rabbit smooth HC (exons 3–6) and rat sarcomeric HC (exons 5–7) in the ATP binding area and 25/50-kDa domain. Not surprisingly, there is marked exon conservation in the ATP and actin binding domains (Babij *et al.*, 1991).

A fascinating disruption of human chromosome 16 results in the expression of the carboxyl-terminal sequence of the human smooth muscle HC fused to the β-subunit of the transcription factor, core binding factor (CBFβ). The resulting gene product, CBFβ-smooth muscle HC, is postulated to be a causative agent of acute myeloid leukemia. Four different breakpoints have been reported for the human smooth muscle HC gene. Each of the chimeric transcripts resulting from these different breakpoints maintains the correct reading frame of the terminal 400 amino acids of the smooth muscle HC (Liu *et al.*, 1993).

In a study analyzing the transforming capabilities of the CBFβ/smooth muscle HC fusion protein, deletion analysis of the fused constructs indicated a role in transformation for both the CBFβ sequences and smooth muscle HC sequences (Hajra *et al.*, 1995). Of

note is that the smooth muscle sequences found to be important were those previously shown to participate in formation of multimeric filaments (Hodge *et al.*, 1992). The requirement of the smooth muscle HC sequences for the transforming properties of CBFβ-smooth muscle HC is the first known instance of a muscle-specific protein being indispensable to the function of an oncoprotein.

B. Alternative Splicing of Smooth Muscle Heavy Chain: Carboxyl Terminal

1. Alternative Splicing to Generate 204- and 200-kDa Isoforms

It is now clear from studies analyzing both the smooth muscle HC gene and its products that the pre-mRNA undergoes alternative splicing in at least two different locations. The first is located at the 3′ end of the coding sequence and results in HCs with two different MWs, 204,000 and 200,000 on SDS–PAGE (see Fig. 2) (Nagai *et al.*, 1989; Babij and Periasamy, 1989).

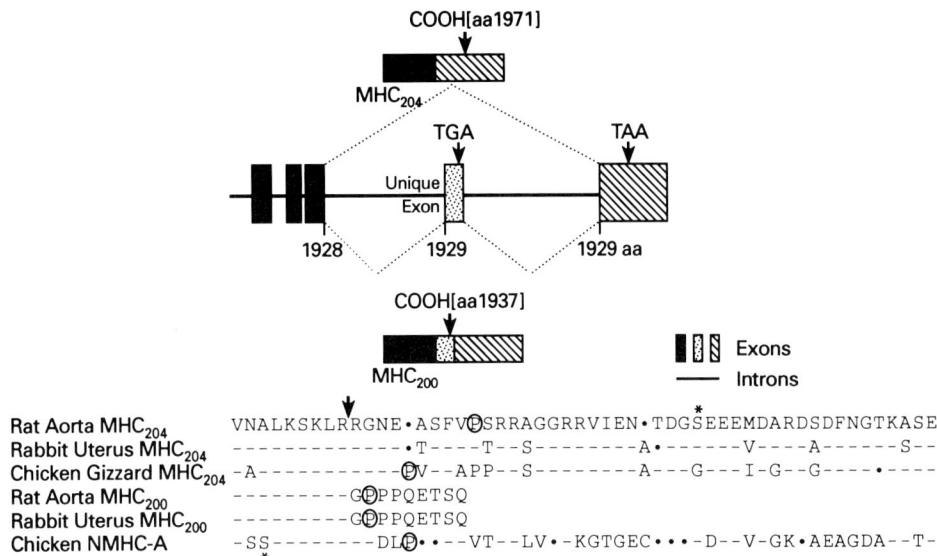

FIGURE 6 Alternative splicing of pre-mRNA encoding smooth muscle myosin to yield HC204 or HC200. The upper part of the figure depicts the splicing of the exons used to generate the two isoforms (modified from Babij and Periasamy, 1989). Below the diagram are the carboxyl-terminal amino acid sequences for smooth muscle HC204 and HC200 from a number of tissues as well as from chicken nonmuscle HC-A. The amino acid sequence numbers correspond to the rabbit smooth muscle numbers in Fig. 3. The arrow between the arginine residues indicates the location of the splice at the amino acid level. The asterisk below the serine residue in the chicken nonmuscle myosin sequence indicates that this residue can be phosphorylated by protein kinase C (Conti et al., 1991). Note that none of the smooth muscle HCs contains this site. The circled P indicates the proline residues that initiate the nonhelical part of the HC tail. The asterisk above the serine in the rat aorta HC204 indicates the site that can be phosphorylated by casein kinase II. Note that the same residue is present in rabbit uterus HC204, but is replaced by a glycine residue in chicken gizzard HC204. HC200 does not contain this site, but chicken nonmuscle HC-A does. The amino acid sequences are from the following sources: rat (Babij and Periasamy, 1989), rabbit (Nagai et al., 1989), chicken gizzard (Yanagisawa et al., 1987), and chicken intestinal epithelial cells (nonmuscle) (Shohet et al., 1989).

As the amino acid sequence in Fig. 6 illustrates, alternative splicing results in two identical isoforms of the smooth muscle HC through amino acid 1928 (the residue numbers in Fig. 6 correspond to the rabbit smooth muscle HC204 sequence shown in Fig. 3). At this point, an exon encoding 9 amino acids is either inserted and terminates the shorter HC200 isoform or the exon is omitted, in which case an exon encoding 43 amino acids (44 for chicken gizzard) terminates the HC204 isoform. The discovery of alternative exons encoding two isoforms that differ only in their carboxyl-terminal amino acid sequence confirmed and extended previous data obtained from analyzing peptide maps of HC204 and HC200, which suggested differences in the primary amino acid sequence (Eddinger and Murphy, 1988). Note that, except for avian gizzard HC204, all isoforms of smooth muscle HC204 contain a serine residue (marked for rat aorta with an asterisk), which can be phosphorylated by casein kinase II in vitro (see Section II). This sequence is not found in the HC200 isoform. Although this site has also been found

to be phosphorylated in intact cells, the function of the phosphorylation is not presently known (Kelley and Adelstein, 1990).

2. Distribution of Isoforms in Tissues

The HC200 and HC204 isoforms are expressed to the same extent at the protein level in a number of smooth muscle cells (Kelley et al., 1991, 1993), but unequal expression of the HC204 and HC200 isoforms does occur in a variety of species and smooth muscle tissues (Cavaillé et al., 1986; Upadhya et al., 1993). For example, cultured rat aorta cells express considerably more HC204 than HC200 (Rovner et al., 1986; Kawamoto and Adelstein, 1987; Babij et al., 1992). Mohammad and Sparrow (1989) demonstrated a ratio of HC204:HC200 of 0.69:1 in both human adult and infant bronchial tissue. This was in contrast to pig airway smooth muscle, where the ratio was found to change during development (Mohammad and Sparrow, 1988). The ratio of the two HC isoforms present in rat uterus was also found to change during pregnancy.

The HC204:HC200 ratio increased from 2.1:1 in nonpregnant rats to 2.6:1 in the pregnant rat. The uterine values contrasted to that found in the rat portal vein, 0.8:1 (Sparrow *et al.*, 1988).

3. Homodimers versus Heterodimers

Because of the dimeric nature of the myosin molecule, the presence of both 204- and 200-kDa isoforms in smooth muscle cells raises the question of whether individual myosin molecules are composed of homodimers of 204- and 200-kDa HCs, heterodimers, or a mixture of homodimers and heterodimers. The observation that some smooth muscles contained equal amounts of HC204 and HC200 was consistent with the possibility that they were composed solely of heterodimers. However, the subsequent finding of nonstoichiometric ratios for the two heavy chains is consistent with the presence of homodimers only, or with a mixture of homodimers and heterodimers.

Evidence that smooth muscle myosin molecules are composed only of homodimers was published by Kelley *et al.*, (1992), who made use of antibodies that were specific for each isoform. As discussed in Section II, immunoaffinity chromatography using isoform-specific antibodies can be used to prepare pure fractions of HC204 and HC200 (see Fig. 2). There was no evidence for heterodimers in either the flow-through or eluted fractions nor were heterodimers found when any remaining proteins bound to the protein G column were electrophoresed in SDS–PAGE. A similar result was obtained when antibodies specific for HC204 were raised against the sequence RRVIENADGSEEEVDAR, and were used to immunoprecipitate a preparation of bovine aorta myosin, similar to that shown in the first lane of Fig. 2. The immunoprecipitated myosin was composed only of dimers of HC204 and the myosin remaining in the supernatant was essentially all HC200.

The ability to purify homodimers of smooth muscle HCs composed solely of HC204 or HC200 permitted a direct comparison of their properties using the *in vitro* motility assay (see Sellers, Chapter 14, this volume). No significant differences were found between the rate at which homodimers of myosin composed solely of HC200 and homodimers of myosin composed solely of HC204 propeled actin filaments. Nor was the rate of actin filament movement found for each of the homodimers significantly different from the rate of movement by unfractionated myosin containing equal amounts of the two isoforms (Kelley *et al.*, 1992).

Using a different, more indirect method, Tsao and Eddinger (1993) concluded that porcine and rabbit smooth muscle myosin molecules from a variety of tissues do contain significant amounts of heterodimer formation between the HC204 and HC200. Their experimental design precluded a direct study of the intact myosin molecule, so they studied the chymotryptic LMM fragment (see Fig. 1). They used $CuCl_2$ oxidation of native LMM to form intramolecular disulfide bonds between the HC fragments. In addition to the expected homodimers of LMM fragments from HC204 or HC200, they found a significant amount of heterodimers. The resolution of these apparently conflicting results with respect to the existence of heterodimers awaits further experimentation.

C. Alternative Splicing of Heavy Chain: Head Region

There is a second region of the pre-mRNA encoding the smooth muscle HC that is also subject to alternative splicing as originally suggested by Hamada *et al.* (1990) (Fig. 7). In this case, splicing of the pre-mRNA results in the insertion or omission of 7 amino acids after amino acid 209, near the ATP binding region (Kelley *et al.*, 1993; Babij, 1993; White *et al.*, 1993). When superimposed on the 3D structure of skeletal muscle HC S-1, this sequence of amino acids is part of what is postulated to be a flexible surface loop that does not appear in the crystal structure and is referred to as loop 1 (see Section III.B).

In general, visceral tissues, such as small bowel, large bowel, and bladder, are highly enriched for, or are composed exclusively of, HC204 and HC200 isoforms that contain the 7-amino-acid insert, whereas the insert is omitted in HCs expressed in vascular tissue. One study comparing avian gizzard smooth muscle myosin with avian aortic smooth muscle myosin permitted a direct functional comparison of two myosins, only one of which has an insert, but both containing equal amounts of HC204 and HC200 (Kelley *et al.*, 1993). Avian gizzard myosin was found to be composed entirely of the inserted isoform based on the reverse transcriptase–polymerase chain reaction (RT–PCR), which showed no evidence for a noninserted isoform. In contrast, none of the HCs that were isolated from avian aorta smooth muscle showed evidence for the presence of the 7-amino-acid insert, either by RT–PCR or by immunoblot analysis using peptide-specific antibodies (Kelley *et al.*, 1993). The gizzard myosin was found to propel actin filaments with a velocity that was 2.5 times greater than the aorta myosin in an *in vitro* motility assay. Since both myosins contained a slightly different complement of LC17 isoforms, the gizzard LC17, which is a single isoform, was exchanged onto the vascular HC. The change in LC17 complement failed to alter the velocity of the aorta myosin in the *in vitro* motility assay, sup-

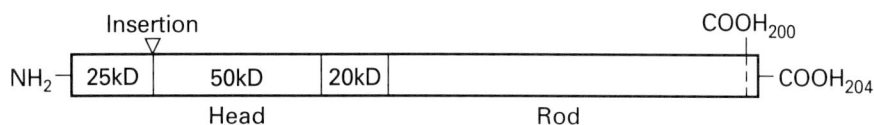

```
                    Insertion                                              COOH₂₀₀
                       ▽                                                      |
         NH₂ ┤  25kD  │    50kD   │  20kD  │                              ┊├─ COOH₂₀₄
                       Head                         Rod
```

```
Chicken Gizzard    HKGKKDTSIT·········QGPSFSYGELEKQLLQA
Rabbit Visceral    ─────────────·········────LA──────────
Rat Visceral       ──────S───·········────A──────────
Chicken Aorta      ──────────·····················──────────
Rabbit Vascular    ──────────·····················──────────
Rat Vascular       ──────S───·····················──────────
Chicken NMHC-B     ───R──HN─PPESPKPVKH─······────R─────
Xenopus NMHC-B     ──────HT─PT────AI───SG─LL─────R─────
```

FIGURE 7 Diagrammatic representation of the smooth muscle HC to show the location of the 7-amino-acid insert. The location of the 7-amino-acid insert in chicken gizzard and rabbit and rat visceral HCs is the same as for the 10- or 16-amino-acid inserts found in chicken and *Xenopus* nonmuscle HC-B, respectively. The amino acids surrounding the inserts are conserved in both smooth muscle and nonmuscle HCs. Whereas the nonmuscle HC contains a consensus sequence for proline-directed kinases (Takahashi *et al.*, 1992; Kelley *et al.*, 1995), this site is not present in the smooth muscle insert. The amino acid sequences are from the following sources: chicken smooth (Kelley *et al.*, 1993; Yanagisawa *et al.*, 1987), rabbit visceral and vascular (Babij, 1993), rat visceral and vascular (White *et al.*, 1993), chicken nonmuscle (Takahashi *et al.*, 1992), and *Xenopus* nonmuscle (Bhatia-Dey *et al.*, 1993).

porting the idea that the 7-amino-acid insertion in the HC was responsible for the increased velocity. The actin-activated MgATPase activity of the inserted myosin was twice that of the noninserted myosin, but this assay was not performed following exchange of the LC17s.

A similar insertion has been found at exactly the same location in the nonmuscle HC-B isoform of chicken brain (Takahashi *et al.*, 1992) and *Xenopus* (Bhatia-Dey *et al.*, 1993). Unlike the smooth muscle insertion, the *Xenopus* insert (see Fig. 7) contains a consensus sequence for cyclin-p34cdc2 kinase and there is evidence that the relevant serine in the sequence becomes phosphorylated during meiosis in *Xenopus* oocytes (Kelley *et al.*, 1995).

As noted, in avian gizzard myosin, all the HC204 and HC200 contain the 7-amino-acid insert. In avian aorta myosin, none of the heavy chains contains the insert. Whether some smooth muscles are composed of mixtures of inserted and noninserted isoforms, as suggested by the experiments of Babij (1993) and White *et al.* (1993), remains to be demonstrated conclusively through the use of peptide-specific antibodies. An alternative explanation for the apparent heterogeneity of some visceral tissues could be the presence of vascular myosin, which lacks the 7-amino-acid insert and which may originate from the blood vessels in the various viscera.

Different smooth muscles vary considerably with respect to whether their resting tone varies in a phasic manner or is maintained in a constant (tonic) manner. The presence of the inserted HC isoform in phasic intestinal smooth muscle and its absence from tonic vascular smooth muscles suggest a possible relation between this property of smooth muscles and the insert. Further study of a large variety of smooth muscles should help to clarify this point.

D. Expression of Smooth Muscle Myosin Fragments

A number of laboratories have made use of the baculovirus expression system to produce active fragments of smooth muscle myosin (Trybus, 1994; Onishi *et al.*, 1995). As expected, expressed HMM containing both LCs was dependent on phosphorylation of the LC20 for actin-activated MgATPase activity, and this activity was enhanced by the addition of tropomyosin (Onishi *et al.*, 1995). Using expressed fragments of the HC and expressed light chains, Trybus (1994) showed that the LC17 is not required to obtain phosphorylation-dependent actin movement in the *in vitro* motility assay. Moreover, a shorter expressed fragment of HMM, with only a 27-nm-long tail, was regulated to a lesser extent by phosphorylation than HMM with a 72-nm-long tail. The tails of the shorter fragments did not fully associate and this resulted in a variable amount of single-headed species, which may have accounted for the apparent lack of regulation. These findings suggest that whereas molecular changes at the active site induced by phosphorylation are independent of the LC17, these changes critically depend on a stable coiled-coil tail that determines how the LC20s interact at the head/rod junction of the HC.

V. KINETICS OF REGULATION

A. Kinetics of ATP Hydrolysis by Actomyosin

When the MgATPase activity of phosphorylated and unphosphorylated HMM was assayed as a function of actin concentration, it was found that the V_{max} of phosphorylated HMM was 25 times larger than that of unphosphorylated HMM, but there was only about a 4-fold difference in the K_m of actin for phosphorylated and unphosphorylated HMM (Table I) (Sellers et al., 1982). Direct measurements of the binding constants of phosphorylated and unphosphorylated HMM to actin in the presence of MgATP confirmed that there was only a 4-fold effect on binding by phosphorylation of the light chains (Sellers et al., 1982). Thus, at 300 μM actin, the actin-activated MgATPase activity of unphosphorylated HMM is only 0.05 s^{-1} compared to 1.75 s^{-1} for that of phosphorylated HMM, yet 60% of the unphosphorylated HMM is bound to actin. These results demonstrate that the unphosphorylated light chain does not act to prevent interaction of myosin with actin, but that phosphorylation exerts its primary effect on some other kinetic step in the hydrolysis of ATP. The fact that the actin-activated MgATPase activity of unphosphorylated HMM was 4% of that of phosphorylated HMM did not reveal the full extent of the phosphorylation-dependent activation. To determine which step in the kinetic cycle was affected by phosphorylation and to determine the full extent of regulation, further transient state kinetic analysis was necessary.

Figure 8 depicts the kinetic pathway for the hydrolysis of MgATP by actomyosin. Most kinetic studies actually employ either HMM or S-1, but the generic term "myosin" will be used in the following discussion. The scheme is similar in detail to that described for skeletal muscle myosin (Adelstein and Eisenberg, 1980; Hibberd and Trentham, 1986; Taylor, 1979). In the absence of actin, one needs to be concerned only with the top portion of the scheme. Myosin actually binds ATP in a two-step process, but only a single step is shown for simplicity, since the first process is very fast compared to the second. Hydrolysis of ATP to ADP·P_i is reversible with an equilibrium constant between 1 and 10 depending on conditions. The products of ATP hydrolysis are released sequentially, with the P_i release preceding ADP release. Marston and Taylor (1980) compared the rate constants for some of these steps using chicken skeletal, cardiac, and smooth muscle myosins under the same conditions. In the absence of actin, there are only small differences between the rate constants of these three myosins.

In the presence of actin, the scheme is necessarily more complex. In the absence of ATP, myosin's affinity for actin is very large and, under most conditions, myosin would be fully complexed with actin. Binding of ATP greatly weakens the affinity of myosin for actin and dissociation of the two proteins rapidly ensues at low protein concentrations. The extent of the ATP-induced dissociation of actin and myosin is a function of the ionic conditions and protein concentration and a rapid equilibrium exists between the two states. Hydrolysis of ATP can occur either by the dissociated myosin or by the actomyosin complex. The net result of ATP binding and hydrolysis is a rapid equilibrium between the four species shown in Fig. 8 in the box marked "weak binding states." The rate of hydrolysis of ATP to give the bound products is fast and is not affected by phosphorylation of HMM (J. R. Sellers, unpublished data).

P_i release also precedes ADP release in the presence of actin. The identification of the rate-limiting step has not been entirely clarified, even for the widely studied skeletal muscle actomyosin system (Hibberd and Trentham, 1986). There are two possibilities that may depend on experimental conditions. One possibility is that P_i release, or an isomerization of AM·ADP·P_i that precedes P_i release, is the rate-limiting step (Stein, 1991). Other studies suggest that, under some conditions, the hydrolysis of ATP may be rate limiting (Tesi et al., 1990). In any case, release of P_i results in an increased affinity of myosin for actin and marks the transition between the weakly and strongly bound states. The affinity of M·ADP for actin is high enough that most of the myosin·ADP complex would be bound to actin. ADP release is fast compared to P_i release and returns myosin to the starting point of the cycle (J. R. Sellers, unpublished data).

TABLE I   Kinetic Parameters of the Interaction of Actin
with Heavy Meromyosin

	K_m (μM)	V_{max} (s^{-1})	At 300 μM actin	
			v (s^{-1})	% bound
Unphosphorylated HMM	130	0.07	0.05	60
Phosphorylated HMM	38	1.89	1.75	90

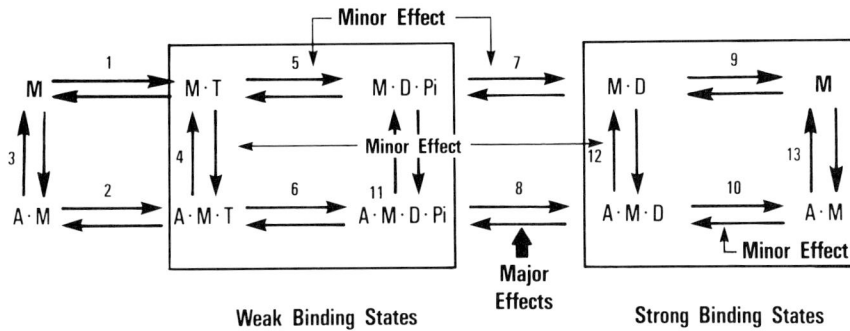

FIGURE 8 Kinetic scheme for the hydrolysis of ATP by smooth muscle actomyosin. M = myosin (or its subfragments); A = actin; T = ATP; D = ADP; P_i = inorganic phosphate.

Measurement of individual rate constants within this cycle demonstrates the kinetic basis for the regulation of actin-activated MgATPase activity of myosin by actin. P_i could potentially be released from either the M·ADP·P_i complex or the AM·ADP·P_i complex. Single turnover experiments of smooth muscle HMM demonstrate that the rate of release of P_i from unphosphorylated HMM occurs at 0.002 s^{-1} in the absence of actin and at an identical rate in the presence of 5 μM actin (Sellers, 1985). Furthermore, the rate of release is not affected by increasing the actin concentration over a large range. This demonstrates that, in the absence of light chain phosphorylation, the regulation of smooth muscle HMM is essentially complete as there is no activation of the rate-limiting step by actin. Phosphorylation must affect the rate of phosphate release by a factor of about 1000 since the V_{max} of the actin-activated MgATPase of phosphorylated HMM under these conditions is about 2 s^{-1} (Sellers et al., 1982). Phosphorylation has only a small effect on the release of P_i in the absence of actin (Sellers, 1985).

Trybus (1989) demonstrated that the same mechanism for regulation operates when intact gizzard myosin is polymerized into filaments. Using a fluorescent ATP analog, formycin triphosphate (FTP), to monitor the dissociation of products, almost identical rate constants were obtained for the rate of P_i release from unphosphorylated myosin filaments, as was obtained using the soluble HMM system described earlier. Furthermore, it was shown that folded, 10S myosin released products even more slowly than unphosphorylated HMM or myosin filaments (Trybus, 1989; Cross et al., 1986, 1988). These experiments dramatically illustrate the importance of single-turnover experiments in the study of myosin-linked regulation, since, as discussed earlier, steady-state measurements may be influenced by a small population of unregulated molecules.

Phosphorylation failed to show any major effect on other steps in the kinetic cycle. Phosphorylation increases the binding constant of M·ADP to actin by a factor of 4–10 (Greene and Sellers, 1987), but does not have a major effect on the release of ADP from AM·ADP (J. R. Sellers, unpublished data). Thus, as shown in Fig. 8, phosphorylation exerts its influence primarily at one step in the kinetic cycle, namely, the release of P_i from an AM·ADP·P_i complex.

Some steady-state kinetic experiments with mammalian smooth muscle myosin came to opposite conclusions about the mechanism of regulation (Wagner and Vu, 1986). These studies showed that phosphorylation had a more profound effect on the binding of actin to myosin than it did on the V_{max} of the MgATPase activity. These studies were typically conducted in the presence of a high concentration of MgCl$_2$ (10 mM) and no single-turnover experiments were performed (Wagner and Vu, 1986). Trybus (1989) found that the MgATPase activity of unphosphorylated myosin filaments, but not phosphorylated myosin filaments, increased in the presence of higher concentrations of Mg^{2+}, which suggested that the high Mg^{2+} concentrations used in the studies with mammalian smooth muscle myosin may have accounted for some of the results.

B. Cooperativity between the Two Heads of Myosin

The two heads of smooth muscle myosin interact cooperatively. Studies with the soluble two-headed HMM fragment clearly demonstrate that, whereas the phosphorylation of the two heads by myosin light chain kinase was random, phosphorylation of both heads was required for the MgATPase activity of either head to be activated by actin (Sellers et al., 1983; Persechini and Hartshorne, 1981). This cooperative activation of the two heads, coupled with the random phosphorylation, meant that at 50% phosphorylation (i.e., 1 mol P_i/mol HMM) there was only 25% of the maximal actin-activated MgATPase activity. Under some conditions, the two heads of a myosin molecule within a filament were phosphorylated in a negatively cooperative manner such that phosphorylation of one head decreased the probability of phosphorylation of the second head of the same molecule. Coupled with the requirement that both heads need to be phosphorylated in order to have activity, this resulted in myosin filaments that were mostly inactive until the level of phosphorylation exceeded 1 mol P_i/mol myosin molecule (Sellers *et al.*, 1983; Persechini and Hartshorne, 1981).

It has been shown that single-headed myosin prepared by proteolytic digestion was not regulated by phosphorylation (Cremo *et al.*, 1995). In this case, one heavy chain of the myosin molecule was full length and complexed with light chains whereas the other heavy chain was cleaved at the S-1/S-2 junction and contained no light chains. The single-headed myosin had an actin-activated MgATPase activity and moved actin filaments regardless of whether the light chain was phosphorylated. The requirement for two heads for regulation is consistent with the fact that S-1 is not regulated even if it contains a phosphorylatable 20-kDa light chain.

VI. PERSPECTIVES

The use of recombinant DNA technology coupled with a new group of assays to quantitate *in vitro* motility and force development between actin filaments and single myosin molecules is increasing the pace with which new information about smooth muscle myosin is being generated. The ability to express active fragments of smooth muscle myosin using the baculovirus expression system and to mutate and delete amino acid residues in the HC and/or LC is providing detailed information about the exact role of each of these polypeptide chains in initiating and regulating smooth muscle contractile activity. A full understanding of this process will require a more exact comprehension of the role of all the contractile proteins present in smooth muscle cells. Many of these are discussed in this volume.

At present, some of the major questions before us with respect to the structure and function of the smooth muscle HC include:

1. How many more smooth muscle HC isoforms remain to be identified? Does each of the HC isoforms have different biological activities, as suggested by the difference between the avian gizzard and avian vascular HC isoforms (Kelley *et al.*, 1993)? How much do these differences in the properties of HC isoforms (and/or LC isoforms) help to explain the substantial differences in the contractile activity between phasic (e.g., intestinal) and tonic (e.g., vascular) smooth muscles?

2. Do the different homodimers formed by the HC play different functional roles? Do they form heterofilaments or do they exist as homofilaments with distinct localizations within different smooth muscle cells?

3. What, if any, is the role of HC phosphorylation which, with the exception of avian gizzard smooth muscle HC204, has been demonstrated to occur both *in vitro* and in intact smooth muscle cells.

4. Finally, the major regulatory elements responsible for initiating and inhibiting smooth muscle HC gene transcription remain to be characterized (for a promising result in *Drosophila*, see Lilly *et al.*, 1995). Defining these regulatory proteins and the relevant DNA elements for vertebrate smooth muscle HC will play a significant role in helping us understand the mechanism underlying smooth muscle cell differentiation.

Acknowledgments

The authors wish to acknowledge Mary Anne Conti, Sachiyo Kawamoto, and Christine A. Kelley for helpful comments on the manuscripts and Catherine S. Magruder for expert editorial assistance. We are grateful to He Jiang for help in preparing some of the figures and would like to thank Ivan Rayment for providing the three-dimensional structural figures of chicken skeletal S-1 that were subsequently labeled.

References

Adelstein, R. S., and Eisenberg, E. (1980). *Annu. Rev. Biochem.* **49,** 921–956.

Babij, P. (1993). *Nucleic Acids Res.* **21,** 1467–1471.

Babij, P., and Periasamy, M. (1989). *J. Mol. Biol.* **210,** 673–679.

Babij, P., Kelly, C., and Periasamy, M. (1991). *Proc. Natl. Acad. Sci. U.S.A.* **88,** 10676–10680.

Babij, P., Kawamoto, S., White, S., Adelstein, R. S., and Periasamy, M. (1992). *Am. J. Physiol.* **262,** C607–C613.

Bhatia-Dey, N., Adelstein, R. S., and Dawid, I. B. (1993). *Proc. Natl. Acad. Sci. U.S.A.* **90,** 2856–2859.

Bonet, A., Mornet, D., Audemard, E., Derancourt, J., Bertrand, R., and Kassab, R. (1987). *J. Biol. Chem.* **262,** 16524–16530.

Cavaillé, F., Janmot, C., Ropert, S., and D'Albis, A. (1986). *Eur. J. Biochem.* **160,** 507–513.

Conti, M. A., Sellers, J. R., Adelstein, R. S., and Elzinga, M. (1991). *Biochemistry* **30,** 966–970.

Cremo, C. R., Sellers, J. R., and Facemyer, K. C. (1995). *J. Biol. Chem.* **270,** 2171–2175.

Cross, R. A., Cross, K. E., and Sobieszek, A. (1986). *EMBO J.* **5,** 2637–2641.

Cross, R. A., Jackson, A. P., Citi, S., Kendrick-Jones, J., and Bagshaw, C. R. (1988). *J. Mol. Biol.* **203,** 173–181.

Deng, Z., Liu, P., Claxton, D. F., Lane, S., Callen, D. F., Collins, F. S., and Siciliano, M. J. (1993). *Genomics* **18**, 156–159.

Eddinger, T. J., and Murphy, R. A. (1988). *Biochemistry* **27**, 3807–3811.

Goldsmith, P., Gierschik, P., Milligan, G., Umson, C. G., Vinitsky, R., Malech, H. L., and Spiegel, A. M. (1987). *J. Biol. Chem.* **262**, 14683–14688.

Greene, L. E., and Sellers, J. R. (1987). *J. Biol. Chem.* **262**, 4177–4181.

Greene, L. E., Sellers, J. R., Eisenberg, E., and Adelstein, R. S. (1983). *Biochemistry* **22**, 530–535.

Hajra, A., Liu, P. P., Wang, Q., Kelley, C. A., Stacy, T., Adelstein, R. S., Speck, N. A., and Collins, F. S. (1995). *Proc. Natl. Acad. Sci. U.S.A.* **92**, 1926–1930.

Hamada, Y., Yanagisawa, M., Katsuragawa, Y., Coleman, J. R., Nagata, S., Matsuda, G., and Masaki, T. (1990). *Biochem. Biophys. Res. Commun.* **170**, 53–58.

Hartshorne, D. J. (1987). *In* "Physiology of the Gastrointestinal Tract" (L. R. Johnson, ed.), 2nd ed., pp. 423–482. Raven Press, New York.

Hibberd, M. G., and Trentham, D. R. (1986). *Annu. Rev. Biophys. Biophys. Chem.* **15**, 119–161.

Hodge, T. P., Cross, R., and Kendrick-Jones, J. (1992). *J. Cell Biol.* **118**, 1085–1095.

Ikebe, M., and Hartshorne, D. J. (1985). *Biochemistry* **24**, 2380–2387.

Katoh, Y., Loukianov, E., Kopras, E., Zilberman, A., and Periasamy, M. (1994). *J. Biol. Chem.* **269**, 30538–30545.

Kawamoto, S. (1994). *J. Biol. Chem.* **269**, 15101–15110.

Kawamoto, S., and Adelstein, R. S. (1987). *J. Biol. Chem.* **262**, 7282–7288.

Kelley, C. A., and Adelstein, R. S. (1990). *J. Biol. Chem.* **265**, 17876–17882.

Kelley, C. A., Kawamoto, S., Conti, M. A., and Adelstein, R. S. (1991). *J. Cell Sci.* **98** (Suppl. 14), 49–54.

Kelley, C. A., Sellers, J. R., Goldsmith, P. K., and Adelstein, R. S. (1992). *J. Biol. Chem.* **267**, 2127–2130.

Kelley, C. A., Takahashi, M., Yu, J. H., and Adelstein, R. S. (1993). *J. Biol. Chem.* **268**, 12848–12854.

Kelley, C. A., Oberman, F., Yisraeli, J. K., and Adelstein, R. S. (1995). *J. Biol. Chem.* **270**, 1395–1401.

Lilly, B., Zhao, B., Ranganayakulu, G., Paterson, B. M., Shulz, R. A., and Olson, E. N. (1995). *Science* **267**, 688–693.

Liu, P., Tarlé, S. A., Hajra, A., Claxton, D. F., Marlton, P., Freedman, M., Siciliano, M. J., and Collins, F. S. (1993). *Science* **261**, 1041–1044.

Maita, T., Yajima, E., Nagata, S., Miyanishi, T., Nakayama, S., and Matsuda, G. (1991). *J. Biochem. (Tokyo)* **110**, 75–87.

Marston, S. B., and Taylor, E. W. (1980). *J. Mol. Biol.* **139**, 573–600.

McLachlan, A. D. (1984). *Annu. Rev. Biophys. Bioeng.* **13**, 167–189.

Mohammad, M. A., and Sparrow, M. P. (1988). *Aust. J. Biol. Sci.* **41**, 409–419.

Mohammad, M. A., and Sparrow, M. P. (1989). *Biochem. J.* **260**, 421–426.

Mornet, D., Bonet, A., Audemard, E., and Bonicel, J. (1989). *J. Muscle Res. Cell Motil.* **10**, 10–24.

Muhlrad, A. (1991). *Biochim. Biophys. Acta Protein Struct. Mol. Enzymol.* **1077**, 308–315.

Murakami, N., Healy-Louie, G., and Elzinga, M. (1990). *J. Biol. Chem.* **265**, 1041–1047.

Nagai, R., Kuro-o, M., Babij, P., and Periasamy, M. (1989). *J. Biol. Chem.* **264**, 9734–9737.

Onishi, H., Maita, T., Matsuda, G., and Fujiwara, K. (1989a). *Biochemistry* **28**, 1898–1904.

Onishi, H., Maita, T., Matsuda, G., and Fujiwara, K. (1989b). *Biochemistry* **28**, 1905–1912.

Onishi, H., Maita, T., Matsuda, G., and Fujiwara, K. (1990). *J. Biol. Chem.* **265**, 19362–19368.

Onishi, H., Maeda, K., Maeda, Y., Inoue, A., and Fujiwara, K. (1995). *Proc. Natl. Acad. Sci. U.S.A.* **92**, 704–708.

Persechini, A., and Hartshorne, D. J. (1981). *Science* **213**, 1383–1385.

Rayment, I., Rypniewski, W. R., Schmidt-Bäse, K., Smith, R., Tomchick, D. R., Benning, M. M., Winkelmann, D. A., Wesenberg, G., and Holden, H. M. (1993a). *Science* **261**, 50–58.

Rayment, I., Holden, H. M., Whittaker, M., Yohn, C. B., Lorenz, M., Holmes, K. C., and Milligan, R. A. (1993b). *Science* **261**, 58–65.

Rovner, A. S., Murphy, R. A., and Owens, G. K. (1986). *J. Biol. Chem.* **261**, 14740–14745.

Sellers, J. R. (1985). *J. Biol. Chem.* **260**, 15815–15819.

Sellers, J. R. (1991). *Curr. Opin. Cell Biol.* **3**, 98–104.

Sellers, J. R., Eisenberg, E., and Adelstein, R. S. (1982). *J. Biol. Chem.* **257**, 13880–13883.

Sellers, J. R., Chock, P. B., and Adelstein, R. S. (1983). *J. Biol. Chem.* **258**, 14181–14188.

Sellers, J. R., Soboeiro, M. S., Faust, K., Bengur, A. R., and Harvey, E. V. (1988). *Biochemistry* **27**, 6977–6982.

Shohet, R. V., Conti, M. A., Kawamoto, S., Preston, Y. A., Brill, D. A., and Adelstein, R. S. (1989). *Proc. Natl. Acad. Sci. U.S.A.* **86**, 7726–7730.

Somlyo, A. P. (1993). *J. Muscle Res. Cell Motil.* **14**, 557–563.

Somlyo, A. P., and Somlyo, A. V. (1994). *Nature (London)* **372**, 231–236.

Sparrow, M. P., Mohammad, M. A., Arner, A., Hellstrand, P., and Rüegg, J. C. (1988). *Pfluegers Arch.* **412**, 624–633.

Spudich, J. A. (1994). *Nature (London)* **372**, 515–518.

Stein, L. A. (1991). *FEBS Lett.* **278**, 131–132.

Stull, J. T., Gallagher, P. J., Herring, B. P., and Kamm, K. E. (1991). *Hypertension* **17**, 723–732.

Takahashi, M., Kawamoto, S., and Adelstein, R. S. (1992). *J. Biol. Chem.* **267**, 17864–17871.

Taylor, E. W. (1979). *CRC Crit. Rev. Biochem.* **6**, 103–164.

Tesi, C., Barman, T., and Travers, F. (1990). *FEBS Lett.* **260**, 229–232.

Trybus, K. M. (1989). *J. Cell Biol.* **109**, 2887–2894.

Trybus, K. M. (1991). *Curr. Opin. Cell Biol.* **3**, 105–111.

Trybus, K. M. (1994). *J. Biol. Chem.* **269**, 20819–20822.

Trybus, K. M., Huiatt, T. W., and Lowey, S. (1982). *Proc. Natl. Acad. Sci. U.S.A.* **79**, 6151–6155.

Tsao, A. E., and Eddinger, T. J. (1993). *Am. J. Physiol.* **264**, H1653–H1662.

Upadhya, A., Samuel, M., Cox, R. H., Bagshaw, R. J., and Chacko, S. (1993). *Hypertension* **21**, 624–631.

Wagner, P. D., and Vu, N. D. (1986). *J. Biol. Chem.* **261**, 7778–7783.

Warrick, H. M., and Spudich, J. A. (1987). *Annu. Rev. Cell Biol.* **3**, 379–421.

Wells, J. A., and Yount, R. G. (1982). *In* "Methods in Enzymology" (D. Frederiksen and L. Cunningham, eds.), vol. 85, pp. 93–116. Academic Press, New York.

White, S., Martin, A. F., and Periasamy, M. (1993). *Am. J. Physiol.* **264**, C1252–C1258.

Yanagisawa, M., Hamada, Y., Katsuragawa, Y., Imamura, M., Mikawa, T., and Masaki, T. (1987). *J. Mol. Biol.* **198**, 143–157.

2

Myosin Light Chains

KATE BÁRÁNY
Department of Physiology and Biophysics
College of Medicine, University of Illinois at Chicago
Chicago, Illinois

MICHAEL BÁRÁNY
Department of Biochemistry
College of Medicine, University of Illinois at Chicago
Chicago, Illinois

I. INTRODUCTION

The myosin molecule contains two types of light chains (LC): regulatory light chains (RLC) and essential light chains (ELC), each myosin head having one RLC and one ELC. In vertebrate smooth muscle myosin, the molecular mass of RLC is 20 kDa (LC20) and that of ELC is 17 kDa (LC17). LC20 is phosphorylatable and its phosphorylation is the key event in regulation of smooth muscle contraction (see Chapters 25–27 this volume), whereas the functional role of LC17 is not known.

In general, all LCs are asymmetric molecules (Stafford and Szent-Györgyi, 1978), which have been localized in the structure of myosin: by the combination of antibody and electron microscopic techniques, RLC or ELC of scallop skeletal myosin (Flicker *et al.*, 1983) and the RLC epitope of chicken pectoralis myosin (Winkelmann *et al.*, 1983) were visualized at the head/rod junction of the myosin molecule. The same site was also suggested from electron microscopic studies for LC20 in gizzard myosin (Craig *et al.*, 1983; Onishi *et al.*, 1983). This morphology was confirmed by Sellers and Harvey (1984), who have demonstrated that LC20 and LC17 are bound to the C-terminal tryptic fragment of gizzard myosin subfragment-1 (S-1). A decade later, X-ray crystallography of chicken pectoralis S-1 provided evidence that RLC is attached to the C-terminal region of the S-1 heavy chain followed by ELC toward the center of the head (Rayment *et al.*, 1993a). It seems likely that the same anatomy holds true for the corresponding regions of smooth muscle S-1.

Vertebrate smooth muscle LCs exist as isoforms with distinct differences in their primary structure (see Figs. 5 and 9). However, the functional significance of the isoforms remains unknown. Most of our knowledge has been accumulated about phosphorylation of LC20, a prerequisite for the activation of the MgATPase activity of smooth muscle myosin by actin and consequently for smooth muscle contraction (Hartshorne, 1987).

In this chapter we review the structure–function relationship of vertebrate smooth muscle myosin LCs. The LCs are the protein cofactors of the smooth muscle contractile machinery, and we will focus on the experiments that investigate the mechanisms of their functional role. We also describe basic methods for the study of the LCs, with special emphasis on the procedures needed for the characterization of the phosphorylated LC20.

II. METHODS

A. Purification of the Light Chains

The key step in LC isolation is the preparation of pure smooth muscle myosin (Hasegawa *et al.*, 1988). This is usually executed by homogenizing the muscle mince with salt solutions (60–100 mM ionic strength) to remove the cytoplasmic proteins and to a certain extent the thin filament proteins. A crude actomyosin is then extracted form the well-washed muscle residue with salt solutions containing ATP and the actin component is removed by ultracentrifugation. The LCs are

OKADAIC ACID

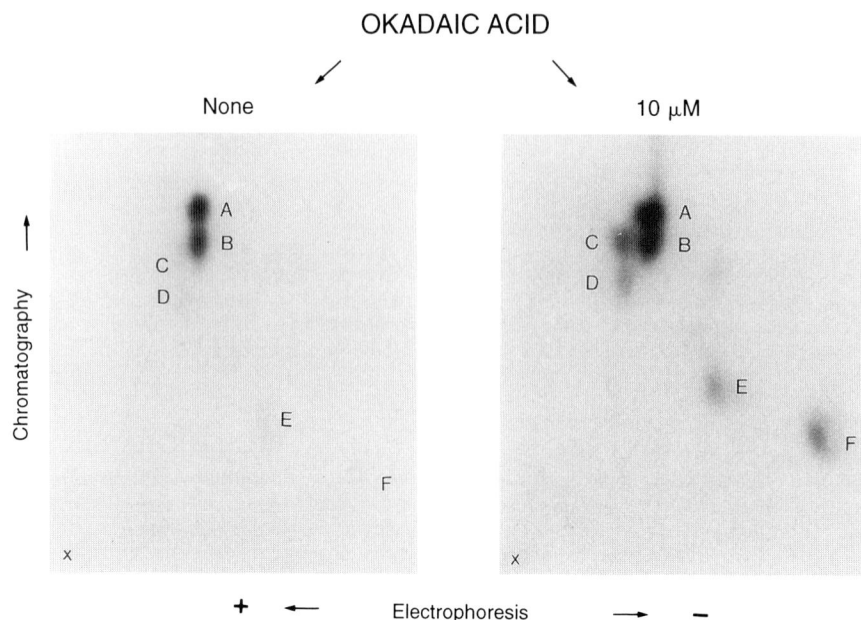

FIGURE 1   Autoradiograms of two-dimensional phosphopeptide maps of LC20 phosphorylated in aortic homogenate. Six phosphopeptides, labeled A through F, are resolved. From
Erdödi *et al.* (1988b, Fig. 2, p. 159).

liberated by denaturing the myosin in 5 *M* guanidine-HCl followed by precipitation of myosin heavy chain (HC) with 50% ethanol (Perrie and Perry, 1970). After removal of ethanol, the LCs are purified by ion-exchange chromatography (Hasegawa *et al.*, 1988; Watanabe *et al.*, 1992).

B. Phosphorylation–Dephosphorylation

The usual conditions for LC20 phosphorylation by myosin light chain kinase (MLCK) or protein kinase C (PKC), and isolation of the phosphorylated light chain, are described by Erdödi *et al.* (1988a). The ^{32}P-labeled LC20 is isolated by 15% polyacrylamide gel electrophoresis (PAGE) in the presence of SDS, localized by autoradiography, and eluted from the gels with 50 m*M* (NH$_4$)HCO$_3$. After centrifugation, the supernatant is dialyzed against water and freeze-dried.

Dephosphorylation of LC20 by myosin light chain phosphatase (MLCP) is described in Chapter 10, this volume. For dephosphorylation of distinct sites of LC20, we refer to Erdödi *et al.* (1989) and Bárány *et al.* (1989).

C. Phosphopeptide Mapping

Phosphopeptide mapping provides qualitative and quantitative information about the sites phosphory-

lated *in vitro* or in the intact muscle. In our laboratory, the ^{32}P-labeled LC20 is isolated by one-dimensional (1D) or two-dimensional (2D) PAGE, digested by trypsin, and the phosphopeptides are separated on Kodak Chromagram 13255 cellulose sheets (Erdödi *et al.*, 1987), resulting in high resolution (Fig. 1). Six peptides, A–F are shown in Fig. 1, and their characterization is described in Section III.D.2. Based on autoradiography, the powder containing the phosphopeptides can be scraped from the cellulose sheets and counted, and the incorporation of [^{32}P]phosphate into peptides of LC20 can be quantified (Rokolya *et al.*, 1991). Alternatively, the phosphopeptides can be eluted from the powder and freeze-dried for phospho-amino acid analysis.

D. Phosphoamino Acid Analysis

The freeze-dried peptide is hydrolyzed in 6 *N* HCl at 110°C for 1 h, HCl is removed by evaporation, and the residue is dissolved in H$_2$O, mixed with unlabeled phosphoamino acid markers, and electrophoresed on the Kodak cellulose sheets (Erdödi *et al.*, 1987). The phosphoamino acids are identified by ninhydrin staining and autoradiography (Fig. 2). The procedure is only qualitative, because of the incomplete hydrolysis of the phosphopeptide and the partial destruction of the liberated phosphoserine and phosphothreonine.

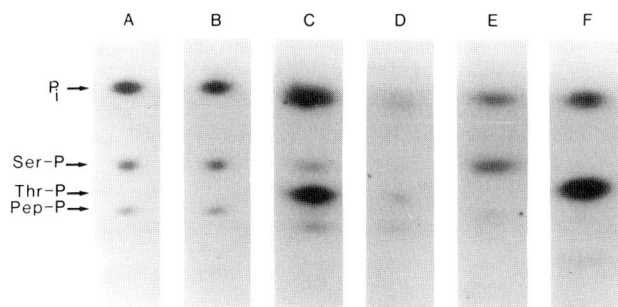

FIGURE 2 Phosphoamino acid analysis of individual phospho-peptides (A–F) from ^{32}P-labeled LC20. P_i, Ser-P, and Thr-P are ^{32}P-labeled inorganic phosphate, serine-phosphate, and threonine-phosphate, respectively, and Pep-P is the unhydrolyzed ^{32}P-labeled phosphate. From Erdödi *et al.* (1988a, Fig. 5, p. 587).

E. Isolation of Phosphopeptides for Sequencing

Affinity chromatography is the first and the major step in phosphopeptide purification. Iminodiacetic acid-epoxy-activated Sepharose 6B (Sigma) complexed with Fe^{3+} absorbs phosphopeptides and phosphoproteins specifically (Muszynska *et al.*, 1986; Murakami *et al.*, 1990). We confirm that this affinity column absorbs only the phosphopeptides from trypsin-digested ^{32}P-labeled LC20, comprising about 15% of the total peptides. We use triethylamine to eluate the ^{32}P-labeled peptides. The eluate is concentrated by the SpeedVac and the phosphopeptides are separated on the Vydac 218TP54 C_{18} column (25 × 0.46 cm) using gradients composed of water/acetonitrile/methanol/trifluoroacetic acid. Pure phosphopeptides, ready for sequencing, can be prepared by this procedure in two or three steps. It is important to know that phosphopeptides are adsorbed onto plasticware and, therefore, only glassware should be used.

F. Antibodies for the Phosphorylated Light Chain

Bennett *et al.* (1988) prepared antibodies that were specific for the phosphorylated form of LC20 using as an antigen a synthetic peptide, containing amino acid residues 6–23 in the LC20 sequence and phosphorylated by MLCK. The antibody obtained did not stain permeabilized smooth muscle cells in the relaxed state. However, when these cells were stimulated to contract, the antibody detected the phosphorylated LC20. The antibody prepared by Sakurada *et al.* (1994) against a shorter phosphorylated synthetic peptide, containing residues 11–22 of LC20, recognized the monophosphorylated LC20 at serine 19 but not the diphosphorylated LC, that is, at Ser-19 and Thr-18.

Such specific antibodies have the potential to follow LC20 phosphorylation in smooth muscle cells under the microscope. Alternatively, they can be used to assay MLCK activity in tissues.

G. Exchange of Light Chains

The structure–function study of LCs requires the exchange of the native LCs in myosin or myofibrils with foreign LCs from a different species, labeled with a fluorescent dye or modified by chemical engineering. Since the LCs are firmly bound to vertebrate smooth muscle myosin, their removal carries the risk of a mild denaturation of the parent myosin molecule. Only LC20 can be taken off from myosin without a harmful effect. LC20 was almost completely exchanged by incubating gizzard myosin (0.5–2 mg/ml) or S-1 (0.25–1 mg/ml) with 10-fold molar excess exogenous LC20 in 0.5 M NaCl, 10 mM EDTA, and 10 mM ATP, pH 7.5, at 40°C. The ATPase activity of myosin or S-1 remained unaltered (Morita *et al.*, 1991). Similar conditions were used for the exchange of recombinant LC20 into smooth muscle myosin (Kamisoyama *et al.*, 1994).

An antibody-affinity column removed more than 80% of the native LC20 from gizzard myosin (Trybus and Lowey, 1988). Trifluoperazine has been introduced for LC20 removal (Trybus *et al.*, 1994); binding of this drug to LC20 appears to weaken its affinity for the HC. Mutant smooth muscle myosin LC20 was also exchanged into chicken pectoralis myofibrils by incubating the fibrils with large excess of exogenous LC20 in 10 mM EDTA at 37°C for 15 min; unbound LC20 was removed by centrifugation of the fibrils at 3000 g (Post *et al.*, 1994).

III. REGULATORY LIGHT CHAIN

A. Isoforms

Multiple forms of LC20 of porcine carotid arterial myosin (Driska *et al.*, 1981), porcine carotid arteries (Ledvora *et al.*, 1983), or rat uterus (Bárány *et al.*, 1987) were observed on 2D gel electrophoretograms. Figure 3 illustrates this for the arteries: four spots of LC20 are resolved by staining (top panel) with estimated isoelectric points of 4.7, 4.8, 4.9, and 5.0 (from left to right), referred to as *Spots* 1, 2, 3, and 4, respectively. The scans (bottom panel) show the changes in the staining intensities of LC20 associated with arterial contraction. The autoradiogram (middle panel) reveals that out of the four spots, three contain phosphory-

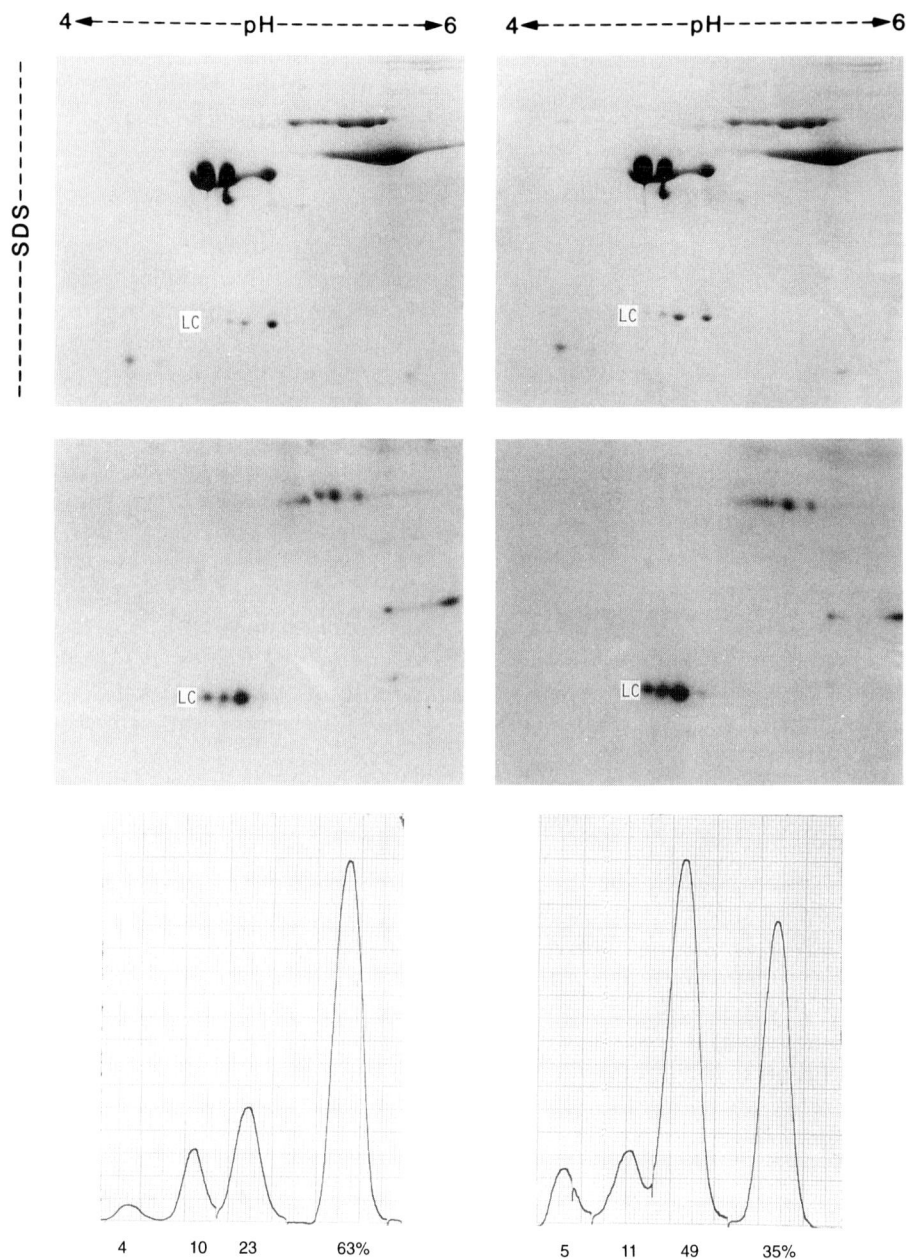

FIGURE 3   The multiple forms of porcine carotid arterial LC20, separated by 2D gel electrophoresis. Upper panel shows the Coomassie blue staining patterns of the arterial proteins (trichloroacetic acid-insoluble residues), middle panel shows the corresponding autoradiograms, and bottom panel shows the densitometric scans of LC20. Left: ^{32}P-labeled arterial muscle was frozen at rest. Right: ^{32}P-labeled muscle frozen 30 sec after 100 mM K$^+$ challenge. LC, phosphorylatable myosin light chain. From Bárány *et al.* (1985b, Fig. 1, p. 204).

lated LC20 and only the fourth most basic spot is nonradioactive. Diverse experiments demonstrated that the four spots of smooth muscle LC20 are not artifacts (Gagelmann *et al.*, 1984; Mougios and Bárány, 1986; Bárány *et al.*, 1987; Csabina *et al.*, 1987).

The experiments of Mougios and Bárány (1986) supported the idea that the multiple LC20 spots origi-

nate from different isoforms. They showed that in fully dephosphorylated porcine arterial muscle, the presence of two LC20 spots persists at proportions of 15 and 85%. The minor and major LC20 isoforms had distinct tryptic peptide maps. Both nonphosphorylated isoforms were recognized by LC20-specific antisera (Erdödi *et al.*, 1987). Furthermore, the antisera reacted with all four spots of LC20 from porcine carotid media (Erdödi et al., 1987; Gaylinn et al., 1989) or rat uterus (Bárány et al., 1987), separated on 2D gels. Both the minor and major isoforms were simultaneously phosphorylated upon contraction of arterial smooth muscle (Erdödi et al., 1987), in cultured rat aortic smooth muscle cells in response to activating agents (Monical et al., 1993), or in aorta actomyosin through the endogenous protein kinases (Mougios and Bárány, 1986; Erdödi et al., 1988a). The isoform concept is further supported by genetic evidence. Inoue et al. (1989) isolated a cDNA clone for the minor LC20 isoform of chicken gizzard myosin with a deduced amino acid sequence that was different in 10 amino acid residues from that of the major LC20 isoform of gizzard myosin. The in vitro transcription/translation product from the cDNA comigrated with the minor isoform of chicken gizzard LC20 in 2D gel electrophoresis, it could be associated with native chicken gizzard myosin, and it was also rapidly phosphorylated by MLCK.

Based on the existence of two nonphosphorylated isoforms that can be mono- or diphosphorylated (Bárány et al., 1985a; Erdödi et al., 1987), the four spots shown in Fig. 3 are explained by the scheme of Fig. 4. As it appears, Spot 2 contains both diphosphorylated and nonphosphorylated isoforms, which is the reason for the four stained and three radioactive spots.

Our laboratory has also investigated the LC20 isoforms of gizzard from chicken, turkey, duck, and goose. Completely dephosphorylated gizzard LC exhibited three spots on 2D gels, with staining intensity distributions of 6–88–6%. The isoelectric point of the major isoform was ~5. Comigration studies of gizzard with porcine aortic LC20 showed that the major isoforms were the same, but the minor ones were different.

B. Amino Acid Sequence of the Isoforms

Zavodny et al. (1990) have isolated two series of cDNAs from a chicken gizzard cDNA library encoding

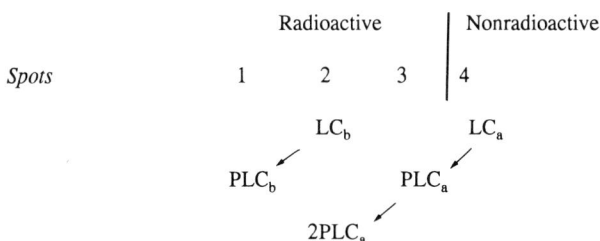

FIGURE 4   Scheme for the explanation of four stained and three radioactive spots on 2D electrophoretogram. LC_a is the major and LC_b is the minor LC20 isoform, and PLC and 2PLC are the mono- and diphosphorylated LC20.

two isoforms of LC20. The amino acid sequence of the isoforms, deduced from the nucleotide sequence, is shown in Fig. 5 (upper two rows) and it is compared with the sequence of LC20 of different smooth muscles (lower three rows). One of the LC20 isoforms is expressed only in adult smooth muscle (LC20-A), whereas the other isoform is expressed in both the smooth and nonmuscle chicken tissues and is designated as cellular RLC (CceRLC). The nucleotide-derived amino acid sequence of LC20-A is the same as that determined by direct chemical sequencing (Maita et al., 1981; Pearson et al., 1984) and it apparently corresponds to the major LC20 isoform. The deduced amino acid sequence of CceRLC differs in 11 residues from that of LC20-A; most substitutions are conservative, but two of them, His-118 → Gln-118 and His-121 → Tyr-121 (LC20-A → CceRLC), should make CceRLC more acidic relative to the major isoform, as this is indeed found for the minor LC20 isoform by 2D gel electrophoresis (Mougios and Bárány, 1986). Furthermore, the difference in 11 amino acids between CceRLC and LC20-A explains the difference in the tryptic peptide maps between the minor and major isoforms of LC20 (Mougios and Bárány, 1986) and, thus, it appears that CceRLC corresponds to the minor LC20 isoform.

Figure 5 shows only a single conservative substitution (K-4 → R-4) between CceRLC and chicken embryonic smooth muscle (LC20-B1), the sequence of which was deduced by Inoue et al. (1989). The overall homology between CceRLC and RamRLC (rat aortic muscle) is 95%, and there is also a close similarity between LC20-A and HsmRLC (human smooth muscle). In contrast, the deduced sequences of RLCs from chicken cardiac and fast-twitch skeletal muscles differ greatly from those of smooth muscles (Fig. 2 in Inoue et al., 1989). Watanabe et al. (1992) sequenced chemically the LC20 of porcine aorta myosin. There was only one amino acid substitution compared with chicken gizzard and two substitutions compared with human umbilical artery smooth muscle LC20. It is noteworthy that the substitutions never take place at positions 18 and 19, the phosphorylation sites for MLCK, or at positions 1, 2, and 9, the phosphorylation sites for PKC.

C. What Is the Functional Significance of LC20 Isoforms?

The finding that the nonmuscle LC20 content in porcine carotid artery (16 ± 3% of the total LC20) matches that of nonmuscle myosin HC content (14 ± 2%) raised the question of a specific function of nonmuscle myosin isoforms (Gaylinn et al., 1989). On the other hand, the proportion of nonmuscle LC20 to

```
                              10           20           30           40
CceRLC   M-S-S-K-A-K-T-K-T-T-K-K-R-P-Q-R-A-T-S-N-V-F-A-M-F-D-Q-S-Q-I-Q-E-F-K-E-A-F-N-M-I-
LC20-A   . . . . R . . A . . . . . . . . . . . . . . . . . . . . . . . . . . . . . . . .-
LC20-B1  . . . . R . . . . . . . . . . . . . . . . . . . . . . . . . . . . . . . . . . .-
RamRLC   . . . . R . . . . . . . . . . . . . . . . . . . . . . . . . . . . . . . . . . .-
HsmRLC   . . . . R . . A . . . . . . . . . . . . . . . . . . . . . . . . . . . . . . . .-

                              50           60           70           80
CceRLC   D-Q-N-R-D-G-F-I-D-K-E-D-L-H-D-M-L-A-S-L-G-K-N-P-T-D-E-Y-L-D-A-M-M-N-E-A-P-G-P-I-
LC20-A   . . . . . . . . . . . . . . . . . . . M . . . . . . . . E G . . S . . . . . .-
LC20-B1  . . . . . . . . . . . . . . . . . . . . . . . . . . . . . . . . . . . . . . .-
RamRLC   . . . . . . . . . . . . . . . . . . . M . . . . . . . . . . . . . . . . . . .-
HsmRLC   . . . . . . . . . . . . . . . . . . . . . . . . . . . . E G . . S . . . . . Y-

                              90          100          110          120
CceRLC   N-F-T-M-F-L-T-M-F-G-E-K-L-N-G-T-D-P-E-D-V-I-R-N-A-F-A-C-F-D-E-E-A-T-G-F-I-Q-E-D-
LC20-A   . . . . . . . . . . . . . . . . . . . . . . . . . . . . . . . . S . . . H . .-
LC20-B1  . . . . . . . . . . . . . . . . . . . . . . . . . . . . . . . . . . . . . . .-
RamRLC   . . . . . . . . . . . . . . . . . . . . . . . . . . . . . . I . T . . . .-
HsmRLC   . . . . . . . . . . . . . . . . . . . . . . . . . . . . S S . . . H . .-

                             130          140          150          160
CceRLC   Y-L-R-E-L-L-T-T-M-G-D-R-F-T-D-E-E-V-D-E-L-Y-R-E-A-P-I-D-K-K-G-N-F-N-Y-I-E-F-T-R-
LC20-A   H . . . . . . . . . . . . . . . . . M . . . . . . . . . . . . . V . . . .-
LC20-B1  . . . . . . . . . . . . . . . . . . . . . . . . . . . . . . . . . . . .-
RamRLC   . . . . . . . . . . . . . . . . . . . . . . . . . . . . . . . . . .-
HsmRLC   H . . . . . . . . . . . . . . . . . M . . . . . . . . . . . . . V . . . .-

                             170
CceRLC   I-L-K-H-G-A-K-D-K-D-D
LC20-A   . . . . . . . . . . .
LC20-B1  . . . . . . . . . . .
RamRLC   . . . . . . . . . . .
HsmRLC   . . . . . . . . . . .
```

FIGURE 5 Comparison of amino acid sequence for myosin regulatory light chain proteins in different species. Cce, chicken cellular; LC20-A, chicken adult smooth muscle; LC20-B1, chicken embryonic smooth muscle; Ram, rat aortic muscle; Hsm, putative human smooth muscle. Amino acid residues are numbered starting at the Ser residue next to the initiator Met. A dot indicates identity of an amino acid residue with that in the top row. Reproduced with permission from Zavodny *et al.* (1990, Fig. 3, p. 937), Copyright 1994, American Heart Association.

muscle LC20 isoforms did not correlate with the proportion of nonmuscle HC to muscle HC isoforms in a variety of smooth muscles (Packer, 1993). Importantly, no evidence was found for differential phosphorylation changes of smooth muscle and nonmuscle LC20 isoforms in response to activating or relaxing agents as expected from a distinct nonmuscle LC20 with a different cellular function and/or anatomical localization (Monical *et al.*, 1993, and results of this laboratory). As a matter of fact, all evidence suggests that the LC20 isoforms are functionally interchangeable, similarly to the smooth muscle actin isoforms (Drew *et al.*, 1991). It appears that in the same smooth muscle cell, several isomyosins exist through the combination of the different HCs and LCs. It is known for some time that the cardiac myosin isoforms, V1 and V3, are simultaneously expressed within a single heart cell (Samuels *et al.*, 1983). The smooth muscle actin, HC and LC isoforms may be expressed as a consequence of a coordinated system of gene regulation but function interchangeably (Gaylinn *et al.*, 1989).

D. Phosphorylation

1. Site

The phosphorylation sites of LC20 are located at the N terminus with the following amino acid sequence (Pearson *et al.*, 1984): acetyl-Ser-Ser-Lys-Arg-Ala-Lys-Ala-Lys-Thr-Thr-Lys-Lys-Arg-Pro-Gln-Arg-Ala-Thr-Ser-Asn-Val-Phe-Ala- (Fig. 5).

Ser-19 is the residue that is phosphorylated by MLCK both *in vitro* (Pearson *et al.*, 1984) and in intact muscle (Colburn *et al.*, 1988; Bárány and Bárány, 1993). At relatively high MLCK concentrations a second site on LC20 is phosphorylated, shown to be Thr-18 (Ikebe *et al.*, 1986). Phosphorylation of Thr-18, in addition to Ser-19, in LC20 of gizzard myosin increases the actin-activated MgATPase activity (Ikebe and Hartshorne, 1985), but phosphorylation at this site had no effect on the velocity of smooth muscle myosin-coated beads in an *in vitro* motility assay (Umemoto *et al.*, 1989). Three residues are phosphorylated by PKC: Ser-1, Ser-2, and Thr-9 (Bengur *et al.*, 1987; Ikebe *et al.*, 1987). The sequential phosphorylation of turkey gizzard heavy meromyosin (HMM) by MLCK and PKC results in a twofold decrease in the actin-activated MgATPase activity of the HMM (Nishikawa *et al.*, 1984). Thr-9 is the major phosphoamino acid resulting from phosphorylation of HMM by PKC.

2. Patterns Characteristic for MLCK and PKC Phosphorylation

Phosphopeptide maps differentiate MLCK-catalyzed LC phosphorylation from that catalyzed by PKC. The two-dimensional map of purified aortic LC20 phosphorylated by MLCK exhibits four peptides: A, B, both containing serine residue (MLCK/Ser), corresponding to the Ser-19 site, and C, D, both containing threonine (MLCK/Thr), most likely corresponding to the Thr-18 site. In addition, peptides C and D also contain trace amounts of serine. When LC20 is phosphorylated by PKC, the map exhibits two peptides: E, containing serine (PCK/Ser), corresponding to the Ser-1 or Ser-2 site, and F, containing threonine (PKC/Thr), corresponding to the Thr-9 site (Figs. 6 and 2). The amino acid sequence of peptides A and B, both containing 18 residues including the phosphorylated Ser-19, was determined (Bárány and Bárány, 1993). Peptide B differed from peptide A by containing Leu in position 12 instead of Gln, and containing Gln in position 16 instead of Glu. The sequence of peptides A and B was very similar to the sequence of residues 17 through 34 of gizzard LC20 (Fig. 4).

Two-dimensional phosphopeptide maps of LC20 from ^{32}P-labeled muscles stimulated with different agents allows one to differentiate between the involvement of MLCK and PKC: the phosphopeptide map of LC20 isolated from K$^+$-stimulated arteries shows predominantly MLCK/Ser peptides (Erdödi *et al.*, 1987), whereas in phorboldibutyrate (PDBu)-treated muscle

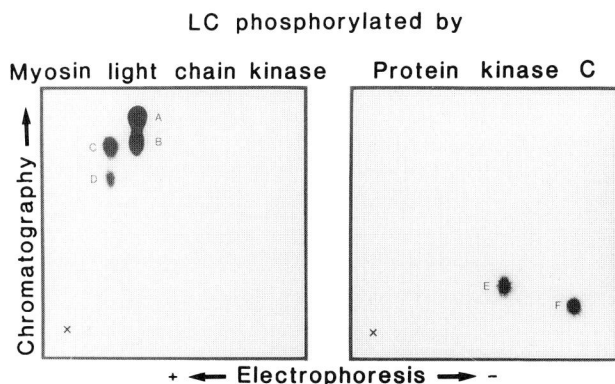

FIGURE 7 Autoradiogram of the phosphopeptide map of LC20 phosphorylated first by PKC, then by MLCK. From Erdödi *et al.* (1988a, Fig. 7, p. 589).

the PKC/Ser peptide is greatly increased and the PKC/Thr peptide is also revealed (Bárány *et al.*, 1990). Upon addition of K$^+$ to a PDBu-contracted artery, the distribution of phosphopeptides shifts toward the MLCK-catalyzed pattern (Bárány *et al.*, 1990). LC20 phosphorylation in smooth muscle homogenate is the *in vitro* correlate of the phosphorylation events taking place in intact muscle. LC20 phosphorylation in aortic homogenate is predominantly due to MLCK, whereas addition of the phosphatase inhibitor okadaic acid (5–10 µM) elicits increased LC20 phosphorylation involving both MLCK and PKC (Fig. 1). Since, in intact muscle, LC20 phosphorylation by PKC is restricted, it is reasonable to assume that a phosphatase inhibits PKC.

Phosphorylation of the distinct sites of LC20 by MLCK and PKC is interrelated. Namely, pre-phosphorylation of LC20 by PKC changes the pattern of MLCK-catalyzed phosphorylation: in isolated LC20, MLCK preferentially phosphorylates MLCK/Ser peptides over those of MLCK/Thr, however, pre-phosphorylation of PKC/Ser and PKC/Thr peptides induces additional phosphorylation of the MLCK/Thr peptides, whereas the relative phosphorylation of the MLCK/Ser peptides decreases (Fig. 7). The interrelation of the MLCK and PKC phosphorylation sites is manifested also in intact muscle. The decrease in LC20 phosphorylation attributable to MLCK in a K$^+$-stimulated artery can be counteracted by addition of PDBu to the K$^+$-stimulated artery through the increase of MLCK/Thr, PKC/Ser, and PKC/Thr peptides (Rokolya *et al.*, 1991).

Proteolysis of PKC resulted in a catalytic fragment, called PKM, that phosphorylated both sets of sites recognized by PKC and MLCK under *in vitro* assay conditions (Nakabayashi *et al.*, 1991).

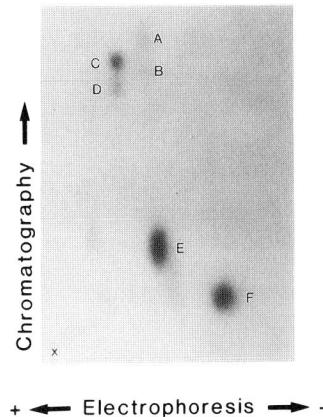

FIGURE 6 Autoradiograms of phosphopeptide maps of isolated LC20 phosphorylated by either MLCK or PKC. From Erdödi *et al.* (1988a, Fig. 4, p. 587).

a. Phosphorylation *in Vitro* versus *in Situ* There are marked differences in LC20 phosphorylation between *in vitro* systems and the intact muscle: (1) The extent of phosphorylation in isolated LC20 or LC20 of actomyosin may be 2 mol P/mol LC20 or higher (Ikebe *et al.*, 1986; Mougios and Bárány, 1986; Erdödi *et al.*, 1988a), whereas in the muscle it is about 1 mol/mol (Mougios and Bárány, 1986) or lower (Csabina *et al.*, 1987). (2) The molar ratio of Ser-P to Thr-P is only 1.3 in LC20 of actomyosin but as high as 11 in K$^+$-stimulated arteries (Mougios and Bárány, 1986), or 6 in oxytocin-stimulated uteri (Csabina *et al.*, 1987). Several factors may influence LC20 phosphorylation: (a) the conformation of the protein (which determines the accessibility of protein kinases and phosphatases to the phosphorylatable residues) is most likely different in the muscle from that in the test tube; (b) the concentration of the cofactors, for example, calmodulin and Ca^{2+}, is variable *in vitro* but regulated in the muscle; and (c) MLCP is virtually absent in the isolated systems, but it is a major factor in the muscle.

3. Region for MLCK Phosphorylation

The importance of Ser-19 phosphorylation for actin-activated smooth muscle myosin MgATPase activity and for contraction of smooth muscle stimulated research to find out which amino acids, surrounding the serine, are required for the phosphorylation. Using synthetic peptide analogs of the native phosphorylation site comprising residues from Lys-11 to Ala-23, it was shown that Arg-16 had a strong influence on the kinetics of peptide phosphorylation (Kemp and Pearson, 1985). The location of Arg-16 in relation to Ser-19, as well as the distance between Arg-13 and Arg-16, was found to be important. Placement of Arg-16 at position 15 caused a complete switch in specificity from the natural Ser-19 phosphorylation site to Thr-18. Increasing the number of alanine residues between Arg-13 and Arg-16 in the model peptide also influenced the kinetics and site specificity of peptide phosphorylation. On the carboxyl side of Ser-19, Val-21 and Phe-22 influenced the V_{max} of peptide phosphorylation, whereas Ala-23 was found not to be essential (Pearson *et al.*, 1986).

From proteolysis of LC20 in turkey gizzard HMM it was also concluded that the four-residue segment, Arg^{13}-Pro^{14}-Glu^{15}-Arg^{16}, is essential for the activation of acto-HMM ATPase activity by the MLCK-induced phosphorylation of Ser-19 (Ikebe and Morita, 1991). Site-directed and chimeric mutations of LC20 suggested that a region distant from the consensus phosphorylation sequence in LC20 is also an important substrate determinant for MLCK (Zhi *et al.*, 1994).

Mutant LC20 with cysteine at position 18, instead of threonine, was prepared by genetic engineering and the cysteine was reacted with the fluorophore acrylodan (Post *et al.*, 1994). When this labeled LC20 was exchanged into turkey gizzard myosin it exhibited nearly control levels of the rate of phosphorylation, K$^+$- and EDTA-activated ATPase activity, and *in vitro* motility, indicating that threonine at position 18 is not essential for the native properties of LC20. Furthermore, the fluorophore acrylodan responded to phosphorylation of Ser-19 with a fluorescence emission quenching and a 28-nm red shift, providing an optical biosensor for monitoring the phosphorylation of LC20.

4. Electrostatic Effect

Mutagenesis of the Ser-19 phosphorylation site was used to elucidate the effect of the negative charge on actomyosin ATPase activity, conformation, and filament formation of gizzard myosin (Kamisoyama *et al.*, 1994). Three mutant LCs were produced; two of them contained Ser-19 substituted by either Asp or Ala and the third contained Asp substituted for both Thr-18 and Ser-19. These mutant LC20s were introduced into myosin or HMM by subunit exchange. The incorporation of the Asp mutant LC20 activated actomyosin ATPase activity but the extent of activation was much lower than that obtained by Ser-19 phosphorylation in the wild-type LC20. The incorporation of the Asp/Asp mutant further activated actomyosin ATPase activity, but the activity still remained below that produced by the phosphorylated LC20. However, the incorporation of Ala mutant did not activate the ATPase, that is, it behaved the same way as dephosphorylated LC20. These mutations have also affected the 10S–6S conformational transition of myosin (see Chapter 3, this volume). The results indicate that a single negative charge at position 19 or two single negative charges at positions 18 and 19 produce an activation of actomyosin ATPase, but do not match the effect of the double negative charge that phosphoryl serine carries at the strategic position. The electrostatic mechanism of phosphorylation was also explored by Sweeney *et al.* (1994), who mutated human smooth muscle LC20 in the region from Lys-11 to Ser-19 and found that decreasing the positively charged amino acid residues in the region mimics aspects of phosphorylation.

5. Regulatory Mechanism

A novel approach was used by Hasegawa *et al.* (1990) to understand the regulatory mechanism of the phosphorylation; these authors focused on the HC rather than on the LC. On the basis of the observation of Rees and Frederiksen (1981) that mild trypsin treatment increases the actin-activated MgATPase activity of porcine aorta myosin, Hasegawa *et al.* subjected porcine aorta smooth muscle myosin to limited proteolysis by *Staphylococcus aureus* protease (V8-protease) in 30 mM KCl, so that the myosin was in the filamentous form. In the presence of F-actin, the junction of the head/rod in the HC was split specifically. Under these conditions, both phosphorylated and unphosphorylated LC20 were resistant to proteolysis. The actin-activated MgATPase activity of the phosphorylated myosin was not affected by the cleavage of the head/rod junction. However, with unphosphorylated myosin the actin-activated MgATPase activity increased as a function of proteolysis time and reached the level of the ATPase activity of the phosphorylated myosin when the digestion became complete. On the basis of these results, Hasegawa *et al.* (1990) suggested that unphosphorylated LC20 suppresses the actin-activated MgATPase activity of smooth muscle myosin and the role of LC20 phosphorylation is to release this suppression. This idea is well supported by the work of Uyeda and Spudich (1993), who isolated recombinant *Dictyostelium* myosin lacking the RLC binding site, and thus did not contain the 18-kDa RLC. The actin-activated MgATPase activity of this myosin was about eight times higher than that of the wild-type myosin containing unphosphorylated LC. This means that not only LC phosphorylation but also removal of the "inhibitory" region from the myosin can increase its ATPase.

6. Thiophosphorylation

Adenosine 5'-O-(3-thiotriphosphate) (ATPγS) is a substrate for MLCK and the resultant thiophosphorylated myosin or LC20 is resistant to hydrolysis by phosphatase (Sherry *et al.*, 1978). Since the mobility of thiophosphorylated LC20 on urea–PAGE is the same as that of phosphorylated LC20, the use of ^{35}S-labeled ATPγS is required to quantify the extent of the reaction (Cassidy *et al.*, 1979). Thiophosphorylation of myosin played a major role in establishing LC20 phosphorylation as a key factor for tension development in smooth muscle (Walsh *et al.*, 1982; Hellstrand and Arner, 1985; Horiuti *et al.*, 1989; Kenney *et al.*, 1990). For details we refer to the review of Hartshorne (1987).

IV. ESSENTIAL LIGHT CHAIN

A. Isoforms

Two isoforms of LC17 have been separated by 2D gel electrophoresis in human uterus (Cavaillé *et al.*, 1986) and porcine aorta (Helper *et al.*, 1988), and were named LC17a and LC17b, "a" for the acidic and "b" for

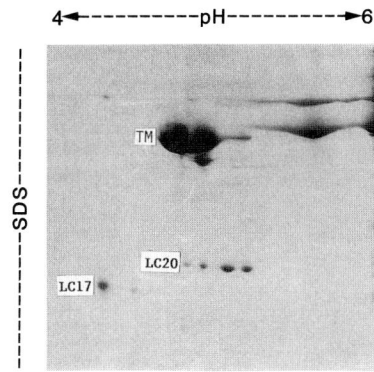

FIGURE 8 Coomassie blue-stained 2D electrophoretogram of porcine carotid arterial proteins. LC17, LC17a, and LC17b isoforms; LC20, multiple forms of LC20; TM, tropomyosin.

the basic isoform, respectively. Figure 8 illustrates the two LC17 isoforms from porcine carotid artery; the percentage distribution of LC17a and LC17b is 85 and 15, and the isoelectric points are approximately 4.2 and 4.3, respectively. The LC17 isoforms were also separated on urea–PAGE, LC17a being the slower and LC17b the faster migrating isoform (Hasegawa *et al.*, 1988). Based on the amino acid sequence, the slower LC17a was named LC17nm (nonmuscle type) and the faster LC17b as LC17gi (gizzard type) (Hasegawa *et al.*, 1992). In this nomenclature, the "nonmuscle" LC17a is the more basic and the "gizzard type" smooth muscle LC17b is the more acidic isoform (Kelley *et al.*, 1993), and therefore the LC17b of urea–PAGE corresponds to the LC17a of the 2D gel and the LC17a of the former corresponds to the LC17b of the latter. In this review, we shall refer to LC17a as the more acidic and LC17b as the more basic isoform, and will make the appropriate corrections in the data presentation of the reverse designations.

The distribution of the LC17a and LC17b in different types of smooth muscles is shown in Table I. In many cases, LC17a is the major isoform and it is the only isoform in most mammalian gastrointestinal muscles and chicken gizzard. In uterus, LC17a gradually increases during pregnancy, whereas in the nonpregnant state LC17b is the predominant isoform. There is some variation in the isoform distribution when the same tissue is analyzed by different laboratories or even by the same laboratory.

B. Amino Acid Sequence of the Isoforms

Hasegawa *et al.* (1992) sequenced LC17a and LC17b of porcine aorta myosin (Fig. 9), both containing 150 amino acid residues with acetylated N-terminal cysteines. Only the C-terminal region differs between the

TABLE I   Distribution of LC17 Isoforms
in Smooth Muscles

Muscle	LC17a (%)	LC17b (%)	Reference[a]
Rat aorta	34	66	(1)
Rat aorta	36	64	(2)
Rat aorta	40	60	(3)
Porcine aorta	60	40	(1)
Porcine aorta	57	43	(2)
Rabbit aorta	69	31	(1)
Rabbit aorta	44	56	(3)
Bovine pulmonary artery	80	20	(2)
Porcine carotid artery	84	16	(2)
Rat portal vein	81	19	(3)
Bovine trachea	60	40	(2)
Bovine trachea	85	15	(2)
Rabbit trachea	95	5	(3)
Rabbit urinary bladder	86	14	(3)
Human uterus nonpregnant	29	71	(4)
Human uterus 21 weeks pregnant	37	63	(4)
Human uterus 32 weeks pregnant	50	50	(4)
Human uterus 40 weeks pregnant	60	40	(4)
Rat uterus nonpregnant	46	54	(5)
Rat uterus 20 days pregnant	65	35	(5)
Porcine jejunum	100	0	(2)
Porcine gastric corpus	100	0	(2)
Rabbit rectococcygeus	100	0	(3)
Guinea-pig taenia coli	79	21	(3)
Chicken gizzard	100	0	(6)

[a]References: (1) Hasegawa *et al.* (1988); (2) Helper *et al.* (1988); (3) Malmqvist and Arner (1991); (4) Cavaillé *et al.* (1986); (5) Morano *et al.* (1993); (6) Kelley *et al.* (1993).

two isoforms, where there are five amino acid substitutions within the last nine residues. The amino acid sequences are the same as those deduced from the nucleotide sequences of cDNA clones isolated from bovine aortic smooth muscle (Lash *et al.*, 1990). The amino acid sequence of LC17a shows 93% and that of LC17b 91% homology with the sequence of chicken gizzard LC17 (Fig. 9). Furthermore, their homologies with rabbit skeletal alkali light chain-2 are 65 and 68%, respectively (Hasegawa *et al.*, 1992).

Nabeshima *et al.* (1987) have isolated two cDNA clones for LC17 mRNA from chicken gizzard and fibroblast cDNA libraries. Sequence analysis of the two

cloned cDNAs revealed that they encode the identical 142-amino-acid sequence with different C terminals of 9 amino acids, each specific for LC17 from gizzard muscle and from fibroblast. DNA blot analysis suggested that the two LC17 mRNAs of gizzard and fibroblast cells are generated from a single gene, probably through alternative RNA splicing mechanisms. Very similar results and conclusions were obtained by Lenz *et al.* (1989) on the generation of LC17 isoforms from cultured human lymphoblasts and heart aorta smooth muscle cells.

C. What Is the Functional Significance of LC17 Isoforms?

The physiological role of LC17 isoforms has been studied by several investigators. The actin-activated MgATPase activity of phosphorylated and nonphosphorylated myosin from porcine gastric corpus (0% LC17b, Table I) was twofold higher that that of myosin from porcine aorta (43% LC17b) (Helper *et al.*, 1988). These authors have also prepared S-1 from three different tissue sources: stomach, pulmonary artery, and aorta, but all of bovine origin. The V_{max} values for the actin-activated MgATPase activity of the S-1 decreased in proportion to the LC17b content (0, 20, and 40%, respectively). This was the first suggestion for modulation of actomyosin ATPase activity by LC17 isoforms. This idea of Helper *et al.* was further elaborated by Hasegawa and Morita (1992), who exchanged the intrinsic LC17 in porcine aorta myosin with externally added isoforms and thereby reconstituted myosins with different content of LC17b, varying from 23 to 81%. The V_{max} of the actin-activated MgATPase activity of the reconstituted myosin decreased with the increase of its LC17b content up to 50%, but remained constant with further increase of LC17b to 81%. The K_m value of the myosin with 81% LC17b content was 20 times lower than that of myosin with 23% LC17b, suggesting that LC17b increases its affinity to actin when reducing the ATPase activity of myosin. Furthermore, they found an isoform-dependent actin binding: isolated LC17b was bound to F-actin with a dissociation constant of 64 μM whereas LC17a was not bound. Malmqvist and Arner (1991) provided the strongest argument for the functional role of the isoforms. They found a correlation between the relative LC17b content of different mammalian smooth muscles and the maximal shortening velocity of skinned fibers prepared form these muscles. An almost inverse relationship was found between shortening velocity and the relative amount of LC17b (Fig. 4 in Malmqvist and Arner, 1991), suggesting that the type of LC17 influences the speed of shortening in smooth

```
                                                              Ac-C-D-F-T-E-
                                                              Ac-. . . . .-
                                                              X -. . . S .-


            10                20                30                40
   D-Q-T-A-E-F-K-E-A-F-Q-L-F-D-R-T-G-D-G-K-I-L-Y-S-Q-C-G-D-V-M-R-A-L-G-Q-N-P-
I  . . . . . . . . . . . . . . . . . . . . . . . . . . . . . . . . . . . . .-
   E . . . . . . . . . . . . . . . . . . . . . . . . . . . . . . . . . . . .-


            50                60                70                80
   T-N-A-E-V-L-K-V-L-G-N-P-K-S-D-E-M-N-V-K-V-L-D-F-E-H-F-L-P-M-L-Q-T-V-A-K-N-K-D-Q-
II . . . . . . . . . . . . . . . . . . . . . . . . . . . . . . . . . . . . . . . .-
   . . . . . M . . . . . . . . . . . . . . . T . N . . Q . . . . . . . . I . . . . .-


            90               100               110
   G-T-Y-E-D-Y-V-E-G-L-R-V-F-D-K-E-G-N-G-T-V-M-G-A-E-I-R-H-V-L-V-T-N-G-E-K-
III . . . . . . . . . . . . . . . . . . . . . . . . . . . . . . . . . . . . .-
   . C F . . . . . . . . . . . . . . . . . . . . . . . . . . . . . . . . . .-


   120              130               140               150
   M-T-E-E-V-E-M-L-V-A-G-H-E-D-S-N-G-C-I-N-Y-E-E-L-V-R-M-V-L-N-G
IV . . . . . . . . . . . . . . . . . . . . . . . . A F . . H I . S .
   . . . . . . . . . . . . . . . . . . . . . . . . . . . . . S .
                 x   y   z   -y  -x    -z
```

FIGURE 9 Comparison of amino acid sequences of smooth muscle LC17. Upper row: LC17a from porcine aorta; middle row: LC17b from porcine aorta; bottom row: chicken gizzard LC17. The four domains (I, II, III, IV) analogous to the EF-hand structure (Kretsinger and Nockolds, 1973) are aligned with each other to produce maximum similarity of the four domains to each other (Collins, 1991). The letters x, y, z, −y, −x, and −z, represent potential Ca-coordinating ligands in the fourth EF-hand region. A dot indicates identity of an amino acid residue with the one immediately above it. Adapted from Hasegawa et al. (1992, p. 800).

muscle. For a comprehensive review of this topic we refer to Somlyo (1993).

Structural studies on the HC component of the smooth muscle myosin molecule did not support the concept that LC17 determines its actin-activated MgATPase activity. Kelley et al. (1993) discovered that chicken gizzard, but not chicken aortic, myosin contains an insert of 7 amino acids in a domain near the ATP binding site in the head region. The presence of the insert in gizzard myosin correlated with a higher velocity of movements of actin filaments sliding over gizzard myosin in the in vitro motility assay and a higher actin-activated MgATPase activity of gizzard myosin, compared with these properties of aortic myosin. Complete exchange of the gizzard myosin light chains (LC20 and LC17a) onto the aortic HC (containing LC20 and both LC17a and LC17b) did not alter the velocity of actin filaments propelled by the reconstituted aortic myosin containing LC20 and LC17a, suggesting that the differences in velocity between gizzard and aorta myosin are most likely due to the 7-amino-acid insert in the gizzard HC and not caused by the difference in the LC17 isoform content of these myosins. Experiments with truncated smooth muscle myosin, containing its native LC20 and either smooth muscle

LC17a or skeletal LC1 or LC3 isoforms, also suggest that the velocity in the in vitro motility assay is determined by HC and not by the isoform of ELC (Trybus, 1994).

For a short time it was believed that LC17 constitutes part of the active site of smooth muscle myosin (Okamoto et al., 1986). However, Grammer and Yount (1993) reversed this suggestion; they concluded that photolabeling of LC17 is not active site related and the active site of gizzard myosin is made up solely of the HC as has been observed in labeling studies of other myosins. Furthermore, through genetic engineering, Uyeda et al. (1994) created chimeric Dictyostelium myosins substituting a 9-amino-acid region at the 50,000/20,000 junction with those from myosin of rabbit skeletal, rat cardiac, or chicken smooth muscles. The actin-activated MgATPase activities of the chimeric Dictyostelium myosins correlated well with the activity of myosin from which the junction region was derived. In other words, smooth muscle myosin enzymatic activity was generated by restructuring Dictyostelium myosin heavy chain. In view of the similarity of the LCs between smooth muscle and Dictyostelium myosins, it appears that LC17 has no role in determining the actin-activated MgATPase activity of smooth muscle myosin. Therefore, the functional significance of the LC17 isoforms in smooth muscle myosin remains uncertain.

V. INTERACTION BETWEEN LIGHT AND HEAVY CHAINS

The three-dimensional structure of chicken skeletal myosin S-1 containing both RLC and ELC (Rayment et al., 1993a) and the structure of the regulatory domain of scallop myosin (Xie et al., 1994) may facilitate understanding the interactions between smooth muscle myosin light and heavy chains. There is considerable analogy in the structure of LCs from various sources, for all possess four homologous regions of the EF-hand structure (Collins, 1991). In the structure of chicken skeletal S-1, the LCs are wound around the long (approximately 85 Å) α-helix of HC that stretches from the thick part of the head containing the active sites to the C terminal of the S-1 heavy chain. The N terminal of RLC wraps around the C terminal of the S-1 heavy chain, followed by the N terminal of ELC. The first 18 amino acid residues in RLC have no electron density, suggesting disorder in this part of the molecule. Consistently, in gizzard actomyosin the N-terminal region of LC20 is readily cleaved by proteolytic enzymes (Jakes et al., 1976), and in intact arterial muscle the phosphoryl group, attached to various residues in the N-terminal region of LC20, is readily exchangeable (M. Bárány et al., 1991; K. Bárány et al., 1992a). In contrast, Cys-108 of LC20 does not react with iodoacetamide in the intact smooth muscle (K. Bárány et al., 1992b), suggesting that Cys-108 is involved in the binding of LC20 to the HC.

In the chicken skeletal model, the structural homology between the RLC and calmodulin was used to delineate the interaction sites (Rayment et al., 1993a). A cluster of hydrophobic residues, including phenylalanine, tryptophan, and methionine, would participate in the binding. The region of Lys-834–Gln-852 of the gizzard smooth muscle HC (Yanagisawa et al., 1987) (corresponding to Asn-825–Lys-842 in the skeletal S-1) and the region of Phe-82–Phe-89 in LC20 could fit these criteria. This is supported by the results of Katoh and Morita (1993), who isolated a 2-kDa peptide from aorta myosin HC, corresponding to residues 835–846 in gizzard HC; this peptide interacted with LC20 isolated from aorta myosin. There is good evidence for the involvement of the C-terminal region of LC20 in the combination with the HC: LC20 mutants in which the Lys-149–Ala-166 segment has been deleted markedly reduced their affinity to HC (Ikebe et al.,

1994), and Lys-845 in the HC could be cross-linked with Asp-168, Asp-170, or Asp-171 in LC20 (Onishi et al., 1992). Site-directed mutagenesis was employed to construct a series of LC20 mutants with successive removal of 3 (Rowe and Kendrick-Jones, 1993) or 6 and more (Trybus et al., 1994) residues from the C terminus and the myosins with LC20 mutants were analyzed for their biological activity. Removal of 12 residues caused a small decrease in velocity in the in vitro motility assay, whereas removal of 26 residues (containing the H helix and the nonfunctional fourth EF-hand, Rayment et al., 1993a) essentially abolished motility (Trybus et al., 1994). It was concluded that the C-terminal portion of LC20 interacts with the Gln-817–Leu-833 region of the HC. This is in accordance with suggestions from previous studies with mutant RLCs that the interactions between the C-terminal domain of LC20 and the HC are primarily responsible for the regulatory capabilities of LC20 (Trybus and Chatman, 1993).

The interaction between LC17 and the HC has not been studied. In the skeletal S-1 structure, ELC interacts with residues Leu-783 to Met-806 in the HC, which would correspond to the Ile-792–Gln-816 segment in smooth muscle HC. LC17 and LC20 could touch each other, similarly to their contact in the scallop myosin regulatory domain (Xie et al., 1994), and it is possible that LC17 is involved in transmitting the conformational coupling signal, initiated by phosphorylation of Ser-19 in LC20, to the actin-activated ATPase site of the smooth muscle HC.

VI. BINDING OF DIVALENT CATIONS BY LIGHT CHAINS

On the basis of amino acid sequence, all myosin LCs contain four homologous regions, I–IV, characteristic for the EF-hand structure (Kretsinger and Nockolds, 1973). Members of this family also include troponin C and calmodulin, key regulatory proteins, which along with the LCs have evolved from a small, ancestral Ca^{2+}-binding protein and possess common structural features (Collins, 1991). The LCs of scallop skeletal muscle myosin are the most studied proteins in this evolutionary system. Removal of the RLC, by washing scallop myosin with EDTA, abolishes the Ca^{2+} sensitivity of myosin–actin interaction and results in a loss of specific Ca^{2+} binding. Readdition of the RLC restores all the Ca^{2+}-dependent functions (Szent-Györgyi et al., 1973; Chantler and Szent-Györgyi, 1980). Similarly, Ca^{2+} dependence of tension generation in skinned scallop fibers requires the presence of RLC (Simons and Szent-Györgyi, 1978, 1985). During evolution, vertebrate smooth muscle became regulated through LC20 phosphorylation rather than by Ca²⁺ binding. Nevertheless, LC20 retained some rudimentary binding of divalent cations involving the amino acid sequence from 41 to 52 (Jakes *et al.*, 1976). For instance, the presence of EDTA is required for substantial LC20 exchange in gizzard myosin (Morita *et al.*, 1991), and LC20 mutants of smooth muscle myosin restore Ca²⁺ regulation to scallop myosin that has been stripped of its native RLC (Rowe and Kendrick-Jones, 1993), or potentiate force in RLC-depleted permeabilized scallop muscle fibers (Sweeney *et al.*, 1994). Hybrid scallop myosin containing gizzard LC20 showed a similar Ca²⁺ binding as native scallop myosin (Kwon *et al.*, 1992). Unexpectedly, the three-dimensional structure of the regulatory domain of scallop myosin revealed that the Ca²⁺-binding site is localized in region 1 of the EF-hand of the ELC (Xie *et al.*, 1994) and not in an EF-hand region of RLC. The Ca²⁺-binding site is stabilized by linkages involving both LCs and the HC. It is of interest that region 1 is the only region of ELC that interacts with RLC, thus the two LCs have a specific contact surface for regulation of scallop myosin function. This may be the reason for having two light chains instead of one.

VII. PERSPECTIVES

At the time of this writing, protein crystallization and structure determination by X-ray diffraction lead the way in understanding the molecular mechanism of muscle contraction (Rayment *et al.*, 1993a,b; Schröder *et al.*, 1993; Xie *et al.*, 1994). It appears that crystallization of smooth muscle S-1, with LC20 and LC17 attached, is the main task ahead. Although there is great enthusiasm for this work, there is an unpredictable time element in crystallization; it may take from several months to many years. For that reason, other avenues are equally important.

Mutations of LC20 coupled with motility assays (see Trybus, Chapter 3 this volume) claim major victories in elucidation of the elementary steps in the contractile interaction between myosin and actin. The scope of this approach is rapidly expanding and new discoveries are expected in the future. Application of these methods to LC17 would be of importance. The first problem to be solved is the reversible removal of LC17 from vertebrate smooth muscle myosin. The question arises whether it is possible to dissect LC17 selectively while LC20 remains bound to the HC.

The finding that in the regulatory domain of scallop myosin there is a contact between RLC and ELC (Xie *et* *al.*, 1994) invites studies to explore an interaction between LC20 and LC17 in vertebrate smooth muscle myosin. Indeed, work on this line has already begun (Trybus, 1994; Higashihara and Ikebe, 1995).

It should be realized that studies on LC interactions in a simplified *in vitro* system may not necessarily correspond to the same interactions in the intact muscle. For instance, the cysteine residue of LC20 is unreactive with iodoacetamide in porcine carotid arteries, whereas the cysteine residues of LC17 are reactive (Bárány *et al.*, 1992b). On the other hand, the lysine amino groups of both LC20 and LC17 are readily available for reductive methylation in porcine uterus, indicating that they are on the surface of the protein (Michael Bárány and Kate Bárány, unpublished). Chemical modification of several other amino acid residues may be explored so that the chemical anatomy of LCs could be established *in situ*.

With so many techniques converging on LC20 and LC17, crystallography, genetic engineering, the combination of molecular biology and fluorescence spectroscopy, and chemical modification, we can expect to see great progress in the near future.

Acknowledgments

We thank Janice Gentry for careful typing of the manuscript. This work was supported by Grants AM 34602 from the National Institutes of Health and CRB 25001 from the University of Illinois at Chicago.

References

Bárány, K., and Bárány, M. (1993). *FASEB J.* **7**, A1078.
Bárány, K., Csabina, S., and Bárány, M. (1985a). *Adv. Protein Phosphatases* **2**, 37–58.
Bárány, K., Ledvora, R. F., and Bárány, M. (1985b). *In* "Calmodulin Antagonists and Cellular Physiology" (H. Hidaka and D. J. Hartshorn, eds.), pp. 199–223. Academic Press, New York.
Bárány, K., Csabina, S., de Lanerolle, P., and Bárány, M. (1987). *Biochim. Biophys. Acta* **911**, 369–371.
Bárány, K., Pato, M. D., Rokolya, A., Erdödi, F., DiSalvo, J., and Bárány, M. (1989). *Adv. Protein Phosphatases* **5**, 517–534.
Bárány, K., Rokolya, A., and Bárány, M. (1990). *Biochim. Biophys. Acta* **1035**, 105–108.
Bárány, K., Polyak, E., and Bárány, M. (1992a). *Biochim. Biophys. Acta* **1134**, 233–241.
Bárány, K., Polyak, E., and Bárány, M. (1992b). *Biochem. Biophys. Res. Commun.* **187**, 847–852.
Bárány, M., Rokolya, A., and Bárány, K. (1991). *Arch. Biochem. Biophys.* **287**, 199–203.
Bengur, A. R., Robinson, E. A., Appella, E., and Sellers, J. R. (1987). *J. Biol. Chem.* **262**, 7613–7617.
Bennett, J. P., Cross, R. A., Kendrick-Jones, J., and Weeds, A. G. (1988). *J. Cell Biol.* **107**, 2623–2629.
Cassidy, P., Hoar, P. E., and Kerrick, W. G. L. (1979). *J. Biol. Chem.* **254**, 11148–11153.
Cavaillé, F., Janmot, C., Ropert, S., and d'Albis, A. (1986). *Eur. J. Biochem.* **160**, 507–513.

Chantler, P. D., and Szent-Györgyi, A. G. (1980). *J. Mol. Biol.* **138,** 473–492.

Colburn, J. C., Michnoff, C. H., Hsu, L. C., Slaughter, C. A., Kamm, K. E., and Stull, J. T. (1988). *J. Biol. Chem.* **263,** 19166–19173.

Collins, J. H. (1991). *J. Muscle Res. Cell Motil.* **12,** 3–25.

Craig, R., Smith, R., and Kendrick-Jones, J. (1983). *Nature (London)* **302,** 436–439.

Csabina, S., Bárány, M., and Bárány, K. (1987). *Comp. Biochem. Physiol. B* **87B,** 271–277.

Drew, J. S., Moos, C., and Murphy, R. A. (1991). *Am. J. Physiol.* **260,** C1332–C1340.

Driska, S. P., Aksoy, M. O., and Murphy, R. A. (1981). *Am. J. Physiol.* **240,** C222–C233.

Erdödi, F., Bárány, M., and Bárány, K. (1987). *Circ. Res.* **61,** 898–903.

Erdödi, F., Rokolya, A., Bárány, M., and Bárány, K. (1988a). *Arch. Biochem. Biophys.* **266,** 583–591.

Erdödi, F., Rokolya, A., DiSalvo, J., Bárány, M., and Bárány, K. (1988b). *Biochem. Biophys. Res. Commun.* **153,** 156–161.

Erdödi, F., Rokolya, A., Bárány, M., and Bárány, K. (1989). *Biochim. Biophys. Acta* **1011,** 67–74.

Flicker, P. F., Wallimann, T., and Vibert, P. (1983). *J. Mol. Biol.* **169,** 723–741.

Gagelmann, M., Rüegg, J. C., and DiSalvo, J. (1984). *Biochem. Biophys. Res. Commun.* **120,** 933–938.

Gaylinn, B. D., Eddinger, T. J., Martino, P. A., Monical, P. L., Hunt, D. F., and Murphy, R. A. (1989). *Am. J. Physiol.* **257,** C997–C1004.

Grammer, J. C., and Yount, R. G. (1993). *Biophys. J.* **64,** A143.

Hartshorne, D. J. (1987). *In* "Physiology of the Gastrointestinal Tract" (L. R. Johnson, ed.), 2nd ed., pp. 423–482. Raven Press, New York.

Hasegawa, Y., and Morita, F. (1992). *J. Biochem. (Tokyo)* **111,** 804–809.

Hasegawa, Y., Ueno, H., Horie, K., and Morita, F. (1988). *J. Biochem. (Tokyo)* **103,** 15–18.

Hasegawa, Y., Tanahashi, K., and Morita, F. (1990). *J. Biochem. (Tokyo)* **108,** 909–913.

Hasegawa, Y., Ueda, Y., Watanabe, M., and Morita, F. (1992). *J. Biochem. (Tokyo)* **111,** 798–803.

Hellstrand, P., and Arner, A. (1985). *Pflüegers Arch.* **405,** 323–328.

Helper, D. J., Lash, J. A., and Hathaway, D. R. (1988). *J. Biol. Chem.* **263,** 15748–15753.

Higashihara, M., and Ikebe, M. (1995). *FEBS Lett.* **363,** 57–60.

Horiuti, K., Somlyo, A. V., Goldman, Y. E., and Somlyo, A. P. (1989). *J. Gen. Physiol.* **94,** 769–781.

Ikebe, M., and Hartshorne, D. J. (1985). *J. Biol. Chem.* **260,** 10027–10031.

Ikebe, M., and Morita, J. I. (1991). *J. Biol. Chem.* **266,** 21339–21342.

Ikebe, M., Hartshorne, D. J., and Elzinga, M. (1986). *J. Biol. Chem.* **261,** 36–39.

Ikebe, M., Harshorne, D. J., and Elzinga, M. (1987). *J. Biol. Chem.* **262,** 9569–9573.

Ikebe, M., Reardon, S., Mitani, Y., Kamisoyama, H., Matsuura, M., and Ikebe, R. (1994). *Proc. Natl. Acad. Sci. U.S.A.* **91,** 9096–9100.

Inoue, A., Yanagisawa, M., Takano-Ohmuro, H., and Masaki, T. (1989). *Eur. J. Biochem.* **183,** 645–651.

Jakes, R., Northrop, F., and Kendrick-Jones, J. (1976). *FEBS Lett.* **70,** 229–234.

Kamisoyama, H., Araki, Y., and Ikebe, M. (1994). *Biochemistry* **33,** 840–847.

Katoh, T., and Morita, F. (1993). *J. Biol. Chem.* **268,** 2380–2388.

Kelley, C. A., Takahashi, M., Yu, J. H., and Adelstein, R. S. (1993). *J. Biol. Chem.* **268,** 12848–12854.

Kemp, B. E., and Pearson, R. B. (1985). *J. Biol. Chem.* **260,** 3355–3359.

Kenney, R. E., Hoar, P. E., and Kerrick, W. G. L. (1990). *J. Biol. Chem.* **264,** 8642–8649.

Kretsinger, R. H., and Nockolds, C. E. (1973). *J. Biol. Chem.* **248,** 3313–3326.

Kwon, H., Melandri, F. D., and Szent-Györgyi, A. G. (1992). *J. Muscle Res. Cell Motil.* **13,** 315–320.

Lash, J. A., Helper, D. J., Klug, M., Nicolozakes, A. W., and Hathaway, D. R. (1990). *Nucleic Acids Res.* **18,** 7176.

Ledvora, R. F., Bárány, K., VanderMeulen, D. L., Barron, J. T., and Bárány, M. (1983). *J. Biol. Chem.* **258,** 14080–14083.

Lenz, S., Lohse, P., Seidel, U., and Arnold, H. H. (1989). *J. Biol. Chem.* **264,** 9009–9015.

Maita, T., Chen, J. I., and Matsuda, G. (1981). *Eur. J. Biochem.* **117,** 417–424.

Malmqvist, U., and Arner, A. (1991). *Pflüegers Arch.* **418,** 523–530.

Monical, P. L., Owens, G. K., and Murphy, R. A. (1993). *Am. J. Physiol.* **264,** C1466–C1472.

Morano, I., Erb, G., and Sogl, B. (1993). *Pflüegers Arch.* **423,** 434–441.

Morita, J. I., Takashi, R., and Ikebe, M. (1991). *Biochemistry* **30,** 9539–9545.

Mougios, V., and Bárány, M. (1986). *Biochim. Biophys. Acta* **872,** 305–308.

Murakami, N., Healy-Louie, G., and Elzinga, M. (1990). *J. Biol. Chem.* **265,** 1041–1047.

Muszynska, G., Anderson, L., and Poráth, J. (1986). *Biochemistry* **25,** 6850–6853.

Nabeshima, Y., Nabeshima, Y. I., Nonomura, Y., and Kuriyama, Y. F. (1987). *J. Biol. Chem.* **262,** 10608–10612.

Nakabayashi, H., Sellers, J. R., and Huang, K. P. (1991). *FEBS Lett.* **294,** 144–148.

Nishikawa, M., Sellers, J. R., Adelstein, R. S., and Hidaka, H. (1984). *J. Biol. Chem.* **259,** 8808–8814.

Okamoto, Y., Sekine, T., Grammer, J. C., and Yount, R. G. (1986). *Nature (London)* **324,** 78–80.

Onishi, H., Wakabayashi, T., Kamata, T., and Watanabe, S. (1983). *J. Biochem. (Tokyo)* **94,** 1147–1154.

Onishi, H., Maita, T., Matsuda, G., and Fujiwara, K. (1992). *Biochemistry* **31,** 1201–1210.

Packer, C. S. (1993). *Biophys. J.* **64,** A32.

Pearson, R. B., Jakes, R., Kendrick-Jones, J., and Kemp, B. E. (1984). *FEBS Lett.* **168,** 108–112.

Pearson, R. B., Misconi, L. Y., and Kemp, B. E. (1986). *J. Biol. Chem.* **261,** 25–27.

Perrie, W. T., and Perry, S. V. (1970). *Biochem. J.* **119,** 31–38.

Post, P. L., Trybus, K. M., and Taylor, D. L. (1994). *J. Biol. Chem.* **269,** 12880–12887.

Rayment, I., Rypniewski, W. R., Schmidt-Bäse, K., Smith, R., Tomchick, D. R., Benning, M. M., Winkelmann, D. A., Wesenberg, G., and Holden, H. M. (1993a). *Science* **261,** 50–58.

Rayment, I., Holden, H. M., Whittaker, M., Yohn, C. B., Lorenz, M., Holmes, K. C., and Milligan, R. A. (1993b). *Science* **261,** 58–65.

Rees, D. D., and Frederiksen, D. W. (1981). *J. Biol. Chem.* **256,** 357–364.

Rokolya, A., Bárány, M., and Bárány, K. (1991). *Biochim. Biophys. Acta* **1057,** 276–280.

Rowe, T., and Kendrick-Jones, J. (1993). *EMBO J.* **12,** 4877–4884.

Sakurada, K., Ikuhara, T., Seto, M., and Sasaki, Y. (1994). *J. Biochem. (Tokyo)* **115,** 18–21.

Samuels, J., Rappaport, L., Mercadier, J., Lompre, A., Sartore, S., Triban, C., Schiaffino, S., and Schwartz, K. (1983). *Circ. Res.* **52,** 200–209.

Schröder, R. R., Manstein, D. J., Jahn, W., Holden, H., Rayment, I.,

Holmes, K. C., and Spudich, J. A. (1993). *Nature (London)* **364,** 171–174.

Sellers, J. R., and Harvey, E. V. (1984). *J. Biol. Chem.* **259,** 14203–14207.

Sherry, J. M. F., Gorecka, A., Aksoy, M. O., Dabrowska, R., and Hartshorne, D. R. (1978). *Biochemistry* **17,** 4411–4418.

Simons, R. M., and Szent-Györgyi, A. G. (1978). *Nature (London)* **273,** 62–64.

Simons, R. M., and Szent-Györgyi, A. G. (1985). *J. Physiol. (London)* **358,** 47–64.

Somlyo, A. P. (1993). *J. Muscle Res. Cell Motil.* **14,** 557–563.

Stafford, W. F., III, and Szent-Györgyi, A. G. (1978). *Biochemistry* **17,** 607–614.

Sweeney, H. L., Yang, Z., Zhi, G., Stull, J. T., and Trybus, K. M. (1994). *Proc. Natl. Acad. Sci. U.S.A.* **91,** 1490–1494.

Szent-Györgyi, A. G., Szentkirályi, E. M., and Kendrick-Jones, J. (1973). *J. Mol. Biol.* **74,** 179–203.

Trybus, K. M. (1994). *J. Biol. Chem.* **269,** 20819–20822.

Trybus, K. M., and Chatman, T. A. (1993). *J. Biol. Chem.* **268,** 4412–4419.

Trybus, K. M., and Lowey, S. (1988). *J. Biol. Chem.* **263,** 16485–16492.

Trybus, K. M., Waller, G. S., and Chatman, T. A. (1994). *J. Cell Biol.* **124,** 963–969.

Umemoto, S., Bengur, A. R., and Sellers, J. R. (1989). *J. Biol. Chem.* **264,** 1431–1436.

Uyeda, T. Q. P., and Spudich, J. A. (1993). *Science* **262,** 1867–1870.

Uyeda, T. Q. P., Ruppel, K. M., and Spudich, J. A. (1994). *Nature (London)* **368,** 567–569.

Walsh, M. P., Bridenbaugh, R., Hartshorne, D. J., and Kerrick, W. G. L. (1982). *J. Biol. Chem.* **257,** 5987–5990.

Watanabe, M., Hasegawa, Y., Katoh, T., and Morita, F. (1992). *J. Biochem. (Tokyo)* **112,** 431–432.

Winkelmann, D. A., Lowey, S., and Press, J. L. (1983). *Cell (Cambridge, Mass.)* **34,** 295–306.

Xie, X., Harrison, D. H., Schlichting, I., Sweet, R. M., Kalabokis, V. N., Szent-Györgyi, A. G., and Cohen, C. (1994). *Nature (London)* **368,** 306–312.

Yanagisawa, M., Hamada, Y., Katsuragawa, Y., Imamura, M., Mikawa, T., and Masaki, T. (1987). *J. Mol. Biol.* **198,** 143–157.

Zavodny, P. J., Petro, M. E., Lonial, H. K., Dailey, S. H., Narula, S. K., Leibowitz, P. J. and Kumar, C. C. (1990). *Circ. Res.* **67,** 933–940.

Zhi, G., Herring, B. P., and Stull, J. T. (1994). *J. Biol. Chem.* **269,** 24723–24727.

3

Myosin Regulation and Assembly

KATHLEEN M. TRYBUS

Rosenstiel Basic Medical Sciences Research Center
Brandeis University
Waltham, Massachusetts

I. INTRODUCTION

Regulatory light chain (RLC) phosphorylation controls smooth muscle myosin's motor and assembly properties. This review describes some structural elements that are necessary to activate myosin's motor properties and to stabilize the folded monomeric conformation. Many key features of both processes have been deduced from the activity and assembly properties of myosin into which mutant RLCs have been incorporated. More recently, expression of mutant smooth muscle myosins using the baculovirus expression system has allowed the role of the heavy chain (HC) in regulation to also be probed. Although light chain (LC) phosphorylation clearly regulates contraction in a smooth muscle cell, there is no strong evidence for a large-scale assembly–disassembly in a smooth muscle cell. This observation raises the interesting question of what factor(s) are involved in filament stabilization *in vivo*. Other unresolved structural issues are how myosin packs into a native filament, and how the nonhelical myosin tailpiece might alter intermolecular myosin interactions.

II. STRUCTURE OF THE LIGHT CHAIN BINDING REGION

Given the pivotal role of the RLC in smooth muscle myosin's function, several important structural features of the LC binding domain and its relationship to myosin's catalytic or motor domain will first be described. The myosin HC, after emerging from the motor domain, forms a long 8.5-nm α-helix that is stabi-

lized by the binding of the essential and the regulatory LCs (ELC and RLC) (Fig. 1). The HC has a gradual bend in the region between the two LCs and one sharp hook at the C terminus, where the two heads join at the head/rod junction (Rayment *et al.*, 1993; Xie *et al.*, 1994). The C-terminal half of the ELC abuts the motor domain and binds to the N-terminal portion of this α-helix, whereas the RLC binds to the C-terminal portion of the HC just below the ELC. The polarity of both LCs is opposite to that of the HC. The N-terminal half of the RLC, which includes the phosphorylatable serine, binds to the HC in the vicinity of the sharp bend, and is stabilized primarily by interactions with hydrophobic and aromatic residues, some of which involve the sequence WQWW found in regulated scallop and smooth muscle myosins.

Despite considerable structural information, the structure of the N-terminal residues of the RLC and the location of the phosphorylatable serine are not known. In the crystal structure of the skeletal muscle myosin head, electron density was not observed for the N terminus of the RLC, suggesting that this region is flexible (Rayment *et al.*, 1993). Density for the N terminus of the scallop myosin RLC begins six residues beyond the serine homologous to Ser 19 in the smooth RLC (Xie *et al.*, 1994). In addition, scallop RLC lacks the N-terminal extension before the serine that contains the basic residues necessary for myosin light chain kinase binding.

How information is transmitted between the neck and the active site remains a key unanswered question. It is noteworthy that the C-terminal "20-kDa" portion of the HC extends throughout the length of the myosin head. It starts at a surface loop that is believed

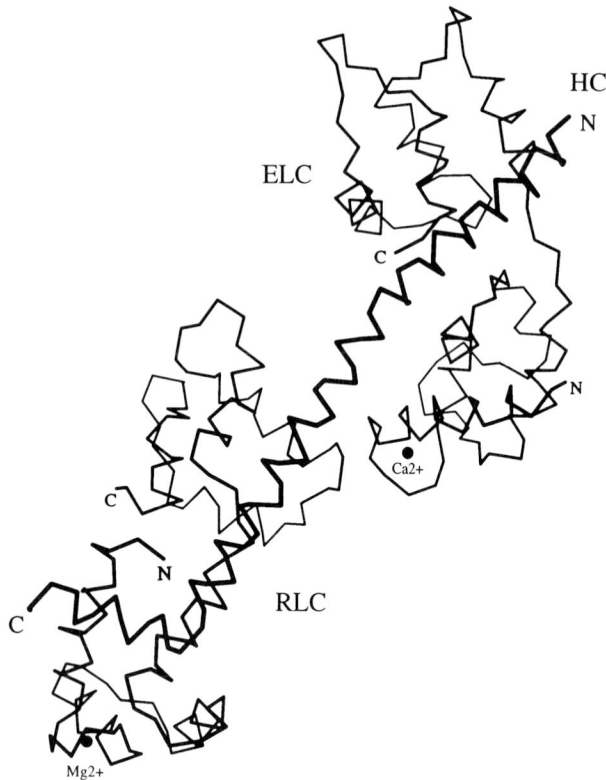

FIGURE 1   Overall fold of the light chain binding region from scallop myosin. The N and C termini of the heavy chain (HC), the regulatory light chain (RLC), and the essential light chain (ELC) are indicated. The α-helical heavy chain (Met 773–Ala 840) has one gradual ≈40° bend in the region between the light chains and one sharp turn at the C terminus, involving the residues Trp-Gln-Trp (824–826). The RLC (Leu 12–Glu 153) winds around the C-terminal half of this HC. Ser 6 of the scallop RLC, which is homologous to the phosphorylatable Ser 19 of the smooth RLC, is six residues before the first amino acid shown here. The ELC (Ser 4–Pro 154) winds around the N-terminal half of this heavy chain. A Mg^{2+} ion is bound to the divalent cation binding site in domain I of the RLC. A Ca^{2+} ion is shown in the calcium-specific binding site that regulates the activity of scallop myosin. See Xie et al. (1994) for details of the scallop regulatory domain structure.

filaments formed from myosin with a dephosphorylated RLC disassemble to a soluble, monomeric conformation in which the rod is bent into thirds (Fig. 2). The monomers reassemble into filaments upon RLC phosphorylation. A similar transition occurs in vertebrate nonmuscle myosins. The bent monomer sediments at a rate of 10S, faster than the typical extended 6S monomer that exists at high ionic strength (>0.3 M KCl). The assembly–disassembly process has thus often been referred to as the 6S–10S transition (Suzuki et al.,

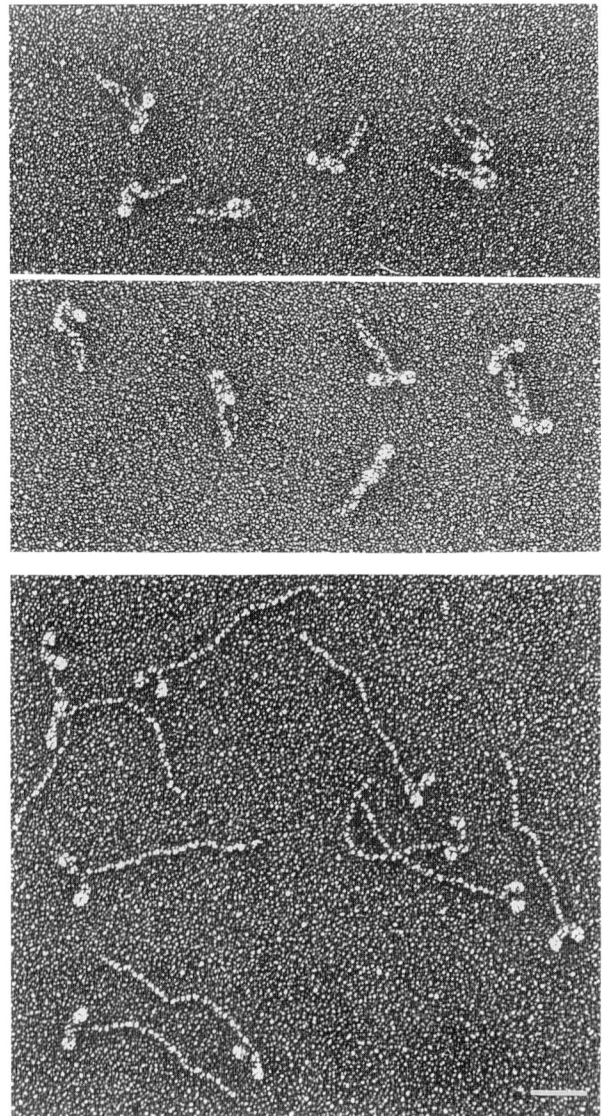

FIGURE 2   Metal-shadowed smooth muscle myosin in the folded and extended conformation. Upper panel: At physiological ionic strength in the presence of MgATP, myosin preferentially forms a folded monomer where the tail is bent into approximately equal thirds. Note the heads often bend down toward the rod. Lower panel: For comparison, extended monomers formed at high ionic strength are shown. Bar-50 nm. Reprinted from Trybus, K. M., and Lowey, S., Journal of Biological Chemistry 259, 8564–8571, 1984.

to be part of the actomyosin interface, continues to cap one part of the nucleotide binding site, and ultimately emerges from the motor domain to form the long α-helical LC binding region (Rayment et al., 1993). It is possible that RLC phosphorylation and actin binding both affect the ATPase site via this pathway.

III. CONTROL OF ASSEMBLY BY LIGHT CHAIN PHOSPHORYLATION

A. The Folded-to-Extended Conformational Transition

A striking feature of smooth muscle myosin is that upon addition of stoichiometric amounts of MgATP,

1978; Trybus *et al.*, 1982; Trybus and Lowey, 1984; Craig *et al.*, 1983). Centrifugation separates the soluble folded monomer from filaments, whereas gel filtration can be used to separate folded from extended monomers because of their difference in Stokes' radius (Trybus and Lowey, 1984, 1988).

The RLC is required to form the folded monomer (Trybus and Lowey, 1988). This observation is consistent with electron microscopy images showing that the tail in the folded monomer is in close proximity to the head/rod junction where the RLC is located (Trybus and Lowey, 1984; Craig *et al.*, 1983; Onishi and Wakabayashi, 1982). The heads in the folded monomer appear to be constrained in a downward orientation against the subragment-2 region of the rod, unlike the numerous orientations of the head relative to the rod observed in the extended monomer.

B. Light Chain Mutants Identify Regions That Stabilize the Folded Monomeric Conformation

1. Exchange of Mutant RLCs into Myosin

The RLC is expressed in high yield in bacteria (>50 mg/liter). The expressed RLC is sequestered into inclusion bodies, which is advantageous for purification since the RLC can be renatured from guanidine hydrochloride (Trybus and Chatman, 1993). Purified expressed RLC mutants can be incorporated into gizzard myosin by two methods. A 10-fold molar excess of RLC incubated with gizzard myosin at elevated temperature (42°C) in the absence of divalent cations results in 90% exchange for RLC mutants whose affinity for the myosin heavy chain is equal to that of the native RLC (Trybus and Chatman, 1993). Alternatively, incubation of gizzard myosin with trifluoperazine, a phenothiazine known to interact with calmodulin, selectivity and completely causes RLC dissociation (Trybus *et al.*, 1994). Subsequent gel filtration yields RLC-deficient myosin. A distinct advantage of this technique is that complete incorporation of mutant RLCs can be achieved, even those with low affinity for the heavy chain (e.g., C-terminal deletion mutants).

2. RLC Residues That Stabilize the Folded Monomer

Mutant RLCs incorporated into gizzard myosin show that two basic residues in the N-terminal region of the RLC are necessary to stabilize the folded monomer (Table I). The folded-to-extended conformational transition is salt dependent (Trybus and Lowey, 1984), and thus ionic interactions are likely to be involved in the stabilization of the bent monomer. Eight of the first 16 residues in the N-terminal region of the RLC are

TABLE I Residues in the N Terminus of the Regulatory Light Chain Required to Form the Folded Monomer[a]

Mutations that mimic phosphorylation's ability to cause the folded monomer to extend and assemble:
R16A, R13A/R16A, S19E, T18E/S19E

```
              13      16   18 19
S S K R A K A K T T K K R P Q R A T S N V F . . .
              ↓       ↓    ↓  ↓
              A       A    E  E
```

Mutations that did not affect formation of the folded monomer:
K11A, K12A

```
                  11 12
S S K R A K A K T T K  K  R P Q R A T S N V F . . .
                  ↓  ↓
                  A  A
```

[a] The first 22 residues of the smooth muscle myosin RLC are shown. Phosphorylation of Ser 19 causes the folded monomer to assemble into filaments. Mutation of the residues shown in the upper part of this table mimic the effect of phosphorylation. Data taken from Sweeney *et al.* (1994).

basic (SSKRAKAKTTKKRPQRATS). Mutation of both R13 and R16 to Ala abolished myosin's ability to fold, with R16 having the larger effect as a single point mutation (Sweeney *et al.*, 1994; Ikebe *et al.*, 1994a). Mutation of K11 or K12, in contrast, had little effect on the monomer–polymer equilibrium. Thus, it is likely that acidic residues in the rod form specific ionic interactions with R13 and R16 in the folded monomer. Consistent with this interpretation, deletion of the N-terminal 16 residues abolished folding (Ikebe *et al.*, 1994a).

Smooth muscle myosin containing a skeletal myosin RLC, which has a Glu residue at the position corresponding to Arg 16 in the smooth muscle RLC sequence, formed filaments. The small amount of soluble material in equilibrium was only partially folded (Trybus and Lowey, 1988). A chimeric RLC with the N-terminal half of smooth RLC and the C-terminal half of skeletal RLC, despite having a completely native N-terminal half, also formed filaments (Trybus and Chatman, 1993). This result implies that the key N-terminal residues must be properly positioned for the rod to bind.

Phosphorylation of S19, by addition of negative charge, must disrupt these ionic interactions and cause unfolding. This interpretation is supported by the observation that the unfolding of the bent monomer by phosphorylation can be mimicked by mutation of T18 and S19 to negatively charged amino acids (D or E) (Sweeney *et al.*, 1994; Kamisoyama *et al.*, 1994).

IV. PHOSPHORYLATION-DEPENDENT CONTROL OF MYOSIN'S MOTOR PROPERTIES

A. ATPase and Motor Properties of Smooth Muscle Myosin

Coupled to the folding of the myosin molecule is a change at the active site that results in "trapping" of nucleotide (Cross *et al.*, 1986). The observed rate of phosphate release from single-turnover experiments, <0.0005 sec^{-1}, likely reflects product release from the small amount of extended myosin that is in equilibrium with the folded monomer (Cross *et al.*, 1988). The structural basis for trapping is not known. It has been proposed that after MgATP is cleaved at the active site, phosphate leaves via a "backdoor" mechanism (Yount *et al.*, 1995). Perhaps in the folded monomer the glycine-rich P-loop is stabilized in a conformation that prevents phosphate release.

Dephosphorylated filamentous myosin, stabilized by binding of antibody, releases nucleotide faster than does dephosphorylated 10S myosin, ≈ 0.002 sec^{-1} (Fig. 3). This rate increases several hundredfold upon RLC phosphorylation in the presence of actin (Trybus, 1989). RLC phosphorylation alone thus effectively regulates filamentous myosin, even though the degree of regulation is increased further if filament disassembly to the folded monomer occurs. The major differences between the folded monomer and an extended dephosphorylated molecule are that the monomer is soluble and inhibits product release more completely.

RLC phosphorylation also causes myosin to move actin filaments in a motility assay. Dephosphorylated extended myosin adhered to a nitrocellulose substratum does not support movement of actin fila-

ments, whereas phosphorylated myosin moves actin at a rate of ≈ 1 μm/sec at 30°C. Dephosphorylated myosin can act as a load to slow the movement of phosphorylated cross-bridges (Warshaw *et al.*, 1990). This observation may provide an explanation for the temporal coupling between RLC phosphorylation and shortening velocity in smooth muscle (Dillon *et al.*, 1981).

B. Structural Elements Required for Phosphorylation-Dependent Activation

1. Mutation of Serine 19

The mechanism to account for phosphorylation-dependent activation of myosin's motor properties is more complex than the interactions that stabilize the folded monomer. Mutation of S19 of the RLC to either D or E did not activate myosin's motor properties, but mutating both T18 and S19 to negatively charged amino acids partially activated motility (Sweeney *et al.*, 1994) and ATPase activity (Kamisoyama *et al.*, 1994). In contrast, rapid actin movement was supported by myosin whose RLC was phosphorylated at either T18 or S19, suggesting that the interaction formed upon activation has specific spatial constraints that are satisfied only by a phosphate moiety (Sweeney *et al.*, 1994).

The unloaded shortening velocity of single smooth muscle cells with mutant RLCs incorporated in the myosin has also been determined (Trybus *et al.*, 1995). The double charge mutants (both T18 and S19 mutated to negatively charged amino acids) caused the single cells to shorten, but at a slower rate than obtained with thiophosphorylated RLC. Simple addition of negative charge thus only partially mimics phosphorylation both *in vitro* and *in situ*.

2. Regulatory Light Chain Removal

RLC removal does not produce a normal "on" state like that obtained upon phosphorylation. RLC-deficient myosin has an approximately 6-fold lower actin-activated ATPase activity that is coupled to a 10-fold decrease in the velocity of actin movement (≈ 0.1 μm/sec versus 1 μm/sec for native myosin, 30°C) (Trybus *et al.*, 1994). RLC removal also increases the basal myosin MgATPase activity 10-fold (from 0.02 sec^{-1} to 0.2 sec^{-1}).

3. Importance of the C-Terminal Half of the Regulatory Light Chain

When skeletal muscle RLC was incorporated into smooth muscle myosin, the hybrid molecule could no longer be activated by phosphorylation (note that skeletal RLC contains a homologous phosphorylatable ser-

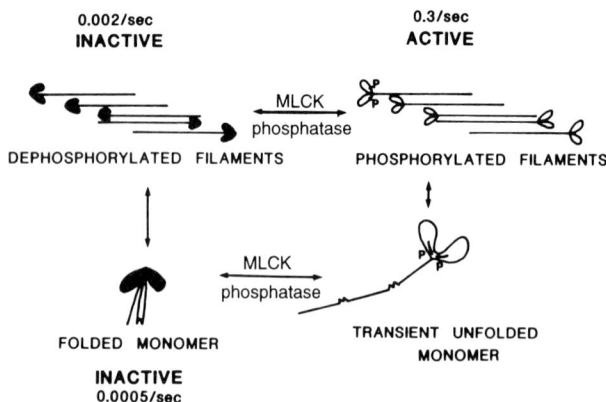

FIGURE 3 Conformation and relative activity of dephosphorylated folded monomers, dephosphorylated filaments, and phosphorylated filaments. Reprinted from Trybus, K. *Cell Motility and the Cytoskeleton* 18, Copyright © 1991 Wiley-Liss, a subsidiary of John Wiley and Sons, Inc.

TABLE II   **Effect of Regulatory Light Chain Mutations on Regulation of Motility**

RLC incorporated	Relative motility	
	Dephosphorylated	Phosphorylated
Smooth	0	100
N-skeletal/C-smooth[a]	0	100
N-smooth/C-skeletal	0	0
Skeletal	0	0
No RLC	10	—
C-terminal deletions		
C▲6	0	100
C▲12	0	75
C▲26	10	10

[a]Smooth and skeletal RLC were interchanged around the common sequence PEDVI (residues 98–102 of the smooth RLC). Data taken from Trybus and Chatman (1993) and Trybus et al. (1994).

ine residue) (Table II). Myosin containing a chimeric RLC with the N-terminal half from smooth muscle RLC and the C-terminal half from skeletal RLC was likewise locked in the "off" state. Replacement of V156–K163 of the smooth RLC (171 amino acids total) with the corresponding skeletal muscle RLC sequence likewise showed very little actin-activated ATPase activity (Ikebe et al., 1994b). In contrast, myosin with an N-skeletal/C-smooth chimeric RLC showed phosphorylation-dependent motility like wild-type RLC (Trybus and Chatman, 1993). Thus the minimum requirements for activation are some smooth muscle specific sequences in the C-terminal half of the RLC and a phosphorylatable serine at the N terminus.

The requirement for a native C-terminal half of the RLC to obtain regulation could reflect the fact that the phosphorylated serine specifically bonds to Arg residues that are found only in smooth muscle RLC. An example of such a mechanism of activation is found with the enzyme glycogen phosphorylase. Activation involves ion-pair interactions between a phosphorylated serine that exists at the highly basic N-terminal region of the molecule and two arginines (Sprang et al., 1988). A preliminary report consistent with a similar mechanism in smooth muscle myosin showed that mutation of six residues in the C terminus of the skeletal RLC to those found in smooth muscle RLC (which includes four Arg residues) restores phosphorylation-dependent regulation (Yang and Sweeney, 1995).

C-terminal deletion mutants of the RLC further show that interactions involving this region of the RLC are important. Removal of 6 residues from the C ter-

minus had no effect on regulation. Removal of 12 residues slowed movement by phosphorylated myosin, whereas deletion of 26 residues abolished phosphorylation-dependent activation (Trybus et al., 1994). The C▲26 mutant surprisingly showed the same depression of motility and activity as did complete RLC removal. Myosin containing the C▲26 RLC showed no tendency to aggregate as did RLC-deficient myosin, thus these changes in activity are due to changes transmitted to the active site (Trybus et al., 1994). How the C-terminal half of the RLC interacts with the HC appears to be critical to obtain regulated release of nucleotide from the active site.

4. The Essential Light Chain Is Not Essential for Regulation

A recent striking finding is that the ELC is not critical for smooth muscle myosin regulation. The interface between the RLC and the ELC was shown to be critical for calcium-dependent regulation of scallop myosin, since the regulatory calcium binding site is stabilized by residues from the ELC, the RLC, and the heavy chain (Xie et al., 1994). In contrast, the motility of an expressed heavy meromyosin (HMM) lacking an ELC was regulated by phosphorylation (Trybus, 1994). Thus the importance of the C-terminal half of the smooth RLC is not due to its interaction with the ELC. If a common mechanism of regulation exists, the switch must depend predominantly on changes in RLC/HC interactions, whether this induces a change in the gradual bend in the α-helix between the two LCs as speculated by Xie et al., (1994), or whether the change is propagated to the motor domain by another mechanism.

C. What Is the Role of the Rod in Regulation?

Two observations led one to believe that any two-headed species was necessary and sufficient to obtain regulation. The single-headed subfragment of myosin, as well as single-headed myosin (Cremo et al., 1995), is unregulated and in the "on" state. In contrast, double-headed myosin or heavy meromyosin prepared by enzymatic digestion requires RLC phosphorylation for activity. An exception to this rule was found because of the success in expressing soluble active fragments of smooth muscle myosin using the baculovirus expression system (Trybus, 1994). Surprisingly, an expressed smooth muscle HMM with a short 27-nm tail was poorly regulated by phosphorylation (Trybus, 1994). Two-headed molecules were seen by electron microscopy, but native gels revealed that this species showed some tendency to dissociate into a monomeric species. The poor regulation of this fragment rules out the possibility of obtaining a single-headed fragment that is well regulated. These results further suggest that to achieve the "off" state, the two heads and RLCs may need to have a precise orientation relative to each other, which is fixed by their attachment to a stable dimeric tail. Longer expressed constructs with tail lengths of 50 and 72 nm were both well regulated by phosphorylation and showed no tendency to dissociate into monomers.

The unusually large number of residues that are necessary to form a stable-coiled-coil region (>180) raises the possibility that a "loose" subfragment-2 region of the molecule may be important for some function of myosin. If all myosin rods show this instability, then a possible role may be to introduce enough flexibility at the region where the heads join so that both heads of myosin can bind equivalently to actin.

Can perturbations to the rod affect the activity of a phosphorylated myosin head, which would normally be active? A monoclonal antibody that binds very near to the C terminus of the myosin rod and depolymerizes myosin filaments caused a loss in motility and actin-activated ATPase activity (Horowitz and Trybus, 1992). These observations could be accounted for by a dramatic decrease in the affinity of the myosin head for actin in the presence of MgATP, but would also suggest that perturbations to the rod can be propagated over a distance to the myosin head.

A related question is whether myosin rods from different muscle types can be interchanged without loss of function. The same phylogenetic tree can be constructed whether myosin head or myosin rod sequences are compared (Goodson and Spudich, 1993). This result suggests that the changes in the head and rod have been coupled throughout evolution, presumably for a functional reason. Chimeric head/rod constructs expressed in the baculovirus insect cell system are being used to test whether phosphorylation-dependent regulation of smooth muscle myosin is retained if the head is attached to the rod of an unregulated myosin (K. M. Trybus and H. L. Sweeney, unpublished data).

V. FILAMENT STRUCTURE

A. Requirements for in Vitro Assembly

An interesting feature of the rod is that only a very limited region is responsible for myosin's insolubility at low ionic strength. The sequence required to confer assembly properties to the smooth muscle myosin rod

has been localized to an approximately 150-residue C-terminal fragment of the rod (Cross and Vandekerckhove, 1986). Consistent with the C-terminal localization of the assembly domain, a monoclonal antibody whose epitope is located very near the end of the myosin rod (but not in the nonhelical tailpiece, see the following) depolymerizes smooth muscle myosin filaments (Trybus and Henry, 1989). The most specific localization of an assembly domain has been shown in a nonmuscle myosin from *Dictyostelium*. The critical sequence for assembly was pinpointed to 35 amino acids that are located approximately 300 residues from the C terminus of that rod (Lee *et al.*, 1994). Because the *Dictyostelium* rod is longer than the smooth muscle myosin rod, these two assembly regions occupy similar positions relative to the head/rod junction. (For a detailed review that contrasts the assembly properties of smooth and nonmuscle myosins from *Dictyostelium* and *Acanthamoeba,* see Trybus, 1991a.)

Kinetic studies have shown that filament initiation is more difficult than subsequent elongation (Cross *et al.,* 1991). In a system where assembly–disassembly might play a large role, for example, in nonmuscle vertebrate cells, this property predicts that the rate at which monomers become available for polymerization could alter both the number and length of myosin filaments that are formed. Thus control of kinase activity, which controls the number of assembly competent extended monomers, could be a factor in determining subsequent polymerization.

B. Polarity of Filaments

In vitro, it is clear that smooth muscle myosin filaments can assemble into side-polar filaments that are never seen with skeletal muscle myosin (Craig and Megerman, 1977) (Fig. 4). The key feature of a side-polar filament is that all myosin heads have the same polarity along one edge of the filament, and the opposite polarity on the other edge, with bare zones on either end of the filament (Fig. 5). The heads of myosin in a bipolar filament, in contrast, reverse polarity at the filament center, thus producing a central bare zone.

The intermolecular interactions in a side-polar and bipolar filament differ. Only antiparallel overlaps are needed to generate a side-polar filament. Such interactions in a bipolar filament are limited to the bare zone. *In vitro* studies clearly show that smooth muscle myosin has a preference for antiparallel interactions. Bipolar smooth muscle myosin filaments greater than 0.5 μm long are generally not formed. Growth beyond this length would require purely parallel interactions between molecules. Short, homogeneous bipolar fila-

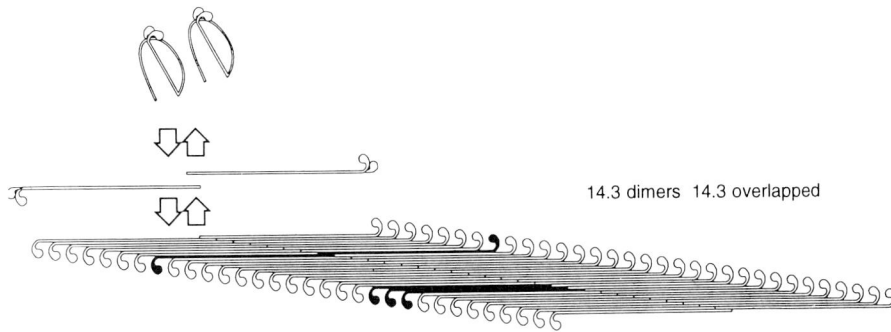

14.3 dimers 14.3 overlapped

FIGURE 4 Diagram of monomer packing in a side-polar filament. This model suggests that the antiparallel overlap between molecules is 14.3 nm. For clarity, only one of myosin's two heads is shown. Reprinted from Cross et al., Embo. J. **10**, 747–756, 1991, by permission of Oxford University Press.

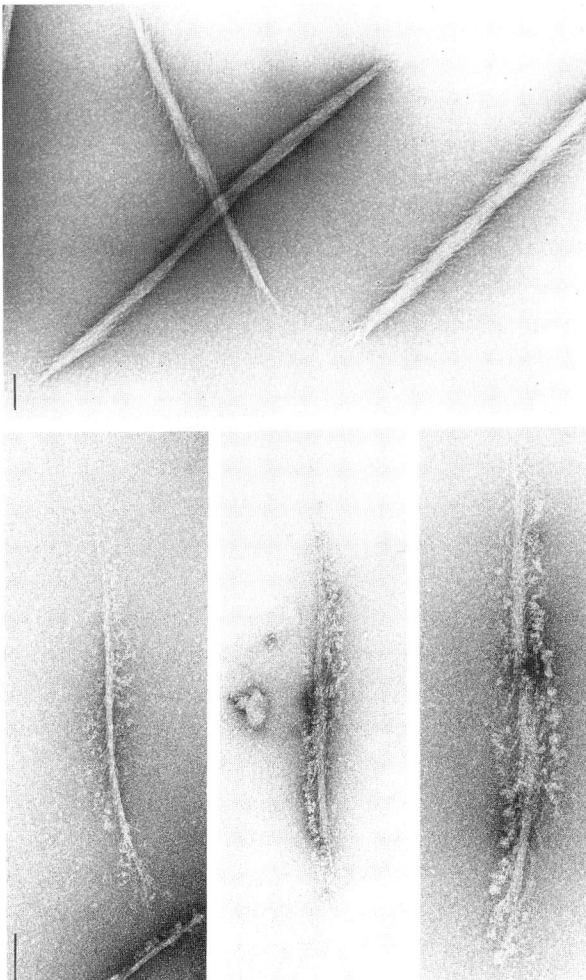

FIGURE 5 Electron micrographs of negatively stained smooth muscle myosin filaments. Upper panel: Smooth muscle myosin rod forms filaments with a distinct side-polar morphology. Lower three panels: Myosin filaments formed by addition of 0.1 M KCl to myosin minifilaments also show a side-polar appearance. Bar-0.1 μm. Reproduced from the Journal of Cell Biology, 1987, **105**, 3007–3019 by copyright permission of The Rockefeller University Press.

ments, called "minifilaments" (Fig. 6), grow into longer side-polar filaments upon addition of salt (see Fig. 5).

The length of the antiparallel overlap between molecules in a side-polar filament is not firmly established. A 43-nm antiparallel overlap was unique to segments formed from smooth muscle myosin rods by precipitation with divalent cations (Kendrick-Jones et al., 1971). Folded dimers with an approximately 40- to 50-nm antiparallel overlap are formed at salt concentrations <50 mM KCl (Trybus and Lowey, 1984). Both of these observations suggest that the molecule favors a 43-nm overlap. Measurements obtained by scanning transmission electron microscopy, however, show that filaments formed in vitro have one antiparallel dimer per 14.3 nm of filament length, a value that favors a model where adjacent molecules have a 14.3-nm antiparallel overlap (Cross and Engel, 1991) (Fig. 4). The 14.3-nm repeat arises from the charge periodicity of the rod, where peaks of positive charge are 14 residues away from peaks of negative charge. By half staggering repeating units of 196 amino acids, a 14.3-nm stagger arises (McLachlan and Karn, 1982).

Although all in vitro studies strongly suggest that the packing arrangements of smooth and skeletal myosin differ, the polarity of molecules within the native filament has been a difficult issue to resolve. This is in large part due to the concern that filaments may disassemble and reassemble during the isolation procedure. Under conditions designed to maximize filament stability, native filaments isolated from skinned amphibian cells showed a continuous 14-nm axial repeat of cross-bridges with no central bare zone, consistent with a side-polar arrangement (Cooke et al., 1989). The physiological advantage of a side-polar filament

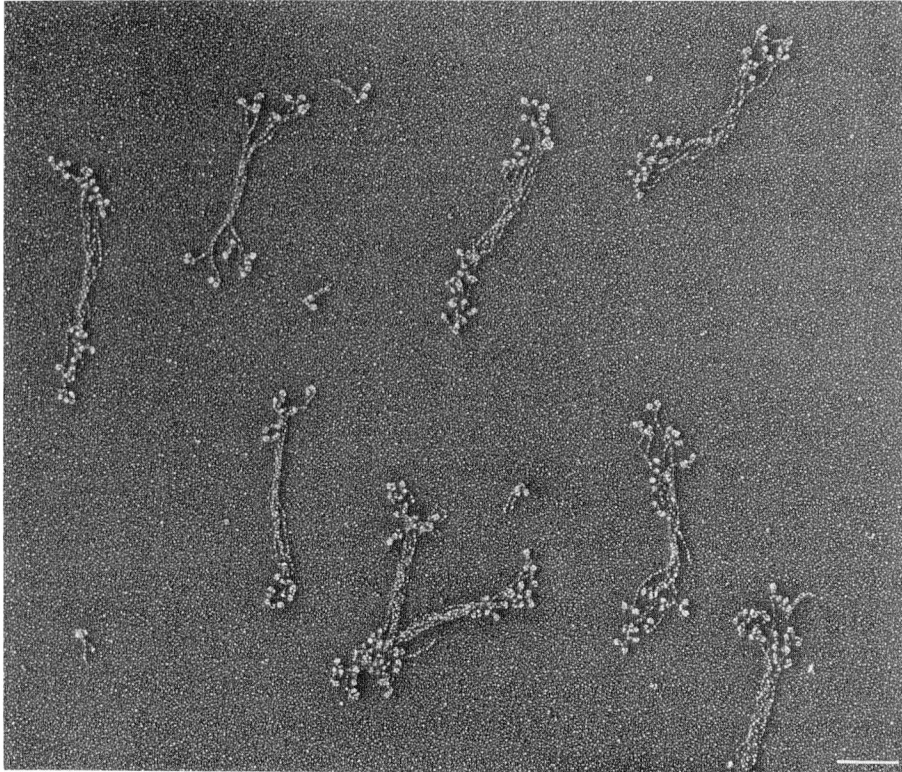

FIGURE 6 Electron micrographs of metal-shadowed minifilaments. These homogeneous short bipolar filaments are composed of 13–14 molecules, with an average filament length of 393 ± 33 nm (5 mM pyrophosphate at pH 7.5). Addition of salt to minifilaments causes them to grow into longer side-polar filaments (see Fig. 5). Note the folded monomers in equilibrium with polymer. Bar-0.1 μm. Reproduced from the *Journal of Cell Biology,* 1987, **105**, 3007–3019 by copyright permission of The Rockefeller University Press.

could be to maximize cross-bridge interactions over a wide range of cell lengths.

C. Role of the Nonhelical Tailpiece

Smooth and nonmuscle myosin rods terminate in a nonhelical tailpiece that is not present in sarcomeric muscle myosins. Two different-length isoforms of this tailpiece have been identified (SM1 and SM2). Since the strongest interactions necessary for assembly have been localized to the C terminus of the rod, there have been suggestions that the two tailpieces may differentially influence filament stability or packing. Moreover, some smooth muscle myosins, such as that from aorta, contain a serine residue in the nonhelical tailpiece that can be phosphorylated in cells by casein kinase II (Kelley and Adelstein, 1990). This site is present only in the longer of the two tailpieces, and it is notably absent from gizzard myosin, where the homologous phosphorylatable Ser residue is Gly.

Consistent with the tailpiece playing a role in assembly, cleavage of this region from gizzard myosin by limited proteolysis increased myosin's ability to polymerize (Ikebe *et al.,* 1991). In contrast, expressed pieces of the rod from a vertebrate nonmuscle myosin from chicken epithelium showed that assembly was markedly decreased upon deletion of the tailpiece (Hodge *et al.,* 1992). A twofold reduction in the size of the tailpiece had little effect on assembly compared with a full-length tailpiece. Both the proteolytic and expression studies suggest that the "critical concentration" for assembly has been altered by the absence of the tailpiece, but the studies disagree on whether the tailpiece enhances or decreases assembly.

Two mechanisms for the role of the tailpiece in assembly have been proposed. One possibility is that the nonhelical tailpiece engages in a specific interaction with a sequence found in the rod of a neighboring molecule. This pathway would then specify a preferred overlap between molecules within a filament. Since Hodge *et al.* (1992) found no evidence of a specific interaction between a tailpiece and the rod of another molecule, they proposed instead that the tailpiece prohibits dominant but unproductive interactions, and by default thus specifies an alternative mode of packing.

D. Filament Assembly–Disassembly in Vivo

In vitro experiments clearly show that purified dephosphorylated myosin in the presence of MgATP preferentially folds into a soluble folded monomer, with intramolecular interactions between portions of the rod. Because of this lability, it is not too surprising that early attempts to visualize filaments in relaxed smooth muscle cells failed. Subsequently, electron micrographs of rapidly frozen, relaxed smooth muscle cells unequivocally showed numerous native thick filaments composed of dephosphorylated myosin (Somlyo *et al.*, 1981). This study did not rule out the possibility that a pool of folded monomers could be recruited to assemble when phosphorylated, thus forming part of smooth muscle myosin's activation pathway. To test this hypothesis, monoclonal antibodies that preferentially react with monomeric myosin were used to probe relaxed and contracted tissue. Quantitative immunofluorescence using these antibodies showed only low levels of monomeric myosin in both the relaxed and contracted states in gizzard muscle (Horowitz *et al.*, 1994). Both results are consistent with little assembly–disassembly during the smooth muscle contractile cycle. An independent study led to a different conclusion, namely, that there was a 1.6-fold increase in filament density upon contraction of the rat anococcygeus muscle (Gillis *et al.*, 1988). It is not known if this difference reflects a real difference in the magnitude of the assembly–disassembly process in different types of smooth muscle cells.

What factor(s) could be responsible for the stabilization of dephosphorylated filaments *in vivo*? One possibility is that the high intracellular concentration of myosin exceeds the critical concentration for assembly even in the dephosphorylated state. Even *in vitro*, dephosphorylated filaments can be formed in the presence of MgATP if the myosin concentration exceeds the critical concentration for assembly (Kendrick-Jones *et al.*, 1987). Second, there are data that suggest that folded dephosphorylated monomers can be induced to assemble by F-actin (Applegate and Pardee, 1992; Rosenfeld *et al.*, 1994). This result suggests that within a smooth muscle cell that has numerous actin filaments, assembled myosin would be favored. Third, a protein identical to the C-terminal portion of myosin light chain kinase, which has been referred to alternately as telokin (Ito *et al.*, 1989) or kinase-related protein, also appears to partially stabilize dephosphorylated filaments to disassembly by MgATP (Shirinsky *et*

al., 1993). Any or all of these factors may be sufficient to explain the presence of apparently stable dephosphorylated filaments in the smooth muscle cell. These explanations also raise the possibility that in a nonmuscle myosin cell, where the myosin concentration is lower and filament-stabilizing proteins may not be present, assembly–disassembly could play a major role in the regulation of myosin activity.

References

Applegate, D., and Pardee, J. D. (1992). *J. Cell Biol.* **117**, 1223–1230.

Cooke, P. H., Fay, F. S., and Craig, R. (1989). *J. Muscle Res. Cell Motil.* **10**, 206–220.

Craig, R., and Megerman, J. (1977). *J. Cell Biol.* **75**, 990–996.

Craig, R., Smith, R., and Kendrick-Jones, J. (1983). *Nature (London)* **302**, 436–439.

Cremo, C. R., Sellers, J. R., and Facemyer, K. C. (1995). *J. Biol. Chem.* **270**, 2171–2175.

Cross, R. A., and Engel, A. (1991). *J. Mol. Biol.* **222**, 455–458.

Cross, R. A., and Vandekerckhove, J. (1986). *FEBS Lett.* **200**, 355–360.

Cross, R. A., Cross, K. E., Sobieszek, A. (1986). *EMBO J.* **5**, 2637–2641.

Cross, R. A., Jackson, A. P., Citi, S., Kendrick-Jones, J., and Bagshaw, C. R. (1988). *J. Mol. Biol.* **203**, 173–181.

Cross, R. A., Geeves, M. A., and Kendrick-Jones, J. (1991). *EMBO J.* **10**, 747–756.

Dillon, P. F., Aksoy, M. O., Driska, S. P., and Murphy, R. A. (1981). *Science* **211**, 495–497.

Gillis, J. M., Cao, M. L., and Godfraind-DeBecker, A. (1988). *J. Muscle Res. Cell Motil.* **9**, 18–28.

Goodson, H. V., and Spudich, J. A. (1993). *Proc. Natl. Acad. Sci. U.S.A.* **90**, 659–663.

Hodge, T. P., Cross, R., and Kendrick-Jones, J. (1992). *J. Cell Biol.* **118**, 1085–1095.

Horowitz, A., and Trybus, K. M. (1992). *J. Biol. Chem.* **267**, 26091–26096.

Horowitz, A., Trybus, K. M., Bowman, D. S., and Fay, F. S. (1994). *J. Cell Biol.* **126**, 1195–1200.

Ikebe, M., Hewett, T. E., Martin, A. F., Chen, M., and Hartshorne, D. J. (1991). *J. Biol. Chem.* **266**, 7030–7036.

Ikebe, M., Ikebe, R., Kamisoyama, H., Reardon, S., Schwonek, J. P., Sanders, C. R., II, and Matsuura, M. (1994a). *J. Biol. Chem.* **269**, 28173–28180.

Ikebe, M., Reardon, S., Mitani, Y., Kamisoyama, H., Matsuura, M., and Ikebe, R. (1994b). *Proc. Natl. Acad. Sci. U.S.A.* **91**, 9096–9100.

Ito, M., Dabrowska, R., Guerriero, V., Jr., and Hartshorne, D. J. (1989). *J. Biol. Chem.* **264**, 13971–13974.

Kamisoyama, H., Araki, Y., and Ikebe, M. (1994). *Biochemistry* **33**, 840–847.

Kelley, C. A., and Adelstein, R. S. (1990). *J. Biol. Chem.* **265**, 17876–17882.

Kendrick-Jones, J., Szent-Györgyi, A. G., and Cohen, C. (1971). *J. Mol. Bol.* **59**, 527–529.

Kendrick-Jones, J., Smith, R. C., Craig, R., and Citi, S. (1987). *J. Mol. Biol.* **198**, 241–252.

Lee, R. J., Egelhoff, T. T., and Spudich, J. A. (1994). *J. Cell. Sci.* **107**, 2875–2886.

McLachlan, A. D., and Karn, J. (1982). *Nature (London)* **299**, 226–231.

Onishi, H., and Wakabayashi, T. (1982). *J. Biochem. (Tokyo)* **92**, 871–879.

Rayment, I., Rypniewski, W. R., Schmidt-Base, K., Smith, R., Tomchick, D. R., Benning, M. M., Winkelmann, D. A., Wesenberg, G., and Holden, H. M. (1993). *Science, 261,* 50–58.

Rosenfeld, S. S., Xing, J., Rener, B., Lebowitz, J., Kar, S., and Cheung, H. C. (1994). *J. Biol. Chem.* **269,** 30187–30194.

Shirinsky, V. P., Vorotnikov, A. V., Birukov, K. G., Nanaev, A. K., Collinge, M., Lukas, T. J., Sellers, J. R., and Watterson, D. M. (1993). *J. Biol. Chem.* **268,** 16578–16583.

Somlyo, A. V., Butler, T. M., Bond, M., and Somlyo, A. P. (1981). *Nature (London)* **294,** 567–569.

Sprang, S. R., Acharya, K. R., Goldsmith, E. J., Stuart, D. I., Varvill, K., Fletterick, R. J., Madsen, N. B., and Johnson, L. N. (1988). *Nature (London)* **336,** 215–221.

Suzuki, H., Onishi, H., Takahashi, K., and Watanabe, S. (1978). *J. Biochem. (Tokyo)* **84,** 1529–1542.

Sweeney, H. L., Yang, Z., Zhi., G., Stull, J. T., and Trybus, K. M. (1994). *Proc. Natl. Acad. Sci. U.S.A.* **91,** 1490–1494.

Trybus, K. M. (1989). *J. Cell Biol.* **109,** 2887–2894.

Trybus, K. M. (1991a). *Curr. Opin. Cell Biol.* **3,** 105–111.

Trybus, K. M. (1991b). *Cell Motil. Cytoskel.* **18,** 81–85.

Trybus, K. M. (1994). *J. Biol. Chem.* **269,** 20819–20822.

Trybus, K. M., and Chatman, T. A. (1993). *J. Biol. Chem.* **268,** 4412–4419.

Trybus, K. M., and Henry, L. (1989). *J. Cell Biol.* **109,** 2879–2886.

Trybus, K. M., and Lowey, S. (1984). *J. Biol. Chem.* **259,** 8564–8571.

Trybus, K. M., and Lowey, S. (1987). *J. Cell Biol.* **105,** 3007–3019.

Trybus, K. M., and Lowey, S. (1988). *J. Biol. Chem.* **263,** 16485–16492.

Trybus, K. M., Huiatt, T. W., and Lowey, S. (1982). *Proc. Natl. Acad. Sci. U.S.A.* **79,** 6151–6155.

Trybus, K. M., Waller, G. S., and Chatman, T. (1994). *J. Cell Biol.* **124,** 963–969.

Trybus, K. M., Carmichael, J., Yagi, S., Sweeney, H. L., and Fay, F. S. (1995). *Biophys. J.* **68,** A62.

Warshaw, D. M., Desrosiers, J. M., Work, S. S., and Trybus, K. M. (1990). *J. Cell Biol.* **111,** 453–463.

Xie, X., Harrison, D. H., Schlichting, I., Sweet, R. M., Kalabokis, V. N., Szent-Gyorgyi, A. G., and Cohen, C. (1994). *Nature (London)* **368,** 306–312.

Yang, Z., and Sweeney, H. L. (1995). *Biophys. J.* **68,** A62.

Yount, R. G., Lawson, J. D., and Rayment, I. (1995). *Biophys. J.* **68,** 44s–49s.

4

Actin and the Structure of Smooth Muscle Thin Filaments

WILLIAM LEHMAN
Department of Physiology,
Boston University School of Medicine,
Boston, Massachusetts

PETER VIBERT
Rosenstiel Basic Medical Sciences Research Center,
Brandeis University,
Waltham, Massachusetts

ROGER CRAIG
Department of Cell Biology,
University of Massachusetts Medical School,
Worcester, Massachusetts

MICHAEL BÁRÁNY
Department of Biochemistry,
College of Medicine,
University of Illinois at Chicago,
Chicago, Illinois

I. INTRODUCTION

Actin plays a vital role in numerous motile and cytoskeletal processes. The diversity of actin isoforms in different muscle types and in nonmuscle tissues is consistent with a protein that displays varied function. Despite the occurrence of multiple isoforms, however, actin amino acid sequences, monomer–monomer interactions, and therefore capacity to form F-actin filaments are highly conserved, and the significance of the isoform distribution is not clear. The high degree of isoform homology is illustrated by the virtual identity of the α-carbon paths taken by amino acids within the atomic structures of α-(muscle) (Kabsch *et al.*, 1990; Holmes and Kabsch, 1991) and β-(cytoplasmic) actin (Schutt *et al.*, 1993). What largely determines and distinguishes tissue-specific thin filament function is not the F-actin assembly per se but its interaction with a host of unique actin-binding proteins. Actin monomers and their associations in F-actin are flexible enough to allow interactions with different binding proteins, and this presumably accounts for such functional diversity (Orlova and Egelman, 1993). It is no doubt also significant that the limited differences that do exist among actin isoforms are largely confined to

N-terminal amino acids lying at the periphery of the molecules. Although not directly involved in filament formation, they are sites for accommodating interactions with varied actin-binding proteins (Vandekerckhove and Weber, 1978; Schutt *et al.*, 1993).

In Section II, we will concentrate on characterizing smooth muscle actin and its isoforms, primarily from a biochemical perspective. In Section III, we will present our current understanding of the structure of native actin assemblies, that is, of thin filaments, which is based largely on analysis of electron microscope images.

II. SMOOTH MUSCLE ACTIN

Different smooth muscles all contain large quantities of actin ranging from about 30 to 50 mg/g wet weight, corresponding to approximately 0.9–1.6 mM *in situ* (Murphy *et al.*, 1977; Hartshorne, 1987). Actin consequently accounts for between 30 and 50% of the total noncollagenous proteins in smooth muscle. As with skeletal muscle, smooth muscle actin can exist *in vitro* in two forms, namely, monomeric-"globular" G-actin and polymeric-filamentous F-actin. The molecular weight of the G-actin monomer is 42,000 (Hart-shorne, 1987); the length and hence mass of F-actin polymers in smooth muscle are uncertain.

A. Methods

1. Preparation of Smooth Muscle Actin

In spite of the high actin content of smooth muscle, preparation of smooth muscle actin is not straightforward. Existing procedures vary from dehydrating muscle with acetone as a first step (Carsten, 1965; Vandekerckhove and Weber, 1979a) to extraction with low-salt (Elce *et al.*, 1981; Cavadore *et al.*, 1985; Ebashi, 1985; Strzelecka-Golaszewska *et al.*, 1985a) or high-salt solutions (Strzelecka-Golaszewska *et al.*, 1980) as a first step prior to acetone treatment. In one of our laboratories (M.B.), actin is prepared from fresh muscle mince that has been extracted with 300 mM KCl, 50 mM histidine buffer (pH 6.9), 1 mM EDTA, 0.5 mM phenylmethylsulfonyl fluoride, and 10 μg leupeptin/ml during brief stirring at 4°C. After centrifugation, the supernatant containing extracted myosin, tropomyosin, calponin, caldesmon, soluble sarcoplasmic proteins, and relatively small amounts of actin is discarded. The pelleted protein containing residual actin is washed with 0.4% NaHCO$_3$, followed by two distilled water washes, and is air-dried after acetone treatment. G-actin is extracted from the resulting acetone powder with 0.2 mM ATP, 0.2 mM CaCl$_2$ at pH 7.0 and 23–25°C for 1 hr and, following filtration through paper pulp, is polymerized into F-actin by addition of 50 mM KCl. Traces of bound tropomyosin are solubilized by the addition of KCl to 600 mM and pure F-actin collected by centrifugation at 176,000 g for 1 hr. The actin pellet is rinsed with distilled water to remove excess KCl and stored at 4°C with small amounts of appropriate buffer added. Before use, the pellet is dispersed in the desired solution. Alternatively, F-actin can be easily and directly prepared from native thin filaments by sedimentation at high salt concentration using the method of Marston and Smith (1984). Actin prepared by these procedures migrates as a single band on one-dimensional SDS–PAGE. Two-dimensional gel electrophoresis, however, demonstrates that smooth muscle actin is a mixture of two or more isoforms.

2. Preparation of Antibodies to Smooth Muscle Actin

Considering the ubiquitous distribution of actin and its highly conserved amino acid sequence, it is not surprising that actin is a poor antigen, regardless of its source. Cavadore *et al.* (1987), however, were able

to prepare antibodies to actin that were specific for the different smooth muscle isoforms. Using glutaraldehyde-stabilized actin as antigen, immune sera were purified by affinity chromatography using isoform-specific N-terminal actin fragments. This procedure yielded several sets of oligoclonal antibodies, including those specific for either α- or γ-smooth muscle actins as well as those for the β- or γ-cytoplasmic actins present in smooth muscle.

3. Determination of Actin Content of Smooth Muscle

Actin in smooth muscle is principally polymerized and restricted to thin filaments; little is present as G-actin (Lehman *et al.*, 1989). The F-actin component of thin filaments is the only known muscle constituent that contains stoichiometric amounts of strongly bound and inexchangeable nucleotide (namely, ADP) as a ligand ($K_m \sim 10^{-10}$ M). Measurement of bound ADP therefore is very specific for assessing actin content in any muscle (Bárány *et al.*, 1992). To make this determination, analytically weighed homogenized muscle samples are washed by sedimentation and resuspension in 0.04 M NaCl to remove free nucleotides. The pelleted residue containing the actin-bound ADP then is extracted with 3.5% perchloric acid. The ADP thereby recovered is quantified spectrophotometrically at 257 nm ($\varepsilon = 14.9 \times 10^3$) and then related to muscle weight. Alternatively, actin content can also be easily estimated by quantitative densitometry of SDS-gel electrophoretograms of total muscle homogenates (Haeberle *et al.*, 1992; Lehman *et al.*, 1993).

B. Characteristic Properties

1. Polymerization

In vitro, the presence of physiological salt concentrations causes G-actin to polymerize to form the F-actin helix, which in native thin filaments represents the backbone of the filament. Polymerization of G-actin into F-actin can be followed by measuring light-scattering or by viscometry measurements. As in skeletal muscle, polymerization of smooth muscle actin depends on actin concentration, temperature, and ionic conditions (Strzelecka-Golaszewska *et al.*, 1980). For both gizzard and bovine aortic muscle actin, the polymerization rate is optimal at 100 mM KCl and 1 mM MgCl$_2$ (Cavadore *et al.*, 1985).

Electron microscopy of negatively stained F-actin preparations shows large variation in filament length (Strzelecka-Golaszewska *et al.*, 1980; Marston and Smith, 1984; Cavadore *et al.*, 1985), which is not surprising since F-actin is easily severed by mechanical perturbation. However, gelsolinlike actin-severing proteins present in F-actin preparations may also contribute to producing unexpectedly short filaments. There is no obvious connection between the observed length distribution of "synthetic" F-actin and of native filaments.

2. Ligand Binding

Actin contains single adenosine nucleotide and divalent cation binding sites (Estes *et al.*, 1992). When either site is unoccupied, actin is rendered unstable and denatures. ATP is the nucleotide moiety bound to G-actin, but, during F-actin polymerization and concomitant ADP hydrolysis, the ADP formed remains bound. Reports are available that in smooth muscle, *in vivo*, the bound metal could be Ca^{2+} (Strzelecka-Golaszewska *et al.*, 1980; Bárány *et al.*, 1992). *In vitro*, nucleotide and divalent ions (either Mg^{2+} or Ca^{2+}) bound to G-actin are freely exchangeable with exogenous ligand; however, after polymerization, the ligands become virtually inexchangeable and trapped in the interior of the protein (see Fig. 1).

3. Activation of Myosin ATPase

The degree of activation of myosin Mg^{2+}-ATPase activity by F-actin, an inherent and physiologically vital property of actin from all muscle sources, has proven difficult to quantify in smooth muscle preparations. Smooth muscle actin purified by various procedures by different investigators not unexpectedly yielded widely varied levels of activation (Próchniewicz and Strzelecka-Golaszewska, 1980; Cavadore *et al.*, 1985; Strzelecka-Golaszewska and Sobieszek, 1981; Carsten, 1965; Ngai *et al.*, 1986). However, when F-actin from smooth and skeletal muscle sources was prepared very gently by the method of Marston and Smith (1984), no sign of differences was apparent. *In vitro* motility assays also confirmed that little functional difference exists between skeletal and smooth muscle F-actin (Harris and Warshaw, 1993). In these studies, the velocities of smooth and skeletal muscle actin filaments moving over myosin-coated surfaces were the same under comparable conditions and depended only on the type of myosin used. Hence, under very controlled conditions, skeletal and smooth

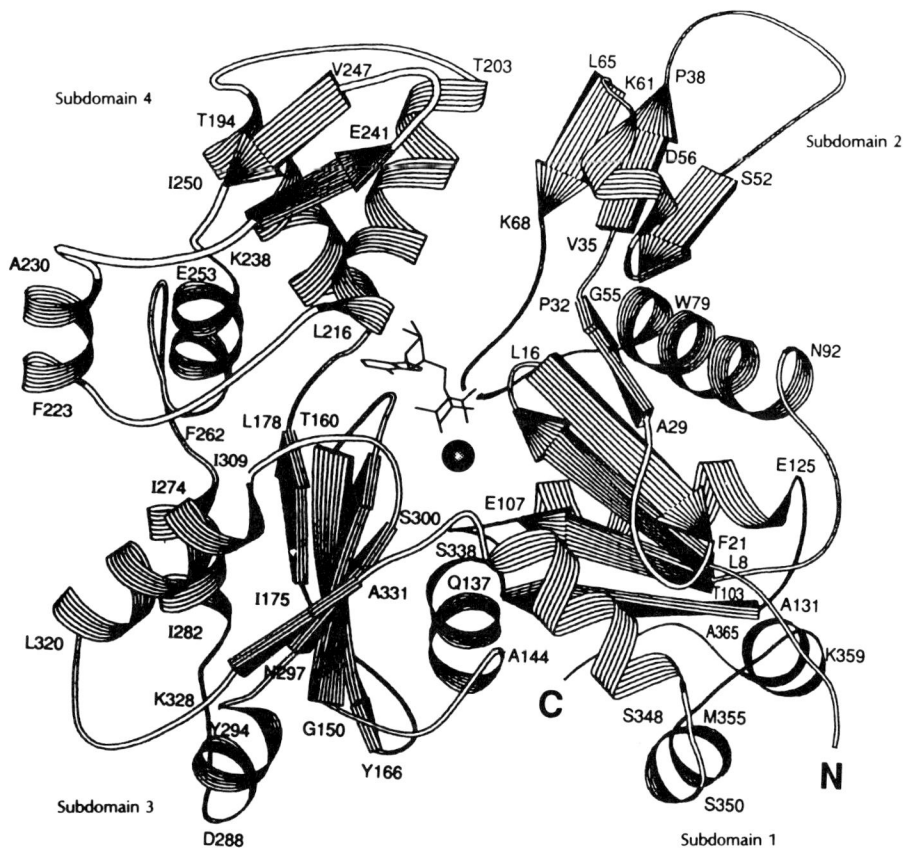

FIGURE 1   Schematic representation of the structure of the actin molecule. ATP and Ca^{2+} are located in the cleft between the major domains. From Kabsch *et al.* (1990, Fig. 1, p. 38). Reprinted with permission from *Nature* **347,** 37–44. Copyright 1990 Macmillan Magazines Limited.

muscle F-actin seemed functionally equivalent and therefore effectively interchangeable.

In both smooth and skeletal muscles, the strong-"rigor" interaction of F-actin and myosin occurring in the absence of ATP involves F-actin binding to double-headed myosin with an expected 2:1 stoichiometry in both smooth and skeletal muscles (Próchniewicz and Strzelecka-Golaszewska, 1980).

C. Isoforms

Actin is a highly conserved protein, and actin variants display approximately 95% amino acid sequence homology (Pollard and Cooper, 1986). In vertebrate tissues, six different actin isoforms have been identified on the basis of electrophoretic mobility and sequence data, and each is encoded by a distinct gene (Vandekerckhove and Weber, 1978; Reddy *et al.*, 1990). α, β, and γ species can be easily separated by isoelectric focusing; the α-isoform is the most acidic and the γ-isoform the least (Vandekerckhove and Weber, 1979a). These variants can be further subdivided on the basis of their amino acid sequence and include three α-isoforms (α-skeletal, α-cardiac, and α-vascular), two γ-isoforms (γ-enteric and γ-cytoplasmic), and one β-isoform (β-cytoplasmic).

Expression of the different actin genes is tissue and not species specific. In visceral smooth muscle (e.g., in esophagus of cat, opossum, and swine and in turkey gizzard), the γ-actin content is relatively high, whereas in vascular smooth muscle (as in carotid artery of swine and dog), α-actin content is high (Fatigati and Murphy, 1984; Hartshorne, 1987). Apparently, elevated levels of α-actin are associated with tissues displaying relatively high tonic activity, whereas γ-actin is characteristic of more phasic contractility. The reason for this divergence is not known. Some smooth muscle tissues contain mixtures of both α- and γ-isoforms, with one form predominating, whereas others display one of the two exclusively (Hartshorne, 1987). The cytoplasmic β-isoform is a significant component in all smooth muscle tissues, and may be part of the smooth muscle cytoskeleton (North *et al.*, 1994b; Gimona and Small, Chapter 7, this volume).

During the development of smooth muscle cells there is a gradual reduction in the relative amount of cytoplasmic β-actin and increase in either α- or γ-"contractile" actin. In tonic aortic muscle of rats, for example, α-actin levels increase over 1.5 times from Day 3 (postpartum) to adult (Day 21), whereas other isoforms decrease (Eddinger and Murphy, 1991). In the case of developing chicken gizzard, containing phasic muscle, γ-actin increases whereas β-actin decreases (Hirai and Hirabayashi, 1983). Corresponding changes occur in uterine muscle during pregnancy (Cavaillé *et al.*, 1986). Hypertrophy of rat portal vein interestingly is associated with an increase in γ-actin and a decrease in α-actin (Malmqvist and Arner, 1990).

D. Amino Acid Sequence

Vandekerckhove and Weber (1979a) determined the complete amino acid sequence of smooth muscle and cytoplasmic actin isoforms (see Fig. 2 for the sequence of bovine aortic muscle α-actin). Comparison of α-smooth muscle actin with skeletal muscle actin reveals great homology with only eight amino acid substitutions noted (cf. Collins and Elzinga, 1975). Strzelecka-Golaszewska *et al.* (1985b) considered that substitution of Cys-17 for Val-17 and/or Ser-89 for Thr-89 may be responsible for decreased stability of the native conformation of purified smooth muscle G-actin, which may partly explain the variability in previously mentioned ATPase data obtained with synthetic F-actin. Comparison of the bovine aortic muscle α-actin sequence (Fig. 2) with that of chicken gizzard γ-actin indicates three additional conservative substitutions, all located at the amino-terminal end of the molecules (Vandekerckhove and Weber, 1979b). β-cytoplasmic actins show another set of differences involving 22–23 residues (Vandekerckhove and Weber, 1979a). The latter presumably are responsible for defining certain structural distinctions observed by Schutt *et al.* (1993).

Table I illustrates the variability of the amino-terminal residues of several different actin isoforms. Mutagenesis in this region can result in partial or complete inhibition of F-actin motility (Sutoh, 1993), consistent with the site being a likely initial target for myosin during the cross-bridge cycle (Rayment *et al.*, 1993b). In fact, numerous studies also have identified several domains of the actin molecule representing clusters of amino acids specifically involved in monomer–monomer interactions (Holmes *et al.*, 1990; Hennessey *et al.*, 1993; Khaitlina *et al.*, 1993; Labbé *et al.*, 1994), actin–myosin interactions (Holmes and Kabsch, 1991; Hennessey *et al.*, 1993; Schröder *et al.*, 1993), actin–tropomyosin interactions (Holmes and Kabsch, 1991; Poole *et al.*, 1994), and actin–caldesmon interactions (Crosbie *et al.*, 1991; Graceffa and Jansco, 1991; Graceffa *et al.*, 1993).

E. Interaction with Thin Filament Proteins

The characteristic stoichiometry of tropomyosin binding, in which one molecule of tropomyosin interacts with approximately 7 actin monomers of F-actin, is the same in skeletal and smooth muscles (Matsumura

```
              10                    20                    30                    40
X-E-E-E-D-S-T-A-L-V-C-D-N-G-S-G-L-C-K-A-G-F-A-G-D-D-A-P-R-A-V-F-P-S-I-V-G-R-P-R-H-

              50                    60                    70    Me              80
Q-G-V-M-V-G-M-G-Q-K-D-S-Y-V-G-D-E-A-Q-S-K-R-G-I-L-T-L-K-Y-P-I-E-H-G-I-I-T-N-W-D-

              90                    100                   110                   120
D-M-E-K-I-W-H-H-S-F-Y-N-E-L-R-V-A-P-E-E-H-P-T-L-L-T-E-A-P-L-N-P-K-A-N-R-E-K-M-T-

              130                   140                   150                   160
Q-I-M-F-E-T-F-N-V-P-A-M-Y-V-A-I-Q-A-V-L-S-L-Y-A-S-G-R-T-T-G-I-V-L-D-S-G-D-G-V-T-

              170                   180                   190                   200
H-N-V-P-I-Y-E-G-Y-A-L-P-H-A-I-M-R-L-D-L-A-G-R-D-L-T-D-Y-L-M-K-I-L-T-E-R-G-Y-S-F-

              210                   220                   230                   240
V-T-T-A-E-R-E-I-V-R-D-I-K-E-K-L-C-Y-V-A-L-D-F-E-N-E-M-A-T-A-A-S-S-S-S-L-E-K-S-Y-

              250                   260                   270                   280
E-L-P-D-G-Q-V-I-T-I-G-N-E-R-F-R-C-P-E-T-L-F-Q-P-S-F-I-G-M-E-S-A-G-I-H-E-T-T-Y-N-

              290                   300                   310                   320
S-I-M-K-C-D-I-D-I-R-K-D-L-Y-A-N-N-V-L-S-G-G-T-T-M-Y-P-G-I-A-D-R-M-Q-K-E-I-T-A-L-

              330                   340                   350                   360
A-P-S-T-M-K-I-K-I-I-A-P-P-E-R-K-Y-S-V-W-I-G-G-S-I-L-A-S-L-S-T-F-Q-Q-M-W-I-S-K-Q-

              370         375
E-Y-D-E-A-G-P-S-L-V-H-R-K-C-F
```

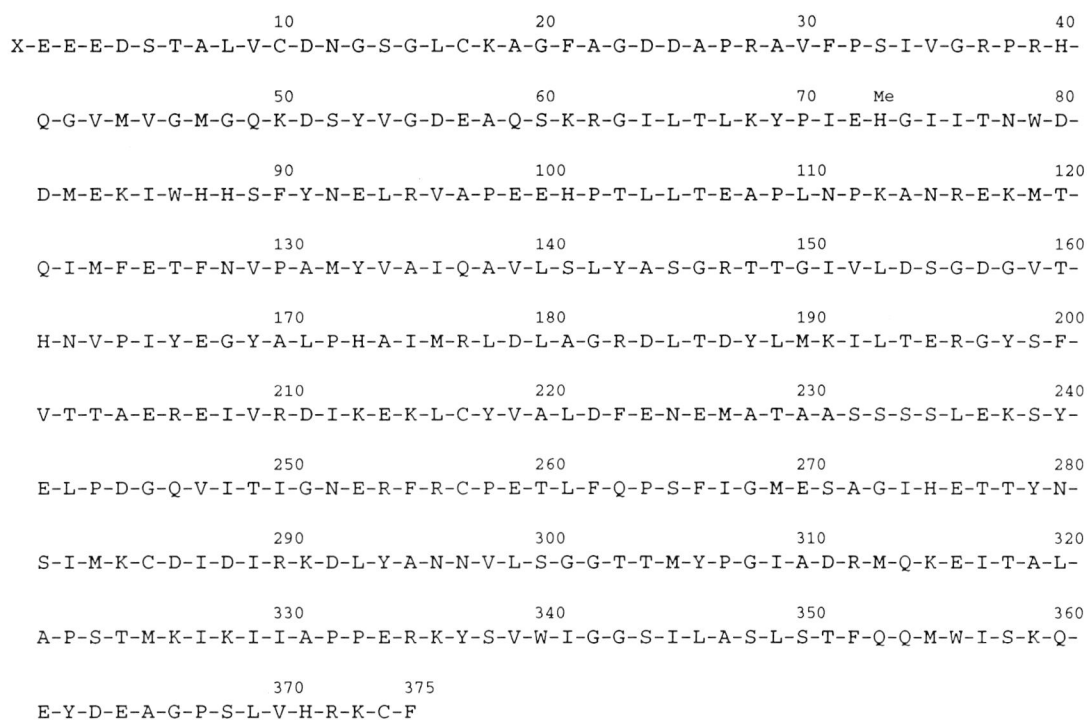

FIGURE 2   Amino acid sequence of bovine aortic muscle α-actin. Residue 73 is 3-methylhistidine. The 234a serine residue in the original sequence has been changed to serine 235, thus the sequence has a total of 375 residues. X refers to a blocked amino terminus, most likely an acetyl group. Adapted from Vandekerckhove and Weber (1979a, Fig. 2, p. 125). Reprinted with permission of Springer-Verlag.

and Lin, 1982; Hartshorne, 1987). Stoichiometries of the smooth muscle-specific actin-binding proteins, caldesmon and calponin, are not as certain. Estimates reported for caldesmon binding to actin and tropomyosin vary (Hartshorne, 1987; Marston and Redwood, 1991), although the probable value is 1 cal-

desmon: 2 tropomyosins: 14 actins (mol:mol:mol) on caldesmon-specific thin filaments (Lehman et al., 1989; Marston, 1990) and 1 caldesmon: 5–6 tropomyosins: 35–40 actins per total cell actin (Lehman et al., 1993). The molar ratio of calponin:tropomyosin:actin in vivo appears to be closer to 1:1:7 in a variety of smooth muscles (Takahashi et al., 1986; Lehman, 1991; Bárány and Bárány, 1993).

F. Modulation of Polymerization

Protein factors that modulate actin polymerization have been isolated from various smooth muscles [e.g., from porcine stomach (Hinssen et al., 1984), bovine aorta (Strzelecka-Golaszewska et al., 1985a), and chicken gizzard (Schröer and Wegner, 1985; Ruhnau et al., 1989)]. As is evident from viscometry and electron microscopy, these so-called "modulator" proteins inhibit G-actin polymerization and also depolymerize assembled F-actin. The modulator proteins isolated from stomach and aorta required Ca^{2+} for their activity and formed a 1:2 complex with actin; they apparently could interact along the length of F-actin filaments. In contrast, activity of the gizzard protein(s) was independent of the Ca^{2+} concentration and appeared to

TABLE I   Comparison of the Amino-Terminal Amino Acids in Smooth and Nonmuscle Actins[a]

	Smooth muscle		Nonmuscle	
	Aorta	Gizzard		
Residue	α	β	β	γ
1	E	—	—	—
2	E	E	D	E
3	E	E	D	E
4	D	E	D	E
5	S	T	I	I
6	T	T	A	A

[a]Adapted from Vandekerckhove and Weber (1979a). Note that residue 1 is absent in gizzard muscle and in nonmuscle actins, which explains their higher isoelectric points relative to aortic muscle α-actin. The blocked amino terminus is not indicated here.

bind to and depolymerize F-actin filaments only from their ends. The porcine stomach protein was purified to homogeneity and migrated as a single band of 85 kDa on SDS-PAGE; the gizzard protein(s) were heterogeneous (20 to 80 kDa).

Modulation of smooth muscle filament length and associations may be very complex and additional proteins such as insertin may be involved in treadmilling of filaments and/or introducing actin molecules between smooth muscle membranes and filaments (Ruhnau *et al.*, 1989). Phosphorylation of talin, another actin-binding protein, may also modulate the attachment of actin filaments to specific sites on plasma membranes (Pavalko *et al.*, 1994). Another actin-binding protein of 25 kDa that was isolated from bovine aorta may also be involved in membrane–cytoskeleton linkage (Kobayashi *et al.*, 1994). If any of these interactions are dynamic, they could indirectly regulate force generation.

III. THIN FILAMENT STRUCTURE

In this section, we will examine the structural impact and function of one class of actin-binding proteins that characterize and distinguish smooth and skeletal muscle thin filaments. These so-called regulatory proteins, namely, tropomyosin and troponin in skeletal muscle, and tropomyosin, caldesmon, and possibly calponin in smooth muscle, form muscle-type specific complexes with F-actin and modulate actomyosin ATPase. The control of contractile activity by troponin–tropomyosin in skeletal muscle has long been acknowledged; thin filament regulation in smooth muscle acting in concert with the well-recognized myosin–phosphorylation control, however, is not as universally accepted.

A. F-actin and Subfragment-1-Decorated F-actin Structure

It is useful to first consider the structure of muscle's mechanochemical transduction elements, actin and the subfragment-1 (S-1) portion of myosin, before analyzing the effects of regulatory proteins that modulate interactions of the two and have the potential therefore to control force generation. Molecular details of the structure of skeletal muscle actin (Kabsch *et al.*, 1990; Holmes and Kabsch, 1991; Lorenz *et al.*, 1993) and S-1 (Rayment *et al.*, 1993a) are known at the atomic level (Fig. 1), and given the general similarity of F-actin and S-1 in smooth and in skeletal muscle, it is reasonable to assume that they are closely related in the two

tissues. Crystallographic information on the actin monomer confirmed that it consists of two domains (inner and outer), as previously indicated by three-dimensional reconstruction of electron micrographs of F-actin (Milligan *et al.*, 1990), and also showed that both inner and outer domains could be further divided into subdomains (3 and 4, 1 and 2, respectively; see Fig. 1). An atomic model of actin connectivity in thin filaments constructed by precisely fitting atomic coordinates of the actin molecule to lower-resolution diffraction data on F-actin (Holmes *et al.*, 1990; Rayment *et al.*, 1993b) again is consistent with the structure of F-actin visualized by three-dimensional reconstruction of electron micrograph images (Milligan *et al.*, 1990). In this model, the well-known helical features of the actin filament are obtained, namely, a right-handed long-pitch double helix of actin monomers with two strands crossing at ~36 nm and a left-handed genetic helix of 5.9-nm pitch with 13 subunits per ~36 nm. Since actin molecules are now known to be asymmetric and certainly are not spherical, traditional models of F-actin depicting double helices of spheres and generating well-defined "grooves" are inaccurate and often misleading.

Atomic models of S-1-decorated F-actin constructed by a process of fitting the atomic structures of both actin and S-1 into the envelope formed by three-dimensional reconstructions of decorated actin (Rayment *et al.*, 1993b) provide important constraints when considering the operation of the S-1-cross-bridge cycle on actin during contraction (Rayment *et al.*, 1993b). It is currently thought that S-1 makes an initial weak contact on the subdomain-1 of the outer domain of actin near the N terminus followed by a strong stereospecific interaction closer to the junction of the inner and outer domains between subdomains 1 and 3; it is the latter tight bond that then leads to or is part of force generation (Rayment *et al.*, 1993b). These considerations are of particular relevance here and provide a structural reference and mechanistic framework for understanding the impact of regulatory proteins on actomyosin interaction in both skeletal and smooth muscle. Our progress in elucidating the impact of thin filament-linked regulatory proteins has benefitted greatly from comparing and contrasting control systems in different tissues. As a means of understanding the role and structural effect of smooth muscle actin-linked proteins, this comparative approach will be taken here in the following discussion. Cogent aspects of the well-characterized troponin–tropomyosin system will be summarized to serve as a foundation for discussion of possible thin filament-linked mechanisms in smooth muscle.

B. Thin Filament Structure in Skeletal Muscle

1. Distribution of Regulatory Proteins on Thin Filaments

Our present understanding of thin filament regulation in general and troponin–tropomyosin action in particular has been achieved in great part because of the elegant structural studies performed on isolated and reconstituted thin filaments and on purified troponin and tropomyosin. Tropomyosin is found on thin filaments in all muscles and in many nonmuscle tissues, but troponin, at least in the case of vertebrates, is limited to striated muscle. Tropomyosin in all muscles is a fibrous protein (Cohen and Szent-Györgyi, 1957) and lies on thin filaments along the long-pitch double-helical array of actin monomers (Moore et al., 1970; O'Brien et al., 1971; Spudich et al., 1972; Milligan et al., 1990). Tropomyosin molecules associate together as a strand and bind in an end-to-end and continuous fashion on thin filaments. Each tropomyosin molecule spans approximately seven actin molecules along the filaments (~38 nm). These features are common to thin filaments in all muscles. In troponin-regulated muscles, each ternary troponin complex binds to a specific region of each tropomyosin molecule along filaments (Ohtsuki et al., 1967; Ebashi and Endo, 1968; Cohen et al., 1972; Ohtsuki, 1974; Flicker et al., 1982); troponin therefore assumes the tropomyosin periodicity of 38 nm. The length of a single troponin complex (~26.5 nm) is shorter than tropomyosin and each extends over about two-thirds of a single tropomyosin molecule (ibid.) Troponin subunit-T is a fairly long asymmetric molecule (~19 nm long), whereas troponin-C and -I are more globular and in the ternary complex bind at the C-terminal end of troponin-T (Flicker et al., 1982). Troponin-T is thought to bind alongside of tropomyosin on thin filaments, bridging the head-to-tail joint between adjacent tropomyosin molecules and then extending over and interacting with the C-terminal half of tropomyosin where troponin-C and -I become localized (Fig. 3). This interaction may impose constraints on tropomyosin position on actin, which are released after Ca^{2+} binding to troponin-C following muscle activation (Phillips et al., 1986).

2. The Steric-Blocking Model

X-ray diffraction studies on intact muscle suggest that the binding of Ca^{2+} to troponin induces changes in the relationship of tropomyosin and actin on thin filaments (Haselgrove, 1972; Huxley, 1972; Parry and Squire, 1973; Kress et al., 1986). These studies are the basis for the well-known steric-blocking model of

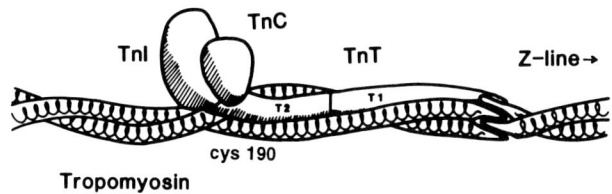

FIGURE 3 A diagram of the tropomyosin–troponin complex in vertebrate skeletal muscle. TnC, TnI, and TnT denote troponin-C, -I, and -T, respectively; T1 and T2 denote the N- and C-terminal parts of TnT. From Flicker et al. (1982, Fig. 3, p. 499). Reprinted with permission of Academic Press.

muscle regulation in which tropomyosin strands running along thin filaments move away from myosin-binding sites on actin when muscle is activated, thereby allowing cross-bridge interaction and contraction to proceed. Though analysis of such X-ray patterns can suggest models of filament structure, unambiguous interpretation is not possible because phase information is lost. In contrast, three-dimensional reconstruction of thin filament electron micrographs offers a means of direct confirmation of the tropomyosin movement. Over the past 20 years, however, attempts to reveal tropomyosin movement by three-dimensional image reconstruction have resulted in a large and often confusing literature (Cohen and Vibert, 1987). The latest three-dimensional reconstructions of skeletal muscle thin filaments indicate that Ca^{2+} binding to troponin fixes the position of tropomyosin along filaments (Milligan et al., 1990); in the absence of Ca^{2+}, however, tropomyosin was not resolved (R. A. Milligan, personal communication) and therefore steric blocking could not be visualized. This shortcoming has been overcome using troponin–tropomyosin-regulated thin filaments from the arthropod *Limulus*, which can be isolated in native form. These thin filaments confer full Ca^{2+} sensitivity on actomyosin ATPase, and the constituent regulatory proteins are functionally interchangeable with their vertebrate counterparts (Lehman et al., 1994).

Maps of *Limulus* thin filaments reveal actin monomers whose bilobed, two-domain shape and monomer–monomer connectivity are typical of F-actin (Lehman et al., 1994). In addition, longitudinally continuous strands of density, presumably tropomyosin, follow long-pitch actin helices and are evident in reconstructions carried out on filaments *both* in the absence and in the presence of Ca^{2+}. Filaments in Ca^{2+}-free buffer show the tropomyosin strands following successive actin monomers in contact with the extreme inner edge of their *outer* domains, whereas in Ca^{2+} the strands are closely associated with the *inner* domain of

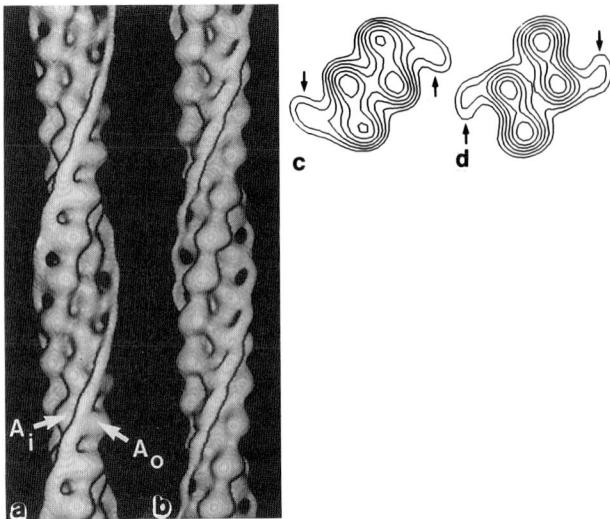

FIGURE 4 Surface views of reconstructed densities of *Limulus* thin filaments (a) in EGTA and (b) in Ca²⁺. The inner and outer domains of actin are marked. Note the difference in position of the tropomyosin strand running along A_o in (a) and A_i in (b). Helical projections (c,d) formed by projecting the above densities along their helical tracks onto a plane perpendicular to the filament axis; thin filaments (c) in EGTA and (d) in Ca²⁺. The tropomyosin strand makes contact with A_o in (c) and A_i in (d). From Lehman *et al.* (1994, Figs. 2 and 3, p. 66). Reprinted with permission from *Nature* **368,** 65–67. Copyright 1994 Macmillan Magazine Limited.

actin (Fig. 4). The effect of Ca²⁺ is to cause an approximately 25° azimuthal shift of tropomyosin strands about the actin helix axis (Lehman *et al.*, 1994). The localization of tropomyosin (and possibly extended parts of troponin, e.g., troponin-T), in both the off- and on-states, to positions originally *predicted* in the steric-blocking model of regulation (Huxley, 1972; Haselgrove, 1972; Parry and Squire, 1973; Milligan and Flicker, 1987) and proposed again on the basis of modeling (Holmes and Kabsch, 1991; Squire *et al.*, 1993; Poole *et al.*, 1994) provides direct structural support for the steric hypothesis of regulation. Reconstructions of negatively stained vertebrate skeletal muscle thin filaments in Ca²⁺-free buffer are identical to corresponding *Limulus* reconstructions and therefore evidence of tropomyosin position in the vertebrate skeletal muscle off-state, consistent with steric blocking, is now also available (Lehman *et al.*, 1995).

The position adopted by tropomyosin in the off-state coincides with a myosin contact site on actin, which as mentioned is thought to be involved in stereospecific and strong actomyosin interaction at actin subdomains 1 and 3 (Rayment *et al.*, 1993b). Results on troponin-regulated muscles therefore are compatible with a steric mechanism in which tropomyosin in relaxed muscle could interfere with the transition from

an initial weak to a strong myosin–cross-bridge association on actin, thereby inhibiting ATPase and cross-bridge cycling. The effect therefore is not to block the binding of myosin and actin entirely as first envisioned in the steric-blocking model (cf. Chalovich *et al.*, 1981; Chalovich and Eisenberg, 1982), but to hinder a part, albeit an essential part, of the binding sequence (cf. Reedy *et al.*, 1994).

C. Thin Filament Structure in Smooth Muscle

1. Thin Filament-Linked Regulation in Smooth Muscle

It has long been known that F-actin structure in thin filaments of vertebrate smooth and skeletal muscle is virtually indistinguishable (Hanson and Lowy, 1963) and that both filament types are saturated with tropomyosin (Sobieszek and Bremel, 1975). Although muscle-specific tropomyosin isoforms exist, the length of tropomyosin molecules and their periodicity on thin filaments are the same in striated and smooth muscles (Matsumura and Lin, 1982). The strength of end-to-end tropomyosin bonding, however, important in determining the equilibrium position of tropomyosin on F-actin, may be stronger in smooth muscle (Sanders and Smillie, 1984). In addition to tropomyosin, the actin-binding proteins, α-actinin and filamin, are present in both muscles; the former is part of dense bodies in smooth and Z-discs in skeletal muscle and the latter is part of the cytoskeletal scaffold in each. We will not be concerned with these cytoskeletal proteins here. In the past 10 years, it has become increasingly apparent that additional actin-binding proteins, possibly analogous to troponin, may also be present in smooth muscle and it is these that we will discuss. The possibility of thin filament-linked regulation in smooth muscle is highlighted by several key findings: (1) the isolation from smooth muscle of caldesmon, a protein with the biochemical characteristics of a thin filament modulator (Sobue *et al.*, 1981); (2) demonstration that this protein is a component of isolated *native* thin filaments capable of regulating actomyosin ATPase (Marston and Lehman, 1985; Marston *et al.*, 1988); (3) analysis showing that caldesmon affects thin filament structure, imposing constraints on tropomyosin by a mechanism that appears to differ from that of troponin in skeletal muscle (Vibert *et al.*, 1993); (4) findings that another smooth muscle protein, calponin, might also be a regulatory factor (Takahashi *et al.*, 1986, 1988a,b); and (5) evidence that one class of smooth muscle thin filaments is complexed with caldesmon and another with calponin and that these are localized separately in different cellular domains (North *et al.*, 1994a).

2. Caldesmon Distribution on Thin Filaments

Caldesmon, an extremely long (~75 nm) and flexible molecule, is found in appreciable quantity in a variety of smooth muscles and also in nonmuscle tissues (Sobue et al., 1981; Ngai et al., 1984; Marston and Lehman, 1985; Bryan and Lee, 1991; Hayashi et al., 1991; Lehman et al., 1993). By interacting with tropomyosin, caldesmon is thought to inhibit actomyosin ATPase at low Ca^{2+}-calmodulin concentration or when unphosphorylated (see Marston and Huber, Chapter 6, this volume, for details). For caldesmon to influence actin in vivo, it would be expected to be in physical contact with the greatest possible number of actin and/or tropomyosin molecules along the length of the filament. By analogy to the assembly of troponin in the skeletal muscle thin filament, its distribution might additionally be expected to be periodic. Indeed, immunolocalization studies (Fig. 5) suggest that C- and N-terminal domains of caldesmon are periodically distributed on smooth muscle thin filaments and that the entire elongated caldesmon molecule is continuously bound (Lehman et al., 1989; Moody et al., 1990). Moreover, the characteristic 38-nm repeat displayed by caldesmon molecules on thin filaments suggests that the protein's location is defined by the underlying tropomyosin periodicity. In vitro work on reconstituted preparations (Bretscher, 1984; Dabrowska et al., 1985; Moody et al., 1985) suggested that in addition to modulating actomyosin ATPase activity, caldesmon, under certain conditions, is capable of cross-linking actin filaments together. A filament-bundling function was considered, without knowing the in vitro susceptibility of caldesmon to sulfhydryl oxidation, which leads artifactually to the bundling observed (Lynch et al., 1987); hypotheses based on intermolecular caldesmon–caldesmon cross-linking are now disregarded. Close examination of native smooth muscle thin filaments by electron microscopy, in fact, reveals no strandlike projections that might represent possible free ends of caldesmon and a potential for the cross-linking function (Moody et al., 1990; Mabuchi et al., 1993; Vibert et al., 1993). In marked contrast, electron microscopy of reconstituted actin–caldesmon shows that although the C-terminal end of caldesmon is bound on F-actin, the N-terminal region projects away freely from the filament shaft (Mabuchi et al., 1993; also Moody et al., 1990). The lack of association of N-terminal portions of caldesmon in reconstituted systems could be due to the absence of tropomyosin or attributed to differences in molecular structure and/or assembly of purified and native caldesmon. Conclusions based on in vitro studies in which N-terminal caldesmon projections might additionally tether or

latch onto myosin therefore should also be regarded with caution (cf. Lash et al., 1986; Hemric and Chalovich, 1988, 1990; Ikebe and Reardon, 1988; Riseman et al., 1989; Sutherland and Walsh, 1989; Marston et al., 1990; Velaz et al., 1990; Katayama and Ikebe, 1995).

If 75-nm-long caldesmon molecules (roughly two times the length of tropomyosin) are present on thin filaments in a molar stoichiometry of approximately 1 caldesmon: 2 tropomyosins: 14 actins (Lehman et al., 1989; 1993; Marston, 1990), a simple model for thin filaments comes to mind. Here caldesmon molecules would run continuously and end-to-end with the position of each defined by contacts with two successive tropomyosin molecules on each of the long-pitch actin helices. A model of this arrangement is presented in Fig. 3 by Marston and Huber in Chapter 6, this volume, which is based on earlier depictions (Lehman et al., 1990; Marston and Redwood, 1991, and which is discussed in detail along with other possibilities in Lehman et al., 1989). Despite the superficial similarity in structural design of smooth and skeletal muscle thin filament components and the comparable effects on actomyosin ATPase by caldesmon and troponin, a homologous mechanism of action by the two regulatory proteins is not as yet evident or requisite. More detailed structural studies on smooth muscle thin filaments, analogous to those performed on skeletal muscle thin filaments in both on- and off-states, are required. Such work has been initiated by Vibert et al. (1993) and by Popp and Holmes (1992).

3. Electron Microscopy and Three-Dimensional Reconstruction of Smooth Muscle Thin Filament Images

Negative staining of thin filament samples reveals the typical double-helical array of actin monomers and, in addition, narrow asymmetric strands running helically along the filament axis (Moody et al., 1990; Vibert et al., 1993). These strands presumably represent tropomyosin or an association of tropomyosin and caldesmon. Density maps calculated from the averaged layer line data in Fourier transforms of images of filaments either isolated and maintained in Ca^{2+}-free buffer or subsequently treated with Ca^{2+}-calmodulin to dissociate caldesmon reveal bilobed actin monomers connected along both the filament's right-handed 36-nm-long-pitch helices and its left-handed 5.9-nm genetic helix (Fig. 6). The actin monomer and connectivity are indistinguishable from those in reconstructions of skeletal muscle thin filaments (Milligan et al., 1990). In addition to actin monomers, maps of filaments that had been maintained in Ca^{2+}-free buffer also display the continuous strands of den-

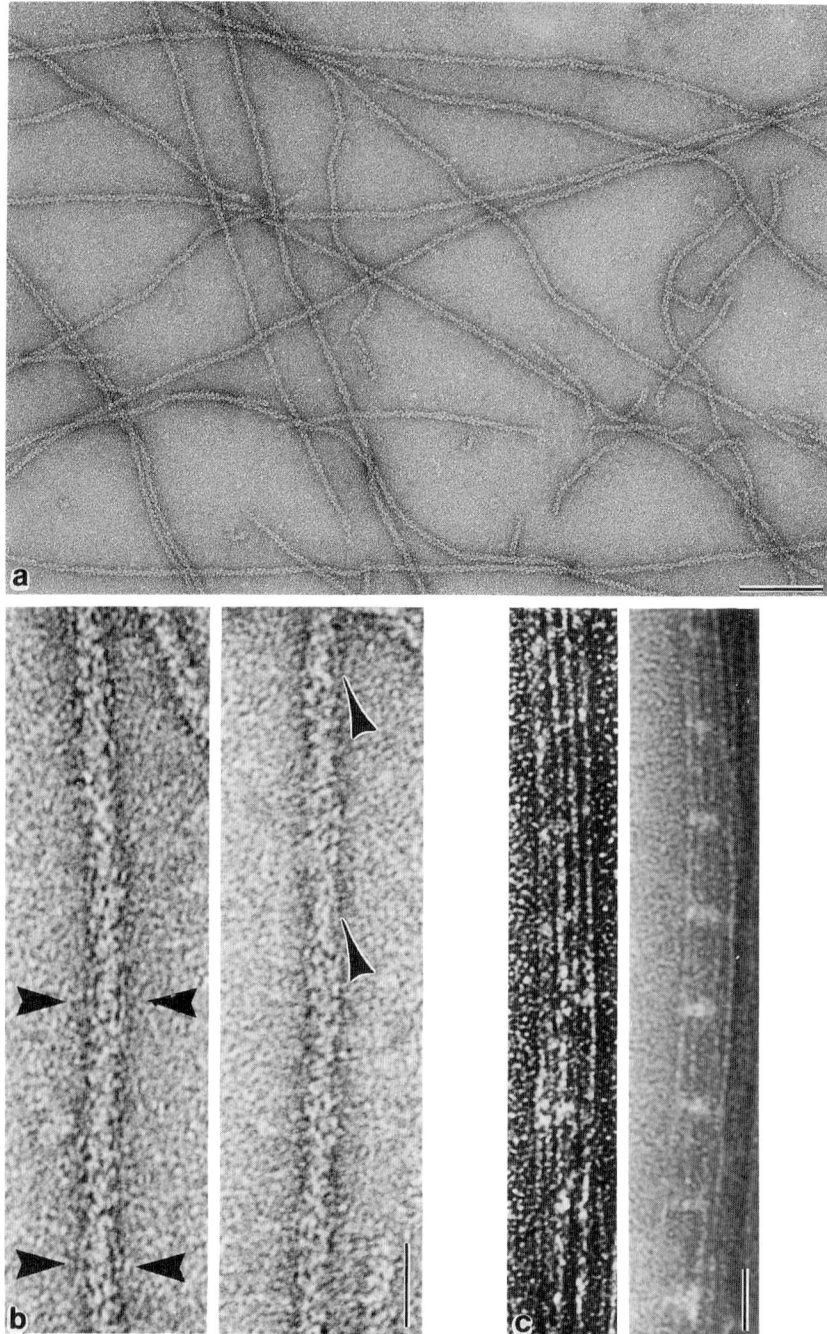

FIGURE 5   (a) Survey of native chicken gizzard thin filaments that were negative stained and photographed using minimal dose electron microscopy. Bar = 100 nm. (b) Higher-magnification electron micrographs of two filaments in (a) in which strands running along the actin helix are visible. Bar = 25 nm. (c) Electron micrographs of chicken gizzard thin filaments precipitated by antibodies to the 40,000 MW C-terminal (left) and the 23,500 MW N-terminal (right) caldesmon fragments. Note the presence of transverse striations at 36- to 40-nm intervals reflecting the site of antibody–antigen interaction. Bar = 25 nm. From Lehman *et al.* (1989, Fig. 3, p. 105) and from Moody *et al.* (1990, Fig. 7, p. 183). Reprinted with permission of Chapman & Hall.

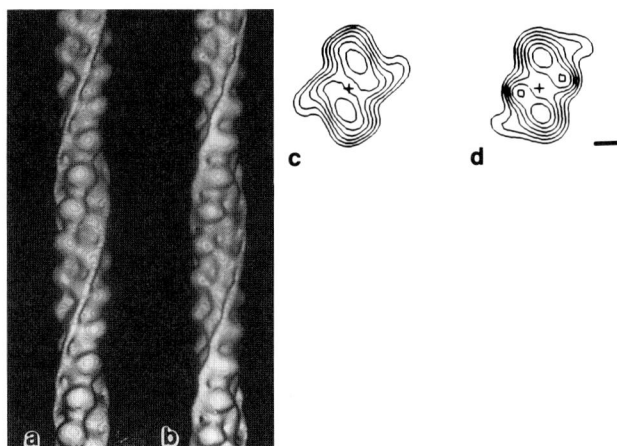

FIGURE 6 Surface views of reconstructed densities of chicken gizzard thin filaments (a) in EGTA and (b) in Ca^{2+}-calmodulin. The inner and outer domains of actin as in Fig. 4. Note the difference in position of the tropomyosin strand running along A_i in (a) and A_o in (b). Helical projections (c,d) formed by projecting the above densities along their helical tracks onto a plane perpendicular to the filament axis; thin filaments (c) in EGTA and (d) in Ca^{2+}-calmodulin. The tropomyosin strand makes contact with A_i in (c) and A_o in (d). Bar = 20 Å. From Vibert *et al.* (1993, Figs. 4 and 5, pp. 317, 318). Reprinted from the *Journal of Cell Biology*, 1993, **123**, 313–321, by copyright permission of the Rockefeller University Press.

sity seen in micrographs. These strands are in contact with the *inner* actin domains and smoothly follow the 36-nm-pitch helix. In striking contrast, strands in the Ca^{2+}-calmodulin-treated filaments appear to be in a different position relative to the actin monomers, closer to the *outer* domain (Fig. 6). This position is consistent with that derived from X-ray diffraction data of actin–tropomyosin gels (Popp and Holmes, 1992). Comparison of the two maps show that the strand position differs by an azimuthal rotation of ~20–25°, representing ~15–20 Å (Vibert *et al.*, 1993). Statistical analysis of the two maps demonstrates that the strand movement is highly significant, whereas actin itself is unchanged at least at 25-Å resolution and ~10-Å positional accuracy. The ability to resolve a strand in reconstructions of both native and caldesmon-deficient filaments, each of which contains tropomyosin, confirms that a major portion of the strand density in thin filaments must be contributed by tropomyosin. Moreover, the observation that the strand occupies different positions in the presence and absence of caldesmon demonstrates that this protein can affect tropomyosin position.

4. Proposed Mechanism of Caldesmon Action

Caldesmon appears to constrain tropomyosin in a position away from the main myosin-binding site known to be on the outer domain of actin (Milligan *et al.*, 1990; Rayment *et al.*, 1993b). *The effect of caldesmon on tropomyosin therefore appears to differ from that of troponin*, which constrains tropomyosin (in the absence of Ca^{2+}) to adopt a position on the outer actin domain, where it may interfere sterically with myosin binding. Hence, in the off-state, tropomyosin in smooth muscle is unlikely to participate in steric hindrance of myosin binding. However, since caldesmon and tropomyosin have not been resolved from each other, caldesmon, in principle, might follow a different path from tropomyosin and function independently, although there is no structural evidence for this. Although the evidence suggests that in the on-state smooth muscle tropomyosin is localized in a position that seemingly would compete with myosin binding, the fact that competition is not observed indicates that tropomyosin needs to be stabilized at this site by a component such as troponin for steric interference to occur. Weakly bound caldesmon in the on-state presumably is not such a determinant.

The possible differences in on- and off-positions of tropomyosin in smooth and skeletal muscle thin filaments may relate to significant differences in enzymatic behavior of the two systems. The well-known observation that tropomyosin activates the smooth muscle actomyosin ATPase (Chacko *et al.*, 1977) and also accelerates the motility of smooth muscle preparations *in vitro* (Shirinsky *et al.*, 1992) should be taken into account when evaluating caldesmon function. Indeed, the degree of actomyosin ATPase potentiation by tropomyosin in the smooth muscle system is considerably greater than that in skeletal muscle preparations (Chacko *et al.*, 1977; Sobieszek and Small, 1977; Chacko, 1981; Lehrer and Morris, 1984; Williams *et al.*, 1984). Tropomyosin potentiation may require a particular steric relationship between actin, tropomyosin, and myosin close to the myosin binding site on the outer actin domain (cf. Phillips *et al.*, 1986; Reedy *et al.*, 1994). Possible caldesmon-induced movement of tropomyosin away from this site or competition with tropomyosin for the site may prevent potentiation from occurring. The work of Chacko and associates (Horiuchi and Chacko, 1989; Horiuchi *et al.*, 1991), in fact, implies that caldesmon may modulate the magnitude of tropomyosin activation, and therefore may control actomyosin ATPase by inhibiting tropomyosin potentiation. This view would explain how a single caldesmon molecule could influence the reactivity of many actin molecules along thin filaments. Such a mechanism of modulation implies that caldesmon may fine-tune contractile activity but not act as an on–off switch per se. This view also fits with results of the elegant experiments of Fay and his collaborators (Itoh *et al.*, 1989), who showed that myosin phosphorylation is necessary and sufficient for the initiation of smooth muscle contraction and concluded that the phosphorylation process itself is the sole switch for smooth muscle activation. Hence, the structural data at present are consistent with a hypothesis that both caldesmon and troponin regulate thin filament interaction with myosin by a unique and distinguishable steric effect on tropomyosin position, which in turn influences the actomyosin ATPase. These distinctions may not be obvious when function is assessed biochemically.

Constraints imposed by caldesmon that disrupt potentiation may not block myosin docking on actin. This process of modulating ATPase could in turn lead to or permit the development of the latch-state of tension maintenance displayed by tonic smooth muscles and observed at low Ca^{2+} concentration. This phenomenon probably involves stable actin–myosin binding and is associated with low actomyosin ATPase activity (Hai and Murphy, 1988; McDaniel *et al.*, 1990). As envisioned, such tension maintenance would be incompatible with a troponin–tropomyosin form of regulation since tropomyosin would in that case block myosin docking at low Ca^{2+} concentrations. Hence, the caldesmon–tropomyosin system may be adapted for muscles that enter a latch-state.

5. Are There Calponin-Specific Thin Filaments?

Calponin, like caldesmon, is an actin-binding protein found in appreciable quantity in a variety of smooth muscles and some nonmuscle tissues (Takahashi *et al.*, 1986, 1988a,b; Lehman and Kaminer, 1984; Winder and Walsh, 1990). Evidence for a role in regulating actomyosin ATPase, although attractive, remains inconclusive at best. Interestingly, it appears likely that calponin and caldesmon are organized in separate classes of thin filaments (Lehman, 1991; Makuch *et al.*, 1991). Immunocytochemical studies by Small and associates (North *et al.*, 1994a) tend to confirm this view and are discussed in Chapter 7 by Gimona and Small (this volume). The occurrence of functionally diverse thin filament domains in smooth muscle may partly explain why smooth muscle actin content is so high. Unfortunately, methods of isolation of native calponin-specific thin filaments (i.e., not involving dissociation and reassociation of constituents as in Nishida *et al.*, 1990) have not been described nor have reconstitution studies produced filaments with native stoichiometry and well-localized proteins (Winder and Walsh, 1990; Makuch *et al.*, 1991; Horiuchi and Chacko, 1991). Hence, meaningful structural studies have not yet been possible on this interesting system.

IV. PERSPECTIVES

Our understanding of smooth muscle contractility and its regulation has frequently lagged behind that of skeletal and cardiac muscle. Although this situation is improving, fundamental questions regarding the mechanism of smooth muscle function remain unresolved. In this chapter, some have been highlighted and include:

1. What is the reason for actin isoform diversity in smooth muscle?
2. How is the pattern of appearance of actin isoforms during smooth muscle development related to cell function?
3. How are actin isoforms segregated in smooth muscle cells?
4. Are subtle differences in actin isoform structure related to functional differences?
5. What determines the binding specificity of actin-associated proteins in smooth muscle?
6. What is the structure of smooth muscle contractile filaments at high resolution?
7. Can caldesmon be resolved from tropomyosin on thin filaments, and do the two proteins follow parallel and contiguous paths?
8. What are the mechanism and function of caldesmon-imposed constraints on tropomyosin location, and how do they relate to smooth muscle regulation?
9. What is the three-dimensional distribution of thin filaments in smooth muscle?
10. What is the real function of calponin?
11. What are the structural consequences of calponin binding on F-actin?

A wealth of new information awaits us.

Acknowledgments

We thank J. Gentry for typing part of the manuscript and M. Picard Craig for photography. Our own work was supported by Grants HL 36153 (W.L.), AR 34711 (R.C.), HL 47530 (R.C.), and AM 34602 (M.B.) from the National Institutes of Health and MCB90-04746 (P.V.) from the National Science Foundation.

References

Bárány, M., and Bárány K. (1993). *Biochim. Biophys. Acta* **1179**, 229–233.
Bárány, K., Polyák, E., and Bárány, M. (1992). *Biochim. Biophys. Acta* **1134**, 233–241.
Bretscher, A. (1984). *J. Biol. Chem.* **259**, 12873–12880.
Bryan, J., and Lee, R. (1991). *J. Muscle Res. Cell Motil.* **12**, 372–375.
Carsten, M. E. (1965). *Biochemistry* **4**, 1049–1054.

Cavadore, J. C., Axelrud–Cavadore, C., Berta, P., Harricane, M. C., and Haiech, J. (1985). *Biochem. J.* **228**, 433–441.

Cavadore, J. C., Martin, F., Calas, B., Mery, J., Berta, P., Benyamin, Y., and Roustan, C. (1987). *Biochem. J.* **242**, 51–54.

Cavaillé, F., Janmot, C., Ropert, S., and d'Albis, A. (1986). *Eur. J. Biochem.* **160**, 507–513.

Chacko, S. (1981). *Biochemistry* **20**, 702–707.

Chacko, S., Conti, M. A., and Adelstein, R. S. (1977). *Proc. Natl. Acad. Sci. U.S.A.* **74**, 129–133.

Chalovich, J. M., and Eisenberg, E. (1982). *J. Biol. Chem.* **257**, 2432–2437.

Chalovich, J. M., Chock, P. B., and Eisenberg, E. J. (1981). *J. Biol. Chem.* **256**, 575–587.

Cohen, C., and Szent-Györgyi, A. G. (1957). *J. Am. Chem. Soc.* **79**, 248.

Cohen, C., and Vibert, P. (1987). *In* "Fibrous Protein Structure" (J. M. Squire and P. Vibert, eds.), pp. 283–306. Academic Press, London.

Cohen, C., Caspar, D. L. D., Johnson, J. P., Nauss, K., Margossian, S. S., and Parry, D. A. D. (1972). *Cold Spring Harbor Symp. Quant. Biol.* **37**, 287–297.

Collins, J. H., and Elzinga, M. (1975). *J. Biol. Chem.* **250**, 5915–5920.

Crosbie, R., Adams, S., Chalovich, J. M., and Reisler, E. (1991). *J. Biol. Chem.* **266**, 20001–20006.

Dabrowska, R., Goch, A., Galazkiewicz, B., and Osinka, H. (1985). *Biochim. Biophys. Acta* **842**, 70–75.

Ebashi, S. (1985). *J. Biochem. (Tokyo)* **97**, 693–695.

Ebashi, S., and Endo, M. (1968). *Prog. Biophys. Mol. Biol.* **28**, 123–183.

Eddinger, T. J., and Murphy, R. A. (1991). *Arch. Biochem. Biophys.* **284**, 232–237.

Elce, J. S., Elbrecht, A. S. U., Middlestadt, M. U., McIntyre, E. J., and Anderson, P. J. (1981). *Biochem. J.* **193**, 891–898.

Estes, J. E., Selden, L. A., Kinosian, H. J., and Gershman, L. C. (1992). *J. Muscle Res. Cell Motil.* **13**, 272–284.

Fatigati, V., and Murphy, R. A. (1984). *J. Biol. Chem.* **259**, 14383–14388.

Flicker, P. F., Phillips, G. N., and Cohen, C. (1982). *J. Mol. Biol.* **162**, 495–501.

Graceffa, P., and Jansco, A. (1991). *J. Biol. Chem.* **266**, 20205–20210.

Graceffa, P., Adam, L. P., and Lehman, W. (1993). *Biochem. J.* **294**, 63–67.

Haeberle, J. R., Hathaway, D. R., and Smith, C. L. (1992). *J. Muscle Res. Cell Motil.* **13**, 81–89.

Hai, C.-M., and Murphy, R. A. (1988). *Am. J. Physiol.* **254**, C99–C106.

Hanson, J., and Lowy, J. (1963). *J. Mol. Biol.* **6**, 46–50.

Harris, D. E., and Warshaw, D. M. (1993). *Circ. Res.* **72**, 219–224.

Hartshorne, D. J. (1987). *In* "Physiology of the Gastrointestinal Tract" (L. R. Johnson, ed.), 2nd ed., pp. 423–482. Raven Press, New York.

Haselgrove, J. C. (1972). *Cold Spring Harbor Symp. Quant. Biol.* **37**, 225–234.

Hayashi, K., Fujio, Y., Kato, I., and Sobue, K. (1991). *J. Biol. Chem.* **266**, 355–361.

Hemric, M. E., and Chalovich, J. M. (1988). *J. Biol. Chem.* **262**, 1878–1885.

Hemric, M. E., and Chalovich, J. M. (1990). *J. Biol. Chem.* **265**, 19672–19678.

Hennessey, E. S., Drummond, D. R., and Sparrow, J. C. (1993). *Biochem. J.* **282**, 657–671.

Hinssen, H., Small, J. V., and Sobieszek, A. (1984). *FEBS Lett.* **166**, 90–95.

Hirai, S., and Hirabayashi, T. (1983). *Dev. Biol.* **97**, 483–493.

Holmes, K. C., and Kabsch, W. (1991). *Curr. Opin. Struct. Biol.* **1**, 270–280.

Holmes, K. C., Popp, D., Gebhard, W., and Kabsch, W. (1990). *Nature (London)* **347**, 44–49.

Horiuchi, K. Y., and Chacko, S. (1989). *Biochemistry* **28**, 9111–9116.

Horiuchi, K. Y., and Chacko, S. (1991). *Biochem. Biophys. Res. Commun.* **176**, 1487–1493.

Horiuchi, K. Y., Samuel, M., and Chacko, S. (1991). *Biochemistry* **30**, 712–716.

Huxley, H. E. (1972). *Cold Spring Harbor Symp. Quant. Biol.* **37**, 361–376.

Ikebe, M., and Reardon, S. (1988). *J. Biol. Chem.* **263**, 3055–3058.

Itoh, T., Ikebe, M., Kargacin, G. J., Hartshorne, D. J., Kemp, B. E., and Fay, F. S. (1989). *Nature (London)* **338**, 164–167.

Kabsch, W., Mannherz, H. G., Suck, D., Pai, E. F., and Holmes, K. C. (1990). *Nature (London)* **347**, 37–44.

Katayama, E., and Ikebe, M. (1995) *Biophys. J.* **68**, 2419–2428.

Khaitlina, S. Y., Moraczewska, J., and Strzelecka-Golaszewska, H. (1993). *Eur. J. Biochem.* **218**, 911–920.

Kobayashi, R., Kubota, T., and Hidaka, H. (1994). *Biochem. Biophys. Res. Commun.* **198**, 1275–1278.

Kress, M., Huxley, H. E., Faruqi, A. R., and Hendrix, J. (1986). *J. Mol. Biol.* **188**, 325–342.

Labbé, J. P., Lelievre, S., Boyer, M., and Benyamin, Y. (1994). *Biochem. J.* **299**, 875–879.

Lash, J., Sellers, J., and Hathaway, D. (1986). *J. Biol. Chem.* **261**, 16155–16160.

Lehman, W. (1991). *J. Muscle Res. Cell Motil.* **12**, 221–224.

Lehman, W., and Kaminer, B. (1984). *Biophys. J.* **45**, 109a.

Lehman, W., Craig, R., Lui, J., and Moody, C. (1989). *J. Muscle Res. Cell Motil.* **10**, 101–112.

Lehman, W., Moody, C., and Craig, R. (1990). *Ann. N.Y. Acad. Sci.* **599**, 75–84.

Lehman, W., Denault, D., and Marston, S. (1993). *Biochim. Biophys. Acta* **1203**, 53–59.

Lehman, W., Craig, R., and Vibert, P. (1994). *Nature (London)* **368**, 65–67.

Lehman, W., Craig, R., and Vibert, P. (1995) *Biophys. J.* (in press).

Lehrer, S. S., and Morris, E. P. (1984). *J. Biol. Chem.* **259**, 2070–2072.

Lorenz, M., Popp, D., and Holmes, K. C. (1993). *J. Mol. Biol.* **234**, 826–836.

Lynch, W. P., Riseman, V. M., and Bretscher, A. (1987). *J. Biol. Chem.* **262**, 7429–7437.

Mabuchi, K., Lin, J.J.-C., and Wang, C.-L. (1993). *J. Muscle Res. Cell Motil.* **14**, 53–64.

Makuch, R., Birukov, K., Shirinsky, V., and Dabrowska, R. (1991). *Biochem. J.* **280**, 33–38.

Malmqvist, U., and Arner, A. (1990). *Circ. Res.* **66**, 832–845.

Marston, S. B. (1990). *Biochem. J.* **272**, 305–310.

Marston, S. B., and Lehman, W. (1985). *Biochem. J.* **231**, 517–522.

Marston, S. B., and Redwood, C. S. (1991). *Biochem. J.* **279**, 1–16.

Marston, S. B., and Smith, C. W. J. (1984). *J. Muscle Res. Cell Motil.* **5**, 559–575.

Marston, S. B., Redwood, C. S., and Lehman, W. (1988). *Biochem. Biophys. Res. Commun.* **155**, 197–202.

Marston, S. B., Redwood, C., and Bennett, P. (1990). *Proc. Eur. Conf. Muscle Motil., 19th*, Brussels, Vol. 2, pp. 351–357.

Matsumura, F., and Lin, J.J.-C. (1982). *J. Mol. Biol.* **157**, 163–171.

McDaniel, N. L., Rembold, C. M., and Murphy, R. A. (1990). *Ann. N.Y. Acad. Sci.* **599**, 66–74.

Milligan, R. A., and Flicker, P. F. (1987). *J. Cell Biol.* **105**, 29–39.

Milligan, R. A., Whittaker, M., and Safer, D. (1990). *Nature (London)* **348**, 217–221.

Moody, C. J., Marston, S. B., and Smith, C. W. J. (1985). *FEBS Lett.* **191**, 107–112.

Moody, C., Lehman, W., and Craig, R. (1990). *J. Muscle Res. Cell Motil.* **11**, 176–185.

Moore, P. B., Huxley, H. E., and DeRosier, D. J. (1970). *J. Mol. Biol.* **50**, 279–285.

Murphy, R. A., Driska, S. P., and Cohen, D. M. (1977). *In* "Excitation–Contraction Coupling in Smooth Muscles" (R. Casteels, T. Godfraind, and J. C. Rüegg, eds.), pp. 417–424. Elsevier-North Holland Biomed. Press, Amsterdam.

Ngai, P. K., Carruthers, C. A., and Walsh, M. P. (1984). *Biochem. J.* **218**, 863–870.

Ngai, P. K., Gröschel-Stewart, U., and Walsh, M. P. (1986). *Biochem. Int.* **12**, 89–93.

Nishida, W., Abe, M., Takahashi, K., and Hiwada, K. (1990). *FEBS Lett.* **268**, 165–168.

North, A. J., Gimona, M., Cross, R. A., and Small, J. V. (1994a). *J. Cell Sci.* **107**, 437–444.

North, A. J., Gimona, M., Lando, Z., and Small, J. V. (1994b). *J. Cell Sci.* **107**, 445–455.

O'Brien, E. J., Bennett, P. M., and Hanson, J. (1971). *Philos. Trans. R. Soc. London, Ser. B* **261**, 201–208.

Ohtsuki, I. (1974). *J. Biochem. (Tokyo)* **75**, 753–765.

Ohtsuki, I., Masaki, T., Nonomura, Y., and Ebashi, S. (1967). *J. Biochem. (Tokyo)* **61**, 817–819.

Orlova, A., and Egelman, E. H. (1993). *J. Mol. Biol.* **232**, 334–341.

Parry, D. A. D., and Squire, J. M. (1973). *J. Mol. Biol.* **75**, 33–55.

Pavalko, F. M., Adam, L., Wu, M. F., Walker, T. L., and Gunst, S. J. (1994). *Biophys. J.* **66**, A411.

Phillips, G. N., Fillers, J. P., and Cohen, C. (1986). *J. Mol. Biol.* **75**, 33–55.

Pollard, T. D., and Cooper, J. A. (1986). *Annu. Rev. Biochem.* **55**, 987–1035.

Poole, K. J. V., Lorenz, M., Evans, G., Rosenbaum, G., and Holmes, K. C. (1994). *Biophys. J.* **66**, 347a.

Popp, D., and Holmes, K. C. (1992). *J. Mol. Biol.* **224**, 65–76.

Próchniewicz, E., and Strzelecka-Golaszewska, H. (1980). *Eur. J. Biochem.* **106**, 305–312.

Rayment, I., Rypniewski, W. R., Schmidt-Bäse, K., Smith, R., Tomchick, D. R., Benning, M. M., Winkelmann, D. A., Wesenberg, G., and Holden, H. (1993a). *Science* **261**, 50–58.

Rayment, I., Holden, H. M., Whittaker, M., Yohn, C. B., Lorenz, M., Holmes, K. C., and Milligan, R. A. (1993b). *Science* **261** 58–65.

Reddy, S., Ozgur, K., Lu, M., Chang, W., Mohan, S. R., Kumar, C. C., and Ruley, H. E. (1990). *J. Biol. Chem.* **265**, 1683–1687.

Reedy, M. K., Reedy, M. C., and Schachat, F. (1994). *Curr. Biol.* **4**, 624–626.

Riseman, V. M., Lynch, W. P., Nefsky, B., and Bretscher, A. (1989). *J. Biol. Chem.* **264**, 2869–2875.

Ruhnau, K., Gaertner, A., and Wegner, A. (1989). *J. Mol. Biol.* **210**, 141–148.

Sanders, C., and Smillie, L. B. (1984). *Can. J. Biochem. Cell Biol.* **62**, 443–448.

Schröder, R. R., Manstein, D. J., Jahn, W., Holden, H., Rayment, I., Holmes, K. C., and Spudich, J. A. (1993). *Nature (London)* **364**, 171–174.

Schröer, E., and Wegner, A. (1985). *Eur. J. Biochem.* **153**, 515–520.

Schutt, C. E., Myslik, J. C., Rozycki, M. D., Goonesekere, N. C. V., and Lindberg, U. (1993). *Nature (London)* **365**, 810–816.

Shirinsky, V. P., Birukov, K. G., Hettash, J. M., and Sellers, J. R. (1992). *J. Biol. Chem.* **267**, 15886–15892.

Sobieszek, A., and Bremel, R. D. (1975). *Eur. J. Biochem.* **55**, 49–60.

Sobieszek, A., and Small, J. V. (1977). *J. Mol. Biol.* **112**, 559–576.

Sobue, K., Muramoto, Y., Fujita, M., and Kakiuchi, S. (1981). *Proc. Natl. Acad. Sci. U.S.A.* **78**, 5652–5655.

Spudich, J. A., Huxley, H. E., and Finch, J. T. (1972). *J. Mol. Biol.* **72**, 619–632.

Squire, J. M., Al-Khayat, H. A., and Yagi, N. (1993). *J. Chem. Soc., Faraday Trans.* **89**, 2717–2726.

Strzelecka-Golaszewska, H., and Sobieszek, A. (1981). *FEBS Lett.* **134**, 197–202.

Strzelecka-Golaszewska, H. Próchniewicz, E., Nowak, E. Zmorzynski, S., and Drabikowski, W. (1980). *Eur. J. Biochem.* **104**, 41–52.

Strzelecka-Golaszewska, H., Zmorzynski, S., and Mossakowska, M. (1985a). *Biochim. Biophys. Acta* **828**, 13–21.

Strzelecka-Golaszewska, H., Venyaminov, S. Y., Zmorzynski, S., and Mossakowska, M. (1985b). *Eur. J. Biochem.* **147**, 331–342.

Sutherland, C., and Walsh, M. P. (1989). *J. Biol. Chem.* **264**, 578–583.

Sutoh, H. (1993). *In* "Mechanism of Myofilament Sliding in Muscle Contraction" (H. Sugi and G. H. Pollack, eds.), pp. 241–244. Plenum, New York.

Takahashi, K., Hiwada, K., and Kokubu, T. (1986). *Biochem. Biophys. Res. Commun.* **141**, 20–26.

Takahashi, K., Abe, M., Hiwada, K., and Kokubu, T. (1988a). *J. Hypertens.* **6**, S40–S43.

Takahashi, K., Hiwada, K., and Kokubu, T. (1988b). *Hypertension* **11**, 620–626.

Vandekerckhove, J., and Weber, K. (1978). *J. Mol. Biol.* **126**, 783–802.

Vandekerckhove, J., and Weber, K. (1979a). *Differentiation (Berlin)* **14**, 123–133.

Vandekerckhove, J., and Weber, K. (1979b). *FEBS Lett.* **102**, 219–222.

Velaz, L., Ingraham, R. H., and Chalovich, J. M. (1990). *J. Biol. Chem.* **265**, 2929–2934.

Vibert, P., Craig, R., and Lehman, W. (1993). *J. Cell Biol.* **123**, 313–321.

Williams, D. L., Greene, L. E., and Eisenberg, E. (1984). *Biochemistry* **23**, 4150–4155.

Winder, S. J., and Walsh, M. P. (1990). *J. Biol. Chem.* **265**, 10148–10155.

THIN FILAMENT AND CALCIUM BINDING PROTEINS

Tropomyosin

LAWRENCE B. SMILLIE

MRC Group in Protein Structure and Function
Department of Biochemistry
University of Alberta
Edmonton, Canada

I. INTRODUCTION

The tropomyosins are widely distributed in nature and found in virtually all eukaryotic cells. Together with actin, myosin, and ancillary proteins, they function as a part of the contractile apparatus and thin filament assemblies of both muscle and nonmuscle cells and tissues. Evidence accumulated in the past two decades has shown them to be remarkably diverse, arising from both multiple genes as well as alternative processing of mRNA. This diversity is undoubtedly a reflection of their varying functions in different tissues and cell types, as well as their location and roles within a single cell type. In spite of this diversity, their basic molecular architecture is the same; two α-helices in parallel and in register arranged like a two-stranded rope as a coiled-coil. Through head-to-tail overlap and association at their ends, they form an extended filament interacting with multiple actin monomers on opposite sides of the F-actin structure. Though tropomyosins (TM) possess many features in common, including regions of highly homologous amino acid sequence, the structural differences in the various TM isoforms can be expected to affect the flexibility and stability of the coiled-coil along its length, extent of head-to-tail association, the strength of interaction with F-actin, and the interaction of myosin heads and of ancillary and/or other regulatory proteins with the F-actin–TM complex.

The TMs occur as two distinct groups in vertebrate species. In differentiated adult muscle, either skeletal, cardiac, or smooth, all TM polypeptide chains are of the same length, 284 amino acid residues. These can be referred to as high-M_r isoforms. In nonmuscle tissues, a variety of lower-M_r isoforms are observed in addition to higher-M_r forms. The polypeptide chain lengths of these vary from 245 to 251 amino acid residues. In their interactions with F-actin, the higher-M_r and lower-M_r forms span 7 and 6 actin monomers, respectively, on each strand of F-actin.

In striated muscles (skeletal and cardiac), TM, together with the troponin complex, provides the calcium switch for the turning on/off of the actomyosin ATPase activity associated with contraction/relaxation. The troponin complex is composed of three components, troponins, I, C, and T, so designated for their abilities to inhibit actomyosin ATPase, to bind Ca^{2+}, and to anchor the whole troponin complex to TM, respectively. Conformational changes produced by the binding of calcium to troponin C are transmitted through the troponin I and T components to TM–actin. The steric relationship of TM to F-actin is altered, facilitating the interaction of myosin heads with the thin filament complex and the generation of ATPase activity and contraction. The process is reversed by the sequestration of calcium in the sarcoplasmic reticulum (Potter and Johnson, 1982; Leavis and Gergely, 1984; Zot and Potter, 1987; Grabarek *et al.*, 1992; Farah and Reinach, 1995). In smooth muscles and nonmuscle tissues, the role of TM is less well understood. Although calcium regulation by phosphorylation/dephosphorylation of myosin light chains is well established, control by calcium at the level of the thin filaments is currently an area of intensive investigation (see Chapters 6, 13, and 24, this volume). Since smooth muscle TM undoubtedly plays a central role in such processes a knowledge of its structural features and interaction properties with other thin filament proteins is obviously crucial to our understanding of the system.

The great bulk of our knowledge of the structure/function relationships of TMs is based on studies with the skeletal fast-twitch muscle proteins, and so it is appropriate in what follows to describe the characteristics of the smooth muscle TMs in a comparative way with those of its skeletal muscle counterparts.

II. TROPOMYOSIN LEVELS IN SMOOTH MUSCLES, PURIFICATION, AND GENERAL PROPERTIES

Estimates of the levels of the major protein components of smooth and skeletal muscles have been compiled by Hartshorne (1987). The concentrations of myosin, actin, and TM for arterial and nonarterial smooth muscles are estimated to be approximately 56 μM, 1.6 mM, 0.27 mM and 56 μM, 0.87 mM, 0.15 mM, respectively. In skeletal muscle, the corresponding concentrations are approximately 0.18, 0.7, and 0.1 mM. Thus in smooth muscles the myosin concentration is significantly reduced, whereas the actin and TM concentrations are elevated relative to those of skeletal muscle. In the present context the important point is that the molar ratio (~1.2:7) of TM to actin monomers is more than adequate to fully saturate the available TM binding sites on F-actin (one per seven actin monomers).

Procedures for the isolation and purification of TMs from muscle tissues are essentially modifications and extensions of those first described by Bailey (1948) and detailed by Smillie (1982). A convenient starting material is an acetone powder of the minced muscle. Alternatively, in the case of skeletal and cardiac tissue, the TM can be prepared as a side product of a troponin preparation from fresh or frozen and thawed minced muscle (see Potter, 1982). The high-ionic-strength extract is subjected to several rounds of isoelectric precipitation (pH 4.6) and $(NH_4)_2SO_4$ fractionation. Further purification can be achieved by chromatography on hydroxylapatite. Purity is assessed by sodium dodecyl sulfate–polyacrylamide gel electrophoresis (SDS–PAGE) in the absence and/or presence of 6 M urea (Sender, 1971). Though not normally a problem with skeletal muscle TM, nucleic acids can be a contaminant of preparations of smooth muscle and nonmuscle TMs. This can be assessed by the ratio of absorbances at 280 and 260 nm and corrected by further purification by column chromatography on DEAE-cellulose or other suitable anion exchanger (Sanders and Smillie, 1984). When prepared under these nondenaturing

conditions, the purified TM can be expected to retain its native conformation and subunit composition. Under denaturing conditions (6–8 M urea, 6 M guanidine HCl, high temperature), the molecules are dissociated into their constituent polypeptide chains. Upon renaturation and depending on the conditions, the subunits may or may not reassemble into their naturally occurring state and dimeric isoform distribution (see the following). High-temperature treatment for the removal of contaminating proteins as a purification step should be avoided. Such treatment can lead to irreversible chemical modification (e.g., possible deamidation or oxidation of Met or Cys residues) and alteration of properties (Graceffa, 1992).

TMs from various sources have very similar physical and solubility properties consistent with their highly asymmetric shape (Tsao et al., 1951; McCubbin et al., 1967) and high charge density of amino acid side chains. With an excess of acidic residues (Asp plus Glu $\cong 28\%$) over basic amino acids (~19%), their isoelectric points are in the range of 4–5. As pointed out originally by Crick (1953), their content of nonpolar residues is consistent with the predictions of a coiled-coil structure. Avian and mammalian TMs are devoid of Pro and Trp.

III. ISOFORM DIVERSITY OF SKELETAL AND SMOOTH MUSCLE TROPOMYOSINS

The first definitive evidence for skeletal muscle TM heterogeneity came from amino acid sequence analysis of the cysteine peptides of unfractionated TM preparations from the back and hind leg muscles of the rabbit (Hodges and Smillie, 1970, 1972). Based on the analytical data of peptides derived from the region of cysteine 190 with multiple substitutions at residue 199, it was concluded that a minimum of four TM polypeptide chains were present in this unfractionated preparation (Sodek et al., 1978). Cummins and Perry (1973) were the first to describe a preparative procedure for the separation of rabbit TM preparation into two fractions, α and β, on CM-cellulose under denaturing conditions (8 M urea). The α (faster) and β (slower) components had different mobilities on SDS–PAGE gels even though complete sequence analyses showed their chain lengths (284 residues) to be the same (Stone and Smillie, 1978; Mak et al., 1979, 1980). Subsequent two-dimensional gel electrophoresis of TMs from a variety of isolated skeletal muscle types showed that the α band of one-dimensional SDS gels could be separated into as many as three components, redesignated α, γ, and δ (Heeley et al., 1983, 1985). The γ and δ

components were shown to be typical of slow-twitch muscles. The proportions of the four components are variable depending on the stage of development and muscle type (fast twitch, slow twitch, mixed). The ratios are altered by cross-innervation, denervation/regeneration, and thyroidectomy (Heeley *et al.*, 1985). The amino acid sequences for the β and α components of rabbit skeletal muscle TMs (Mak *et al.*, 1979; Stone and Smillie, 1978) would correspond to the β and redesignated α components, respectively. In the chicken and rabbit, an α component has been detected in slow muscles and can be designated slow-twitch α-TM (Montarras *et al.*, 1981; Bronson and Schachat, 1982). Similar extra components have been detected in the cat (Steinbach *et al.*, 1980), human (Billeter *et al.*, 1981), and rat (Heeley *et al.*, 1985). These slow-twitch α-TM components probably correspond to one or other of the γ and δ isoforms reported by Heeley *et al.* (1983, 1985).

Aside from correlations of their relative abundance in different muscle types and during development, almost nothing is known of the functional and interaction properties of these various TM isoforms. Although an early study (Leger *et al.*, 1976) reported no obvious functional differences in the α and β isoforms, a preliminary report (Thomas and Smillie, 1994) indicates differences in the head-to-tail polymerization, binding to troponin T, and effects on actomyosin subfragment-1 ATPase of αα-, αβ-, and ββ-TM dimers. In addition, Schachat and his colleagues have provided evidence for a correlation between the expression of different isoforms of fast muscle troponin T, TM, and α-actinin. Thus certain combinations of isoforms of these three proteins appear to be co-expressed in at least three distinct programs. Each program of expression has been reported to have distinctive effects on the calcium sensitivity and apparent cooperativity of tension development in fibers (Schachat *et al.*, 1990). An extension of these observations to *in vitro* characterization of the purified isoforms (or recombinant forms thereof) and their interactions would be of great interest.

The isoform diversity in striated muscles and their electrophoretic gel patterns are further complicated by partial phosphorylation of serine 283, the penultimate COOH-terminal residue (Ribolow and Bárány, 1977; Mak *et al.*, 1978; Montarras *et al.*, 1981, 1982; Heeley *et al.*, 1982, 1985). The extent of phosphorylation is highest for the α isoform in embryonic and neonatal tissue, dropping to much lower levels in adult muscle. Heeley *et al.* (1982, 1989) and Heeley (1994) have compared the phosphorylated and nonphosphorylated forms of rabbit skeletal αα TM and observed significant differences

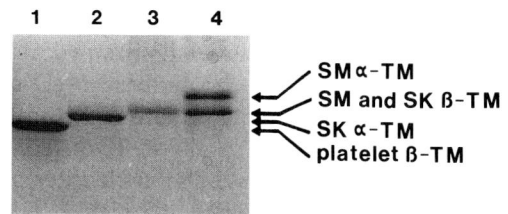

FIGURE 1 Sodium dodecyl sulfate–polyacrylamide gel electrophoresis of equine platelet ββ-TM (lane 1), rabbit skeletal muscle αα-TM (lane 2), rabbit skeletal muscle ββ-TM (lane 3), and chicken gizzard TM (lane 4). Note that the chicken gizzard TM is composed of two subunits α and β in a molar ratio of 1:1. This figure is adapted from Fig. 1b of Lau *et al.*, (1985), in which the α-TM chain was inappropriately designated γ-TM.

in their polymerizability, interactions with troponin T, and effects on reconstituted actomyosin subfragment-1 ATPase.

In smooth muscle the isoform diversity appears to be less complex. In adult chicken gizzard, the TM migrates as two major bands on SDS electrophoretic gels (Cummins and Perry, 1974; Lau *et al.*, 1985; Fig. 1). The faster component was designated β since it had a mobility similar to that of the β band of rabbit skeletal muscle TM. The second band, migrating more slowly than any of the skeletal TMs, was inappropriately designated γ. Protein amino acid sequence analyses (Lau *et al.*, 1985) and the elucidation of the TM genes (see reviews by MacLeod, 1987; Lees-Miller and Helfman, 1991; Pittenger *et al.*, 1994; Helfman, 1994) have demonstrated that this γ component is in fact derived from the α gene by alternative mRNA processing. It therefore should be referred to as α-TM and will be so designated subsequently in this manuscript. Both the chicken gizzard α and β components are present in close to equal amounts in adult tissue, have the same chain lengths (284 residues) as their skeletal counterparts, and are not phosphorylated (O'Connor *et al.*, 1979; Montarras *et al.*, 1981; Sanders and Smillie, 1985). In fact, Ser-283, the phosphorylated residue in skeletal and cardiac TMs, is replaced by Asn in both the α and β smooth muscle isoforms (Sanders and Smillie, 1985). There have been no reports of post-translational modifications of smooth muscle TMs.

The TM isoform content of a variety of smooth muscle tissues has been examined by one- and two-dimensional gel electrophoresis, including aorta, pulmonary artery, carotid, trachealis, esophagus, duodenum, jejenum, taenia coli, rectum, and others (Fatigati and Murphy, 1984; Yamaguchi *et al.*, 1984; Xie *et al.*, 1991). In all cases, two major components were observed to be present in close to equal amounts. The designation of these as α and β was made on the basis of relative mobilities in the SDS gel system and may require alteration depending on the species (as with chicken gizzard in Fig. 1). Although differences were observed in the positions of these components on two-dimensional gels, the two TM isoforms from different tissues of the same species appeared to be identical as assessed by this criterion. In an examination of eight digestive organs of the adult chicken, three high-M_r and four low-M_r isoforms were observed in addition to the usual α and β components (Xie *et al.*, 1991). These extra TMs were attributed to the nonmuscle contamination in the samples. Hosoya *et al.* (1989a,b) have followed TM isoform diversity and total TM content in developing chicken gizzard from 7-day embryos to the adult. In embryos, four high-M_r and five low-M_r isoforms were observed in addition to α and β. In early embryonic tissue the amounts of these additional isoforms were greater than those of the α and β isoforms. The former were progressively replaced by α and β, until at the time of hatching, and subsequently, only the α and β components were present. The data indicated that the time course of expression of the α component (designated herein as α as described earlier) preceded that of β and was higher than β during the period of Day 10 to hatching. The observations also indicated that the ratio of total TM (all isoforms) to actin was lower during the embryonic stages. These developmental changes in TMs presumably correspond to the maturation of mesenchymal and myoblasts into mature and fully differentiated smooth muscle cells. To this author's knowledge, no reports of the effects of denervation or altered hormone balance on the TM isoform distribution of smooth muscle cells have appeared in this literature.

As indicated in the foregoing, since the relative proportions of the α and β components in skeletal muscle are variable, the TM molecule can be made up of αα, αβ, or ββ dimers. This would also be true of smooth muscle TM, even though the ratio of the two chains is close to 1:1. Substantial evidence now exists that for both muscle types the most thermodynamically stable form is the αβ dimer. Thus in the case of skeletal muscle in which the α:β ratio is >1:1, the TM species present would be a mixture of αα and αβ with minimal amounts of ββ. In smooth muscles in which α:β is 1:1, the predominant species is the αβ dimer (Bronson and Shachat, 1982; Eisenberg and Kielley, 1974; Lehrer, 1975; Holtzer *et al.*, 1984; Brown and Schachat, 1985; Lehrer *et al.*, 1989; Sanders *et al.*, 1986; Graceffa, 1989; Jancso and Graceffa, 1991; Lehrer and Stafford, 1991). Upon denaturation and renaturation, the distribution of the α and β chains in the dimeric species is dependent on temperature and ionic strength of renaturation (see, e.g., Lehrer and Stafford, 1991). Under appropriate conditions the subunits will assume their *in vivo* distribution by chain exchange.

IV. TROPOMYOSIN GENES, ORIGIN OF ISOFORM DIVERSITY, AND AMINO ACID SEQUENCES

TM diversity of isoforms is now explicable in terms of multiple genes, alternative transcriptional promoters, and alternative splicing of primary gene transcripts (for reviews, see Lees-Miller and Helfman, 1991; Pittenger *et al.*, 1994; Helfman, 1994). In the human, four TM genes have been identified to date and have been designated *TPM1* to *TPM4* in the human genome (Cuticchia and Pearson, 1993). *TPM1*, more commonly known as the α gene, contains 15 exons, of which 3, 4, 5, 7, and 8 are expressed in all known isoforms. The other 10 exons can be alternatively spliced to produce at least 9 different isoforms in different mammal and avian species. The chain lengths of these isoforms can be classified into long or high-M_r (281–284 amino acid residues) and short or low-M_r (245–251 residues) forms. In adult smooth and striated muscles, only the high-M_r isoforms are observed. Both low-and high-M_r chains have been demonstrated in nonmuscle tissues. This gene gives rise to the α-TM isoform in smooth muscle and the fast-twitch skeletal muscle α-TM in vertebrates (Fig. 2) and to cardiac α-TM in mammals.

TPM2, also known as the β gene, contains 9 exons in the rat, of which exons 3, 4, 5, 7, and 8 are expressed in all known isoforms. Identified isoforms include skeletal muscle β-TM, smooth muscle β-TM, and fibroblast TM-1, all 284 amino acid residues in length. In the chicken, this gene is somewhat different and gives rise also to a low-M_r fibroblast, TM-3b. The *TPM3* gene in humans has previously been known as hTM_{nm}, where the subscript refers to nonmuscle (Clayton *et al.*, 1988). However, in addition to encoding a human nonmuscle fibroblast TM_{30nm} (or TM-5) of 248 amino acids, the so-called slow-twitch skeletal muscle α-TM is derived from this gene. This latter gene product presumably corresponds to one of the γ and δ components observed in slow-twitch muscles referred to in the previous section (Heeley *et al.*, 1983, 1985). No smooth muscle TM isoform has been identified as arising from this gene. Gene *TPM4*, also designated *TM4* and containing 8 exons in the rat (Lees-Miller, *et al.*, 1990), has been shown to encode only one isoform to date (fibroblast TM-4, 248 amino acid residues long). It corresponds to a human cDNA defining a gene encoding TM_{30pl}, a human fibroblast TM (MacLeod *et al.*, 1987). The latter is virtually identical in amino acid

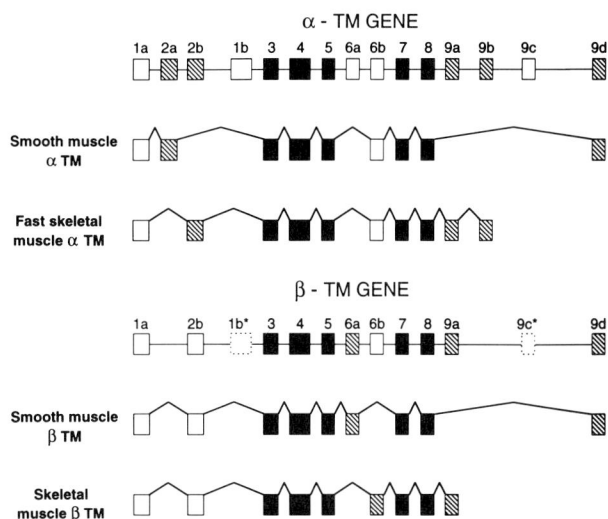

FIGURE 2   Intron–exon organization of the α- and β-TM genes of mammals. Those of the chicken and quail are believed to be similar. Smooth and fast skeletal muscle TM chains are encoded by both genes. In the human, these genes are designated *TPM1* and *TPM2*. For the α gene, alternative splicing of exons 2a/2b and 9d/9ab produces smooth and fast-twitch skeletal muscle α-TM chains, respectively. For the β-gene, codons 6a/6b and 9d/9a are alternatively spliced to produce the smooth muscle and fast-twitch skeletal muscle β-TM chains. Usage of other exons in the expression of the smooth and skeletal muscle TM chains is the same for both genes. β-TM gene exons lb* and 9c* have been detected in chicken but not in rat. A variety of other nonmuscle isoforms of TM chains are produced by different combinatorial patterns of exon expression (for reviews, see Lees-Miller and Helfman, 1991; Pittenger *et al.*, 1994; Helfman, 1994). This figure is adapted from Fig. 1 of Helfman (1994) and Fig. 2 of Lees-Miller and Helfman (1991).

sequence to equine platelet TM (Lewis *et al.*, 1983), whose properties and interaction with F-actin and troponin have been examined in some detail (Côté and Smillie, 1981a,b,c; Pearlstone and Smillie, 1982).

Of these TM genes, *TPM1* (α gene) and *TPM2* (β gene) are of particular significance in the present context since the former encodes the skeletal fast-twitch α-TM and smooth muscle α-TM isoforms whereas the latter encodes the skeletal and smooth muscle β-TMs, respectively. The exon/intron organization of the two genes is illustrated in Fig. 2. Alternative splicing of the RNA transcript of the α gene for exons 2a/2b and 9d/9ab produces the smooth muscle and skeletal fast-twitch α-TM isoforms, respectively. Similarly, exons 6a/6b and 9d/9a in the β gene are alternatively spliced to produce the smooth and skeletal β isoforms.

The corresponding amino acid sequences for the chicken gizzard α- and β-TM chains are shown in Fig. 3. Exon boundaries corresponding to the splice positions in the α and β genes of Fig. 2 are indicated as vertical lines. The two chains are seen to be highly homologous and of the same length (284 residues),

suggesting very similar structural properties. Amino acid substitutions (a total of 71) are distributed throughout, although not randomly, consistent with their origins from separate genes. As can be noted, a high proportion of these substitutions, 29 of a total of 71 (or 41%), are concentrated between amino acid residues 39 to 80, corresponding to exons 2a and 2b of the α and β genes, respectively. The functional significance of these differences between the smooth muscle α and β chain is presently unknown.

Since the α and β genes each give rise to both smooth muscle and skeletal TM isoforms, large regions of the amino acid sequences of the smooth and skeletal muscle TM isoforms arising from each gene are identical. The sequences of the smooth and skeletal muscle TMs differ only in those regions that have arisen from alternative splicing. This is shown schematically in Fig. 4, where the differences between the smooth and skeletal α-TMs are highlighted, as are the differences in the corresponding β-TMs. The details of these amino acid sequence differences are compiled in Table I. It will be noted that in any attempts to correlate functional and interactive properties of smooth and skeletal αβ-TM dimers with structural differences, three regions of amino acid sequence dissimilarity must be considered. These include residues 39–81 in the α chain, 189–213 in the β chain, and 258–284 in both chains. The functional/structural correlations of these are discussed in the next section.

V. AMINO ACID SEQUENCE AND COILED-COIL STRUCTURE

The demonstration of a repeating pattern of nonpolar and polar amino acids in the amino acid sequences of the α and β chains of rabbit skeletal muscle TM was the first confirmation of predictions for a coiled-coil structure made by Francis Crick in a classic paper in 1953. In such a structure, two α-helices interact along their length by knobs-into-holes packing of nonpolar residues to form a hydrophobic core. Polar and ionic side chains are directed toward the exterior of the two-stranded ropelike arrangement, where they can interact with solvent and/or other protein molecules. Because of the nonintegral nature of the α-helix, nonpolar core residues are expected to occur at an average interval of 3.5 residues. The sequence analyses demonstrated the presence of such a regular pattern of hydrophobic residues that extends throughout the entire lengths of their 284-residue polypeptide chains (Hodges *et al.*, 1972; Sodek *et al.*, 1972; Stone and Smillie, 1978; Mak *et al.*, 1980). Considering the sequence as a repeating pseudoheptapeptide in which

```
                                                                              1a│2a
         a b c d e f g a b c d e f g a b c d e f g a b c d e f g a b c d e f g a b c│d e
SMα      X-M-D-A-I-K-K-K-M-Q-M-L-K-L-D-K-E-N-A-L-D-R-A-E-Q-A-E-A-D-K-K-A-A-E-E-R-S-K-Q│L-E-
SMβ      - - E - - - - - - - - - - - - - I - - - - - - - - - - Q - - D - C - - │- -
         1                 10                  20                  30           1a│2b  40

         f g a b c d e f g a b c d e f g a b c d e f g a b c d e f g a b c d e f g a b c
SMα      D-D-I-V-Q-L-E-K-Q-L-R-V-T-E-D-S-R-D-Q-V-L-E-E-L-H-K-S-E-D-S-L-L-S-A-E-E-N-A-A-K-
SMβ      E-E-Q-Q-G - Q - K - K-G - - - E-V-E-K-Y-S - S-V-K-E-A-Q-E-K - E-Q - - K-K - T-D-
                       50                  60                  70                  80

      2a│3
         d e f g a b c d e f g a b c d e f g a b c d e f g a b c d e f g a b c d e f g a
SMα      A-E-S-E-V-A-S-L-N-R-R-I-Q-L-V-E-E-E-L-D-R-A-Q-E-R-L-A-T-A-L-Q-K-L-E-E-A-E-K-A-A-
SMβ      - - A - - - - - - - - - - - - - - - - - - - - - - - - - - - - - - - - - - - - -
      2b│3                    90                 100                 110                120

                         3│4
         b c d e f g a b c d e f g a b c d e f g a b c d e f g a b c d e f g a b c d e f
SMα      D-E-S-E-R─G-M-K-V-I-E-N-R-A-Q-K-D-E-E-K-M-E-I-Q-E-I-Q-L-K-E-A-K-H-I-A-E-E-A-D-R-
SMβ      - - - - - - - - - - - - M - - - - - L - - M - - - - - - - - - - - -
                         3│4      130                 140                 150           160

                 4│5                                           5│6b
         g a b c│d e f g a b c d e f g a b c d e f g a b c d e f│g a b c d e f g a b c d
SMα      K-Y-E-E│V-A-R-K-L-V-I-I-E-G-D-L-E-R-A-E-E-R-A-E-L-S-E-S│K-C-A-E-L-E-E-E-L-K-T-V-
SMβ      - - - -│- - - - - - V L - - E - - - S - - - - V-A - -│R-V R Q - - - - R - M-
                 4│5                170                 180         5│6a  190             200

                             6b│7                                       7│8
         e f g a b c d e f g a b c│d e f g a b c d e f g a b c d e f g a b c│d e f g a b
SMα      T-N-N-L-K-S-L-E-A-Q-A-E-K─Y-S-Q-K-E-D-K-Y-E-E-E-I-K-V-L-T-D-K-L-K-E│A-E-T-R-A-E-
SMβ      D-Q-S - - - - I - S-E - E - - T - - - - - - - - - L - G-E - - -│- - - - -
                             6a│7  210                 220                 230     7│8  240

                                     8│9d
         c d e f g a b c d e f g a b c d e│f g a b c d e f g a b c d e f g a b c d e f g
SMα      F-A-E-R-S-V-T-K-L-E-K-S-I-D-D-L-E│E-K-V-A-H-A-K-E-E-N-L-N-M-H-Q-M-L-D-Q-T-L-L-E-
SMβ      - - - - - A - - - T - - - -│- S-L - S - - - - - V-G-I - - V - - - - - -
                                 8│9d     250                 260                 270      280

         a b c d
SMα      L-N-N-M
SMβ      - - - L
                 284
```

FIGURE 3   Amino acid sequences of α- and β-TMs of chicken smooth muscle, products of two separate genes. The smooth muscle α-TM (SMα) sequence is as detailed by Lau *et al.*, (1985) for the chicken gizzard protein with corrections at positions 172, 199, 203, and 204 based on the cDNA sequence for chicken skeletal muscle α-TM reported by Gooding *et al.* (1987). The β-TM sequence is as reported by Sanders and Smillie (1985) for the gizzard protein and corroborated by cDNA (Helfman *et al.*, 1984) and genomic DNA sequencing (Libri *et al.*, 1989). Exon boundaries are indicated by the vertical lines.

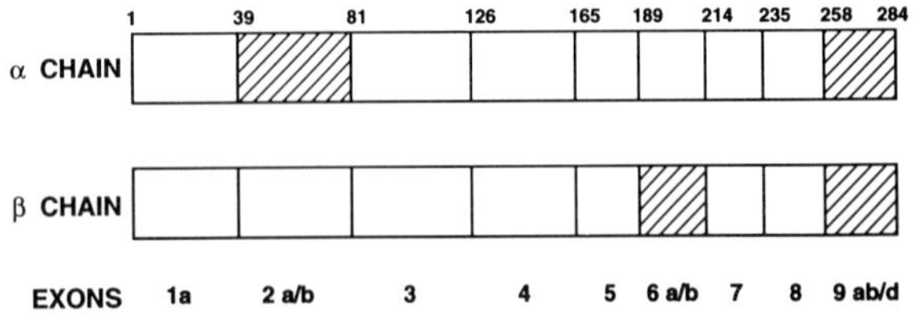

FIGURE 4   Comparison of skeletal and smooth muscle TM primary structures. Shaded areas represent alternatively spliced exons, which are different in the smooth muscle α- and β-TM chains when compared with those in skeletal muscle. Nonshaded areas are identical in smooth muscle and skeletal fast-twitch muscle α-TM chains. This is true also of the smooth and skeletal muscle β-TM chains. Exon boundaries for the α gene are for the highly homologous quail genomic sequence (Lindquester *et al.*, 1989). These are at identical amino acid positions as for the product of the β gene (Libri *et al.*, 1989; Forry-Shardies *et al.*, 1990).

the amino acid residues are labeled *a* to *g*, then positions *a* and *d* are largely occupied by nonpolar residues, whereas the great majority occupying positions *b*, *c*, *e*, *f*, and *g* are polar or ionic in nature. This repeating pattern in the case of the smooth muscle α and β chains is shown in Fig. 3 and applies to all known TM sequences as well as many other proteins. The spatial relationships of these residues in a schematic representation of a cross section of a coiled-coil is shown in Fig. 5.

In addition to the nonpolar interactions involving *a* and *d* positions of both chains, ionic interactions undoubtedly play a role in the stabilization of the structure. Thus, for example, a high proportion of residues occupying positions *e* are Glu residues whereas *g* positions are often basic in nature. In the coiled-coil with

TABLE I **Amino Acid Sequences of Alternatively Spliced Exons in Chicken α and β TM Genes as Expressed in Smooth and Skeletal Muscle**

α Gene[a]

Exon	Expressed in:	Sequence
		40 — 50 — 60 — 70 — 80
2a	Smooth muscle	L- E- D- D- I- V- Q- L- E- K- Q- L- R- V- T- E- D- S- R- D- Q- V- L- E- E- L- H- K- S- E- D- S- L- L- S- A- E- E- N- A- A- K
2b	Skeletal muscle	- - - E- L - A - Q - K - K- G - - - E- L - K- Y- S - S - K- D- A- Q- E- K - E- L - D- K- K - T- D
		258 — 270 — 284
9ab	Skeletal muscle	D- E- L- Y- A- Q- K- L- K- Y- K- A- I- S- E- E- L- D- H- A- L- N- D- M- T- S- I
9d	Smooth muscle	E- K- L- A- H- A - E- E- N- L- N- M- H- Q- M - - A- T - L- E- L- N- N- M

β Gene[b]

Exon	Expressed in:	Sequence
		189 — 200 — 210 — 213
6a	Smooth muscle	R- V- R- Q- L- E- E- E- L- R- T- M- D- Q- S- L- K- S- L- I- A- S- E- E- E
6b	Skeletal muscle	K- C- G- D - - - - - K- I- V- T- N- N - - - - E - Q- A- D- K
		258 — 270 — 280 — 284
9a	Skeletal muscle	D- E- V- Y- A- Q- K- M- K- Y- K- A- I- S- E- E- L- D- N- A- L- N- D- I- T- S- L
9d	Smooth muscle	E- S- L- A- S- A - E- E- N- V- G - E- Q- V - - Q- T - L- E- L- N- N- L

[a] Smooth muscle sequences are as reported by Lau *et al.* (1985) for smooth muscle α TM. Skeletal sequences are as in Gooding *et al.* (1987). Exon boundaries are as in the quail α gene (Lindquester *et al.*, 1989).

[b] Smooth muscle sequences are as reported by Sanders and Smillie (1985); skeletal sequences are as detailed in Libri *et al.* (1989); exon boundaries as in Libri *et al.*, (1989) and Forry-Shaudies *et al.* (1990).

the α-helices running in the same direction (i.e., parallel arrangement), position *e* in one helix is in a suitably located position to interact with position *g* in the preceding heptapeptide of the second helix. Other ionic interactions involving other combinations of acid and basic residues in the TM coiled-coil may also contrib-

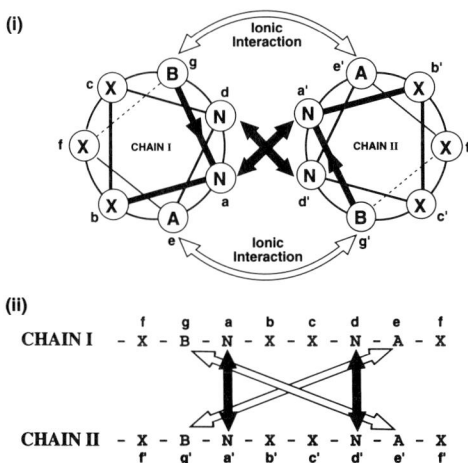

FIGURE 5 Representation of the two coiled-coil polypeptide chains of TM in cross section (i) and as the linear amino acid sequence (ii). Letters *a* to *g* represent the amino acid positions in the pseudoheptapeptide repeat. The *a, a'* and *d, d'* positions are nonpolar (N) and interact to form the core of the two-stranded structure. Acidic (A) and basic (B) amino acids occur frequently in *e, e'* and *g, g'* positions. These can form ionic links. The outer *b, c, f* and *b', c', f'* positions are usually but not always ionic or polar.

ute to its stability. The coiled-coil is now a well-established protein-structured motif present in a large number of documented cases. Factors such as the effect of different amino acids at different positions in the repeating heptapeptide on the stability of the structure are an area of active research. The reader is referred to reviews and references cited therein for further information (Adamson *et al.*, 1993; Cohen and Parry, 1990; Zhou *et al.*, 1994). These studies have not yet reached the point where stability differences between different TM isoforms or between different regions of their coiled-coils are explicable in terms of their sequences.

Using a variety of experimental approaches, evidence has been obtained by various workers (Woods, 1969, 1976; Lehrer, 1978; Pato *et al.*, 1981; Potekhin and Privalov, 1982; Holtzer *et al.*, 1983; Betteridge and Lehrer, 1983; Graceffa and Lehrer, 1980, 1984; Ueno, 1984) for variations in stability of the striated muscle TM coiled-coil along its length. Thermal unfolding as monitored by far-UV circular dichroism measurements as a function of increasing temperature shows a so-called broad pretransition (~20% of total ellipticity change) followed by a single transition at higher temperature. The pretransition has been ascribed to a less stable central region of the structure (~ residues 132–182) and the main transition to the simultaneous unfolding of the NH$_2$-terminal (residues 1–131) and COOH-terminal (~ residues 182–284) regions (Ishii *et al.*, 1992). These interpretations are in general agreement with earlier proteolytic studies that demonstrated relatively trypsin-resistant cores encompassing residues 13–125 and 183–284 (Pato *et al.*, 1981; see also Ueno, 1984). Thermal stability studies showed a lesser degree of stability for those fragments derived from the COOH-terminal region (residues 183–284) than from the NH$_2$ terminus (13–125). These latter observations are consistent with the TM crystal structural studies of Phillips *et al.* (1986), whose data suggest that the COOH-terminal half of the striated muscle TM ($\alpha\alpha$ or $\beta\beta$) is a less stable and more flexible coiled-coil than the NH$_2$-terminal portion. Similar but less extensive studies on the native $\alpha\beta$ and renatured $\alpha\alpha$ and $\beta\beta$ dimers of chicken gizzard TM (Woods, 1976; Lehrer *et al.*, 1984; Sanders *et al.*, 1986; Lehrer and Stafford, 1991; Wrabl *et al.*, 1994) have indicated a single reversible transition without a pretransition. Relative stabilities of these dimers appeared to be consistent with the numbers of putative favorable ionic *e* to *g* interactions as detailed by Sanders and Smillie (1985). The absence of a pretransition in these unfolding profiles appears to be inconsistent, however, with its assignment to the central region ($\sim$ residues 132–182) of the skeletal muscle TM. The amino acid sequences of the chicken skeletal muscle α- and β-TM chains over this region are identical to that of smooth muscle (see Fig. 4). It seems unlikely that this phenomenon is explicable in terms of species variation, since interspecies differences in TM sequences are minimal and highly conservative in this region of the chains (e.g., see Fig. 2 of Sanders and Smillie, 1985). Either the pretransition phenomenon must be reassigned to regions of the structure that do differ significantly in sequence or one must invoke a long-range effect of these on the stability of the central region. This latter possibility seems unlikely. The structural basis of this stability difference between skeletal and smooth muscle TMs is therefore presently unclear but obviously must be ultimately related to the differences in their primary structures.

Both electron microscopic and X-ray diffraction studies of paracrystalline fibrous arrays (see McLachlan and Stewart, 1976a; Stewart, 1981, and references therein) and of different crystal forms of TM have provided vital information on how molecules are organized in these structures and on their interaction and molecular parameters (Bailey, 1948; Phillips *et al.*, 1979, 1980, 1986; Cabral-Lilly *et al.*, 1991; Whitby *et al.*, 1992; Chacko and Phillips, 1992; Miegel *et al.*, 1993; Xie *et al.*, 1994). Because of the very high water content of the crystals (>95%), diffraction data are weak and high-resolution data are not yet available. However, a molecular structure at 15 Å has been reported (Phillips

et al., 1986), and data analyzed from a spermine-induced TM crystal form at 9 Å (Whitby *et al.*, 1992). These analyses have shown the pitch of the coiled-coil to be close to 140 Å and the effective molecular length to be 406 Å. In its interaction with actin, this value for the pitch enables each turn of the TM coiled-coil to make equivalent interactions with each actin monomer on the thin filament. Both in the crystal and on the F-actin strands, the TM is supercoiled (a coiled coiled-coil) with a right-handed sense. The radius of the supercoil in the crystal lattice is smaller (10 to 23 Å) than on the F-actin ($\sim$40 Å). This reduces the projected molecular length on the actin to about 385 Å. This degree of supercoiling on actin leads to seven half turns of the coiled-coil per molecular length. These relationships may be visualized in Fig. 6. Other features of the TM structure, as they relate to head-to-tail polymerization and binding properties, are considered in the following.

VI. ACTIN BINDING AND HEAD-TO-TAIL POLYMERIZATION

Depending on its length, a high- or low-M_r TM coiled-coil molecule spans 7 or 6 monomeric actin molecules, respectively, on each of the two strands of F-actin. Each TM molecule is linked by interaction and overlap of its NH$_2$- and COOH-terminal ends to contiguous molecules to form an extended filament. The binding of skeletal muscle TM to F-actin has been shown to be dependent on ionic strength and especially the Mg^{2+} ion concentration. Binding isotherms are highly cooperative and TM is believe to be largely responsible for the cooperative features of the skeletal muscle thin filament assembly in its interaction with the myosin head and in the effects of Ca^{2+} ion concentration. The spanning of 7 actin monomers on each strand (or 14 on both strands) of F-actin by a muscle TM molecule predicts a repeating pattern (7- or 14-fold) in the structural features of the TM coiled-coil. The likelihood that this would be reflected in the amino acid sequence was investigated by several groups (Stone *et al.*, 1975; Parry, 1975; McLachlan and Stewart, 1975). Although no repeats of identical amino acids were observed, a quasi-equivalent repeat of charged and nonpolar residues was detected. McLachlan and Stewart (1976b) concluded that there was a significant 14-fold periodicity of acidic and outer nonpolar residues (excluding those in the core *a* and *d* positions). Basic residues are more randomly distributed. Statistically the repeat was about 19.66 residues, and since the effective length of each chain of the molecule in a head-to-tail association of TM molecules may be taken

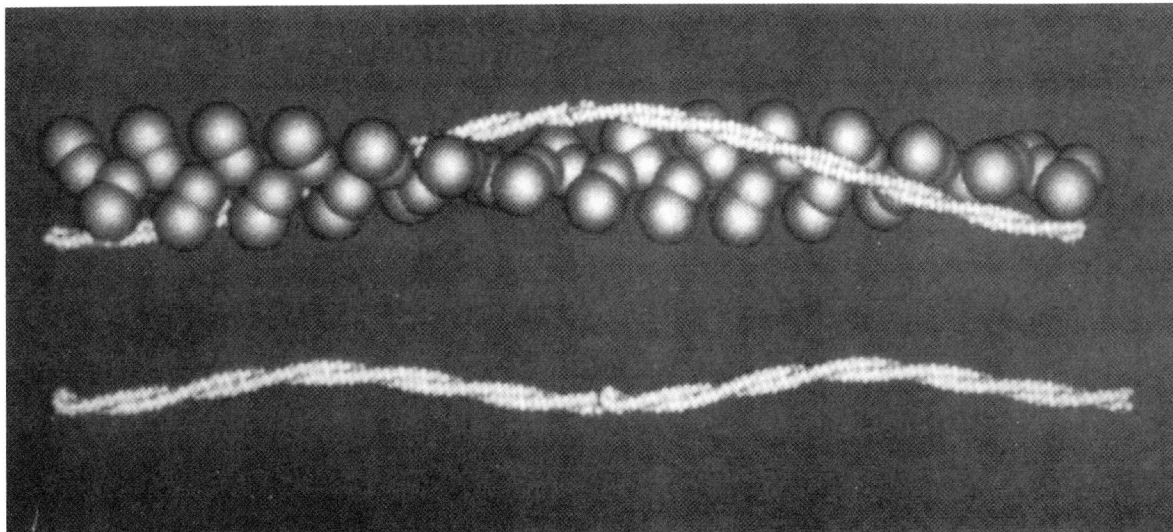

FIGURE 6   Two TM molecules are shown linked head-to-tail as observed in the X-ray crystal structure. Note the right-handed supercoil with a small variable radius. When bound to actin, each TM is believed to be supercoiled with a radius of about 40 Å. The projected molecular length is reduced to ~385 Å. With a pitch of ~140 Å, each half turn of the TM coiled-coil can make equivalent interactions with each of seven actin monomers. The structure of the head-to-tail joint is unknown but is believed to be a globular domain. This representation is reproduced from Fig. 11 of Phillips *et al.* *J. Mol. Biol.* **192**, 111–131 (1986) by permission of Academic Press.

as 275 residues, this repeat pattern (19.66 × 14 = 275) would be continuous in passing from one TM molecule to another along the F-actin structure. Each of the 14 periods would consist of a negative region of about 11 residues and a positive region (relatively devoid of acidic residues) of about 9 residues. The nonpolar residues (those not found in *a* or *d* positions) are found almost exclusively in a band that encompasses the positive zone and overlaps into the negative zone by about 4 residues [see Fig. 2 of McLachlan and Stewart (1976b) or of Smillie (1979)]. In a detailed hand analysis of the sequence, McLachlan and Stewart (1976b) suggested that the 14 periods could be divided into two sets of 7, α and β, which alternate throughout the sequence. Although the general features of the α and β alternating zones are alike in general, the α bands were concluded to be more regular in their repeat pattern. The authors suggested that the α and β zones represented different binding sites for actin monomers, the α zones interacting with monomers on one strand and the β zones with monomers on the other strand.

Phillips has reexamined this question (Phillips *et al.*, 1986) by taking into consideration the molecular parameters of the coiled-coil as determined in the crystal structure, the azimuthal positions of the amino acid side chains, and how these could be presented to each actin monomer along the F-actin strand (see Figs. 12 and 13 of Phillips *et al.*, 1986). Each of the α zones was

observed to display a characteristic constellation of side chains, including a prominent stripe of electronegative oxygen atoms, a flanking array of electropositive nitrogen atoms, and a region of nonpolar residues between the positive and negative regions. In contrast, inspection of the β zones failed to reveal any strong homologies except for broad zones of charge. Phillips suggests that although the β zones could be involved in long-range aspects of actin binding, it is likely that the more detailed similarity of the α sites indicates their significance for specific interaction with actin. By implication, the TM would interact with only one strand of F-actin.

In an elegant series of experiments, Hitchcock-DeGregori and Varnell (1990) attempted to address the question of 7 or 14 actin binding sites by preparing mutants of chicken skeletal α-TM chains in which one-half, two-thirds, and one binding site (on the assumption of 7 such sites) were deleted. In actin binding assays only the mutant with one site deleted bound to actin. However, with troponin present, both the one-half and one binding site deletion mutants bound to actin and behaved normally in a reconstituted regulated actomyosin S-1 ATPase system. The two-thirds deletion mutant was inactive in all assays. For various reasons discussed by the authors, the results were inconclusive in distinguishing between a 14-fold versus 7-fold periodicity for actin binding, but did confirm the importance of such a periodicity.

Because of the quasi-equivalent nature of the repeating pattern noted here, the TM binding affinity at individual actin monomers is likely to be variable along its length. The structural departures in the regularity of the coiled-coil and differences in disorder (stability) noted in the crystal structure (Phillips *et al.*, 1986) are also indicative of this. Of particular importance to F-actin binding affinity are the ends of the molecules involved in the head-to-tail overlap. This association of TM molecules by interaction and overlap of their NH$_2$- and COOH-terminal ends involving some 8–9 amino acid residues is reflected in large increases in viscosity and apparent M_r at low ionic strength. Removal of several residues at the COOH terminus, including Met-281 by carboxypeptidase treatment as well as acetylation of Lys-7 at the NH$_2$ terminus, markedly reduces this viscosity (Johnson and Smillie, 1977). More exhaustive treatment with carboxypeptidase to quantitatively remove residues 274–284 produces a nonpolymerizable TM that no longer binds to F-actin at the protein concentrations and ionic conditions normally employed in such assays (Mak and Smillie, 1981a,b). Recombinant skeletal muscle TM produced in *E. coli*, differing from muscle TM only in lacking the acetylated α-NH$_2$ group, also fails to polymerize head-to-tail and to bind to F-actin (Hitchcock-DeGregori and Heald, 1987). These properties are restored when an acetylated recombinant TM is expressed in insect SF9 cells using the baculovirus expression vector system (Urbancikova and Hitchcock-DeGregori, 1994). Normal head-to-tail polymerization and actin binding are apparently either fully or partially restored by extension of the NH$_2$ terminus in fusion TMs by 2, 3, or 17 amino acids (Monteiro *et al.*, 1994; Urbancikova and Hitchcock-DeGregori, 1994). The importance of NH$_2$-terminal residues 1–10 of skeletal muscle TM is further illustrated by loss of F-actin binding and head-to-tail polymerizability when these residues are deleted in recombinant TM (Cho *et al.*, 1990). This deletion mutant, unlike carboxypeptidase-treated TM and nonacetylated recombinant TM, is not induced to bind to F-actin by whole troponin and has no regulatory function. Interestingly, the sequence of residues 2–7 of TM (DAIKKK) has been detected in cofilin, an actin binding protein that binds stoichiometrically to F-actin (Matsuzaki *et al.*, 1988).

In the case of smooth muscle TM, these NH$_2$-terminal modifications can be expected to exert similar but not identical effects since, although their NH$_2$-terminal sequences are the same, their COOH-terminal sequences are different (Fig. 4 and Table I). When chicken gizzard TM was assessed for head-to-tail polymerizability in comparison with the skeletal protein by both viscosity and analytical centrifugation, a considerably increased head-to-tail interaction was indicated by both criteria for the smooth muscle protein (Sanders and Smillie, 1984). The stoichiometry and cooperativity of binding to F-actin, however, were the same for the two proteins, as were the effects of [KCl] and [Mg^{2+}]. Cho and Hitchcock-DeGregori (1991) have reported similar viscosity results for the two proteins isolated from the respective muscles, but have observed that the nonacetylated recombinant skeletal and smooth muscle αα-TMs differed significantly in their affinity for F-actin. Using chimeric forms of the two proteins, they concluded that nonacetylated TM with smooth muscle exon 9 bound with at least five-fold greater affinity than the same TM but substituted with skeletal type exon 9. This apparent discrepancy in actin binding results of the two laboratories may arise from the use of NH$_2$-terminal acetylated TMs isolated from muscle in the one case and nonacetylated recombinant proteins in the other. The development of procedures for the preparation of fully functional recombinant TMs (either acetylated or with two- to three-amino acid NH$_2$-terminal extensions) (Urbancikova and Hitchcock-DeGregori, 1994; Monteiro *et al.*, 1994) should help to remove the uncertainties associated with the interpretation of results using the nonacetylated recombinant TMs.

From the observations described here and other reports in the literature (Fowler and Bennett, 1984; Broschat and Burgess, 1986; Heald and Hitchcock-DeGregori, 1988), there does not appear to be a direct cause-and-effect relationship between propensity for head-to-tail polymerization and affinity of actin binding. There may be several reasons for this, including a less than satisfactory experimental method for measuring head-to-tail interaction. Viscosity measurements can only be considered a qualitative tool for this purpose and are subject to major uncertainties in their experimental application. Clearly a more reliable experimental method is needed. The methodology by which the TM is prepared may also have significant effects on the polymerization properties of the protein. Methionines, of which there are three in residues 1–10 of the muscle TMs, are susceptible to oxidation to methionine sulfoxide. Such a chemical modification could significantly alter the interaction properties of this region of the structure. The complexity of the head-to-tail polymerization and actin binding relationship may also result from the interplay of interactions between the several components, particularly if residues 1–10 of TM contribute disproportionately to the overall binding affinity for actin, as is suggested by some of the observations described in the foregoing. Thus the structural surface of this overlap region presented to the binding site on actin may well be affected by the extent of the head-to-tail interaction. It could also be affected by the binding of other regulatory proteins. In the skeletal system, troponin T is known to bind to the overlap region and significantly increase head-to-tail polymerization. In smooth muscle, caldesmon and calponin may play an analogous role, perhaps by long-range effects transmitted through the TM coiled-coil. The major binding sites for these two proteins have been reported to span residues 142–227 of TM and not likely to involve the head-to-tail overlap directly (Watson et al., 1990a,b; Childs et al., 1992). Unfortunately, although the TM head-to-tail overlap of the thin filament assembly is of critical importance to our understanding of both systems, the nature of its structural organization in unknown. Originally postulated to involve a side-by-side abutment of the ends of the two coiled-coils, the crystallographic studies indicate a globular domain (Phillips et al., 1986). In light of their different sequences this domain is likely to show significant folding differences in the skeletal and smooth muscle isoforms.

VII. PERSPECTIVES

Although our knowledge of the genetic and structural basis of TM isoform diversity has advanced very significantly in recent years, the interactive, functional, and biological roles of these isoforms during development and in adult muscle tissue are only poorly if at all understood. Because of their very similar physicochemical properties, this can be ascribed in part to the uncertainties and difficulties in the isolation of the individual isoforms from muscle tissues in pure form and in amounts adequate for studies of their interactive properties with the other thin filament proteins. Since the differences in their effects on ATPase activity and tension development in reconstituted systems are likely to be subtle, present assay procedures may well be inadequate for distinguishing what may be small but important differences in the context of the intact tissue. The application of modern molecular biological approaches for the expression and preparation of fully functional TM isoforms and mutated varieties thereof in bacterial and eukaryotic systems, an approach now well under way in many laboratories, will undoubtedly lead to new insights into their functional roles. Studies of interaction properties with other purified thin filament regulatory protein isoforms and their incorporation into reconstituted thin filament assemblies and skinned fibers for ATPase and tension measurements, as well as into the newly developed motility assay systems, should provide a wealth of new information. Overexpression in transgenic mice or in mammalian cells infected with appropriate viral vectors carrying the TM isoforms also promises to be a fruitful approach. Substantial progress can be anticipated in the delineation of the sites of interaction of TM isoforms with the several regulatory proteins in smooth and skeletal muscles and in the details of their molecular structures. Such information will come from improvements in the resolution obtainable from a variety of technical approaches, including X-ray crystallography, multinuclear multidimensional NMR, cryo-electron microscopy, and others. Developments in the next several years will be both exciting and informative.

Acknowledgments

The author is indebted to Dawn Lockwood and Lorne Burke for their help and patience in the preparation of this manuscript and illustrations. Acknowledgment is also made to the MRC of Canada for financial assistance.

References

Adamson, J. G., Zhou, N. E., and Hodges, R. S. (1993). Curr. Opin. Biotechnol. 4, 428–437.

Bailey, K. (1948). Biochem. J. 43, 271–279.

Betteridge, D. R., and Lehrer, S. S. (1983). J. Mol. Biol. 167, 481–496.

Billeter, R., Heizmann, C. W., Reist, U., Howald, H., and Jenny, E. (1981). FEBS Lett. 132, 133–136.

Bronson, D. D., and Schachat, F. H. (1982). J. Biol. Chem. 257, 3937–3944.

Broschat, K. O., and Burgess, D. R. (1986). J. Biol. Chem. 261, 13350–13359.

Brown, H. R., and Schachat, F. H. (1985). Proc. Natl. Acad. Sci. U.S.A. 82, 2359–2363.

Cabral-Lilly, D., Phillips, G. N., Jr., Sosinsky, G. E., Melanson, L., Chacko, S., and Cohen, C. (1991). Biophys. J. 59, 805–814.

Chacko, S., and Phillips, G. N., Jr. (1992). Biophys. J. 61, 1256–1266.

Childs, T. J., Watson, M. H., Novy, R. E., Lin, J.J.-C., and Mak, A. S. (1992). Biochim. Biophys. Acta 1121, 41–46.

Cho, Y.-J., and Hitchcock-DeGregori, S. E. (1991). Proc. Natl. Acad. Sci. U.S.A. 88, 10153–10157.

Cho, Y.-J., Liu, J., and Hitchcock-DeGregori, S. E. (1990). J. Biol. Chem. 265, 538–545.

Clayton, L., Reinach, F. C., Chumbley, G. M., and MacLeod, A. R. (1988). J. Mol. Biol. 201, 507–515.

Cohen, C., and Parry, D. A. D. (1990). Proteins: Struct., Funct. Genet. 7, 1–15.

Côté, G. P., and Smillie, L. B. (1981a). J. Biol. Chem. 256, 7257-7261.

Côté, G. P., and Smillie, L. B. (1981b). J. Biol. Chem. 256, 11004–11010.

Côté, G. P., and Smillie, L. B. (1981c). J. Biol. Chem. 256, 11999–12004.

Crick, F. H. C. (1953). Acta Crystallogr. 6, 689–697.

Cummins, P., and Perry, S. V. (1973). Biochem. J. 133, 765–777.

Cummins, P., and Perry, S. V. (1974). Biochem. J. 141, 43–49.

Cuticchia, A. J., and Pearson, P. L. (1993). "Human Gene Mapping: A Compendium." Johns Hopkins Univ. Press, Baltimore.

Eisenberg, E., and Kielley, W. W. (1974). J. Biol. Chem. 249, 4742–4748.

Farah, C. S., and Reinach, F. C. (1995). *FASEB J.* **9**, 755–767.

Fatigati, V., and Murphy, R. A. (1984). *J. Biol. Chem.* **259**, 14383–14388.

Forry-Shardies, S., Maihle, N. J., and Hughes, S. H. (1990). *J. Mol. Biol.* **211**, 321–330.

Fowler, V. M., and Bennett, V. (1984). *J. Biol. Chem.* **259**, 10896–10903.

Gooding, C., Reinach, F. C., and MacLeod, A. R. (1987). *Nucleic Acids/Res.* **15**, 8105.

Grabarek, Z., Tao, T., and Gergely, J. (1992). *J. Muscle Res. Cell Motil.* **13**, 383–393.

Graceffa, P. (1989). *Biochemistry* **28**, 1282–1287.

Graceffa, P. (1992). *Biochim. Biophys. Acta* **1120**, 205–207.

Graceffa, P., and Lehrer, S. S. (1980). *J. Biol. Chem.* **255**, 11296–11300.

Graceffa, P., and Lehrer, S. S. (1984). *Biochemistry* **23**, 2606–2612.

Hartshorne, D. J. (1987). *In* "Physiology of the Gastrointestinal Tract" (L. R. Johnson, ed.), 2nd ed., pp. 423–482. Raven Press, New York.

Heald, R. W., and Hitchcock-DeGregori, S. E. (1988). *J. Biol. Chem.* **263**, 5254–5259.

Heeley, D. H. (1994). *Eur. J. Biochem.* **221**, 129–137.

Heeley, D. H., Moir, A. J. G., and Perry, S. V. (1982). *FEBS Lett.* **146**, 115–118.

Heeley, D. H., Dhoot, G. K., Frearson, N. Perry, S. V., and Vrbova, G. (1983). *FEBS Lett.* **152**, 282–286.

Heeley, D. H., Dhoot, G. K., and Perry, S. V. (1985). *Biochem. J.* **226**, 461–468.

Heeley, D. H., Watson, M. H., Mak, A. S., Dubord, P., and Smillie, L. B. (1989). *J. Biol. Chem.* **264**, 2424–2430.

Helfman, D. M. (1994). *Soc. Gen. Physiol. Ser.* **49**, 105–115.

Helfman, D. M., Feramisco, J. R., Ricci, W. M., and Hughes, S. H. (1984). *J. Biol. Chem.* **259**, 14136–14143.

Hitchcock-DeGregori, S. E., and Varnell, T. A. (1990). *J. Mol. Biol.* **214**, 885–896.

Hitchcock-DeGregori, S. E., and Heald, R. W. (1987). *J. Biol. Chem.* **262**, 9730–9735.

Hodges, R. S., and Smillie, L. B. (1970). *Biochem. Biophys. Res. Commun.* **41**, 987–994.

Hodges, R. S., and Smillie, L. B. (1972). *Can. J. Biochem.* **50**, 312–329.

Hodges, R. S., Sodek, J., Smillie, L. B., and Jurasek, L. (1972). *Cold Spring Harbor Symp. Quant. Biol.* **37**, 299–310.

Holtzer, M. E., Holtzer, A., and Skolnick, J. (1983). *Macromolecules* **16**, 173–180.

Holtzer, M. E., Breiner, T., and Holtzer, A. (1984). *Biopolymers* **23**, 1811–1833.

Hosoya, M., Miyazaki, J.-I., and Hirabayashi, T. (1989a). *J. Biochem. (Tokyo)* **105**, 712–717.

Hosoya, M., Miyazaki, J.-I., and Hirabayashi, T. (1989b). *J. Biochem. (Tokyo)* **106**, 998–1002.

Ishii, Y., Hitchcock-DeGregori, S., Mabuchi, K., and Lehrer, S. S. (1992). *Protein Sci.* **1**, 1319–1325.

Jancso, A., and Graceffa, P. (1991). *J. Biol. Chem.* **266**, 5891–5897.

Johnson, P., and Smillie, L. B. (1977). *Biochemistry* **16**, 2264–2269.

Lau, S. Y. M., Sanders, C., and Smillie, L. B. (1985). *J. Biol. Chem.* **260**, 7257–7263.

Leavis, P. C., and Gergely, J. (1984). *CRC Crit. Rev. Biochem.* **16**, 235–305.

Lees-Miller, J. P., and Helfman, D. M. (1991). *BioEssays* **13**, 429–437.

Lees-Miller, J. P., Yan, A., and Helfman, D. M. (1990). *J. Mol. Biol.* **23**, 399–405.

Leger, J., Bouveret, P., Schwartz, K., and Swynghedauw, B. (1976). *Pfluegers Arch.* **362**, 271–277.

Lehrer, S. S. (1975). *Proc. Natl. Acad. Sci. U.S.A.* **72**, 3377–3381.

Lehrer, S. S. (1978). *J. Mol. Biol.* **118**, 209–226.

Lehrer, S. S., and Stafford, W. F. (1991). *Biochemistry* **30**, 5682–5688.

Lehrer, S. S., Betteridge, D., Graceffa, P., Wong, S., and Seidel, J. (1984). *Biochemistry* **23**, 1591–1595.

Lehrer, S. S., Qian, Y., and Hvidt, S. (1989). *Science* **246**, 926–928.

Lewis, W., Côté, G. P., Mak, A. S., and Smillie, L. B. (1983). *FEBS Lett.* **156**, 269–273.

Libri, D., Lemonnier, M., Meinnel, T., and Fiszman, M. Y. (1989). *J. Biol. Chem.* **264**, 2935–2944.

Lindquester, G. J., Flach, J. E., Fleenor, D. E., Hickman, K. H., and Devlin, R. B. (1989). *Nucleic Acids Res.* **17**, 2099–2118.

MacLeod, A. R. (1987). *BioEssays* **6**, 208–212.

MacLeod, A. R., Talbot, K., Smillie, L. B., and Houlker, C. (1987). *J. Mol. Biol.* **194**, 1–10.

Mak, A. S., and Smillie, L. B. (1981a). *Biochem. Biophys. Res. Commun.* **101**, 208–214.

Mak, A. S., and Smillie, L. B. (1981b). *J. Mol. Biol.* **149**, 541–550.

Mak, A. S., Smillie, L. B., and Bárány, M. (1978). *Proc. Natl. Acad. Sci. U.S.A.* **75**, 3588–3592.

Mak, A. S., Lewis, W. G., and Smillie, L. B. (1979). *FEBS Lett.* **105**, 232–234.

Mak, A. S., Smillie, L. B., and Stewart, G. R. (1980). *J. Biol. Chem.* **255**, 3647–3655.

Matsuzaki, F., Matsumoto, S., Yahara, I., Yonezawa, N., Nishida, E., and Sakai, H. (1988). *J. Biol. Chem.* **263**, 11564–11568.

McCubbin, W. D., Kouba, R. F., and Kay, C. M. (1967). *Biochemistry* **6**, 2417–2424.

McLachlan, A. D., and Stewart, M. (1975). *J. Mol. Biol.* **98**, 293–304.

McLachlan, A. D., and Stewart, M. (1976a). *J. Mol. Biol.* **103**, 251–269.

McLachlan, A. D., and Stewart, M. (1976b). *J. Mol. Biol.* **103**, 271–298.

Miegel, A., Lee, L., Dauter, Z., and Maeda, Y. (1993). *Adv. Exp. Med. Biol.* **332**, 25–32.

Montarras, D., Fiszman, M. Y., and Gros, F. (1981). *J. Biol. Chem.* **256**, 4081–4086.

Montarras, D., Fiszman, M. Y., and Gros, F. (1982). *J. Biol. Chem.* **257**, 545–548.

Monteiro, P. B., Lataro, R. C., Ferro, J. A., and Reinach, F. C. (1994). *J. Biol. Chem.* **269**, 10461–10466.

O'Connor, C. M., Balzer, D. R., and Lazarides, E. (1979). *Proc. Natl. Acad. Sci. U.S.A.* **76**, 819–823.

Parry, D. A. D. (1975). *J. Mol. Biol.* **98** 519–535.

Pato, M. D., Mak, A. S., and Smillie, L. B. (1981). *J. Biol. Chem.* **256**, 593–601.

Pearlstone, J. R., and Smillie, L. B. (1982). *J. Biol. Chem.* **257**, 10587–10592.

Phillips, G. N., Jr., Lattman, E. E., Cummins, P., Lee, K. Y., and Cohen, C. (1979). *Nature (London)* **278**, 413–417.

Phillips, G. N., Jr., Fillers, J. P., and Cohen, C. (1980). *Biophys. J.* **32**, 485–502.

Phillips, G. N., Jr., Fillers, J. P., and Cohen, C. (1986). *J. Mol. Biol.* **192**, 111–131.

Pittenger, M. F., Kazzaz, J. A., and Helfman, D. M. (1994). *Curr. Opin. Cell Biol.* **6**, 96–104.

Potekhin, S. A., and Privalov, P. L. (1982). *J. Mol. Biol.* **159**, 519–535.

Potter, J. D. (1982). *In* "Methods in Enzymology" (S. P. Colowick and N. B. Kaplan, eds.), Vol. 85, pp. 241–263. Academic Press, New York.

Potter, J. D., and Johnson, J. D. (1982). *Calcium Cell Funct.* **2**, 145–173.

Ribolow, H., and Bárány, M. (1977). *Arch. Biochem. Biphys.* **179**, 718–720.

Sanders, C., and Smillie, L. B. (1984). *Can. J. Biochem. Cell Biol.* **62**, 443–448.

Sanders, C., and Smillie, L. B. (1985). *J. Biol. Chem.* **260**, 7264–7275.

Sanders, C., Burtnick, L. D., and Smillie, L. B. (1986). *J. Biol. Chem.* **261**, 12774–12778.

Schachat, F., Briggs, M. M., Williamson, E. K., McGinnis, H., Diamond, M. S., and Brandt, P. W. (1990). *In* "The Dynamic State of Muscle Fibers" (D. Pette, ed.), pp. 279–291. de Gruyter, Berlin.

Sender, P. M. (1971). *FEBS Lett.* **17**, 106–110.

Smillie, L. B. (1979). *Trends Biochem. Sci.* **4**, 151–155.

Smillie, L. B. (1982). *In* "Methods in Enzymology" (S. P. Colowick and N. B. Kaplan, eds.), Vol. 85, pp. 234–241. Academic Press, New York.

Sodek, J., Hodges, R. S., Smillie, L. B., and Jurasek, L. (1972). *Proc. Natl. Acad. Sci. U.S.A.* **69**, 3800–3804.

Sodek, J., Hodges, R. S., and Smillie, L. B. (1978). *J. Biol. Chem.* **253**, 1129–1136.

Steinbach, J. H., Schubert, D., and Eldridge, L. (1980). *Exp. Neurol.* **67**, 655–669.

Stewart, M. (1981). *J. Mol. Biol.* **148**, 411–425.

Stewart, M., and McLachlan, A. D. (1976). *J. Mol. Biol.* **103**, 251–269.

Stone, D., and Smillie, L. B. (1978). *J. Biol. Chem.* **253**, 1137–1148.

Stone, D., Sodek, J., Johnson, P. J., and Smillie, L. B. (1975). *Proc. FEBS Meet.* **31**, 125–136.

Thomas, L., and Smillie, L. B. (1994). *Biophys. J.* **66**, A310.

Tsao, T.-C., Bailey, K., and Adair, G. S. (1951). *Biochem. J.* **49**, 27–36.

Ueno, H. (1984). *Biochemistry* **23**, 4791–4798.

Urbancikova, M., and Hitchcock-DeGregori, S. E. (1994). *J. Biol. Chem.* **269**, 24310–24315.

Watson, M. H., Kuhn, A. E., and Mak, A. S. (1990a). *Biochim. Biophys. Acta* **1054**, 103–113.

Watson, M. H., Kuhn, A. E., Novy, R. E., Lin, J.J.-C., and Mak, A. S. (1990b). *J. Biol. Chem.* **265**, 18860–18866.

Whitby, F. G., Kent, H., Stewart, F., Stewart, M., Xie, X., Hatch, V., Cohen, C., and Phillips, G. N., Jr. (1992). *J. Mol. Biol.* **227**, 441–452.

Woods, E. F. (1969). *Int. J. Protein Res.* **1**, 29–45.

Woods, E. F. (1976). *Aust. J. Biol. Sci.* **29**, 405–418.

Wrabl, J., Holtzer, M. E., and Holtzer, A. (1994). *Biopolymers* **34**, 1659–1667.

Xie, L., Miyazaki, J.-I., and Hirabayashi, T. (1991). *J. Biochem.* **109**, 872–878.

Xie, X., Rao, S., Walian, P., Hatch, V., Phillips, G. N., Jr., and Cohen, C. (1994). *J. Mol. Biol.* **236**, 1212–1226.

Yamaguchi, M., Ver, A., Carlos, A., and Seidel, J. C. (1984). *Biochemistry* **23**, 774–779.

Zhoú, N. E., Kay, C. M., and Hodges, R. S. (1994). *J. Mol. Biol.* **237**, 500–512.

Zot, A. S., and Potter, J. D. (1987). *Annu. Rev. Biophys. Biophys. Chem.* **16**, 535–559.

6

Caldesmon

STEVEN B. MARSTON and PIA A. J. HUBER

Department of Cardiac Medicine
National Heart and Lung Institute
London, England

I. INTRODUCTION

Caldesmon (CD) is a ubiquitous protein in smooth muscle cells. The smooth muscle isoform has a sequence-derived molecular mass of 89–93 kDa, but migrates on SDS gels at 120–150 kDa. Within the smooth muscle cell, CD is localized within the contractile apparatus (Furst *et al.*, 1986) and when the contractile filaments are isolated it is found to be tightly bound to the thin filaments (Marston and Lehman, 1985). It is suggested that its *in vivo* function involves regulation of thin filament activity and possibly a role in the assembly and stabilization of thick and thin filaments.

A single CD gene has been identified and sequenced in human and chicken, consisting of at least 14 exons (Fig. 1) (Hayashi *et al.*, 1992; Haruna *et al.*, 1993). All the smooth muscle CD cDNAs so far sequenced correspond to the transcript 1',2,3a,3b,4,5,6–13 and code for a protein of 793 amino acids in human (Humphrey *et al.*, 1992; Hayashi *et al.*, 1992) and 771 amino acids in chicken (Hayashi *et al.*, 1989; Haruna *et al.*, 1993). This isoform is known as CD*h*. In this chapter we have adopted the amino acid numbering originally given for human CD by Humphrey *et al.* (1992), since all the human sequence determinations agree (Humphrey *et al.*, 1992; Hayashi *et al.*, 1992). The original cDNA sequence of chicken gizzard CD (Bryan *et al.*, 1989) omitted 15 amino acids compared to the genomic sequence (Haruna *et al.*, 1993) but is widely used in the literature, therefore we will also give the chicken sequence numbers according to Bryan *et al.* (1989) in brackets.

Although a single splicing pattern has been found for all the smooth muscle CD cDNAs sequenced to date, the genomic sequence permits several isoforms. Messenger RNAs and protein with and without exon 4 have been detected and heterogeneity has been found in exon 1 (Hayashi *et al.*, 1992; Haruna *et al.*, 1993), which is apparently responsible for the formation of two bands of CD*h* commonly seen on SDS gels (Payne *et al.*, 1995) (see Fig. 1).

A second class of CD isoforms (CD*l*) is expressed in nonmuscle tissues with 538 amino acids in human (Novy *et al.*, 1991; Humphrey *et al.*, 1992) and 524 or 517 in chicken (Hayashi *et al.*, 1991; Bryan and Lee, 1991). The length is reduced by the splicing out of exons 3b and 4 from the primary transcript (Hayashi *et al.*, 1992; Haruna *et al.*, 1993).

II. PREPARATION

Caldesmon is heat stable and acid soluble but extremely sensitive to proteolytic degradation. Most preparations take advantage of these properties (Lynch and Bretscher, 1986; Bretscher, 1984; Smith *et al.*, 1987). The initial step is to homogenize 50–100 g smooth muscle tissue in 5 volumes of 300 mM KCl, 1mM EGTA, 0.5 mM MgCl$_2$, 50 mM imidazole-HCl, pH 6.9 at 4°C, plus 2 μg/ml each of chymostatin and leupeptin to stop proteolysis (conventional protease inhibitors such as PMSF and TLCK are ineffective). Smooth muscle is tough and elastic and mincing fol-

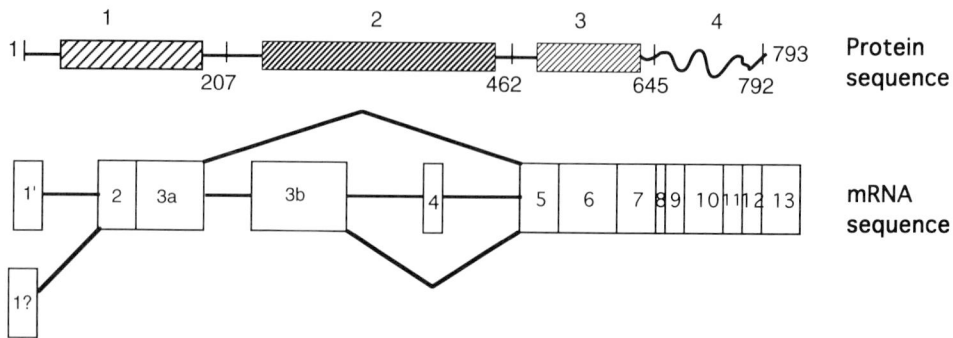

FIGURE 1   The four-domain structure of human CD*h* related to the exon structure of the human CD gene, based on the genomic sequence published by Hayashi *et al.* (1992). Amino acid numbers defining the beginning of each domain refer to the human CD cDNA sequence (Humphrey *et al.*, 1992). Variants of CD*h* can arise from splicing exon 4 in or out and from using different first exons, presumably with different promoters. The nonmuscle isoform message is produced by excluding exons 3b and 4 (Payne *et al.*, 1995).

lowed by a Polytron homogenizer is required. This procedure has been successful with chicken gizzard, rabbit stomach, uterus, and aorta, and sheep and pig aortas. Calf and bovine aortas are too tough and should be avoided. The mixture is rapidly heated to 95°C (300–500 ml in a 600-W microwave oven for 8–10 min), maintained for 2 min, and then rapidly cooled to 4°C. After centrifugation (10 min at 18,000g), the supernatant contains mainly CD, tropomyosin (TM), and calmodulin. Reducing pH to 2.9 with 1 M HCl precipitates most of the TM, leaving CD in solution. After centrifugation (10 min at 18,000g), the supernatant is readjusted to pH 7.0. Subsequent ammonium sulfate precipitation at 30–50% saturation yields up to 90% pure CD. The pH 2.9 step may be omitted if caldesmon is subsequently column purified, thus increasing the yield. In chicken gizzard, all CD is the CD*h* isoform, whereas in blood vessels (e.g., sheep aorta), 20% is the CD*l* isoform.

Final purification and separation of CD*l* from CD*h* may be achieved by ion-exchange chromatography. We use Q-Sepharose Fast Flow in 20 mM Tris-HCl, pH 7.5, 2.5 mM DTT with a NaCl gradient of 0–400 mM. CD*l* is eluted at 140–170 mM NaCl and CD*h* is eluted at 180–200 mM NaCl. It is advisable to concentrate the pure protein by dialysis against 20% polyethylene glycol 20,000 since ammonium sulfate precipitation or lyophilization can produce insoluble aggregates. The pure protein can be stored frozen for at least 6 months. CD migrates anomalously on SDS gels owing to its high content of glutamic acid, thus CD*h* (true M_r 92,000) migrates at 120,000–150,000 and CD*l* (true M_r 63,000) migrates at 70,000–80,000 (Bryan, 1989; Redwood and Marston, 1993).

III. STRUCTURE OF CALDESMON IN SOLUTION

CD is an extremely long and thin molecule. A length of 75–80 nm has been estimated by direct observation in the electron microscope (Mabuchi and Wang, 1991). Observation by electron microscopy and nuclear magnetic resonance (nmr) spectroscopy suggests that it is made up of several rigid rod sections joined by flexible connections. One rod section, extending over 30–40 nm in the center of the molecule, is present only in CD*h* (Stafford *et al.*, 1990; Mabuchi and Wang, 1991). It corresponds to the sequence coded for by exons 3b and 4 (Fig. 1).

A four-domain model was proposed for CD from analysis of the amino acid sequence, predicted secondary structure, and proteolytic digestion patterns (Marston and Redwood, 1991). This model is compatible with the recent genomic sequence and with functional analysis.

Domain 1 stretches from amino acids 1 to 207 (198 in chicken) and includes exons 1,2, and 3a. This domain is extended, may contain some α-helix, and contains binding sites for myosin and TM. Domain 2 stretches from 208 to 462 (431 in chicken), corresponding to exons 3b and 4, and binds to TM. This domain is absent from all CD*l*.

Domain 2 is predominantly α-helical and almost entirely made up of acidic and basic amino acids; in chicken CD it consists of a motif of 15 amino acids based on the sequence EEE(R/K)KAAEERERAKA, which is repeated 10 times (Bryan *et al.*, 1989; Hayashi *et al.*, 1989). This motif is considerably less well conserved in the human CD*h* sequence (Humphrey *et al.*, 1992), how-

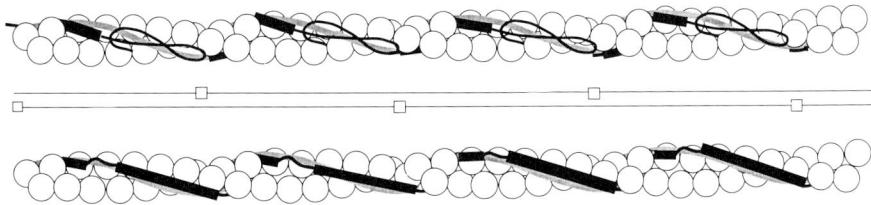

FIGURE 2   A model for the incorporation of an extended caldesmon molecule into the thin filaments of smooth muscle. To produce a 38-nm repeat of structure, each CD is placed in register with a TM and extends for 76 nm, the length of two TMs. The CD molecules alongside TM in the actin helix are staggered by the length of one TM relative to each other as is shown schematically by the pair of lines with blocks. A consequence of this arrangement, besides producing the 38-nm repeat, is that the filament has no radial symmetry and consequently any particular part of CD is on the same side of the filament. We have therefore shown the filament in two views that are rotated by 180 degrees. One shows the domains 3 and 4 of one CD and the beginning of domain 1 of the next. The other shows the rest of domain 1 and the central helix domain 2. Reproduced from Marston and Redwood (1991), with permission.

ever, in both species there is a pattern of acidic and basic amino acids that enables the α-helix to be stabilized by salt-bridge interactions of the side chains. Domain 2 contains an α-helix of at least 55 turns, which is stable in solution (Wang et al., 1991b).

Domain 3 stretches from amino acids 463 (432) to 645 (589) and corresponds to exons 5, 6, and 7. The core of domain 3 is again a single α-helix stabilized by intrahelix salt bridges. The sequence from 565 (509) to 621 (565) is 43% identical to the amino acids 90–146 of rabbit skeletal muscle troponin T, which is a TM binding site in troponin T. Domain 4 stretches from 646 (590) to the C terminus and corresponds to exons 8–13. This is an extended and stably folded region that contains little regular secondary structure (Levine et al., 1990). It is, however, the most highly conserved region and contains all the important regulatory interaction sites with actin and calmodulin.

IV. CALDESMON AS A COMPONENT OF THE THIN FILAMENTS

In the smooth muscle cell, CD is incorporated into the thin filaments in the "contractile domain" of the cell (Furst et al., 1986; North et al., 1994a). Ultrastructural studies presented in Chapter 4 (this volume) have shown that CD is located in the thin filament in an extended form beside TM along the axis of the actin double helix. The model (Fig. 2) places CD in potential contact with actin and TM throughout its length and allows a possible end-to-end interaction. These structural arrangements form the basis of caldesmon function in the thin filament.

Native thin filaments extracted from smooth

muscles contain CD, TM, and actin. The measured CD: actin ratio is 1:16 (Marston, 1990; Lehman et al., 1993), which is close to the 1:14 ratio predicted in the model. Smooth muscle thin filaments are Ca^{2+} regulated and there is good evidence that CD plays a key role in this regulation:

1. CD content and Ca^{2+} sensitivity in thin filament preparations are correlated (Marston and Smith, 1984; Gusev et al., 1994).
2. Antibodies to CD interfere with the Ca^{2+} sensitivity of thin filaments (Marston et al., 1988b).
3. A Ca^{2+}-regulated thin filament can be reconstituted from pure actin, TM, CD, and calmodulin (Sobue et al., 1981; Dabrowska et al., 1985; Ngai and Walsh, 1985; Smith et al., 1987; Pritchard and Marston, 1989).

An additional property of CD in thin filaments is to cross-link thick and thin filaments owing to its ability to bind to actin and myosin. This suggests an additional role of CD in directing and stabilizing filament assembly and possibly in latch (see Section IX).

V. CALDESMON–ACTIN–TROPOMYOSIN INTERACTIONS

The basis of the proposed regulatory function of caldesmon is its ability to inhibit the interaction of myosin with the thin filament. A prerequisite for this is that CD binds to actin. In the native thin filament, TM is also present, and it is found that the presence of TM has effects on CD binding to actin and a very profound amplifying effect on CD inhibition. We therefore have to consider the interaction of CD with actin and TM together.

Smith *et al.* (1987) measured CD binding to actin–TM over a very wide range of CD concentration and concluded that there were two classes of binding sites: high-affinity sites with a stoichiometry around 0.06 CD/actin–TM with $K_1 = 4 \times 10^7 M^{-1}$ at $I = 0.09 M$ and low-affinity sites with a stoichiometry of 0.5–1 CD/actin–TM, $K_2 = 5 \times 10^5 M^{-1}$. The two classes of binding sites have also been observed by Yamakita *et al.* (1992) and in recombinant CD fragments by Redwood and Marston (1993) and Redwood *et al.* (1993).

If TM is not present, the tight binding is barely detectable, but the weaker class of actin–CD binding sites is clearly observed, with a two- to three-fold diminution in affinity. Thus it seems likely that the tight binding sites are TM dependent. Unfortunately, the majority of CD–actin binding curves have not been measured over a wide enough concentration range to distinguish the two sites; they are commonly analyzed as single-site binding, which tends to yield a stoichiometry of around 0.15 mol CD/mol actin and affinities in the range of 10^6–$10^7 M^{-1}$ (Tanaka *et al.*, 1990; Velaz *et al.*, 1989; Riseman *et al.*, 1989; Dabrowska *et al.*, 1985; Lash *et al.*, 1986), with little difference between actin and actin–TM.

Functional studies clearly show the presence of high-affinity, TM-dependent sites linked to inhibition. Thus CD inhibition of actin–TM correlates with 1 CD bound per 14 actin (Smith *et al.*, 1987; Marston and Redwood, 1992, 1993; Velaz *et al.*, 1989), whereas 1 CD for 1–2 actin is required to inhibit actin filaments in the absence of TM (see Fig. 6) (Marston and Redwood, 1993). In addition, we have observed that CD switches off the actin–TM filament movement in the motility assay with a half-maximal effect at 3 nM CD, corresponding to an affinity of $>10^8 M^{-1}$ ($I = 0.09 M$) (Fraser and Marston, 1995); if TM is absent, no effect is observed at low concentrations.

These experiments with reconstituted thin filaments agree with the stoichiometry and stability of CD present in native thin filaments extracted from smooth muscles (Marston, 1990). The stoichiometry is 1 CD : 16 actin and the estimated affinity of CD for the thin filament is $2 \times 10^7 M^{-1}$ ($I = 0.09 M$). Thus a complex of 1 CD : 2 TM : 14 actin, which resembles the native thin filament, may be reconstituted in the test tube. Actin activation of myosin MgATPase and motility is switched off in this complex and a Ca^{2+}-regulated thin filament can be reconstituted by the further addition of calmodulin (Smith *et al.*, 1987; Marston and Smith, 1985). At high CD concentrations, more CD binds to the actin or actin–TM filament up to a maximum of one per actin; this can only be achieved in the test tube and has hardly any physiological significance.

A. The Caldesmon–Tropomyosin Interaction

CD binds to TM with an affinity in the region of 10^6 M^{-1}. The interaction is strongly dependent on ionic strength. Affinities ranging from $2 \times 10^7 M^{-1}$ at very low salt concentrations to $10^5 M^{-1}$ at $I = 0.1 M$ have been recorded (Horiuchi and Chacko, 1988; Watson *et al.*, 1990; Smith *et al.*, 1987; Redwood *et al.*, 1990). The affinity is fourfold greater when the TM is already bound to the actin filament (Horiuchi and Chacko, 1988). None of the binding studies has shown any cooperativity in CD–TM binding, whereas the binding of TM to actin is highly cooperative (Sanders and Smillie, 1984). Experiments using fragments of CD and TM indicate that there are specific sites in both molecules. CD has strong TM binding sites in domains 2 and 4 (Hayashi *et al.*, 1991; Redwood and Marston, 1993; Fraser *et al.*, 1994; Huber *et al.*, 1994) and a slightly weaker site in domain 1 (Redwood and Marston, 1993; Bogatcheva *et al.*, 1993). It is surprising to find that little or no TM binding has been demonstrated in domain 3 of CD (Fraser *et al.*, 1994; Huber *et al.*, 1995c; Hayashi *et al.*, 1991), since this has a region of strong homology with a TM binding domain, T1, of skeletal muscle troponin T. However, it has been reported that the T1 domain has little affinity for smooth muscle TM (Pearlstone and Smillie, 1982).

The location of the TM binding site within domain 1 seems to be toward its C terminus since longer fragments, for example, 1–175 (Bogatcheva *et al.*, 1993; Vorotnikov *et al.*, 1993), bind significantly tighter than the shorter fragment N128, residues 1–128 (Redwood and Marston, 1993). Within domain 4, the TM binding is probably restricted to the N-terminal region, 663–714 (606–657). Fragments containing part of this sequence do bind to TM (e.g., H4, H2), whereas C-terminal fragments such as H9, H8, and 658C do not (Redwood and Marston, 1993; Fraser *et al.*, 1994; Huber *et al.*, 1995c) (see Fig. 4).

A major site on TM has been identified between 142 and 227. There is also a suggestion of a weaker site in the N terminus of TM (11–127) (Watson *et al.*, 1990). These results can be reconciled with the model of thin filament structure (Figs. 2 and 3). CD and TM are both extended and can make multiple binding contacts along their length. CD is double the length of TM, hence the strong site (142–227) in adjacent TMs can probably interact with domain 2 and the N terminus of domain 4, whereas the weaker site (11–127) could interact with domains 1 and 3. We have no evidence for the relative polarity of CD and TM; Fig. 3 shows an

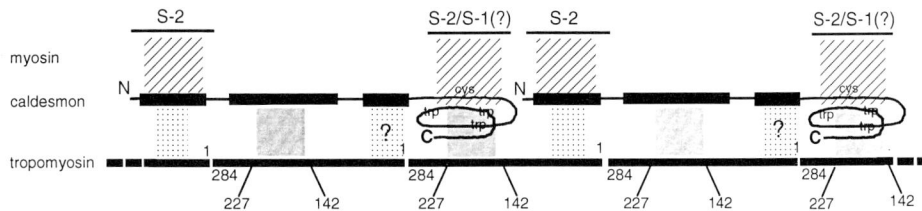

FIGURE 3 Location of CD–tropomyosin and CD–myosin binding sites. CD and TM are shown bound in an antiparallel configuration. One CD extends over parts of three TMs such that the strong binding site in TM is opposite domain 2 and the N terminus of domain 4. Myosin subfragment-2 is indicated to be able to bind to extended segments in domains 1 and 3–4.

antiparallel configuration by analogy with troponin T binding to TM (Watson *et al.*, 1990). However, a parallel configuration would also fit the data.

B. The Caldesmon–Actin Interaction

Actin binding sites are confined to the C terminus of CD. Many workers have demonstrated that domains 3 and 4, whether prepared by proteolytic digestion or expressed as recombinant proteins, are indistinguishable from whole CD (Redwood *et al.*, 1990; C.-L.A. Wang *et al.*, 1991a; Hayashi *et al.*, 1991; Z. Wang *et al.*, 1994). Although there is evidence for a weak interaction between domain 3 and actin (Leszyk *et al.*, 1989; Levine *et al.*, 1990; Hayashi *et al.*, 1991), recombinant fragments containing domains 1,2 and 3 did not bind to actin in quantitative assays (Redwood *et al.*, 1990; Wang *et al.*, 1994). This sets a limit of less than 4×10^3 M^{-1} for domain 3 binding to actin.

Domain 4 contains the high-affinity actin binding sites that are involved in inhibition of the actin filament's activity. CD fragment 606C (Marston and Redwood 1992) (Fig. 4), the equivalent chymotryptic fragments ("18K" and "20K") (Szpacenko and Dabrowska, 1986), and recombinants such as "22K" (Wang *et al.*, 1991a) and C178H (Hayashi et al., 1991) all bind strongly to actin and are highly potent TM-dependent inhibitors of actin-activated ATPase. The binding of 606C to actin and actin–TM has been demonstrated to be the same as for the whole CD molecule (Redwood and Marston, 1993).

There are a number of actin binding sequences in domain 4. Fragments from the N terminus (the 7.3-kDa fragment, 69 amino acids long, Fig. 4), the middle (H2, 85 amino acids), or the C terminus (658C, 10K, and H9, 99 and 67 amino acids, respectively, Fig. 4) all bind actin with affinities of about 10^5 M^{-1} at low ionic strength, which is about one-tenth of the affinity of native CD or 606C. The shortest actin-binding fragment studied is LW30, which binds actin at one-fifth of the affinity of the longer sequences.

The binding studies do not indicate any localized actin-binding sites: affinity is roughly proportional to peptide length and does not depend on the peptide's position within domain 4. Some subsites may be distinguished on a functional basis, although the analysis is complicated because of different measuring conditions in different laboratories. Most domain 4 fragments that have been studied can inhibit actin-activated ATPase activity. Where it has been measured (658C, 7.3K, LW30) it appears that 100% inhibition corresponds to 0.5 to 1 CD fragment bound per actin—the same as whole CD (Marston and Redwood, 1993; Mezgueldi *et al.*, 1994; Chalovich *et al.*, 1992; Bartegi *et al.*, 1990).

The sequence which would appear to be required for TM dependent effects of CD is represented by amino acids 726-767 (669-710 in chicken). It is interesting to note that a peptide segment corresponding to exon 11 in the CD genomic sequence (amino acids 734-765 (677-708)), is present in all the inhibitory fragments and thus seems to be essential for tropomyosin enhanced inhibition of the ATPase. The C-terminal fragments 658C and H9 are TM-enhanced inhibitors. In the case of 658C it has been demonstrated that full inhibition corresponds to 1 658C bound per 14 actin (Redwood and Marston, 1993), and a high-affinity, low-stoichiometry, TM-dependent actin binding component has also been observed similar to whole CD and 606C. The domain 4 N-terminal 7.3K fragment and the short C-terminal fragment LW30 show little or no enhancement of inhibition in the presence of TM, whereas H7 (N terminus and middle of domain 4) does inhibit in a TM-enhanced way.

If these results can be taken at face value, the middle of domain 4 [726–767 (669–710 in chicken)] is involved in the TM enhancement of CD inhibition and binding to actin. Since this sequence is well separated from the putative TM binding sequence in domain 4, the effects of TM are probably indirect. The C-terminal end of this sequence is particularly rich in proline, a motif known to be present in many protein–protein binding sites (Williamson, 1994). Amino acids 752–771 (695–714),

tropomyosin binding

tropomyosin -linked actin binding

643
(587)
GSSLKIEERA EFLNKSVQKS SKVKSTHQAA IVSKIDSRLE QYTSAIEGTK SAKPTKPAAS DLPVPAEGVR NIKSMWEKGN VFSSPTRAGT PNKETAGLKV GVSSRINEWL TKTPDGNKSP APKPSKLLRG DVSSKRNLWE KQSVDKVTSP TKV

(A)(-) (M.PA.TT. V)   (VV.N) (A) (A)   (GGT)   (E)   (G)

650   656   793(737)
        (600)   (756)
        PKC   (PAASSSKVTATGKKSETNG...)

H4

622 (566)   681 (624)   793

715 (658)   726 (669)

H2

622 (566)   683 (626)   767 (710)   793

H7

715 (658)   767 (710)

H9

H8

714 (657)   730 (675)   743 (686)   749 (692)   759 (702)   779 (722)   783 (726)
715 (659)   PKC
PKC   cdc2   PKC   MAPK   PKC

GS17C
GVTNIKSMWEKGN VFSC
calmodulin binding peptide
(Zhan et al. 1992)

M73
SMWEKGN VFSSPGF
calmodulin binding peptide

NK21
NKETAGLKV GVSSRINEWL TK
calmodulin binding peptide
(Mezgueldi et al. 1994)

LW30
L TKTPDGNKSP APKPDLRPGDVSSKRNLW
actin binding and
inhibitory peptide
(Mezgueldi et al. 1994)

V12T
V GVSSRINEWL TKT
calmodulin binding peptide

G12S
G DVSSKRNLWE KQS
calmodulin binding peptide

7.3 kDa (Chalovich et al 1992)

606C   715 (658)   658C   (756)

663 (606)   715 (658)   (756)

(=10K fragment, Bartegi et al. 1990)

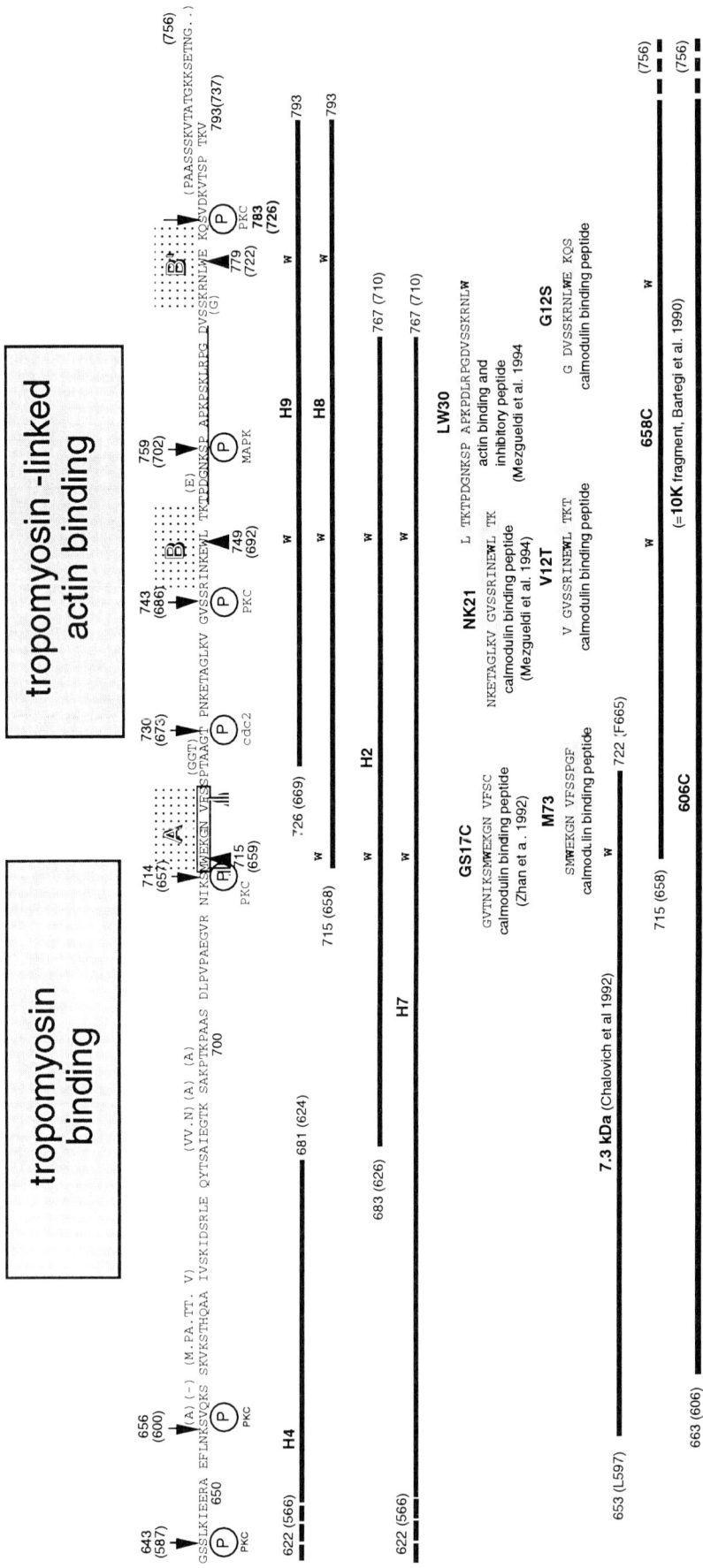

FIGURE 4   Sequence and binding sites in domain 4 of human caldesmon. The amino acids of chicken CD, which differ from human, are shown in brackets above the main sequence. Numbering is for human CDh with chicken CDh in brackets. The human recombinants H2,H7, and H9 (Marston et al., 1994a), chicken recombinants 606C and 658C (Redwood and Marston, 1993), and the 7.3K chymotryptic fragment (Chalovich et al., 1992) are shown as lines. Peptides M73 (Marston et al., 1994a), GS17C (Zhan et al., 1991), V12T (Huber et al., 1995b), LW30, and NK21 (Mezgueldi et al., 1994) are indicated and the locations of the tryptophans and phosphorylation sites are shown on the sequence. The proposed position of the TM binding region and the sequence required for TM-dependent inhibition are indicated schematically. The sequence of site A is boxed.

underlined in Fig. 4, show a striking similarity to a sequence in smooth muscle myosin light chain kinase, [41]PKTPVPEKVPPPKPATPDFRSVL[63], which is believed to be an actin binding sequence (Kobayashi *et al.*, 1992). The importance of this part of the sequence to CD function is supported by three lines of experiments.

1. Serine 759 (702) is the main site of phosphorylation by MAP kinase (see Adam, Chapter 13, this volume), and phosphorylation at this site, or modification of serine to aspartic acid by site-directed mutagenesis, has a distinct effect: TM-dependent inhibition is diminished and TM-dependent high-affinity actin binding is absent (Redwood *et al.*, 1993; Childs *et al.*, 1992).

2. The peptide LW30 is part of this sequence and is an actin-binding inhibitor, whereas peptides containing only the C terminus or N terminus are nonfunctional (Mezgueldi *et al.*, 1994).

3. The recombinant peptide H2 contains a segment of this sequence. H2 is an inhibitor of actin-activated ATPase but has the opposite effect, that is, it activates actin–TM activity (Marston *et al.*, 1994b). H2 is a potent antagonist of the TM-dependent inhibition owing to whole CD. Thus H2 contains a sequence that is essential but not sufficient for TM-dependent inhibition; it seems that flanking actin binding sites, on either the N-terminal or the C-terminal ends (e.g., H7,H9) but preferably both (606C), are also required for effective inhibition.

Most of the interactions between CD and actin have been mapped to subdomain 1 of actin. This is the region of actin that in the three-dimensional structure contains both the N and C termini and important myosin and troponin I binding sites (Sheterline and Sparrow, 1994). Removal of three amino acids from the C terminus of actin weakens its interaction with both CD and troponin I (Makuch *et al.*, 1992), and reduction of the number of acidic amino acids in the N-terminal five amino acids (substitution or deletion mutants) decreases inhibition of actin-activated ATPase by CD but does not alter its binding affinity (Crosbie *et al.*, 1994).

Nuclear magnetic resonance studies show structural detail of this interaction (Mornet *et al.*, 1995; Levine *et al.*, 1990). Amino acids 1–7 of actin interact with domain 4 but not with the C-terminal fragment 658C, thus actin 1–7 probably binds to the N terminus of domain 4, whereas the C-terminal half binds elsewhere on actin, currently not identified. A paramagnetic label attached to cysteine 636 (580) perturbs the nmr signals of histidine 667 (610), tyrosine 682 (625), and at least one of the tryptophans 716,749, or 779 (659,692, or 722), indicating that all of these residues are within 1.5 nm of the cysteine 636 (580). This means

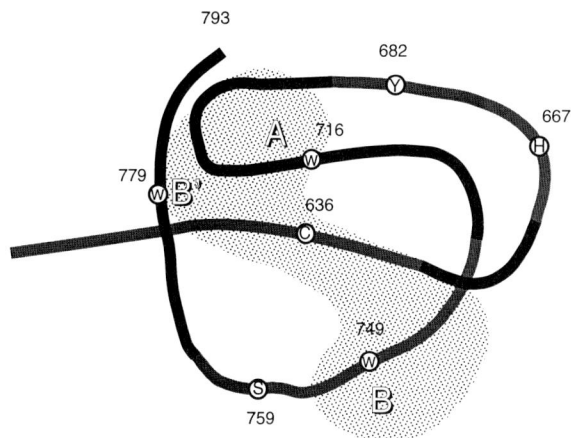

FIGURE 5 A model of the two-dimensional structure of domain 4 derived from structural and functional studies. This hypothetical structure shows domain 4 of CD bound to Ca²⁺·calmodulin. Sites A,B, and B' are indicated as including both lobes of calmodulin bound in an extended conformation, together with CD tryptophans 716,749, and 779. The peptide chain is looped so that the three tryptophans and tyrosine 682 and histidine 667 are within 1.5 nm of cysteine 636. The exons are indicated by alternating shading of the peptide chain. Exon 11 is the loop at the bottom, which includes tryptophan 749 and the MAP kinase site serine 759.

that the peptide chain of domain 4 must fold into a comparatively compact structure, such as that shown in Fig. 5. Such a configuration would enable several peptide segments in domain 4 to be presented close to subdomain 1 of actin.

VI. CALDESMON INHIBITION OF ACTOMYOSIN ATPase ACTIVITY

The enzymatic cycle of actomyosin MgATPase involves equilibrium binding of myosin·ADP·P_i to actin to form an initial "weak" binding complex ($K = 10^4$ M^{-1}) followed by a rate-limiting transition to a "strong" binding complex ($K = 10^8 M^{-1}$)—the "power stroke" of the cross-bridge cycle—with subsequent release of P_i and ADP. Binding and hydrolysis of ATP at the active site of myosin breaks the "strong" complexes and the myosin·ADP·P_i can subsequently rebind as a weak complex to another actin monomer. Inhibition of this cycle could occur either by reducing the affinity of weak complexes for actin or by reducing the rate of the transition from weak to strong complexes.

It is abundantly clear that in the absence of tropomyosin, CD inhibits actin activation of myosin MgATPase by reducing the affinity of weak myosin complexes for actin. This has been demonstrated directly since CD decreases the affinity of S-1·ADP·P_i for

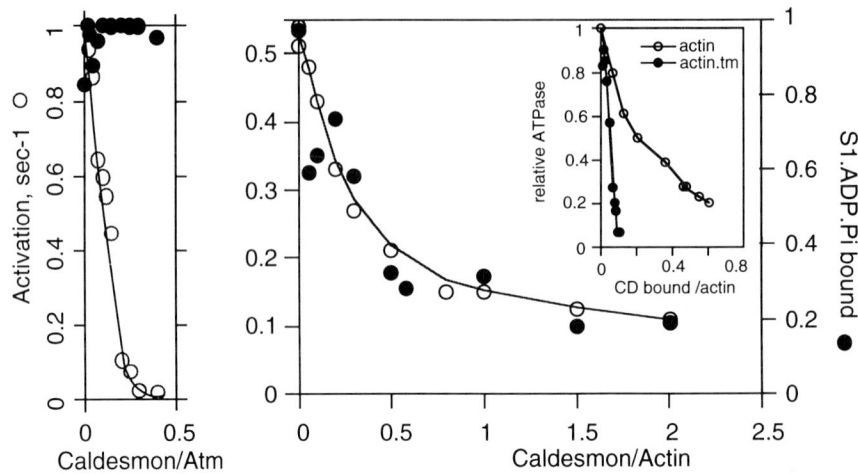

FIGURE 6   The mechanism of caldesmon inhibition of actin and actin–tropomyosin activation of myosin MgATPase. The left-hand plot shows that inhibition of actin–TM activation is potent and not associated with S-1·ADP·P$_i$ displacement from actin. The right-hand plot shows that inhibition of actin activation is associated with displacement of S-1·ADP·P$_i$. Conditions: 6 μM skeletal muscle S-1 and 4 μM actin or actin–TM (0.3 mg aorta TM/mg actin) and 0 to 10 μM CD were assayed for activation of S-1 MgATPase activity (open circle), the fraction of S-1·ADP·P$_i$ bound to actin (expressed relative to the amount bound in the absence of CD, which was 5–10%) (solid circle) and CD binding (see inset graph) at 25°C, pH 7.1, I = 0.057. The inset graph shows the relationship between inhibition and CD bound to actin in the presence and absence of smooth muscle TM. Reproduced from Marston and Redwood (1993) with permission.

actin and it is also implied by the effects of CD on the kinetics of actin-activated myosin ATPase; that is, CD increases K_m but V_m is unchanged (Horiuchi *et al.*, 1991; Chalovich *et al.*, 1987). Kinetically this effect appears to be a direct competition between CD and myosin·ADP·P$_i$ for a site on actin. Thus CD displaces S-1·ADP·P$_i$ from actin and the displacement correlates with inhibition (Fig. 6); conversely, S-1, S-1·AMP·PNP (strong binding complexes), and S-1·ADP·P$_i$ (weak binding complex) displace CD from actin filaments under the appropriate conditions (Marston and Redwood, 1992, 1993; Velaz *et al.*, 1990; Chalovich *et al.*, 1987). In each case, the stoichiometry of the competition is 1 CD: 1 actin: 1 S-1.

Direct competition is structurally possible since domain 4 of CD binds to sites in or near subdomain 1 of actin, which is also known to be the site where the myosin S-1 binds (Rayment *et al.*, 1993). S-1, troponin I, and CD contacts in this subdomain have been studied by nmr spectroscopy (Levine *et al.*, 1990; Mornet *et al.*, 1995), which reveals that the molecular contacts of the three proteins with amino acids in actin subdomain 1 are not the same. Nevertheless, the troponin I inhibitory peptide displaced 658C from actin, indicating that 658C is located sufficiently close to actin 1–7 that it sterically blocks this site and may thus com-

pete with S-1·ADP·P$_i$ for binding and activation of the ATPase (Huber *et al.* 1995b).

CD inhibition of actin-activated myosin ATPase at 1 : 1 ratio is observable only in the laboratory. Its significance as a regulatory mechanism is severely limited because the thin filaments of smooth muscle contain only 1 CD per 14 actin (Fig. 2), so that no more than 7% inhibition is possible by this mechanism. On the other hand, it has been demonstrated in numerous laboratories that CD: actin ratios similar to those *in vivo* are potently inhibitory when tropomyosin is present.

VII. TROPOMYOSIN-DEPENDENT INHIBITION OF ACTOMYOSIN ATPase BY CALDESMON

The key feature of the TM-dependent inhibition of actin activity by CD is that it takes place at low ratios of CD to actin (see Fig. 6). This high degree of cooperativity is critically dependent on the TM isoform. Smooth muscle TM permits maximal inhibition at a 1:14 ratio, but filaments containing skeletal muscle TM need 1 CD : 7 actin for equivalent inhibition (Marston and Redwood, 1993). When CD inhibits actin–TM-activated ATPase at physiological ratios of CD : actin, there is no detectable change in the binding of S-1

·ADP·P$_i$ to actin–TM (Fig. 6) (Marston and Redwood, 1992, 1993). Inhibition of actin–TM does not operate by controlling the weak interaction, instead it involves a control of the rate-limiting step of the ATPase.

How could CD control a rate process at ratios of 1 CD : 2 TM : 14 actin molecules? In striated muscles, troponin and tropomyosin control a rate process that is propagated over 7 actin by a well-understood mechanism (Lehrer, 1994). Striated muscle troponin I and caldesmon domain 4 have no amino acid sequence similarities but the functional similarities are striking, and we have so far not been able to find any important characteristics that they do not share (Marston and Redwood, 1993; Marston et al., 1994b). Consequently, we have proposed a mechanism for TM-dependent CD regulation based on the troponin mechanism and illustrated in Fig. 7. The basis of the model is that CD is an allosteric effector switching actin–TM between two states, "on" and "off." The cooperativity derives from the fact that TM changes state as a unit, acting on up to 14 actin, even though the effector protein (CD) binds only to a single actin. This binding site corresponds to the low-stoichiometry, high-affinity, TM-dependent actin binding site already identified.

We have tested the specific predictions of the proposed mechanism by showing that when actin–TM filaments are switched off by CD, the binding of strong myosin complexes (S-1·ADP, S-1·AMP·PNP) is blocked (Marston et al., 1994b). Moreover, in the *in vitro* motility assay, CD domain 4 switches motile actin–TM filaments off without affecting the velocity of movement of the remaining motile filaments or displacing filaments from myosin. This form of behavior is compatible with the model and indicates that the actin–TM filaments form a cooperative unit. The behavior of domain 4 and troponin I is identical in the *in vitro* motility assay, thus further supporting the idea of a common regulatory mechanism in striated and smooth muscle thin filaments (Fraser and Marston, 1995a,b).

VIII. Ca^{2+} CONTROL OF CALDESMON INHIBITION

Since CD was first isolated as a Ca^{2+}·calmodulin binding protein (Sobue et al., 1981), it has been supposed that Ca^{2+}-dependent regulation of the thin filaments is due to Ca^{2+} binding to calmodulin (CM), thereby reversing inhibition (Smith et al., 1987; Marston and Smith, 1985; Sobue et al., 1982). This may readily be incorporated into the regulatory model (Fig. 7). However, there are some uncertainties about Ca^{2+} control of CD that have not been fully resolved.

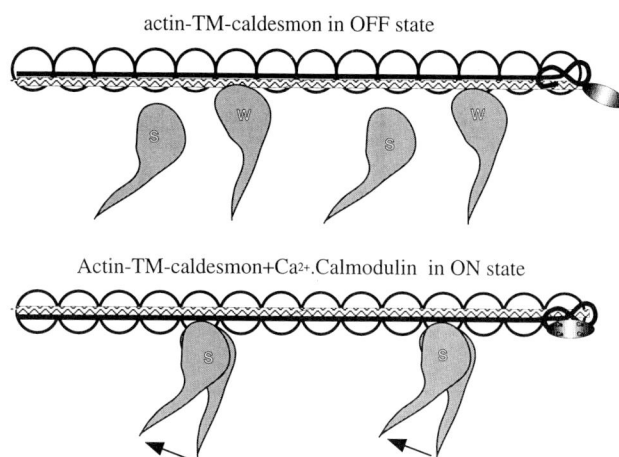

FIGURE 7 A model for caldesmon regulation of the actin–tropomyosin filament. CD acts as an allosteric effector that controls the state of the actin–TM filament. Actin–TM exists in two states, "off" and "on," in equilibrium. In the "off" state the strand of TM is located over the outer edge of the actin monomer in such a way that it prevents binding of myosin to actin in the "strong" conformation but allows binding in the "weak" conformation. CD binds preferentially to actin–TM when it is in the "off" state, thus the actin–TM unit to which it binds is unable to interact cyclically with myosin. The function of the Ca^{2+} binding protein is to cause a conformational change that alters the preference of CD from binding to actin–TM in the "off" state to binding to actin–TM in the "on" state. Ca^{2+} activation thus proceeds without dissociation of CD from actin. In the "on" state, "strong" and "weak" myosin complexes can bind and cross-bridges can cycle. The effective unit regulated by one CD spans at least 14 actin monomers.

Calmodulin is a prime candidate for the role of the Ca^{2+}-sensitizing protein since it is the most abundant calcium binding protein in smooth muscles, with a total concentration in the same range as CD. Under particular reaction conditions (37°C/120 mM KCl), Ca^{2+}-regulated synthetic thin filaments may be reconstituted using only actin, TM, CD, and CM. In contrast, native thin filaments are Ca^{2+} sensitive over a wide range of conditions; and in conditions optimal for Ca^{2+} sensitivity in native thin filaments, the synthetic thin filaments are usually completely Ca^{2+} insensitive (Smith et al., 1987; Pritchard and Marston, 1989; Marston et al., 1988a).

We and others have tested other Ca^{2+} binding proteins that might give regulation of CD inhibition that closer resembles that of native thin filaments. In general it has been found that the Ca^{2+}-sensitizing capacities of S-100, troponin C, SMCaBP-11, and caltropin (see Mani and Kay, Chapter 8, this volume) are just the same as CM when compared under the same conditions (Pritchard and Marston, 1991; Skripnikova and Gusev, 1989; Mani et al., 1992; Fujii et al., 1990). Since none of these proteins is as abundant as CM in smooth muscle, there is currently no proven alternative candidate for the Ca^{2+}-sensitizing component of smooth muscle thin filaments.

Pritchard and Marston (1993) suggested that there could be another protein mediating between CD and CM in native thin filaments and isolated a crude preparation that had the desired property. Gusev et al. (1994) noted that native thin filament preparations contain a number of additional proteins, previously thought to be impurities, which, when added in physiological ratios, could influence Ca^{2+} regulation to make the synthetic filament more like the native system. These factors included filamin, gelsolin, and rigor myosin heads, and it was suggested that they acted by changing the TM equilibrium toward the "on" state (see Fig. 7). Since increased ionic strength and temperature and myosin rigor heads, which are well known to favor the "on" state of actin–TM, also favor Ca^{2+} sensitivity in reconstituted thin filaments, it seems possible that the anomalies between reconstituted and native systems may be simply due to subtle differences in the on/off equilibrium constant.

When $Ca^{2+} \cdot CM$ binds to CD attached to actin–TM, it reverses CD inhibition; since CM does not bind in the absence of Ca^{2+}, this mechanism makes a Ca^{2+}-dependent switch (Smith et al., 1987; Sobue et al., 1982; Dabrowska et al., 1985). $Ca^{2+} \cdot CM$ binding to CD affects its interaction with actin–TM. Under many experimental conditions this is manifested as a weakening of the affinity of CD for actin and TM and in extreme circumstances CD dissociates from actin. However, in most test tube situations, effective Ca^{2+}-dependent regulation involves the formation of a $Ca^{2+} \cdot CM$–CD–actin–TM complex (Smith et al., 1987; Sobue et al., 1982; Pritchard and Marston, 1989; Ngai and Walsh, 1985).

$Ca^{2+} \cdot CM$ binds to CD with an affinity in the range 10^6–$10^7 M^{-1}$, which is largely unaffected by temperature and ionic strength (Mills et al., 1988; Shirinsky et al., 1988; Smith et al., 1987; Pritchard and Marston, 1989); in the absence of Ca^{2+}, CM binding to CD is at least 100-fold weaker. $Ca^{2+} \cdot CM$ appears to bind to CD in an extended conformation rather like troponin C binding to troponin I (Mabuchi et al., 1995).

$Ca^{2+} \cdot CM$ binds to CD at a number of discrete sites that may be no longer than nine amino acids. From studies with recombinant fragments and synthetic peptides, two short calmodulin binding sequences with affinities in the $10^6 M^{-1}$ range have been clearly identified in domain 4 (Zhan et al., 1991; Mezgueldi et al., 1994; Marston et al., 1994a; Huber et al., 1995a). In Fig. 4 these are shown using the notation of Marston et al. (1994a) as sites A ([715]MWEKGNVFS[723]) and B ([744]SRINEWLTK[752]). There may be a third site B'

([773]SSKRNLWEK[781]). The involvement of tryptophans 716 (659), 749 (692), and 779 (722) in the interaction with CM has been determined by measurements of the intrinsic tryptophan fluorescence (Shirinsky et al., 1988; Zhan et al., 1991; Mezgueldi et al., 1994; Marston et al., 1994a) and by nmr spectroscopy (Fraser et al., 1994; Mornet et al., 1995) and supports these assignments. Site A, which corresponds to the core sequence of peptides GS17C and M73 (Figs. 4 and 5), is not involved in regulation, whereas $Ca^{2+} \cdot CM$ binding at site B is coupled to reversal of CD inhibition (Marston et al., 1994a). Site B is close to the important TM-dependent inhibitory sequence of CD located in exon 11 but by itself has no actin affinity or inhibitory potency.

Mutations in either lobe of $Ca^{2+} \cdot CM$ reduce or even abolish reversal of inhibition (Huber et al., 1995a), consequently we have proposed a two-sited interaction on both molecules with both lobes of CM involved. The weaker binding of Ca^{2+} in the N-terminal domain of CM is the main switch of the Ca^{2+}-dependent reversal of inhibition, however, all the sites need to be intact for functional coupling. Figure 5 shows a possible conformation of the CD–$Ca^{2+} \cdot CM$ complex incorporating all the currently known structural features. It is believed that actin–TM is in the "on" state when this complex is formed (Fig. 7), presumably because $Ca^{2+} \cdot CM$ binding induces a conformational change in the CD–actin interface.

IX. CALDESMON–MYOSIN INTERACTION

Caldesmon binds to smooth muscle myosin in vitro. The interaction is very dependent on the ionic strength, reaching $>10^6 M^{-1}$ at ionic strengths below $0.03 M$. Most measurements of stoichiometry give two to three CDs bound per myosin molecule, although some reports suggest the stoichiometry is 1:1 in the presence of ATP. The binding site on myosin appears to be in the subfragment-2 region since proteolytically derived myosin rod and heavy meromyosin bind as well as whole myosin, whereas S-1 and light meromyosin bind very poorly (Ikebe and Reardon, 1988; Marston et al., 1992; Hemric and Chalovich, 1990; Hemric et al., 1993). Binding is largely specific to smooth muscle myosin since skeletal muscle myosin binds to CD much weaker or not at all. Electron micrographs of a complex of smooth muscle myosin rod filaments show bound CD coating the S-2 portion of the myosin, giving the appearance of whiskers (Marston et al., 1992).

A myosin binding site on CD has been identified in domain 1 and has been studied in some detail (Velaz et al., 1990; Bogatcheva et al., 1993; Redwood and Marston, 1993), however, N-terminal CD fragments show reduced binding compared with full-length CD, and an additional site has been demonstrated in domains 3 and 4 (Yamakita et al., 1992; Huber et al., 1993). Attempts at locating the C-terminal site more precisely have not been very successful and suggest that myosin binds at an extended site, most likely in the N-terminal half on domain 3 and extending into domain 4 (Huber et al., 1995c; Fraser et al., 1994). The N- and C-terminal myosin binding sites seem to work independently and are not cooperative.

The CD–myosin S-2 binding interaction has some interesting consequences for test tube experimentation that may be of physiological relevance. Thick and thin filaments can be cross-linked through CD in an interaction that is independent of the actin–myosin interaction which produces movement. This was directly demonstrated in the electron microscope with filaments assembled from myosin rod (i.e., no S-1), which were observed to be cross-linked to thin filaments by CD (Marston et al., 1992). When CD-containing actin filaments activate smooth muscle heavy meromyosin (HMM), MgATPase, the HMM is found to bind 40 times tighter to the actin filament than the "weak" actomyosin binding due to the cross-linking effect of CD (Lash et al., 1986; Marston, 1989a). This "tight binding" is independent of myosin phosphorylation and apparently independent of the acto-HMM ATPase. The cross-linking interaction can also promote myosin filament polymerization by stabilizing the filaments formed (Ikebe and Reardon, 1988; Hemric et al., 1994).

In the in vitro motility assay, two effects can be identified. Addition of CD promotes the interaction and movement of actin filaments over smooth muscle myosin, presumably because the myosin–CD–actin interaction tethers thick and thin filaments together, permitting interaction (Haeberle et al., 1992b). At higher CD concentrations, CD tethering appears to exert a drag upon movement of actin filaments over myosin that slows down the speed of filament movement (Horiuchi and Chacko, 1995).

The physiological role of CD binding to myosin in smooth muscles is quite uncertain. It has been hypothesized that CD cross-linking might be responsible for tension maintenance (latch) in smooth muscles (Marston, 1989b; Walsh and Sutherland, 1989), however, alternative explanations not involving CD are reasonably satisfactory (see Strauss and Murphy, Chapter 26, this volume), and it is not known whether CD cross-links are able to bear any significant load. A more likely function of the CD-myosin interaction is the organization of the filaments (Yamashiro and Mat-

sumura, 1991) as has been demonstrated by Hemric et al. (1994). Any of these functions requires that CD cross-linking is a regulated process and therefore considerable attention as been paid to the modulation of this interaction.

$Ca^{2+} \cdot CM$ was found to reduce the binding of CD to myosin with the same effectiveness as the binding of CD to actin (Hemric et al., 1993). However, it was ineffective in antagonizing the binding of the purified N-terminal myosin binding region of CD. As relatively high concentrations of CM are needed for this, it is not clear whether myosin binding is regulated by $Ca^{2} \cdot CM$.

Phosphorylation of CD has been described to abolish myosin binding (Sutherland and Walsh, 1989). Casein kinase II has been identified as the major endogenous CD kinase in sheep aorta that reduced the interaction of CD with smooth muscle myosin and TM. It was found to phosphorylate only Ser-73 and Thr-83 (Bogatcheva et al., 1993; Sutherland et al., 1994; Vorotnikov et al., 1993; Wawrzynow et al., 1991). A significant reduction of myosin binding after phosphorylation of CD has also been reported with CM-dependent protein kinase II (PK II), which often copurifies with CD (Hemric et al., 1993; Sutherland and Walsh, 1989). Up to 8 mol of P_i are incorporated into sites on caldesmon, including Ser-73 (Ikebe and Reardon, 1990). The mechanism of reversibility of the CD–myosin interaction in these interactions is of high interest and awaits future investigations.

X. THE ROLE OF CALDESMON IN REGULATING SMOOTH MUSCLE CONTRACTILITY

Direct evidence for a role of CD in regulating smooth muscle contraction is at present rather unsatisfactory. Several workers have shown that CD affects actin filament movement in an in vitro motility assay (Shirinsky et al., 1992; Haeberle et al., 1992b; Okagaki et al., 1991). The interpretations of the mechanism of the CD effect have been contradictory, however, a more thorough analysis (Fraser and Marston, 1995b) has given clear evidence that CD primarily inhibits actin–TM filament motility by switching off filaments as a unit, rather than by reducing their velocity or displacing filaments from myosin. This finding is compatible with the behavior of smooth muscle thin filaments in the test tube and indicates a similarity between the CD effect and troponin.

Caldesmon has been shown to reduce isometric tension in chemically skinned smooth or skeletal muscle fibers, thus clearly suggesting that it can influence contractility (Szpacenko et al., 1985; Pfitzer et al.,

1993; Brenner *et al.*, 1991). However, these experiments involved adding CD to a muscle that already contained a full complement of native regulatory proteins, so it is not possible to conclude that these effects are the same as the effects of CD in its native environment. Rather more convincing is the experiment of Taggart and Marston (1988), which used skeletal muscle fibers that had been treated to remove their regulatory protein and showed that CD was effectively inhibiting tension production at low concentrations.

Indirect evidence for a role of CD in intact tissue may be obtained by considering the quantity and location of CD in smooth muscles. We have assessed all the published data on the content of smooth muscle contractile proteins, and we evaluated protein concentrations reported in absolute values and ratios. Using the ratios of cell volume : tissue volume given by Murphy *et al.* (1977), the actin concentration in chicken gizzard smooth muscle cells was 724–973 μM, the content of TM was 100–121 μM, and that of CD was 23–33 μM, yielding a consensus of 1 CD:29 actin (four studies) (Yamazaki *et al.*, 1987; Lehman *et al.*, 1993; Bretscher, 1984; Murakami and Uchida, 1985). In vascular smooth muscle cells, actin concentration was 299–524 μM, TM was 43–69 μM, and CD was 10–16 μM (four studies), yielding a mean ratio of 1 CD : 32 actin (Yamazaki *et al.*, 1987; Lehman *et al.*, 1993; Reckless *et al.*, 1994; Clark *et al.*, 1986). Measurements of CD content in gut and uterine muscle also fall into this range (Haeberle *et al.*, 1992a; Lehman *et al.*, 1993). In general, therefore, the CD content of smooth muscles is of the order of 1 CD for 29–32 actin molecules. One study seems to be the odd one out; Haeberle *et al.* (1992a) obtained quite a low actin content and a *very* low CD content in bovine aorta (but see Section II for comments on bovine aorta).

The location of CD in the smooth muscle cell has been determined by immunofluorescence microscopy. A number of studies have shown the presence of two populations of thin filaments in smooth muscles; one population is mainly β-actin and associated with the cytoskeleton, filamin, and calponin, whereas the remainder (about 60% of the total in chicken gizzard) is mainly α-actin and is associated with contractile proteins, myosin, TM, and CD (Furst *et al.*, 1986; North *et al.*, 1994a,b; Lehman *et al.*, 1987).

The location of CD and its cellular concentration are compatible with its proposed function. A CD:actin ratio of 1:30 in the whole muscle cell corresponds to 1:18 if CD is confined to the contractile domain. This is close to the CD:actin ratio obtained when thin filaments are extracted from actomyosin (1:16; Marston, 1990) and is also compatible with the model structure of the thin filament in smooth muscle (Fig. 2). Thus CD

meets the criteria of quantity, location, and functionality that would be necessary for it to play a part in regulating smooth muscle contraction.

Although myosin phosphorylation is undoubtedly necessary for initiation and maintenance of contractility, there is evidence that a second regulatory system is involved in relaxation under certain conditions. A number of laboratories have demonstrated relaxation of smooth muscle following a contraction where Ca^{2+} concentration falls without a change in the level of myosin phosphorylation using a variety of agents. This is described in detail by Bárány and Bárány (Chapter 25, this volume). A Ca^{2+}-dependent switching off of thin filaments by CD could account for these observations. CD has been experimentally linked to such relaxation since activation of MAP kinase specifically phosphorylates CD and leads to suppression of the relaxing phenomenon (Khalil and Morgan, 1993; Khalil *et al.*, 1995). Therefore, on balance, we believe CD does play a physiological role in controlling smooth muscle contractility by inactivation of the thin filament at low Ca^{2+}.

XI. PERSPECTIVES

In vitro experimentation has shown that caldesmon is an integral component of smooth muscle thin filaments and plays a central role in their Ca^{2+}-dependent regulation. A mechanism analogous to that of troponin has been proposed; this now requires extensive testing. The structure of the regulatory domain of caldesmon is not well defined and we look forward to being able to describe this in three dimensions and in combination with its physiological partners actin, tropomyosin, and calmodulin.

The greatest challenge ahead, because it is the most difficult experimentally, is to determine the physiological role of caldesmon. *In vitro* experiments show how caldesmon might regulate smooth muscle contractility in concert with myosin phosphorylation, but they can never demonstrate that it does. For this we need new tools that can manipulate caldesmon within the intact cell. It is to be hoped that modern antisense RNA and transgenic techniques could provide the answer.

References

Bartegi, A., Fattoum, A., Derancourt, J., and Kassab, R. (1990). *J. Biol. Chem.* **265**, 15231–15238.
Bogatcheva, N. V., Vorotnikov, A. V., Birukov, K. G., Shirinsky, V. P., and Gusev, N. B. (1993). *Biochem. J.* **290**, 437–442.
Brenner, B., Yu, L. C., and Chalovich, J. M. (1991). *Proc. Natl. Acad. Sci. U.S.A.* **88**, 5739–5743.

Bretscher, A. (1984). *J. Biol. Chem.* **259**, 12873–12880.

Bryan, J. (1989). *J. Muscle Res. Cell Motil.* **10**, 95–96.

Bryan, J., and Lee, R. (1991). *J. Muscle Res. Cell Motil.* **12**, 372–375.

Bryan, J., Imai, M., Lee, R., Moore, P. Cook, R. G., and Lin, W. G. (1989). *J. Biol. Chem.* **264**, 13873–13879.

Chalovich, J. M., Cornelius, P., and Benson, C. E. (1987). *J. Biol. Chem.* **262**, 5711–5716.

Chalovich, J. M., Bryan, J., Benson, C. E., and Velaz, L. (1992). *J. Biol. Chem.* **267**, 16644–16650.

Childs, T. J., Watson, M. H., Sanghera, J. S., Campbell, D. L., Pelech, S. L., and Mak, A. S. (1992). *J. Biol. Chem.* **267**, 22853–22859.

Clark, T., Ngai, P. K., Sutherland, C., Groschel-Stewart, U., and Walsh, M. P. (1986). *J. Biol. Chem.* **261**, 8028–8035.

Crosbie, R. H., Miller, C., Chalovich, J., Rubenstein, P. A., and Reisler, E. (1994). *Biochemistry* **33**, 3210–3216.

Dabrowska, R., Goch, A., Galazkiewicz, B., and Osinska, H. (1985). *Biochim. Biophys. Acta* **842**, 70–75.

Fraser, I. D. C. and Marston, S. B. (1995a). *J. Biol. Chem.* **270**, 7836–7841.

Fraser, I. D. C., and Marston, S. B. (1995b). *J. Biol. Chem.* (in press).

Fraser, I. D. C., Huber, P. A. J., Torok, K., Slatter, D. A., Gusev, N. B., and Marston, S. B. (1994). *J. Muscle Res. Cell Motil.* **15**, 218.

Fujii, T., Machino, K., Andoh, H., Satoh, T., and Kondo, Y. (1990). *J. Biochem. (Tokyo)* **107**, 133–137.

Furst, D. O., Cross, R. A., De Mey, J., and Small, J. V. (1986). *EMBO. J.* **5**, 251–257.

Gusev, N. B., Pritchard, K. P., Hodgkinson, J. L., and Marston, S. B. (1994). *J. Muscle Res. Cell Motil.* **15**, 672–681.

Haeberle, J. R., Hathaway, D. R., and Smith, C. L. (1992a). *J. Muscle Res. Cell Motil.* **13**, 81–89.

Haeberle, J. R., Trybus, K. M., Hemric, M. E., and Warshaw, D. M. (1992b). *J. Biol. Chem.* **267**, 23001–23006.

Haruna, M., Hayashi, K., Yano, H., Takeuchi, O., and Sobue, K. (1993). *Biochem. Biophys. Res. Commun.* **197**, 146–153.

Hayashi, K., Kanda, K., Kimizuka, F., Kato, I., and Sobue, K. (1989). *Biochem. Biophys. Res. Commun.* **164**, 503–511.

Hayashi, K., Fujio, Y., Kato, I., and Sobue, K. (1991). *J. Biol. Chem.* **266**, 355–361.

Hayashi, K., Yano, H., Hashida, T., Takeuchi, T., Takeda, O., Asada, K., Takahashi, E., Kato, I., and Sobue, K. (1992). *Proc. Natl. Acad. Sci. U.S.A.* **89**, 12122–12126.

Hemric, M. E., and Chalovich, J. M. (1990). *J. Biol. Chem.* **265**, 19672–19678.

Hemric, M. E., Lu, F. W. M., Shrager, R., Carey, J., and Chalovich, J. M. (1993). *J. Biol. Chem.* **268**, 15305–15311.

Hemric, M. E., Tracy, P. B., and Haeberle, J. R. (1994). *J. Biol. Chem.* **269**, 4125–4128.

Horiuchi, K. Y., and Chacko, S. (1988). *Biochemistry* **27**, 8388–8393.

Horiuchi, K. Y., and Chacko, S. (1995). *J. Muscle Res. Cell Motil.* **16** 11–19.

Horiuchi, K. Y., Samuel, M., and Chacko, S. (1991). *Biochemistry* **30**, 712–717.

Huber, P. A. J., Redwood, C. S., Avent, N. D., Tanner, M. J. A., and Marston, S. B. (1993). *J. Muscle Res. Cell Motil.* **14**, 385–391.

Huber, P. A. J., Grabarek, Z., Slatter, D. A., Levine, B. A., and Marston, S. B. (1995a). *Biophys. J.* **68**, A59.

Huber, P. A. J., Fraser, I. D. C., Marston, S. B., Levine, B. A. (1995b). *J. Muscle Res. Cell. Motil.* **16**, (in press).

Huber, P. A. J., Fraser, I. D. C., and Marston, S. B. (1995c). *Biochem. J.* (in press).

Humphrey, M. B., Herrera-Sosa, H., Gonzalez, G., Lee, R., and Bryan, J. (1992). *Gene* **112**, 197–205.

Ikebe, M., and Reardon, S. (1988). *J. Biol. Chem.* **263**, 3055–3058.

Ikebe, M., and Reardon, S. (1990). *J. Biol. Chem.* **265**, 17607–17612.

Khalil, R. A., and Morgan, K. G. (1993). *Am. J. Physiol.* **265**, C406–C411.

Khalil, R. A., Wang, C.-L.A., and Morgan, K. G. (1995). *Biophys. J.* **68**, A75.

Kobayashi, H., Inoue, A., Mikawa, T., Kuwayama, H., Hotta, Y., Masaki, T., and Ebashi, S. (1992). *J. Biochem. (Tokyo)* **112**, 786–791.

Lash, J. A., Sellers, J. R., and Hathaway, D. R. (1986). *J. Biol. Chem.* **261**, 16155–16160.

Lehman, W., Sheldon, A., and Madonia, W. (1987). *Biochim. Biophys. Acta* **914**, 35–39.

Lehman, W., Denault, D., and Marston, S. B. (1993). *Biochim. Biophys. Acta* **1203**, 53–59.

Lehrer, S. S. (1994). *J. Muscle Res. Cell Motil.* **15**, 232–236.

Leszyk, J., Mornet, D., Audemard, E., and Collins, J. H. (1989). *Biochem. Biophys. Res. Commun.* **160**, 210–216.

Levine, B. A., Moir, A. J. G., Audemard, E., Mornet, D., Patchell, V. B., and Perry, S. V. (1990). *Eur. J. Biochem.* **193**, 6987–696.

Lynch, W., and Bretscher, A. (1986). *In* "Methods in Enzymology" (R. B. Vallee, ed.), Vol. 134, pp. 37–42. Academic Press, Orlando, FL.

Mabuchi, K., and Wang, C.-L.A. (1991). *J. Muscle Res. Cell Motil.* **13**, 146–151.

Mabuchi, Y., Wang, C.-L.A., and Grabarek, Z. (1995). *Biophys. J.* **68**, A359.

Makuch, R., Kolakowski, J., and Dabrowska, R. (1992). *FEBS Lett.* **297**, 237–240.

Mani, R. S., McCubbin, W. D., and Kay, C. M. (1992). *Biochemistry* **31**, 11896–11901.

Marston, S. B. (1989a). *Biochem. J.* **259**, 303–306.

Marston, S. B. (1989b). *J. Muscle Res. Cell Motil.* **10**, 97–100.

Marston, S. B. (1990). *Biochem. J.* **272**, 305–310.

Marston, S. B., and Lehman, W. (1985). *Biochem. J.* **231**, 517–522.

Marston, S. B., and Redwood, C. S. (1991). *Biochem J* **279**, 1–16.

Marston, S. B., and Redwood, C. S. (1992). *J. Biol. Chem.* **267**, 16796–16800.

Marston, S. B., and Redwood, C. S. (1993). *J. Biol. Chem.* **268**, 12317–12320.

Marston, S. B., and Smith, C. W. J. (1984). *J. Muscle Res. Cell Motil.* **5**, 559–575.

Marston, S. B., and Smith, C. W. (1985). *J. Muscle Res. Cell Motil.* **6**, 669–708.

Marston, S. B., Pritchard, K., Redwood, C. S., and Taggart, M. J. (1988a). *Biochem. Soc. Trans.* **16**, 494–497.

Marston, S. B., Redwood, C. S., and Lehman, W. (1988b). *Biochem. Biophys. Res. Commun.* **155**, 197–202.

Marston, S. B. Pinter, K., and Bennett, P. M. (1992). *J. Muscle Res. Cell Motil.* **13**, 206–218.

Marston, S. B., Fraser, I. D. C., Huber, P. A. J., Pritchard, K., Gusev, N. B., and Torok, K. (1994a). *J. Biol. Chem.* **269**, 8134–8139.

Marston, S. B., Fraser, I. D. C., and Huber, P. A. J. (1994b). *J. Biol. Chem.* **269**, 32104–32109.

Mezgueldi, M., Derancourt, J., Callas, B., Kassab, R., and Fattoum, A. (1994). *J. Biol. Chem.* **269**, 12824–12832.

Mills, J. S., Walsh, M. P., Nemcek, K., and Johnson, J. D. (1988). *Biochemistry* **27**, 991–996.

Mornet, D., Bonet-Kerrache, A., Strasburg, G. M., Patchell, V. B., Perry, S. V., Huber, P. A. J., Marston, S. B., Slatter, D. A., Evans, J. S., and Levine, B. A (1995). *Biochemistry* **34**, 1893–1901.

Murakami, U., and Uchida, K. (1985). *J. Biochem. (Tokyo)* **98**, 187–197.

Murphy, R. A., Driska, S. P., and Cohen, D. M. (1977). *In* "Excitation–Contraction Coupling in Smooth Muscle" (R. Casteels, T. Godfraind, and J. C. Rüegg, eds.), pp. 417–424. Elsevier-North-Holland Biomed. Press, Amsterdam.

Ngai, P. K., and Walsh, M. P. (1985). *Biochem. J.* **230**, 695–707.

North, J. A., Gimona, M., Cross, R. A., and Small, J. V. (1994a). *J. Cell Sci.* **107**, 437–444.

North, A. J., Gimona, M., Lando, Z., and Small, J. V. (1994b). *J. Cell Sci.* **107**, 445–455.

Novy, R. E., Lin, J.L.-C., and Lin, J.J.-C. (1991). *J. Biol. Chem.* **266**, 16917–16924.

Okagaki, T., Higahi-Fujime, S., Ishikawa, R., Takano-Ohmuro, H., and Kohama, K. (1991). *J. Biochem. (Tokyo)* **109**, 858–866.

Payne, A. M., Yue, P., Pritchard, K., and Marston, S. B. (1995). *Biochem. J.* **305**, 445–450.

Pearlstone, J. R., and Smillie, L. B. (1982). *J. Biol. Chem.* **257**, 10587–10592.

Pfitzer, G., Zeugner, C., Trotschka, M., and Chalovich, J. M. (1993). *Proc. Natl. Acad. Sci. U.S.A.* **90**, 5904–5908.

Pritchard, K., and Marston, S. B. (1989). *Biochem. J.* **257**, 839–843.

Pritchard, K., and Marston, S. B. (1991). *Biochem. J.* **277**, 819–824.

Pritchard, K. P., and Marston, S. B. (1993). *Biochem. Biophys. Res. Commun.* **190**, 668–673.

Rayment, I., Holden, H. M., Whittaker, M., Yohn, C. B., Lorenz, M., Holmes, K. C., and Milligan, R. A. (1993). *Science* **261**, 58–65.

Reckless, J., Pritchard, K., Marston, S. B., Fleetwood, G., and Tilling, L. (1994). *Atheroscler. Thromb.* **14**, 1837–1845.

Redwood, C. S., and Marston, S. B. (1993). *J. Biol. Chem.* **268**, 10969–10976.

Redwood, C. S., Marston, S. B., and Gusev, N. K. (1993). *FEBS Lett.* **327**, 85–89.

Redwood, C. S., Marston, S. B., Bryan, J., Cross, R. A., and Kendrick-Jones, J. (1990). *FEBS Lett.* **270**, 53–56.

Riseman, V. M., Lynch, W. P., Nefsky, B., and Bretscher, A. (1989). *J. Biol. Chem.* **264**, 2869–2875.

Sanders, C., and Smillie, L. B. (1984). *Can. J. Biochem. Cell Biol.* **62**, 443–448.

Sheterline, P., and Sparrow, J. C. (1994). *Protein Profile* **1**, 1–121.

Shirinsky, V. P., Bushueva, T. L., and Frolova, S. I. (1988). *Biochem. J.* **255**, 203–208.

Shirinsky, V., Birukov, K. G., Hettasch, J. M., and Sellers, J. R. (1992). *J. Biol. Chem.* **267**, 15886–15892.

Skripnikova, E. V., and Gusev, N. B. (1989). *FEBS Lett.* **257**, 380–382.

Smith, C. W., Pritchard, K., and Marston, S. B. (1987). *J. Biol. Chem.* **262**, 116–122.

Sobue, K., Muramoto, Y., Fujita, M., and Kakiuchi, S. (1981). *Proc. Natl. Acad. Sci. U.S.A.* **78**, 5652–5655.

Sobue, K., Morimoto, K., Inui, M., Kanda, K., and Kakiuchi, S. (1982). *Biomed. Res.* **3**, 188–196.

Stafford, W. F., Jancso, A., and Graceffa, P. (1990). *Arch. Biochem. Biophys.* **281**, 66–69.

Sutherland, C., and Walsh, M. P. (1989). *J. Biol. Chem.* **264**, 578–583.

Sutherland, C., Renaux, B. S., McKay, D., and Walsh, M. P. (1994). *J. Muscle Res. Cell Motil.* **15**, 440–456.

Szpacenko, A., and Dabrowska, R. (1986). *FEBS Lett.* **202**, 182–186.

Szpacenko, A., Wagner, J., Dabrowska, R., and Ruegg, J. C. (1985). *FEBS Lett.* **192**, 9–12.

Taggart, M. J., and Marston, S. B. (1988). *FEBS Lett.* **242**, 171–174.

Tanaka, T., Ohta, H., Kanda, K., Hidaka, H., and Sobue, K. (1990). *Eur. J. Biochem.* **188**, 495–500.

Velaz, L., Hemric, M. E., Benson, C. E., and Chalovich, J. M. (1989). *J. Biol. Chem.* **264**, 9602–9610.

Velaz, L., Ingraham, R. H., and Chalovich, J. M. (1990). *J. Biol. Chem.* **265**, 2929–2934.

Vorotnikov, A. V., Gusev, N. B., Hua, S., Collins, J. H., Redwood, C. S., and Marston, S. B. (1993). *FEBS Lett.* **334**, 18–22.

Walsh, M. P., and Sutherland, C. (1989). *Adv. Exp. Med. Biol.* **255**, 337–346.

Wang, C.-L.A., Wang, L.-W.C., Xu, S., Lu, R. C., Saavedra-Alanis, V., and Bryan, J. (1991a). *J. Biol. Chem.* **266**, 9166–9172.

Wang, C.-L.A., Chalovich, J. M,. Graceffa, P., Lu, R. C., Mabuchi, K., and Stafford, W. F. (1991b). *J. Biol. Chem.* **266**, 13958–13963.

Wang, Z., Horiuchi, K. Y., Jacob, S. S., Gopalakurup, S., and Chacko, S. (1994). *J. Muscle Res. Cell Motil.* **15**, 646–658.

Watson, M. H., Kuhn, A. E., Novy, R. E., Lin, J.J.-C., and Mak, A. S. (1990). *J. Biol. Chem.* **265**, 18860–18866.

Wawrzynow, A., Collins, J. H., Bogatcheva, N. V., Vorotnikov, A. V., and Gusev, N. B. (1991). *FEBS Lett.* **289**, 213–216.

Williamson, M. P. (1994). *Biochem. J.* **297**, 249–260.

Yamakita, Y., Yamashiro, S., and Matsumura, F. (1992). *J. Biol. Chem.* **267**, 12022–12029.

Yamashiro, S., and Matsumura, F. (1991). *BioEssays* **13**, 563–568.

Yamazaki, K., Itoh, K., Sobue, K., Mori, T., and Shibata, N. (1987). *J. Biochem. (Tokyo)* **101**, 1–9.

Zhan, Q., Wong, S. S., and Wang, C.-L.A. (1991). *J. Biol. Chem.* **266**, 21810–21814.

7

Calponin

MARIO GIMONA

Cold Spring Harbor Laboratory
Cold Spring Harbor, New York

J. VICTOR SMALL

Department of Physics
Institute of Molecular Biology
Austrian Academy of Sciences
Salzburg, Austria

I. INTRODUCTION

In the early 1970s when a consensus had been reached about the nature of the Ca^{2+}-regulatory complexes in the contractile apparatus of cross-striated muscle (see, e.g., Cold Spring Harbor Symp. 46, 1972), smooth muscle biochemistry was only in its infancy. Now, more than 20 years later, we have a clearer picture of the targets of Ca^{2+} action on the thin and the thick filaments of smooth muscle, but for a number of reasons agreement is still lacking about how proteins associated with the actin-containing thin filaments may influence the contractile activity of this muscle type *in vivo*. One reason for this dilemma is that thin filaments from smooth muscle vary in their composition depending on the conditions employed for their isolation (see Small and Sobieszek, 1980; Nishida *et al.*, 1990; Marston, 1991; Lehman, 1991), indicating that the actin-associated components are not tenaciously bound in a single, homogeneous complex. The situation is further complicated by the presence in smooth muscle tissue of a mixture of actin isotypes, including nonmuscle-type actins (Vandekerckhove and Weber, 1981), which are segregated in different compartments of the cell—in the contractile apparatus and in the cytoskeleton (North *et al.*, 1994b).

Despite these complications, it is generally agreed that the thin filaments of the smooth muscle cell contain (on average) actin complexed with at least three proteins: tropomyosin (TM), caldesmon (CD), and calponin (CP). In the present chapter we aim to review what is currently known about CP and ideas about its possible function. Brief mention of CD will also be made in the context of properties that distinguish it from CP; a more detailed treatment of CD is found in Marston and Huber, Chapter 6 (this volume).

II. PURIFICATION AND PHYSICOCHEMICAL PROPERTIES

Calponin was first isolated from chicken gizzard (Takahashi *et al.*, 1986) and bovine aorta (Takahashi *et al.*, 1988a,b) on the basis of the protein's heat stability (Takahashi *et al.*, 1986). In Takahashi's method, chicken gizzard smooth muscle is boiled in a water bath and a protein extract made from the subsequently chilled mince at high ionic strength (300 mM KCl). CP is purified from this extract by ammonium sulfate precipitation at 30% saturation followed by dialysis and column separation by cation-exchange chromatography on SP-Sephadex C-50 and gel filtration on Ultrogel AcA 44 (both in the presence of 6 M urea). The yield is around 40 mg per 100 g of minced muscle (Takahashi *et al.*, 1986). This basic purification scheme has been generally adopted, although several groups have introduced modifications, mostly by using different chromatography media (Winder and Walsh, 1990c; Marston, 1991; Abe *et al.*, 1990a; Bárány *et al.*, 1991). CP can also be purified in its native form in the absence of 6 M urea (Vancompernolle *et al.*, 1990) by dissolving and dialyzing the ammonium sulfate pellet after extraction in a low pH (pH 5.4) buffer (Vancompernolle *et al.*, 1990; Abe *et al.*, 1990b). By this method, purified

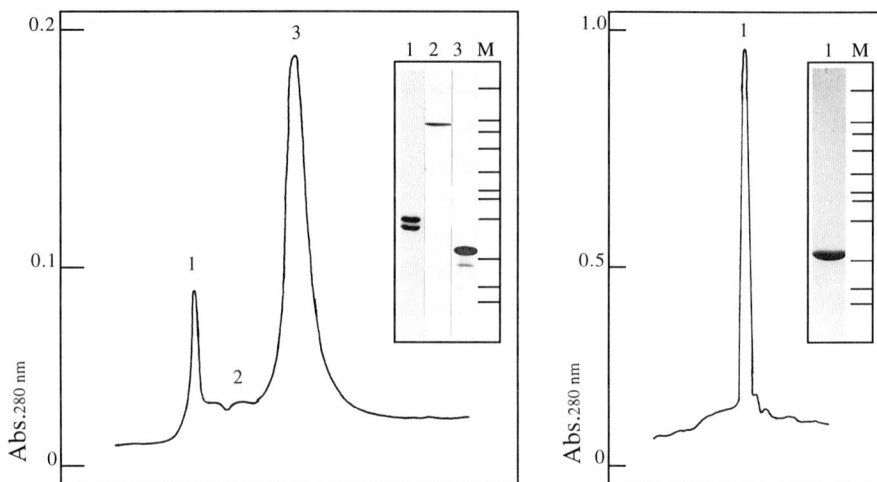

FIGURE 1   Purification of native smooth muscle CP using S-Sepharose fast flow cation-exchange (left) and FPLC Superose 12 gel filtration media (right). Insets show the gel patterns for the specific elution of TM (1), CD (2), and CP (3). M represents molecular weight markers. From Vancompernolle et al. (1990, Fig. 1, p. 147).

CP is obtained in two steps using S-Sepharose fast flow and FPLC Superose 12-column chromatography media (Fig. 1). In addition to gizzard, CP has been purified from hog and toad stomach (Gimona et al., 1990; Winder et al., 1993b), as well as from bovine and sheep aorta (Takahashi et al., 1988a,b; Marston, 1991). We may note that the either heat alone or heat and urea denaturation are most commonly used in the purification for CP, but little is known about the unfolding/refolding abilities of CP and the influence of this treatment on the properties of the molecule.

In solution, CP is a monomer as determined by gel filtration under nondenaturing conditions (Takahashi et al., 1986) and migrates as a 35-kDa polypeptide on denaturing polyacrylamide gels (Takahashi et al., 1988a; Winder et al., 1990). The extinction coefficient ($E_{1\%}^{277}$) was estimated as 11.3 for avian CP (Winder et al., 1990, 1991) and 8.9 for porcine CP (Wills et al., 1995). The Stokes radii for aorta and gizzard CP are 27.8 and 27.1 Å, respectively, as determined by gel chromatography, and the sedimentation coefficient $S_{20,w}^{0}$ for gizzard CP is 3.16 (Takahashi et al., 1988a).

Other work has centered on the production of recombinant CP in bacteria for functional studies (Gong et al., 1993; Wills et al., 1995). Engineered subdomains of CP have been purified to homogeneity from the soluble E. coli fractions by column chromatography on hydroxylapatite, Mono-S, and Superose 12 (Wills et al., 1995). The use of these subdomains in both binding and in vitro motility assays is now allowing assignment of binding and inhibitory activities to different regions of the molecule (see Section VI).

III. ISOFORM DIVERSITY

CP from all species analyzed so far occurs as multiple basic isoforms with isoelectric points around 9 (Takahashi et al., 1988a). Using two-dimensional (2D) gel electrophoresis it has been established that CP is expressed in up to three major isoelectric variants (α, β, γ; see Fig. 2) of similar molecular mass, ranging from pI's of 9.9 to 9.4 (Gimona et al., 1992). The isoforms appear progressively with the differentiation in chicken gizzard, porcine stomach (Gimona et al., 1992), and human uterine smooth muscle (Draeger et al., 1991, 1993) starting with the most basic (α) isoform. The exact nature and significance of these isoforms is unknown, but in vitro translation experiments (Gimona et al., 1992) have pointed to the involvement of post-translational modifications other than phosphorylation. Along with the down-regulation of CP expression in cultured smooth muscle cells, the relative amounts of the CP isoforms were observed to decrease in the reverse order of their appearance during differentiation (Draeger et al., 1991).

Takahashi and Nadal-Ginard (1991) identified, in addition to the 34-kDa CP, a chicken cDNA encoding a low-molecular-mass CP variant (28 kDa) that lacked an internal 40-amino acid stretch in the carboxyl-terminal part of the molecule. A protein that could correspond to this variant from chicken gizzard was first detected in human uterus by Draeger et al. (1991) by Western blotting, but its expression was strictly limited to the urogenital tract (Draeger et al., 1993). This low-molecular-weight isoform was also absent in malig-

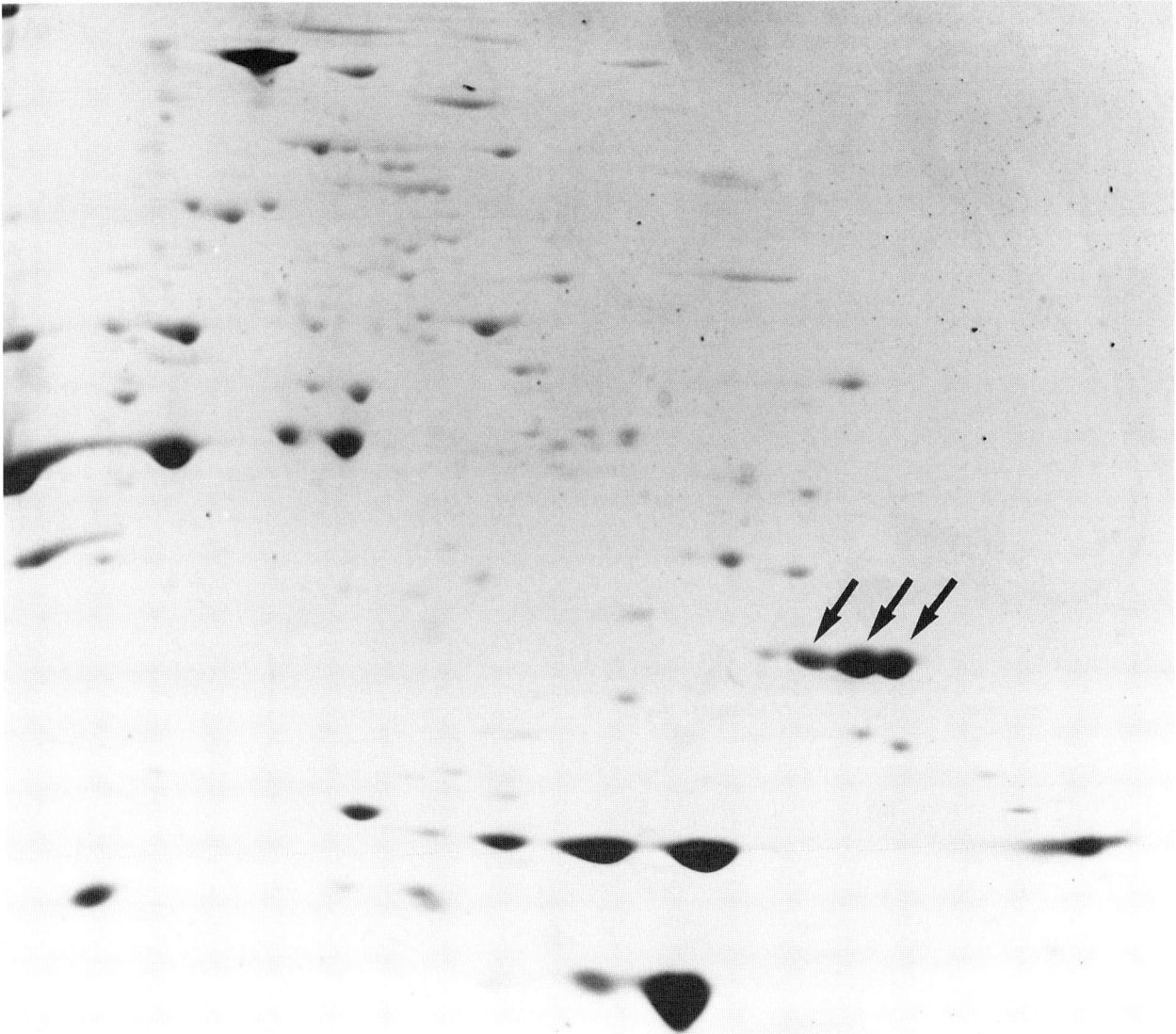

FIGURE 2 Silver-stained mini 2D gel of chicken gizzard smooth muscle showing the three isoelectric variants (α, β, γ from right to left) of CP expressed in this tissue type.

nant leiomyomas of uterine smooth muscle tissue and it was immediately down-regulated in cultures of normal myometrium.

IV. PRIMARY STRUCTURE AND GENETIC VARIANTS

The first partial protein sequence of turkey gizzard CP (Vancompernolle *et al.*, 1990) and the complete chicken sequence derived from the cDNA (Takahashi and Nadal-Ginard, 1991) revealed a novel primary se-

quence exhibiting homologies to another smooth muscle-specific but poorly characterized protein, SM22 (Lees-Miller *et al.*, 1987; Pearlstone *et al.*, 1987). Although Takahashi and Nadal-Ginard (1991) originally noted strong homologies between CP and troponin T (TnT), as well as to parts of CD and the *ras* gene-encoded protein p21, these claims were subsequently withdrawn (Takahashi and Nadal-Ginard, 1993). Complete sequences of rat aorta (Nishida *et al.*, 1993) as well as of pig and mouse stomach CP (Strasser *et al.*, 1993) have revealed a high degree of conservation between species (Fig. 3 and Table I). A notable feature of the C-terminal third of CP is the presence of

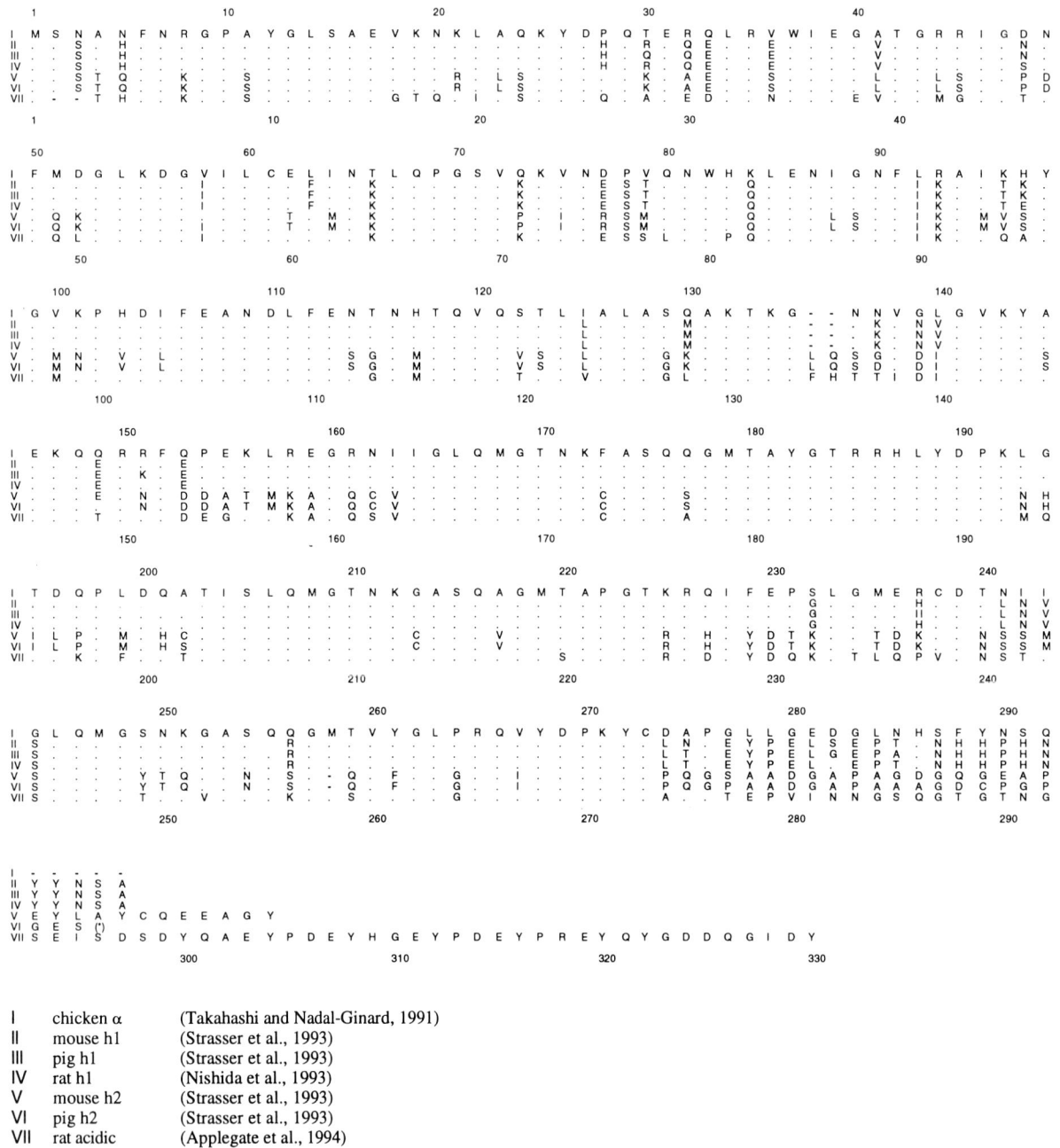

FIGURE 3　Comparison of the amino acid sequences of CPs from different species. The amino acid positions are numbered according to the chicken alpha (upper lane) and rat acidic (lower lane) sequences. Dots represent identical residues, horizontal bars indicate gaps introduced into the sequences for maximal alignment, and asterisks represent incomplete sequences.

I	chicken α	(Takahashi and Nadal-Ginard, 1991)
II	mouse h1	(Strasser et al., 1993)
III	pig h1	(Strasser et al., 1993)
IV	rat h1	(Nishida et al., 1993)
V	mouse h2	(Strasser et al., 1993)
VI	pig h2	(Strasser et al., 1993)
VII	rat acidic	(Applegate et al., 1994)

three homologous repeats of around 30 amino acids each (Takahashi and Nadal-Ginard, 1991). Most of this C-terminal part can be removed without affecting CP's inhibitory activity, however (Winder and Walsh, 1990a), and appears to have no significant influence on the secondary structure of the rest of the molecule (M. Mezgueldi, P. Strasser, M. Jaritz, A. Fattoum, and M. Gimona, unpublished). Other reports have revealed the existence of more acidic CP variants at the cDNA level (Strasser et al., 1993; Applegate et al., 1994) (Fig. 3), but neither of these groups was able to detect the shorter cDNAs corresponding to that described earlier for chicken gizzard (Takahashi and Nadal-Ginard, 1991). A protein of 45 kDa showing around 40% se-

TABLE I Molecular Masses, Isoelectric Points, and Number of Amino Acid Residues for Sequenced Calponins

Source	Molecular mass (Da)	pI	Residues	Reference[a]
Chicken α	32,333	9.91	292	(1)
Mouse h1	33,335	9.68	297	(2)
Pig h1	33,203	9.93	297	(2)
Rat h1	33,342	8.74	297	(3)
Mouse h2	33,104	7.82	305	(2)
Pig h2[b]	31,981	8.33	296[b]	(2)
Rat acidic	36,377	5.2	330	(4)
Chicken β	28,127	9.95	252	(1)

[a]References: (1) Takahashi and Nadal-Ginard, 1991; (2) Strasser et al., 1993; (3) Nishida et al., 1993; (4) Applegate et al., 1994.

[b]Incomplete sequence.

quence identity to CP in a 150-amino acid overlap region has been described in the parasite *Onchocerca volvulus* (Irvine *et al.*, 1994).

Although the C-terminal 100 amino acids of CP are apparently unimportant for the inhibition of actomyosin ATPase (e.g., Mezgueldi *et al.*, 1992), it is this part of the molecule that shows close sequence homology to a number of other muscle proteins (Vancompernolle *et al.*, 1990), making it worthy of closer attention. Specifically, the three 30-amino acid repeat sequences show homology to parts of the smooth muscle protein SM22, to the *Drosophila* muscle protein mp20 (Ayme-Southgate *et al.*, 1989), and to the rat neuronal protein np25 (see Goetinck and Waterston, 1994b). Perhaps most interesting is the finding of an even closer homology of these repeats to the unc-87 protein of the *Caenorhabdites elegans* obliquely striated body wall muscle (Goetinck and Waterston, 1994b). The ~40-kDa protein product of the *unc-87* gene exhibits 7 repeats of 25 amino acids (Goetinck and Waterston, 1994b), distributed throughout the molecule, that show up to 70% homology to those found in CP (Fig. 4). The unc-87 protein also associates with actin filaments, independently of TM, and appears to be essential for the maintenance of the sarcomere lattice integrity (Goetinck and Waterston, 1994a). We may thus entertain the possibility that CP performs a structural as well as the regulatory role that is suggested by its *in vitro* properties (see the following section). The phylogenetic relationship between CP and SM22-like proteins as derived from the sequence homologies is schematically represented in Fig. 5.

```
CP consensus I      I  G  L  Q  M  G  T  N  K  x  A  S  Q  x  G  M  T  A  Y  G  T  R  R
SM 22 consensus     ψ  G  L  Q  M  G  S  N  φ  G  A  S  Q  A  G  M  T  G  Y  G  R  P  R
np25                I  G  L  Q  M  G  S  N  K  G  A  S  Q  A  G  M  T  G  Y  G  M  P  G
mp20                V  G  L  Q  A  G  S  N  K  G  A  T  Q  A  G  -  -  N  L  G  A  G  R
unc87 consensus     φ  P  L  Q  x  G  T  N  K  x  x  S  Q  K  G  M  T  G  F  G  T  x  R


CP consensus II     I  S  L  Q  M  G  T  N  K  μ  A  S  Q  φ  G  M  T  A  P  G  T  φ  R
SM 22 consensus     φ  G  L  Q  M  G  S  N  φ  G  A  S  Q  A  G  M  T  G  Y  G  R  P  R
np25                I  G  L  Q  M  G  S  N  K  G  A  S  Q  A  G  M  T  G  Y  G  M  P  G
mp20                V  G  L  Q  A  G  S  N  K  G  A  T  Q  A  G  -  -  N  L  G  A  G  R
unc 87 consensus    φ  P  L  Q  x  G  T  N  K  x  x  S  Q  K  G  M  T  G  F  G  T  x  R


CP consensus III    φ  S  L  Q  M  G  μ  N  K  G  A  μ  Q  X  G  M  μ  V  ψ  G  L  x  R
SM 22 consensus     φ  G  L  Q  M  G  S  N  φ  G  A  S  Q  A  G  M  T  G  Y  G  R  P  R
np25                I  G  L  Q  M  G  S  N  K  G  A  S  Q  A  G  M  T  G  Y  G  M  P  G
mp20                V  G  L  Q  A  G  S  N  K  G  A  T  Q  A  G  -  -  N  L  G  A  G  R
unc87 consensus     φ  P  L  Q  x  G  T  N  K  x  x  S  Q  K  G  M  T  G  F  G  T  x  R


consensus           φ  x  L  Q  x  G  μ  N  φ  x  x  μ  Q  x  G  M  T  x  x  G  x  x  x
```

x = any amino acid;
φ = Ile (I),Val (V),Met (M), Val (V) = hydrophobic;
ϕ = Lys (K) or Arg (R) = positively charged;
μ = Asn (N), Cys (C), Gln (Q), Gly (G), Ser (S), Thr (T), Tyr (Y) = uncharged polar;
ψ = Tyr (Y) or Phe (F);

FIGURE 4 Comparison of the C-terminal sequences of CP with those of homologous proteins. The consensus sequences of the first, second, and third repeats of CP (each top lane) are compared to the homologous regions in SM22 (lane 2), np25 (lane 3), mp20 (lane 4), and the consensus sequence determined for the seven homologous repeats found in *unc87* gene product (lane 5).

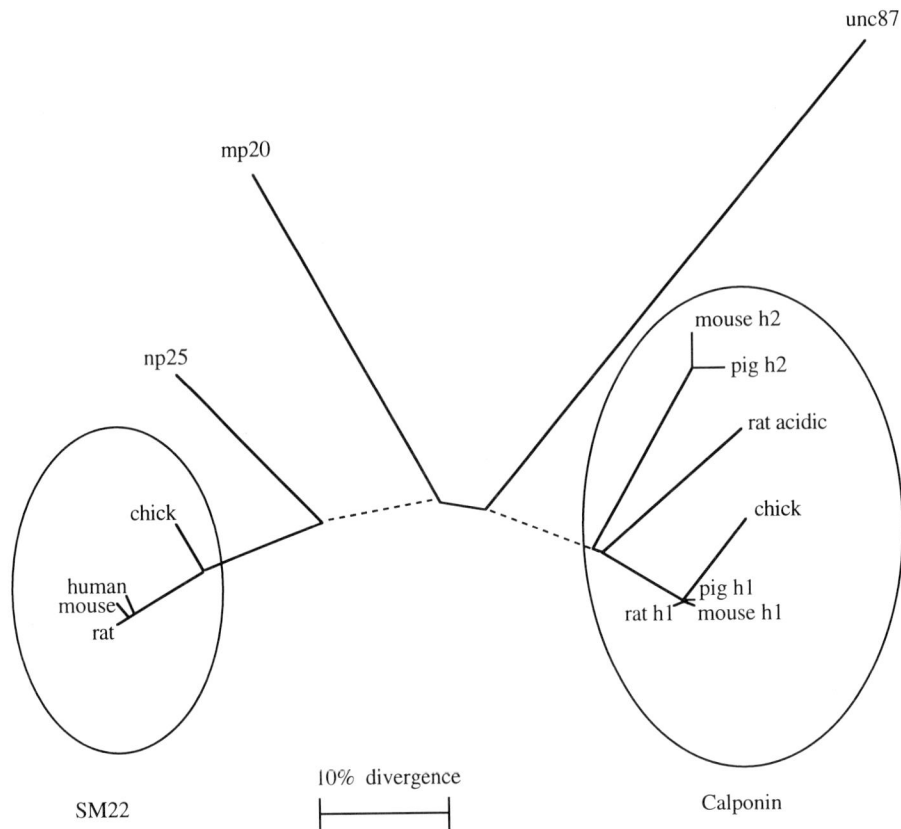

FIGURE 5 Schematic representation of the phylogenetic interrelationship between CPs and proteins related in sequence to CP. (Figure kindly provided by Dr. S. J. Winder.)

V. *IN VITRO* ACTIVITY

A. *Interaction of Calponin with Other Proteins*

As originally shown by Takahashi and colleagues (1986), CP binds to and sediments with F-actin, the binding reaching saturation *in vitro* at a stoichiometry of CP : actin of 1:1 (Makuch *et al.*, 1991; Horiuchi and Chacko, 1991). Binding to actin is independent of the presence of TM (Table II) despite the fact that CP can bind TM independently (Takahashi *et al.*, 1987; Vancompernolle *et al.*, 1990; Nakamura *et al.*, 1993). Notably, the source of F-actin greatly influences the strength of CP binding, the affinity for smooth muscle actin being seven- to eightfold greater than for actin from skeletal muscle (Winder *et al.*, 1991) (Table II).

Reports that CP directly binds Ca^{2+} (Winder *et al.*, 1991) have been disputed (Wills *et al.*, 1993), but there is general agreement that Ca^{2+} mediates the binding of Ca^{2+} receptor molecules to CP. Wills and coworkers (1993, 1994) have analyzed in parallel the binding of

four calcium binding proteins to CP: calmodulin (CaM), caltropin (CaT), S-100 b, and troponin C (TnC). Their findings show that CaT and S-100 b (from brain) bind with higher affinity to CP than does CaM (Table II) and that TnC binds only weakly. Other studies ascribe even weaker affinities for the binding of CaM to CP than those given by Wills (Nakamura *et al.*, 1993; Fujii *et al.*, 1994). Since CaT occurs in amounts comparable to CaM in smooth muscle (CaT : CaM estimated as 1:5; Mani and Kay, 1990), Wills and coworkers argue that this protein more likely targets CP *in vivo* (see also Mani and Kay, Chapter 8, this volume). The associations of all these Ca^{2+}-receptor molecules to CP are, however, of relatively low affinity compared to, for example, the binding affinity of CaM to myosin light chain kinase (MLCK) ($K_d \sim 3$ nM; Malencik and Anderson, 1986). Additional data suggest that CP binds also to myosin (Szymanski and Tao, 1993a,b; Lin *et al.*, 1993) and to CD (Vancompernolle *et al.*, 1990), but these associations are apparently too weak to be relevant to CP function.

TABLE II Binding Characteristics of Calponin

Ligand	Source[a]	K_d (μM)	Stoichiometry of ligand: calponin	Reference
Actin ± TM	Sm	0.045	~0.3	Winder *et al.* (1991)
Actin ± TM	Sk	0.35	~0.3	Winder *et al.* (1991)
TM	Sm	1.8		Nakamura *et al.* (1993)
Calmodulin	Br	0.64		Nakamura *et al.* (1993)
Calmodulin	Br	0.080		Winder *et al.* (1993a)
Calmodulin	Br	0.22[b]	2	Wills *et al.* (1993)
S-100 b	Br	0.03[b]	2	Wills *et al.* (1993)
S-100	Br	0.70	1	Fujii *et al.* (1994)
Caltropin	Sm	0.13[b]	2	Wills *et al.* (1993)

[a]Abbreviations: Sm, smooth muscle; Sk, skeletal muscle; Br, brain; TM, tropomyosin.

[b]Value for binding of the first ligand.

B. Inhibition of Actomyosin ATPase and in Vitro Motility by Calponin

CP causes an inhibition of actomyosin ATPase (Table III) as well as an arrest of actin filament movement in *in vitro* motility assays (Shirinsky *et al.*, 1992; Haeberle, 1994; K. Anderson, M. Gimona, and J. V. Small, unpublished results), in a dose-dependent fashion. The degree of inhibition and dose dependence that has been reported is rather variable (Table III) and stems in part from the use of different preparations of myosin and actin, purified from either smooth or skeletal muscle (Winder *et al.*, 1992). The data nevertheless show a constant inhibition of MgATPase, up to as much as 80% (Table III). According to Takahashi *et al.* (1986), CP is as abundant as TM in smooth muscle, suggesting a stoichiometry of CP to actin *in vivo* of around 1:7. This value is slightly higher than that deduced from measurements of the protein content of isolated thin filaments for which stoichiometric values of CP to actin range from 1:10 (Nishida *et al.*, 1990) to 1:16 (Marston, 1991). Since Winder and Walsh (1990b,c) observed, *in vitro*, half-maximal inhibition of MgATPase by CP at a molar ratio to actin of around 1:5 (Fig. 6), we may conclude that if CP is homogeneously distributed throughout the thin filament population in smooth muscle (but see the following) it has the potential of inhibiting actomyosin ATPase no more than about 30%.

The inhibitory effect of CP on actomyosin MgATPase can be partially reversed in the presence of Ca^{2+} by the Ca^{2+} binding proteins CaM, CaT, and S-100, the latter two proteins being more effective at equimolar ratios to CP (Wills *et al.*, 1994; Fujii *et al.*, 1994). In line with their relatively low-affinity associations with CP (Table II), complete reversal of inhibition could be achieved only with a high molar excess of each Ca^{2+} binding protein: 6-fold for CaT and S-100 (Wills *et al.*, 1994) and 10-fold for CaM (Makuch *et al.*, 1991). The same effect was observed in motility assays, where restoration of movement of actin over myosin required high concentrations (10 μM) of CaM (Shirinsky *et al.*, 1992). CP can also be phosphorylated *in vitro* and this phosphorylation abolishes its inhibitory activity (Winder and Walsh, 1990c). CP phosphorylation is considered in more detail in Section VI.

C. The Relationship of Calponin to Caldesmon

We have already noted that smooth muscle thin filaments contain both CD and CP. Analysis of isolated smooth muscle thin filaments has revealed, on average, similar amounts of CP and CD with a composition, on a molar basis of actin: TM : CD : CP, of 7:0.9–1:0.5–0.6 : 0.45–0.7 (Marston, 1991; Nishida *et al.*, 1990). Since CD alone also inhibits actomyosin ATPase and this inhibition is released by Ca^{2+}/CaM (see Marston and Huber, Chapter 6, this volume), the relative roles of CD and CP have inevitably been brought into question. Depending on the report (Makuch *et al.*, 1991; Marston, 1991; Winder *et al.*, 1992), either CD or CP has a more potent inhibitory effect on actomyosin MgATPase activity. It is not yet clear why two apparently similar actin regulator proteins are required in smooth muscle, but differences in their properties suggest that they perform distinct functions. Some of these differences are highlighted in the following. Sobue *et al.* (1985) showed originally that CD inhibits primarily the two- to threefold activation of ATPase that is induced by TM in smooth muscle actomyosin (see Small and Sobieszek, 1980), whereas the inhibition by CP is, as already noted, unaffected by TM (Winder and Walsh, 1990c). When mixed together with F-actin in saturating amounts, CP and CD compete for binding to actin, CP being more effective at displacing CD than vice versa (Makuch *et al.*, 1991). This result indicates that CP and CD do not form a mutual complex on F-actin and may in fact reside on different populations of thin filaments *in vivo* (Makuch *et al.*, 1991) for which some independent evidence has been obtained (Lehman, 1991; see also the following).

TABLE III   *In Vitro* Inhibition of Actomyosin ATPase or Actin Filament Motility by Calponin

Calponin[a]	Actin	Myosin	Tropomyosin	V_{max} (%)	K_m (%)	Molar ratio CP/A	Arrest of motility (µM CP)	Reference
Ch.g.	Ch.g.	Ch.g.	Ch.g.	78		0.33		Winder and Walsh (1990c)
Ch.g.	Ch.g.	Ch.g.		25		0.5		Abe et al. (1990a)
Ch.g.	Ch.g.	Ch.g.	Ch.g.	40		0.5		Abe et al. (1990b)
Ch.g.	Ch.g.	Ch.g.	Ch.g.	33[b]	0	0.1[b]		Nishida et al. (1990)
Ch.g.	Ch.g.	Ch.g. (HMM)	Ch.g.	80	50	0.6		Horiuchi and Chacko (1991)
Ch.g.	Ch.g.	Ch.g.		90	50	0.6		Horiuchi and Chacko (1991)
Ch.g.	Rsk	Rsk (HMM)		60		0.65		Miki et al. (1992)
Ch.g.	Rsk	Rsk		80		2.0		Makuch et al. (1991)
T.g.	Rsk	Rsk		50		1.0		Mezgueldi et al. (1992)
T.g.	Rsk	Rsk (S-1)		48		0.5		Wills et al. (1994)
Ch.g.	Rsk	Rsk		75				Winder et al. (1992)
Ch.g.	Rsk	Rsk	Rsk	35				Winder et al. (1992)
Ch.g.	Rsk	Rsk	Rsk + TN	35				Winder et al. (1992)
T.g.	Rsk	Rsk (S-1)		50		1.0		Mezgueldi et al. (1992)
T.g.	Rsk	T.g.					0.5–1.0	Shirinsky et al. (1992)
Ch.g.	Rsk	Ch.g.					1–2	Haeberle (1994)
T.g.	Rsk	Rsk					1–2	Shirinsky et al. (1992)
T.g.	T.g.	Rsk (HMM)					0.4	K. Anderson (unpublished)

[a] Abbreviations: Ch.g., chicken gizzard; T.g., turkey gizzard; Rsk, rabbit skeletal muscle; HMM, heavy meromyosin; S-1, myosin subfragment-1; TN, troponin.

[b] Native thin filaments including TM and caldesmon.

Further differences between CP and CD relate to their binding sites on the actin molecule: CP binds to the C terminus and CD to the N terminus of actin (Mezgueldi et al., 1992; Miki et al., 1992), consistent with different mechanisms of their action at the actin–myosin interface (Miki et al., 1992). Noda et al. (1992) have observed the induction of structural changes in actin upon binding to CP.

Current data suggest that CD exerts its inhibition by displacing weak-binding cross-bridges from actin (Chalovich, 1992) in line with its sharing a binding site on actin with the myosin head (Bartegi et al., 1990). By virtue of its association with TM, it may alternatively, or in parallel, diminish the proportion of TM in the strong, activating state (Horiuchi and Chacko, 1989; Marston and Redwood, 1993). In the latter respect, Marston and Redwood (1993) have drawn an analogy between CD and the complex of TnI and TnT. CP, on the other hand, does not require TM for inhibition nor does it block the myosin binding site (Miki et al., 1992)

and may exert its effect indirectly by affecting the strongly bound myosin heads (Mezgueldi et al., 1992). This would be consistent with CP having little effect on myosin binding, while slowing significantly the catalytic step (Horiuchi and Chacko, 1991). Results from a novel modification of the *in vitro* motility assay have been interpreted as showing an increased binding of actin to myosin heads in the presence of CP and a concomitant decrease in filament velocity (Haeberle, 1994). Accordingly, it has been suggested that CP contributes to the development of the latch state of smooth muscle that is characterized by tension maintenance with slowly cycling cross-bridges (Haeberle, 1994). All the *in vitro* assays have so far been designed to test the effect of CP on actomyosin ATPase. However, CP could just as well influence the binding of proteins to actin that share a common site of association at the actin C terminus. We shall discuss this possibility further when elaborating on the localization of CP in smooth muscle.

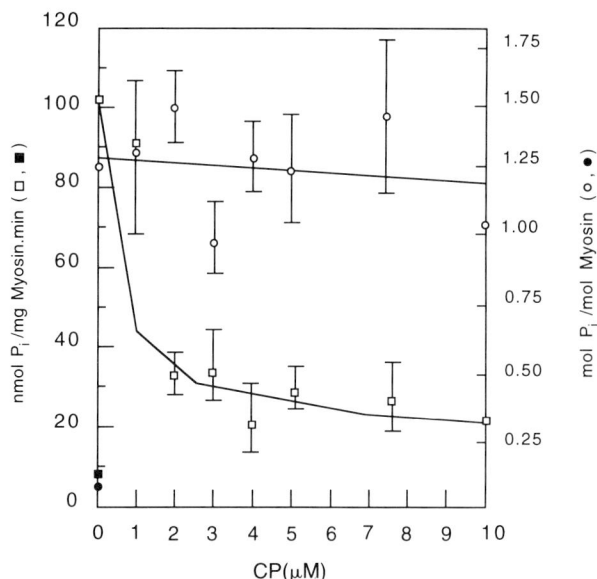

FIGURE 6   Inhibition of the actin-activated myosin MgATPase by CP. Actomyosin ATPase rates in the presence (□) and the absence (■) of Ca^{2+}·myosin phosphorylation levels were quantified in the presence (○) and the absence (●) of Ca^{2+}. Actin concentration was 6 μM. Reproduced with permission from Winder and Walsh (1990c), Fig. 1, p. 10150, Copyright 1990, The American Society for Biochemistry and Molecular Biology, Inc.

VI. MOLECULAR DOMAIN ORGANIZATION AND BINDING MOTIFS

The different *in vitro* properties of CP, namely, inhibition of actomyosin ATPase as well as binding to F-actin, TM, and Ca^{2+} binding proteins, derive from separate or combined subdomains of the molecule. NTCB cleavage at cysteine residues in positions 61 and 238 of chicken gizzard CP yields two N-terminal fragments of 30 and 21 kDa (Winder and Walsh, 1990a; Mezgueldi *et al.*, 1992), both of which retain the basic *in vitro* properties of native CP (F-actin, TM, and CaM binding and ATPase inhibition). Chymotryptic digestion of turkey gizzard CP produces a 22-kDa N-terminal fragment that is retained on columns carrying immobilized TM (Vancompernolle *et al.*, 1990) and contains the TM binding site. From viscosity measurements, light scattering, electron microscopy, and affinity chromatography, Childs *et al.* (1992) delineated the region between residues 142 and 227 in CP and the region around Cys 190 in TM as the domains of interaction between these two proteins.

In another, elegant study, Mezgueldi and colleagues (1992) demonstrated that the amino-terminal region from residues 7–182 is capable of performing the entire functional repertoire of CP. Deductions from chemical cross-linking and the binding properties of proteolytic subfragments of this part of the mol-

ecule led to the assignment of the region embracing residues 52–144 for CaM binding and residues 145–182 for actin binding (M. Mezgueldi, P. Strasser, M. Jaritz, A.Fattoum, and M. Gimona, unpublished results). A second binding site for Ca^{2+} binding protein(s) at the very N-terminal region spanning residues 1–45 has also been identified using recombinant mouse CP fragments (Wills *et al.*, 1995). Recombinant full-length chicken gizzard CP has been shown to be functionally indistinguishable from CP purified from muscle (Gong *et al.*, 1993). Additional physicochemical and biochemical measurements on both native and recombinant CP subfragments (M. Mezgueldi *et al.*, unpublished) support earlier assumptions (Mezgueldi *et al.*, 1992; Strasser *et al.*, 1993) that the N terminus of CP is more structurally conserved than the C terminus. A detailed map of the functional domains identified so far is shown in Fig. 7.

CP can be phosphorylated *in vitro* (Nakamura *et al.*, 1993; Winder *et al.*, 1993a) and this phosphorylation reverses CP's inhibitory activity (Naka *et al.*, 1990; Nakamura *et al.*, 1993; Winder *et al.*, 1993a). Accordingly, it has been suggested that CP phosphorylation may be pivotal in regulation of CP function. CP is phosphorylated at multiple sites (Fig. 8) by the two serine/threonine kinases, protein kinase c (PKC) and Ca/CaM-dependent kinase II (Nakamura *et al.*, 1993; Winder *et al.*, 1993a). Whereas Winder *et al.* (1993a) specifically revealed serines as targets for the kinases, Naka *et al.* (1990) demonstrated a preference of PKC for threonine over serine phosphorylation. Nakamura *et al.* (1993) identified Thr 184 as the preferred site of PKC action *in vitro* (Fig. 8). Since CP is not readily phosphorylated *in vivo*, however, the biological significance of CP phosphorylation has been brought into question (see Bárány and Bárány, Chapter 25, this volume). In the latter respect, it is notable that CP phosphorylation by PKC *in vitro* is suppressed partially (Nakamura *et al.*, 1993) or completely (Nagumo *et al.*,

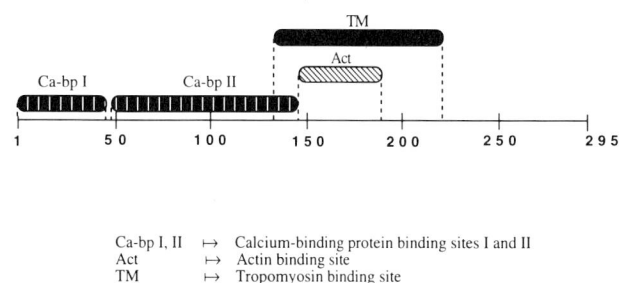

Ca-bp I, II ↦ Calcium-binding protein binding sites I and II
Act ↦ Actin binding site
TM ↦ Tropomyosin binding site

FIGURE 7   Schematic representation of the functional binding domains identified in CP. The N terminus harbors two separate sites for Ca^{2+} binding protein(s), whereas the middle of the molecule contains the binding sites for TM and F-actin, which are likely to overlap with each other.

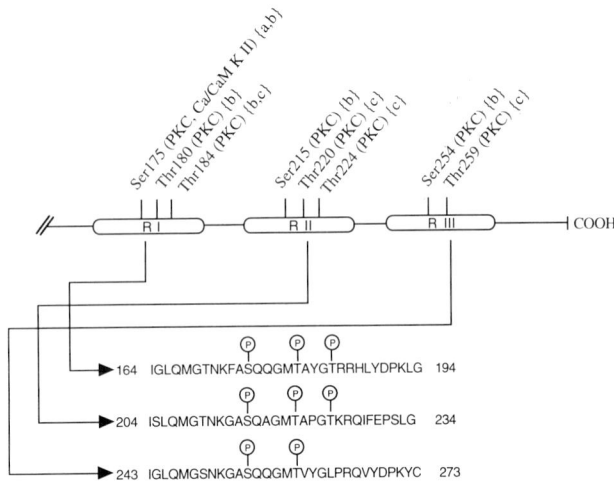

FIGURE 8 Summary of the phosphorylation sites identified in CP, showing the three internal 30-amino acid repeats (RI, RII, RIII). References: (a) Winder *et al.* (1993b); (b) Winder and Walsh (1990a); (c) Nakamura *et al.* (1993).

1994) when CP is bound to F-actin or to isolated thin filaments. Inhibition of phosphorylation is complete at a molar ratio of actin : CP of 10 : 1.

VII. TISSUE SPECIFICITY AND LOCALIZATION

The results of immunoblotting with antibodies against CP indicate that CP is more or less exclusively expressed in smooth muscle *in vivo* (Takahashi *et al.*, 1987; Gimona *et al.*, 1990). It is not found in skeletal or cardiac muscle and is absent in nonmuscle tissues, including brain (Takahashi *et al.*, 1987), kidney, liver, and spleen (Gimona *et al.*, 1990). But there are detectable amounts in blood platelets (Takeuchi *et al.*, 1991), cells that express other proteins typical for smooth muscle (Turner and Burridge, 1989), and it has also been reported to occur in adrenal medulla (Takahashi *et al.*, 1987). In the latter case, however, the investment of vascular tissue makes it difficult to exclude the possibility of the presence of contaminating smooth muscle.

Studies using immunofluorescence microscopy of tissue sections and cultured cells confirm the immunoblotting data and reveal additional aspects of CP distribution. During development in the chick, CP is exclusively restricted to smooth muscle and its putative precursor cells (Duband *et al.*, 1993). In humans it is also found in myoepithelial cells and in myofibroblasts of mammary tissue (Lazard *et al.*, 1993). Both of these cell types express other smooth muscle markers, including smooth muscle actin and, variably, smooth muscle myosin (Lazard *et al.*, 1993), and have

been generally considered as related to smooth muscle (Franke *et al.*, 1980; Schmitt-Gräff *et al.*, 1994). Current data suggest that there is a temporal shift in the expression of smooth muscle proteins during differentiation, with CP and heavy CD appearing later than other phenotypic markers: desmin, smooth muscle α-actin, smooth muscle myosin, and MLCK (Duband *et al.*, 1993; Frid *et al.*, 1992). Myoepithelial cells as well as myofibroblasts can express CP without CD, however, suggesting that CP and CD may be independently regulated at the gene level (Lazard *et al.*, 1993). Consistent with its smooth muscle specificity, CP expression is down-regulated in smooth muscle cells cultured *in vitro* (Gimona *et al.*, 1990; Birukov *et al.*, 1991; Durand-Arczynska *et al.*, 1993). Before it is down-regulated CP is found on the actin filament bundles of cultured cells (Gimona *et al.*, 1990; Birukov *et al.*, 1991) in line with the *in vitro* binding of CP to F-actin. Interestingly, and in association with its down-regulation, CP is not uniformly localized on the stress fiber system of cultured smooth muscle cells, but is restricted to the more central regions of the actin bundles, away from their sites of anchorage to the cell membrane (Gimona *et al.*, 1990; Birukov *et al.*, 1991). This localization may reflect the association of CP with a specific actin isoform (Birukov *et al.*, 1991) or may result, for example, from a coordinate degradation of CP along with a dynamic remodeling of the actin cytoskeleton (Wang, 1987). Weak labeling for CP has been reported on the actin bundles of fibroblasts (Takeuchi *et al.*, 1991), but the antibody used in this study cross-reacted also with a 23-kDa polypeptide, which could correspond to the sequence-related smooth muscle protein SM22 that is not down-regulated in culture (Gimona *et al.*, 1990) and has been described in primary fibroblasts (Santarèn *et al.*, 1987).

The localization of CP in differentiated smooth muscle *in vivo* is most relevant to discussion about this protein's function. In differentiated, isolated smooth muscle cells, CP codistributes with longitudinally oriented bundles that also contain actin (Walsh *et al.*, 1993; North *et al.*, 1994b). There is a drawback with isolated cells, however, in that they often become distorted on isolation and can suffer degradation of their cytoskeleton through the penetration into the cells of the enzymes employed for cell separation (Small *et al.*, 1990; North *et al.*, 1994a). This problem can be obviated by using intact tissue. In this regard, immunofluorescence and immunoelectron microscopy of thin sections of intact tissue has shown that the smooth muscle cell exhibits two structural domains, a contractile domain and a cytoskeleton domain, each characterized by a specific subset of components (Small *et al.*, 1986; North *et al.*, 1994a). The contractile domain con-

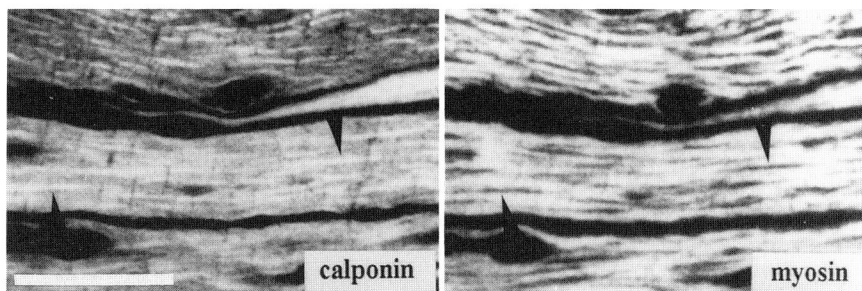

FIGURE 9 Longitudinal, 0.25-μm section of chicken gizzard smooth muscle double-labeled with antibodies against CP (left) and myosin (right). The unstained channels in B, corresponding to the cytoskeletal domain, are strongly stained by the CP antibody, which also labels the contractile domain. Reproduced with permission from North *et al.* (1994a), Fig. 4a,b, Copyright 1994, the Company of Biologists LTD.

tains myosin, smooth muscle actin, and CD and the cytoskeletal domain contains desmin, filamin, and the cytoplasmic β-actin isoform (North *et al.*, 1994b). A perhaps surprising finding was that CP is present in both domains (Fig. 9), in clear contrast to the situation with CD (Fürst *et al.*, 1986), and in fact appears to occur in highest concentrations in the cytoskeleton (North *et al.*, 1994a). This result implies that CP may indeed be segregated in different thin filaments from those containing CD, as also suggested by *in vitro* studies (see Marston and Huber, Chapter 6, this volume), and may regulate functions of the cytoskeleton that remain to be defined. For example, CP could modulate the interaction of filamin and actin in the cytoskeleton and thereby modify its viscoelastic properties. More work is needed to establish the contribution of the cytoskeleton of smooth muscle to the mechanical properties of this tissue.

VIII. PERSPECTIVES

We are still much in the dark regarding CP function. Does it perform both structural and regulatory roles, and if so what are they? Are these roles carried out in the cytoskeleton or in the contractile apparatus, or both? Does CP influence or complement CD function? And if Ca^{2+} binding proteins are the primary regulators, which one(s) operate on CP? To determine the function of CP *in vivo* will require new approaches. In part because of the presence of multiple targets for Ca^{2+} binding proteins in smooth muscle (including MLCK, CD, and CP), it has so far proved difficult to define the relative roles that myosin and the actin-associated proteins play in smooth muscle. Here specifically, we will want to know what roles CD and CP perform in the process of tension maintenance

at low energy cost (tone) that is characteristic for smooth muscle tissue. A current barrier to further progress is the lack of a model skinned muscle system that mimics tonic contraction and for which molecular components can be readily exchanged. The only way out may be the analysis of tissues from transgenic mice bearing deletions or mutations in the molecules of interest.

In the meantime, the ability to generate large quantities of purified recombinant CP and its subdomains opens the way for the production of crystals for the determination of the molecular structure via X-ray crystallography. This should provide the basis for analyzing in more detail the interaction between CP and its binding partners. We can also anticipate more insight into the *in vitro* function of CP subdomains from binding studies and *in vitro* motility assays.

Acknowledgments

We are grateful to Dr. S. J. Winder for the permission to reproduce the phylogenetic tree for calponin prior to publication and for helpful discussions. We also thank Dr. D. M. Helfman for his support. The studies of the authors were supported in part by grants from the Austrian Science Research Council.

References

Abe, M., Takahashi, K., and Hiwada, K. (1990a). *J. Biochem. (Tokyo)* **107**, 507–509.

Abe, M., Takahashi, K., and Hiwada, K. (1990b). *J. Biochem. (Tokyo)* **108**, 835–838.

Applegate, D., Feng, W., Green, R. S., and Taubman, M. B. (1994). *J. Biol. Chem.* **269**, 10683–10690.

Ayme-Southgate, A., Laski, P., French, C., and Pardue, M. L. (1989). *J. Cell Biol.* **108**, 521–531.

Bárány, M., Rokolya, A., and Bárány, K. (1991). *FEBS Lett.* **279**, 65–68.

Bartegi, A., Fattoum, A., and Kassab, R. (1990). *J. Biol. Chem.* **265,** 2231–2237.

Birukov, K. G., Stepanova, O. V., Nanaev, A. K., and Shirinsky, V. P. (1991). *Cell Tissue Res.* **266,** 579–584.

Chalovich, J. M. (1992). *Pharmacol. Ther.* **55,** 95–148.

Childs, T. J., Watson, M. H., Novy, R. E., Lin, J. J., and Mak, A. S. (1992). *Biochim. Biophys. Acta* **1121,** 41–46.

Draeger, A., Gimona, M., Stuckert, A., Celis, J. E., and Small, J. V. (1991). *FEBS Lett.* **291,** 24–28.

Draeger, A., Graf, A.-H., Staudach, A., North, A. J., and Small, J. V. (1993). *Virchows Arch. B.* **64,** 21–27.

Duband, J.-L., Gimona, M., Scatena, M., Sartore, S., and Small, J. V. (1993). *Differentiation (Berlin)* **55,** 1–11.

Durand-Arczynska, W., Marmy, N., and Durand, J. (1993). *Histochemistry* **100,** 465–471.

Franke, W. W., Schmid, E., Freudenstein, C., Appelhans, B., Osborn, M., Weber, K., and Keenan, K. W. (1980). *J. Cell Biol.* **84,** 633–654.

Frid, M. G., Shekhonin, B. V., Koteliansky, V. E., and Glukhova, M. A. (1992). *Dev. Biol.* **153,** 185–193.

Fujii, T., Oomatsuzawa, A., Kuzunicki, N., and Kondo, Y. (1994). *J. Biochem. (Tokyo)* **116,** 121–127.

Fürst, D. O., Cross, R. A., De Mey, J., and Small, J. V. (1986). *EMBO J.* **5,** 251–257.

Gimona, M., Herzog, M., Vandekerckhove, J., and Small, J. V. (1990). *FEBS Lett.* **274,** 159–162.

Gimona, M., Sparrow, M. P., Strasser, P., Herzog, M., and Small, J. V. (1992). *Eur. J. Biochem.* **205,** 1067–1075.

Goetinck, S., and Waterston, R. H. (1994a). *J. Cell Biol.* **127,** 71–78.

Goetinck, S., and Waterston, R. H. (1994b). *J. Cell Biol.* **127,** 79–93.

Gong, B. J., Mabuchi, K., Takahashi, K., Nadal-Ginard, B., and Tao, T. (1993). *J. Biochem. (Tokyo)* **114,** 453–456.

Haeberle, J. R. (1994). *J. Biol. Chem.* **269,** 12424–12431.

Horiuchi, K. Y., and Chacko, S. (1989). *Biochemistry* **28,** 9111–9116.

Horiuchi, K. Y., and Chacko, S. (1991). *Biochem. Biophys. Res. Commun.* **176,** 1487–1493.

Irvine, M., Huima, T., Prince, A. M., and Lustigman, S. (1994). *Mol. Biochem. Parasitol.* **65,** 135–146.

Lazard, D., Sastre, X., Frid, M. G., Glukhova, M. A., Thiery, J. P., and Koteliansky, V. E. (1993). *Proc. Natl. Acad. Sci. U.S.A.* **90,** 999–1003.

Lees-Miller, J. P., Heeley, D. H., Smillie, L. B., and Kay, C. M. (1987). *J. Biol. Chem.* **262,** 2988–2993.

Lehman, W. (1991). *J. Muscle Res. Cell Motil.* **12,** 221–224.

Lin, Y., Ye, L. H., Ishikawa, R., Fujita, K., and Kohama, K. (1993). *J. Biochem. (Tokyo)* **113,** 643–645.

Makuch, R., Birukov, K., Shirinsky, V., and Dabrowska, R. (1991). *Biochem. J.* **280,** 33–38.

Malencik, D. A., and Anderson, S. R. (1986). *Biochemistry* **25,** 709–721.

Mani, R. S., and Kay, C. M. (1990). *Biochemistry* **29,** 1398–1404.

Marston, S. B. (1991). *FEBS Lett.* **292,** 179–182.

Marston, S. B., and Redwood, C. S. (1993). *J. Biol. Chem.* **268,** 12137–12320.

Mezgueldi, M., Fattoum, A., Derancourt, J., and Kassab, R. (1992). *J. Biol. Chem.* **267,** 15943–15951.

Miki, M., Walsh, M. P., and Hartshorne, D. J. (1992). *Biochem. Biophys. Res. Commun.* **187,** 867–871.

Nagumo, H., Sakurada, K., Seto, M., and Sasaki, Y. (1994). *Biochem. Biophys. Res. Commun.* **203,** 1502–1507.

Naka, M., Kureishi, Y., Muroga, Y., Takahashi, K., Ito, M., and Tanaka, T. (1990). *Biochem. Biophys. Res. Commun.* **171,** 933–937.

Nakamura, F., Mino, T., Yamamoto, J., Naka, M., and Tanaka, T. (1993). *J. Biol. Chem.* **268,** 6194–6201.

Nishida, W., Abe, M., Takahashi, K., and Hiwada, K. (1990). *FEBS Lett.* **268,** 165–168.

Nishida, W., Kitami, Y., and Hiwada, K. (1993). *Gene* **130,** 297–302.

Noda, S., Ito, M., Watanabe, S., Takahashi, and Maruyama, K. (1992). *Biochem. Biophys. Res. Commun.* **185,** 481–487.

North, A. J., Gimona, M., Cross, R. A., and Small, J. V. (1994a). *J. Cell Sci.* **107,** 437–444.

North, A. J., Gimona, M., Lando, Z., and Small, J. V. (1994b). *J. Cell Sci.* **107,** 445–455.

Pearlstone, J. R., Weber, M., Lees-Miller, J. P., Carpenter, M. R., and Smillie, L. B. (1987). *J. Biol. Chem.* **262,** 5985–5991.

Santarèn, J. F., Blüthmann, H., Macdonald-Bravo, H., and Bravo, R. (1987). *Exp. Cell Res.* **173,** 341–348.

Schmitt-Gräff, A., Desmouliére, A., and Gabbiani, G. (1994). *Virchows Arch.* **524,** 3–24.

Shirinsky, V. P., Biryukov, K. G., Hettasch, J. M., and Sellers, J. R. (1992). *J. Biol. Chem.* **267,** 15886–15892.

Small, J. V., and Sobieszek, A. (1980). *Int. Rev. Cytol.* **64,** 241–306.

Small, J. V., Fürst, D. O., and DeMey, J. (1986). *J. Cell Biol.* **102,** 210–220.

Small, J. V., Herzog, M., Barth, M., and Draeger, A. (1990). *J. Cell Biol.* **111,** 2451–2461.

Sobue, K., Takahashi, K., and Wakabayshi, I. (1985). *Biochem. Biophys. Res. Commun.* **132,** 645–651.

Strasser, P., Gimona, M., Mössler, H., Herzog, M., and Small, J. V. (1993). *FEBS Lett.* **330,** 13–18.

Szymanski, P. T., and Tao, T. (1993a). *FEBS Lett.* **331,** 256–259.

Szymanski, P. T., and Tao, T. (1993b). *FEBS Lett.* **334,** 379–382.

Takahashi, K., and Nadal-Ginard, B. (1991). *J. Biol. Chem.* **266,** 13284–13288.

Takahashi, K., and Nadal-Ginard, B. (1993). *J. Biol. Chem.* **267,** 26198.

Takahashi, K., Hiwada, K., and Kokubu, T. (1986). *Biochem. Biophys. Res. Commun.* **141,** 20–26.

Takahashi, K., Hiwada, K., and Kokubu, T. (1987). *Life Sci.* **41,** 291–296.

Takahashi, K., Hiwada, K., and Kokubu, T. (1988a). *Hypertension* **11,** 620–626.

Takahashi, K., Abe, M., Hiwada, K., and Kokubu, T. (1988b). *J. Hypertens.* **6**(suppl. 4), S40–S43.

Takeuchi, K., Takahashi, K., Abe, M., Nishida, W., Hiwada, K., Nabeya, T., and Maruyama, K. (1991). *J. Biochem. (Tokyo)* **109,** 311–316.

Turner, C. E., and Burridge, K. (1989). *Eur. J. Cell Biol.* **49,** 202–206.

Vancompernolle, K. Gimona, M., Herzog, M., Van Damme, J., Vandekerckhove, J., and Small, J. V. (1990). *FEBS Lett.* **274,** 146–152.

Vandekerckhove, J., and Weber, K. (1981). *Eur. J. Biochem.* **113,** 595–603.

Walsh, M. P., Carmichael, J. D., and Kargacin, G. J. (1993). *Am. J. Physiol.* **265,** 1371–1378.

Wang, Y.-L. (1987). *J. Cell Biol.* **105,** 2811–2816.

Wills, F. L., McCubbin, W. D., and Kay, C. M. (1993). *Biochemistry* **32,** 2321–2328.

Wills, F. L., McCubbin, W. D., and Kay, C. M. (1994). *Biochemistry* **33,** 5562–5569.

Wills, F. L., McCubbin, W. D., Gimona, M., Strasser, P., and Kay, C. M. (1995). *Protein Sci.* **3,** 2311–2321.

Winder, S. J., and Walsh, M. P. (1990a). *Biochem. Int.* **22,** 335–341.

Winder, S. J., and Walsh, M. P. (1990b). *Prog. Clin. Biol. Res.* **327,** 141–148.

Winder, S. J., and Walsh, M. P. (1990c). *J. Biol. Chem.* **265,** 10148–10155.

Winder, S. J., Sutherland, C., and Walsh, M. P. (1990). *In* "Regulation of Smooth Muscle: Progress in Solving the Puzzle" (R. S. Moreland, ed.) Plenum, New York. Pp. 37–51.

Winder, S. J., Sutherland, C., and Walsh, M. P. (1991). *Adv. Exp. Med. Biol.* **304,** 37–51.

Winder, S. J., Sutherland, C., and Walsh, M. P. (1992). *Biochem. J.* **288,** 733–739.

Winder, S. J., Walsh, M. P., Vasulka, C., and Johnson, J. D. (1993a). *Biochemistry* **32,** 13327–13333.

Winder, S. J., Allen, B. G., Fraser, E. D., Kang, H.-M., Kargacin, G. J., and Walsh, M. P. (1993b). *Biochem. J.* **296,** 827–836.

8

Calcium Binding Proteins

RAJAM S. MANI and CYRIL M. KAY

MRC Group in Protein Structure and Function
Department of Biochemistry,
University of Alberta,
Edmonton, Alberta, Canada

I. INTRODUCTION

In smooth muscle, calcium exerts its control via mechanisms based on both the thick filaments (Sobieszek and Bremel, 1975) and the thin actin-based filaments (Sobue and Sellers, 1991; Chalovich, 1993). The calcium binding protein calmodulin (CaM) is considered to be the Ca^{2+}-dependent regulatory component of the myosin-linked system. The thin filament-linked pathway involves actin-mediated regulation by caldesmon (CaD), a protein that is bound to the thin filaments (Fürst *et al.*, 1986). Native smooth muscle thin filaments contain actin–tropomyosin and CaD in molar ratios of 14:2:1 (Marston *et al.*, 1992). When purified CaD binds to actin–tropomyosin, it inhibits actomyosin ATPase activity (Moody *et al.*, 1985; Sobue *et al.*, 1985). This inhibition is not Ca^{2+} sensitive but can be rendered Ca^{2+} dependent by a suitable Ca^{2+} binding regulatory protein that can bind to CaD and thus reverse its inhibitory effect on the actomyosin ATPase activity. CaM in the presence of Ca^{2+} can reverse the inhibitory effect of CaD. However, the affinity of CaM for CaD ($K \sim 1.6 \times 10^6 M^{-1}$) is rather weak, and for this reason, a very large molar excess of CaM over CaD ($\approx$25-fold higher) is required to regulate CaD (Shirinsky *et al.*, 1988; Pritchard and Marston, 1989). Hence, it is conceivable that the native thin filament is regulated not by CaM but by a different calcium binding protein.

In our laboratory, we have developed a new protocol for isolating four calcium binding proteins from smooth muscle. These include 67-kDa calcium binding protein (67 calcimedin), calmodulin, 12-kDa calcium binding protein, and an 11-kDa smooth muscle calcium binding protein referred to as caltropin (CaT) by drawing analogy with troponin C, the calcium binding subunit of the troponin complex. The purpose of this chapter is to describe the calcium binding properties of these calcium binding proteins with special emphasis on CaT. With CaT we will demonstrate how it can regulate CaD in a calcium-dependent manner. In this context, the effect of CaT on the CaD–myosin and the CaD–actin interactions in the presence of Ca^{2+} will be discussed.

II. INSOLATION OF CALCIUM BINDING PROTEINS

Calcium binding proteins (CaBP) in general bind hydrophobic resins in a calcium-dependent manner (Moore and Dedman, 1982), and this property was utilized in the isolation scheme developed for purifying calcium binding proteins from smooth muscle (Mani and Kay, 1990). About 350 g of chicken gizzard was homogenized in a Waring blender using 1 liter of buffer consisting of 40 mM Tris, pH 7.5, 80 mM, NaCl, 2 mM EDTA, 0.05% sodium azide, 1 mM phenylmethanesulfonyl fluoride, and 0.5 ml aprotinin. The homogenate was centrifuged in a Beckman J2-21 centrifuge at 10,000 rpm for 75 min. Proteins in the supernatant were concentrated by ammonium sulfate addition (95%, 700 g/liter). Precipitated protein was dissolved in 40 mM Tris, pH 7.5, containing 80 mM NaCl, and subjected to centrifugation at 14,000 rpm for 45 min. After adjusting the free calcium concentra-

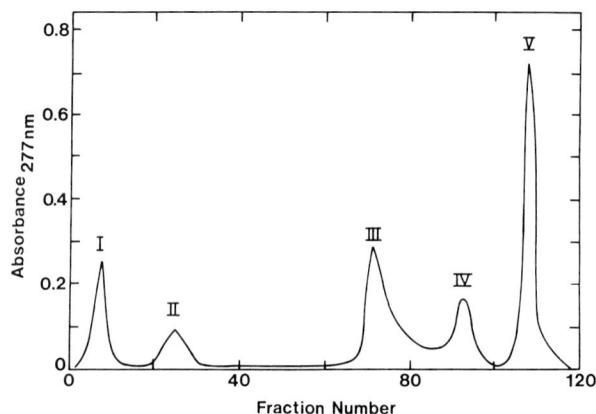

FIGURE 1   Elution profile of calcium binding proteins from a DEAE-cellulose column. Solvent system used was 20 mM imidazole (pH 6.2), 1 mM EDTA, containing 1000 ml of a linear NaCl gradient from 0 to 0.25 M. Reprinted with permission from *Biochemistry* (29). Copyright 1990 American Chemical Society.

tion in the supernatant to 2 mM, it was applied to a phenyl Sepharose column equilibrated with 40 mM Tris, pH 7.5, 80 mM NaCl, and 2 mM CaCl$_2$. Calcium binding proteins were eluted from this column by using 4 mM EGTA. Eluted proteins were next applied to a DEAE-cellulose column in the presence of 20 mM imidazole, pH 6.2, and 1 mM EDTA. Calcium binding proteins were eluted by using a linear NaCl gradient. Under these conditions, most of the 12-kDa CaBP eluted in the void volume along with some 35-kDa calcimedin. Caltropin (11-kDa CaBP) eluted around 0.15 M NaCl in peak III as indicated in Fig. 1 and was essentially homogeneous (>90% pure) (Mani and Kay, 1990).

Peak III material, when analyzed by SDS–polyacrylamide gel electrophoresis, revealed CaT as the major protein along with some 67-kDa calcimedin. These two proteins could easily be separated on an Ultragel AcA44 (LKB) gel filtration column equilibrated with 50 mM Tris, pH 7.5, containing 200 mM NaCl, 2 mM EDTA, and 0.05% NAN$_3$. The first peak corresponded to 67-kDa calcimedin and the protein isolated under peak II was a homogeneous preparation of CaT. The major proteins in peaks IV and V from the DEAE-cellulose column were 67-kDa calcimedin and calmodulin; 67-kDa calcimedin in a pure form was isolated by applying peak IV material on to an AcA44 gel filtration column. Calmodulin, which elutes at the end of the gradient, was essentially homogeneous. In peak 1, 12-kDa CaBP could easily be separated from 35-kDa calcimedin on an Ultragel AcA44 gel filtration column; 12-kDa CaBP can also be separated from other CaBP's with a Waters Model 625 HPLC system using a 1 × 30-cm AP-1 glass column packed with Waters Accell plus QMA anion-exchange media as described in our paper (Mani and Kay, 1992). When the isolated 12-kDa CaBP

was run in 20% SDS gels, it migrated with a molecular weight of 12,000 and was homogeneous. The gels were calibrated with standard proteins using the Sigma gel kit protein samples.

III. AMINO ACID COMPOSITION OF SMOOTH MUSCLE CALCIUM BINDING PROTEINS

A comparison of the amino acid analysis of our preparation of CaBP's with other known low-molecular-mass calcium binding proteins is summarized in Table I. Caltropin, for instance, is distinct from other proteins listed in the table, especially when one considers phe : tyr ratios. Other proteins have a low content of tyrosine with the result that the phe:tyr ratio is anywhere from 3 to 8, whereas caltropin has a high content of tyr (4 residues) and a phe:tyr ratio of 1. Twelve-kDa CaBP also isolated from the same source is distinct from CaT since the phe:tyr ratio is 6 for this protein. It is characterized by a high content of methionine and arginine residues. Other proteins that are similar to 12-kDa CaBP include the parvalbumins. Again, in this instance 12-kDa CaBP is different since it has 4 arginine residues, whereas parvalbumins are characterized by a single arginine residue. The amino acid composition of 67-kDa protein is very similar to that of 67-kDa calcimedin prepared by Moore (1986).

IV. CALCIUM BINDING PROPERTIES OF 12-kDa CALCIUM BINDING PROTEIN

A calcium binding assay using Arsenazo 111 (Sigma) indicated that the protein binds 1 mol of Ca^{2+}/mol of protein (Mani and Kay, 1992). Upon binding Ca^{2+}, the protein undergoes a conformational change as demonstrated by UV difference spectroscopy, a powerful technique to study conformational changes in proteins upon ligand binding. The difference in the absorption properties of 12-kDa CaBP between 250 and 300 nm, induced by Ca^{2+}, is shown in Fig. 2.

The dominant difference peaks at 287 and 280 nm in the presence of Ca^2 arise from the perturbation of the single tyrosyl chromophore. The sign of the tyrosyl difference peaks (i.e., a red shift) suggests that the chromophore is in a less polar environment in the presence of Ca^{2+} (Donovan, 1969). In addition, a sharpening of the fine structure of the absorption bands below 270 nm is observed, suggesting that one or more of the phenylalanine residues are perturbed in the presence of Ca^{2+}. Another spectroscopic technique that has proved useful in our hands to study protein–ligand and protein–protein interaction is circular dichroism (CD) measurements (Mani and Kay,

TABLE I   Amino Acid Composition of Calcium Binding Proteins

Amino Acid	CaT[a]	12 kDa[b]	CaM[c]	S-100b[a]	Parvalbumin[a]	67 kDa[d]	Calcimedin
Aspartate	8	17	23	9	15	10	10
Threonine	3	8	12	3	5	4	4
Serine	4	3	4	5	11	7	9
Glutamate	17	20	27	19	9	14	15
Proline	3	2	2	0	0	3	3
Glycine	10	8	11	4	9	11	12
Alanine	11	8	11	5	11	9	10
Cysteine	0	0	0	0	0	—	—
Valine	4	4	7	6	5	7	4
Methionine	2	6	9	3	3	2	2
Isoleucine	4	6	8	4	6	4	3
Leucine	12	8	9	8	9	8	6
Tyrosine	4	1	2	1	0	3	3
Phenylalanine	5	6	8	7	8	3	4
Tryptophan	0	0	0	0	—	1	1
Histidine	2	1	1	5	2	2	4
Lysine	11	6	7	8	15	8	9
Trimethyllysine	0	0	1	0	0	0	0
Arginine	2	4	6	1	1	4	2

[a]Mani and Kay (1990).
[b]Mani and Kay (1992).
[c]Watterson et al. (1980).
[d]Mani and Kay (1989).

1987). In proteins, near-UV–CD measurements will furnish information regarding the local environments of aromatic rings, whereas far-UV studies will yield information regarding secondary structure. Addition of Ca^{2+} caused a slight increase in ellipticity around 222 nm. Analysis of the CD data according to the method of Chen et al. (1974) indicated an increase in apparent α-helical content from 35 to 39% in the presence of Ca^{2+}. There was no significant change in the β-structure (~20%), whereas the random coil structure decreased from 47 to 41% in the presence of Ca^{2+}. The protein seems to acquire a more ordered secondary structure in the presence of Ca^{2+}. The aromatic CD spectrum of 12-kDa CaBP in the absence and in the presence of Ca^{2+} is shown in Fig. 3. The contribution appears to be due largely to phenylalanine with ellipticity bands at 262 and 268 nm, since their sign and position are in complete agreement with the CD spectrum of acetyl-L-phenylalanine methyl ester (Goodman and Toniolo, 1968).

Fluorescence spectroscopy is another powerful technique that can be utilized to study protein conformation in solution. Fluorescent probes like TNS [6-(p-toluidino)naphthalene-2-sulfonic acid], a hydrophobic probe, and a sulfhydryl probe such as acrylodan [6-(acryloyl-2-dimethyl aminonaphthalene] are useful for studying protein conformation. Twelve-kDa

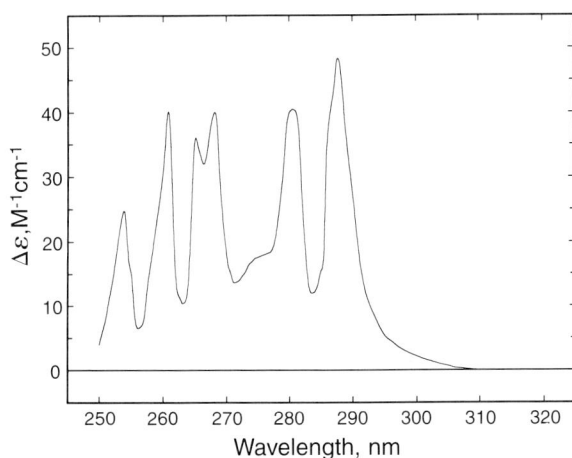

FIGURE 2   UV difference spectrum of 12-kDa CaBP in 0.1 M Tris (pH 7.5), 0.5 mM $CaCl_2$. The units are expressed as differences in molar absorption, $\Delta\epsilon$. The protein concentration was 2 mg/ml. Reprinted with permission from Arch. Biochem. Biophys. (296). Copyright 1992 Academic Press, Inc.

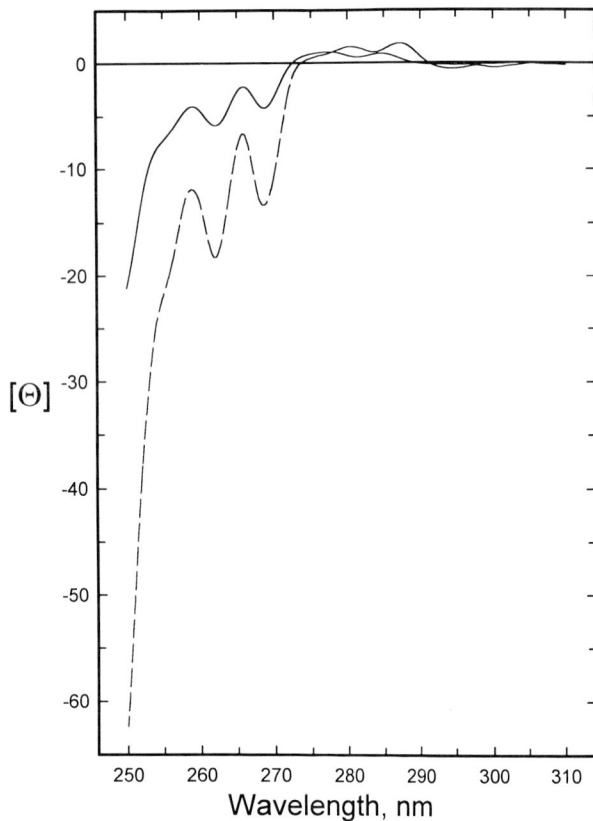

FIGURE 3   Aromatic CD spectra of 12-kDa CaBP in 50 mM Mops, pH 7.2, 50 mM NaCl, and 1 mM EGTA (solid line) and in 50 mM Mops, pH 7.2, 50 mM NaCl, and 2 mM CaCl$_2$ (dashed line). Reprinted with permission from *Arch. Biochem. Biophys.* (296). Copyright 1992 Academic Press, Inc.

CaBP binds to a phenyl-Sepharose column in the presence of Ca^{2+}, which suggests that this protein exposes a hydrophobic region when it binds Ca^{2+}. For this reason the hydrophobic probe TNS was used to label the protein. TNS-labeled apoprotein when excited at 345 nm had an emission maximum at 463 nm. Addition of Ca^{2+} to TNS-labeled protein resulted in a sevenfold enhancement in fluorescent intensity and this was accompanied by a blue shift in the emission maximum to 445 nm, implying that the probe in the presence of Ca^{2+} occupies a more hydrophobic environment (McClure and Edelman, 1966; Mani and Kay, 1992). A K_d value of 1.1×10^{-6} M was estimated for calcium binding from the fluorescence titration data. Twelve-kDa CaBP binds 1 mol of Ca^{2+}/mol of protein at pH 7.2. Hence, the protein has a single high-affinity Ca^{2+} binding site.

At present, the exact biological function of this protein is not known. When tested for its ability to reverse caldesmon's inhibition of the actin-activated myosin ATPase activity, 12-kDa CaBP had no significant effect on the ATPase activity when the mole ratio of 12-kDa

CaBP to caldesmon used was 1:1. This would imply that 12-kDa CaBP probably does not bind to CaD. Hence its function must involve binding to some other protein in smooth muscle. One could resort to an affinity column of 12-kDa CaBP to isolate its receptor in smooth muscle as a first step in understanding its possible functional role in this tissue.

V. PURIFICATION AND CHARACTERIZATION OF 67-kDa CALCIMEDIN

The 67-kDa calcimedin, isolated by using a phenyl-Sepharose affinity column followed by DEAE-cellulose and gel filtration chromatographies, was homogeneous by the criterion of SDS–polyacrylamide gel electrophoresis. In 15% SDS gels the protein migrated with a M_r value of 67,000 (Mani and Kay, 1989). In non-SDS gels the protein moved faster in the presence of EDTA. A decrease in mobility in the presence of Ca^{2+} could be due to a decrease in negative charge on the protein resulting from binding Ca^{2+}. In this respect, 67-kDa calcimedin behaves similarly to calmodulin and S-100b protein (Mani *et al.*, 1982). The amino acid composition of 67-kDa protein is presented in Table I together with that of the 67-kDa calcimedin (Moore, 1986). Good agreement exists between these two proteins. The latter is the largest member of a family of Ca^{2+} binding proteins that associate with membranes and phospholipids in a Ca^{2+}-dependent manner (for a review, see Crompton *et al.*, 1988). So far, seven distinct members of this novel family, called annexins I–VII (Geisow, 1986; Kaetzel and Dedman, 1989), have been identified. These are related through high degrees of sequence similarities and appear to be structurally and functionally distinct from the calmodulin/troponin C family. These calcium/phospholipid binding proteins are composed of conserved repeating domains of about 70 residues. The EF-hand structure is not predicted from primary sequence data (Crompton *et al.*, 1988).

A. Ca^{2+} Binding Properties of 67-kDa Calcimedin

A calcium binding assay revealed the binding of 4 mol of Ca^{2+}/mol of protein at pH 7.5 (Mani and Kay, 1989). Ca^{2+} binding affects the environments of the tryptophan and the tyrosine residues. Figure 4 shows the Ca^{2+}-induced difference spectrum of 67-kDa protein.

The difference peak at 292 nm is characteristic of a red shift of the tryptophan absorption band. The dominant difference peaks at 287 and 280 nm arise from the perturbation of one or more tyrosine residues. The

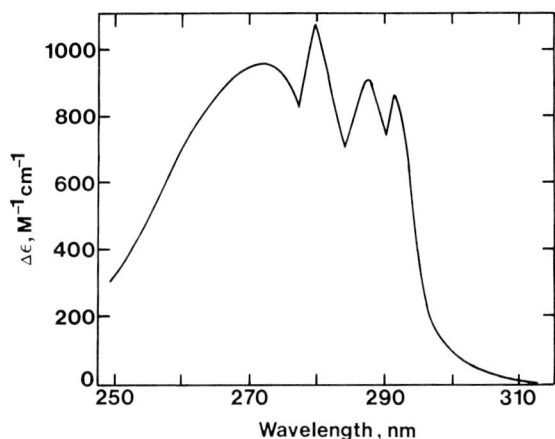

FIGURE 4 UV difference spectrum of 67-kDa calcimedin in 0.1 M Tris (pH 7.5) and 0.5 mM CaCl$_2$. Reprinted with permission from *Biochem. J.* (259). Copyright 1989. The Biochemical Society and Portland Press.

sign of the tryptophan and tyrosine difference peaks (i.e., a "red shift") suggests that the chromophores are in a less polar environment in the presence of Ca^{2+} (Mani and Kay, 1989). Far-UV–CD measurements revealed only a small change in the overall secondary structure of the protein upon Ca^{2+} binding. The $[\theta]_{222}$ nm value decreased by only 10%. Mg^{2+} addition, on the other hand, produced no significant change in the $[\theta]_{222}$ nm value, suggesting that the observed effect with Ca^{2+}, though small, is specific for this metal ion (Mani and Kay, 1989). Sixty-seven-kDa calcimedin is believed to expose a hydrophobic patch upon Ca^{2+} binding. For this reason, 67-kDa protein was labeled with the hydrophobic probe TNS. Addition of Ca^{2+} to TNS-labeled protein resulted in a 25% enhancement in fluorescence intensity and this was accompanied by a blue shift of the emission maximum, implying that the probe in the presence of Ca^{2+} occupies a more hydrophobic environment. Fluorescence titration of labeled 67-kDa calcimedin with Ca^{2+} yielded a K_d value of 2 × 10^{-5} M and this corresponds to the Ca^{2+} concentration required to produce 50% of the observed increase in fluorescence intensity (Mani and Kay, 1989). The protein binds 4 mol of Ca^{2+}/mol of protein, but fluorescence titration with Ca^{2+} indicated only one class of binding site in the presence of the probe. One possible interpretation would be that all four calcium binding sites have the same affinity for calcium or, alternatively, we may be monitoring only one or more calcium binding sites that are located close to the TNS binding site and, as a consequence, the observed K_d value for Ca^{2+} corresponds only to those sites that are in the near vicinity of the probe.

The cellular function of 67-kDa calcimedin is still uncertain. Calcimedins are known to have a broad tissue distribution. Since higher levels of calcimedins are found in the liver, it has been suggested that these may be involved in the secretory process. Relative amounts of calcimedins vary from tissue to tissue and from species to species. For instance, bovine liver is estimated to contain 93% of the total calcimedin in the 67-kDa form, whereas rat liver has only 63% of calcimedin as 67-kDa protein (Matthew *et al.*, 1986). Further studies on the functional difference of livers in the two species may shed some light on the possible functional role of these proteins. It has also been suggested that 67-kDa protein may be involved in events involving the plasma membrane calcium channels. Investigations on other functional properties of the annexins, such as binding to elements of the cytoskeleton (Crompton *et al.*, 1988) or participation in exocytosis (Ali *et al.*, 1989), will hopefully provide more detailed insight into the physiological role of this interesting protein family.

VI. CALMODULIN

Calmodulin, the ubiquitous and multifunctional Ca^{2+} binding protein, mediates many of the regulatory effects of Ca^{2+}, including the contractile state of smooth muscle. Calmodulin was discovered as an activator of cyclic nucleotide phosphodiesterase (PDE) (Cheung, 1970; Kakiuchi *et al.*, 1970). Smooth muscle calmodulin contains 148 amino acid residues (Grand and Perry, 1978) and its amino acid sequence is almost identical to that of the well-characterized brain protein (Watterson *et al.*, 1980). Its structure has been highly conserved throughout evolution. CaM is a monomeric, globular protein of $M_r = 16,680$. It is highly acidic (pI $\sim$ 4) owing to its high content ($\sim$35%) of aspartate and glutamate residues. Most CaM's contain a single ϵ-N-trimethyllysine (at position 115) as a result of post-translational modification.

Calmodulin belongs to a family of structurally related calcium binding proteins often referred to as "EF-hand Ca^{2+}-binding proteins" based on the crystallographic structure of a Ca^{2+} binding site in carp parvalbumin (Kretsinger and Nockolds, 1973). An EF-hand consists of a helix–loop–helix structure composed of $\sim$30 amino acids, the two α-helices lying nearly perpendicular and flanking the 12-residue calcium binding loop. CaM has four EF-hands consistent with the presence of four calcium binding sites. Ca^{2+} binding induces a marked conformational change in CaM as detected by spectroscopic measurements (Wang *et al.*, 1975; Walsh *et al.*, 1979; Seamon, 1980; Ikura *et al.*, 1984). Binding of Ca^{2+} gives CaM its functional configuration, which in turn recognizes a variety of protein targets such as phosphodiesterase, protein kinases, NAD kinase, and calcium pumps, as well as motility proteins (Manalan and Klee, 1984; Means, 1988; O'Neil and DeGrado, 1990). X-ray crystallographic structure determination of Ca^{2+}–CaM revealed a dumbbell structure of overall length 65 Å with N- and C-terminal globular domains (each containing two Ca^{2+} binding sites) connected by a central seven-turn α-helix (Babu et al., 1985, 1988; Chattopadhyaya et al., 1992). The flexibility of the central helix in solution allows the globular ends of the molecule to come together when CaM binds to target proteins or peptides.

A. Calmodulin and Smooth Muscle Contraction

In smooth muscle contraction, two major pathways in the interaction between thin and thick filaments have been generally accepted in the regulatory system following an increase in the intracellular Ca^{2+} concentration. Myosin filament activity involves the phosphorylation of myosin catalyzed by myosin light chain kinase (MLCK) (Adelstein and Eisenberg, 1980; Sommerville and Hartshorne, 1986). CaM in the presence of Ca^{2+} activates MLCK, which transfers the phosphate of ATP to the 20-kDa light chain of myosin and the phosphorylation triggers cross-bridge cycling and the development of force or contraction of the muscle. For this reason, the interaction between CaM and MLCK has received a great deal of attention. The solution structure of a complex of Ca^{2+}–CaM with the 26-residue M13 peptide derived from the CaM binding region of MLCK has been determined by multidimensional nuclear magnetic resonance (NMR) (Ikura et al., 1990, 1992). The crystal structure of the complex of Ca^{2+}–CaM with a 20-residue CaM binding peptide derived from smooth muscle MLCK has also been determined (Meador et al., 1992). These complexes assume a compact globular shape with a flexible linker that enables the two domains to clamp the respective target peptide, which itself adopts an α-helical conformation. This complex formation between CaM and M13 peptide results in stabilizing CaM as shown by amide proton exchange experiments (Spera et al., 1991). The increased stability of the ternary complex (Ca^{2+}–CaM–M13), largely due to the hydrophobic interaction, produces a significant increase in the affinity of the complex for Ca^{2+} ($pCa_{50\%} \sim 7.4$; Yagi et al., 1989). Since CaM binds Ca^{2+} with a $pCa_{50\%}$ value around 5.5, MLCK exists as an inactive (or almost inactive) apoenzyme in the resting smooth muscle with a free Ca^{2+} concentration ($[Ca^{2+}]_i$) of 0.1–0.2 μM (Williams and Fay, 1986). Upon stimulation of the cell, $[Ca^{2+}]_i$ rises to ~0.5–0.7 μM. As a result Ca^{2+} binds to CaM, inducing a conformational change resulting in the exposure of (hydrophobic) sites of interaction with MLCK. The resultant ternary complex (Ca^{2+}–CaM–

MLCK) is fully active due to a conformational change induced in the kinase by the binding of CaM (Olwin et al., 1984).

Smooth muscle also has a Ca^{2+}-dependent regulatory mechanism associated with the thin filaments, made up of actin, tropomyosin, caldesmon, and a calcium-sensitizing factor related to CaM (Marston and Smith, 1985). The regulation of thin filaments is mediated by caldesmon, which is a potent inhibitor of actin–tropomyosin activation of myosin MgATPase activity (Marston and Redwood, 1991, 1992). This inhibition is not Ca^{2+} sensitive but can be rendered Ca^{2+} dependent by a suitable Ca^{2+} binding regulatory protein that can bind to caldesmon and thus reverse its inhibitory effect on the actomyosin ATPase activity. In vitro, CaM in the presence of Ca^{2+} can reverse the inhibitory effect of CaD on the actomyosin ATPase activity. However, the affinity of CaM for CaD ($K \sim 1.6 \times 10^6\ M^{-1}$) is rather weak, and for this reason, a very large molar excess of CaM over CaD ($\approx$25-fold higher) is required to regulate CaD at 25°C (Shirinsky et al., 1988; Pritchard and Marston, 1989). Hence, it is conceivable that the native thin filament is regulated not by CaM but by a different calcium binding protein.

We have isolated a novel calcium binding protein from smooth muscle that is referred to as caltropin, with a subunit molecular weight of 11,000. CaT was much more potent than CaM in reversing the inhibitory effect of CaD in the presence of Ca^{2+} (see the following), making it a potential Ca^{2+} factor in regulating CaD in smooth muscle. Hence, we believe the principal function of CaM in smooth muscle is to activate cross-bridge cycling and the development of force in response to a $[Ca^{2+}]_i$ transient via the activation of myosin light chain kinase and phosphorylation of myosin, that is, it operates at the level of them myosin thick filament.

VII. ISOLATION AND CHARACTERIZATION OF CALTROPIN

Caltropin, isolated using a phenyl-Sepharose affinity column followed by ion-exchange and gel filtration chromatographies, was homogeneous and the protein migrated with a molecular weight value of 11,000 on SDS gels (Mani and Kay, 1990). In non-SDS gels the protein moved faster in the presence of EDTA. A decrease in mobility in the presence of Ca^{2+} could be due to a decrease in negative charge on the protein resulting from binding the cation. Caltropin behaves similarly to CaM but different from muscle troponin C. Low-speed sedimentation equilibrium studies under denaturing conditions in 6 M guanidine hydrochloride suggested a subunit molecular weight of 10,500 ± 500 (Mani and Kay, 1990), in excellent agreement with the SDS gel result. Sedimentation equilibrium runs in benign medium consisting of 0.1 M Tris, pH 7.5, and 0.1 M NaCl yielded a molecular weight of 21,000 ± 500, suggesting that the protein exists as a dimer of 21,000 in native solvents. In this respect, CaT behaves similarly to brain S-100 proteins, which are also known to exist as dimers (ββ or αβ) of 21,000 molecular weight (Mani et al., 1982; Mani and Kay, 1983).

A. Ca^{2+} Binding Properties of Caltropin

Caltropin binds 2 mol of calcium per subunit (Fig. 5). Direct calcium binding assays were carried out using Arsenazo 111 as described by Mani and Kay (1990). The binding curve was fitted by the Hill equation describing the binding of ligand (Ca^{2+}) to the protein (Edsall and Gutfreund, 1983). Best fit was observed when the Hill coefficient (n) was 1.3 and the calculated apparent dissociation constant (K_d) for Ca^{2+} was $3 \times 10^{-6} M$. The observed K_d value probably represents a composite value for the two sites since the value of n obtained is greater than one. Calcium-induced UV difference spectra of CaT are shown in Fig. 6. The dominant peaks at 287 and 280 nm arise from the perturbation of one or more tyrosine residues. The sign of the tyrosyl difference peaks suggests that the chromophore is in a less polar environment. The change in absorption of the protein at 287 nm as a function of Ca^{2+} concentration is shown in the inset graph (Fig. 6).

From the titration data, it is possible to identify two

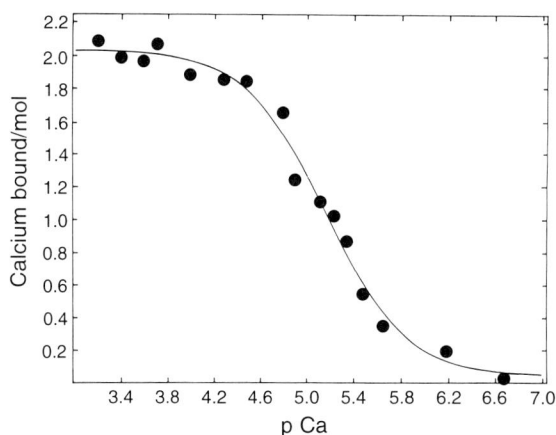

FIGURE 6 Change in molar extinction coefficient (Δε), obtained by UV difference spectroscopy, when caltropin was mixed with 1 (solid line) or 2 (dashed line) molar equivalents of Ca^{2+}/mol of protein. In both experiments, the concentration of CaT used was 0.4 mg/ml in 0.1 M Tris, pH 7.5, containing 0.1 mM EGTA. Inset: the changes in absorption (Abs) at 287 nm, measured at increasing free Ca^{2+} concentrations, are expressed as a percentage of the maximum change. Reprinted with permission from *Biochemistry* (29). Copyright 1990 American Chemical Society.

classes of calcium binding sites with K_d values of 8×10^{-7} and $6 \times 10^{-5} M$. Both far- and near-UV–CD measurements also revealed that CaT undergoes a conformational change upon Ca^{2+} binding (Mani and Kay, 1990). Addition of Ca^{2+} resulted in a nearly 15% decrease in the observed tyrosine fluorescence intensity. A plot of the relative fluorescence intensity versus Ca^{2+} added indicated the presence of two sets of Ca^{2+} binding sites on CaT with K_d values of 2×10^{-7} and $8 \times 10^{-5} M$ and these values are in good agreement with UV difference spectroscopy results.

B. Caldesmon Regulation by Caltropin

Caldesmon is a major actin binding protein associated with thin filament of smooth muscle (Sobue et al., 1981) and nonmuscle cells (Owada and Kakiuchi, 1984). When purified CaD binds to actin–tropomyosin, it inhibits actomyosin ATPase activity. CaM in the presence of Ca^{2+} can reverse caldesmon's inhibition of the actin-activated myosin ATPase. However, the affinity of CaM for CaD is rather low ($K_{aff} = 10^6 M^{-1}$), with the result that a large molar excess of CaM (~25-fold) at 25°C in 60 mM KCl is required to reverse caldesmon's inhibition (Pritchard and Marston, 1989; Shirinsky et al., 1988). On the other hand, CaT in the presence of 0.2 mM Ca^{2+} is very effective in releasing caldesmon's inhibition at 25°C. Complete recovery in the ATPase rate is achieved when 1 mol of CaT is added per mol of CaD (Mani et al., 1992). In fact, most of the inhibition was released (~90%) by the time 0.5 mol of CaT was added per mol of CaD. Thus CaT is

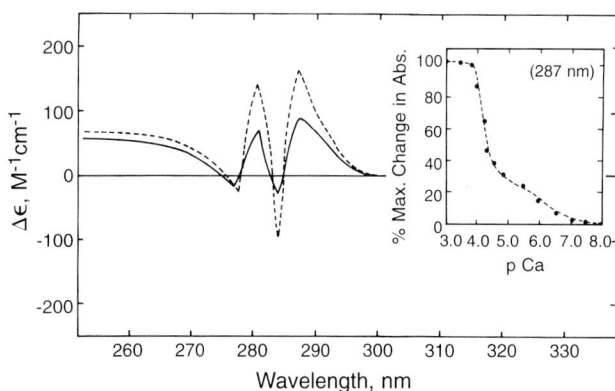

FIGURE 5 Ca^{2+} binding to caltropin. The binding assay was carried out using the calcium indicator Arsenazo III (Sigma). The solid circles represent the results from the assay. The solvent system used for the assay was 25 mM HEPES buffer, pH 7.5. The solid line represents the best fit of the data. Reprinted with permission from *Biochemistry* (29). Copyright 1990 American Chemical Society.

very potent compared to CaM. Evidence for direct binding of CaT to CaD was obtained using affinity column (Mani *et al.*, 1992). When CaD was applied to a CaT-Sepharose 4B affinity column in the presence of 1 mM Ca²⁺, most of the protein was retained by the column. The bound CaD was eluted from the column by adding 2 mM EGTA in the buffer. SDS gel electrophoresis showed that CaD alone was present in the bound fraction. These results clearly indicate that CaT–CaD interaction is calcium dependent since CaD could be eluted from this affinity column only in the presence of EGTA.

Spectroscopic techniques as well have been used in demonstrating an interaction between CaD and CaT (Mani and Kay, 1990). For fluorescence studies, the sulfhydryl groups in CaD were labeled with acrylodan. Addition of CaT to labeled CaD in the presence of Ca²⁺ produced nearly 50% increase in fluorescence intensity and the emission maximum shifted to 492 nm from 504 nm. Labeled CaD was titrated with CaT. Analysis of the titration curve by curve fitting showed that the best fit was obtained when the molar ratio of CaT to CaD in the complex at saturation was 1:1. A K_d value of $(9 \pm 2) \times 10^{-8} M$ was obtained from the titration. However, there was no evidence of any interaction between CaT and CaD in the absence of Ca²⁺.

Caltropin, which is modulated by Ca²⁺, binds to CaD in a Ca²⁺-dependent manner with high affinity. It is also very potent in regulating caldesmon's inhibition in a calcium-dependent manner and could conceivably be the Ca²⁺ factor that regulates CaD in smooth muscle.

C. Regulation of Caldesmon and Heavy Meromyosin Interaction by Caltropin

Current studies suggest a vital role for CaD in the regulation of smooth muscle contraction since it inhibits the actin-activated ATPase activity of myosin (Dabrowska *et al.*, 1985; Smith and Marston, 1985; Moody *et al.*, 1985; Sobue *et al.*, 1985) and its subfragments (Lash *et al.*, 1986; Chalovich *et al.*, 1987). However, defining the precise role of CaD in cells requires an understanding of its interaction with several key proteins, namely, actin, myosin, calmodulin, and caltropin.

CaD binds to actin and myosin tightly with affinities near 10⁷ and 10⁶ M^{-1}, respectively (Velaz *et al.*, 1989; Hemric and Chalovich, 1990). The myosin binding activity is localized in the N-terminal "27K" chymotryptic fragment of CaD (Velaz *et al.*, 1990), whereas the actin binding region(s) are largely confined to a C-terminal "35K" fragment (Szpacenko and Dabrowska, 1986; Fujii *et al.*, 1987). According to Chalovich *et al.* (1987), caldesmon acts as a competitive inhibitor of the

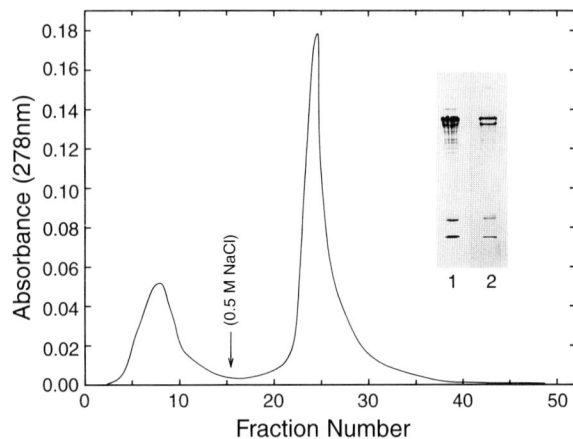

FIGURE 7 CaD-Sepharose affinity chromatography of heavy meromyosin. HMM (1.5 mg) was applied to a CaD affinity column equilibrated with buffer containing 20 mM Tris (pH 7.5), 0.5 mM DTT, 40 mM NaCl, and 0.2 mM CaCl₂. The column was washed with the buffer, and at the point indicated by the arrow the bound protein was eluted with the same buffer containing 0.5 M NaCl. The inset shows the SDS gel electrophoretic profile of control HMM (lane 1) and HMM eluted from the column with 0.5 M NaCl (lane 2). Reprinted with permission from *Biochemistry* (32). Copyright 1993 American Chemical Society.

binding of myosin to actin. Even though the myosin binding site is thought to be on the N-terminal end of CaD, the stoichiometry and the affinity with which it binds to CaD are not well established. Quantitation of the binding is important in determining if this association is specific and whether this interaction could occur under *in vivo* conditions on the basis of the affinity with which it binds to CaD. We studied the interaction between smooth muscle heavy meromyosin (HMM) and CaD using a CaD affinity column and spectroscopic techniques. Direct binding of HMM to CaD was established using affinity chromatography (Fig. 7). When HMM was applied to a CaD-Sepharose 4B column equilibrated with 20 mM Tris, pH 7.5, 40 mM NaCl, 1 mM DTT, and 0.2 mM CaCl₂, most of the protein was retained by the column. Bound HMM was eluted by increasing the NaCl concentration to 0.5 M. The SDS gel electrophoretic pattern of the protein eluted with 0.5 M NaCl is shown in the Fig. 7 inset. Also included in the figure is the gel profile obtained for the isolated HMM prior to affinity chromatography, used as a control. On the basis of the peptide composition seen on the gel, the bound protein was identified as HMM and it retained its full biological activity (Mani and Kay, 1993).

1. Circular Dichroism Studies

HMM and CaD were mixed in a molar ratio of 0.5:1.0, respectively, since two CaD molecules can bind to a molecule of HMM (Hemric and Chalovich,

1990). The experimentally observed ellipticity values deviated from the theoretical values particularly in the 222- and 207-nm-wavelength regions, suggesting that the interaction has produced a conformational change (Mani and Kay, 1993). For instance, at 222 nm the difference between the observed and the theoretical ellipticity value was nearly 1300 deg. cm^2/dmol, whereas the experimental error in these measurements is only ±300 deg. cm^2/dmol, thus clearly indicating that the interaction has produced a conformational change. For the ternary complex, CaT, CaD, and HMM were mixed in molar ratios of 1:1:0.5, respectively. Formation of the ternary complex also produced a conformational change since the observed ellipticity at 222 nm differed from the theoretical value by 2300 deg. cm^2/dmol. However, when the three proteins were mixed in the absence of Ca^{2+}, that is, in 1 mM EGTA, there was no evidence for any ternary complex formation since the observed ellipticity at 222 nm was in excellent agreement with the theoretical value.

2. Fluorescence Studies

Addition of HMM to CaD labeled with acrylodan resulted in a nearly 20% decrease in fluorescence intensity with no significant change in the emission maximum, which was centered at 517 nm (Mani and Kay, 1993). When fluorescence titration of labeled CaD was carried out with HMM, more than 70% of the fluorescence change was completed by the time 0.5 mol of HMM was added per mol of CaD (Mani and Kay, 1993). At equimolar ratio, the observed fluorescence change was greater than 80%. When Scatchard analysis was carried out, a best fit was obtained when the ratio of CaD to HMM was 2:1. This indicates the existence of two CaD binding sites on HMM, in agreement with an earlier observation of Hemric and Chalovich (1990). The slope term of the Scatchard plot yielded a K_a value of $(4.5 \pm 0.5) \times 10^7 \, M^{-1}$, suggesting that HMM binds to CaD with high affinity. Titration of HMM with the CaD–CaT complex was also carried out. HMM binds to the binary complex with lower affinity and the observed K_a value obtained was $(1.5 \pm 0.5) \times 10^7 \, M^{-1}$. The presence of CaT has lowered the affinity of HMM for CaD by nearly threefold. CaT, which is very potent in reversing the inhibitory effect of CaD in the presence of Ca^{2+}, is also able to modulate the interaction between CaD and HMM in a Ca^{2+}-dependent manner.

The CaT effect may be summarized as follows. As CaD can bind to both actin and myosin with high affinity, it is able to inhibit the actin-activated myosin ATPase activity in interfering in the actin–myosin interaction. In the presence of Ca^{2+}–CaT, the affinity of CaD for myosin is lowered nearly threefold, with the result that myosin is able to interact effectively with actin

thereby restoring the actin-activated myosin ATPase activity. Alternatively, CaD in the absence of CaT acts as a competitive inhibitor of myosin binding to actin, resulting in a lowered ATPase level since CaD can bind to actin at or near the myosin binding site, thereby blocking myosin from binding to actin. CaD, which binds to CaT in the presence of Ca^{2+} with high affinity (Mani *et al.*, 1992), undergoes a conformational change with the result that it no longer is able to complete with myosin for binding to actin, resulting in the release of the inhibition.

D. Effect of Caltropin on the Caldesmon–Actin Interaction

CaD, by virtue of its ability to interact with both myosin and actin, is able to inhibit the actin-activated myosin ATPase activity by interfering in the actin–myosin interaction. Any proposed mechanism to account for the effect of CaT on CaD should include not only the CaD–myosin interaction but also the CaD–actin interaction. Cosedimentation studies were carried out to study the binding of CaD to actin. When a CaD–actin mixture was subjected to high-speed centrifugation (100,000g) for 30 min, most of the CaD cosedimentated with F-actin whether Ca^{2+} was present or not (Mani and Kay, 1995). Addition of Ca^{2+}–CaT to actin prior to the addition of CaD had a pronounced effect on caldesmon's ability to interact with actin. Under these conditions, most of the CaD did not cosediment with actin, that is, CaD was now present in the supernatant along with caltropin. However, in the minus calcium state, that is, in the presence of 1 nM EDTA, most of the CaD pelleted along with actin, suggesting that CaT had no significant effect on CaD–actin interaction in the minus calcium state. This finding is consistent with bioassay results (Mani *et al.*, 1992) in which CaT was effective in releasing caldesmon's inhibition only in the presence of calcium.

1. Fluorescence Studies

Acrylodan-labeled CaD, when excited at 375 nm, had an emission maximum at 515 ± 2 nm. Addition of actin resulted in a blue shift in the emission maximum, which was not centered at 505 ± 2 nm. There was no significant increase in fluorescence intensity around the emission maximum, but a significant increase in relative fluorescence intensity (nearly 70%) was observed at 470 nm. During the titration of labeled CaD with actin, the fluorescence intensity at 470 nm was monitored. Maximum increase in fluorescence intensity was achieved by the time 7 mol of actin were added per mol of CaD (Fig. 8).

Also shown in Fig. 8 is the titration of actin with the CaD–CaT complex. In this instance, the maximum

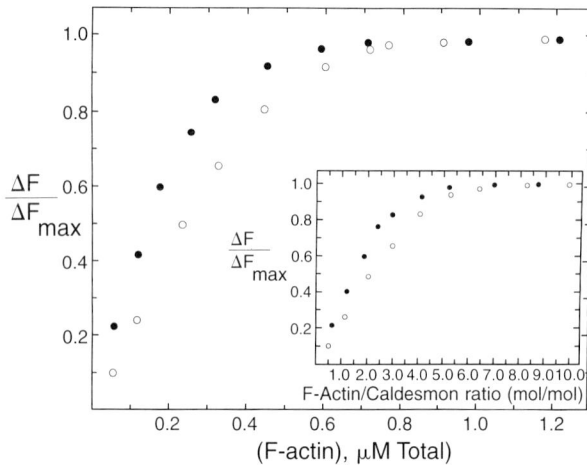

FIGURE 8 Influence of actin on acrylodan–CaD (solid circle) and acrylodan CaD–CaT fluorescence (open circle). The initial CaD concentration was $1.0 \times 10^7\,M^{-1}$. Measurements were carried out in 25 mM Tris (pH 7.5), 42 mM NaCl, 2 mM $MgCl_2$, 0.2 mM $CaCl_2$, and 1 mM DTT at 20°C. Relative change in fluorescence intensities ($\Delta F/\Delta F_{max}$) at 470 nm are plotted as a function of actin concentration. The excitation wavelength was 375 nm. Reprinted with permission from *J. Biol. Chem.* (270). Copyright 1995. The American Society for Biochemistry and Molecular Biology.

change was reached when 7 mol of actin were added. However, actin binds to the CaD–CaT complex with lower affinity. For example, for a 50% increase in fluorescence intensity, the amount of actin required for CaD and the CaD–CaT complex corresponded to 0.15 and 0.25 μM, respectively. The slope of the titration curve suggests the existence of more than one class of binding site (Mani and Kay, 1995). Analysis of the titration curve suggested the presence of three high-affinity actin binding sites, that is, out of seven actins that are required for saturation, only three are binding with high affinity. From the slope term of the Scatchard plot (Mani and Kay, 1995), a K_a value of $(6.0 \pm 0.5) \times 10^{-7}\,M^{-1}$ was calculated from three sets of titration data, suggesting that CaD binds to actin with high affinity. Actin binds to the binary complex with lower affinity [$K_a \sim (2.5 \pm 0.5) \times 10^7\,M^{-1}$]. The presence of

CaT has lowered the affinity of CaD for actin by more than twofold. CaD has been implicated to act as a competitive inhibitor of myosin binding to actin. CaD binds to actin with strong affinity at or near the myosin binding region (i.e., at the N-terminal end of actin), thus blocking myosin from binding to actin, resulting in lowered ATPase activity (Chalovich *et al.*, 1987). CaT, which binds to CaD with high affinity, undergoes a conformational change; as a consequence we believe it binds to actin with lower affinity and is no longer able to compete with myosin for binding to actin, resulting in the recovery of the ATPase activity. Graceffa and Jancso (1991) have also proposed a similar mechanism to explain the effect of Ca^{2+}–CaM on the CaD–actin interaction.

VIII. CONCLUSION

A possible fine-tuning mechanism for cooperative CaD regulation of the thin filament is shown in Fig. 9. The regulatory properties of CaD are similar to troponin in most details (Marston and Redwood, 1993). The C-terminal region of CaD, comprising amino acid residues from 658 to 756 and named domain "4b," is closely analogous to the inhibitory peptide of troponin (TN-I). The domain "3" of CaD, that is, residues from 508 to 565, is homologous to the CB_2 region of TN-T, that is, residues 71–151, which are implicated in tropomyosin binding. For this reason, in the model we have indicated region 4b and domain 3 of CaD to bind to actin and tropomyosin, respectively. TN-C (the Ca^{2+} binding subunit) is part of the troponin complex, whereas in smooth muscle the Ca^{2+} factor is not a part of CaD. CaD can bind tightly to myosin as well as to actin, and CaT has no effect on these interactions in the absence of Ca^{2+} as shown in the figure. CaM and CaT in the presence of Ca^{2+} can release caldesmon's inhibition without displacing CaD from the actin filament (Smith *et al.*, 1987; Marston and Smith, 1985; Mani *et al.*, 1992). According to this model, in the presence of

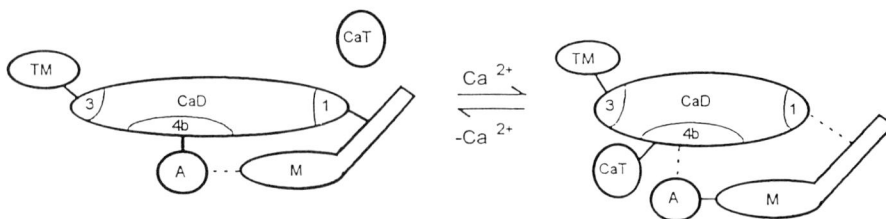

FIGURE 9 A troponinlike model for cooperative caldesmon regulation of the thin filament caldesmon (CaD), tropomyosin (TM), actin (A), and myosin (M). Domain 1 corresponds to the N-terminal region of caldesmon; domains 3 and 4b represent C-terminal amino acid residues from 508 to 565 and from 658 to 756, respectively. Caltropin (CaT). Reprinted with permission from *J. Biol. Chem.* (270). Copyright 1995. The American Society for Biochemistry and Molecular Biology.

Ca^{2+}, CaT binds to CaD with high affinity and induces a conformational change in CaD with the result that CaD binds to actin with lower affinity (~threefold) and no longer is able to compete with myosin for binding to actin, resulting in the release of the inhibition. The mode of interaction between this cooperative unit consisting of CaD–TM–actin plays a key role in regulating the smooth muscle thin filament. Even though both CaM and CaT can bind to CaD in a Ca^{2+}-dependent manner, their mode of interaction with CaD must be different, since CaT is much more potent than CaM in releasing caldesmon's inhibition. In addition, CaM binds to CaD at the C-terminal end (Zhan et al., 1991) at two sites (site A, from Met-658 to Ser-666, and site B, from Asn-675 to Lys-695) and perturbs the tryptophan residue(s) that are located in this region (trp 659 and 692). CaT is also believed to bind at the C-terminal end of CaD (Mani et al., 1992), but the binding site must be different because addition of CaT had no effect on the tryptophan fluorescence. Thus, although the two Ca^{2+} binding proteins behave similarly, there are differences in their interaction with CaD. Future studies with CaT should be directed toward identifying the binding sites on CaD, which may shed more light in understanding its mode of interaction.

References

Adelstein, R. S., and Eisenberg, E. (1980). Annu. Rev. Biochem. 49, 921–956.

Ali, S. M., Geisow, M. J., and Burgoyne, R. D. (1989). Nature (London) 340, 313–315.

Babu, Y. S., Sack, J. S., and Cook, W. J. (1985). Nature London) 315, 37–40.

Babu, Y. S., Bugg, C. E., and Cook, W. J. (1988). J. Mol. Biol. 204, 191–204.

Chalovich, J. M. (1993). Pharmacol. Ther. 55, 95–148.

Chalovich, J. M., Cornelius, P., and Benson, C. E. (1987). J. Biol. Chem. 262, 5711–5716.

Chattopadhyaya, R., Meador, W. E., Means, A. R., and Quiocho, F. A. (1992). J. Mol. Biol. 228, 1177–1192.

Chen, Y., Yang, J. T., and Chan, K. H. (1974). Biochemistry 13, 3350–3359.

Cheung, W. Y. (1970). Biochem. Biophys. Res. Commun. 38, 533–538.

Crompton, M. R., Moss, S. E., and Crompton, M. J. (1988). Cell (Cambridge, Mass.) 55, 1–3.

Dabrowska, R., Goch, A., Galazkiewicz, B., and Osinska, H. (1985). Biochim. Biophys. Acta 842, 70–75.

Donovan, J. W. (1969). In "Physical Principles and Techniques of Protein Chemistry" (S. J. Leach, ed.), pp. 101–170. Academic Press, New York.

Edsall, J. T., and Gutfreund, H. (1983). "Biothermodynamics," pp. 157–207. Wiley, New York.

Fujii, T., Imai, M., Rosenfeld, G. C., and Bryan, J. (1987). J. Biol. Chem. 262, 16155–16160.

Fürst, D. O., Cross, R. A., Demey, R. A., and Small, J. V. (1986). EMBO J. 5, 251-257.

Geisow, M. J. (1986). FEBS Lett. 203, 99–103.

Goodman, M., and Toniolo, C. (1968). Biopolymers 6, 1673–1689.

Graceffa, P., and Jancso, A. (1991). J. Biol. Chem. 266, 20305–20310.

Grand, R. J. A., and Perry, S. V. (1978). FEBS Lett. 92, 137–142.

Hemric, M. E., and Chalovich, J. M. (1990). J. Biol. Chem. 265, 19672–19678.

Ikura, M., Hiraoki, T., Hikichi, K., Minowa, O., Yamaguchi, H., Yazawa, M., and Yagi, K. (1984). Biochemistry 23, 3124–3128.

Ikura, M., Kay, L. E., and Bax, A. (1990). Biochemistry 29, 4659–4667.

Ikura, M., Clore, G. M., Gronenborn, A. M., Zhu, G., Klee, C. B., and Bax, A. (1992). Science 256, 632–638.

Kaetzel, M. A., and Dedman, J. R. (1989). J. Biol. Chem. 264, 14463–14470.

Kakiuchi, S., Yamazaki, R., and Nakajima, H. (1970). Proc. Jpn. Acad. 46, 589–594.

Kretsinger, R. H., and Nockolds, C. E. (1973). J. Biol. Chem. 248, 3313–3326.

Lash, J. A., Sellers, J. R., and Hathaway, D. R. (1986). J. Biol. Chem. 261, 16155–16160.

Manalan, A. S., and Klee, C. B. (1984). Adv. Cyclic Nucleotide Protein Phosphorylation Res. 18, 227–278.

Mani, R. S., and Kay, C. M. (1983). biochemistry 22, 3902–3907.

Mani, R. S., and Kay, C. M. (1987). In "Methods in Enzymology" (A. Means and P. Conn, eds.), Vol. 139, pp. 168–187. Academic Press, Orlando, FL.

Mani, R. S., and Kay, C. M. (1989). Biochem. J. 259, 799–804.

Mani, R. S., and Kay, C. M. (1990). Biochemistry 29, 1398–1404.

Mani, R. S., and Kay, C. M. (1992). Arch. Biochem. Biophys. 296, 442–449.

Mani, R. S., and Kay, C. M. (1993). Biochemistry 32, 11217–11223.

Mani, R. S., and Kay, C. M. (1995). J. Biol. Chem. 270, 6658–6663.

Mani, R. S., Boyes, B. E., and Kay, C. M. (1982). Biochemistry 21, 2607–2612.

Mani, R. S., McCubbin, W. D., and Kay, C. M. (1992). Biochemistry 31, 11896–11901.

Marston, S. B., and Redwood, C. S. (1991). Biochem. J. 231, 5167–522.

Marston, S. B., and Redwood, C. S. (1992). J. Biol. Chem. 267, 16796–16800.

Marston, S. B., and Redwood, C. S. (1993). J. Biol. Chem. 268, 12317–12320.

Marston, S. B., and Smith, C. W. J. (1985). J. Muscle Res. Cell Motil. 6, 669–708.

Marston, S. B., Pinter, K., and Bennett, P. (1992). J. Muscle Res. Cell Motil. 13, 206–218.

Matthew, J. R., Krokak, J. M., and Dedman, J. R. (1986). J. Cell. Biochem. 32, 223–234.

McClure, W. O., and Edelman, G. M. (1966). Biochemistry 5, 1908–1919.

Meador, W. E., Means, A. R., and Quiocho, F. A. (1992). Science 257, 1251–1255.

Means, A. R. (1988). Recent Prog. Horm. Res. 44, 223–286.

Moody, C. J., Marston, S. B., and Smith, C. W. J. (1985). FEBS Lett. 191, 107–112.

Moore, P. B. (1986). Biochem. J. 238, 49–54.

Moore, P. B., and Dedman, J. R. (1982). J. Biol. Chem. 257, 9663–9667.

Olwin, B. B., Edelman, A. M., Krebs, E. G., and Storm, D. R. (1984). J. Biol. Chem. 259, 10949–10955.

O'Neil, K. T., and DeGrado, W. F. (1990). Trends Biochem. Sci. 15, 59–64.

Owada, M. K., and Kakiuchi, S. (1984). Proc. Natl. Acad. Sci. U.S.A. 81, 3133–3137.

Pritchard, K., and Marston, S. B. (1989). Biochem. J. 257, 839–843.

Seamon, K. B. (1980). Biochemistry 19, 207–215.

Shirinsky, V. P., Bushueva, T. C., and Frolova, S. I. (1988). Biochem. J. 255, 203–208.

Smith, C. W., and Marston, S. B. (1985). *FEBS Lett.* **184,** 115–119.

Smith, C. W., Pritchard, K., and Marston, S. B. (1987). *J. Biol. Chem.* **262,** 116–122.

Sobieszek, A., and Bremel, R. D. (1975). *Eur. J. Biochem.* **55,** 49–60.

Sobue, K., and Sellers, J. R. (1991). *J. Biol. Chem.* **266,** 12115–12118.

Sobue, K., Muramoto, Y., Fujita, M., and Kakiuchi, S. (1981). *Proc. Natl. Acad. Sci. U.S.A.* **78,** 5652–5655.

Sobue, K., Takahashi, K., and Wakabayashi, I. (1985). *Biochem. Biophys. Res. Commun.* **132,** 645–651.

Sommerville, L. E., and Hartshorne, D. J. (1986). *Cell Calcium* **7,** 353–364.

Spera, S., Ikura, M., and Bax, A. (1991). *J. Bimol. NMR* **1,** 155–165.

Szpacenko, A., and Dabrowska, R. (1986). *FEBS Lett.* **202,** 182–186.

Velaz, L., Hemric, M. E., Benson, C. E., and Chalovich, J. M. (1989). *J. Biol. Chem.* **264,** 9602–9610.

Velaz, L., Ingraham, R. H., and Chalovich, J. M. (1990). *J. Biol. Chem.* **265,** 2929–2934.

Walsh, M. P., Stevens, F. C., Oikawa, K., and Kay, C. M. (1979). *J. Biochem. (Tokyo)* **57,** 267–278.

Wang, J. H., Teo, T. S., Ho, H. C., and Stevens, F. C. (1975). *Adv. Cyclic Nucleotide Res.* **5,** 179–194.

Watterson, D. M., Sharief, F., and Vanaman, T. C. (1980). *J. Biol. Chem.* **255,** 962–975.

Williams, D. A., and Fay, F. S. (1986). *Am. J. Physiol.* **250,** C779–C779.

Yagi, K., Yazawa, M., Ikura, M., and Hikichi, K. (1989). *In* "Calcium Protein Signalling" (H. Hidaka, E. Carafoli, A. R. Means, and T. Tanaka, eds.), pp. 147–154. Plenum, New York.

Zhan, Q., Wong, S. S., and Wang, C.-L.A. (1991). *J. Biol. Chem.* **266,** 21810–21814.

ENZYMES OF PROTEIN PHOSPHORYLATION– DEPHOSPHORYLATION

Myosin Light Chain Kinase

JAMES T. STULL, JOANNA K. KRUEGER, KRISTINE E. KAMM,
ZHONG-HUA GAO, GANG ZHI, and ROANNA PADRE

Department of Physiology
The University of Texas Southwestern Medical Center at Dallas
Dallas, Texas

I. INTRODUCTION

Ca²⁺ is a key second messenger involved in many different signaling pathways in eukaryotic cells. Increases in cytosolic Ca^{2+} concentrations occur principally by Ca^{2+} influx through plasma membrane Ca^{2+} channels or through release from internal storage compartments (Fig. 1). The binding of Ca^{2+} to a calmodulin, a modulator protein found in all eukaryotic cells, leads to a conformation capable of binding to an inactive myosin light chain kinase (MLCK). This binding results in isomerization of MLCK to an active form, which subsequently phosphorylates serine 19 of the myosin regulatory light chain (RLC). For myosin from smooth and nonmuscle cells, RLC phosphorylation markedly increases actin-activated myosin MgATPase activity. Myosin RLC phosphorylation plays a central role in many cellular processes, including initiation of smooth muscle contraction (Kamm and Stull, 1985a; Hartshorne, 1987), platelet aggregation (Sellers and Adelstein, 1987), endothelial cell retraction (Sheldon *et al.*, 1993), exocytosis (Choi *et al.*, 1994), and lymphocyte receptor capping (Kerrick and Bourguignon, 1984). In skeletal and cardiac muscle, the actin-activated myosin MgATPase activity is maximal, even in the absence of RLC phosphorylation. However, RLC phosphorylation by a unique MLCK appears to move the cross-bridge away from the thick filament backbone, leading to potentiation of skeletal muscle contraction (Sweeney *et al.*, 1993).

Decreases in cytosolic Ca^{2+} concentrations result in the dissociation of calmodulin from MLCK and con-version of the kinase to an inactive enzyme. With kinase inactivation, RLC is dephosphorylated by a myosin light chain phosphatase, protein phosphatase-1M (MLCP), localized to the contractile elements (Shimizu *et al.*, 1994; Shirazi *et al.*, 1994).

Although this simplified scheme predicts a unique Ca^{2+}-RLC phosphorylation relationship, it is clear that other modulatory influences alter the Ca^{2+} sensitivity of RLC phosphorylation in smooth muscle by affecting the activities of MLCK and MLCP, respectively (Stull *et al.*, 1993; Somlyo and Somlyo, 1994; also see Kamm and Grange, Chapter 27, this volume).

This chapter will focus on the biochemical properties of MLCK in relation to its activation in smooth muscle and to recent insights into the molecular structure of the catalytic core and its regulation by an autoinhibitory region and calmodulin binding domain.

II. MYOSIN LIGHT CHAIN KINASE ACTIVITY *IN VIVO*

The amount of MLCK in smooth muscle tissues is similar to or greater than that found in skeletal muscle, which is considerably greater than that found in nonmuscle cells and cardiac muscle (Stull *et al.*, 1986). The average value is approximately 3–4 μM, which is significantly less than 70–80 μM myosin RLC, its only physiological substrate (Stull *et al.*, 1986; Hartshorne, 1987). It is predicted from enzyme kinetics and substrate concentrations that there is sufficient MLCK for the rapid phosphorylation of myosin RLC and, hence,

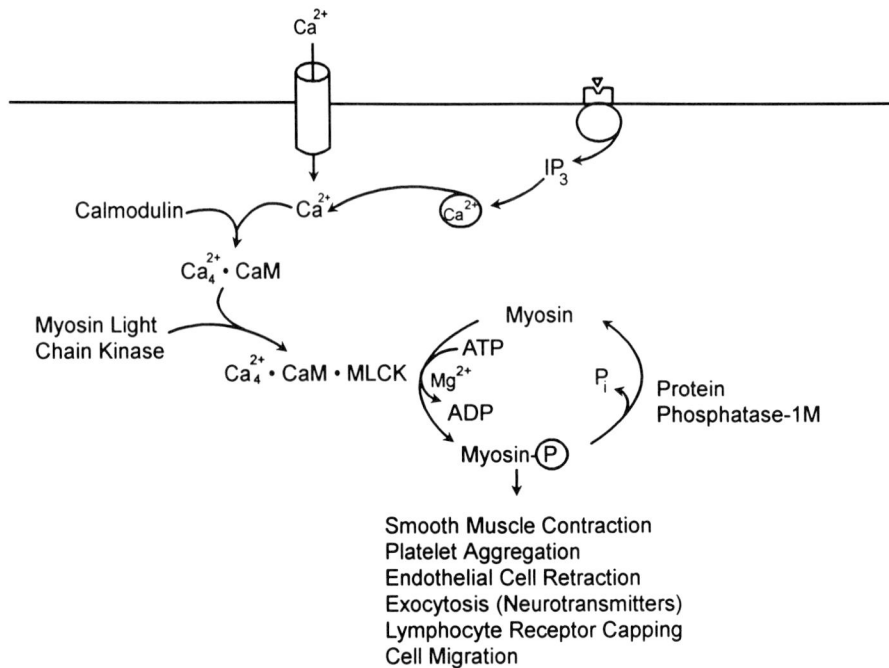

FIGURE 1   Scheme for activation of MLCK in cells.

force development in smooth muscle (Kamm and Stull, 1985a).

A. *Activation Kinetics*

We have developed a preparation of smooth muscle that is sufficient for studying kinetic events in the activation of contraction (Kamm and Stull, 1985b). Myosin RLC phosphorylation and force development are rapid and nearly maximal when tracheal smooth muscle strips are stimulated by acetylcholine release during selective electric field stimulation of intrinsic cholinergic nerves. The rate of the contraction is independent of tissue cross-sectional area, indicating that field stimulation avoids most of the diffusion problems delaying activation of cells in agonist-stimulated tissues. An automated, rapid-release electronic freezing device, coupled with the ability to simultaneously measure $[Ca^{2+}]_i$ and contraction, allows temporal resolution of cellular processes involved in initiating contraction in the millisecond time range (Kamm and Stull, 1986; Word *et al.*, 1994).

Stimulation of acetylcholine release results in small but significant increases in cytosolic Ca^{2+} concentrations within 100 ms (Fig. 2). The cytosolic Ca^{2+} concentration continues to increase for 3 s of continuous stimulation, reaching values of 225 nM relative to the resting value of 115 nM (Word *et al.*, 1994). The

increase in cytosolic Ca^{2+} concentration results in calmodulin (CaM) binding and activation of the calmodulin-dependent enzyme phosphodiesterase by 500 ms (Miller-Hance *et al.*, 1988). For technical reasons, it was not possible to assess the fractional activation of MLCK. Since MLCK and phosphodiesterase have a similar affinity for calmodulin ($K_{CaM} = 1$ nM), the fractional activation of phosphodiesterase may reflect the activation properties of MLCK. The activation of phosphodiesterase occurs after small increases in cytosolic Ca^{2+} concentrations and remains at its maximal extent from 500 ms to 3 s (Fig. 2). Biochemical measurements show that calmodulin binds MLCK rapidly ($10^7 M^{-1} s^{-1}$), which is close to the limiting value for a diffusion-controlled association (Bowman *et al.*, 1992; Kasturi *et al.*, 1993) and is consistent with the proposal that activation precedes RLC phosphorylation.

Not until 500 ms following neurostimulation is there a significant increase in RLC phosphorylation, coincident with the latency in force development (Fig. 2). However, after this latency, myosin RLC phosphorylation and stiffness both increase more rapidly than either force or cytosolic Ca^{2+} concentrations (Fig. 2). Interestingly, the kinetics of myosin RLC phosphorylation are described as a pseudo-first-order rate of 1 s^{-1}, showing no evidence of an ordered or cooperative phosphorylation of myosin RLC (Kamm and Stull,

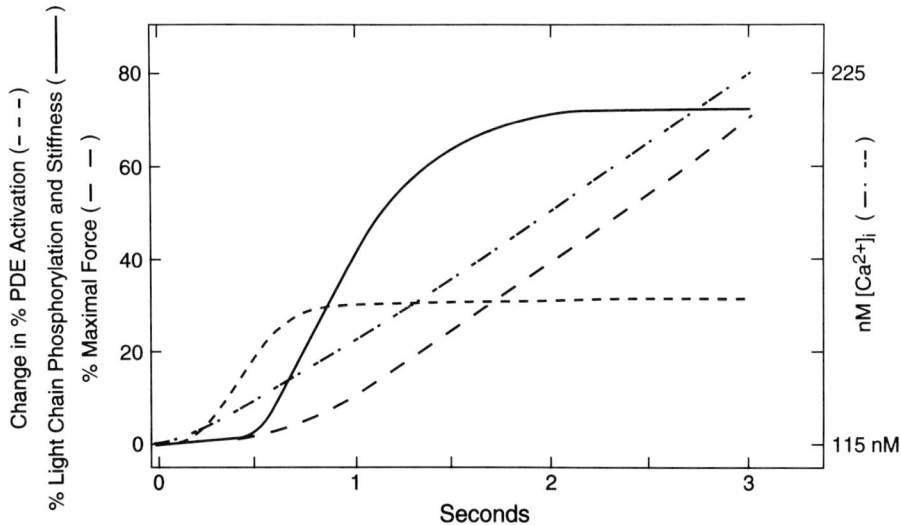

FIGURE 2 Kinetic properties of cellular processes leading to MLCK activation in neurally stimulated tracheal smooth muscle.

1986). The rate of phosphorylation is greater than the estimated rate of dephosphorylation, $0.25\ s^{-1}$ (Kamm and Stull, 1985b). Measurements of nonphosphorylated, monophosphorylated, and diphosphorylated forms of myosin in tracheal smooth muscle tissue provide direct evidence that myosin RLC phosphorylation occurs as a random rather than an ordered or cooperative phosphorylation process (Persechini *et al.*, 1986). These results are consistent with biochemical measurements (Sellers *et al.*, 1983; Trybus and Lowey, 1985).

Both RLC phosphorylation and active stiffness increase more rapidly than isometric force during the initiation of the contraction (Kamm and Stull, 1986). These observations suggest that phosphorylation of myosin RLC allows cross-bridge attachment to actin. The delay in force development may result from cooperative effects of phosphorylation on activation whereby force depends on formation of doubly phosphorylated myosin (Persechini and Hartshorne, 1981; Sellers *et al.*, 1983); however, other contributions, including a delay in the expression of force through series elastic element in the tissue, cannot be excluded (Aksoy *et al.*, 1983).

Although purified smooth muscle myosin RLC may be phosphorylated by other protein kinases *in vitro* (Kamm and Stull, 1989), direct measurements of the sites of phosphorylation in myosin RLC during contractions in tracheal smooth muscle tissues show that there is only one prominent site of phosphorylation, serine 19 (Colburn *et al.*, 1988; Kamm *et al.*, 1989). The multifunctional Ca^{2+}/calmodulin-dependent protein kinase II may also phosphorylate serine 19; however, inhibition of this kinase activity in smooth muscle cells does not diminish RLC phosphorylation (Tansey *et al.*, 1992, 1994).

B. Myosin Light Chain Kinase Inhibitors

The identification and development of specific inhibitors of MLCK could provide valuable experimental tools for exploring physiological functions of MLCK and myosin phosphorylation in smooth muscle and nonmuscle cells. Although inhibitors that act by binding calmodulin have been used experimentally, their usefulness is limited because they inhibit other calmodulin-dependent enzymes in tissues and cells (Asano and Stull, 1985; Nakanishi *et al.*, 1992). Therefore, attempts have been made to develop novel reagents that inhibit MLCK activity directly. ML-9 [1-(5-chloronaphthalenesulfonyl)-1*H*-hexahydro-1, 4-diazepine] inhibits purified smooth muscle MLCK competitively with respect to ATP (Saitoh *et al.*, 1987). Predictably, it inhibits RLC phosphorylation in actomyosin, skinned fibers, and smooth muscle strips (Ishikawa *et al.*, 1988). However, ML-9 also inhibits other protein kinases, so its specificity under different experimental conditions needs to be established.

New experimental strategies for inhibiting kinases have included the use of autoinhibitory domains from specific protein kinases (Kemp *et al.*, 1991). The autoinhibitory sequences are similar in amino acid composition to the substrates of the kinase and block catalytic activity in the presence of Ca^{2+}/calmodulin. Such peptides have been useful in examining MLCK phosphorylation in skinned fibers (Kargacin *et al.,* 1990) and intact cells (Ito *et al.,* 1989). Microinjection of a constitutively active, proteolyzed form of MLCK causes contraction of isolated smooth muscle cells with no change in the Ca^{2+} concentration (Ito *et al.,* 1989). This kinase is inhibited with a pseudosubstrate inhibitor peptide. The autoinhibitory domain peptides contain basic residues resembling the natural substrate recognition sequence for the kinase without the phosphorylatable serine. One potential problem, however, is that many protein kinases share some of the same basic amino acid determinants for substrate recognition, which could limit specificity. For example, Smith *et al.* (1990) found that the autoinhibitory domain of smooth muscle MLCK inhibited both protein kinase C and MLCK with similar potency. These observations emphasize the need to establish specificity of inhibition with appropriate control experiments even with peptide inhibitors.

Wortmannin is produced by the fungal strain *Talaromyces wortmannin* KY12420, which inhibits smooth muscle MLCK competitively and irreversibly with respect to ATP (Nakanishi *et al.,* 1992). At concentrations less than 1 μM, MLCK activity was inhibited, whereas at a 10-fold higher concentration there was no effect on the activities of other protein kinases. Wortmannin inhibited myosin RLC phosphorylation and contraction in rat aorta. These initial experiments demonstrated the potential usefulness of wortmannin as an experimental reagent that selectively inhibits MLCK and not other protein kinases. Unfortunately, it also inhibits other enzymes, including, in particular, phosphatidylinositol 3-kinase (Nakanishi *et al.,* 1994; Kimura *et al.,* 1994).

III. PURIFICATION AND ASSAY

Because of MLCK's central role in smooth muscle contraction and various nonmuscle cell processes, many investigations have focused on the biochemical and molecular properties of purified MLCK. Purification schemes for MLCK from mammalian smooth muscle tissues generally involve modifications of the procedure originally described for avian smooth muscle MLCK (Adelstein and Klee, 1981). The kinase has been purified from many types of mammalian smooth muscles, including trachea, aorta, myometrium, and stomach (Stull *et al.,* 1986).

The following procedure describes a purification scheme for MLCK from bovine tracheal smooth muscle (Stull *et al.,* 1990). All purification procedures are performed at 4°C. Bovine tracheal smooth muscle is dissected, frozen in liquid nitrogen, and stored at −60°C. Frozen tissues (500 g) are homogenized in 2 liters of 40 mM 3-(N-morpholino)propanesulfonic acid (MOPS), 5 mM ethylene-bis(oxyethylenenitrilo)tetraacetic acid (EGTA), 50 mM MgCl$_2$, 1 mM dithiothreitol, 0.1 mM phenylmethylsulfonyl fluoride, 0.1 mM N$^\alpha$-*p*-tosyl-L-lysine chloromethyl ketone, 0.1 nM L-1-tosylamido-2-phenylethyl chloromethyl ketone, 8 mg/liter lima bean trypsin inhibitor, and 10 μM leupeptin at pH 7.2. After centrifugation at 10,000× *g* for 40 min, the supernatant fraction is diluted with 20 mM MOPS, 1 mM EGTA, 1mM EDTA, 1 mM EGTA, and 1 mM dithiothreitol at pH 7.0 to less than 2.5 mS. Diethylaminoethyl cellulose (DE-52) equilibrated in the same buffer (150 g dry wet) is added and stirred for 1 hr. After washing the resin on a scintered glass funnel, the kinase is eluted by bringing the conductivity of the resin in 100 ml buffer to 15 mS with solid NaCl. The solution is collected, diluted to a conductivity of less than 2.5 mS, and added to DE-52 (50 g dry) equilibrated in 20 mM MOPS, 1 mM EDTA, and 1 mM dithiothreitol at pH 7.0 and containing protease inhibitors as listed in the foregoing. After stirring for 1 hr, the resin is drained and washed two times on a scintered glass funnel and poured into a column. The kinase is eluted with a 1.5-liter gradient containing 0–300 mM NaCl in the equilibration buffer. Fractions containing kinase activity are pooled, brought to 500 mM NaCl by the addition of solid NaCl, and concentrated in an Amicon Concentrator (PM-30 filter). MLCK activity is measured as described by Blumenthal and Stull (1980). The concentrate is brought to 10% sucrose and is applied to a Biogel A-1.5-m column (5 × 90 cm) that is equilibrated with 20 mM MOPS, 0.5 M NaCl, 1 mM EDTA, 1 mM dithiothreitol, 0.1 mM EGTA, 0.1 mM N$^\alpha$-*p*-tosyl-L-lysine chloromethyl ketone, and 10 μM leupeptin at pH 7.0. Active kinase fractions are pooled and dialyzed against 20 mM MOPS, 1 mM EDTA, 1 mM EGTA, 1 mM dithiothreitol, and 50 mM NaCl at pH 7.0. The dialyzed pool is applied to a phosphocellulose column (5 × 4 cm) equilibrated in the same buffer. MLCK is eluted with 400 ml equilibration buffer plus 0.4 M NaCl.

The kinase fraction is brought to 2 mM CaCl$_2$, 2 mM magnesium acetate, and 1 mM dithiothreitol at pH 7.0 and applied to a calmodulin-Sepharose column (2.5 × 10 cm). The column is washed extensively with 10 mM MOPS, 1 mM CaCl$_2$, 2 mM magnesium acetate, 0.2 M NaCl, 1 mM dithiothreitol, 10 μM leupeptin, and 0.1 mM phenylmethylsulfonyl fluoride at pH 7.0. MLCK is eluted with wash buffer containing 10 mM EGTA and dialyzed against 10 mM potassium phosphate and

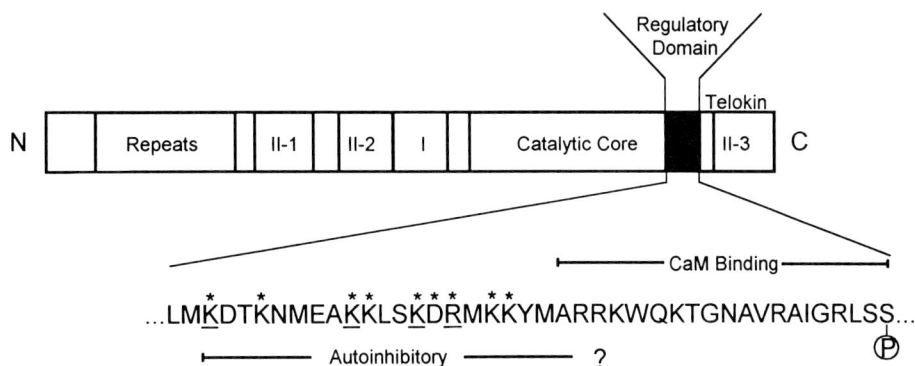

FIGURE 3 Domain organization of mammalian smooth/nonmuscle MLCK. Residues in the regulatory domain have been examined individually by mutagenesis for involvement in autoinhibition (asterisk). Four basic residues (underlined) may contribute to autoinhibition by binding to acidic residues in the catalytic core. The sequence of the calmodulin binding domain is noted. The original pseudosubstrate sequence (. . . RRKWQKTGNAV . . .) is in the N terminus of the calmodulin binding domain. The serine phosphorylated *in vivo* by the multifunctional Ca^{2+}/calmodulin-dependent protein kinase II is identified with the circled "P." The phosphorylation desensitizes MLCK to activation by Ca^{2+}/calmodulin.

1 m*M* dithiothreitol at pH 6.8. The dialyzed sample is applied to a hydroxylapatite column (1 × 4 cm). After washing the column in the same buffer, the kinase is eluted with a 10 to 300 m*M* potassium phosphate gradient containing 1 m*M* dithiothreitol at pH 6.8 in a total volume of 120 ml. The fractions containing MLCK activity are brought to 10% glycerol and stored at −60°C.

IV. BIOCHEMICAL AND MOLECULAR PROPERTIES OF MYOSIN LIGHT CHAIN KINASE

Ca^{2+}/calmodulin-dependent MLCKs have been cloned and sequenced from chicken (Olson *et al.*, 1990), rabbit (Gallagher *et al.*, 1991), and bovine (Kobayashi *et al.*, 1992) smooth muscles, in addition to rabbit (Herring *et al.*, 1992) and chicken (Leachman *et al.*, 1992) skeletal muscles. Structural analyses as well as differences in enzymatic properties demonstrate that there are at least two distinct genes for MLCK, represented by the smooth/nonmuscle versus the skeletal muscle MLCK. The domain organization of the smooth muscle MLCK is similar, although not identical. The first 80 residues are highly conserved in all three smooth muscle kinases (Fig. 3) and contain an actin binding domain (Kanoh *et al.*, 1993). The chemical identification of the N-terminal sequence of the chicken smooth muscle MLCK is consistent with the predicted start site for expression of all three kinases (Faux *et al.*, 1993). Furthermore, the expression of

these clones results in protein with the same molecular mass as that found in smooth muscle tissues and nonmuscle cells (Gallagher *et al.*, 1991). These results are not consistent with the report that the chicken nonmuscle MLCK contains a 286-residue extension at the N terminus beyond the highly conserved sequence (Shoemaker *et al.*, 1990). The calculated mass of this form of MLCK is 165 kDa, which is considerably larger than the 130-kDa molecular mass measured by SDS–PAGE for the avian smooth and nonmuscle MLCK (Gallagher *et al.*, 1991).

Following the N-terminal 80 residues, there is a tandem repeat of 12 amino acids in the rabbit and bovine MLCKs that is not present in chicken smooth muscle MLCK (Fig. 3). There are some differences in the size of the repeat in the two mammalian enzymes. Furthermore, the identity of the amino acid residues in this region is about 70%, which is the least conserved portion in smooth muscle MLCKs. The higher molecular masses of the mammalian MLCKs compared to the avian smooth muscle MLCK are primarily due to the presence of these tandem repeat sequences.

Following the tandem repeat sequences, four unc structural motifs are present in all three smooth muscle MLCKs but not in the skeletal muscle enzymes. The unc motifs are repeat structures of approximately 100 amino acids found in twitchin, the *Caenorhabditis elegans* unc-22 gene product. Motif I is related to the type III module of fibronectin (Campbell and Spitzfaden, 1994) and is N-terminal of the catalytic core (Fig. 3). The general topology of this fibronectin module contains seven antiparallel β-strands arranged in two sheets, enclosing a core of highly conserved hydrophobic residues. Motif II belongs to the C II set of the immunoglobulin superfamily. It is repeated three times in the smooth muscle MLCK, two of which are between the tandem repeat sequences and motif I (Fig. 3). The third motif II is located at the C terminus of MLCK. Interestingly, motif II-3 is expressed as an independent protein, telokin, in some smooth muscle tissues (Ito *et al.*, 1989; Gallagher and Herring, 1991). The expression of telokin is due to an alternate promotor in the MLCK gene (Gallagher and Herring, 1991). Telokin is an acidic protein containing a total of 154 amino acid residues, 103 visible in the crystal structure consisting of seven strands of antiparallel β-pleated sheets that form a barrel (Holden *et al.*, 1992). The function of telokin has not been elucidated, although this motif may be responsible for the binding of the kinase to myosin filaments (Shirinsky *et al.*, 1993).

MLCK contains a catalytic core in the central portion of the enzyme that is highly homologous to other protein kinases (Figs. 3 and 4). The crystal structure of several protein kinases has been elucidated, including cAMP-dependent protein kinase (Knighton *et al.*, 1991), cyclin-dependent kinase II (DeBondt *et al.*, 1993), MAP kinase (Zhang *et al.*, 1994), and twitchin (Hu *et al.*, 1994). The catalytic cores of these kinases contain two lobes with the smaller lobe binding MgATP, leaving the γ-phosphate positioned for catalytic transfer to the RLC. Phosphorylation occurs at a cleft between the two lobes, with the larger lobe providing binding sites for protein substrates.

C-terminal of the catalytic core is the autoregulatory domain which contains a sequence connecting the catalytic core to the calmodulin binding domain (Fig. 3). This portion of MLCK appears to fold back onto the catalytic core, thereby blocking substrate phosphorylation (Kemp *et al.*, 1994). The inactive MLCK is converted to an active form upon Ca^{2+}/calmodulin binding, presumably due to the specific displacement of this regulatory region allowing RLC binding and phosphorylation. This type of intramolecular autoinhibition has been referred to as intrasteric regulation (Kemp *et al.*, 1994). A more detailed description of molecular structures involved in autoregulation and catalysis follows.

A. Calmodulin Binding Domain

Calmodulin is a 16.8-kDa protein with a dumbbell-shaped structure in which a pair of EF Ca^{2+} binding sites are contained in each lobe (Babu *et al.*, 1988). In the crystal structure the overall length of calmodulin is 65 Å, and the two lobes are connected by a seven-turn

α-helix. The central helix tethers the two lobes and its flexibility allows them to wrap around the amphiphilic peptide analog of the calmodulin binding region of smooth muscle MLCK (Fig. 5) (Meador *et al.*, 1992). The structure is more compact than calmodulin alone and has the shape of an ellipsoid. The bound calmodulin forms a tunnel diagonal to its long axis that contains the helical peptide. The hydrophobic side of the amphiphilic α-helical peptide is in close contact to the hydrophobic regions of calmodulin in the two lobes of calmodulin. Acidic residues at the binding surface extremities interact strongly with the basic residues in the peptide (Afshar *et al.*, 1994). Thus, both electrostatic and van der Waals' interactions contribute to the high-affinity binding of 185 contacts between the calmodulin binding domain peptide and calmodulin, in which 80% are van der Waals' contacts and 15% are hydrogen bonds (Meador *et al.*, 1992). All seven basic residues in the calmodulin binding peptide form salt bridges with calmodulin. The arginine at the C terminus of the peptide interacts by hydrogen bonding and charge coupling with the bend in the central helix as well as with helices preceding and following the bend. These interactions stabilize the bent structure and probably explain why this arginine is crucial for calmodulin binding (Bagchi *et al.*, 1992a).

Evidence obtained with tryptic fragments of calmodulin suggests that the calmodulin lobes make distinct contributions to binding and activation of smooth muscle MLCK (Persechini *et al.*, 1994). Furthermore, the tethering function of the central helix can be mimicked by high concentrations of both fragments. It is important to note that MLCK activation involves more than simply calmodulin binding. Mutation of a few residues in domains 1 and 3 convert calmodulin from an activator to a competitive inhibitor (Van Berkum and Means, 1991; Su *et al.*, 1994).

Specific hydrophobic and electrostatic interactions are essential for calmodulin binding and activation in smooth muscle MLCK, including, in particular, the Trp at the N terminus and the most C-terminal Ile (Figs. 3 and 5) (Shoemaker *et al.*, 1990; Bagchi *et al.*, 1992a; Fitzsimons *et al.*, 1992; Matsushima *et al.*, 1994). These residues are underlined in the following sequence for the calmodulin binding region: ARRKWQKTGHAVRAIGRL. As noted by Bagchi *et al.* (1992a), more amino acids toward the C-terminal end of the calmodulin binding domain are critical for the calmodulin-dependent activation of the kinase than the ones present toward the N-terminal region. Mutations in the residues marked with an asterisk allow activation, but the concentration of Ca^{2+}/calmodulin required for activation is increased 15- to 45-fold (Fitzsimons *et al.*, 1992).

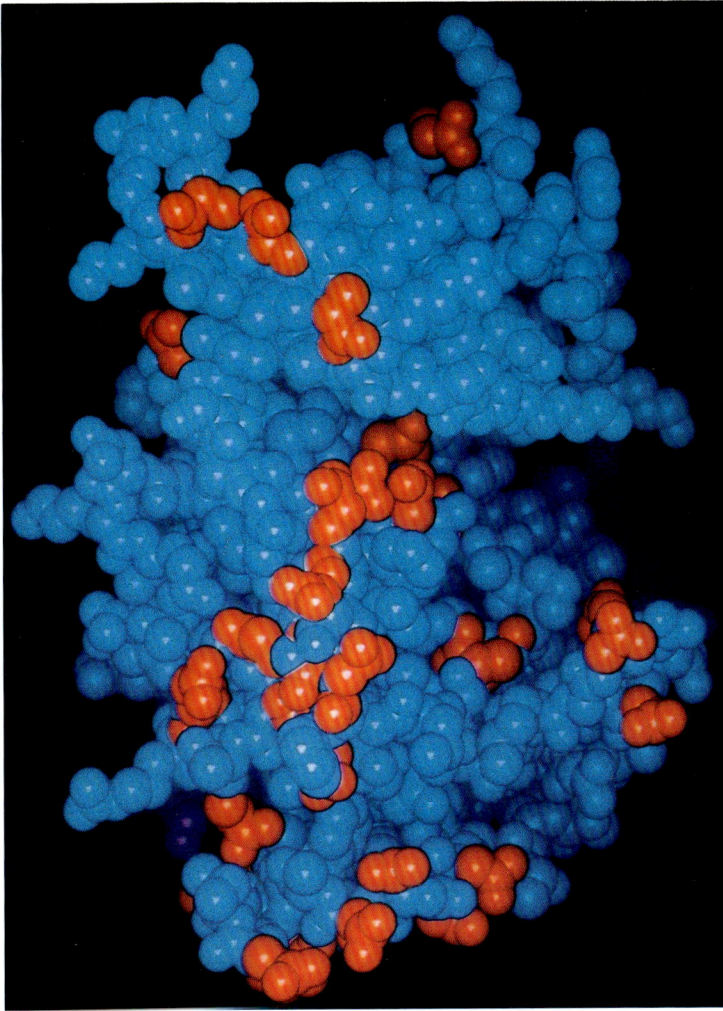

FIGURE 4 Space-filling model of the catalytic core of rabbit smooth/nonmuscle MLCK. The model displayed is derived from the chicken smooth muscle catalytic core model (Knighton *et al.*, 1992). Acidic residues in the smaller ATP binding lobe and larger light chain binding lobe are red.

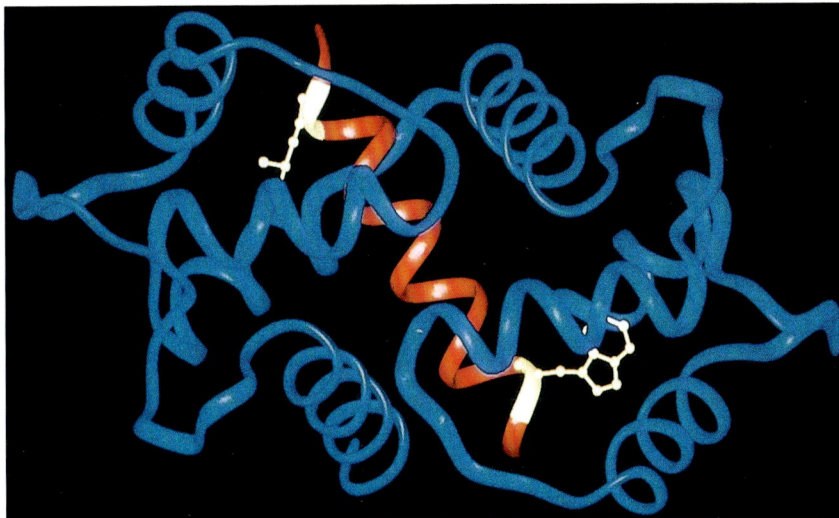

FIGURE 5 View of the α-carbon backbone of calmodulin (blue) bound to calmodulin binding domain peptide (red) of smooth muscle MLCK. This view is derived from the crystal structure (Meador *et al.*, 1992) as deposited in the Brookhaven Protein Database (reference number 1CDL). The peptide includes hydrophobic residues (yellow), Trp and Leu, at its N and C termini, respectively.

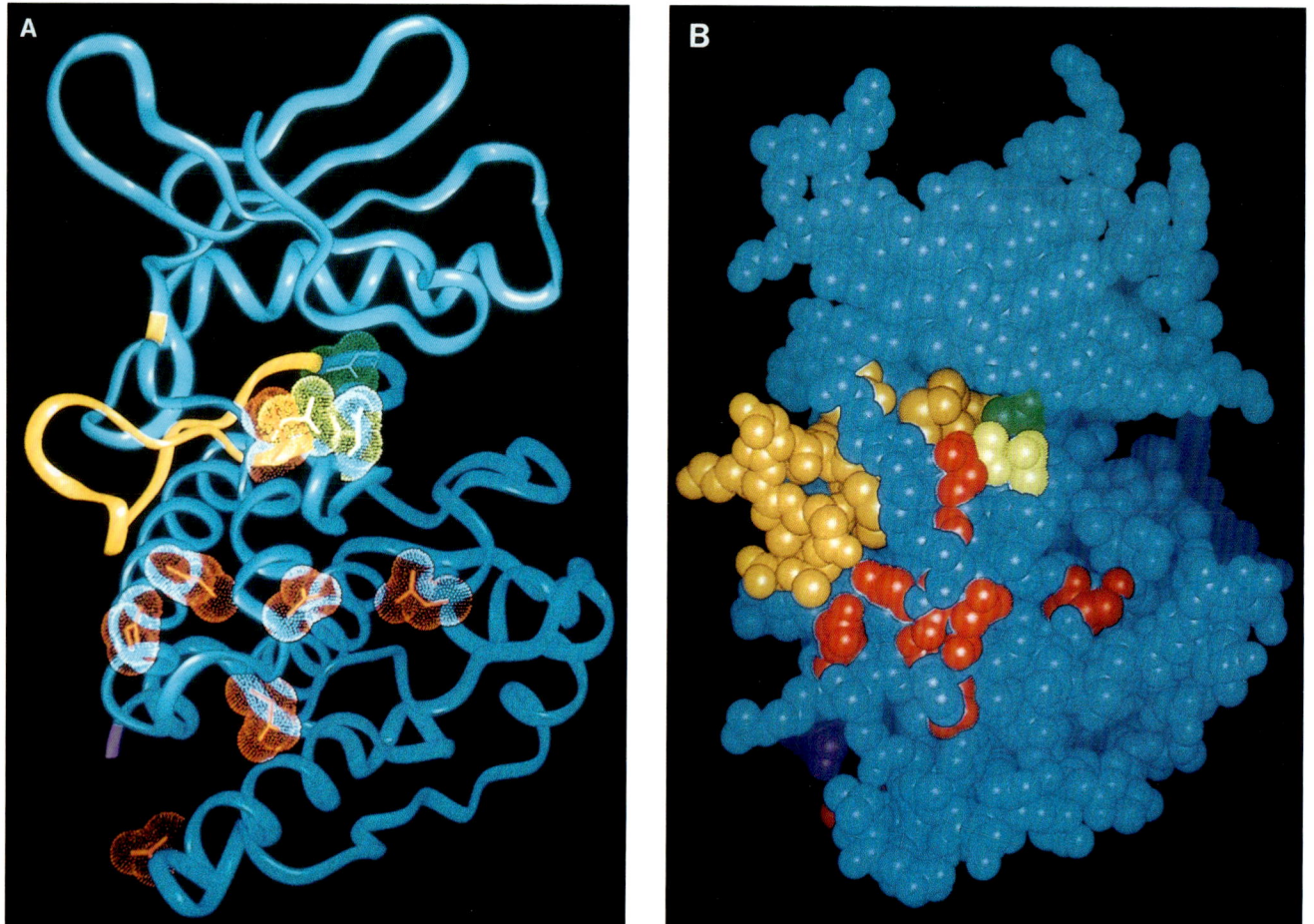

FIGURE 6   Model structure of MLCK. (A) Ribbon diagram of a model of the catalytic core of rabbit smooth/nonmuscle MLCK. The model is derived from the chicken smooth muscle MLCK model (Knighton *et al.*, 1992). The C-terminal tail of the catalytic core is purple. The side chains of the two acidic residues Glu^{777} and Glu^{821} that appear to bind the autoinhibitory region and arginine at P-3 in RLC are yellow. The side chains of the seven acidic residues that may bind the autoinhibitory region, but not light chain, are red. A synthetic peptide substrate with a cross-linking amino acid at P-9 labels a hydrophobic residue in the connection between the small and large lobes as well as β-strands 7 and 8 (orange). The side chain of the catalytic Asp is shown in green. The distant orientations of the substrate in the cleft and the autoinhibitory domain across the surface of the catalytic core converge to the active site of MLCK. The G α-helix is in the lower right corner of the structure. (B) Space-filling model of the catalytic core of MLCK with important residues colored as described in part A.

B. Catalysis

The consensus sequence of smooth muscle myosin RLC (**KKRXXRXXSNVF**) for phosphorylation by MLCK was originally defined with synthetic peptide substrates (Kemp and Pearson, 1990; Kennelly and Krebs, 1991). Results from these studies showed two groups of substrate determinants involving basic residues N-terminal and hydrophobic residues C-terminal of the phosphoacceptor serine. Furthermore, the spatial arrangement of these basic residues was crucial. Arginine at P-3 and the three basic residues at P-6, P-7, and P-8 were important for low K_m values. For example, changing any one of the three basic residues to an alanine increased the K_m values 25- to 50-fold. Deletion of the hydrophobic residues in synthetic peptide substrates resulted in marked decreases in the V_{max} values.

The V_{max} value of a synthetic peptide containing residues 11–23 has a K_m value that is similar to that of the intact regulatory RLC; however, its V_{max} value is less than 10% of the RLC. Thus, there appear to be substrate determinants missing in the synthetic peptide. This problem was explored by mutation analyses of the intact RLC (Zhi et al., 1994). Additionally, the importance of individual residues in the consensus phosphorylation sequence was reexamined in the context of the intact RLC. Surprisingly, individual alanine substitutions for basic residues at P-6 to P-8 resulted in modest changes in K_m values. The greatest effect was noted with arginine at P-3, where substitution of an alanine resulted in an increase in K_m and a decrease in V_{max} values, respectively, with an overall change in the V_{max}/K_m ratio of 34-fold (Table I). A charge reversal mutation at this position resulted in a 340-fold decrease in the V_{max}/K_m ratio. These results show that significant substrate determinants are different in synthetic peptides compared to the intact RLC in residues N-terminal of the phosphoacceptor serine. Additional experiments showed the importance of hydrophobic residues at P + 1 to P + 3 (Table I).

An additional substrate recognition motif appears to be contained somewhere within subdomains 1 and 2 of RLC (Zhi et al., 1994). Skeletal muscle myosin RLC is a poor substrate for the smooth muscle MLCK (Table I). Substitution of glutamate at the P-3 position for an arginine improves the ability of the skeletal muscle myosin RLC to be phosphorylated by the smooth muscle kinase; however, the V_{max}/K_m ratio is only 0.06 compared to 3.4 for the wild-type smooth muscle RLC. Thus, a basic residue at the P-3 position is important, but is not the sole determinant for substrate specificity. This fact is further illustrated by the poor kinetic properties for phosphorylation of a chimera RLC contain-

TABLE I Phosphorylation Properties of Mutated Regulatory Light Chains [a]

RLC	K_m (μM)	V_{max} (μ min^{-1} mg^{-1})	V_{max}/K_m
N-terminal mutations			
Wild-type smooth	8	28	3.50
R16A	71	8	0.11
R16E	99	1	0.01
Del Q20	36	6	0.17
Del V21	46	4	0.09
Del F22	63	6	0.01
Chimeras			
Wild-type skeletal	94	3	0.03
Sm N/Sk 1–4	23	6	0.24
Sm N, 1–2/Sk 3–4	9	27	3.00

[a] Mutated and wild-type RLCs were expressed and purified (Zhi et al., 1994). Individual substitutions or deletions (Del) of residues in the N terminus (residues 1–29) are noted. Chimeras of skeletal and smooth muscle RLCs were made with the respective N termini as well as with subdomains 1–4.

ing 29 N-terminal residues of the smooth muscle RLC and subdomains I–IV of the skeletal muscle RLC (Table I). If the chimera RLC contains the N terminus of the smooth muscle RLC plus subdomains 1 and 2, the kinetic properties are similar to those of the wild-type smooth muscle RLC. Thus, some structure within subdomains 1 and 2 is important for substrate recognition by smooth muscle MLCK.

C. Intrasteric Regulation

Many protein kinases are maintained in an inhibited state by an autoinhibitory domain within the protein or as a separate inhibitory subunit (Kemp et al., 1994). Activation thus involves exposure of the catalytic site so that the protein substrate may bind and be phosphorylated. A number of autoinhibitory sequences have structural similarities to the consensus phosphorylation site in the respective protein kinase substrates. Thus, the enzyme's active site may be sterically blocked by the autoinhibitory domain via a pseudosubstrate mechanism, where residues in the catalytic core bind to residues in the autoinhibitory region or substrate.

In the case of smooth muscle MLCK, a pseudosubstrate structure with similarity to the consensus phosphorylation sequence defined by synthetic peptide substrates was recognized in the N terminus of the calmodulin binding region (Kemp et al., 1987). The pseudosubstrate structure includes an identical number and spatial arrangement of basic residues found to be important substrate determinants in synthetic peptides for myosin RLC, including basic residues at P-3, P-6, P-7, and P-8 (Fig. 3). Synthetic peptides containing the pseudophosphorylation sequence act as substrate antagonists (Kemp *et al.*, 1987). Additionally, removal of the pseudosubstrate region with limited tryptic digestion results in a constitutively active fragment of MLCK that is independent of calmodulin (Olson *et al.*, 1990; Pearson *et al.*, 1988). However, analyses of the cleavage sites have given different interpretations to which residues are important for autoinhibition. Ikebe *et al.* (1989) suggested that the sequence N-terminal of the calmodulin binding domain (DTKNMEAK-KLSKDRMKK) was sufficient for autoinhibition (Fig. 3). Furthermore, the primary determinants for calmodulin binding were contained in the more C-terminal sequence (ARRKWQKTGHAVRAIGRL). Conflicting results have been obtained by Pearson *et al.* (1991) with limited digestion of MLCK by different proteases in addition to inhibition studies with synthetic peptides. They concluded that basic residues in the sequence NMEAKKLSKDRMKKYMARRKWQ were responsible for autoinhibition, which were inaccessible to proteolysis in the absence of calmodulin. Furthermore, the inhibitory properties of a synthetic peptide (DTKNMEAKKLSKDRMKK) corresponding to the autoinhibitory region proposed by Ikebe *et al.* (1989) was three orders of magnitude less potent than a pseudosubstrate peptide (SKDRMKKYMARRK-WRKTGHAV) proposed by Pearson *et al.* (1991).

There may be different interactions between autoinhibitory region and substrate peptides with the catalytic core of smooth muscle MLCK. For example, as noted earlier, the cluster of three basic residues N-terminal of the phosphoacceptor serine (P-6, P-7, and P-8) are not important substrate determinants in the intact RLC. Additionally, a serine-containing analog of the pseudosubstrate peptide RRKWQKTGSAV was not phosphorylated by MLCK (Pearson *et al.*, 1991). This raises a concern that an autoinhibitory region may interact differently as part of the enzyme compared to when it is a free peptide.

Mutation analyses have provided additional insights into the intrasteric regulation of MLCK. A series of point mutations in the autoinhibitory region of MLCK show that charge reversal mutations of the three adjacent basic residues RRK in the pseudosubstrate region to EEE did not activate the enzyme (Shoemaker *et al.*, 1990; Fitzsimons *et al.*, 1992). However, charge reversal mutations of six basic residues N-terminal of these three basic residues did result in a constitutively active kinase (Shoemaker *et al.*, 1990). Single

mutations were not made, so the individual contributions of the six residues for converting the enzyme to a constitutively active form were not available. Additionally, deletion of most of the connecting region and calmodulin binding domain resulted in a constitutively active kinase (Shoemaker *et al.*, 1990; Ito *et al.*, 1991).

The effects of individual substitutions of residues in the connecting region from the catalytic core up to the calmodulin binding domain implicate four of the eight basic residues in autoregulation (Fig. 3) (Fitzsimons *et al.*, 1992; Krueger *et al.*, 1994). It is significant to note that some of these basic residues are well beyond the defined pseudosubstrate sequence, which suggests that most of the connecting region is involved in intrasteric autoregulation. This idea is further supported by the demonstration that deletion of the calmodulin binding domain at the N-terminal boundary of the tribasic residues defined in the pseudosubstrate sequence results in the expression of a kinase that is not activated by Ca^{2+}/calmodulin (Yano *et al.*, 1993). However, limited proteolysis of this recombinant MLCK produces Ca^{2+}/calmodulin-independent activity. These observations support the hypothesis that a significant portion of the autoregulatory sequence lies N-terminal of the calmodulin binding domain.

On the basis of mutation analysis, it is not possible to determine whether a specific residue within the calmodulin binding domain also binds to the catalytic core if the mutation affects calmodulin binding. As discussed earlier, mutations of the four basic residues central to the pseudosubstrate hypothesis do not result in a constitutively active kinase (Shoemaker *et al.*, 1990; Bagchi *et al.*, 1992a; Fitzsimons *et al.*, 1992). However, these mutations increase K_{CaM}, presumably due to interference with electrostatic interactions between calmodulin and the calmodulin binding domain of the kinase (Fitzsimons *et al.*, 1992). Contrary to the results of Yano *et al.* (1993), Ito *et al.* (1991) found that a deletion near the N terminus of the calmodulin binding domain resulted in kinase that was not activated by Ca^{2+}/calmodulin or by limited proteolysis. However, the enzyme activity was low and Yano *et al.* (1993) suggested improper folding of the kinase so that the activity of the autoinhibitory region was attenuated.

Although there may be disagreements over the precise boundaries of the autoinhibitory region, it can bind to the active site of the enzyme with appropriate mutations. When the tribasic residues were changed to alanine and a major portion of the N terminus of the calmodulin binding domain was mutated to residues found in the RLC substrate (. . . RRKWQKTGHAV . . . mutated to . . . AAAWQRATSNV . . .), the mutated enzyme was rapidly autophosphorylated intramolecularly in the absence of Ca^{2+}/calmodulin, presumably at the introduced serine (Bagchi et al., 1992b). Autophosphorylation activated the enzyme in the absence of Ca^{2+}/calmodulin so it could phosphorylate RLC, presumably by displacing the autoinhibitory calmodulin binding domain from the active site. Surprisingly, the simple introduction of serine at the predicted phosphorylation site in the calmodulin binding domain does not result in autophosphorylation. However, these results indicate that a portion of the autoinhibitory region is near the active site.

If an acidic residue in the catalytic core of smooth muscle MLCK binds to a basic residue in the autoinhibitory domain, a charge reversal mutation of the acidic residue should weaken their electrostatic interactions (Gallagher et al., 1993; Krueger et al., 1994). We have examined acidic residues predicted to be on the surface of the catalytic core of smooth muscle MLCK for interactions with the autoinhibitory region, RLC, or both (Fig. 4). From a total of 20 acidic residues predicted to be on the surface of the larger lobe of the catalytic core, only 7 are implicated in binding directly or indirectly to the autoinhibitory domain. These mutations resulted in significant decreases in K_{CaM} values with no significant effect on K_m or V_{max} values for the regulatory RLC. It is proposed that a lowered K_{CaM} value reflects a weakened binding of the inhibitory region to the catalytic core. The placement of these acidic residues defines a binding pathway that primarily traverses the D α-helix (Fig. 6). This binding pathway is distinct from that proposed in a molecule model where the autoinhibitory domain was placed on the catalytic core of MLCK in a position similar to the PKI peptide on the catalytic core of the cAMP-dependent protein kinase (Knighton et al., 1992). This model proposed that both the pseudosubstrate sequence and basic residues forming the consensus phosphorylation sequence in synthetic peptides made contacts with residues in the G α-helix (Knighton et al., 1992; Kemp et al., 1994). However, none of the acidic residues in or near the G α-helix was implicated as binding the autoinhibitory domain.

In the proposed model these residues were also thought to bind the RLC at the respective three basic residues at P-6 to P-8. However, there were no or modest changes in the K_m values even with charge reversal mutations. For example, the decrease in the V_{max}/K_m ratio for RLC varied from 1- to 2-fold. These modest changes are in contrast to the marked changes observed in the V_{max}/K_m ratios for mutations at the arginine at P-3 and the hydrophobic residues in the RLC (Zhi et al., 1994). With a charge reversal mutation at P-3, the V_{max}/K_m ratio decreased 340-fold. Similarly, charge reversal mutations at two acidic residues pre-

dicted to bind to arginine at P-3 in the RLC decreased the V_{max}/K_m ratio 138- and 72-fold, respectively (Herring et al., 1992). These residues are in the cleft between the two lobes of the kinase near the catalytic site (Fig. 6). Furthermore, these two acid residues also appear to bind to or near the autoinhibitory domain (Gallagher et al., 1993). Their exposure upon calmodulin binding may be sufficient for binding of arginine at P-3 in the RLC substrate and phosphorylation at the phosphoacceptor serine.

The crystal structure of a related kinase, twitchin, has been reported (Hu et al., 1994). The portion of the kinase C-terminal to the catalytic core extends across the surface of the large lobe, similar to the distinct pathway proposed for the autoinhibitory sequence of MLCK (Fig. 6). It then extends through the active site between the large and small lobes. The extensive contacts with the catalytic core may account for autoinhibition. The physiological activator and substrate for twitchin are yet to be identified. Although we do not have a crystal structure of MLCK, twitchin kinase provides an elegant structural example of intrasteric regulation. It is expected that MLCK will reveal a similar, but not identical, structural mechanism for autoregulation.

Additional information about substrate binding to MLCK was recently obtained. The substrate binding properties near the active site of skeletal muscle MLCK were investigated with a synthetic peptide containing a photoreactive amino acid, p-benzoylphenylalanine (Bpa), incorporated N-terminal of the phosphoacceptor serine at the P-9 position (BpaKKRAARATSNVFA). The structures of the catalytic cores of smooth and skeletal MLCK are highly homologous and the peptide is a substrate for both kinases. Upon photolysis, peptide was cross-linked to the kinase in a Ca^{2+}/calmodulin-dependent manner (Gao et al., 1995). The extent of incorporation was saturable at 1 mol peptide/mol kinase, and the concentration of peptide required for half-maximal labeling was similar to the K_m value (8 μM). The incorporation of peptide into kinase resulted in a proportional loss of enzyme activity. Furthermore, incorporation was inhibited by peptide or protein substrates, and the incorporated peptide was phosphorylated. These data are consistent with the specific incorporation of the peptide into the substrate binding site of the kinase. Twenty percent of the total incorporation was in a portion of the autoinhibitory domain, supporting the hypothesis that it is near the catalytic site (Bagchi et al., 1992b). This placement is consistent with an intrasteric mechanism of regulation (Hu et al., 1994).

The identity of the cross-linked sites showed that the major labeled site (66%) was a hydrophobic residue in the connecting sequence between the small and large lobes of the catalytic core (Fig. 6). Another 14% was contained in a peptide that forms β-strands 7 and 8. These strands are close to the labeled residue in the connecting sequence between the lobes (Fig. 6).

If the phosphoacceptor serine of the peptide is positioned near the γ-phosphate of ATP in the postulated catalytic base, the cross-linked peptide lies along the cleft between the two lobes rather than oriented toward the proposed substrate binding groove and G α-helix (Fig. 6). The orientation of the peptide is similar to the substrate recognition fragment of PKI bound to the catalytic subunit of the cAMP-dependent protein kinase (Knighton *et al.*, 1991). However, the more extended α-helical portion of PKI is oriented perpendicular to the substrate recognition fragment and binds down in the proposed substrate binding groove near the G α-helix. The high-affinity binding of this portion of PKI is due to Phe[10], which is at the P-11 position relative to the pseudophosphoacceptor alanine. This placement appears to be unique to PKI. A recent kinetic analysis of synthetic peptide substrates derived from PKI led to the conclusion that a peptide containing only a phosphoacceptor serine in place of the pseudophosphoacceptor alanine was the most effective substrate with two key residues essential for substrate activity at P-3 and P + 1. Most interestingly, Phe[10] at P-11 did not affect the kinetic properties for phosphorylation, although its importance in the inhibitory peptide containing alanine in the place of serine was recognized (Mitchell *et al.*, 1995). It was proposed that the hydrophobic binding pocket that recognized Phe[10] may act to contribute specificity for the interaction between the catalytic subunit of cAMP-dependent protein kinase and PKI but not substrates. These results are consistent with the orientation of the synthetic peptide substrate on the surface of MLCK.

V. SUMMARY AND PERSPECTIVES

MLCK plays a central role in the initiation of smooth muscle contraction and many nonmuscle motile processes owing to its Ca^{2+}/calmodulin-dependent phosphorylation of myosin RLC. Physiological experiments with rapid, synchronous activation of smooth muscle cells demonstrate that the significant 500-ms latency for RLC phosphorylation and force development is due to the time required for increases in cytosolic Ca^{2+} concentration and activation of MLCK. Once activated, MLCK rapidly ($1 s^{-1}$) phosphorylates RLC in a random mechanism leading to the rapid attachment of cross-bridges and subsequent force development. Although the general properties of this cellu-

lar activation pathway are identified, additional information is needed to apply our understanding of biochemical mechanisms to smooth muscle cells. For example, with the demonstration that there is little free calmodulin in smooth muscle cells (Tansey *et al.*, 1994), all of the MLCK may not be activated by Ca^{2+}/calmodulin; that is, the availability of calmodulin may be limiting. What are the dynamic properties of calmodulin binding and activation of MLCK in cells? Can the calmodulin that is bound in other cellular compartments rapidly exchange with the calmodulin that activates MLCK? The biochemical observations that MLCK binds to actin and to myosin do not provide sufficient insights into MLCK localization *in vivo*. Is it bound to actin, myosin, or both? Is the binding regulated? New experimental approaches involving combinations of molecular biology, protein biochemistry, and cellular biophysics should provide answers in the near future to these and other important issues related to the regulation of myosin light chain kinase activity in smooth muscle.

Recent progress by a number of investigators provides new insights into the structural properties of MLCK important for autoregulation and catalysis. The high-resolution structure of calmodulin bound to a peptide of the calmodulin binding domain of smooth muscle MLCK explains the role of key residues identified by mutagenesis analysis. However, the mechanism of activation can be dissociated from calmodulin binding and new experimental approaches are needed to explain these observations. MLCK is inhibited by an intrasteric mechanism involving the folding of a C-terminal extension of the kinase back onto the catalytic core. Recombinant DNA techniques combined with protein biochemistry have provided a general perspective on the mechanism. However, controversy remains about the structural details that probably will not be resolved until the atomic-resolution structures of the kinase with and without calmodulin are resolved.

Finally, new information about the mechanism of RLC phosphorylation shows the importance of structure in RLC subdomains 1 and 2, well beyond the phosphoacceptor serine. How does this structural determinant work? It is important for the structure of the N terminus of the light chain or does it bind directly to the kinase? Additional investigations combining biochemical and molecular biological approaches may soon provide important information.

References

Adelstein, R. S., and Klee, C. B. (1981). *J. Biol Chem.* **256,** 7501–7509.

Afshar, M., Caves, L. S. D., Guimard, L., Hubbard, R. E., Calas, B., Grassy, G., and Haiech, J. (1994). *J. Mol. Biol.* **244,** 554–571.

Aksoy, M. O., Mras, S., Kamm, K. E., and Murphy, R. A. (1983). *Am. J. Physiol.* **245,** C255–C270.

Asano, M., and Stull, J. T. (1985). *In* "Calmodulin Antagonists and Cellular Physiology" (H. Hidaka, and D. J. Hartshorne, eds.), pp. 225–260. Academic Press, Orlando, FL.

Babu, Y. S., Bugg, C. E., and Cook, W. J. (1988). *J. Mol. Biol.* **204,** 191–204.

Bagchi, I. C, Huang, Q., and Means, A. R. (1992a). *J. Biol. Chem.* **267,** 3024–3029.

Bagchi, I. C., Kemp, B. E., and Means, A. R. (1992b). *Mol. Endocrinol.* **6,** 621–626.

Blumenthal, D. K., and Stull, J. T. (1980). *Biochemistry* **19,** 5608–5614.

Bowman, B. F., Peterson, J. A., and Stull, J. T. (1992). *J. Biol. Chem.* **267,** 5346–5354.

Campbell, I. D., and Spitzfaden, C. (1994). *Structure* **2,** 333–337.

Choi, O. H., Adelstein, R. S., and Beaven, M. A. (1994). *J. Biol. Chem.* **269**(1), 536–541.

Colburn, J. C., Michnoff, C. H., Hsu, L., Slaughter, C. A., Kamm, K. E., and Stull, J. T. (1988). *J. Biol. Chem.* **263,** 19166–19173.

De Bondt, H. L., Rosenblatt, J., Jancarik, J., Jones, H. D., Morgan, D. O., and Kim, S.-H. (1993). *Nature (London)* **363,** 595–602.

Faux, M. C., Mitchelhill, K. I., Katsis, F., Wettenhall, R. E. H., and Kemp, B. E. (1993). *Mol. Cell. Biochem.* **127/128,** 81–91.

Fitzsimons, D. P., Herring, B. P., Stull, J. T., and Gallagher, P. J. (1992). *J. Biol. Chem.* **267,** 23903–23909.

Gallagher, P. J., and Herring, B. P. (1991). *J. Biol. Chem.* **266,** 23945–23952.

Gallagher, P. J., Herring, B. P., Griffin, S. A., and Stull, J. T. (1991). *J. Biol. Chem.* **266,** 23936–23944.

Gallagher, P. J., Herring, B. P., Trafny, A., Sowadski, J., and Stull, J. T. (1993). *J. Biol. Chem.* **268,** 26578–26582.

Gao, Z.-H., Zhi, G., Herring, B. P., Moomaw, C., Deogny, L., Slaughter, C. A., and Stull, J. T. (1995). *J. Biol. Chem.* **270,** 10125 10135.

Hartshorne, D. J. (1987). *In* "Physiology of the Gastrointestinal Tract" (L. R. Johnson, ed.), pp. 423–482. Raven Press, New York.

Herring, B. P., Gallagher, P. J., and Stull, J. T. (1992). *J. Biol. Chem.* **267,** 25945–25950.

Holden, H. M., Ito, M., Hartshorne, D. J., and Rayment, I. (1992). *J. Mol. Biol.* **227,** 840–851.

Hu. S.-H., Parker, M. W., Lel, J. Y., Wilce, M. C. J., Benian, G. M., and Kemp, B. E. (1994). *Nature (London)* **369,** 581–584.

Ikebe, M., Maruta, S., and Reardon, S. (1989). *J. Biol. Chem.* **264,** 6967–6971.

Ishikawa, T., Chijiwa, T., Hagiwara, M., Mamiya, S., Saitoh, M., and Hidaka, H. (1988). *Mol. Pharmacol.* **33,** 598–603.

Ito, M., Dabrowska, R., Guerriero, V., Jr., and Hartshorne, D. J. (1989). *J. Biol. Chem.* **264,** 13971–13974.

Ito, M. Guerriero, V., Jr., Chen, X., and Hartshorne, D. J. (1991). *Biochemistry* **30,** 3498–3503.

Kamm, K. E., and Stull, J. T. (1985a). *Annu. Rev. Pharmacol. Toxicol.* **25,** 593–620.

Kamm, K. E., and Stull, J. T. (1985b). *Am. J. Physiol.* **249,** C238–C247.

Kamm, K. E., and Stull, J. T. (1986). *Science* **232,** 80–82.

Kamm, K. E., and Stull, J. T. (1989). *Annu. Rev. Physiol.* **51,** 299–313.

Kamm, K. E., Hsu, L.-C., Kubota, Y., and Stull, J. T. (1989). *J. Biol. Chem.* **264,** 21223–21229.

Kanoh, S., Ito, M., Niwa, E., Kawano, Y., and Hartshorne, D. (1993). *Biochemistry* **32,** 8902–8907.

Kargacin, G. J., Ikebe, M., and Fay, S. F. (1990). *Am. J. Physiol.* **259,** C315–C324.

Kasturi, R., Vasulka, C., and Johnson, J. D. (1993). *J. Biol. Chem.* **268,** 7958–7964.

Kemp, B. E., and Pearson, R. B. (1990). *Trends Biochem. Sci.* **15,** 342–346.

Kemp, B. E., Pearson, R. B., Guerriero, V., Jr., Bagchi, I. C., and Means, A. R. (1987). *J. Biol. Chem.* **262,** 2542–2548.

Kemp, B. E., Pearson, R. B., and House, C. M. (1991). *In* "Methods in Enzymology" (T. Hunter and B. Sefton, eds.), Vol. **201,** pp. 287–304.

Kemp, B. E., Parker, M. W., Hu, S., Tiganis, T., and House, C. (1994). *Trends Biochem. Sci.* **19,** 440–444.

Kennelly, P. J., and Krebs, E. G. (1991). *J. Biol. Chem.* **266,** 15555–15558.

Kerrick, W. G. L., and Bourguignon, L. Y. W. (1984). *Proc. Natl. Acad. Sci. U.S.A.* **81,** 165–169.

Kimura, K., Hattori, S., Kabuyama, Y., Shizawa, Y., Takayanagi, J., Nakamura, S., Toki, S., Matsuda, Y., Onondera, K., and Fukui, Y. (1994). *J. Biol. Chem.* **269,** 18961–18970.

Knighton, D. R., Zheng, J., Ten Eyck, L. F., Ashford, V. A., Xuong, N.-H., Taylor, S. S., and Sowadski, J. M. (1991). *Science* **253,** 353–484.

Knighton, D. R., Pearson, R. B., Sowadski, J. M., Means, A. R., Ten Eyck, L. F., Taylor, S. S., and Kemp, B. E. (1992). *Science* **258,** 130–135.

Kobayashi, H., Inoue, A., Mikawa, T., Kuwayama, H., Hotta, Y., Masaki, T., and Ebashi, S. (1992). *J. Biochem. (Tokyo)* **112,** 786–791.

Krueger, J. K., Gallagher, P. J., and Stull, J. T. (1994). *Biophys. J.* **66,** A31.

Leachman, S. A., Gallagher, P. J., Herring, B. P., McPhaul, M. J., and Stull, J. T. (1992). *J. Biol. Chem.* **267,** 4930–4938.

Matsushima, S., Huang, Y.-P., Dudas, C. V., Guerriero, V., Jr., and Hartshorne, D. J. (1994). *Biochem. Biophys. Res. Commun.* **202,** 1329–1336.

Meador, W. E., Means, A. R., and Quiocho, F. A. (1992). *Science* **257,** 1251–1255.

Miller-Hance, W. C., Miller, J. R., Wells, J. N., Stull, J. T., and Kamm, K. E. (1988). *J. Biol. Chem.* **263,** 13979–13982.

Mitchell, R. D., Glass, D. B, Wong, C.-W., Angelos, K. L., and Walsh, D. A. (1995). *Biochemistry* **34,** 528–534.

Nakanishi, S., Kakita, S., Takahashi, I., Kawahara, K., Tsukuda, E., Sano, T., Yamada, K., Yoshida, M., Kase, H., and Matsuda, Y. (1992). *J. Biol. Chem.* **267,** 2157–2163.

Nakanishi, S., Catt, K. J., and Balla, T. (1994). *J. Biol. Chem.* **269,** 6528–6535.

Olson, N. J., Pearson, R. B., Needleman, D. S., Hurwitz, M. Y., Kemp, B. E., and Means, A. R. (1990). *Proc. Natl. Acad. Sci. U.S.A.* **87,** 2284–2288.

Pearson, R. B., Wettenhall, R. E. H., Means, A. R., Hartshorne, D. J., and Kemp, B. E. (1988). *Science* **241,** 970–973.

Pearson, R. B., Ito, M., Morrice, N. A., Smith, A. J., Condron, R., Wettenhall, R. E. H., Kemp, B. E., and Hartshorne, D. J. (1991). *Eur. J. Biochem.* **200,** 723–730.

Persechini, A., and Hartshorne, D. J. (1981). *Science* **213,** 1383–1385.

Persechini, A., Kamm, K. E., and Stull, J. T. (1986). *J. Biol. Chem.* **261,** 6293–6299.

Persechini, A., McMillan, K., and Leakey, P. (1994). *J. Biol. Chem.* **269,** 16148–16154.

Saitoh, M., Ishikawa, T., Matsushima, S., Naka, M., and Hidaka, H. (1987). *J. Biol. Chem.* **262,** 7796–7801.

Sellers, J. R., and Adelstein, R. S. (1987). *In* "The Enzymes" (P. D. Boyer and E. G. Krebes, eds.), 3rd ed., p. 382. Academic Press, Orlando, FL.

Sellers, J. R., Chock, P. B., and Adelstein, R. S. (1983). *J. Biol. Chem.* **258,** 14181–14188.

Sheldon, R., Moy, A., Lindsley, K., Shasby, S., and Shasby, D. M. (1993). *Am. J. Physiol.* **265,** L606–L612.

Shimizu, H., Ito, M., Miyahara, M., Ichikawa, K., Okubo, S., Konishi, T., Naka, M., Tanaka, T., Hirano, K., Hartshorne, D. J., and Nakano, T. (1994). *J. Biol. Chem.* **269**, 30407–30411.

Shirazi, A., Iizuka, K., Fadden, P., Mosse, C., Somlyo, A. P., Somlyo, A. V., and Haystead, T. A. J. (1994). *J. Biol. Chem.* **269**, 31598–31606.

Shirinsky, V. P., Vorotnikov, A. V., Birukov, K. G., Nanaev, A. K., Collinge, M., Lukas, T. J., Sellers, J. R., and Watterson, D. M. (1993). *J. Biol. Chem.* **268**, 16578–16583.

Shoemaker, M. O., Lau, W., Shattuck, R. L., Kwiatkowski, A. P., Matrisian, P. E., Guerra-Santos, L., Wilson, E., Lukas, T. J., Van Eldik, L. J., and Watterson, D. M. (1990). *J. Cell Biol.* **111**, 1107–1125.

Smith, M. K., Colbran, R. J., and Soderling, T. R. (1990). *J. Biol. Chem.* **265**, 1837–1840.

Somlyo, A. P., and Somlyo, A. V. (1994). *Nature (London)* **372**, 231–236.

Stull, J. T., Nunnally, M. H., and Michnoff, C. H. (1986). *In* "The Enzymes" (E. G. Krebs and P. D. Boyer, eds.), pp. 113–166. Academic Press, Orlando, FL.

Stull, J. T., Hsu, L.-C., Tansey, M. G., and Kamm, K. E. (1980). *J. Biol. Chem.* **265**, 16683–16690.

Stull, J. T., Tansey, M. G., Tang, D.-C., Word, R. A., and Kamm, K. E. (1993). *Mol. Cell. Biochem.* **127/128**, 229–237.

Su, Z., Fan, D., and Georges, S. E. (1994). *J. Biol. Chem.* **269**, 16761–16765.

Sweeney, H. L., Bowman, B. F., and Stull, J. T. (1993). *Am. J. Physiol.* **264**, C1085–C1095.

Tansey, M. G., Word, R. A., Hidaka, H., Singer, H., Schworer, C. M., Kamm, K. E., and Stull, J. T. (1992). *J. Biol. Chem.* **267**, 12511–12516.

Tansey, M. G., Luby-Phelps, K., Kamm, K. E., and Stull, J. T. (1994). *J. Biol. Chem.* **269**, 9912–9920.

Trybus, K. M., and Lowey, S. (1985). *J. Biol. Chem.* **260**, 15988–15995.

Van Berkum, M. F. A., and Means, A. R. (1991). *J. Biol. Chem.* **266**, 21488–21495.

Word, R. A., Tang, D.-C., and Kamm, K. E. (1994). *J. Biol. Chem.* **269**, 21596–21602.

Yano, K., Araki, Y., Hales, S. J., Tanaka, M., and Ikebe, M. (1993). *Biochemistry* **32**, 12054–12061.

Zhang, F., Strand, A., Robbins, D., Cobb, M. H., and Goldsmith, E. J. (1994). *Nature (London)* **367**, 704–710.

Zhi, G., Herring, B. P., and Stull, J. T. (1994). *J. Biol. Chem.* **269**, 24723–24727.

10

Myosin Light Chain Phosphatase

FERENC ERDÖDI
Department of Medical Chemistry
University Medical School of Debrecen
Debrecen, Hungary

MASAAKI ITO
1st Department of Internal Medicine
Mie University School of Medicine
Tsu, Mie, Japan

DAVID J. HARTSHORNE
Muscle Biology Group
Shantz Building
University of Arizona
Tucson, Arizona

I. INTRODUCTION

A. Importance of Myosin Phosphorylation in Smooth Muscle

It has been established for several years that the major mechanism for regulation of contraction in smooth muscle is myosin phosphorylation (Hartshorne, 1987). Phosphorylation of the two 20,000-dalton light chains of myosin (LC20) activates the actin-dependent ATPase activity of myosin and this initiates the contractile response. Dephosphorylated myosin is associated with relaxed muscle. In this scheme there are two key enzymes: the myosin light chain kinase (MLCK) and the myosin light chain phosphatase (MLCP). Obviously a balance of these two activities determines the level of myosin phosphorylation.

Much more information is available for MLCK. Several cDNAs have been isolated and cloned and some of the functional regions of the molecule have been identified. This is not the case for MLCP. Until recently there was no consensus about the identity of the phosphatase involved in myosin dephosphorylation and since there are no strict sequence requirements (in terms of primary structure) for most phosphatases, there was the possibility that several phosphatases could be involved.

An impetus to smooth muscle phosphatase research was provided in the last few years. With the increased application of fluorescent dyes to measure internal Ca^{2+} concentration it was realized that the force/Ca^{2+} ratio could vary for different methods of stimulation (Somlyo and Somlyo, 1994). This implied that the balance of the two key enzymes ("on–off" activities) was altered and evidence was presented to suggest an inhibition of phosphatase activity. Later it was proposed that this inhibition resulted from a G-protein-linked mechanism (Somlyo and Somlyo, 1994), although the second messenger involved has not been unequivocally identified. The importance of these results was that for the first time the regulation of myosin phosphatase activity became a realistic possibility and this regulatory process could form a critical component of the integrated smooth muscle regulatory mechanism. Spurred by these intriguing results, several laboratories renewed efforts to isolate MLCP and particularly to search for possible regulatory subunits. Three groups have isolated a multisubunit phosphatase from gizzard (Alessi *et al.*, 1992; Shimizu *et al.*, 1994) and pig bladder (Shirazi *et al.*, 1994). The presumed holoenzyme is trimeric and consists of 130-, 38-, and 20-kDa subunits. This phosphatase will be documented further in this chapter. To provide a historical perspective, some of the previous phosphatase bibliography will be reviewed.

B. Classification of the Ser/Thr-Specific Protein Phosphatases

Protein phosphatases (PP) are divided into phosphoserine/threonine (Ser/Thr)-specific and phosphotyrosine (Tyr)-specific enzymes based on the distinct specificity toward the phosphorylated residues. Smooth muscle myosin is phosphorylated at Ser 19 (and sometimes Thr 18) at the N-terminal end of LC20. Thus, the Ser/Thr phosphatases are those pertinent to this discussion. Ser/Thr protein phosphatases were first classified by Ingebritsen and Cohen (1983) according to their relative substrate preference and their sensitivity to inhibition by two acid and heat-stable proteins, called inhibitor-1 (I1) and inhibitor-2 (I2). Type 1 protein phosphatase (PP1) preferentially dephosphorylates the β-subunit of phosphorylase kinase, and is often sensitive to inhibition by nanomolar concentrations of I1 or I2. There is one puzzling feature that should be mentioned at this stage. The inhibition of PP1 by I2 is not always a consistent feature of the smooth muscle phosphatases. Reduced sensitivity to I2 has been observed for several myosin phosphatases (Pato and Kerc, 1985; Erdödi et al., 1989; Mitsui et al., 1992; Tulloch and Pato, 1991). The molecular basis for this reduced sensitivity is not known. In contrast, type 2 (PP2) protein phosphatase dephosphorylates the α-subunit of phosphorylase kinase to a greater extent than that of the β-subunit, and it is insensitive to inhibition by either I1 or I2. The family of PP2 includes subtypes, called PP2A, PP2B, and PP2C. PP2A has significant phosphorylase phosphatase activity and it is active in the absence of Ca^{2+} and Mg^{2+}. Both PP2B and PP2C have very low phosphorylase phosphatase activity and they require Ca^{2+}–calmodulin or Mg^{2+} for activity, respectively. It was suggested that these four classes of PP catalytic subunits (PP1, PP2A, PP2B, and PP2C) could account for most, if not all, of the dephosphorylation processes on Ser/Thr residues in mammalian cells. In support of this view, the catalytic subunits of the four PP classes have been identified in many mammalian cells and their regulatory roles in distinct biochemical pathways have also been demonstrated (Cohen, 1989; Walter and Mumby, 1993; Bollen and Stalmans, 1992).

Several other criteria have also been introduced to distinguish PP1 and PP2A. These procedures are based on the effects of polyanionic and polycationic macromolecules on the activity of PP1 and PP2A. Heparin inhibits the phosphorylase phosphatase activity of PP1, whereas it is without effect on the activity of PP2A (Gergely et al., 1984). Heparin, coupled to activated Sepharose, is also useful in the chromatographic separation of PP1 from PP2A, since PP1 binds tightly to the matrix, whereas PP2A flows through or binds only weakly, eluting at low ionic strength (Gergely et al., 1984; Erdödi et al., 1985). Polycationic macromolecules (such as histone H1, protamine, polylysine, and polybrene) stimulate the phosphorylase phosphatase activity of PP2A, whereas they inhibit (or have no influence) the activity of PP1 (Erdödi et al., 1985; Pelech and Cohen, 1985; DiSalvo et

al., 1985). The application of heparin or the polycationic macromolecules for the differentiation of PP1 and PP1A is limited, since their effects can alter depending on the particular substrate used.

Another useful diagnostic aid is okadaic acid. This marine toxin is a potent inhibitor of phosphatases (Takai et al., 1987) and its different sensitivity to PP1 and PP2A is the critical feature. PP2A is approximately 4000-fold more sensitive than PP1 (Takai et al., 1995). The K_i value for PP2A is approximately 30 pM (Takai et al., 1995). PP2B is inhibited only in the micromolar range and PP2C is not affected by okadaic acid (Cohen, 1991).

The introduction of a classification system for PP has been useful in characterization of the most abundant enzymatic forms from various tissues. However, cloning of cDNA for the PP catalytic subunits has led to the conclusion that PP is an expanding family, and other types of Ser/Thr phosphatases (PPX, PPZ, etc.) exist and are distinct from PP1 and PP2 (Cohen and Cohen, 1989). Recognition of the dual specificity (Ser/Thr and Tyr) has been a significant development and the Tyr phosphatases also represent a new class (Guan et al., 1991). Thus the application of a rigid classification may not be appropriate and the use of a more flexible system may be imposed by future research.

C. Subunit Structure and Regulation of Protein Phosphatases

The catalytic subunits of PP can be complexed with various regulatory subunit(s) in cells, resulting in heterogeneity of the enzymes within the same class (Cohen, 1989; Walter and Mumby, 1993; Bollen and Stalmans, 1992; Mumby and Walter, 1993; Mayer-Jaekel and Hemmings, 1994). This diversity has become even more apparent with the identification of distinct isoforms of both the catalytic and regulatory subunits (Arino et al., 1988; Sasaki et al., 1990; Hendrix et al., 1993). The associating regulatory subunits in the distinct PP forms are believed to play crucial roles in mediating the substrate specificity as well as sensitivity to inhibitors or activators of PP. They have also been implicated in directing the enzymes to specific cellular compartments. In these regards, the structure and regulation of PP1 and PP2A have been intensively studied for the past few years and data have accumulated for the functions associated with the holoenzyme forms.

PP2A has been isolated in various dimeric and trimeric forms. It appears that the core structure of PP2A consists of a 36-kDa catalytic and a 65-kDa regulatory subunit and this dimer can associate with a third variable subunit ranging from 54 to 130 kDa (Walter and

Mumby, 1993; Mayer-Jaekel and Hemmings, 1994). The substitution of this third subunit by the transforming small T and middle T antigen of SV40 and polyoma viruses results in a change of enzymatic activity and specificity (Walter and Mumby, 1993). This displacement of the variable subunit has been implicated in the regulation of cell transformation. PP2A is considered to be largely cytosolic in the cell and the function of the regulatory subunit with regards to localization with subcellular compartments has yet to be elucidated.

PP1 enzymatic forms are present in the cytosolic and particulate fractions of the cell. PP1 was purified from the cytosolic fraction of rabbit skeletal muscle in an inactive form and called the MgATP-dependent phosphatase, since it could be activated by incubation with MgATP plus a protein factor identified later as glycogen synthase kinase-3 (Bollen and Stalmans, 1992). In the MgATP-dependent PP1 the catalytic subunit is complexed with I2, and activation of the phosphatase requires phosphorylation of I2. PP1 has been shown to associate with many cellular organelles, including glycogen particles, sarcoplasmic reticulum, myofibrillar proteins, and nucleus. Several lines of evidence suggest that the regulatory subunits that associate with the catalytic subunit play an important role in targeting PP1 to the various cell compartments or organelles. The glycogen-bound form of PP1 consists of the catalytic subunit (37kDa) and a 161-kDa glycogen binding subunit (Hubbard and Cohen, 1989). The same enzyme targets the sarcoplasmic reticulum of skeletal muscle, and this is probably due to the presence of a hydrophobic membrane-spanning domain in the 161-kDa subunit, revealed by the protein sequence deduced from the cloned cDNA (Tang et al., 1991). It has also been demonstrated that the association of PP1 with glycogen is under hormonal control. The 161-kDa subunit is phosphorylated in vivo in response to adrenalin and this phosphorylation weakens the interaction between the catalytic and glycogen binding subunit, resulting in the release of catalytic subunit to the cytosol (Hubbard and Cohen, 1989). This translocation leads to inhibition of the dephosphorylation processes, ensuring enhanced activation of the glycogenolytic enzymes. The glycogen binding subunit is also phosphorylated in response to insulin on a distinct site from that phosphorylated in response to adrenalin. This insulin-dependent phosphorylation increases the affinity of the protein to the catalytic subunit, causing the latter to rebind to the glycogen particles (Dent et al., 1990). The targeting subunits that direct PP1 to the nucleus and to skeletal or smooth muscle myofibrils are distinct from the glycogen binding subunit. It is assumed that several low-molecular-mass inhibitory proteins might be responsible for lo-

calization of PP1 to the nucleus (Beullens et al., 1992). The two proteins targeting the phosphatase to the skeletal and smooth muscle contractile apparatus also are distinct proteins (Dent et al., 1992; Alessi et al., 1992).

Thus there is ample precedence for regulation of phosphatase activity. The challenge for future research is to determine if the smooth muscle myosin phosphatase is regulated, and if so, how this mechanism integrates with the overall contraction–relaxation cycle of smooth muscle.

II. SMOOTH MUSCLE MYOSIN PHOSPHATASE

A. Assay and Preparation

The phosphorylated form of the isolated LC20 of myosin serves as an excellent substrate for the various forms of PP1, PP2A, and PP2C (Pato and Adelstein, 1983). However, since myosin is the native substrate, it is important to consider only those phosphatases that can dephosphorylate myosin. When LC20 associates with the heavy chains, its availability for dephosphorylation by many MLCP decreases dramatically. Furthermore, subunit composition of the phosphatase might also hinder interaction with intact myosin (Pato and Kerc, 1986). Besides this primary requirement that the MLCP should be effective with the native substrate, binding studies with phosphorylated and dephosphorylated myosins are also useful to indicate interaction with myosin (Sellers and Pato, 1984). Assay of MLCP with both P-LC20 and P-myosin [or heavy meromyosin (HMM)] during the course of preparation is essential to distinguish myosin phosphatases from other phosphatases. Use of P-myosin is more difficult because of its low solubility but myosin can be substituted for the soluble HMM, since the latter carries all the phosphatase binding properties associated with intact myosin (Mitsui et al., 1992). Table I summarizes protein phosphatases that have been purified from smooth muscle and qualified as myosin phosphatase based on the foregoing criteria. An ongoing controversy has been whether myosin phosphatase is PP1 or PP2A, or a "third type" of enzyme similar to but not identical with PP1 or PP2A. Application of improved analytical (e.g., sequence analysis) as well as immunologic techniques to classify the smooth muscle protein phosphatases has established that PP1 and PP2A might be the two major PP classes involved in the dephosphorylation of myosin (Alessi et al., 1992; Okubo et al., 1993) and other proteins such as caldesmon and calponin (Winder et al., 1992; Ichikawa et al., 1993). For these reasons, we attempted to assign

TABLE I   Myosin Phosphatases from Smooth Muscle

Enzyme	Tissue fraction isolated	Subunit component(s) (MW)	Type classified	Reference
Myosin phosphatase	Bovine aorta cytosol	67,000 38,000	PP2A	Werth *et al.* (1982)
Myosin light chain phosphatase	Chicken gizzard cytosol	67,000 54,000 34,000	PP2A	Onishi *et al.* (1982)
Polycation-modulated phosphatase	Bovine aorta cytosol	72,000 53,000 35,000	PP2A	DiSalvo *et al.* (1985)
Smooth muscle phosphatase-IV	Turkey gizzard cytosol	58,000 40,000	PP1	Pato and Kerc (1985)
Smooth muscle phosphatase-III	Turkey gizzard cytosol	40,000 (multimeric)	PP1	Tulloch and Pato (1991)
Myosin light chain phosphatase	Chicken gizzard cytosol	37,000 (multimeric)	??	Yoshida and Yagi (1988)
Myosin-associated phosphatase	Chicken gizzard myosin	34,000 (multimeric)	PP1	Mitsui *et al.* (1992)
Avian smooth muscle phosphatase-I	Chicken gizzard myofibrils	130,000 37,000 20,000	PP1	Alessi *et al.* (1992)
Myosin-bound phosphatase	Chicken gizzard actomyosin	58,000 38,000	PP1	Okubo *et al.* (1993)
Myofibrillar form of light chain phosphatase	Turkey gizzard myofibrils	35,000	PP1	Nowak *et al.* (1993)
Mammalian myosin light chain phosphatase	Pig bladder	130,000 37,000 20,000	PP1	Shirazi *et al.* (1994)

the phosphatases listed in Table I as either PP1 or PP2A based on their subunit structure, substrate specificity, immunologic cross-reactivity, and responses to known effectors of PP1 and PP2A, where applicable.

Myosin phosphatases are prepared from both the soluble (cytosolic) and the myofibrillar fraction of smooth muscle. If the source of phosphatase is the soluble fraction, then chromatographic procedures are usually preceded by fractionation with ammonium sulfate. Myosin phosphatase from myofibrils, crude actomyosin, or myosin is solubilized at high ionic strength by 0.6 NaCl plus detergent (Alessi *et al.*, 1992), or a high concentration of $MgCl_2$, 50 to 80 m*M* (Mitsui *et al.*, 1992; Okubo *et al.*, 1993). Although the sequence may be different, most of the purification procedures include ion-exchange chromatography steps (anion and cation exchange) and gel filtration. Cation-exchange column is replaced by heparin-Sepharose in some cases (Tulloch and Pato, 1991; Mitsui *et al.*, 1992; Okubo *et al.*, 1993). Immobilized myosin or HMM (in unphosphorylated or phosphorylated

form) has also been used to absorb myosin phosphatase and to separate contaminating proteins (Pato and Kerc, 1985; Tulloch and Pato, 1991). The use of thiophosphorylated LC20 in an affinity column has been used in many procedures to improve homogeneity (Onishi *et al.*, 1982; Pato and Kerc, 1985; Mitsui *et al.*, 1992; Okubo *et al.*, 1993; Nowak *et al.*, 1993; Shirazi *et al.*, 1994). Myosin phosphatase is sensitive to cleavage by various proteinases, therefore, the application of a proteinase inhibitor "cocktail" is strongly recommended during the course of preparation. Many of the inhibitors used are added for insurance since the specific cell proteases that cleave MLCP are not identified. A likely candidate is the Ca^{2+}-dependent protease, and inhibitors to this enzyme should be used, for example, leupeptin. As a matter of fact we also use diisopropylfluorophosphate. Similar methods that do not minimize proteolysis may result in MLCPs with distinct structure, and this often can be explained by loss or partial cleavage of the phosphatase subunit(s).

B. Properties of Phosphatases Isolated from Smooth Muscles

Many phosphatases have been prepared from several smooth muscle sources as shown in Table I. One of the earlier concerns was whether the myosin phosphatase is a type 1 or 2A enzyme. The myosin phosphatase that was purified by Onishi *et al.* (1982) resembles PP2A based on its subunit composition. This phosphatase affected the superprecipitation of actomyosin via dephosphorylation of LC20 and was assumed to act on intact myosin. Pato and Kerc (1986) purified a phosphatase from gizzard with very similar subunit structure and classified it as PP2A. This phosphatase could not dephosphorylate or bind to intact myosin, suggesting that trimeric PP2A might not be involved in the dephosphorylation of myosin *in vivo*. This controversy is not resolved, but it should be noted that different substrates were used and in the procedure of Onishi *et al.* (1982) no chromatographic separation of PP2A and PP1 was attempted. The myosin phosphatase obtained from the cytosol of bovine aorta (Werth *et al.*, 1982), although not classified at the time of preparation, shows obvious similarities to the dimeric PP2A regarding its subunit structure. It has high myosin phosphatase activity in the presence of metal ions (Co^{2+} or Mn^{2+}) and this enzyme was separated from a trimeric phosphatase of low myosin phosphatase activity (possibly trimeric PP2A) during ionic-exchange chromatography. The high myosin phosphatase activity of a dimeric PP2A is consistent with the results of Pato and Kerc (1986), who reported that proteolytic digestion of the trimeric PP2A from smooth muscle led to the loss of the 55-kDa subunit and this increased the ability of PP2A to dephosphorylate HMM. The polycation-modulated phosphatase isolated from bovine and aortic media (DiSalvo *et al.*, 1985) proved to be a trimeric form of PP2A, and this enzyme dephosphorylated intact myosin and affected smooth muscle contractility in skinned fiber preparations (Bialogjan *et al.*, 1987), suggesting that certain types of trimeric PP2A might be involved in the dephosphorylation of myosin *in vivo*.

Pato and coworkers (Pato and Adelstein, 1983; Pato and Kerc, 1985, 1986) separated and purified four smooth muscle protein phosphatases, identified as SMP-I, SMP-II, SMP-III, and SMP-IV. SMP-I and SMP-II dephosphorylated isolated LC20, and these two enzymes were classified as PP2A and PP2C, respectively (Pato and Adelstein, 1983). SMP-III and SMP-IV dephosphorylated both isolated LC20 and intact myosin, and these enzymes were purified and characterized (Pato and Kerc, 1985; Tulloch and Pato, 1991). SMP-IV was composed of 40- and 58-kDa subunits and eluted as a 150-kDa entity from a gel filtration column. It bound to both dephosphorylated and phosphorylated myosin (Sellers and Pato, 1984), preferentially dephosphorylated the β-subunit of phosphorylase kinase, but was not inhibited by I2. The properties of this phosphatase were similar but not identical to those of PP1. SMP-III was also similar to PP1 in respect that it bound tightly to heparin-Sepharose, but was insensitive to inhibition by I2 and did not dephosphorylate either the α-subunit or the β-subunit of phosphorylase kinase (Tulloch and Pato, 1991). SMP-III appeared as a 390-kDa species of multimeric 40-kDa subunits on gel filtration. Myosin phosphatase with similar multimeric structure (dimer and tetramer involving 37-kDa subunits) was isolated by Yoshida and Yagi (1988), however, no experiments were carried out to classify this enzyme. A PP2B has also been found in chicken gizzard and suggested to dephosphorylate calponin (Fraser and Walsh, 1995). Earlier a PP2A was shown to be involved in the dephosphorylation of calponin (Winder *et al.*, 1992).

From this summary it is clear that several phosphatases may dephosphorylate myosin and there is no obvious way to identify the *in vivo* activity. One approach was to characterize the phosphatase that is associated with myosin (or actomyosin) on the assumption that binding to myosin is a property expected for MLCP. Using okadaic acid as a probe it was found that the major phosphatase in gizzard actomyosin preparations was PP1 (Ishihara *et al.*, 1989). Subsequently several groups isolated MLCP using myosin or actomyosin as a source. The preparation of Mitsui *et al.* (1992) showed a single band of 34 kDa on SDS–PAGE, but a molecular mass of 125 kDa under nondissociating conditions, suggesting a tetramer. This phosphatase preferentially dephosphorylated the β-subunit, was inhibited by I1 as is PP1, but was only partially inhibited by I2. The purified preparation did not bind to myosin, implying that an important "protein factor" responsible for myosin binding was lost during the preparation procedure. An avian smooth muscle PP1, prepared from an actomyosin-rich fraction, was composed of 130-, 37-, and 20-kDa subunits and gel-permeation chromatography indicated a molecular mass of 230 kDa (Alessi *et al.*, 1992). Patterns of inhibition by okadaic acid and I2 suggested a type 1 enzyme and the 38-kDa subunit was found to be the β-isoform of PP1 [i.e., the δ-isoform using the nomenclature of Sasaki *et al.*, (1990)]. The concentration of the phosphatase in gizzard was estimated as 0.7 μM (compared to about 4 μM for MLCK and 80 to 100 μM for LC20). The activity of the holoenzyme toward myosin as substrate was higher than that of the catalytic subunit and it was suggested that this reflected the binding of 130 kDa to myosin, that is, the 130 kDa is a targeting subunit. It was also shown that the catalytic subunit bound to the 130-kDa component. Subsequently, Okubo *et al.* (1993) isolated a myosin-bound phosphatase from chicken gizzard with a subunit composition of 58 and 38 kDa (similar to SMP-IV). The 38-kDa subunit was the δ-isoform of the PP1 catalytic subunit (PP1cδ). However, the 58-kDa subunit was later shown to be a proteolytic fragment of the 130-kDa subunit (Okubo *et al.*, 1994). But it was useful in that the 58-kDa subunit bound to both myosin and PP1cδ and also was used to generate a monoclonal antibody (Okubo *et al.*, 1993, 1994). Two other trimeric phosphatases with subunit compositions similar to that of Alessi *et al.* (1992) have been isolated, one from chicken gizzard (Shimizu *et al.*, 1994) and one from pig bladder (Shirazi *et al.*, 1994). In addition it was suggested, based on experiments carried out with Triton X-100-skinned portal vein, that the trimeric enzyme is the native MLCP, and that the targeting subunit plays an important role (Shirazi *et al.*, 1994). From these three preparations the consensus is that the 38-kDa subunit is the catalytic subunit and the 130-kDa subunit is a target molecule for myosin and PP1c. The function of the 20-kDa subunit is not known.

A "myofibrillar" form of MLCP distinct from those just described was isolated from turkey gizzards by Nowak *et al.* (1993). This consisted of a monomeric 35-kDa component that bound to unphosphorylated myosin. It was suggested that *in vivo* this protein is complexed with MLCK and a 63-kDa subunit. The relationship of this MLCP to the trimeric phosphatases is not known. In contrast, several other groups found that PP1c did not bind to unphosphorylated myosin (Alessi *et al.*, 1992; Mitsui *et al.*, 1992; Okubo *et al.*, 1993).

The foregoing discussion has been limited to dephosphorylation of LC20 at Ser 19. However, LC20 has several potential phosphorylation sites: Ser 1, Ser 2, and Thr 9 [by protein kinase C (Ikebe *et al.*, 1987)] and Thr 18 and Ser 19 (by MLCK). The obvious question is whether one or more phosphatases are involved for each site. Several of the phosphatases listed in Table I (SMP-I through -IV and the polycation-modulated phosphatase) were more effective with Ser 19, Thr 18, and Thr 9 compared to Ser 1/2 (Bárány *et al.*, 1989). Yamakita *et al.* (1994) have demonstrated the importance of the phosphorylation of myosin on Ser 1/2 in mitotically arrested cells. They also showed that release of mitotic arrest resulted in dephosphorylation of Ser 1/2 with a concomitant increase of the phosphorylation level of Ser 19. This inverse change in the phosphorylation levels of Ser 1/2 and Ser 19 suggests the involvement of distinct phosphatases. Since PP1, PP2A, and PP2C are not effective with Ser 1/2 (Bárány

et al., 1989; Erdödi *et al.*, 1989), the phosphatase responsible for the dephosphorylation of this site remains to be identified.

C. Phosphatases of Different Smooth Muscles

A systematic evaluation of phosphatase types in different smooth muscles has not been carried out. Most of the biochemical characterizations have used gizzard as a source (for practical considerations) and only a few other smooth muscles have been studied (see Table I). A larger variety of muscle have been studied using more physiological approaches, either intact or skinned fibers, and these have frequently demonstrated existence of myosin phosphatase but generally have not identified the phosphatase involved.

Smooth muscles can be divided into phasic (most intestinal muscle, gizzard, and uterus) and tonic (major arteries) depending on physiological characteristics. The tonic muscles have a slower relaxation rate and Gong *et al.* (1992a) showed that the phosphatase level in femoral artery was lower than that in phasic muscle (ileum and portal vein). The use of phosphatase inhibitors suggested a type 1 enzyme. In addition, MLCK activity was lower in the tonic muscle (Gong *et al.*, 1992a). The molecular basis for the lower activity in tonic muscle is not established but could reflect a lower concentration, a different phosphatase, or different regulatory mechanisms.

Two aortic phosphatases, prepared from the cytosolic fraction, appear to be PP2A enzymes (Table I). However, evidence is accumulating to suggest that PP1 might be the major myosin phosphatase. This is well documented for the gizzard enzyme, but may also apply to vascular muscle (Gong *et al.*, 1992a). A partially purified phosphatase from aorta resembled SMP-IV and had properties of the type 1 enzyme (Erdödi *et al.*, 1989). This was suggested as the major myosin phosphatase in aorta muscle. The PP1cδ isoform has been detected in cDNA libraries of rat and human aortas (M. Ito, unpublished observations). Even though PP1 may be the major myosin phosphatase of both tonic and phasic smooth muscles, some differences would not be unexpected. For example, in preliminary experiments it was shown (F. Erdödi, unpublished observations) that the amount of PP1 bound to actomyosin was considerably less for aorta compared to gizzard. Thus myosin binding properties may vary.

D. Isoforms of the Type 1 Catalytic Subunit

Several laboratories have demonstrated the existence of PP1c isoforms. (Dombradi *et al.*, 1990; Sasaki *et al.*, 1990; Wadzinski *et al.*, 1990). We will use the no-

```
δ   MADGE-LNVDSLITRLLEVRGCRPGKIVQMTEAEVRGLCIKSREIFLSQP   49
α   •S•S•K••L••I•G•••••Q•S••••N••L••N•I••••L•••••••••   50
γ1  •••IDK••I••I•Q••••••••SK•••N••LQ•N•I••••L•••••••••   50
γ2  •••IDK••I••I•Q••••••••SK•••N••LQ•N•I••••L•••••••••   50

δ   ILLELEAPLKICGDIHGQYTDLLRLFEYGGFPPEANYLFLGDYVDRGKQS   99
α   ••••••••••••••••••Y•••••••••••••••S••••••••••••••  100
γ1  ••••••••••••••••••Y•••••••••••••••S••••••••••••••  100
γ2  ••••••••••••••••••Y•••••••••••••••S••••••••••••••  100

δ   LETICLLLAYKIKYPENFFLLRGNHECASINRIYGFYDECKRRFNIKLWK  149
α   •••••••••••••••••••••••••••••••••••••••••Y•••••••  150
γ1  •••••••••••••••••••••••••••••••••••••••••Y•••••••  150
γ2  •••••••••••••••••••••••••••••••••••••••••Y•••••••  150

δ   TFTDCFNCLPIAAIVDEKIFCCHGGLSPDLQSMEQIRRIMRPTDVPDTGL  199
α   •••••••••••••••••••••••••••••••••••••••••••••Q••  200
γ1  •••••••••••••••••••••••••••••••••••••••••••••Q••  200
γ2  •••••••••••••••••••••••••••••••••••••••••••••Q••  200

δ   LCDLLWSDPDKDVQGWGENDRGVSFTFGADVVSKFLNRHDLDLICRAHQV  249
α   •••••••••••••••••••••••••••E••A•••HK••••••••••••  250
γ1  •••••••••••••F••••••••••••••E••A•••HK••••••••••••  250
γ2  •••••••••••••L••••••••••••••E••A•••HK••••••••••••  250

δ   VEDGYEFFAKRQLVTLFSAPNYCGEFDNAGGMMSVDETLMCSFQILKPSE  299
α   ••••••••••••••••••••••••••A•••••••••••••••••AD   300
γ1  ••••••••••••••••••••••••••A••••••••••••••••••A•   300
γ2  ••••••••••••••••••••••••••A••••••••••••••••••A•   300

δ   K-KAKY-QYGGLNSG-RPVTPPR---T--A-NPPK-K----R         327
α   •N•G••G•FS•••P•G••I••••---NS-•-KA-•-•              330
γ1  •-•-•P------•AT-•••••••GMI•KQ•-----•-•             323
γ2  •-•-•P------•AT-•••••••---VGSGL••SIQ•ASNY•NNTVLYE   337
```

FIGURE 1 Comparison of the sequences of the isoforms of the PP1 catalytic subunit (Sasaki *et al.*, 1990). Identical residues are indicated by dots; lack of aligning residues is indicated by bar.

menclature of Sasaki *et al.* (1990). Four cDNA clones were isolated from rat libraries and these corresponded to isoforms α, γ$_1$, γ$_2$, δ. Their molecular masses are 38, 36, 39, and 37 kDa, respectively. These four isoforms are thought to be products of three genes, with the γ-isoform produced by alternative splicing (see also Dombradi *et al.*, 1990). The deduced amino acid sequences showed about 90% identity with differences mainly in the C-terminal regions. The C-terminal differences allow the preparation of selective antibodies (Shima *et al.*, 1993). The sequences of the four isoforms are given in Fig. 1 and the differences are indicated.

The PP1cδ isoform is associated with the trimeric phosphatase from chicken and turkey gizzard (Okubo *et al.*, 1993). Alessi *et al.* (1992) also showed the presence of this isoform (in their nomenclature it is the β-isoform). The same isoform is associated with skeletal muscle myosin (Dent *et al.*, 1992). It is reasonable to expect that each isoform has distinctive properties, although this has not yet been demonstrated. Zhang *et al.* (1992) developed an expression system for the PP1c subunit and found that the four isoforms were similar with respect to inhibition by I2 and okadaic acid (Zhang *et al.*, 1993). They required Mn^{2+} for full activity. The PP1c has also been expressed in the baculo-

virus system (Cohen and Berndt, 1991) and here Mn^{2+} was used in the renaturation procedure. [It is interesting that the MLCP isolated by Okubo *et al.* (1993) was dependent on Co^{2+} and Mn^{2+} for activity.] Mutagenesis of residues 274–277 (GEFD, identical for all four isoforms) to YRCG (from PP2A) resulted in an increased sensitivity to okadaic acid, suggesting that this region of the molecule is involved in binding to the phosphatase inhibitors (Zhang *et al.*, 1994). There was also a decrease in the sensitivity to I2, indicating that the binding sites for I2 and okadaic acid may be close.

The distribution of the isoforms in different tissue has not been analyzed in detail. The information that does exist suggests that various isoforms are enriched in some tissues. For example, PP1cγ$_1$ was high in brain and PP1cδ was enriched in brain, lung, and intestine. Both were low in liver and skeletal muscle (Shima *et al.*, 1993). PP1cγ$_2$ was shown to be high in testis and may interact with a member of the 70-kDa heat-shock protein family (Chun *et al.*, 1994). Whether there is a temporal distribution, for example, during the cell cycle, is not known.

The PP1cδ isoform is present in smooth muscles and was detected bound to gizzard myosin (Okubo *et al.*, 1994). In addition, cDNA clones for PP1cδ have been isolated from cDNA libraries of chicken gizzard

(Shimizu *et al.*, 1994) and rat and human aorta (M. Ito, unpublished observations). The nucleotide sequences of the δ-isoforms of gizzard and human aorta (in open reading frames) were 86 and 94% identical to the rat isoform, respectively. The deduced amino acid sequences for the δ-isoform from these three sources were identical (Shimizu *et al.*, 1994; M. Ito, unpublished observations). PP1cα also was detected in the gizzard cDNA library (M. Ito, unpublished observations).

Consideration of other catalytic subunits (PP2A, etc.) is beyond the scope of this article and the reader is referred to Cohen (1991).

E. *Targeting Subunits*

Hundreds of different kinases have been identified but there are relatively few phosphatases. Obviously the phosphatases are coordinated with the kinases and one mechanism that could increase substrate specificity and possibly play a regulatory role involves targeting subunits. This concept was developed by Cohen and colleagues and the classic example is the 161-kDa glycogen-binding subunit (see previous discussion; Dent *et al.*, 1990) associated with PP1. It has been suggested that the 130-kDa subunit of the trimeric phosphatase is a targeting molecule in that it couples the substrate, myosin, and the PP1c subunit (Alessi *et al.*, 1992). The evidence in support of this theory is: the trimeric MLCP has higher activity toward HMM than the isolated catalytic subunit (Alessi *et al.*, 1992; Shirazi *et al.*, 1994); the holoenzyme is more effective in promoting relaxation, compared to the catalytic subunit (Shirazi *et al.*, 1994); arachidonic acid dissociates the holoenzyme into subunits and slows relaxation and LC20 dephosphorylation (Gong *et al.*, 1992b); the 58-kDa fragment of the 130-kDa subunit (Okubo *et al.*, 1993) and the intact 130-kDa subunit bind to myosin (Shimizu *et al.*, 1994); and the 130-kDa subunit binds the catalytic subunit (Alessi *et al.*, 1992; Shimizu *et al.*, 1994).

Using the monoclonal antibody to the 58-kDa fragment, it was found, by Western blots, that a protein similar or identical to the 130-kDa subunit was present in a wide range of tissues (Okubo *et al.*, 1994). A cross-reacting species was present in each of the smooth muscles examined (stomach, gizzard, aorta, oviduct, and small intestine) and was also detected in brain, spleen, kidney, lung, and cardiac muscle. Slightly different apparent molecular masses were observed in many tissues, for example, in cardiac muscle a doublet of 143 and 137 kDa was observed. It is not known if these different immunoreactive species are different isoforms. Northern blots confirmed these data, and

using a labeled cDNA probe, mRNA at 5.7 kilobases was detected in those tissues but not in liver or skeletal muscle (Shimizu *et al.*, 1994). These results suggest that the 130-kDa subunit may be involved in cell function other than contraction. Possibilities include cell motility or cytoskeletal reorganization. It is not known whether myosin is an obligatory partner.

The large subunit of the trimeric MLCP, thought to be a targeting subunit, has been characterized from two sources. cDNA clones were isolated from a chicken gizzard cDNA library and overlapping clones indicated the presence of two isoforms. These have been called M130 and M133 (the actual molecular masses are 106.7 and 111.6 kDa, respectively). The insert of the larger isoform is residues 512–552. The major isoform in gizzard is M130. The N-terminal third of the molecule is composed of eight repeat sequences between residues 39 and 295. Five of the eight repeats contained 33 residues, with the fourth, fifth, and eighth containing 29, 31, and 32 residues, respectively. Twenty of the residues in these repeat sequences are conserved and correspond to the sequence motif for cdc10/SW16 or ankyrin repeat. The ankyrin repeat is found in several other proteins and these are listed by Shimizu *et al.* (1994). Several of these proteins are involved in tissue differentiation or regulation of the cell cycle. It is also possible that the ankyrin repeat may be involved with the cytoskeleton (cf. the role of ankyrin in the red blood cell membrane). Other features of the molecule include an acidic cluster (326–372); two ionic clusters (719–755 and 814–848 for M133); and a Ser/Thr-rich region (770–793). Overall the molecule is hydrophilic. It has no significant homology to other PP1 binding proteins. Several potential phosphorylation sites exist for the cAMP-dependent protein kinase, protein kinase C, p34[cdc2], and glycogen synthase kinase-3. The 58-kDa fragment, obtained in the preparation of Okubo *et al.* (1993), is the N-terminal fragment of M130 (or M133). This fragment binds to myosin and to the catalytic subunit, but probably not to the 20-kDa subunit.

In another study two subunits were characterized: the small subunit (M_{21}) from chicken gizzard and the large subunit (M_{110}) from rat aorta (Chen *et al.*, 1994). Overall matching identity of the M130 and M_{110} is 79%. The N-terminal regions of the two molecules are 96% homologous and only 9 of 257 residues are different. The degrees of similarity for different regions of the molecules are shown in Fig. 2. Seven ankyrin repeats were identified in M_{110}. The region 171 to 197 was not considered by Chen *et al.* (1994) as an ankyrin repeat, but was included by Shimizu *et al.* (1994). The regions of the two molecules that are markedly different are 553–610 (present in M130/M133, not in M_{110})

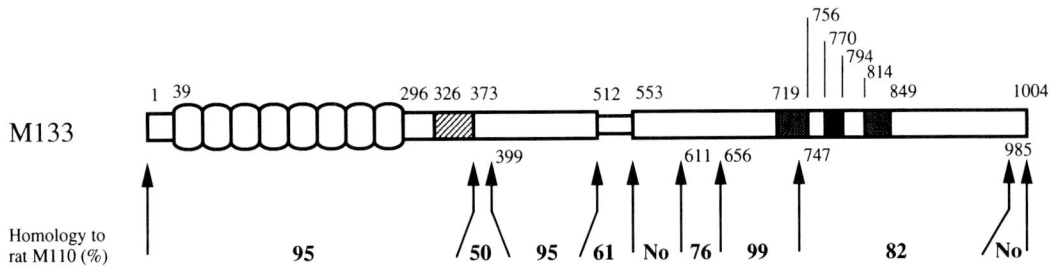

FIGURE 2 Comparison of the gizzard M133 subunit with the Rat M110 subunit. The domain map for the M133 subunit is shown diagrammatically. Various regions are indicated: the 8 ankyrin repeats are shown between residues 39 and 295; an acidic cluster lies at 326–372; ionic clusters (acidic and basic) at 719–755 and 814–848; insert region (not present in M130) at 512–552; and the Ser/Thr-rich cluster at 770–793. For different segments of M133 the homology of M110 is shown (%).

and the C-terminal regions of the two molecules. In the rat isoform, a four-7 unit leucine zipper sequence is found that is not obvious in M130. It is interesting that M_{21} also contains the leucine zipper sequence and these may be important for interaction of the two subunits or with other proteins, for example, myosin (Chen *et al.*, 1994).

III. PHOSPHATASE INHIBITORS

Several phosphatase inhibitors are available and have been useful in many aspects of smooth muscle research. Okadaic acid was the first inhibitor to be widely used and inhibition of phosphatase activity by this compound was demonstrated using smooth muscle (Hartshorne *et al.*, 1989). Calyculin A was also used in several earlier studies. Since 1990 other compounds have been described as phosphatase inhibitors. These are remarkable in that despite considerable differences in structure, they are relatively specific as phosphatase inhibitors. These inhibitors include: tautomycin, as antifungal antibiotic produced by *Streptomyces* (MacKintosh and Klumpp, 1990; Hori *et al.*, 1991); the cyclic peptide hepatoxins, microcystins, and nodularin (Eriksson *et al.*, 1990; MacKintosh *et al.*, 1990) produced by cyanobacteria; and the cantharidin derivatives (Li and Casida, 1992; Honkanen, 1993), which are widely used in agriculture as herbicides or insecticides. Cantharidin is the active ingredient of blister beetle extract. [This is only a partial listing; for a more complete review, see Fujiki and Suganuma (1993)].

All of these compounds are effective inhibitors of PP1 and PP2A when used *in vitro*. Okadaic acid shows a marked difference between the two classes with a much more potent inhibition of type 2A. Using phosphorylated LC20 as substrate, K_i values for

microcystin-LR, calyculin A, and tautomycin were approximately 0.05, 1, and 0.48 nM for PP1 and 0.013, 0.1, and 29 nM for PP2A, respectively (Takai *et al.*, 1995). Of these inhibitors only tautomycin is more potent for PP1. K_i values for cantharidin are 1.11 and 0.19 μM for PP1 and PP2A, respectively (Honkanen, 1993), using phosphohistone as substrate.

When applied to intact cells, different factors should be considered. The microcystins, cantharidin, and endothal are not permeable across the plasmalemma but may be taken up by hepatocytes via the bile acid transport system (Eriksson *et al.*, 1990). A derivative of endothal, endothal thioanhydride, is permeable across the cell membrane (M. Hirano and F. Erdödi, unpublished observations). Thus permeability across lipid bilayers is one consideration. In addition, the potency of inhibitors used with intact cells always appears less than the *in vitro* assays. The potency with intact cells is often 10- to 100-fold less sensitive. For example, with 3T3 fibroblasts, external concentrations above 10 nM were required to elicit shape changes (Chartier *et al.*, 1991). This difference in dose dependence could be due to inefficient uptake by the cells or preferential localization of the inhibitor with lipids. Cohen *et al.* (1989) suggested that the higher concentrations required with intact cells reflect the intracellular concentration of the targeted phosphatase.

The cell-permeable inhibitors induce dramatic changes in cell morphology. These have been documented in several cell types (see Hirano *et al.*, 1992). In 3T3 fibroblasts the changes involve a reorganization of F-actin and its gradual aggregation into a ball-like structure that is connected to the nucleus via cables of intermediate filaments. Several proteins are phosphorylated, but of the major cytoskeletal proteins only vimentin and myosin were identified (Chartier *et al.*, 1991; Hirano and Hartshorne, 1993). It is not known if these extensive changes occur in smooth muscle cells, but since this is a possibility, the mechanics of the contractile process induced by phosphatase inhibitors should be interpreted with caution.

In spite of these words of caution, the phosphatase inhibitors still represent a powerful tool for smooth muscle research. They are specific and effectively inhibit PP1 and PP2A activities in either the intact cell or extracts and have been used to stabilize phosphorylation levels during preparative procedures. In addition, they may be used in affinity columns, for example, the use of a microcystin-LR derivative for the isolation of PP1 (Moorhead *et al.*, 1994).

IV. SUMMARY

Smooth muscle contains many protein phosphatases, as expected for any eukaryotic cell, and several of these have been shown to use phosphorylated myosin as substrate. Whether myosin is a native substrate is not known, but the possibility that more than one phosphatase acts on myosin cannot be excluded. However, if the criterion is that the relevant phosphatase should be associated with, that is, bind to, the contractile apparatus, then the possibilities are more restricted. The consensus is that the myosin-bound phosphatase is a type 1 enzyme and the catalytic subunit has been identified as the δ-isoform. In addition, several groups have shown that a trimeric enzyme is associated with myosin and the possibility of targeting subunits has been raised. The molecular masses of the trimer are (approximately) 130, 38, and 20 kDa, and the 38-kDa subunit is the catalytic subunit. Thus there are two subunits that could be involved in targeting the catalytic subunit to the substrate or in regulation of phosphatase activity. Some evidence, certainly not conclusive, has indicated that a fragment of the larger subunit and the large subunit itself can bind to myosin. Thus there is experimental support for the idea that the large subunit is a targeting subunit and in this capacity would bind to the substrate, myosin, and also to the catalytic subunit. The small subunit could also be a targeting molecule, and its role with respect to the myosin-bound PP1 activity has yet to be established. It is interesting that the 20-kDa subunit from chicken gizzard has a leucine zipper sequence at its C-terminal end and thus its interaction with some other component would be predicted. Presumably its partner would also contain the leucine zipper motif and this might include the large subunit from the rat trimeric phosphatase, that is, the M_{110} subunit. Details of the interactions, including the identification of various binding sites, remain to be determined.

The activity of the trimeric phosphatase toward P-myosin is higher than that of the isolated catalytic subunit. Thus dissociation of the trimeric phosphatase could be a regulatory mechanism. The factors involved in dissociation of PP1c are not established, although there is the suggestion that arachidonic acid is implicated. It is possible that dissociation of the trimeric phosphatase is the end result of the G-protein-linked inhibition of phosphatase activity, observed in several systems. Another possibility is that the trimeric phosphatase binds only to phosphorylated myosin and has a lower affinity for the dephosphorylated form. Whatever the scenario imagined, it is clear that much remains to be learned about regulation of myosin phosphatase activity.

Another intriguing point concerns the stoichiometry of myosin heads (or LC20) to the phosphatase. Assuming a myosin concentration of 50 μM, and a phosphorylation level of 50% (of total myosin), the P-LC20 concentration would also be 50 μM. This is much higher than the phosphatase concentration (about 0.7 μM) and thus there is the enigma of accessibility. If the phosphatase is bound to the thick filaments, how does it access the many phosphorylated myosin heads? Is there another mechanism involved in the "on–off" binding to myosin? There are many points to be resolved and an emphasis of future research will be to identify regulatory mechanisms involved in the dephosphorylation of smooth muscle myosin.

Acknowledgments

This work was supported in part by grants from the National Institutes of Health, HL 23615 and HL 20984 (D.J.H.), by a grant from OTKA (F.E.), and by Grants-in-Aid (M.I.) for Scientific Research and for Scientific Research on priority areas from the Ministry of Education, Science and Culture, Japan.

References

Alessi, D., MacDougall, L. D., Sola, M. M., Ikebe, M., and Cohen, P. (1992). *Eur. J. Biochem.* **210**, 1023–1035.

Arino, J., Woon, C. W., Brautigan, D. L., Miller, T. B., and Johnson, G. L. (1988). *Proc. Natl. Acad. Sci. U.S.A.* **85**, 4252–4256.

Bárány, K., Pato, M., Rokolya, A., Erdödi, F., DiSalvo, J., and Bárány, M. (1989). *Adv. Protein Phosphatases* **5**, 517–534.

Beullens, M., Van Eynde, A., Stalmans, W., and Bollen, M. (1992). *J. Biol. Chem.* **267**, 16538–16544.

Bialojan C., Rüegg, J. C., and DiSalvo, J. (1987). *Pfuegers Arch.* **410**, 304–312.

Bollen, M., and Stalmans, W. (1992). *Crit. Rev. Biochem. Mol. Biol.* **27**, 227–281.

Chartier, L., Rankin, L. L., Allen, R. E., Kato, Y., Fusetani, N., Karaki, H., Watabe, S., and Harshorne, D. J. (1991). *Cell Motil. Cytoskel.* **18**, 26–40.

Chen, Y. H., Chen, M. X., Alessi, D. R., Campbell, D. G., Shanahan, C., Cohen, P., and Cohen, P. T. W. (1994). *FEBS Lett.* **356**, 51–55.

Chun, Y.-S., Shima, H., Nagasaki, K., Sugimura, T., and Nagao, M. (1994). *Proc. Natl. Acad. Sci. U.S.A.* **91**, 3319–3323.

Cohen, P. (1989). *Annu. Rev. Biochem.* **58**, 453–508.

Cohen, P. (1991). *In* "Methods in Enzymology" (T. Hunter and B. Sefton, eds.), Vol. 201, pp. 389–398. Academic Press, San Diego, CA.

Cohen, P., and Cohen, P. T. W. (1989). *J. Biol. Chem.* **201**, 21435–21438.

Cohen, P., Klumpp, S., and Schelling, D. L. (1989). *FEBS Lett.* **250**, 596–600.

Cohen, P. T. W., and Berndt, N. (1991). *In* "Methods in Enzymology" (T. Hunter and B. Sefton, eds.), Vol. 201, pp. 408–414. Academic Press, San Diego, CA.

Dent, P., Lavoinne, A., Nakielny, S., Caudwell, F. B., Watt, P., and Cohen, P. (1990). *Nature (London)* **348**, 302–308.

Dent, P., MacDougall, L. K., MacKintosh, C., Campbell, D. G., and Cohen, P. (1992). *Eur. J. Biochem.* **210**, 1037–1044.

DiSalvo, J., Gifford, D., and Kokkinakis, A. (1985). *Adv. Protein Phosphatases* **1**, 327–345.

Dombradi, V., Axton, J. M., Brewis, N. D., DaCruze Silva, E. F., Alphey, L., and Cohen, P. T. W. (1990). *Eur. J. Biochem.* **194**, 739–745.

Erdödi, F., Csortos, C., Bot, G., and Gergely, P. (1985). *Biochem. Biophys. Res. Commun.* **128**, 705–712.

Erdödi, F., Rokolya, A., Bárány, M., and Bárány, K. (1989). *Biochim. Biophys. Acta* **1011**, 67–74.

Eriksson, J. E., Toivola, D., Meriluoto, J. A. O., Karaki, H., Han, Y.-G., and Hartshorne, D. J. (1990). *Biochem. Biophys. Res. Commun.* **173**, 1347–1353.

Fraser, E. D., and Walsh, M. P. (1995). *Biophys. J.* **68**, 162a.

Fujiki, H., and Suganuma, M. (1993). *Adv. Cancer Res.* **61**, 143–194.

Gergely, P., Erdödi, F., and Bot, G. (1984). *FEBS Lett.* **169**, 45–48.

Gong, M. C., Cohen, P., Kitazawa, T., Ikebe, M., Masua, M., Somlyo, A. P., and Somlyo, A. V. (1992a). *J. Biol. Chem.* **267**, 14662–14668.

Gong, M. C., Fuglsang, A., Alessi, D., Kobayashi, S., Cohen, P., Somlyo, A. V., and Somlyo, A. P. (1992b). *J. Biol. Chem.* **267**, 21492–21498.

Guan, K., Broyles, S. S., and Dixon, J. E. (1991). *Nature (London)* **350**, 359–362.

Hartshorne, D. J. (1987). *In* "Physiology of the Gastrointestinal Tract" (L. R. Johnson, ed.), 2nd ed., pp. 423–482. Raven Press, New York.

Hartshorne, D. J., Ishihara, H., Karaki, H., Ozaki, H., Sato, K., Hori, M., and Watabe, S. (1989). *Adv. Protein Phosphatases* **5**, 219–231.

Hendrix, P., Turowski, P., Mayer-Jaekel, R. E., Goris, J., Hofsteenge, J., Merlevede, W., and Hemmings, B. A. (1993). *J. Biol. Chem.* **268**, 7330–7337.

Hirano, K., and Hartshorne, D. J. (1993). *Eur. J. Cell Biol.* **62**, 59–65.

Hirano, K., Chartier, L., Taylor, R. G., Allen, R. E., Fusetani, N., Karaki, H., and Hartshorne, D. J. (1992). *J. Muscle Res. Cell Motil.* **13**, 341–353.

Honkanen, R. E. (1993). *FEBS Lett.* **330**, 283–386.

Hori, M., Magae, J., Han, Y.-G., Hartshorne, D. J., and Karaki, H. (1991). *FEBS Lett.* **285**, 145–148.

Hubbard, M. J., and Cohen, P. (1989). *Eur. J. Biochem.* **180**, 457–465.

Ichikawa, K., Ito, M., Okubo, S., Konishi, T., Nakano, T., Mino, T., Nakamura, F., Naka, M., and Tanaka, T. (1993). *Biochem. Biophys. Res. Commun.* **193**, 828–833.

Ikebe, M., Hartshorne, D. J., and Elzinga, M. (1987). *J. Biol. Chem.* **262**, 9569–9573.

Ingebritsen, T., and Cohen, P. (1983). *Science* **221**, 331–338.

Ishihara, H., Martin, B. L., Brautigan, D. L., Karaki, H., Ozaki, H., Kato, Y., Fusetani, N., Watabe, S., Hashimoto, K., Uemura, D., and Hartshorne, D. J. (1989). *Biochem. Biophys. Res. Commun.* **159**, 871–877.

Li, Y.-M., and Casida, J. E. (1992). *Proc. Natl. Acad. Sci. U.S.A.* **89**, 11867–11870.

MacKintosh, C., and Klumpp, S. (1990). *FEBS Lett.* **277**, 137–140.

MacKintosh, C., Beattie, K. A., Klumpp, S., Cohen, P., and Codd, G. A. (1990). *FEBS Lett.* **264**, 187–192.

Mayer-Jaekel, R. E., and Hemmings, B. A. (1994). *Trends Cell Biol.* **4**, 287–291.

Mitsui, T., Inagaki, M., and Ikebe, M. (1992). *J. Biol. Chem.* **267**, 16727–16735.

Moorhead, G., MacKintosh, R. W., Morrice, N., Gallagher, T., and MacKintosh, C. (1994). *FEBS Lett.* **356**, 46–50.

Mumby, M. C., and Walter, G. (1993). *Physiol. Rev.* **73**, 673–699.

Nowak, G., Rainer, F., and Sobieszek, A. (1993). *Biochim. Biophys. Acta* **1203**, 230–235.

Okubo, S., Erdödi, F., Ito, M., Ichikawa, K., Konishi, T., Nakano, T., Kawamura, T., Brautigan, D. L., and Hartshorne, D. J. (1993). *Adv. Protein Phosphatases* **8**, 795–314.

Okubo, S., Ito, M., Takashiba, Y., Ichikawa, K., Miyahara, M., Shimizu, H., Konishi, T., Shima, H., Nagao, M., Hartshorne, D. J., and Nakano, T. (1994). *Biochem. Biophys. Res. Commun.* **200**, 429–434.

Onishi, H., Umeda, J., Uchiwa, H., and Watanabe, S. (1982). *J. Biochem. (Tokyo)* **91**, 265–271.

Pato, M. D., and Adelstein, R. S. (1983). *J. Biol. Chem.* **258**, 7047–7054.

Pato, M. D., and Kerc, E. (1985). *J. Biol. Chem.* **260**, 12359–12366.

Pato, M. D., and Kerc, E. (1986). *J. Biol. Chem.* **261**, 3770–3774.

Pelech, S., and Cohen, P. (1985). *Eur. J. Biochem.* **128**, 245–251.

Sasaki, K., Shima, H., Kitagawa, Y., Irino, S., Sugimura, T., and Nagao, M. (1990). *Jpn. J. Cancer Res.* **81**, 1272–1280.

Sellers, J. R., and Pato, M. D. (1984). *J. Biol. Chem.* **259**, 7740–7746.

Shima, H., Hatano, Y., Chun, Y.-S., Sugimura, T., Zhang, Z., Lee, E. Y. C., and Nagao, M. (1993). *Biochem. Biophys. Res. Commun.* **192**, 1289–1296.

Shimizu, H., Ito, M., Miyahara, M., Ichikawa, K., Okubo, S., Konishi, T., Naka, M., Tanaka, T., Hirano, K., Hartshorne, D. J., and Nakano, T. (1994). *J. Biol. Chem.* **269**, 30407–30411.

Shirazi, A., Iizuka, K., Fadden, P., Mosse, C., Somlyo, A. P., Somlyo, A. V., and Haystead, T. A. J. (1994). *J. Biol. Chem.* **269**, 31598–31606.

Somlyo, A. P., and Somlyo, A. V. (1994). *Nature (London)* **372**, 231–236.

Takai, A., Bialojan, C., Troschka, M., and Rüegg, J. C. (1987). *FEBS Lett.* **217**, 81–84.

Takai, A., Sasaki, K., Nagai, H., Mieskes, G., Isobe, M., Isono, K., and Yasumoto, T. (1995). *Biochem. J.* **306**, 657–665.

Tang, P. M., Bondor, J. A., Swiderek, K. M., and DePaoli-Roach, A. A. (1991). *J. Biol. Chem.* **266**, 15782–15789.

Tulloch, A. G., and Pato, M. D. (1991). *J. Biol. Chem.* **266**, 20168–20174.

Walter, G., and Mumby, M. C. (1993). *Biochim. Biophys. Acta* **1155**, 207–226.

Wadzinski, B. E., Heasley, L. E., and Johnson, G. L. (1990). *J. Biol. Chem.* **265**, 21504–21508.

Werth, D. K., Haeberle, J. R., and Hathaway, D. R. (1982). *J. Biol. Chem.* **257**, 7306–7309.

Winder, S. J., Pato, M. D., and Walsh, M. P. (1992). *Biochem. J.* **286**, 197–203.

Yamakita, Y., Yamashiro, S., and Matsumura, F. (1994). *J. Cell Biol.* **124,** 129–137.

Yoshida, M., and Yagi, K. (1988). *J. Biochem. (Tokyo)* **103,** 380–385.

Zhang, Z., Bai, G., Deans-Zirattu, S., Browner, M. F., and Lee, E. Y. C. (1992). *J. Biol. Chem.* **267,** 1484–1490.

Zhang, Z., Bai, G., Shima, M., Zhao, S., Nagao, M., and Lee, E. Y. C. (1993). *Arch. Biochem. Biophys.* **303,** 402–406.

Zhang, Z., Zhao, S., Long, F., Zhang, L., Bai, G., Shima, H. Nagao, M., and Lee, E. Y. C. (1994). *J. Biol. Chem.* **269,** 16997–17000.

11

Calcium/Calmodulin-Dependent Protein Kinase II

HAROLD A. SINGER, S. THOMAS ABRAHAM,
and CHARLES M. SCHWORER

Sigfried and Janet Weis Center for Research
Geisinger Clinic
Danville, Pennsylvania

I. INTRODUCTION

Ca^{2+}/calmodulin (Ca^{2+}/CaM)-dependent protein kinases are important mediators of signal transduction events triggered by stimuli that increase intracellular levels of free $[Ca^{2+}]$. Phosphorylase kinase and myosin light chain kinase (MLCK) are two examples of Ca^{2+}/CaM-dependent kinases that have specific substrates and well-defined functions. Several Ca^{2+}/CaM-dependent kinases have been identified that phosphorylate multiple substrates *in vitro* and are likely to have diverse cellular functions analogous to other multifunctional kinases, such as the cyclic nucleotide-dependent protein kinases and protein kinase C (PKC). Ca^{2+}/calmodulin-dependent protein kinase II, referred to here as CaM-kinase II, is the best known of these multifunctional CaM-kinases. Much of the current state of knowledge relating to this protein kinase has resulted from the availability of readily purified kinase from brain, where it accounts for as much as 2% of the total protein in certain regions (Erondu and Kennedy, 1985). Interest in brain CaM-kinase II has been fueled by its implicated involvement in such critical functions as synaptic transmission and memory (Hanson and Schulman, 1992). However, CaM-kinase II is ubiquitous, and as insight into its unique structural and autoregulatory features has grown, investigators have become interested in its functional roles in other tissues. In this chapter we will review these structural and regulatory properties with emphasis on CaM-kinase II and its cellular activation in smooth muscle.

II. STRUCTURAL PROPERTIES OF CALMODULIN-KINASE II

A. Holoenzyme and Subunit Structure

CaM-kinase II has been purified from mammalian brain and a number of peripheral tissues such as liver and heart (see Colbran and Soderling, 1990, for review). In each case it has been described as a large multimeric protein of approximately 300–600 kDa composed of 50- to 60-kDa kinase subunits. Four primary subunits (α, β, δ, γ) were first identified and cloned from rat brain cDNA libraries (Lin *et al.*, 1987; Bennett and Kennedy, 1987; Tobimatsu *et al.*, 1988; Tobimatsu and Fugisawa, 1989). The domain structures of the known subunits and their calculated molecular masses are shown in Fig. 1A. Multiple alternatively spliced CaM-kinase II subunits that are closely related to the rat α-subunit (77% amino acid identity) have also been identified in *Drosophila* (Cho *et al.*, 1991; Ohsako *et al.*, 1993), indicating the kinase is strongly conserved across species. The first 315 amino acids (AA) are highly conserved among the subunits and comprise the "catalytic/regulatory" domain containing a consensus ATP binding sequence, catalytic, and regulatory regions. A variable region separates the catalytic/regulatory domain from a C-terminal "association" domain, which is approximately 150 AA in length and also highly conserved among subunits. The δ-subunit originally cloned from brain is distinguished structurally from the other subunits by a unique 21-AA carboxyl terminus (Tobimatsu and Fu-

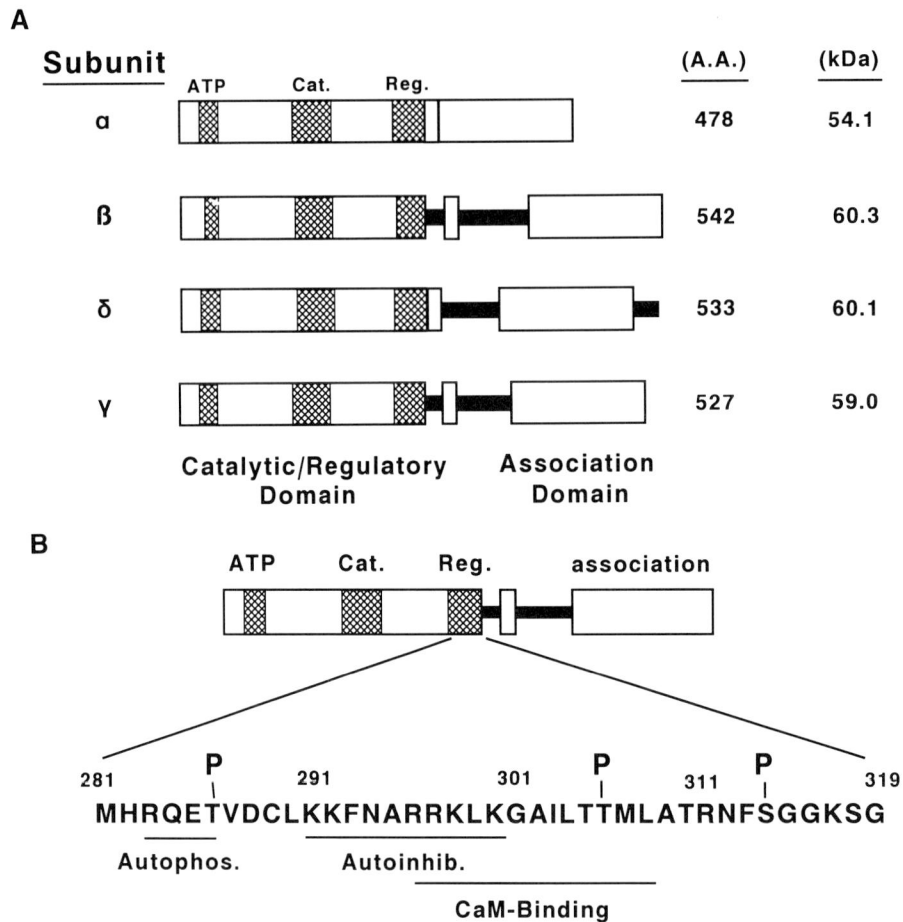

FIGURE 1   (A) CaM-kinase II subunit structures. Open boxes indicate conserved domains and solid boxes variable domains. Shaded areas designate conserved ATP binding (ATP), catalytic (Cat.), and regulatory (Reg.) domains. The conserved association domain is required for the formation of the multimeric holoenzyme. Sizes of the full-length subunits are indicated by the number of amino acids in the sequence (A.A.) and calculated molecular mass (kDa). (B) CaM-kinase II regulatory domain. Overlapping autoinhibitory (Autoinhib.) and calmodulin (CaM) binding domains and known autophosphorylation sites are indicated. Thr[286] has been identified as the primary autophosphorylation site (Autophos.) conferring Ca^{2+}/CaM-independent activity to the kinase. Secondary autophosphorylation events include Thr[306], which *in vitro* results in a decreased binding affinity for CaM.

jisawa, 1989). Proteolytic cleavage of subunits in the regulatory region (Colbran *et al.*, 1988; Yamagata *et al.*, 1991) or expression of mutant kinase subunits truncated in the regulatory region (Cruzalegui *et al.*, 1992) generates a constitutively active monomeric kinase activity, hence the name for the conserved "association" domain, which is inferred to be required for formation of the holoenzyme. Electron microscopic imaging of purified brain CaM-kinase II molecules suggests that the holoenzyme has the shape of a rosette (similar to that depicted in Fig. 4), consistent with the association of 8–10 "dumbbell"-shaped kinase subunits (Kanaseki *et al.*, 1991).

An additional level of structural complexity within the CaM-kinase II subunits has become apparent with the identification of a number of subunit variants. Alternatively spliced forms of the β-subunit (β, β') (Bulleit *et al.*, 1988), α-subunit (α, α-33) (Benson *et al.*, 1991), δ-subunit (δ_{1-4}) (Schworer *et al.*, 1993), and γ-subunit (γ_{A-C}) (Nghiem *et al.*, 1993; Zhou and Ikebe, 1994), differing in variable domain nucleic acid sequence, have been identified. In addition, Mayer *et al.* (1993) have described δ-subunit variants lacking the δ-specific C-terminal sequence. Figure 2 depicts the domain structure of δ-isoforms deduced from rat nucleic acid sequences identified by reverse transcriptase–polymerase chain reaction (RT–PCR) cloning of mRNA from various rat tissues (Schworer *et al.*, 1993;

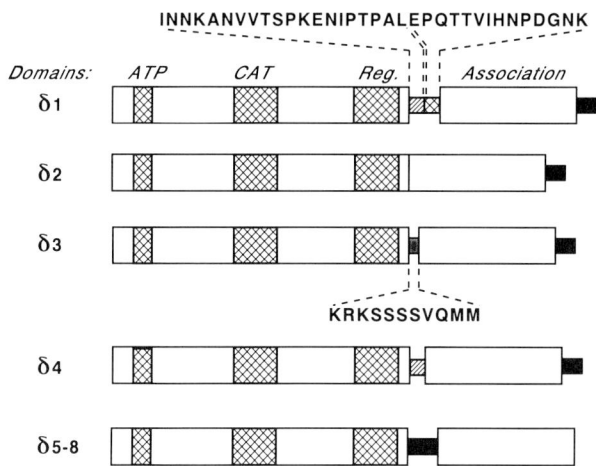

FIGURE 2   Structure of δ-subunit variants deduced from PCR-amplified target sequences of δ-subunit cDNA from rat brain, VSM, and skeletal muscle (Schworer et al., 1993). The deduced amino acid sequence in the variable domain is indicated. The variable domain is deleted in δ$_2$, substituted in δ$_3$, and truncated in δ$_4$ by deletion of the δ$_1$ variable domain sequence EPQ...GNK. δ$_{5-8}$ signifies δ-subunit variants equivalent to δ$_{1-4}$ but lacking the unique C terminus (Mayer et al., 1993).

Mayer et al., 1993). There are interesting similarities between these and the other reported subunit splice variants, indicating some conservation of structure in the variable region. For example, α-subunits and the δ$_2$-subunit completely lack variable region sequence, and a 15-AA sequence (-E-P-...-N-K-) at the end of the variable domain in δ$_1$ is conserved in the β- and γ-subunits, but missing in α, β', and δs$_{2-4}$.

The length of the subunit variable region may determine or influence the number of subunits comprising the holoenzyme. CaM-kinase II purified from rat forebrain has an α/β subunit stoichiometry of 3:1 and occurs primarily as homomultimers of 10 α-subunits, whereas CaM-kinase II from cerebellum, which is enriched in the larger β-subunit (α/β = 1:4), occurs as homomultimers averaging 8 β-subunits (Kanaseki et al., 1991). In other brain regions and in peripheral tissues where the expression of multiple subunits is more balanced, it is likely that heteromultimers also form (Yamauchi et al., 1989). The number and composition of subunits could affect certain kinetic properties of a CaM-kinase II holoenzyme, such as the cooperative intersubunit autophosphorylation described in Section III.B.

In many tissues such as brain, heart, and liver, CaM-kinase II is mainly cytosolic, whereas in others a variable amount has been reported to be associated with membranes or particulate fractions (Colbran and Soderling, 1990). CaM-kinase II has been localized in junctional sarcoplasmic reticulum from skeletal

muscle (Chu et al., 1990), brush border membranes from ileum (Cohen et al., 1990), and cytoskeletons from postsynaptic densities (Sahyoun et al., 1985), ileal enterocytes (Matovcik et al., 1993), and adrenal cortical cells (Papadopoulus et al., 1990). The structural features in the holoenzyme or the individual subunits that are responsible for directing the localization are only beginning to be considered experimentally. Interestingly, three CaM-kinase II subunits (α-33, δ$_3$, and γ$_A$) share a conserved 11-AA variable domain insert (K-R-K-S-S-S-S-V-Q/H-L/M-M), which recently has been shown to be involved in nuclear targeting (Srinivasan et al., 1994). Using cotransfection of CaM-kinase II subunit mutants containing or lacking this sequence, it was demonstrated that the amount of CaM-kinase II localizing in the nucleus of overexpressing COS cells and cardiomyocytes was dependent on the relative proportion of the transfected subunits containing this sequence. Though it remains to be shown that this sequence is involved in controlling nuclear localization of CaM-kinase II in normal cells containing physiological concentrations of CaM-kinase II, this study provides direct evidence that the CaM-kinase II subunit variable domain and holoenzyme subunit composition are important factors in targeting the kinase to specific subcellular localizations and substrates.

B. Tissue Distribution of Calmodulin-Kinase II Subunits

The subunit structure of native CaM-kinase II isozymes and the expression of subunits at the level of mRNA have not been studied in detail in tissues other than brain, where the α- and β-subunit mRNA is enriched compared to δ- and γ-subunit mRNA (Tobimatsu and Fujisawa, 1989). In brain, α-, β-, and β'-subunits are differentially expressed both regionally and developmentally (Kelly et al., 1987; Tobimatsu and Fujisawa, 1989). Based on Northern analysis, skeletal muscle has been reported to express primarily β-, δ-, and γ-subunit mRNA, whereas cardiac muscle, smooth muscles, and other peripheral organs appear to express δ- and γ-subunits (Tobimatsu and Fujisawa, 1989). RT–PCR analysis of mRNA from various tissue sources indicates that the full-sized δ-subunit is expressed only in brain (Schworer et al., 1993). Therefore, the reported distribution of the δ-subunit mRNA in peripheral tissues reflects the differential distribution of the smaller δ$_2$, δ$_3$ and δ$_4$ variants. For example, the RT–PCR analysis indicates that aortic smooth muscle expresses δ$_2$ and δ$_3$, cardiomyocytes primarily δ$_3$, and skeletal muscle the δ$_4$-isoforms (Schworer et al., 1993).

Confirmation of the subunit composition in a given

CaM-kinase II holoenzyme requires that the subunits and alternatively spliced variants be identified using specific antibodies. Given the high degree of homology between the subunits and variants, production of such reagents can be expected to be difficult. To date, γ-subunit-specific antibodies have not been described. However, monoclonal antibodies specific for α- and β-subunits are available and have been used to identify these subunits in brain holoenzymes (Baitinger *et al.*, 1990). These antibodies fail to cross-react with CaM-kinase II from rat aortic vascular smooth muscle or swine cartoid artery, consistent with a lack of α- and β-subunit mRNA expression in these tissues (Singer, H. A., unpublished observations). Expression of varying sized δ-subunit variants has been documented at the protein level in rat brain and aortic medial smooth muscle extracts by immunoblotting with an antipeptide antibody (CK2-Delta) that specifically recognizes δ-subunit variants containing the unique 21-AA C terminus (Schworer *et al.*, 1993). A 53-kDa δ-subunit (δ_2) and approximately 80% of the total CaM-kinase II activity can be immunoprecipitated from lysates of cultured rat aortic vascular smooth muscle (VSM) cells with the CK2-Delta antiserum (Fig. 3). In contrast, only a few percent of the total CaM-kinase II activity and multiple subunits (50, 53, 58 kDa) are immunoprecipitated from rat brain lysates with CK2-Delta. Fifty-kDa α-subunits and 60-kDa β-subunits are detected by ^{125}I-labeled calmodulin overlay in preparations of purified rat brain CaM-kinase II (first lane) and in immune complexes from rat brain lysates formed with a subunit nonspecific antipeptide antibody (CK2-CAT)

directed to a conserved sequence in the catalytic domain (last lane). The low efficiency of immunoprecipitation from brain compared to VSM cell extracts with CK2-Delta is consistent with the bulk of the brain CaM-kinase II being composed of α- and β-subunits. The appearance of multiple subunits in the immune complexes from rat brain indicates either the expression of multiple δ-subunits (i.e., δ_1 and δ_2) and/or immunoprecipitation of hetermultimers containing both δ- and non-δ-subunits.

C. Related Calmodulin-Kinases (Calmodulin-Kinases I,III,IV,V)

Several other Ca^{2+}/CaM-dependent serine/threonine kinases have been identified that can be distinguished from CaM-kinase II on the basis of native size, substrate specificity, and autophosphorylation properties. One of these is CaM-kinase Gr or IV, which is a multifunctional kinase found primarily in brain (Means *et al.*, 1991). CaM-kinase IV has a peptide substrate specificity that is similar to that of CaM-kinase II, and it also undergoes Ca^{2+}/CaM-dependent autophosphorylation on a serine residue (Frangakis *et al.*, 1991). However, autophosphorylation of CaM-kinase IV expressed in insect cells does not result in the generation of Ca^{2+}/CaM-independent activity (Cruzalegui and Means, 1993). CaM-kinase IV is only 30–40% identical to a CaM-kinase II subunit and is monomeric with an estimated molecular mass of 61–67 kDa by SDS–PAGE depending on its source (Means *et al.*, 1991; Cruzalegui and Means, 1993). Two other classes of CaM-dependent kinases, CaM-kinases Ia and Ib (De-Remer *et al.*, 1992; Picciotto *et al.*, 1993) and CaM-kinase III (Nairn *et al.*, 1985), can readily be distinguished from CaM-kinase II by their native size and restricted substrate specificity. Although these protein kinases also autophosphorylate, this is not associated with a transition to a Ca^{2+}/CaM-independent activity. A monomeric (37–41 kDa) Ca^{2+}/calmodulin-dependent kinase activity designated CaM-kinase V, which is similar catalytically to CaM-kinase II, has been isolated from rat cerebrum (Mochizuki *et al.*, 1993). CaM-kinase V undergoes Ca^{2+}/CaM-dependent autophosphorylation, but this has not yet been shown to lead to formation of autonomous activity.

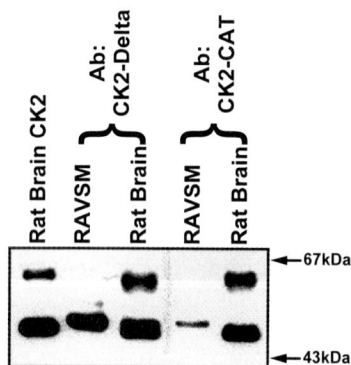

FIGURE 3 Immunoprecipitation of CaM-kinase II from rat brain and cultured rat aortic VSM cell (RAVSM) lysates with antibodies specific to the δ-subunit C terminus (CK2-Delta) or a conserved sequence in the catalytic/regulatory domain (CK2-CAT). The immune complexes were subjected to SDS–PAGE and transferred to immobilon membranes. CaM-kinase II subunits were detected by an overlay with ^{125}I-labeled calmodulin and autoradiography. The left lane is a control containing partially purified rat brain CaM-kinase II (CK2), which was not immunoprecipitated.

III. Regulation of Calmodulin-Kinase II Activity

A. Activation by Ca^{2+}/Calmodulin

The K_m of purified CaM-kinase II for calmodulin has been reported to be in the range of 20–100 nM at saturating $[Ca^{2+}]$ with one calmodulin bound per subunit (Colbran and Soderling, 1990; Schulman and Hanson, 1993). This calmodulin affinity is about a factor of 10 lower than that for other calmodulin-activated kinases like MLCK (Kamm and Stull, 1989). Analysis of α- and β-subunits expressed in mammalian cells indicates that calmodulin affinity varies somewhat between these subunits (Yamauchi *et al.*, 1989). However, no significant differences were found in another study comparing expressed α- and δ-subunits (Edman and Schulman, 1994). Based on homology to the calmodulin binding domains in phosphorylase kinase and MLCK, a putative regulatory region in CaM-kinase II was identified (Fig. 1B) and peptides based on AA sequence in this region were designed and assessed for their ability to inhibit calmodulin binding and activation of CaM-kinase II (Payne *et al.*, 1988). The calmodulin binding domain has been localized to amino acids 296–309 and a peptide corresponding to this sequence has been shown to inhibit Ca^{2+}/CaM-dependent activation of CaM-kinase II with an IC_{50} of 57 nM and CaM-activated phosphodiesterase with an IC_{50} of 3.4 nM. Similar studies evaluating the ability of peptides to competitively inhibit the activity of autophosphorylated Ca^{2+}/CaM-*independent* CaM-kinase II identified an autoinhibitory domain corresponding to amino acids 291–302 (Payne *et al.*, 1988; Colbran and Soderling, 1990). Additional inhibitory potency was gained by extending the peptides to amino acid 281, though inhibition became noncompetitive and involved inhibition of ATP binding (Colbran and Soderling, 1990). Other studies using truncated mutants, site-directed mutagenesis, and molecular modeling approaches (Cruzalegui *et al.*, 1992) have generally confirmed the C-terminal boundary of the calmodulin binding domain and position of the autoinhibitory domain as indicated in Fig. 1B.

B. Autophosphorylation of Calmodulin-Kinase II

Early studies of CaM-kinase II indicated that the kinase subunits autophosphorylated with stoichiometries estimated to be as high as 4 mol P_i/mol subunit (Ahmad *et al.*, 1982). Ca^{2+}/CaM-dependent autophosphorylation was subsequently shown to result in the generation of activator-independent kinase activity to a maximum level of 40–70% of total Ca^{2+}/CaM-stimulated activity (Miller and Kennedy, 1986; Schworer *et al.*, 1986; Lou *et al.*, 1986; Lai *et al.*, 1986). Once CaM-kinase II is activator independent or "autonomous," additional autophosphorylation can take place in the absence of Ca^{2+}/CaM (Miller and Kennedy, 1986; Hashimoto *et al.*, 1987). These properties distinguish CaM-kinase II from other Ca^{2+}/CaM-

dependent protein kinases and could be of physiological importance, for example, by effecting prolonged intracellular responses to transient Ca^{2+} signals. Therefore, the mechanisms, sites, and functional consequences of CaM-kinase II autophosphorylation events have been intensively investigated using peptide mapping and sequencing approaches and more recently site-directed mutagenesis.

It is now known that generation of activator-independent CaM-kinase II activity requires autophosphorylation of a specific threonine residue (Thr^{286}) in the autoinhibitory domain (Schworer *et al.*, 1988; Thiel *et al.*, 1988; Hanson *et al.*, 1989). Autophosphorylation on this residue is via a highly cooperative intraholoenzyme intersubunit reaction that occurs only between activated (Ca^{2+}/CaM bound) subunits (Hanson *et al.*, 1994; Mukherji and Soderling, 1994). In addition to generating autonomous activity, autophosphorylation of CaM-kinase II on Thr^{286} has been shown to result in calmodulin "trapping" as a consequence of nearly a thousandfold decrease in calmodulin off-rate (Meyer *et al.*, 1992). This activation scheme is summarized in Fig. 4. Autophosphorylation-dependent calmodulin trapping and enzymatic activity in the absence of Ca^{2+}/CaM provide mechanisms for prolonging kinase activity beyond the dura-

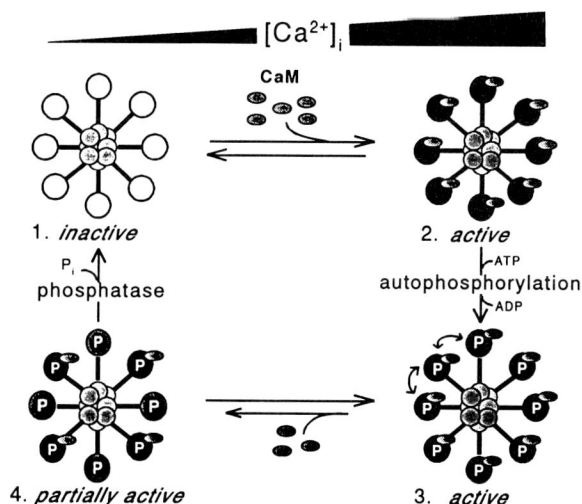

FIGURE 4   Model for activation and autophosphorylation of a CaM-kinase II holoenzyme based on Hanson *et al.* (1994). 1–2: Multimers of CaM-kinase II subunits are activated by increasing concentrations of free intracellular Ca^{2+} [(Ca^{2+})$_i$] and association of Ca^{2+}/CaM complexes. 2–3: Activated subunits (solid) with bound Ca^{2+}/CaM are autophosphorylated on Thr^{286} (P) in a cooperative intraholoenzyme intersubunit reaction. 3–4: Phosphorylated subunits have an increased affinity for CaM and are partially active (shaded circles) in the absence of bound Ca^{2+}/CaM. 4–1: Dissociation of Ca^{2+}/CaM and the action of phosphoprotein phosphatases return autophosphorylated subunits to the inactive state.

tion of the Ca^{2+} signal and for responding in a frequency-dependent manner to pulsatile changes in free intracellular Ca^{2+} (Hanson et al., 1994).

Once subunits are autophosphorylated on Thr^{286} in vitro, Ca^{2+} chelation and dissociation of CaM result in the autophosphorylation of additional serine and threonine residues, including Thr^{305} and/or Thr^{306} in the calmodulin binding domain (Miller and Kennedy, 1986; Hashimoto et al., 1987; Patton et al., 1990). Phosphorylation of $Thr^{305/306}$ blocks rebinding of CaM and this is reflected by a loss of total Ca^{2+}/CaM-dependent activity to a level equal to the activator-independent activity. Autophosphorylation of Thr^{306} in this manner also appears to be an intraholoenzyme intersubunit reaction (Mukherji and Soderling, 1994). Under certain experimental conditions it has been possible to demonstrate the slow autophosphorylation of Thr^{306} and concomitant loss of CaM-kinase II activity without addition of activators and prior autophosphorylation of Thr^{286} (Colbran, 1993). In this case, Thr^{306} autophosphorylation appears to be catalyzed by an intrasubunit reaction (Mukherji and Soderling, 1994).

C. Substrate Specificity

Analysis of the sequences around phosphorylated residues and analysis of the phosphorylation kinetics using a series of small peptide substrates (Pearson et al., 1985) indicate that a minimal consensus phosphorylation sequence for CaM-kinase II is -R-X-X-S/T-. However, there are exceptions to this, including the sequence around the Thr^{306} autophosphorylation site. As would be expected given this relatively loose consensus phosphorylation sequence, numerous proteins have been shown to be substrates for CaM-kinase II in vitro (Colbran and Soderling, 1990; Schulman and Hanson, 1993), including smooth muscle regulatory proteins such as LC20, caldesmon, calponin, and MLCK (see Section VI). Few of these potential substrates have been shown to be phosphorylated specifically by CaM-kinase II in vivo. The difficulties in determining in vivo substrates for CaM-kinase II are not unique for this protein kinase and include a lack of specific tools and approaches for manipulating kinase activity in situ, the requirement for assessing site-specific phosphorylation of the target proteins, and the presence of other multifunctional kinases that are Ca^{2+} dependent and share substrate specificity, such as CaM-kinase IV and PKC. Tools that have been used to identify physiological substrates and functions for CaM-kinase II include inhibitory peptides based on AA sequences in the regulatory domain (Payne et al., 1988) and specific chemical inhibitors such as KN-62 and KN-93. KN-62 inhibits CaM-kinase II with a K_i of

$0.9 \mu M$ with no significant inhibitory effects on MLCK, cAMP-dependent protein kinase, or PKC (Tokumitsu et al., 1990). The mechanism of inhibition by KN-62, and a more water-soluble derivative KN-93 (Sumi et al., 1991), appears to be by competitive inhibition of calmodulin binding. Consequently, the autophosphorylated form of the kinase is not affected by the inhibitor (Tokumitsu et al., 1990). KN-62 has been used in intact PC-12 cells to inhibit CaM-kinase II autophosphorylation and generation of autonomous activity following stimulation by ionophores or KC1 depolarization (Tokumitsu et al., 1990).

IV. ACTIVATION AND AUTOPHOSPHORYLATION OF CALMODULIN-KINASE II in Situ

There is substantial evidence that CaM-kinase II autophosphorylation events, which have been characterized in vitro, also occur in situ following activation of intact cells with stimuli that increase intracellular free Ca^{2+}. Increases in activator-independent CaM-kinase II activity, reflecting activation of CaM-kinase II and autophosphorylation of Thr^{286}, have been demonstrated in cerebellar granule cells (Fukunaga et al., 1989), GH_3 cells (Jefferson et al., 1991), and PC-12 cells (MacNicol et al., 1990) stimulated by agonists or ionophores. CaM-kinase II δ-subunits undergo phosphorylation in intact cultured VSM cells stimulated with the Ca^{2+} ionophore ionomycin, an effect that can be potentiated by preincubating the cells with the protein phosphatase inhibitor okadaic acid (Fig. 5A). Increases in Ca^{2+}/CaM-independent CaM-kinase II activity, reflecting activation and autophosphorylation of CaM-kinase II in situ, can also be demonstrated in lysates from ionomycin-stimulated VSM cells in response to increases in free intracellular calcium ($[Ca^{2+}]_i$), which are well within a physiological range (i.e., 150–400 nM) (Fig. 5B). Physiological stimuli such as angiotensin II, vasopressin, ATP, and platelet-derived growth factor (PDGF) also stimulate variable increases in autonomous CaM-kinase II activity in parallel with their abilities to increase $[Ca^{2+}]_i$ over this range (Abraham, S. T., unpublished data). In these experiments, CaM-kinase II activity was assayed using a synthetic peptide substrate (autocamtide-2; KKALRRQETVDAL) that corresponds in part to the conserved autophosphorylation site in the known CaM-kinase II subunits. Autocamtide-2 has proven to be a useful substrate for measuring CaM-kinase II activity since it is not a substrate for cAMP-dependent protein kinase, smooth muscle MLCK, or PKC under the activation conditions used (Hanson et al., 1989; Schworer, C. M., unpublished data).

B

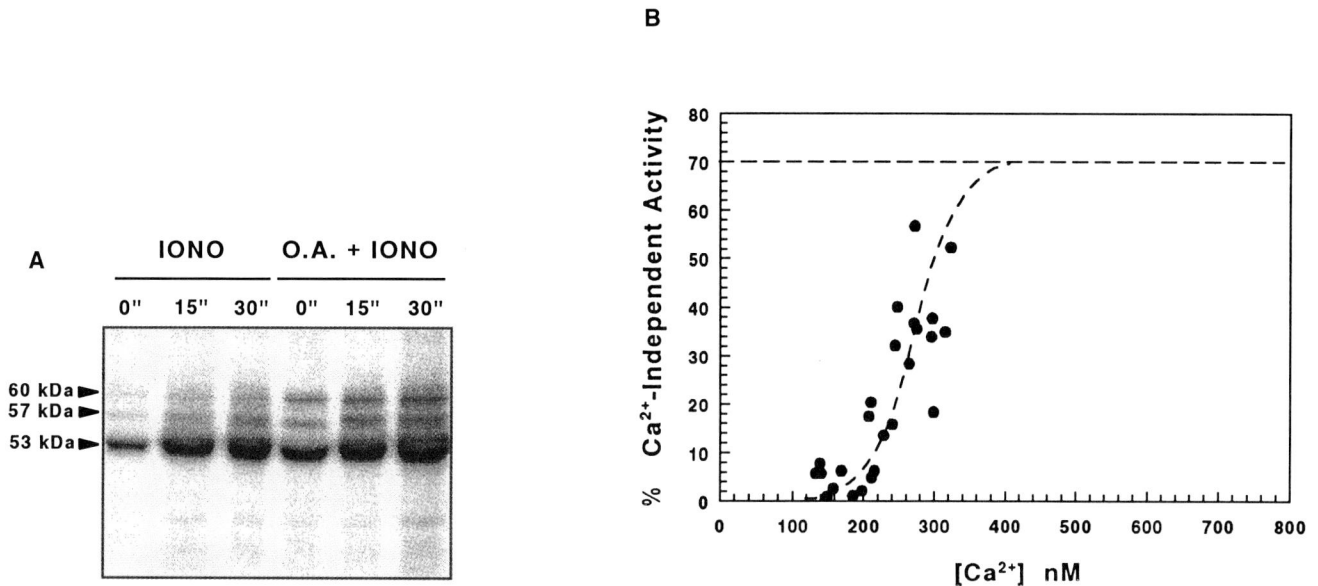

FIGURE 5 (A) Autophosphorylation of CaM-kinase II δ-subunits in cultured rat aortic VSM cells. CaM-kinase II was immunoprecipitated with a δ-subunit-specific antibody (CK2-Delta) from lysates of ^{32}P-labeled cells following stimulation of the intact cells for the indicated times with ionomycin (1 μM). Pretreatment of the cells with the protein phosphatase inhibitor okadaic acid (O.A.; 1 μM) potentiated the phosphorylation response. (B) Correlation between [Ca^{2+}]$_i$ (measured by fura-2 fluorescence) and autophosphorylation-dependent generation of Ca^{2+}/calmodulin-independent kinase activity in suspensions of cultured rat aortic VSM cells stimulated by varying concentrations of ionomycin.

V. CALMODULIN-KINASE II ISOZYMES IN SMOOTH MUSCLE

A. Caldesmon Kinase

Previous reports describing CaM-kinase II in smooth muscle have related primarily to a Ca^{2+}/CaM-dependent kinase activity that copurifies with smooth muscle caldesmon (Ngai and Walsh, 1985; Scott-Woo and Walsh, 1988; Abougou et al., 1989). Evidence that the caldesmon kinase activity was related to CaM-kinase II included close similarities in activator dependencies and substrate specificity using caldesmon (Abougou et al., 1989) and synapsin as substrates (Scott-Woo et al., 1990). Protein bands in the 50-kDa range, consistent with known CaM-kinase II subunit sizes, were identified on Western blots of purified caldesmon with an antipeptide antibody to a conserved sequence in CaM-kinase II subunits (Abougou et al., 1989). Dissociation of the endogenous Ca^{2+}/CaM-dependent kinase activity and caldesmon was ultimately achieved by treatment of a myofibrillar fraction (from avian gizzard) with 30 mM Mg^{2+} and resolution by anion-exchange chromatography (Ikebe et al., 1990). The purified kinase was concluded to be a smooth muscle isozyme of CaM-kinase II based on its catalytic properties (using caldesmon as the primary substrate), its ability to autophosphorylate resulting in generation of Ca^{2+}/CaM-independent kinase activity, and some physical properties, such as the apparent subunit molecular weight on SDS–PAGE gels (Ikebe et al., 1990). However, the native size of this kinase, the structural basis for its apparent tight binding to caldesmon, and its relationship to CaM-kinase II holoenzymes in other smooth muscle cellular compartments have not been determined.

B. Arterial Smooth Muscle Calmodulin-Kinase II

Ca^{2+}/CaM-dependent kinase activity is also readily measured in low-ionic-strength extracts from rat aortic (Fig. 5B) (Schworer et al., 1993) and swine carotid arterial smooth muscle (Fig. 6) using the peptide substrate autocamtide-2, consistent with a cytosolic distribution. The specific activity of the soluble kinase is approximately one-tenth of that in an analogous fraction of rat or swine forebrain (Schworer et al., 1993), where the average concentration of CaM-kinase II subunits has been estimated to be 1% of total brain protein. On the basis of a native size of 550–650 kDa estimated by gel filtration chromatography, peptide substrate specificity, autophosphorylation properties, and cross-reactivity with subunit nonselective anti-CaM-kinase

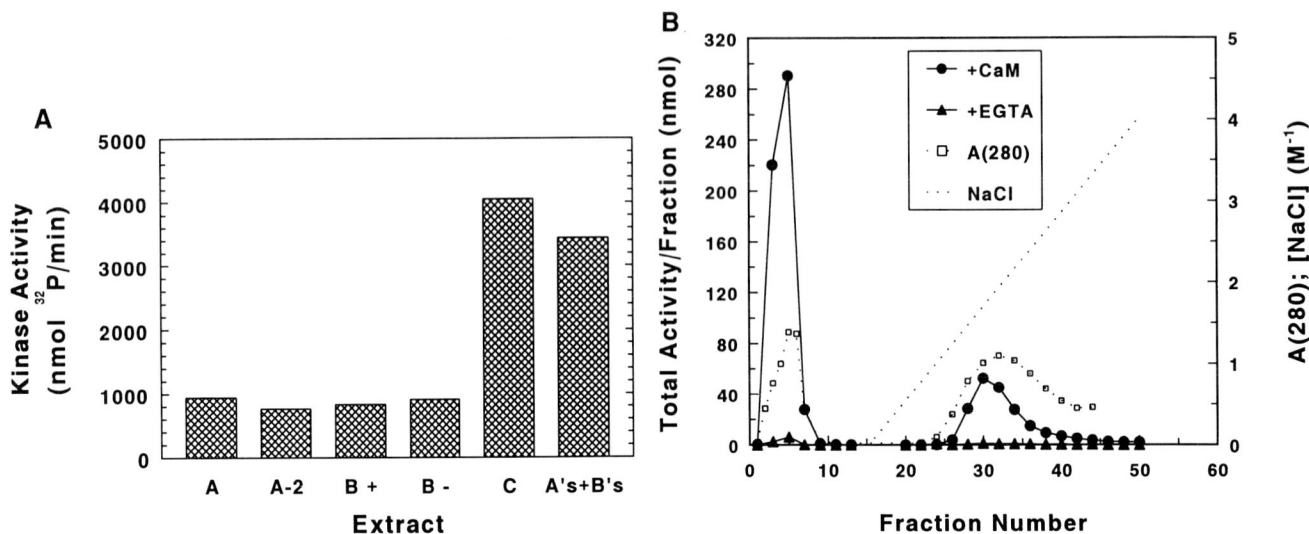

FIGURE 6   (A) Sequential extraction of Ca^{2+}/CaM-dependent autocamtide-2 kinase activity from swine carotid artery medial smooth muscle. Twenty-five grams of tissue were homogenized in low-ionic-strength buffer (50 mM NaCl, 4 mM EDTA, 1 mM EGTA, 2 mM DTT, 5 mg/l leupeptin, 20 mM Tris, pH 7.5, 4°C). The 10,000 g supernatant is denoted Extract A. The pellet was re-extracted with this buffer (Extract A-2), buffer "B" containing 40 mM KCl, 1 mM $MgCl_2$, 1 mM EGTA, 1 mM DTT, 0.2 mM PMSF, 0.05% Triton X-100, 20 mM Tris (Extract B+), buffer "B" minus Triton (Extract B−), and finally buffer B containing 80 mM KCl and 30 mM $MgCl_2$ (Extract C). Total Ca^{2+}/CaM-dependent autocamtide-2 kinase activity in each extract is plotted. (B) Anion-exchange chromatography of Extract C on DEAE-Sephacel. Autocamtide-2 kinase activity was assayed in each fraction plus (solid circles) and minus (solid triangles) Ca^{2+}/CaM. Also plotted is absorbance (open squares) and the NaCl gradient (dotted line).

II antibodies, we have identified the kinase from swine carotid arterial smooth muscle as an isozyme(s) of CaM-kinase II composed of 54- and 58-kDa subunits (Singer *et al.*, 1993). Northern analysis and Western blotting with the δ-subunit-specific antibody (CK2-Delta) are consistent with the interpretation that the 54- and 58-kDa subunits are forms of the δ- and γ-subunits, respectively (Singer, H. A., unpublished data). Thus, this readily solubilized smooth muscle CaM-kinase II is similar to brain, heart, and liver CaM-kinase II, which are large multimers composed of 54- to 60-kDa kinase subunits (Colbran and Soderling, 1990).

Following repeated extraction with low-ionic-strength buffers, a substantial pool of Ca^{2+}/CaM-dependent kinase activity (approximately half of the total) can be recovered by extracting a myofibril-rich 10,000 g pellet from swine carotid arterial smooth muscle with a buffer containing 30 mM $MgCl_2$ as the key ingredient (Fig. 6A). This extraction is similar to that which was shown to dissociate caldesmon and its associated CaM-kinase activity (Ikebe *et al.*, 1990). The Ca^{2+}/CaM-dependent kinase activity in this extract resolves into two fractions on an anion-exchange column, with a large fraction flowing through, which is atypical for CaM-kinase II (Fig. 6B). Further analysis of these fractions by gel filtration chromatography and immunoblotting indicates that the salt-eluted activity

is similar to the readily solubilized CaM-kinase II in extracts A and B in that it has a native size of 550–650 kDa and is composed of 54- and 58-kDa subunits. In contrast, the fraction not retained by the anion-exchange column is monomeric with an estimated size of 35 kDa. On the basis of a comparison of auto-phosphorylation kinetics, catalytic properties, and cross-reactivity with anti-CaM-kinase II antibodies, we have concluded that the low-molecular-weight kinase represents the catalytic/regulatory domain of a CaM-kinase II subunit, or a closely related kinase (Schworer *et al.*, 1994). The physiological significance, if any, of this CaM-kinase II-related activity, or its relationship to the previously described CaM-kinase activity that co-purifies with caldesmon, is not clear at this time.

VI. POTENTIAL FUNCTIONS OF CALMODULIN-KINASE II IN SMOOTH MUSCLE

Phosphorylation of caldesmon *in vitro* by CaM-kinase II (or PKC) has been shown to interfere with caldesmon binding to F-actin, and results in a reversal of the caldesmon inhibitory effect on actin-activated myosin ATPase activity (Ngai and Walsh, 1987). This has led to speculation that these kinases may be involved in regulating smooth muscle contractile activity by directly phosphorylating caldesmon. The fact that CaM-kinase II or a similar kinase activity copurifies with caldesmon and phosphorylates it *in vitro* to high stoichiometries is consistent with this possibility. However, this may be misleading since the specific sites phosphorylated in caldesmon by CaM-kinase II and PKC *in vitro* (Ikebe and Reardon, 1990b) have been shown to be different from the sites phosphorylated in intact arterial smooth muscle in response to addition of contractile stimuli (Adam *et al.*, 1989, 1992). In these studies, the sites phosphorylated in caldesmon from intact tissues were the same as those phosphorylated *in vitro* by mitogen-activated protein kinase (MAP-kinase). Thus, if CaM-kinase II is involved in regulating caldesmon phosphorylation and function *in vivo*, it may be indirect via effects on the MAP-kinase signaling pathway. Calponin is another smooth muscle thin filament-associated protein that inhibits actin-activated smooth muscle myosin ATPase *in vitro* (Takahashi *et al.*, 1988). Like caldesmon, phosphorylation of calponin *in vitro* by CaM-kinase II interferes with its inhibitory actions on actin-activated myosin ATPase activity (Winder and Walsh, 1990). Because calponin has not yet been shown to be phosphorylated to a significant extent in intact smooth muscle (Bárány *et al.*, 1991), the functional significance of the *in vitro* phosphorylations catalyzed by CaM-kinase II or PKC remains speculative.

A strong case can be made for a physiologically important role for CaM-kinase II in phosphorylating and modulating the activity of MLCK in smooth muscle. MLCK is phosphorylated by CaM-kinase II *in vitro* (Hashimoto and Soderling, 1990; Ikebe and Reardon, 1990a) on a site near the calmodulin binding domain, resulting in decreased affinity of MLCK for calmodulin (Stull *et al.*, 1990). Phosphorylation of this site in MLCK and a subsequent decrease in MLCK affinity for CaM occurs *in vivo* in cultured tracheal smooth muscle cells, in response to addition of contractile stimuli (Stull *et al.*, 1990). Treatment of intact tracheal smooth muscle cells with the CaM-kinase II inhibitor KN-62 inhibits phosphorylation of MLCK and *potentiates* LC20 phosphorylation in response to Ca^{2+}-mobilizing stimuli (Tansey *et al.*, 1992). Similar effects have been demonstrated in permeabilized cells using KN-62 or a peptide inhibitor corresponding to amino acids 281–309 in the regulatory domain of CaM-kinase II (Tansey *et al.*, 1994). The calcium dependence for phosphorylating MLCK in tracheal cells has been reported to be half-maximal at a concentration of free intracellular Ca^{2+} equal to 500 nM (Tansey *et al.*, 1994), consistent with the *in situ* Ca^{2+} dependence for activating CaM-kinase II shown in Fig. 5B. Thus, CaM-kinase II ap-

pears to negatively modulate MLCK activity, resulting in the attenuation of LC20 phosphorylation, during activation of smooth muscle by contractile stimuli.

CaM-kinase II may also be involved in control of smooth muscle contractile activity by regulating free intracellular $[Ca^{2+}]$. CaM-kinase II and other multifunctional protein kinases have been reported to phosphorylate dihydropyridine receptors (Ca^{2+} channels) *in vitro* (Jahn *et al.*, 1988). Peptide inhibitors of CaM-kinase II specifically block Ca^{2+}-induced enhancement of L-type Ca^{2+} current in isolated toad stomach smooth muscle cells (McCarron *et al.*, 1992) or rabbit heart ventricular myocytes (Anderson *et al.*, 1994). Thus CaM-kinase II has been implicated in the positive feedback control of voltage-regulated Ca^{2+} channels. CaM-kinase II may also regulate Ca^{2+} uptake into sarcoplasmic reticulum (SR). The cardiac SR Ca^{2+}-ATPase (SERCA2) is under tonic inhibitory control by the SR-associated protein, phospholamban. Phosphorylation of phospholamban by cAMP-dependent protein kinase or CaM-kinase II relieves its inhibitory effect on SERCA2 activity, resulting in enhanced Ca^{2+} uptake into SR (LePeuch *et al.*, 1979). In addition, SERCA2 can be directly phosphorylated on a specific serine residue (Ser^{38}) by an endogenous SR Ca^{2+}/CaM-dependent kinase or exogenous CaM-kinase II, resulting in SERCA2 activation independently of phospholamban effects (Xu *et al.*, 1993; Toyofuku *et al.*, 1994). Other membrane channels, including anion channels from bovine tracheal epithelial cells that are phosphorylated and activated *in vitro* by CaM-kinase II (Fuller *et al.*, 1994), may be physiologically relevant substrates for the kinase.

Given the ubiquity of CaM-kinase II and the universal importance of Ca^{2+} as a second messenger, it is not surprising that CaM-kinase II has been implicated in the control of other basic cellular processes that are not unique to smooth muscle. For example, there is increasing evidence that regulation of certain transcriptional responses by Ca^{2+}-dependent pathways is mediated by CaM-kinase II and/or CaM-kinase IV (Sun *et al.*, 1994; Enslen and Soderling, 1994). Expression of a constitutively active fragment of CaM-kinase II acts to increase the rate of transcription of several genes (Kapiloff *et al.*, 1991) by phosphorylating a transcription factor that binds to cAMP response elements (CREB) (Sheng *et al.*, 1991; Sun *et al.*, 1994). CaM-kinase II has also been implicated in the fundamental process of cell cycle control (Means, 1994). In nonmammalian models such as sea urchin (Baitinger *et al.*, 1990) and frog (Lorca *et al.*, 1993) oocytes, CaM-kinase II inhibitor peptides have been shown to block progression of mitosis. Transient expression of a truncated constitutively active form of CaM-kinase II in stable mammalian cell lines results in a complete arrest of cell cycle progression in the G2 phase (Planas-Silva and Means, 1992). Other data supporting proposed roles for CaM-kinase II in cell cycle control include an immunofluorescence study of several mammalian cell lines where CaM-kinase II was shown to be localized in the nuclear matrix and nucleoli during interphase and in the mitotic apparatus during mitosis (Ohta *et al.*, 1990). Studies establishing a structural basis for the localization of CaM-kinase II in the nucleus (Srinivasan *et al.*, 1994) support the concept that CaM-kinase II is involved in the regulation of fundamental nuclear processes associated with gene transcription and cell division.

VII. SUMMARY

Even though CaM-kinase II is a ubiquitous multifunctional protein kinase controlled by a universal second messenger, relatively little is known about its specific cellular functions in tissues other than brain. This is true in smooth muscle, where there has been substantial speculation regarding roles for CaM-kinase II in controlling or modulating contractile function. Studies using relatively specific pharmacological inhibitors of CaM-kinase II have hinted at the importance of the kinase in controlling contractile function by modulating availability of activator Ca^{2+} and by phosphorylating and desensitizing MLCK. Molecular approaches have provided evidence that CaM-kinase II may be involved in controlling basic cellular processes such as gene transcription and cell growth. The unique autoregulatory properties of CaM-kinase II, which have been defined *in vitro*, have physiological implications that are only just beginning to be evaluated *in vivo*. Much additional work is required to assess the significance *in vivo* of specific autophosphorylation events that modify CaM-kinase II activator requirements or total activity. Biochemical and molecular approaches are starting to provide detailed knowledge of CaM-kinase II holoenzyme and subunit structure in smooth muscle. This information, coupled with a better basic understanding of how CaM-kinase II structure influences subcellular distribution, substrate specificity, and autoregulatory properties, should ultimately provide a basis for the development of specific tools and approaches aimed at establishing functional roles for the kinase.

References

Abougou, J.-C., Hagiwara, M., Hachiya, T., Terasawa, M., Hidaka, H., and Hartshorne, D. J. (1989). *FEBS Lett.* **257**, 408–410.
Adam, L. P., Gapinski, C. J., and Hathaway, D. R. (1992). *FEBS Lett.* **302**, 223–226.
Adam, L. P., Haeberle, J. R., and Hathaway, D. R. (1989). *J. Biol. Chem.* **264**, 7698–7703.
Ahmad, Z., DePaoli-Roach, A. A., and Roach, P. J. (1982). *J. Biol. Chem.* **257**, 8348–8355.
Anderson, M. E., Braun, A. P., Schulman, H., and Premack, B. A. (1994). *Circ. Res.* **75**, 854–861.
Baitinger, C., Alderton, H., Poenie, M., Schulman, H., and Steinhardt, R. A. (1990). *J. Cell Biol.* **111**, 1763–1773.
Bárány, M., Rokolya, A., and Bárány, K. (1991). *FEBS Lett.* **279**, 65–68.
Bennett, M. K., and Kennedy, M. B. (1987). *Proc. Natl. Acad. Sci. U.S.A.* **84**, 1794–1798.
Benson, D. L., Isackson, P. J., Gail, C. M., and Jones, E. G. (1991). *J. Neurosci.* **11**, 31–47.
Bulleit, R. E., Bennett, M. K., Molfoy, S. S., Hurley, J. B., and Kennedy, M. B. (1988). *Neuron* **1**, 63–72.
Cho, K. O., Wall, J. B., Pugh, P. C., Ito, M., Mueller, S. A., and Kennedy, M. B. (1991). *Neuron* **7**, 439–450.
Chu, A., Sumbilla, C., Inesi, G., Jay, S. D., and Campbell, K. P. (1990). *Biochemistry* **29**, 5899–5905.
Cohen, M. E., Reinlib, L., Watson, A. J., Gorelick, F., Rys-Sikore, K., Tse, M., Rood, R. P., Czernik, A. J., Sharp, G. W., and Donowitz, M. (1990). *Proc. Natl. Acad. Sci. U.S.A.* **87**, 8990–8994.
Colbran, R. J. (1993). *J. Biol. Chem.* **268**, 7163–7170.
Colbran, R. J., and Soderling, T. R. (1990). *Curr. Top. Cell. Regul.* **31**, 181–215.
Colbran, R. J., Fong, Y.-L., Schworer, C. M., and Soderling, T. R. (1988). *J. Biol. Chem.* **263**, 18145–18151.
Cruzalegui, F. H., and Means, A. R. (1993). *J. Biol. Chem.* **268**, 26171–26178.
Cruzalegui, F. H., Kapiloff, M. S., Morfin, J.-P., Kemp, B. E., Rosenfeld, M. G., and Means, A. R. (1992). *Proc. Natl. Acad. Sci. U.S.A.* **89**, 12127–12131.
DeRemer, M. F., Saeli, R. J., and Edelman, A. M. (1992). *J. Biol. Chem.* **267**, 13460–13465.
Edman, C. F., and Schulman, H. (1994). *Biochim. Biophys. Acta* **1221**, 89–101.
Enslen, H., and Soderling, T. R. (1994). *J. Biol. Chem.* **269**, 20872–20877.
Erondu, N. E., and Kennedy, M. B. (1985). *J. Neurosci.* **5**, 3270–3277.
Frangakis, M. V., Ohmstede, C. A., and Sahyoun, N. (1991). *J. Biol. Chem.* **266**, 11309–11316.
Fukunaga, K., Rich, D. P., and Soderling, T. R. (1989). *J. Biol. Chem.* **264**, 21830–21836.
Fuller, C. M., Ismailov, I. I., Keeton, D. A., and Benos, D. J. (1994). *J. Biol. Chem.* **269**, 26642–26650.
Hanson, P. I., and Schulman, H. (1992). *Annu. Rev. Biochem.* **61**, 559–601.
Hanson, P. I., Kapiloff, M. S., and Lou, L. L. (1989). *Neuron* **3**, 59–70.
Hanson, P. I., Meyer, T., Stryer, L., and Schulman, H. (1994). *Neuron* **12**, 943–956.
Hashimoto, Y., and Soderling, T. R. (1990). *Arch. Biochem. Biophys.* **278**, 41–45.
Hashimoto, Y., Schworer, C. M., Colbran, R. J., and Soderling, T. R. (1987). *J. Biol. Chem.* **262**, 8051–8055.
Ikebe, M., and Reardon, S. (1990a). *J. Biol. Chem.* **265**, 8975–8978.
Ikebe, M., and Reardon, S. (1990b). *J. Biol. Chem.* **265**, 17607–17612.
Ikebe, M., Reardon, S., Scott-Woo, G. C., Zhou, Z., and Koda, Y. (1990). *Biochemistry* **29**, 11242–11248.
Jahn, H., Nastainczyk, W., Rohrkasten, A., Schneider, T., and Hofmann, F. (1988). *Eur. J. Biochem.* **178**, 535–542.
Jefferson, A. B., Travis, S. M., and Schulman, H. (1991). *J. Biol. Chem.* **266**, 1484–1490.
Kamm, K. E., and Stull, J. T. (1989). *Annu. Rev. Physiol.* **51**, 299–313.
Kanaseki, T., Ikeuchi, Y., Sugiura, H., and Yamauchi, T. (1991). *J. Cell Biol.* **115**, 1049–1060.

Kapiloff, M. S., Mathis, J. M., Nelson, C. A., Lin, C. R., and Rosenfeld, M. G. (1991). *Proc. Natl. Acad. Sci. U.S.A.* **88,** 3710–3714.

Kelly, P. T., Shields, S., Conway, K., Yip, R., and Burgin, K. (1987). *J. Neurochem.* **49,** 1927–1940.

Lai, Y., Nairn, A. C., and Greengard, P. (1986). *Proc. Natl. Acad. Sci. U.S.A.* **83,** 4253–4257.

LePeuch, C. J., Haiech, J., and Demaille, J. G. (1979). *Biochemistry* **18,** 5150–5157.

Lin, C. R., Kapiloff, M. S., Durgerian, S., Tatemoto, K., Russo, A. F., Hanson, P., Schulman, H., and Rosenfeld, M. G. (1987). *Proc. Natl. Acad. Sci. U.S.A.* **84,** 5962–5966.

Lorca, T., Cruzalegui, F. H., Fesquet, D., Cavadore, J.-C., Mery, J., Means, A. R., and Dorée, M. (1993). *Nature (London)* **366,** 270–273.

Lou, L. L., Lloyd, S. J., and Schulman, H. (1986). *Proc. Natl. Acad. Sci. U.S.A.* **83,** 9497–9501.

MacNichol, M., Jefferson, A. B., and Schulman, H. (1990). *J. Biol. Chem.* **265,** 18055–18058.

Matovcik, L. M., Haimowitz, B., Goldenring, J. R., Czernik, A. J., and Gorelick, F. S. (1989). *Am. J. Physiol.* **244,** C1029–C1036.

Mayer, P., Mohlig, M., Schatz, H., and Pfeiffer, A. (1993). *FEBS Lett.* **333,** 315–318.

McCarron, J. G., McGeown, J. G., Reardon, S., Ikebe, M., Fay, F. S., and Walsh, J. V. (1992). *Nature (London)* **357,** 74–77.

Means, A. R. (1994). *FEBS Lett.* **347,** 1–4.

Means, A. R., Cruzalegui, F., LeMagueresse, B., Needleman, D. S., Slaughter, G. R., and Ono, T. (1991). *Mol. Cell. Biol.* **11,** 3960–3971.

Meyer, T., Hanson, P. I., Stryer, L., and Schulman, H. (1992). *Science* **256,** 1199–1202.

Miller, S. G., and Kennedy, M. B. (1986). *Cell (Cambridge, Mass.)* **44,** 861–870.

Mochizuki, H., Ito, T., and Hidaka, H. (1993). *J. Biol. Chem.* **268,** 9143–9147

Mukherji, S., and Soderling, T. R. (1994). *J. Biol. Chem.* **269,** 13744–13747.

Nairn, A. C., Bhagat, B., and Palfrey, C. (1985). *Proc. Natl. Acad. Sci. U.S.A.* **79,** 7939–7943.

Ngai, P. K., and Walsh, M. P. (1985). *Biochem. J.* **230,** 695–707.

Ngai, P. K., and Walsh, M. P. (1987). *Biochem. J.* **244,** 417–425.

Nghiem, P., Saati, S. M., Martens, C. L., Gardner, P., and Schulman, H. (1993). *J. Biol. Chem.* **268,** 5471–5479.

Ohsako, S., Nishida, Y., Ryo, H., and Yamauchi, T. (1993). *J. Biol. Chem.* **268,** 2052–2062.

Ohta, Y., Ohba, T., and Miyamoto, E. (1990). *Proc. Natl. Acad. Sci. U.S.A.* **87,** 5341–5345.

Papadopoulus, V., Brown, A. S., and Hall, P. F. (1990). *Mol. Cell. Endocrinol.* **74,** 109–123.

Patton, B. L., Miller, S. G., and Kennedy, M. B. (1990). *J. Biol. Chem.* **265,** 11204–11212.

Payne, M. E., Fong, Y.-L., Ono, T., Colbran, R. J., Kemp, B. E., Soderling, T. R., and Means, A. R. (1988). *J. Biol. Chem.* **263,** 7190–7195.

Pearson, R. B., Woodgett, J. R., Cohen, P., and Kemp, B. E. (1985). *J. Biol. Chem.* **260,** 14471–14476.

Picciotto, M. R., Czernik, A. J., and Nairn, A. C. (1993). *J. Biol. Chem.* **268,** 26512–26521.

Planas-Silva, M. D., and Means, A. R. (1992). *EMBO J.* **11,** 507–517.

Sahyoun, N., Levine, H., Bronson, D., Siegel-Greenstein, F., and Cuatrecasas, P. (1985). *J. Biol. Chem.* **260,** 1230–1237.

Schulman, H., and Hanson, P. I. (1993). *Neurochem. Res.* **18,** 65–77.

Schworer, C. M., Colbran, R. J., and Soderling, T. R. (1986). *J. Biol. Chem.* **261,** 8581–8584.

Schworer, C. M., Colbran, R. J., Keefer, J. R., and Soderling, T. R. (1988). *J. Biol. Chem.* **263,** 13486–13489.

Schworer, C. M., Rothblum, L. I., Thekkumkara, T., and Singer, H. A. (1993). *J. Biol. Chem.* **268,** 14443–14449.

Schworer, C. M., Sweeley, C. S., and Singer, H. A. (1994). *FASEB J.* **8,** A393.

Scott-Woo, G. C., and Walsh, M. P. (1988). *Biochem. J.* **255,** 817–824.

Scott-Woo, G. C., Sutherland, C., and Walsh, M. P. (1990). *Biochem. J.* **268,** 367–370.

Sheng, M., Thompson, M. A., and Greenberg, M. E. (1991). *Science* **252,** 1427–1430.

Singer, H. A., Sweeley, C., and Schworer, C. M. (1993). *Biophys. J.* **64,** A33.

Srinivasan, M., Edman, C. F., and Schulman, H. (1994). *J. Cell Biol.* **126,** 839–852.

Stull, J. T., Hsu, L. C., Tansey, M. G., and Kamm, K. E. (1990). *J. Biol. Chem.* **265,** 16683–16690.

Sumi, M., Kiuchi, K., Ishikawa, T., Ishii, A., Hagiwara, M., Nagatsu, T., and Hidaka, H. (1991). *Biochem. Biophys. Res. Commun.* **181,** 968–975.

Sun, P., Enslen, H., Myung, P. S., and Maurer, R. A. (1994). *Genes Dev.* **8,** 2527–2539.

Takahashi, K., Hiwada, K., and Kokubu, T. (1988). *Hypertension* **11,** 620–626.

Tansey, M. G., Word, R. A., Hidaka, H., Singer, H. A., Schworer, C. M., Kamm, K. E., and Stull, J. T., (1992). *J. Biol. Chem.* **267,** 12511–12516.

Tansey, M. G., Luby-Phelps, K., Kamm, K. E., and Stull, J. T. (1994). *J. Biol. Chem.* **269,** 9912–9920.

Thiel, G., Czernik, A. J., Gorelik, F., Nairn, A. C., and Greengard, P. (1988). *Proc. Natl. Acad. Sci. U.S.A.* **85,** 6337–6341.

Tobimatsu, T., and Fujisawa, H. (1989). *J. Biol. Chem.* **264,** 17907–17912.

Tobimatsu, T., Kameshita, I., and Fujisawa, H. (1988). *J. Biol. Chem.* **263,** 16082–16086.

Tokumitsu, H., Chijwa, T., Hagiwara, M., Mizutani, A., Terasawa, T., and Hidaka, H. (1990). *J. Biol. Chem.* **265,** 4315–4320.

Toyofuku, T., Kurzydlowski, K., Narayanan, N., and MacLennan, D. H. (1994). *J. Biol. Chem.* **269,** 26492–26496.

Winder, S. J., and Walsh, M. P. (1990). *J. Biol. Chem.* **265,** 10148–10155.

Xu, A., Hawkins, C., and Narayanan, N. (1993). *J. Biol. Chem.* **268,** 8394–8397.

Yamagata, Y., Czernik, A. J., and Greengard, P. (1991). *J. Biol. Chem.* **266,** 15391–15397.

Yamauchi, T., Ohsaka, S., and Deguchi, T. (1989). *J. Biol. Chem.* **264,** 19108–19116.

Zhou, Z. L., and Ikebe, M. (1994). *Biochem. J.* **299,** 489–495.

12

Protein Kinase C

HAROLD A. SINGER

Sigfried and Janet Weis Center for Research
Geisinger Clinic
Danville, Pennsylvania

I. INTRODUCTION

Protein kinase C (PKC) was originally isolated from brain as a proenzyme that was activated by proteolysis to form a stable, multifunctional, serine/threonine protein kinase activity called protein kinase M (PKM) (Inoue *et al.*, 1977). Interest in the enzyme was stimulated by the discovery that physiological concentrations of free Ca^{2+} and membrane phospholipids, later identified as diacylglycerol (DAG) and phosphatidylserine (PS), could synergistically activate the kinase independently of any proteolytic events (Kishimoto *et al.*, 1980). It is currently recognized that phospholipase C activation and subsequent hydrolysis of membrane phosphoinositides and phosphatidylcholine are proximal events in the signal transduction process triggered by a variety of extracellular stimuli, including smooth muscle contractile and proliferative stimuli (Lee and Severson, 1994). Products of the phospholipase C-catalyzed reaction include inositol 1,4,5-trisphosphate, a soluble second messenger that stimulates release of intracellular Ca^{2+}, and membrane DAG, the endogenous lipid activator of PKC. Thus, PKC has assumed a position of central importance in current models of intracellular signaling and it has been the subject of intense research.

Most insights into PKC function in smooth muscle have been inferred from the cellular responses to phorbol ester activators of PKC and putative PKC inhibitors, and from assays of PKC activation based on content of membrane-bound Ca^{2+}/phospholipid-dependent kinase activity. The current recognition that PKC is a large family of related protein kinases requires consideration of the distribution and functional roles of specific PKC isozymes within a tissue. This monograph will focus on the occurrence, properties, and possible functions of PKC isozymes in smooth muscle with particular emphasis on an unresolved role for PKC in the regulation of smooth muscle contractile function.

II. PROPERTIES OF PROTEIN KINASE C ISOZYMES

A. Structural Properties

An understanding of how PKC is involved in cellular regulation is currently complicated by the fact that as many as 11 distinct PKC isozymes representing 10 separate gene products have been identified by molecular cloning techniques (Asaoka *et al.*, 1992; Hug and Sarre, 1993). Based on deduced amino acid sequences, the PKC isozymes are all single polypeptides of 67–115 kDa and can be categorized according to their regulatory domain structure and activator requirements (Fig. 1). The C-terminal half of the protein comprises the catalytic domain. A consensus ATP binding site ($xGxGx_2Gx_{16}Kx$) in the C3 region and the catalytic core of the kinase (C4) are conserved among the isozymes and have homology with other protein kinases in general (Hanks and Quinn, 1991). The first PKC activities isolated required both Ca^{2+} and phospholipid activators (DAG and PS) for activity. This group is composed of the "conventional" or cPKCs, α, βI, βII, and γ, with the two β-isozymes arising by alternative gene splicing. A conserved N-terminal reg-

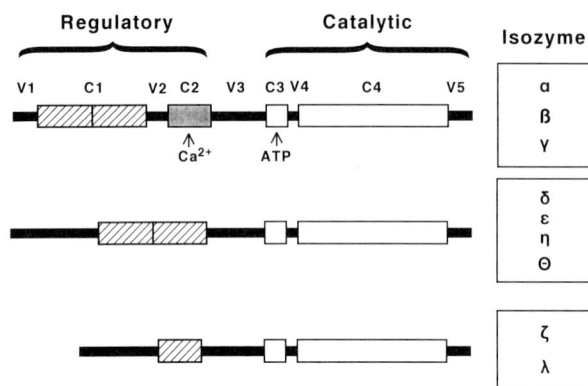

FIGURE 1 Structure of PKC isozymes. Conserved regions among the isozymes are indicated by the large rectangles (C1–4). Variable domains are indicated by the solid bars (V1–5). Structure and activator requirements define three classes of PKC. Conserved Ca^{2+} binding (C2) and diacylglycerol or phorbol ester binding (C1) regions are found in the first group, the second and third groups lack C2 and the requirement of Ca^{2+} for activity, and the third group also lacks a complete C1 region and members of this group are not activated by phorbol esters.

ulatory domain (C1 region) has been identified in these isozymes that is composed of two cysteine-rich regions forming consensus Zn^{2+} binding domains. The C1 domain appears to be involved in binding DAG and is also the binding site for phorbol esters that substitute for DAG and directly activate PKC (Bell and Burns, 1991; Castagna et al., 1982). Although Zn^{2+} binding has been shown to affect kinase activity and phorbol ester binding in vitro (Csermely et al., 1988), the significance of this in vivo is not known. A conserved region designated C2 is also found in the regulatory domain of the cPKC isozymes and appears to be involved in Ca^{2+} binding (Bell and Burns, 1991). The second group in the PKC family is composed of the "new" or nPKCs, δ, ε, η, and θ. These isozymes contain the C1 domain and are activated by DAG or phorbol esters, but lack the Ca^{2+} binding domain (C2) and do not require Ca^{2+} for activity. The third group of isozymes is composed of "atypical" or aPKCs, ζ and λ, which lack both the Ca^{2+} binding domain and a complete phorbol ester binding domain. Recombinant PKCζ and λ do not require Ca^{2+} or DAG for activity and do not bind phorbol esters with high affinity. PKCμ is the most recent isozyme identified and is structurally similar to aPKCs (Johannes et al., 1994; Valverde et al., 1994). Like other aPKCs, expressed PKCμ only weakly binds to and is not activated by phorbol ester (Johannes et al., 1994).

B. Regulation of Protein Kinase C Activity

Proteolytic cleavage of PKC isozymes in the variable region (V3) that separates the regulatory and catalytic

domains (Kishimoto et al., 1983), or expression of mutant cDNA encoding only the catalytic domain of an isozyme (James and Olson, 1992), generates a constitutively active kinase activity analogous to PKM. These observations are consistent with a model of PKC activation where the regulatory domain, in the absence of activators, interacts with and inhibits the catalytic site in the kinase. Inhibition of the catalytic site has been shown by mutational analysis to be via a pseudosubstrate sequence (xRxxAxR/Kx) that is located at the beginning of the C1 domain and is conserved in all the isozymes (House and Kemp, 1987). This motif, with serine or threonine substituted for alanine, forms a consensus phosphorylation sequence found in a number of PKC substrates (House et al., 1987). Association of the inactive kinase with membrane PS provides access of the kinase to activator DAG, resulting in a conformational change that exposes the active site in the kinase (Bell and Burns, 1991). In cPKC isozymes, Ca^{2+} acts synergistically with PS to promote localization of the kinase to the membrane (Newton and Koshland, 1989). In vitro studies using preparations of purified PKC isozymes in mixed micellar assays, which allow control of lipid composition and stoichiometries, indicated that the requirements for PS and DAG are quite specific and that a single molecule of DAG is required for activation of the kinase (Hannun et al., 1986; Newton and Koshland, 1989). However, based on the sensitivity of the complex protein/lipid/metal ion interactions to specific in vitro assay conditions, and variability in binding affinities of the various recombinant PKC isozymes for cofactors (Burns et al., 1990; Liyanage et al., 1992; Kazanietz et al., 1993), this general model may not fully explain the activation of all PKC isozymes in vivo. For example, studies indicate that PKCζ can be activated in vitro by phosphatidic acid in the absence of membrane phospholipid or Ca^{2+}, consistent with a role for this acidic phospholipid as a soluble activator of intracellular PKCζ in vivo (Limatola et al., 1994).

In addition to allosteric regulation of PKC activity by membrane phospholipids, it has become apparent that phosphorylation of the kinase itself is an important regulatory mechanism. Using site-directed mutagenesis approaches, phosphorylation of PKCα on Thr^{497} (Cazaubon et al., 1994) or PKCβII on Thr^{500} (Orr and Newton, 1994) has been shown to be essential for catalytic activity. The catalytic subunit of protein phosphatase 1 specifically dephosphorylates this site, resulting in an inactive kinase that has no intrinsic capacity to rephosphorylate and activate when phosphatase is removed (Dutil et al., 1994). Based on this, and an analysis of the sequence around this site, it has been hypothesized that transphosphorylation of Thr^{500} is catalyzed by an unidentified proline-directed protein kinase (Dutil *et al.*, 1994). Molecular models of PKC structure indicate that this residue is located on a loop near the entrance to the catalytic site (Orr and Newton, 1994), and because this threonine is conserved among the PKC isozymes, it is predicted that phosphorylation of this site is a fundamental requirement for activity in all PKC isozymes. There is similar evidence that phosphorylation of Thr[642] in the V5 region of PKCβI is required for PKC catalytic activity, though it is not clear if this is a *trans-* or *auto*phosphorylation event (Zhang *et al.*, 1994). Phosphorylation of this residue appears to decrease the affinity of PKC for membranes. This site, which is conserved in all PKC isozymes, is flanked by a proline residue and may also be the target of a proline-directed protein kinase. Several activation-dependent *auto*phosphorylation sites in PKC isozymes have been mapped to serine or threonine residues in the variable regions V1, V2, and V5 (Flint *et al.*, 1990). Although the functional consequences of these phosphorylations are not yet clear, some investigators have taken advantage of the phenomenon to provide an index of isozyme-specific activation by measuring phosphorylation of immunoprecipitated kinase (e.g., Mitchell *et al.*, 1989). Future work in these areas can be expected to provide important insights into the regulation of PKC isozyme activity and the integration of this signaling system with pathways involving proline-directed protein kinases.

C. Tissue Distribution

As new PKC isozymes have been cloned, their tissue distribution has been surveyed by Northern analysis of mRNA expression using cDNA probes and by Western blotting using antipeptide antibodies raised to deduced amino acid sequences in variable domains (usually V5) to assess expression at the protein level. A general finding has been that the PKC isozymes are differentially expressed as a function of cell type, and in most cells at least one member of each group is represented (Asaoka *et al.*, 1992; Hug and Sarre, 1993). PKCα is expressed in most tissues, PKCβI and βII appear to be variably expressed, and PKCγ is largely restricted to neural tissues (Yoshida *et al.*, 1988). In the group of nPKCs, δ or ε is widely expressed (Wetsel *et al.*, 1992), PKCη is restricted primarily to epithelial tissues (Greif *et al.*, 1992), and PKCθ to skeletal muscle (Osada *et al.*, 1992). The aPKC isozymes λ, ζ, and PKCμ all have wide but distinct tissue distributions (Akimoto *et al.*, 1994; Johannes *et al.*, 1994; Valverde *et al.*, 1994). These types of data, along with evidence for isozyme-specific subcellular localization, activation properties, and *in vitro* substrate specificity, support the generalization that individual PKC isozymes are likely to phosphorylate and regulate specific substrates *in situ*.

D. Subcellular Localization

PKC can be found in cytosolic and membrane fractions in virtually all cells examined, and based on its phospholipid requirements for activation, it has been argued that active PKC and its protein substrates should be restricted to the membrane compartment. However, a large number of intracellular and even nuclear proteins are phosphorylated following stimulation of cells or tissues with phorbol ester activators of PKC. To account for this, either active forms of PKC must localize to intracellular sites, soluble activators must be generated that activate intracellular pools of PKC, or membrane PKC must trigger another cytosolic kinase(s) that then results in the phosphorylation response. Consistent with the latter possibility, PKC has now been shown in a number of cell types, including smooth muscle, to activate a signaling pathway involving a cascade of intracellular protein kinases that includes mitogen-activated protein kinase (MAP-kinase), another ubiquitous multifunctional serine/threonine kinase (see Section IV.C). There is evidence that activated PKC can bind to specific intracellular receptors, called RACKs (*r*eceptors for *a*ctivated *C-k*inase) (Mochley-Rosen *et al.*, 1992), and that inhibition of this binding can inhibit the actions of activated PKC (Smith and Mochley-Rosen, 1992). RACKs described so far are proteins of about 30–36 kDa and are tightly bound to a detergent-insoluble particulate fraction (Ron *et al.*, 1994). The conserved C2 region in cPKC appears to be involved in RACK binding, though there is some evidence that multiple RACKs may exist that bind PKC isozymes lacking the C2 region (Mochley-Rosen *et al.*, 1992).

The existence of RACKs may explain the reported association of various PKC isozymes with cytoskeletal elements (Leach *et al.*, 1989; Mochley-Rosen *et al.*, 1990). In adrenal Y-1 cells, activation of cytoskeletal-associated PKC resulted in the phosphorylation of several endogenous substrates, including LC20, a response associated with a shape change in the cytoskeleton (Papadopoulos and Hall, 1989). Stimulation of Swiss 3T3 fibroblasts with serum has been shown to result in the recruitment of PKCδ into newly formed focal adhesions (Barry and Critchley, 1994). Intermediate filament proteins, actin binding proteins such as profilin and filamin, and focal contact-associated proteins, such as talin, vinculin, and integrins, have all been reported to be substrates for PKC *in vitro*. Thus, PKC may play a role in regulating the interaction of these cytoskeletal proteins (Woods and Couchman, 1992; Barry and Critchley, 1994), which are required for

VSM cell motility and for transmitting force between cells.

Phorbol ester (Leach et al., 1989; Rogue et al., 1990) or peptide growth factor stimulation (Fields et al., 1990; Neri et al., 1994) has been reported to increase nuclear PKC activity and/or to result in translocation of PKC isozymes to nuclei. Increased nuclear PKC has been correlated with the phosphorylation of nuclear protein substrates such as lamins and with mitogenic responses (Fields et al., 1990) or cell differentiation (Hocevar and Fields, 1991). There is evidence that diacylglycerol levels may be regulated in nuclear membranes in response to extracellular stimuli, through as yet unidentified pathways, providing a mechanism for recruiting and activating cPKC and nPKC isozymes in this compartment (Divecha et al., 1991; Leach et al., 1992). In vascular smooth muscle (VSM), nuclear lamin phosphorylation in response to angiotensin II stimulation has been linked to activation of PKC (Tsuda and Alexander, 1990). Using immunofluorescence and digital imaging microscopy, PKCζ has been reported to translocate to the nucleus of VSM cells following stimulation of the cells with phenylephrine (Khalil et al., 1992). A dominant kinase-negative mutant of PKCζ has been shown to block PMA-stimulated phosphorylation and activation of the transcription factor NF-κB (Diaz-Meco et al., 1993) and mitogenesis in Xenopus oocytes and rat fibroblasts (Berra et al., 1993). Based on these types of data, nuclear PKC has been proposed to be involved in controlling gene transcription responses required for cell growth and proliferation.

III. EXPRESSION OF PROTEIN KINASE C ISOZYMES IN SMOOTH MUSCLE

A complete understanding of PKC function in smooth muscle requires biochemical, immunological, and/or molecular approaches aimed at defining the content of specific isozymes. Ca^{2+}- and phospholipid-dependent PKC activity has been purified from rabbit iris smooth muscle (Howe and Abdel-Latif, 1988), chicken gizzard smooth muscle (DeVries et al., 1989), bovine aortic VSM (Dell et al., 1988), and swine carotid VSM (Schworer and Singer, 1991). Based on hydroxylapatite chromatography and immunoblotting with the type-specific antibodies, the principal cPKC isozyme expressed in bovine aortic VSM was determined to be PKCα (Watanabe et al., 1989). Two peaks of cPKC activity were resolved chromatographically from rat cerebral microvessels (Kobayashi et al., 1991) and swine carotid arterial smooth muscle extracts (Schworer and Singer, 1991) and identified as PKCα and β isozymes by immunoblotting with type-specific

monoclonal antibodies. Using antipeptide antibodies that distinguish between PKCβI and βII, PKCβII appears to be the form expressed in swine carotid VSM (Singer, H. A., unpublished observations). The specific activity of cPKC in unpurified VSM cell lysates was found in these studies to be in the range of 0.5–4 nmol/min/mg protein using histone IIIS as a substrate. These values are about one-tenth that found in brain but comparable to PKC activities reported in other peripheral tissues.

PKC activity purified from swine carotid artery and assayed with myelin basic protein can be partially stimulated by lipid activators independently of Ca^{2+}, consistent with the activity of an nPKC isozyme(s) (Schworer and Singer, 1991). Because the chromatographic properties of the nPKCs and aPKCs are similar to those of the cPKCs or are not known, there have been few attempts to individually purify and characterize these isozymes from intact tissues. Based on studies of eukaryotically expressed cDNAs encoding the isozymes, histone IIIS and/or myelin basic protein, which are commonly used to assay the cPKCs (Burns et al., 1990), are poor substrates for most of the nPKCs and aPKCs (Kazanietz et al., 1993). Although the PKC isozymes have varying relative affinities for peptide substrates (Karanietz et al., 1993), there is not a series of peptide substrates that can be used in activity assays to unambiguously identify specific isozymes in a mixture. Therefore, identification of the nPKC and aPKC isozymes in tissues, including smooth muscle, has been mainly by immunoblotting with type-specific antipeptide antibodies. In the cases of intact swine carotid artery, rat aorta, and cultured rat aortic VSM extracts, we have identified immunoreactive bands of 76 and 67–76 kDa on Western blots using anti-PKCδ and anti-PKCζ antipeptide antibodies, respectively (Fig. 2). A small amount of PKCε was also identified in partially purified fractions of PKC from carotid VSM (Singer, H. A., unpublished data).

Table I summarizes the result of several studies that surveyed PKC isozyme expression in extracts or homogenates of intact smooth muscle or cultured smooth muscle cells by immunoblotting with type-specific antibodies. A consistent finding is that multiple PKC isozymes are expressed in various smooth muscles, and in most cases one PKC isozyme from each group (usually α, δ, and ζ) is expressed. PKCβ, which is abundant in swine carotid, appears to be expressed in only a few other vascular tissues. Likewise, PKCε from the nPKC group appears to be variably expressed. Considering the diversity in smooth muscle embryological origins (Miano et al., 1994), it might be expected that some of the PKC isozymes, which were noted earlier to be variably expressed in

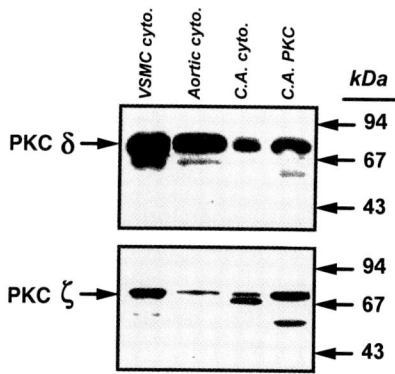

FIGURE 2 Immunoblots of VSM samples with antipeptide antibodies specific for the Ca^{2+}-independent PKC isozymes PKCδ (top) and PKCζ (bottom). Detection was with ^{125}I-labeled IgG and autoradiography. Lanes from left to right are: cultured rat aortic VSM cell cytosol, rat aortic medial smooth muscle cytosol, swine carotid artery (C.A.) cytosol, and purified PKC from C.A. The positions of molecular mass standards are indicated on the right.

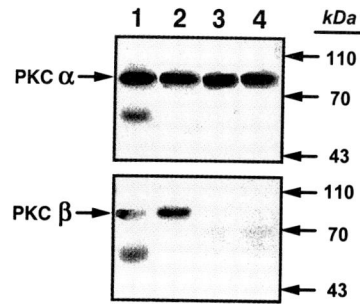

FIGURE 3 Immunoblots of swine VSM and cultured VSM cell extracts with PKCα (top) and PKCβ isozyme-specific monoclonal antibodies. Lane 1: aortic extract; lane 2: extract from acutely dispersed carotid VSM cells; lane 3: extract from carotid VSM cells in 1° culture (Day 13): lane 4: extract from carotid VSM cells second passage (Day 11). The positions of molecular mass standards are indicated on the right.

different tissues, would also be variably expressed in different types of smooth muscle or even VSM from different vascular beds. Another possible explanation for some of the variability in PKC isozyme expression, particularly in cultured cells, is that expression of certain isozymes may be regulated as a function of the differentiated state of the cells. PKCζ expression has been noted to be low and variable in proliferating cultured VSM (Assender et al., 1994). Total content of PKC activity (Barzu et al., 1994) and PKCα expression (Haller et al., 1995) has been reported to increase with differentiation (assessed by α-smooth muscle actin expression) of cultured rat aortic VSM cells. We have consistently observed that PKCβ expression in swine carotid VSM cells is lost in cell culture coincident with loss of the differentiated contractile phenotype (Fig. 3).

Evidence that PKCβ is independently regulated in cells includes the observation that PKCβ expression is specifically increased in rat aorta and heart in response to streptozotocin-induced diabetes (Inoguchi et al., 1992).

Approaches for evaluating PKC isozyme expression at the level of mRNA are required to confirm the specificity of the antibodies used in the immunoblotting studies and to ultimately understand the physiological and pathophysiological control of isozyme expression. Screening for the expression of 10 PKC isozyme mRNAs by Northern blotting is tedious and relatively insensitive, requiring the isolation of large amounts of RNA. To define PKC isozyme expression in VSM, we have used reverse transcription (RT) of total RNA (1 µg) followed by polymerase chain reaction (PCR) amplification of first-strand cDNA with isozyme-specific oligonucleotide primers (18–21 mers)

TABLE I PKC Isozyme Expression in Smooth Muscle

Smooth muscle	PKC isozymes[a]	Method[b]	Reference
Swine carotid VSM	α, βII, δ, ζ, (ε)	IB	Schworer and Singer (1991), Fig. 2
Rat aortic VSM	α, βII, δ, ζ	IB	Inoguchi et al. (1992), Fig. 2
	α, δ, ζ, η, (β), (ε)	RT–PCR	Inoguchi et al. (1992), Fig. 4
Ferret aorta VSM	α, ε, ζ	IF, IB	Khalil et al. (1992)
Human saphenous vein VSM	α, δ, (ζ)	IB, RT–PCR	Assender et al. (1994)
Hamster vas deferens	α, δ, ζ	IB, RT–PCR	Assender et al. (1994)
Human renal artery cultured VSM	α, δ	IB	Assender et al. (1994)
	α, δ, ε	RT–PCR	Assender et al. (1994)
Rat aortic cultured VSM	α, δ, ε, ζ	IB, RT–PCR	Ali et al. (1994)
Rat mesenteric artery cultured VSM	α, δ, ζ, (ε)	IB	Dixon et al. (1994)

[a]Parentheses indicate trace amounts.

[b]IB, immunoblotting; IF, immunofluorescence; RT–PCR, reverse transcriptase–polymerase chain reaction amplification of mRNA.

(Fig. 4). This technique, though not absolutely quantitative, is very sensitive and is a quick method for screening expressed genes. The PCR primers were chosen so that the amplified products included the V3 variable domain. Using rat brain RNA as a control, an expected product for each of the isozymes was resolved on agarose gels. Identities of the ethidium bromide-stained PCR products were verified by Southern blotting using ^{32}P-labeled oligonucleotide probes complementary to V3 isozyme-specific sequences (not shown).

PCR amplification of the PKC isozyme mRNA target sequences from rat aortic medial smooth muscle RNA resulted in products from PKCs α, δ, ζ, and η, confirming our immunoblot analysis of protein expression (with the exception of PKCη, which has not been evaluated by immunoblotting). PKCβ and ε products from rat aortic VSM were not visible by ethidium bromide staining but small amounts could be detected by Southern blotting with high specific activity probes (not shown). The very low level of PKCβ mRNA expression was unexpected based on reports of PKCβ protein expression in rat aorta (Inoguchi *et al.*, 1992). These data raise questions regarding the stability of the PKCβ message in VSM and/or the specificity of β-subtype-specific antibodies, but illustrate the importance of verifying isozyme expression at both the mRNA and protein levels. A similar pattern of PKC isozyme expression (α, δ, ε, ζ,) in *cultured* rat aortic VSM has been reported by Ali *et al.* (1994). Additional

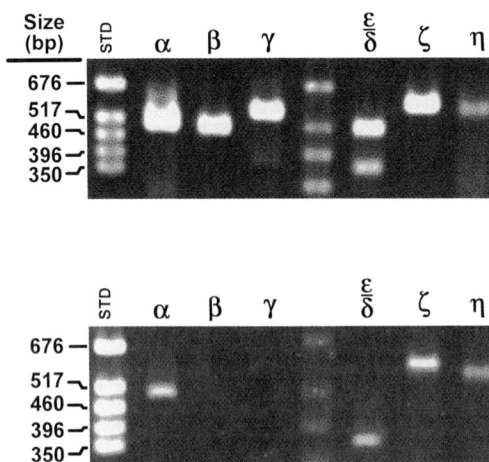

FIGURE 4 Ethidium bromide-stained agarose gel showing size-fractionated RT–PCR products amplified from adult rat brain RNA (top) and rat aortic medial smooth muscle RNA (bottom). Type-specific primer pairs were used to amplify targets of the following expected sizes: PKCα (513 bp), β (501 bp), γ (567 bp), ζ (600 bp), and η (578 bp). A common set of primers was used to amplify the PKCε (503 bp) and δ (357 bp) targets. Target identities were confirmed by Southern blotting using type-specific V3 probes.

studies are required to evaluate the expression in smooth muscle of the recently identified PKC isozymes (θ, η, λ). Once the pattern of PKC isozyme expression has been determined, the RT–PCR approach can be modified so that it is quantitative (Dostal *et al.*, 1994), providing a sensitive approach for assessing the regulation of isozyme gene expression.

IV. ROLE OF PROTEIN KINASE C IN REGULATING SMOOTH MUSCLE CONTRACTION

A. Protein Kinase C Activation in Smooth Muscle

Based on observations that phorbol ester activators of PKC cause slowly developing sustained contractile responses in arterial smooth muscle, a role for PKC in regulating arterial smooth muscle contraction was proposed as early as 1984 (Rasmussen *et al.*, 1984; Danthuluri and Deth, 1984). Since then, phorbol ester-stimulated contractile responses, and potentiation or inhibition of contractile responses to other stimuli, have been characterized in diverse smooth muscle preparations (e.g., Baraban *et al.*, 1985; Mitsui and Karaki, 1993; Langlands and Diamond, 1994). Consistent with the idea that PKC is involved in regulating VSM contraction, putative inhibitors of PKC such as H-7, staurosporine, and calphostin C have been widely reported to inhibit phorbol ester- and physiological stimulus-induced contractile responses (e.g., Singer, 1990a; Ratz, 1990; Shimamoto *et al.*, 1993) and myogenic tone (Henrion and Laher, 1993). PKC downregulation by prolonged exposure to phorbol 12,13-dibutyrate (PDBu) has also been reported to inhibit agonist-stimulated contraction in VSM (Merkel *et al.*, 1991). These pharmacological approaches have proven useful in directing attention to the possibility that PKC may be involved in regulating smooth muscle contraction and several other responses, including the feedback regulation of receptor-coupled phosphoinositide metabolism (Lee and Severson, 1994). However, because phorbol esters cause sustained unregulated activation of PKC and PKC inhibitors have only relative specificity for PKC compared to other kinases, there are clear limitations regarding the interpretation of these types of data in a physiological context.

Direct evidence that a regulatory pathway involving PKC exists in smooth muscle includes the demonstration that physiological contractile stimuli such as angiotensin II, endothelin, and norepinephrine cause sustained increases in the membrane content of the endogenous PKC activator, DAG (Lee and Severson,

1994). In addition, phorbol ester- and contractile stimulus-induced increases in membrane cPKC activity, reflective of PKC translocation and activation *in situ,* have been documented in cultured VSM cells (e.g., Dixon *et al.,* 1994) and intact preparations of arterial (Haller *et al.,* 1990; Singer *et al.,* 1992) and tracheal (Langlands and Diamond, 1994) smooth muscle. By immunoblotting with type-specific monoclonal antibodies, PKCα and β content in particulate fractions from carotid artery VSM was found to correlate temporally with measured increases in cPKC activity and tonic force maintenance following stimulation of intact tissues with physiological stimuli, including KCl depolarization (Singer *et al.,* 1992). This study also indicated that pharmacological activation of the most abundant PKC isozymes (α and β) by PDBu in this tissue was similar (both qualitatively and quantitatively) to activation of these isozymes by physiological contractile stimuli. Using immunofluorescence approaches, PKCα and ε have been reported to translocate to the plasma membrane in isolated ferret aortic VSM cells following stimulation with phenylephrine (Khalil *et al.,* 1992). In Ca^{2+}-depleted cells, translocation of PKCε was correlated with a Ca^{2+}-independent contraction in response to phenylephrine, raising the possibility that this Ca^{2+}-independent PKC isozyme was involved in regulating the contractile response (Khalil *et al.,* 1992). A specific knock-out or inhibition of PKCε will be required to test this hypothesis.

B. Phorbol Ester-Stimulated Contractile Responses

Ca^{2+}/calmodulin-dependent activation of myosin light chain kinase (MLCK) with phosphorylation of the 20-kDa regulatory myosin light chains (LC20) on Ser[19] by MLCK is accepted as the primary cross-bridge regulatory mechanism in smooth muscle (Kamm and Stull, 1989). However, sustained contractile responses in arterial and tracheal smooth muscle are characterized by high levels of cross-bridge activation with low levels of free intracellular [Ca^{2+}], low levels of LC20 phosphorylation, and slow shortening velocities. The temporal dissociation of LC20 phosphorylation and cross-bridge activation (or force production), and the hyperbolic relationship between steady-state force and LC20 phosphorylation, has led to the controversial proposal that there may be alternative regulatory mechanisms controlling smooth muscle actin/myosin interactions that are superimposed on the known regulatory mechanism involving LC20 phosphorylation (see Murphy, 1994, for review).

Given the fundamental importance of a second regulatory system in controlling smooth muscle contrac-

tile function, considerable effort has been directed at defining the mechanistic basis for phorbol ester-induced contractile responses. The first reports describing such responses in intact preparations of VSM showed them to be partially dependent on the availability of extracellular Ca^{2+} (Danthuluri and Deth, 1984; Baraban *et al.,* 1985). Phorbol esters were found to increase the sensitivity of the contractile apparatus to added Ca^{2+} in permeabilized preparations of arterial smooth muscle but could not independently contract the tissue at free [Ca^{2+}] less than 100 nM (Chatterjee and Tejada, 1986; Miller *et al.,* 1986). These observations focused attention on the possibility that at least part of the phorbol ester-stimulated contractile response could be mediated by Ca^{2+}-dependent activation of MLCK and phosphorylation of LC20 on Ser[19]. Consistent with this, LC20 phosphorylation was shown to increase in response to phorbol ester addition in both rabbit aorta (Singer and Baker, 1987) and swine carotid artery (Rembold and Murphy, 1988; Ratz *et al.,* 1989), where the response was correlated with increases in free intracellular [Ca^{2+}] measured by aequorin luminescence (Rembold and Murphy, 1988). However, even after depletion of intracellular Ca^{2+} pools in rabbit aorta (Singer and Baker, 1987) and swine carotid artery (Singer, 1990b; Bárány *et al.,* 1992; Fulginiti *et al.,* 1993) to levels that did not support MLCK activation by normal stimuli, both contractile and LC20 phosphorylation responses were measured in response to PDBu stimulation. This property of phorbol ester activators of PKC to increase free intracellular Ca^{2+} in smooth muscle is apparently tissue specific, since in ferret and rat aorta, phorbol ester-induced contractile responses were not associated with increases in free intracellular [Ca^{2+}] (Jiang and Morgan, 1987).

The role of LC20 phosphorylation in phorbol ester-stimulated contractile responses has been largely resolved by defining the specific phosphorylated residues. *In vitro* studies using purified PKC and smooth muscle LC20 as a substrate identified Thr[9] followed by Ser[1 and/or 2] as the primary PKC phosphorylation sites and phosphorylation of these sites was associated with a decrease in actomyosin ATPase activity (Ikebe *et al.,* 1987). In contrast, MLCK is known to phosphorylate Ser[19] as a primary site and Thr[18] as a secondary site, both of which result in activation of actomyosin ATPase (Ikebe and Hartshorne, 1985). Stimulation of intact [32P]-labeled bovine tracheal strips (Kamm *et al.,* 1989), rabbit aortic rings (Singer *et al.,* 1989), and swine carotid artery strips (Singer, 1990b; Bárány *et al.,* 1990) with high concentrations of phorbol ester (>1 μM) was found to increase LC20 phosphorylation primarily on Ser[1,2] with only low amounts of net Thr[9] phosphorylation reported. Some Ser[19] phosphorylation was observed in response to PDBu in carotid artery (Singer, 1990b), but even the most conservative analysis of the data indicated that full contraction in response to PDBu (100% of KCl-induced responses) was associated with very small increases in Ser[19] monophosphorylation (3–8% of total Ser[19]). PDBu was also reported to stimulate disproportionate increases in LC20 diphosphorylation, primarily by increasing phosphorylation of Ser[1,2], but also by stimulating some phosphorylation of Thr[18] (Singer, 1990b; Rokolya *et al.*, 1991). The mechanism underlying the selective phosphorylation of Thr[18] is not clear, but it suggests the activity of an unidentified kinase. Based on these experiments and studies characterizing the mechanical properties of arterial smooth muscle activated by phorbol esters (Fulginiti *et al.*, 1993), it was concluded that phorbol ester-stimulated contraction in smooth muscle can occur by a mechanism that is independent of LC20 phosphorylation.

Phosphorylation of LC20 in permeabilized gizzard smooth muscle by direct addition of PKC (Sutton and Haeberle, 1990) or its constitutively active proteolytic fragment, PKM (Parente *et al.*, 1992), results in phosphorylation of LC20 mainly on Ser[1,2] with a small amount on Thr[9]. The only mechanical response observed in those studies in response to direct addition of PKC or PKM was an attenuation of Ca^{2+}-induced contraction in permeabilized gizzard smooth muscle fibers (Parente *et al.*, 1992). Phorbol ester-induced attenuation of contractile responses has been observed in ileal and uterine smooth muscle (Baraban *et al.*, 1985) and it was hypothesized that during sustained activation of intact smooth muscle by phorbol esters or physiological stimuli, PKC was proteolyzed at the membrane forming PKM, which then phosphorylated LC20 on Ser[1,2] and Thr[9] and attenuated Ca^{2+}-dependent contractile responses (Andrea and Walsh, 1992). However, we have found no evidence for significant cytosolic accumulation of activator-independent PKC activity (PKM) in response to sustained activation of PKC by PDBu or physiological stimuli in swine carotid artery (Singer *et al.*, 1992). Furthermore, phosphorylation of LC20 in response to physiological stimuli (phenylephrine, carbachol, histamine, KCl) requires Ca^{2+} and has been shown to be restricted to Ser[19] with small amounts on Thr[18] in diphosphorylated LC20, consistent with the activation of MLCK (Singer *et al.*, 1989; Kamm *et al.*, 1989; Singer, 1990b; Bárány *et al.*, 1990). Therefore, the available evidence suggests that phosphorylation of LC20 on Ser[1,2] and Thr[9], either directly by PKC or indirectly by PKM, is not of physiological significance in differentiated VSM. Interestingly, increases in LC20 phosphorylation on Ser[1,2] have been observed in fibroblasts undergoing mitosis

(Yamakita *et al.*, 1994). This phosphorylation event has been attributed to cdc2 kinase and is thought to be important in regulating myosin filament formation during cell division.

To account for activation of arterial smooth muscle independently of LC20 phosphorylation, attention has been focused on the possible roles of the thin filament-associated regulatory proteins, caldesmon and calponin. Both proteins have been localized in the actomyosin domain of the smooth muscle cell and both have been shown to inhibit actin-activated myosin ATPase by interacting with F-actin, tropomyosin, and/or myosin (Clark *et al.*, 1986; Takahashi *et al.*, 1988). Caldesmon has also been shown to inhibit tension development in chemically permeabilized gizzard smooth muscle (Pfitzer *et al.*, 1993). Furthermore, inhibition of caldesmon/F-actin interaction in permeabilized VSM resulted in contractile force generation independently of changes in $[Ca^{2+}]$, supporting the concept that caldesmon may function as a regulator *in situ* independently of LC20 phosphorylation (Katsuyama *et al.*, 1992). Both caldesmon (Adam *et al.*, 1989) and calponin (Winder and Walsh, 1990) can be phosphorylated *in vitro* by PKC, and once phosphorylated they no longer inhibit actomyosin ATPase. Whether or not calponin is phosphorylated and regulated *in vivo* remains controversial (Bárány *et al.*, 1991; Winder and Walsh, 1993). Increased caldesmon phosphorylation has been demonstrated in ^{32}P-labeled carotid artery strips stimulated by either KCl depolarization or PDBu (Adam *et al.*, 1989). However, sequence analysis of the sites phosphorylated *in vivo* indicated that they were not the sites in caldesmon phosphorylated *in vitro* by purified PKC or Ca^{2+}/calmodulin-dependent protein kinase II (Adam *et al.*, 1992). The endogenous phosphorylation sites in caldesmon form consensus sequences for, and are phosphorylated by, proline-directed kinases such as MAP-kinase or p34[cdc2] kinase (Adam *et al.*, 1992; Childs *et al.*, 1992). Given these data, it was hypothesized that smooth muscle caldesmon is regulated by MAP-kinase. Unproven extrapolations of this hypothesis are that MAP-kinases mediate phorbol ester-stimulated, LC20 phosphorylation-independent contractile responses in VSM, and that physiological activation of PKC and control of contractile activity are by a mechanism involving MAP-kinase-dependent caldesmon phosphorylation.

C. Protein Kinase C-Dependent Activation of Mitogen-Activated Protein Kinase

The ability of PKC to activate the MAP-kinase signaling pathway has been well established in numerous cell lines (e.g., L'Allemain *et al.*, 1991), including

VSM, where MAP-kinase activation in response to several physiological stimuli has been shown to be PKC dependent (Tsuda *et al.*, 1992; Kribbon *et al.*, 1993; Khalil and Morgan, 1993). MAP-kinases are ubiquitous, multifunctional, serine/threonine kinases and are themselves activated by phosphorylation catalyzed by a specific kinase, MAP-kinase kinase (MEK). Input of PKC into the MAP-kinase pathway appears to be at the level of a specific MEK kinase or *raf*-kinase, the upstream activators of MEK (Marquardt *et al.*, 1994). Coexpression of *raf-1* kinase and Ca^{2+}- and phospholipid-dependent isozymes of PKC (α,β,γ) but not Ca^{2+}-independent forms of PKC (δ,ζ,η) in insect cells has been shown to result in the phosphorylation and activation of *raf*-kinase activity (Sozeri *et al.*, 1993). Although the direct outcomes of activating this signaling pathway from the standpoint of regulating smooth muscle contraction are not yet understood, it is clear that MAP-kinase is involved in regulating gene transcription (Seth *et al.*, 1992) and protein synthesis (Lin *et al.*, 1994). An attractive hypothesis is that PKC-dependent activation of the MAP-kinase signaling pathway could serve to coordinate differentiated smooth muscle contractile activity with contractile protein gene expression and protein synthesis.

V. SUMMARY

This review has focused on a possible role for PKC in the control of arterial smooth muscle contractile function. However, based on studies in smooth muscle or by analogy with other cell types, several other functions have been suggested for the kinase, ranging from feedback regulation of signal transduction events to control of gene expression and cell growth (Andrea and Walsh, 1992; Lee and Severson, 1994). The identification of multiple PKC isozymes with distinct activation properties, subcellular distributions, and tissue distributions provides a basis for explaining these diverse roles. Future research will require determination of the complement of PKC isozymes expressed in a given smooth muscle cell type and new approaches designed to dissect the functional roles of a specific isozyme. Such approaches will depend heavily on isozyme-specific immunological and molecular probes and a better understanding of how PKC isozymes are regulated by specific lipid activators, autophosphorylation, and intracellular binding proteins.

Studies aimed at defining how phorbol ester activators of PKC stimulate contractile responses in smooth muscle have pointed to the existence of an unidentified mechanism for activating cross-bridges that is independent of LC20 phosphorylation of Ser[19]. The physiological significance of this remains uncertain. A promising area of investigation will be to determine to what extent activation of a signaling pathway involving MAP-kinase contributes to the PKC-dependent control of smooth muscle contraction. Activation of these or similar cytosolic kinases could help explain the diversity of phosphorylation events that are triggered by PKC activators, including phosphorylation of caldesmon, which may be involved in regulating contractile function. Finally, it may be worthwhile to consider if PKC-dependent activation of the MAP-kinase signaling pathway provides a mechanism for coordinating contractile function, gene expression, and protein synthesis in differentiated smooth muscle.

References

Adam, L. P., Haeberle, J. R., and Hathaway, D. R. (1989). *J. Biol. Chem.* **264**, 7698–7703.

Adam, L. P., Gapinski, C. J., and Hathaway, D. R. (1992). *FEBS Lett.* **302**, 223–226.

Akimoto, K., Mizuno, K., Osada, S., Hirai, S., Tanuma, S., Suzuki, K., and Ohno, S. (1994). *J. Biol. Chem.* **269**, 12677–12683.

Ali, S., Becker, M. W., Davis, M. G., and Dorn, G. W., II (1994). *Circ. Res.* **75**, 836–843.

Andrea, J. E., and Walsh, M. P. (1992). *Hypertension* **20**, 585–595.

Asaoka, Y., Nakamura, S., Yoshida, K., and Nishizuka, Y. (1992). *Trends Biochem. Sci.* **17**, 414–417.

Assender, J. W., Kontny, E., and Fredholm, B. B. (1994). *FEBS Lett.* **342**, 76–80.

Baraban, J. M., Gould, R. J., Peroutka, S. J., and Snyder, S. H. (1985). *Proc. Natl. Acad. Sci. U.S.A.* **82**, 604–607.

Bárány, K., Rokolya, A., and Bárány, M. (1990). *Biochim. Biophys. Acta* **1035**, 105–108.

Bárány, K., Rokolya, A., and Bárány, K. (1991). *FEBS Lett.* **279**, 65–68.

Bárány, K., Polyak, E., and Bárány, M. (1992). *Biochim. Biophys. Acta* **1134**, 233–241.

Barry, S. T., and Critchley, D. R. (1994). *J. Cell Sci.* **107**, 2033–2045.

Barzu, T., Herbert, J., Desmouliere, A., Carayon, P., and Pascal, M. (1994). *J. Cell. Physiol.* **160**, 239–248.

Bell, R. M., and Burns, D. J. (1991). *J. Biol. Chem.* **266**, 4661–4664.

Berra, E., Diaz-Meco, M. T., Dominguez, I., Municio, M. M., Sanz, L., Lozano, J., Chapkin, R. S., and Moscat, J. (1993). *Cell (Cambridge, Mass.)* **74**, 555–563.

Burns, D. J., Bloomenthal, J., Lee, M., and Bell, R. M. (1990). *J. Biol. Chem.* **265**, 12044–12051.

Castagna, M., Takai, Y., Kaibuchi, K., Sano, K., Kikkawa, U., and Nishizuka, Y. (1982). *J. Biol. Chem.* **257**, 7847–7851.

Cazaubon, S., Bornancin, F., and Parker, P. J. (1994). *Biochem. J.* **301**, 443–448.

Chatterjee, M., and Tejada, M. (1986). *Am. J. Physiol.* **251**, C356–C361.

Childs, T. J., Watson, M. H., Sanghera, J. S., Campbell, D. L., Pelech, S. L., and Mak, A. S. (1992). *J. Biol. Chem.* **267**, 22853–22859.

Clark, T., Ngai, P. K., Sutherland, T., Groschel-Stewart, U., and Walsh, M. P. (1986). *J. Pharmacol. Exp. Ther.* **261**, 8028–8035.

Csermely, P., Szamel, M., Resch, K., and Somogyi, J. (1988). *J. Biol. Chem.* **263**, 6487–6490.

Danthuluri, N. R., and Deth, R. C. (1984). *Biochem. Biophys. Res. Commun.* **125**, 1103–1109.

Dell, K. R., Walsh, M. P., and Severson, D. L. (1988). *Biochem. J.* **254**, 455–462.

DeVries, G., Fraser, E. D., and Walsh, M. P. (1989). *Biochem. Cell Biol.* **67**, 260–270.

Diaz-Meco, M. T., Berra, E., Municio, M. M., Sanz, L., Lozano, J., Dominguez, I., Diaz-Golpe, V., Lain de Lera, M. T., Alcami, J., Paya, C. V., Arenzana-Seisdedos, F., Virelizier, J., and Moscat, J. (1993). *Mol. Cell. Biol.* **13**, 4770–4775.

Divecha, N., Banfic, H., and Irvine, R. F. (1991). *EMBO J.* **10**, 3207–3214.

Dixon, B. S., Sharma, R. V., Dickerson, T., and Fortune, J. (1994). *Am. J. Physiol.* **266**, C1406–C1420.

Dostal, D. E., Rothblum, K. N., and Baker, K. M. (1994). *Anal. Biochem.* **223**, 239–250.

Dutil, E. M., Keranen, L. M., DePaoli-Roach, A. A., and Newton, A. C. (1994). *J. Biol. Chem.* **269**, 29359–29362.

Fields, A. P., Tyler, G., Kraft, A. S., and May, W. S. (1990). *J. Cell Sci.* **96**, 107–114.

Flint, A. J., Paladini, R. D., and Koshland, D. E. J. (1990). *Science* **249**, 408–411.

Fulginiti, J., III, Singer, H. A., and Moreland, R. S. (1993). *J. Vasc. Res.* **30**, 315–322.

Greif, H., Ben-Chaim, J., Shimon, T., Bechor, E., Eldar, H., and Livneh, E. (1992). *Mol. Cell. Biol.* **12**, 1304–1311.

Haller, H., Smallwood, J. I., and Rasmussen, H. (1990). *Biochem. J.* **270**, 375–381.

Haller, H., Lindschau, C., Quass, P., Distler, A., and Luft, F. C. (1995). *Circ. Res.* **76**, 21–29.

Hanks, S. K., and Quinn, A. M. (1991). *In* "Methods in Enzymology" (T. Hunter and B. Sefton, eds.), Vol. 200, pp. 38–61. Academic Press, San Diego, CA.

Hannun, Y. A., Loomis, C. R., and Bell, R. M. (1986). *J. Pharmacol. Exp. Ther.* **261**, 7184–7190.

Henrion, D., and Laher, I. (1993). *Hypertension* **22**, 78–83.

Hocevar, B. A., and Fields, A. P. (1991). *J. Biol. Chem.* **266**, 28–33.

House, C., and Kemp, B. E. (1987). *Science* **238**, 1726–1728.

House, C., Wettenhall, R. E. H., and Kemp, B. E. (1987). *J. Biol. Chem.* **262**, 772–777.

Howe, P. H., and Abdel-Latif, A. A. (1988). *Biochem. J.* **255**, 423–429.

Hug, H., and Sarre, T. F. (1993). *Biochem. J.* **291**, 329–343.

Ikebe, M., and Hartshorne, D. J. (1985). *J. Biol. Chem.* **260**, 10027–10031.

Ikebe, M., Hartshorne, D. J., and Elzinga, M. (1987). *J. Biol. Chem.* **262**, 9569–9573.

Inoguchi, T., Battan, R., Handler, E., Sportsman, J. R., Health, W., and King, G. I. (1992). *Proc. Natl. Acad. Sci. U.S.A.* **89**, 11059–11063.

Inoue, M., Kishimoto, A., Takai, Y., and Nishizuka, Y. (1977). *J. Biol. Chem.* **252**, 7610–7616.

James, G., and Olson, E. (1992). *J. Cell Biol.* **116**, 863–874.

Jiang, M. J., and Morgan, K. G. (1987). *Am. J. Physiol.* **253**, H1365–H1371.

Johannes, F. J., Prestle, J., Eis, S., Oberhagemann, P., and Pfizenmaier, K. (1994). *J. Biol. Chem.* **269**, 6140–6148.

Kamm, K. E., and Stull, J. T. (1989). *Annu. Rev. Physiol.* **51**, 299–313.

Kamm, K. E., Hsu, L.-C., Kubota, Y., and Stull, J. T. (1989). *J. Biol. Chem.* **264**, 21223–21229.

Katsuyama, H., Wang, C.-L. A., and Morgan, K. G. (1992). *J. Biol. Chem.* **267**, 14555–14588.

Kazanietz, M. G., Areces, L. B., Bahador, A., Mischak, H., Goodnight, J., Mushinski, J. F., and Blumberg, P. M. (1993). *Mol. Pharmacol.* **44**, 298–307.

Khalil, R. A., and Morgan, K. G. (1993). *Am. J. Physiol.* **265**, C404–C411.

Khalil, R. A., Lajoie, C., Resnick, M. S., and Morgan, K. G. (1992). *Am. J. Physiol.* **263**, C714–C719.

Kishimoto, A., Takai, Y., Mori, T., Kikkawa, U., and Nishizuka, Y. (1980). *J. Biol. Chem.* **255**, 2273–2276.

Kishimoto, A., Kijakawa, N., Shiota, M., and Nishizuka, Y. (1983). *J. Biol. Chem.* **258**, 1156–1164.

Kobayashi, H., Mizuki, T., Okazaki, M., Kuroiwa, A., and Izumi, F. (1991). *Experientia* **47**, 245–247.

Kribbon, A., Wieder, E. D., Li, X., Van Putten, V., Granot, Y., Schrier, R. W., and Nemenoff, R. A. (1993). *Am. J. Physiol.* **265**, C939–C945.

L'Allemain, G., Sturgill, T. W., and Weber, M. J. (1991). *Mol. Cell. Biol.* **11**, 1002–1008.

Langlands, J. M., and Diamond, J. (1994). *Eur. J. Pharmacol.* **266**, 229–236.

Leach, K. L., Powers, E. A., Ruff, V. A., Jaken, S., and Kaufmann, S. (1989). *J. Cell Biol.* **109**, 685–695.

Leach, K. L., Ruff, V. A., Jarpe, M. B., Adams, L. D., Fabbro, D., and Raben, D. M. (1992). *J. Biol. Chem.* **267**, 21816–21822.

Lee, M. W., and Severson, D. L. (1994). *Am. J. Physiol.* **267**, C659–C678.

Limatola, C., Schaap, D., Moolenaar, W. H., and van Blitterswijk, W. J. (1994). *Biochem. J.* **304**, 1001–1008.

Lin, T., Kong, X., Haystead, T. A., Pause, A., Belsham, G., Sonenberg, N., and Lawrence, J. C., Jr. (1994). *Science* **266**, 653–656.

Liyanage, M., Frith, D., Livneh, E., and Stabel, S. (1992). *Biochem. J.* **283**, 781–787.

Marquardt, B., Frith, D., and Stabel, S. (1994). *Oncogene* **9**, 3213–3218.

Merkel, L. A., Rivera, L. M., Colussi, J., and Perrone, M. H. (1991). *J. Pharmacol. Exp. Ther.* **257**, 134–140.

Miano, J. M., Cserjesi, P., Ligon, K. L., Periasamy, M., and Olson, E. N. (1994). *Circ. Res.* **75**, 803–812.

Miller, J. R., Hawkins, D. J., and Wells, J. N. (1986). *J. Pharmacol. Exp. Ther.* **239**, 38–42.

Mitchell, F. E., Marais, R. M., and Parker, P. J. (1989). *Biochem. J.* **261**, 131–136.

Mitsui, M., and Karaki, H. (1993). *Br. J. Pharmacol.* **109**, 229–233.

Mochly-Rosen, D., Henrich, C. J., Cheever, L., Khaner, H., and Simpson, P. C. (1990). *Cell Regul.* **1**, 693–706.

Mochly-Rosen, D., Miller, K. G., Scheller, R. H., Khaner, H., Lopez, J., and Smith, B. L. (1992). *Biochemistry* **31**, 8120–8124.

Murphy, R. A. (1994). *FASEB J.* **8**, 311–318.

Neri, L. M., Billi, A. M., Manzoli, L., Rubbini, S., Gilmour, R. S., Cocco, L., and Martelli, A. M. (1994). *FEBS Lett.* **347**, 63–68.

Newton, A. C., and Koshland, D. E. J. (1989). *J. Biol. Chem.* **264**, 14909–14915.

Orr, J. W., and Newton, A. C. (1994). *J. Biol. Chem.* **269**, 27715–27718.

Osada, S., Mizuno, K., Saido, T., Suzuki, K., Kuroki, T., and Ohno, S. (1992). *Mol. Cell. Biol.* **12**, 3930–3938.

Papadopoulos, V., and Hall, P. F. (1989). *J. Cell Biol.* **108**, 553–567.

Parente, J. E., Walsh, M. P., Kerrick, W. G. L., and Hoar, P. E. (1992). *J. Muscle Res. Cell Motil.* **13**, 90–99.

Pfitzer, G., Zeugner, C., Troschka, M., and Chalovich, J. M. (1993). *Proc. Natl. Acad. Sci. U.S.A.* **90**, 5904–5908.

Rasmussen, H., Forder, J., Kojima, I., and Scriabine, A. (1984). *Biochem. Biophys. Res. Commun.* **122**, 776–784.

Ratz, P. H. (1990). *J. Pharmacol. Exp. Ther.* **252**, 253–259.

Ratz, P. H., Hai, C., and Murphy, R. A. (1989). *Am. J. Physiol.* **256**, C96–C100.

Rembold, C. M., and Murphy, R. A. (1988). *Am. J. Physiol.* **255**, C719–C723.

Rogue, P., Labourdette, G., Masmoudi, A., Yoshida, Y., Huang, F. L., Huang, K.-P., Zwiller, J., Vincendon, G., and Malviya, A. N. (1990). *J. Biol. Chem.* **265**, 4161–4165.

Rokolya, A., Bárány, M., and Bárány, K. (1991). *Biochim. Biophys. Acta* **1057**, 276–280.

Ron, D., Chen, C., Caldwell, J., Jamieson, L., Orr, E., and Mochly-Rosen, D. (1994). *Proc. Natl. Acad. Sci. U.S.A.* **91**, 839–843.

Schworer, C. M., and Singer, H. A. (1991). *Adv. Exp. Med. Biol.* **304,** 353–361.

Seth, A., Gonzalez, F. A., Gupta, S., Raden, D. L., and Davis, R. J. (1992). *J. Biol. Chem.* **267,** 24796–24804.

Shimamoto, Y., Shimamoto, H., Kwan, C. Y., and Daniel, E. E. (1993). *Am. J. Physiol.* **264,** H1300–H1306.

Singer, H. A. (1990a). *J. Pharmacol. Exp. Ther.* **252,** 1068–1074.

Singer, H. A. (1990b). *Am. J. Physiol.* **259,** C631–C639.

Singer, H. A., and Baker, K. M. (1987). *J. Pharmacol. Exp. Ther.* **243,** 814–821.

Singer, H. A., Oren, J. W., and Benscoter, H. A. (1989). *J. Biol. Chem.* **264,** 21215–21222.

Singer, H. A., Schworer, C. M., Sweeley, C., and Benscoter, H. A. (1992). *Arch. Biochem. Biophys.* **299,** 320–329.

Smith, B. L., and Mochly-Rosen, D. (1992). *Biochem. Biophys. Res. Commun.* **188,** 1235–1240.

Sozeri, O., Vollmer, K., Liyanage, M., Firth, D., Kour, G., Mark, G. E., and Stabel, S. (1993). *Oncogene* **7,** 2259–2262.

Sutton, T. A., and Haeberle, J. R. (1990). *J. Biol. Chem.* **265,** 2749–2754.

Takahashi, K., Hiwada, K., and Kokubu, T. (1988). *Hypertension* **11,** 620–626.

Tsuda, T., and Alexander, R. W. (1990). *J. Biol. Chem.* **265,** 1165–1170.

Tsuda, T., Kawahara, Y., Ishida, Y., Koide, M., Shii, K., and Yokoyama, M. (1992). *Circ. Res.* **71,** 620–630.

Valverde, A. M., Sinnett-Smith, J., Lint, J. V., and Rozengurt, E. (1994). *Proc. Natl. Acad. Sci. U.S.A.* **91,** 8572–8576.

Watanabe, M., Hachiya, T., Hagiwara, M., and Hidaka, H. (1989). *Arch. Biochem. Biophys.* **273,** 165–169.

Wetsel, W., Khan, W., Merchenthaler, I., Rivera, H., Halpern, A., Phung, H., Negro-Vilar, A., and Hannun, Y. (1992). *J. Cell Biol.* **117,** 121–133.

Winder, S. J., and Walsh, M. P. (1990). *J. Biol. Chem.* **265,** 10148–10155.

Winder, S. J., and Walsh, M. P. (1993). *Cell. Signal.* **5,** 677–686.

Woods, A., and Couchman, J. R. (1992). *J. Cell Sci.* **101,** 277–290.

Yamakita, Y., Yamashiro, S., and Matsumura, F. (1994). *J. Cell Biol.* **124,** 129–137.

Yoshida, Y., Huang, F. L., Nakabayashi, H., and Huang, K. (1988). *J. Biol. Chem.* **263,** 9868–9873.

Zhang, J., Wang, L., Schwartz, J., Bond, R. W., and Bishop, W. R. (1994). *J. Biol. Chem.* **269,** 19578–19584.

Mitogen-Activated Protein Kinase

LEONARD P. ADAM

Medicine and Physiology/Biophysics
Krannert Institute of Cardiology
Indiana University School of Medicine
Indianapolis, Indiana

I. INTRODUCTION

Protein kinases are classified, in general, as either serine/threonine or tyrosine kinases. Several exceptions exist since there are enzymes that phosphorylate histidine residues and at least one kinase phosphorylates both serine/threonine *and* tyrosine residues. Each of these enzymes has its own substrate specificity, that is, a unique substrate amino acid sequence or sequence motif that is required for modification. For one class of serine/threonine protein kinases, the site of phosphorylation is either a serine or threonine residue that is immediately followed by proline. This class is called proline-directed protein kinase (PDPK). There are several members of the PDPK class, including p34^{cdc2} and the mitogen-activated protein kinases (MAPKs). MAPKs include a family of enzymes in the 40- to 45-kDa range that are called extracellular signal-regulated kinases (ERKs), and higher-molecular-weight c-*Jun* N-terminal kinases (JNKs). Certain of the JNKs are activated by cellular stress and because of this are referred to as stress-activated protein kinases (or SAPKs). The SAPKs are distinguished from the other MAPKs on the basis of molecular mass (54 versus 40–45 kDa) and substrate specificity.

Smooth muscle is unique among the three muscle types (cardiac, skeletal, and smooth) because it exists in two distinct phenotypes. The *contractile* phenotype of smooth muscle is responsible for generating tension; the *secretory* or *proliferative* phenotype of smooth muscle undergoes cellular division and has only a limited capacity to undergo shortening. Proliferative smooth muscle is commonly derived, *in vitro*, by placing contractile smooth muscle tissue, typically aortas, in culture and allowing proliferating cells to migrate out of the explant. Proliferating smooth muscle cells are commonly used as model for certain pathological states of smooth muscle proliferation resulting from vascular injury (e.g., restenosis associated with angioplasty). However, in many respects the cells more closely resemble other cultured cell types such as fibroblasts than they do fully differentiated contractile smooth muscle. Because of the difference in phenotype, care will be taken in this discussion to distinguish between studies of MAPK in the two types of smooth muscle. Emphasis will be given to a discussion of MAPK in the contractile phenotype of smooth muscle since this phenotype is physiologically responsible for smooth muscle contractile behavior.

These two phenotypes of smooth muscle, in addition to having marked differences in contractile activity, express different isoforms of several contractile proteins and certain soluble enzymes. In particular, proliferative smooth muscle contains at least three PDPKs; p34^{cdc2}, p42^{MAPK}, and p44^{MAPK}. In the contractile phenotype of smooth muscle, only p42^{MAPK} and p44^{MAPK} have been identified. The precise function and a complete description of the substrates for MAPK in the contractile phenotype of smooth muscle are unknown; however, one substrate that has been identified is the actin and myosin binding protein, caldesmon. Because of the phosphorylation of caldesmon, MAPK may be involved in either smooth muscle contractile regulation or the structural organization of actin filaments within smooth muscle cells.

II. MEASUREMENT OF MITOGEN-ACTIVATED PROTEIN KINASE ACTIVITY IN SMOOTH MUSCLE

There are at least three techniques for measuring MAPK activity in contractile smooth muscle. In the first technique to be discussed, phosphotyrosine content is measured in p42^{MAPK} and p44^{MAPK}. In the second, phosphotransferase activity is measured using an "in gel" protein kinase assay. In the final technique, phosphotransferase activity using MAPK-specific substrates is measured in tissue extracts. The first two methods are relatively simple to perform; however, for technical reasons, comparisons of MAPK activity among different experiments not assayed at the same time may not be quantitative. The third method allows for direct, quantitative comparisons of MAPK activity among successive experiments. For all three techniques, the starting material is finely ground smooth muscle tissue. At appropriate experimental time points, contractile smooth muscle is frozen rapidly, ground to a fine powder under liquid N_2, and stored at $-80°C$ until assayed for kinase activity. This is necessary because the relatively large amount of connective tissue in contractile smooth muscle reduces the ability to extract cytosolic proteins. The extraction of proteins from finely ground powdered tissue is more efficient than from smooth muscle homogenized using a polytron or small tissue homogenizer.

A. Measurement of Phosphotyrosine Content in Mitogen-Activated Protein Kinase

MAPK is active only when phosphorylated on two specific amino acids, one threonine and one tyrosine. Because of this, MAPK activity can be measured by determining the content of phosphotyrosine in the enzyme using highly specific antiphosphotyrosine antibodies. Known amounts of protein are separated by polyacrylamide gel electrophoresis, transferred to sheets of nitrocellulose (or polyvinylidine difluoride), and probed with antiphosphotyrosine antibodies. To increase the signal produced by the detection system, it may be necessary to immunoprecipitate MAPK from known quantities of tissue, or to partially purify MAPK, biochemically, before immunoblot analysis. An increase in specific antiphosphotyrosine antibody binding implies that MAPK activity increases.

This technique has the advantage that it is relatively simple to perform. However, comparisons among successive experiments are dependent on several factors. These factors include the efficiency of protein transfer to nitrocellulose (or polyvinylidine difluoride) and the quantitative nature of the antibody detection system. In addition, if it is necessary to partially purify or immunoprecipitate MAPK, it is important that the kinase is quantitatively recovered at all steps.

B. "In-Gel" Kinase Assay

The principal features of this assay are that (1) MAPK is separated from other cellular proteins by polyacrylamide gel electrophoresis using a gel impregnated with a MAPK substrate (usually myelin basic protein), (2) MAPK is renatured within the gel, and (3) renatured MAPK is used to phosphorylate the substrate within the gel using exogenous ^{32}P-labeled ATP. This method, based on the work of Kameshita and Fujisawa (1989) and Geahlen *et al.* (1986), has the advantage that it is both sensitive and relatively easy to perform. Kinase activities can be quantitated in successive experiments by determining the amount of radiolabel associated with the substrate; however, comparisons among different experiments require that equivalent amounts of substrate are available to the kinase within the gel and that renaturation occurs to the same extent. Care must also be taken when interpreting the data since the assay will detect all kinases that can be renatured and have the ability to phosphorylate the substrate within the gel, not just MAPK. Finally, it is important that the time of incubation with ATP and the amount of protein loaded on the gel be controlled so that the assay is in the linear range each time it is performed.

Known amounts of protein extracts or MAPK immunoprecipitates are separated by electrophoresis on SDS–polyacrylamide gels containing 0.5 mg/ml myelin basic protein (added before polymerization). SDS is removed from the gel by two washes in a solution of 20% 2-propanol, 50 mM Tris, PH 8.0 for 1 hr each, followed by a wash in 50 mM Tris, pH 8.0 containing 5 mM β-mercaptoethanol. MAPK in the gel is denatured in 6 M guanidine for 1 hr and then renatured by a combination of five washes in 50 mM Tris, pH 8.0, 0.04% Tween-40, 5 mM β-mercaptoethanol and one wash with 40 mM HEPES, pH 8.0, 2 mM dithiothreitol, 10 mM MgCl$_2$. Following renaturation and washing, the gel is incubated in 40 mM HEPES, pH 8.0, 0.5 mM EGTA, 10 mM MgCl$_2$, and 0.04 mM ^{32}P-labeled ATP (5 μCi/ml). After incubation for an appropriate length of time, the gel is washed with 5% trichloroacetic acid, 1% sodium pyrophosphate until all free radioactivity is liberated. The gel is then dried and subjected to autoradiography. The amount of phosphate incorporated into the substrate is quantitated by either scanning densitometry of the X-ray film or by liquid scintillation counting of the MAPK bands excised from the gel.

C. Kinase Assay Using a Mitogen-Activated Protein Kinase-Specific Substrate

In this method, phosphotransferase activity is used to measure MAPK activity in extracts of smooth muscle tissue. MAPK-specific activity is detected using a synthetic peptide as a substrate in the phosphotransferase reaction. The use of myelin basic protein will result in the detection of all kinases that phosphorylate this substrate (including MAPK, protein kinase C, and other protein kinases), whereas the use of a synthetic peptide, designed as a proline-directed protein kinase substrate, results in the detection of MAPK-specific activity, exclusively (Adam *et al.*, 1995; Clark-Lewis *et al.*, 1991).

Proteins are extracted from liquid nitrogen-pulverized tissue in 10 volumes of a buffer at 4°C containing 20 mM Tris, pH 7.5, 5 mM EGTA, 1 mM sodium orthovanadate, 20 mM β-glycerophosphate, 10 mM NaF, 1 mM dithiothreitol, and a cocktail of protease inhibitors (1 μg/ml aprotinin and 0.1 mM each of phenylmethylsulfonyl fluoride, N-tosyl-L-phenylalanine chloromethyl ketone, and Nα-p-tosyl-L-lysine chloromethyl ketone). The extract is clarified by centrifugation for 10 min at 100,000 g and the supernatant assayed for protein concentration and phosphotransferase activity. Phosphotransferase activity is measured by incubating 10 μl of the tissue extract in 50 μl (final volume) of a buffer containing 12.5 mM MOPS, pH 7.2, 12.5 mM β-glycerophosphate, 7.5 mM MgCl$_2$, 0.5 mM EGTA, 0.05 mM NaF, 0.5 mM sodium orthovanadate, 2 mM dithiothreitol, 0.25 mM ^{32}P-labeled ATP, and 500 μM peptide (APRTPGGRR). The reaction is terminated by the addition of trichloroacetic acid (10% final weight/volume) and centrifuged for 5 min at 14,000 g to remove the acid-denatured proteins. The supernatant is spotted onto phosphocellulose paper (P81), which is then washed four times in 50 mM H$_3$PO$_4$ at 4°C. The paper is washed briefly in 95% ethanol and the amount of radioactivity determined by liquid scintillation counting. Activities are quantitated in units of picomoles of phosphate transferred to the substrate/minute/mg protein and compared among various experimental interventions. This assay is valid if (1) the peptide detects only MAPK-specific phosphotransferase activity and (2) the activity of MAPK is not altered during the extraction and assay. These validations are confirmed for extracts of porcine carotid arteries using the conditions described here.

III. MITOGEN-ACTIVATED PROTEIN KINASE CASCADE

A. Activation by Insulin, Growth Factors, and Pharmacological Agents

Mitogen-activated protein kinases consist of a family of serine/threonine protein kinases first identified because their activity increased in response to stimulation by a number of mitogenic factors, especially insulin (Clarke, 1994; Davis, 1993; Pelech and Sanghera, 1992; White and Kahn, 1994). Since their discovery in the mid-1980s, MAPKs have been identified in virtually all tissues. MAPKs phosphorylate a number of intracellular substrates and are activated as part of a signaling cascade that transduces extracellular stimuli into a host of physiological responses. Because growth factors and other agents acting on cell-surface membrane receptors regulate MAPKs in the molecular mass range of 40–45 kDa, these particular MAPKs are commonly called extracellular signal-regulated kinases (Boulton *et al.*, 1990).

The most studied and best understood isoforms of MAPK have molecular masses in the range of 40–45 kDa, although larger-molecular-weight species are described. These high-molecular-weight MAPKs include the c-*Jun* N-terminal kinases. The JNKs are themselves a family of kinases in the 50-kDa range (especially p54) and, because of their activation in response to stress, certain members are referred to as stress-activated protein kinases (Kyriakis *et al.*, 1994). To date, the JNKs are not identified in contractile smooth muscle, and the MAPKs that have been identified, p44^{MAPK} and p42^{MAPK}, are essentially ERK1 and ERK2. All MAPKs are inactive when dephosphorylated, and become active only when phosphate is covalently attached to two specific amino acids: one threonine and one tyrosine. The structure of one MAPK, ERK2, is known at the 2.3-Å level (Zhang *et al.*, 1994). The data obtained from this study provide insight into the mechanism for MAPK activation by phosphorylation. Within the amino acid loop containing the protein kinase catalytic domain of ERK2, an additional number of amino acids are present relative to the catalytic domains of other protein kinases. The sites of phosphorylation in ERK2 occur in this stretch called the "lip" region, so called because the catalytic loop is extended owing to the insertion of the extra amino acids. These structural data now provide a molecular basis for understanding how phosphorylation of MAPK leads to its activation. Phosphorylation appears to relieve steric constraints that normally inhibit substrate binding.

In several cultured cell systems, MAP-kinases "turn on" by the activation of growth factor receptors that contain tyrosine kinase domains. The responsible growth factors include, but are not limited to, epidermal growth factor (EGF), platelet-derived growth factor (PDGF), fibroblast growth factor (FGF), nerve growth factor (NGF), and insulin and insulin-like growth factor-I (IGF-I) (Ahn *et al.*, 1991; Chao *et al.*, 1994; Graves *et al.*, 1993; Hoshi *et al.*, 1988; Jaiswal *et al.*, 1993; Lamy *et al.*, 1993). In addition, activation can result by stimulation with vasopressin, angiotensin II, platelet-activating factor (PAF), endothelin-1, phorbol esters, interleukins, growth hormone, and muscarinic agonists (Alessandrini *et al.*, 1992; Campbell *et al.*, 1992; Crespo *et al.*, 1994; Honda *et al.*, 1994; Weber *et al.*, 1994; Yin and Yang, 1994). Many of these agents stimulate MAPK by first activating G-proteins at the cell-surface membrane. However, the precise pathways for activation by the various agents are only partially understood.

B. Mitogen-Activated Protein Kinase Cascade Members, Including the Protein Kinases raf, MEKK, and MEK

MAPK is activated by a unique dual function kinase called MEK, which is an acronym for MAPK and ERK-kinase (Crews *et al.*, 1992; Zheng and Guan, 1993). MEK has also been named MAPK-kinase (or MAPKK) and is unusual because it phosphorylates MAPK on both a threonine *and* a tyrosine amino acid. MEK consists of a family of at least three protein kinase isoforms, each of which shows differential reactivity toward the different members of the MAPK family. Each MEK isoform is activated through phosphorylation mechanisms. When fully dephosphorylated, MEK is inactive. When phosphorylated by either MEK-kinase (MEKK) or the proto-oncogene *raf*, the phosphotransferase activity of MEK is "turned on." Although both *raf* and MEKK can activate MEK, differential activation of either of these kinases can lead to the activation of different downstream signal transduction pathways. Therefore, the roles of *raf* and MEKK in the cell are not limited to MEK and, therefore, MAPK activation.

The proto-oncogene *raf* becomes active in response to *ras* stimulation. However, the mechanisms through which *ras* accomplishes this are only partially understood. *Ras* requires post-translational modification (farnesylation) to effect cellular transformation and the activation of the enzymes, *raf* and MAPK (Itoh *et al.*, 1993). It appears that *ras* functions primarily to localize *raf* to the surface membrane, where additional factors are involved in the subsequent activation of *raf*

(Stokoe *et al.*, 1994). Additional proteins implicated in this process include the 14-3-3 proteins (for references, see Morrison, 1994).

C. G-Protein Activation of Mitogen-Activated Protein Kinase

In the first studies to investigate the surface-membrane protein requirements for MAPK activation by growth factors, MAPK activity correlated with *ras* activity. Further studies have shown that mutated oncogenic *ras*, transfected into cells, leads to the constitutive activation of MAPK, suggesting that *ras* lies upstream of MAPK. The precise mechanisms leading to *ras* activation in untransfected cells are not completely understood. However, *ras* activity increases as a result of growth factor or phorbol ester stimulation, implicating tyrosine kinase-dependent processes and/or protein kinase C. In addition to the involvement of PKC, certain of the intermediates between the activation of insulin receptors and *ras* are known. These intermediates include the insulin receptor substrate-1 (IRS-1) and a complex of two proteins, Grb2 and SOS. Intermediates between other growth factor receptors, or protein kinase C, and *ras* are incompletely understood except that Grb2 and SOS can activate *ras* upon stimulation of the EGF receptor (White and Kahn, 1994).

Not only is MAPK activated as a result of stimulating low-molecular-weight GTP binding proteins (such as *ras*) at the cell-surface membrane, but MAPK activity can increase in response to stimulation of membrane receptors acting through heterotrimeric GTP binding proteins. In particular, the stimulation of muscarinic or adrenergic receptors can lead to the activation of both *ras* and MAPK in a G_i-dependent manner (Alblas *et al.*, 1993; Winitz *et al.*, 1993; Crespo *et al.*, 1994). On the other hand, MAPK is activated in response to certain stimuli as a result of heterotrimeric G-protein stimulation, and stimulation by phorbol esters, in a *ras-independent* fashion (de Vries-Smits *et al.*, 1992; Ahn *et al.*, 1992a,b; Lange-Carter *et al.*, 1993). Thus, at least two known pathways exist leading to the activation of MAPK, both of which initiate at the cell surface. The best characterized of these pathways begins with *ras* (Fig. 1).

D. Substrate Specificity of Mitogen-Activated Protein Kinase

The sites on proteins covalently phosphorylated by MAPK suggest that this kinase has a very specific substrate phosphorylation consensus sequence. One of the first proteins identified as a substrate for MAPK was myelin basic protein (MBP). Phosphorylation of

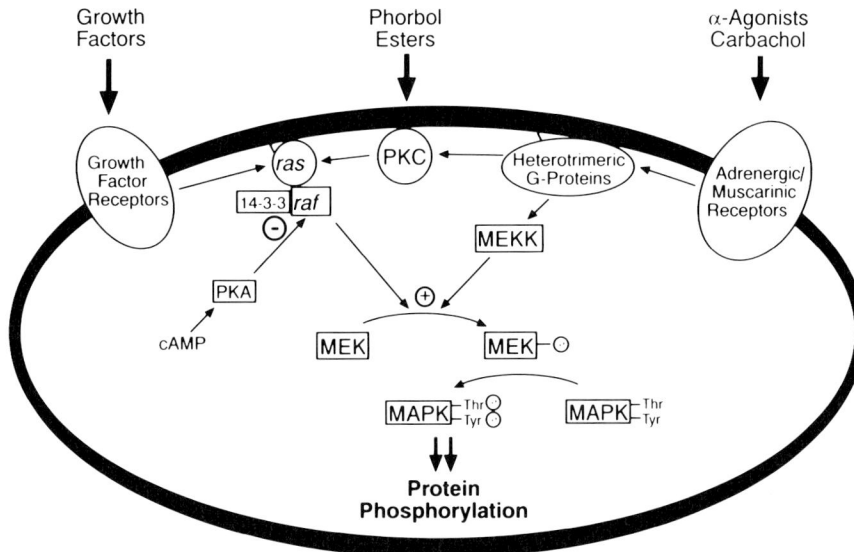

FIGURE 1 A schematic representation of the mechanisms for MAPK activation in cells. The activation of MAPK results from cell stimulation by a number of extracellular stimuli, including growth factors, phorbol esters, and a number of pharmacological agents, such as α-adrenergic agonists and muscarinic agonists (carbachol). Stimulation by these agents leads to enhanced G-protein activity (either heterotrimeric G-proteins or low-molecular-weight G-proteins such as *ras*). *Ras*, in combination with a family of proteins called 14-3-3, activates *raf*. Either *raf* or MEKK activates MEK, ultimately resulting in the phosphorylation of MAPK on a threonine and tyrosine residue. cAMP and cAMP-dependent protein kinase (PKA) inhibit this process by inhibiting *raf*.

MBP occurs on specific threonine residues that are proline directed; that is, the phosphorylated amino acids are immediately followed by proline in the amino acid sequence of the protein. In studies of peptide substrates phosphorylated by MAPK it appears that the sequence Ser/Thr-Pro is a minimum consensus phosphorylation sequence for MAPK. However, the amino acids carboxyl- and amino-terminal to this sequence modify the ability of MAPK to covalently attach phosphate. In particular, the placement of proline at position -2 (relative to the phosphorylated amino acid) increases kinase activity. On the basis of these data, the optimal consensus phosphorylation sequence for MAPK is Pro-X-(Ser/Thr)-Pro (Clark-Lewis *et al.*, 1991). Certain proteins, including caldesmon, are phosphorylated by MAPK on sites that do not match this optimal consensus sequence exactly (Adam and Hathaway, 1993).

Several proteins are substrates for MAPK, both *in vivo* and *in vitro* (for reviews, see Pelech and Sanghera, 1992; Davis, 1993). Physiological candidate substrates include proteins involved in cell proliferation and growth (c-*fos*, c-*myc*, c-*jun*, the 90-kDa ribosomal S6 kinase or p90[rsk], oncoprotein 18, the phosphorylated heat- and acid-stable protein regulated by insulin or PHAS-I), cell signaling (cytoplasmic phospholipase

A2 or cPLA2), and proteins for which the function of phosphorylation is incompletely understood (caldesmon). In addition to cell growth and proliferation, the activation of MAPK is linked to physiological functions such as osmosensing in yeast, stretch sensing in cardiac tissue, and a potential role in smooth muscle contractile or cytoskeletal function. A complete description of the substrates phosphorylated by MAPK under physiological conditions in different cell types, and the responses controlled by those modifications, are areas of intense investigation.

IV. EVIDENCE FOR MITOGEN-ACTIVATED PROTEIN KINASE IN THE CONTRACTILE PHENOTYPE OF SMOOTH MUSCLE

A. Phosphorylation of Caldesmon

Vascular smooth muscle caldesmon is phosphorylated in resting muscle and the stoichiometry of phosphorylation increases in response to pharmacological stimulation (Adam *et al.*, 1989). On the basis of these findings, phosphorylation mechanisms are proposed to modulate the physiological effects of caldesmon on either contractile or cytoskeletal function. There are two sites of phosphorylation on caldesmon in the carboxyl terminus of the molecule. The sequences of these two sites are VTS*PTKV (which are the carboxyl-terminal seven amino acids of the protein) and PDGNKS*PAPK (Adam *et al.*, 1992). These sites are both proline directed, by virtue of the Ser-Pro phosphorylation motif, and are the only *in vitro* sites of phosphorylation by MAPK. These sites are also modified by p34[cdc2], *in vitro*; however, it appears that MAPK is more likely to be the physiologically relevant "caldesmon kinase" since there is no evidence to date for p34[cdc2], or any PDPK other than MAPK, in arterial muscle (Adam and Hathaway, 1993; Adam *et al.*, 1995). With the use of a specific peptide substrate, the 42- and 44-kDa isoforms of MAPK (p42[MAPK] and p44[MAPK]) were found to contain all the proline-directed protein kinase activity in contractile smooth muscle. However, it is possible that PDPKs other than MAPK exist in smooth muscle and that (1) the activity of these are not detected using this peptide substrate and (2) they phosphorylate caldesmon.

Although these sequences are the only two sites in porcine caldesmon phosphorylated by MAPK, *in vitro* and in intact muscle, the sites of phosphorylation by MAPK in gizzard caldesmon are different (Childs *et al.*, 1992; Childs and Mak, 1993). Gizzard caldesmon is phosphorylated by p44[MAPK] on at least one serine and one threonine amino acid, and not on serine residues alone. Also, one of the sequences in mammalian caldesmon, phosphorylated by MAPK, is not present in gizzard caldesmon. In addition to differences in the phosphorylation of avian and mammalian caldesmon by MAPK, different isoforms of MAPK are expressed in fully differentiated gizzard and mammalian smooth muscle. Chicken gizzard contains p42[MAPK], whereas rat aorta and porcine carotid artery contain both p42[MAPK] and p44[MAPK] (Adam and Hathaway, 1993; Childs *et al.*, 1992; Katoch and Moreland, 1995). The reasons for this differential expression of avian and mammalian MAP-kinases, as well as the differential phosphorylation of avian and mammalian caldesmon by MAPK, are unknown.

In a study by Childs *et al.* (1992), the functional effects of gizzard caldesmon phosphorylation have been investigated. The addition of actin and/or calcium/calmodulin (but not tropomyosin) significantly inhibits the phosphorylation of gizzard caldesmon by p44[MAPK]. On the other hand, caldesmon phosphorylation by p44[MAPK] only slightly inhibits its binding to actin. Complementing these experiments, Redwood *et al.* (1993) investigated the functional effects of phosphorylation of a specific 99-amino acid fragment of gizzard caldesmon. This fragment inhibits actomyosin ATPase activity and contains two amino acids that are within preferred phosphorylation sequences for MAPK—Thr[673] and Ser[702]. The sequence containing serine is present in mammalian caldesmon, the threonine sequence is not. Phosphorylation of the serine residue by MAPK inhibits the actin binding capacity of the caldesmon fragment and reverses the ability of this fragment to inhibit actomyosin ATPase activity. Interestingly, MAPK does not phosphorylate the threonine residue even though it is suspected that this site is phosphorylated in holo-gizzard caldesmon. Collectively, these data support a role for MAPK phosphorylation in caldesmon function; in particular, a role for the phosphorylation of Ser[702], common to both avian and mammalian caldesmon, is suggested.

B. Several Elements of the Mitogen-Activated Protein Kinase Cascade Are Present in Contractile Smooth Muscle

In addition to p42[MAPK] and p44[MAPK], other elements of the "MAPK cascade" are present in vascular smooth muscle, including *raf* and MEK (Adam and Hathaway, 1993; Childs and Mak, 1993). Although not implicated in MAPK activation, certain *ras*-related low-molecular-weight GTP binding proteins are also identified. These include members of the *smg* and *rho* families (Kawahara *et al.*, 1990; Kawata *et al.*, 1989). The functions of these GTP binding proteins in contractile smooth muscle are unknown; however, two laboratories (Hirata *et al.*, 1992; Kitazawa *et al.*, 1991) have shown that GTP enhances smooth muscle contractile activity by inhibiting myosin phosphatase. The specific GTP binding proteins modulating this effect remain undetermined and it is unknown if *smg* or *rho* link to MAPK in intact contractile smooth muscle.

In addition to these elements of the "MAPK cascade," contractile smooth muscle contains heterotrimeric GTP binding proteins that may activate MAPK. Adrenergic stimulation of contractile smooth muscle has pronounced effects on contractility that are not explained by a simple alteration in intracellular free calcium (Aburto *et al.*, 1993). Furthermore, it is known that adrenergic stimulation can cause a G_i-dependent increase in MAPK activity in some tissues. It is possible that some of the actions of adrenergic agonists on smooth muscle contractile behavior involve the activation of MAPK, since elements of the pathway linking adrenergic stimulation to MAPK are present in contractile smooth muscle.

C. Mitogen-Activated Protein Kinase Activity in Contractile Smooth Muscle

1. Activities in Unstimulated Smooth Muscle

MAPK is active when isolated from chicken gizzard, rat aorta, and porcine carotid smooth muscles

(Adam *et al.*, 1995; Childs and Mak, 1993; Katoch and Moreland, 1995). In chicken gizzard and rat aorta, immunoprecipitated MAPK exhibits phosphotransferase activity toward myelin basic protein and purified caldesmon. When crudely extracted from porcine carotid arteries, MAPK exhibits phosphotransferase activity that has been quantitated using a peptide substrate (APRTPGGRR). To determine if the peptide kinase activity detected in crude tissue extracts is specific for MAPK (the peptide may detect other protein kinases), proteins in the extracts have been separated on a Mono-Q fast-performance liquid chromatography column as shown in Fig. 2, and further characterized. Only two peaks of phosphotransferase activity

elute from the column corresponding to tyrosine-phosphorylated p42MAPK and p44MAPK. Importantly, these two peaks contain all the phosphotransferase activity present in crude extracts. Thus it appears that MAPK in fully differentiated smooth muscle accounts for all peptide kinase activity and exhibits some of the same characteristics as MAPK in proliferating cells. For example, smooth muscle MAPK becomes activated as a result of tyrosine phosphorylation and phosphorylates similar substrates (peptide and protein).

2. Activation by Mechanical Load

In unstimulated porcine carotid arteries dissected from animals and attached to force transducers, *in vitro*, MAPK levels are relatively high. By comparison, the activity of MAPK in porcine carotid arteries frozen rapidly, *in situ*, is very low (Fig. 3) (Adam *et al.*, 1995). These data suggest that the manipulation of arteries associated with dissection from animals, storage at 4°C, and attachment to force transducers induce MAPK activity. It is possible that factors released upon arterial injury are responsible for this increase in activity. It is also possible that stretch associated with the attachment of arteries to force transducers is responsible. In support of this second possibility, quantitative analysis of MAPK activity in porcine carotid arteries has shown that these activities increase in response to the initial load applied to muscles. In muscles unloaded at 4°C or stretched at 4°C, MAPK activity is expectedly very low. However, when arterial muscle strips are mechanically loaded at physiological temperatures, MAPK activity increases. The level of kinase activity increases in proportion to the amount of initial load applied to the muscle up to a maximum. These data are consistent with observations made in cultured cardiac myocytes (Komuro and Yazaki, 1994; Sadoshima and Izumo, 1993; Yamazaki *et al.*, 1993). In cardiac myocytes, stretching induces the activation of certain proteins and the expression of a number of immediate-early genes, presumably with the ultimate purpose of increasing cell growth leading to hypertrophy. Proteins either activated or induced by stretch include p21ras, p90rsk, protein kinase C, phospholipases C and D, c-*fos*, c-*myc*, and MAPK.

There are several similarities between the activation of MAPK by stretch in cardiac myocytes and in contractile smooth muscle. In cardiac myocytes, MAPK activity increases in response to stretch, reaching a maximum at approximately 10 min, coincident with the increase in S6 kinase activity (p90rsk), and then declines to basal levels in 30–60 min (Sadoshima and Izumo, 1993; Yamazaki *et al.*, 1993). Under the same conditions, protein kinase C activity elevates more rapidly and is maintained for at least 30 min; c-*fos* and

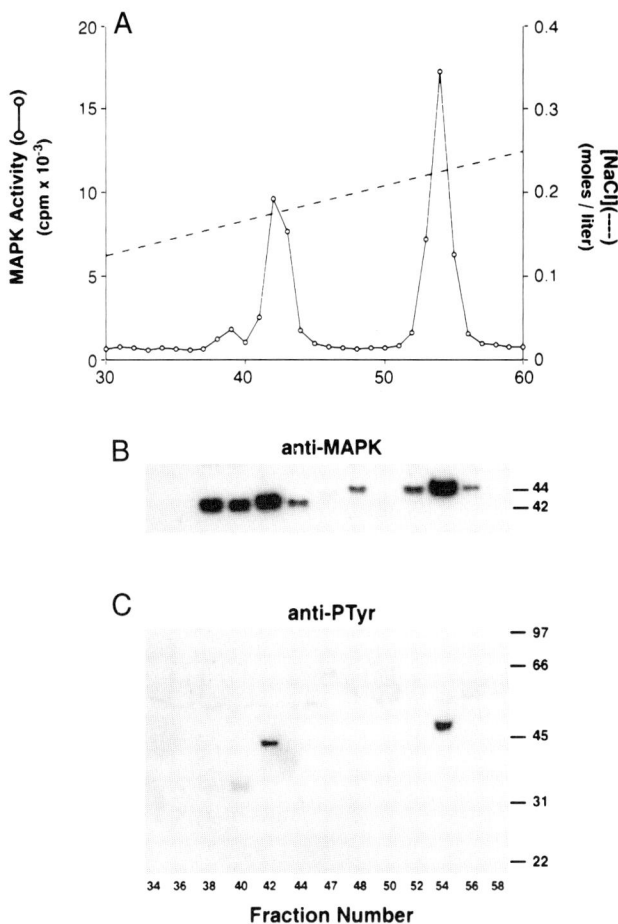

FIGURE 2 Line graph (A) and immunoblots (B,C) showing separation of MAPK isoforms by Mono-Q fast-performance liquid chromatography (Adam *et al.*, 1995). Extracts of porcine carotid arteries were separated on a 1-ml Mono-Q column. Aliquots from each fraction were assayed for MAPK activity (A). Only fractions containing activity above background are presented for clarity. Proteins from specified fractions were separated by SDS–polyacrylamide gel electrophoresis, transferred to nitrocellulose, and assayed by immunoblot for the presence of either MAPK (B) or phosphotyrosine (C). Reprinted with permission. "Circulation Research." Copyright 1995 American Heart Association.

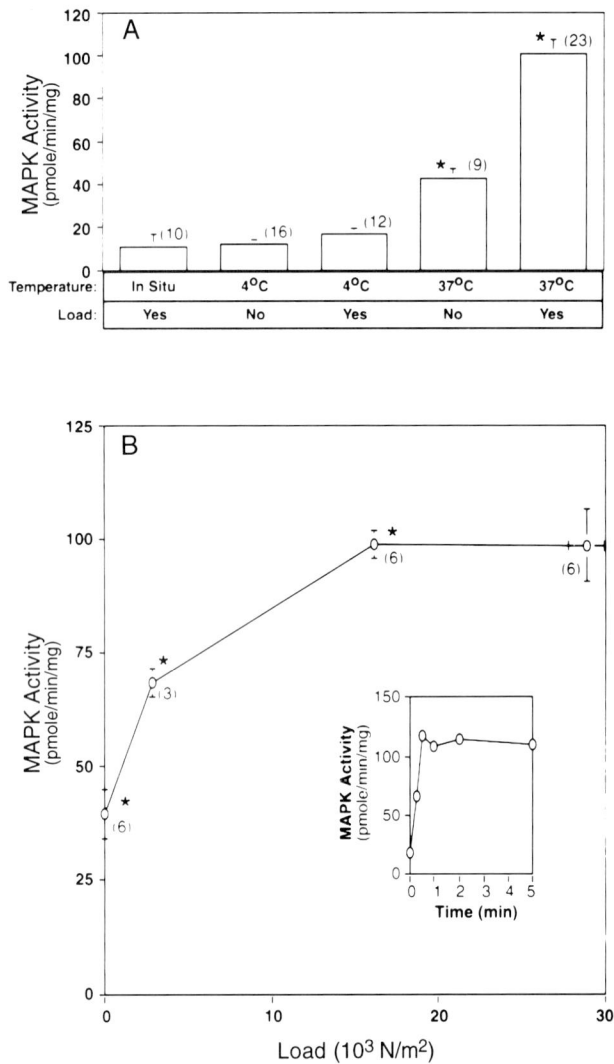

FIGURE 3 Effects of load on MAPK activity in unstimulated porcine carotid arteries (Adam et al., 1995). (A) Bar graph showing MAPK activity measured in arteries frozen immediately on dissection from animals (in situ) or after storage and incubation in physiological saline solution at the indicated temperature. Muscles were either unloaded (no) or had 12.5 g of tension applied (yes). (B) Line graph showing MAPK activity measured in muscles to which various initial loads were applied at 37°C. In the inset, kinase activity was measured under maximal load after various times of incubation at 37°C. Each time point represents the results from three to six experiments. Reprinted with permission. "Circulation Research." Copyright 1995 American Heart Association.

c-myc expression develops more slowly over the time course of hours. These data are consistent with the hypothesis that protein kinase C activates MAPK, ultimately leading to the activation of a number of proto-oncogenes responsible for an increase in cellular growth. In carotid arteries, the application of mechanical load induces an increase in MAPK activity that develops rapidly along the time course seen with cardiac

myocytes, yet is maintained for at least 30 min of stretching. In vascular smooth muscle the function of this increase is unknown; however, MAPK may be a contributing signal for smooth muscle cell growth, dedifferentiation, an alternation and restructuring of the cellular cytoskeleton, or altered contractility.

3. Pharmacological Activation

In addition to activation by stretch, MAPK activity rises in response to pharmacological stimulation in contractile smooth muscle (Adam et al., 1995; Katoch and Moreland, 1995). Porcine carotid arteries stimulated with KCl, phorbol 12,13-dibutyrate (PDBu), or histamine demonstrate elevated levels of active MAPK. The activities of p42[MAPK] and p44[MAPK] increase in tandem under all conditions tested; that is, KCl, PDBu, and histamine do not preferentially activate one MAPK isoform over the other isoform. With KCl and histamine stimulation, MAPK activity reaches a maximum within several minutes. PDBu induces a more slowly developing increase in MAPK activity that is maintained for at least one hour. These data suggest that increases in intracellular free calcium, and/or the stimulation of protein kinase C, in vascular smooth muscle result in MAPK activation. Pharmacological increases in MAPK activity are in addition to the levels that result from the application of mechanical load.

D. Cytochemical Localization of Mitogen-Activated Protein Kinase in Arterial Smooth Muscle

Upon stimulation of quiescent cultured cells with serum, growth factors, or phorbol esters, a translocation of a subfraction of MAPK to the cell-surface membrane occurs. These data are in agreement with reports that show a recruitment by ras of the protein kinase raf to the cell membrane, resulting in raf, and subsequently MAPK, activation. In fully differentiated contractile smooth muscle stimulated with phenylephrine, MAPK transiently moves from the cytoplasm to the surface membrane (Khalil and Morgan, 1993). MAPK translocates to the surface membrane before contraction and then redistributes to the cytoplasm coincident with muscle cell shortening. Protein kinase C inhibitors block these translocations, suggesting a role for PKC in the activation of MAPK. In further support of a role for PKC in this process, PKC translocates to the surface membrane at the same time as MAPK.

Collectively, studies of MAPK activation and translocation in response to mechanical and pharmacological stimulation provide important data to support different roles for the physiology of MAPK in different cell types. In cultured cells, where MAPK contributes to cell growth and proliferation, MAPK translocates to the nucleus (Gonzalez et al., 1993). In fully differentiated smooth muscle, where MAPK is implicated in either contractile regulation or cytoskeletal organization, MAPK translocates to the surface membrane where activation is thought to occur, followed by cytoplasmic redistribution. Once active MAPK redistributes within the cytoplasm of contractile smooth muscle, it is able to phosphorylate certain myofibrillar proteins (e.g., caldesmon) as well as other proteins that have yet to be identified.

V. COMPARISON OF MITOGEN-ACTIVATED PROTEIN KINASE ACTIVATION IN CONTRACTILE VERSUS PROLIFERATIVE SMOOTH MUSCLE

MAP-kinases are implicated in several cell cycle transition checkpoints, including G0/G1, G1/S, and G2/M (Gotoh et al., 1991; Tamemoto et al., 1992; Watson et al., 1993). In the specific case of cultured vascular smooth muscle, Watson et al. (1993) have shown that MAPK is important for G0/G1 transitions. MAPK activity in quiescent cells peaks within minutes of serum stimulation and then declines to near basal levels during progression through the cell cycle. Clearly, MAPK plays a very important role in proliferative smooth muscle physiology.

Consistent with these observations, other investigators have found that essentially all agents that stimulate smooth muscle cell proliferation also stimulate MAPK activity. Included among these agents are (1) arginine vasopressin, PDGF, endothelin-1, angiotensin II, and phorbol 12-myristate 13-acetate in rat aortic smooth muscle cells, (2) thromboxane/prostaglandin mimetics in guinea pig coronary artery smooth muscle cells, and (3) vasopressin and phorbol esters in A7R5 cells (Granot et al., 1993; Graves et al., 1993; Jones et al., 1994; Koide et al., 1992; Kribben et al., 1993; Langan et al., 1994; Li et al., 1994; Morinelli et al., 1994; Tsuda et al., 1992; Weber et al., 1994). Interestingly, one agent that inhibits cell proliferation, heparin, also inhibits MAPK activity in cultured rat aortic cells (Ottlinger et al., 1993). Thus, MAPK activity parallels proliferative capacity in cultured smooth muscle.

Other work has shown that MAPK is subject to regulation by intracellular cAMP. Agents that increase intracellular cAMP inhibit MAPK in fibroblasts, adipocytes, and CHO cells (Hordijk et al., 1994; Sevetson et al., 1993; Wu et al., 1993). In human aortic smooth

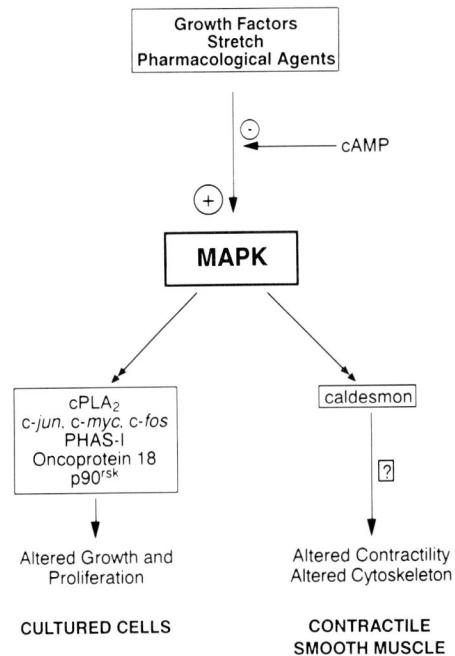

FIGURE 4   Schematic depicting the potential roles for MAPK in contractile versus proliferative (or cultured) smooth muscle. MAPK is activated in response to stimulation by growth factors, stretch, and pharmacological agents. This process can be inhibited by cAMP and cAMP-dependent protein kinase. Once activated, MAPK phosphorylates a number of intracellular proteins (both cytoplasmic and nuclear) that result in an alteration of growth and proliferation in cultured cells. In contractile smooth muscle, MAPK phosphorylation of caldesmon may lead to alterations in muscle contractility or actin filament structure.

muscle cells, forskolin, isoproterenol, and cholera toxin inhibit PDGF-induced activation of MAPK (Graves et al., 1993). These observations are particularly interesting in light of the fact that cAMP inhibits contractility in vascular smooth muscle (Kamm and Stull, 1985). It appears that increases in MAPK activity are stimulatory for proliferation in cultured vascular smooth muscle cells and are stimulatory for contraction in fully differentiated vascular smooth muscle tissue. Conversely, cAMP is inhibitory for contraction in smooth muscle tissue (where its effects on MAPK are unknown) and is inhibitory for growth in proliferative smooth muscle (where it inhibits MAPK) (Fig. 4).

VI. SUMMARY

There are two isoforms of MAPK in contractile, vascular, smooth muscle—p42[MAPK] and p44[MAPK]—and one in gizzard smooth muscle—p42[MAPK]. The level of MAPK activity in porcine carotid arteries is altered in response to pharmacological stimulation and the application of mechanical load in a manner that is consistent with a role for the kinase in muscle contraction. The potential involvement of MAPK in the contractile responsiveness of smooth muscle may result from the phosphorylation of the actin binding protein, caldesmon. Caldesmon is phosphorylated in intact tissue on the same sites that MAPK phosphorylates *in vitro*. Importantly, the sequences in caldesmon that are phosphorylated in intact tissue are "proline directed" and MAPK is the only proline-directed protein kinase identified in contractile smooth muscle to date. Although it is apparent that MAPK can phosphorylate caldesmon under physiological conditions, the possibility remains that there are other caldesmon kinases; and there are certainly substrates for MAPK, other than caldesmon, in fully differentiated smooth muscle.

A role for MAPK in the contractile responsiveness of smooth muscle may result from either of two possible mechanisms, both involving caldesmon phosphorylation. First, caldesmon phosphorylation by MAPK may lead directly to an alteration of actomyosin activity. Caldesmon may exert this effect alone, or in concert with other myofibrillar proteins such as calponin. Second, phosphorylation of caldesmon may alter the dynamics of actin filament organization within the cell. Caldesmon phosphorylation may result in alterations of the cellular cytoskeleton that must occur during prolonged contractions.

Because of the involvement of MAPK in smooth muscle proliferation, growth, and caldesmon phosphorylation, this kinase appears to play a central role in signal transduction for both the contractile and proliferative phenotypes of smooth muscle. Despite different potential roles for MAPK in these two tissue types, there are important similarities in kinase activation and potential physiological function. Agents that increase MAPK activity are stimulatory for contraction and proliferation, whereas cAMP inhibits both processes. In sum, the function of MAPK in the *contractile* phenotype of vascular smooth muscle is unknown, but may be severalfold. MAPK activation may be an initiating factor leading to smooth muscle cell growth, dedifferentiation and/or proliferation. On the other hand, MAPK signaling processes may lead to smooth muscle actin filament restructuring or alterations in contractility.

References

Aburto, T. K., Lajoie, C., and Morgan, K. G. (1993). *Circ. Res.* **72**, 778–785.

Adam, L. P., and Hathaway, D. R. (1993). *FEBS Lett.* **322**, 56–60.

Adam, L. P., Haeberle, J. R., and Hathaway, D. R. (1989). *J. Biol. Chem.* **264**, 7698–7703.

Adam, L. P., Gapinski, C. J., and Hathaway, D. R. (1992). *FEBS Lett.* **3**, 223–226.

Adam, L. P., Franklin, M. T., Raff, G. J., and Hathaway, D. R. (1995). *Circ. Res.* **76**, 183–190.

Ahn, N. D., Seger, R., Bratlien, R. L., Diltz, C. D., Tonks, N. K., and Krebs, E. G. (1991). *J. Biol. Chem.* **266**, 4220–4227.

Ahn, N. D., Seger, R., and Krebs, E. G. (1992a). *Curr. Opin. Cell Biol.* **4**, 992–999.

Ahn, N. D., Robbins, D. J., Haycock, J. W., Seger, R., Cobb, M. H., and Krebs, E. G. (1992b). *J. Neurochem.* **59**, 147–156.

Alblas, J., van Corven, E. J., Hordijk, P. L., Milligan, G., and Moolenaar, W. H. (1993). *J. Biol. Chem.* **268**, 22235–22238.

Alessandrini, A., Crews, C. M., and Erikson, R. L. (1992). *Proc. Natl. Acad. Sci. U.S.A.* **89**, 8200–8204.

Boulton, T. G., Yancopoulos, G. D., Gregory, J. S., Slaughter, C., Moomaw, C., Hsu, J., and Cobb, M. H. (1990). *Science* **249**, 64–67.

Campbell, G. S., Pang, L., Miyasaka, T., Saltiel, A. R., and Carter-Su, C. (1992). *J. Biol. Chem.* **267**, 6074–6080.

Chao, T.-S.O., Foster, D. A., Rapp, U. R., and Rosner, M. R. (1994). *J. Biol. Chem.* **269**, 7337–7341.

Childs, T. J., and Mak, A. S. (1993). *Biochem. J.* **296**, 745–751.

Childs, T. J., Watson, M. H., Sanghera, J. S., Campbell, D. L., Pelech, S. L., and Mak, A. S. (1992). *J. Biol. Chem.* **267**, 22853–22859.

Clark-Lewis, I., Sanghera, J. S., and Pelech, S. L. (1991). *J. Biol. Chem.* **266**, 15180–15184.

Clarke, P. R. (1994). *Curr. Biol.* **4**, 647–650.

Crespo, P., Xu, N., Simonds, W. F., and Gutkind, J. S. (1994). *Nature (London)* **369**, 418–420.

Crews, C. M., Alessandrini, A., and Erikson, R. L. (1992). *Science* **258**, 478–480.

Davis, R. J. (1993). *J. Biol. Chem.* **268**, 14553–14556.

de Vries-Smits, A. M. M., Burgering, B. M. T., Leevers, S. J., Marshall, C. J., and Bos, J. L. (1992). *Nature (London)* **357**, 602–604.

Geahlen, R. L., Anostario, M., Low, P. S., and Harrison, M. L. (1986). *Anal. Biochem.* **153**, 151–158.

Gonzalez, F. A., Seth, A., Raden, D. L., Bowman, D. S., Fay, F. S., and Davis, R. J. (1993). *J. Cell Biol.* **122**, 1089–1101.

Gotoh, Y., Nishida, E., Matsuda, S., Shiina, N., Kosako, H., Shiokawa, K., Akiyama, T., Ohta, K., and Sakai, H. (1991). *Nature (London)* **349**, 251–254.

Granot, Y., Erikson, E., Fridman, H., Van Putten, V., Williams, B., Schrier, R. W., and Maller, J. L. (1993). *J. Biol. Chem.* **268**, 9564–9569.

Graves, L. M., Bornfeldt, K. E., Raines, E. W., Potts, B. C., Macdonald, S. G., Ross, R., and Krebs, E. G. (1993). *Proc. Natl. Acad. Sci. U.S.A.* **90**, 10300–10304.

Hirata, K., Kikuchi, A., Sasaki, T., Kuroda, S., Kaibuchi, K., Matsuura, Y., Seki, H., Saida, K., and Takai, Y. (1992). *J. Biol. Chem.* **267**, 8719–8722.

Honda, Z., Takano, T., Gotoh, Y., Nishida, E., Ito, K., and Shimizu, T. (1994). *J. Biol. Chem.* **269**, 2307–2315.

Hordijk, P. L., Verlaan, I., Jalink, K., van Corven, E. J., and Moolenaar, W. H. (1994). *J. Biol. Chem.* **269**, 3534–3538.

Hoshi, M., Nishida, E., and Sakai, H. (1988). *J. Biol. Chem.* **263**, 5396–5401.

Itoh, T., Kaibuchi, K., Masuda, T., Yamamoto, T., Matsuura, Y., Maeda, A., Shimizu, K., and Takai, Y. (1993). *J. Biol. Chem.* **268**, 3025–3028.

Jaiswal, R. K., Murphy, M. B., and Landreth, G. E. (1993). *J. Biol. Chem.* **268**, 7055–7063.

Jones, L. G., Ella, K. M., Bradshaw, C. D., Gause, K. C., Dey, M., Wisehart-Johnson, A. E., Spivey, E. C., and Meier, K. E. (1994). *J. Biol. Chem.* **269**, 23790–23799.

Kameshita, I., and Fujisawa, H. (1989). *Anal. Biochem.* **183**, 139–143.

Kamm, K. E., and Stull, J. T. (1985). *Annu. Rev. Pharmacol. Toxicol.* **25**, 593–620.

Katoch, S. S., and Moreland, R. S. (1995). *Am. J. Physiol.* **269**, H222–H229.

Kawahara, Y., Kawata, M., Sunako, M., Araki, S., Koide, M., Tsuda, T., Fukuzaki, H., and Takai, Y. (1990). *Biochem. Biophys. Res. Commun.* **170**, 673–683.

Kawata, M., Kawahara, Y., Araki, S., Sunako, M., Tsuda, T., Fukuzaki, H., Mizoguchi, A., and Takai, Y. (1989). *Biochem. Biophys. Res. Commun.* **163**, 1418–1427.

Khalil, R. A., and Morgan, K. G. (1993). *Am. J. Physiol.* **265**, C406–C411.

Kitazawa, T., Masuo, M., and Somlyo, A. P. (1991). *Proc. Natl. Acad. Sci. U.S.A.* **88**, 9307–9310.

Koide, M., Kawahara, Y., Tsuda, T., Ishida, Y., Shii, K., and Yokoyama, M. (1992). *J. Hypertens.* **10**, 1173–1182.

Komuro, I., and Yazaki, Y. (1994). *Trends Cardiovasc. Med.* **4**, 117–121.

Kribben, A., Wieder, E. D., Li, X., Van Putten, V., Granot, Y., Schrier, R. W., and Nemenoff, R. A. (1993). *Am. J. Physiol.* **265**, C939–C945.

Kyriakis, J. M., Banerjee, P., Nikolakaki, E., Dai, T., Rubie, E. A., Ahmad, M. F., Avruch, J., and Woodgett, J. R. (1994). *Nature (London)* **369**, 156–160.

Lamy, F., Wilkin, F., Baptist, M., Posada, J., Roger, P. P., and Dumont, J. E. (1993). *J. Biol. Chem.* **268**, 8398–8401.

Langan, E. M.., Youkey, J. R., Elmore, J. R., Franklin, D. P., and Singer, H. A. (1994). *J. Surg. Res.* **57**, 215–220.

Lange-Carter, C. A., Pleiman, C. M., Gardner, A. M., Blumer, K. J., and Johnson, G. L. (1993). *Science* **260**, 315–319.

Li, X., Kribben, A., Wieder, E. D., Tsai, P., Nemenoff, R. A., and Schrier, R. W. (1994). *Hypertension* **23**, 217–222.

Morinelli, T. A., Zhang, L.-M., Newman, W. H., and Meier, K. E. (1994). *J. Biol. Chem.* **269**, 5693–5698.

Morrison, D. (1994). *Science* **266**, 56–57.

Ottlinger, M. E., Pukac, L. A., and Karnovsky, M. J. (1993). *J. Biol. Chem.* **268**, 19173–19176.

Pelech, S. L., and Sanghera, J. S. (1992). *Science* **257**, 1355–1356.

Redwood, C. S., Marston, S. B., and Gusev, N. B. (1993). *FEBS Lett.* **327**, 85–89.

Sadoshima, J., and Izumo, S. (1993). *EMBO J.* **12**, 1681–1692.

Sevetson, B. R., Kong, X., and Lawrence, J. C. (1993). *Proc. Natl. Acad. Sci. U.S.A.* **90**, 10305–10309.

Stokoe, D., Macdonald, S. G., Cadwallader, K., Symons, M., and Hancock, J. F. (1994). *Science* **264**, 1463–1414.

Tamemoto, H., Kadowaki, T., Tobe, K., Ueki, K., Izumi, T., Chatani, Y., Kohno, M., Kasuga, M., Yazaki, Y., and Akanuma, Y. (1992). *J. Biol. Chem.* **267**, 20293–20297.

Tsuda, T., Kawahara, Y., Ishida, Y., Koide, M., Shii, K., and Yokoyama, M. (1992). *Circ. Res.* **71**, 620–630.

Watson, M. H., Venance, S. L., Pang, S. C., and Mak, A. S. (1993). *Circ. Res.* **73**, 109–117.

Weber, H., Webb, M. L., Serafino, R., Raylor, D. S., Moreland, S., Norman, J., and Molloy, C. J. (1994). *Mol. Endocrinol.* **8**, 148–158.

White, M. F., and Kahn, C. R. (1994). *J. Biol. Chem.* **269**, 1–4.

Winitz, S., Russell, M., Qian, N.-X., Gardner, A., Dwyer, L., and Johnson, G. L. (1993). *J. Biol. Chem.* **268**, 19196–19199.

Wu, J., Dent, P., Jelinek, T., Wolfman, A., Weber, M. J., and Sturgill, T. W. (1993). *Science* **262**, 1065–1069.

Yamazaki, T., Tobe, K., Hoh, E., Maemura, K., Kaida, T., Komuro, I., Tamemoto, H., Kadowaki, T., Nagai, R., and Yazaki, Y. (1993). *J. Biol. Chem.* **268**, 12069–12076.

Yin, T., and Yang, Y.-C. (1994). *J. Biol. Chem.* **269**, 3731–3738.

Zhang, F., Strand, A., Robbins, D., Cobb, M. H., and Goldsmith, E. J. (1994). *Nature (London)* **367**, 704–711.

Zheng, C.-F., and Guan, K.-L. (1993). *J. Biol. Chem.* **268**, 23933–23939.

MOTILE SYSTEMS

14

In Vitro Motility Assays with Smooth Muscle Myosin

JAMES R. SELLERS

Laboratory of Molecular Cardiology
National Heart, Lung, and Blood Institute
National Institutes of Health
Bethesda, Maryland

I. INTRODUCTION

The study of the interaction of actin and myosin was remarkably advanced in 1983 when Sheetz and Spudich published the first *in vitro* motility assay in which the movement of polymer beads coated with myosin filaments could be observed to take place on actin cables exposed by dissection of the green alga *Nitella axillaris*. Up to this point, biochemists could measure the binding of actin to myosin and the actin-activated MgATPase activity of myosin, but measurement of the movement of actin and myosin was left to the realm of physiologists working with intact or skinned muscle fibers. With the advent of the *in vitro* motility assay, biochemists could use video microscopy to quantify the rate of movement of myosin on organized arrays of actin filaments while varying the ionic composition and other parameters of the assay medium. *Nitella*, which has multinucleated cells up to 10 cm in length, was ideally suited for this assay. Cell biologists had studied the cytoplasmic streaming of the membranous organelles that were continuously moving over cables of actin filaments firmly affixed to the array of chloroplasts that lined the boundaries of the cell (Kachar *et al.*, 1987). Sheetz and Spudich (1983) demonstrated that these cables could be exposed by making a longitudinal cut of the cell wall of *Nitella* followed by pinning the flaps to a Sylgar-coated surface.

This assay was very innovative and was the subject of many studies, but had several limitations. The dissection was technically difficult and the exposed actin cables degenerated with time, especially in the presence of Ca^{2+}. Monomeric myosin or heavy mero-

myosin (HMM) did not reproducibly support movement unless coupled to the beads via antibodies against the tail portion of the molecule (Hynes *et al.*, 1987). The system was particularly difficult to use when the rate of movement was slow and required long periods of observation. It involved the use of plant actin, which, although homologous to vertebrate actin, was not identical. In addition, there were concerns that components of the dissected *Nitella* cytoplasm, such as its own myosins, phosphatases, kinases, or proteases, may interfere with the movement of the exogenously added myosin.

The *Nitella* assay served as the inspiration for a second-generation *in vitro* motility assay. In this assay, which was developed by Kron and Spudich (1986), myosin is bound to a glass coverslip, where it moves actin filaments that are free in solution (see Fig. 2). It is a remarkable assay in which the essence of muscle contraction, the relative movement of actin and myosin, can be reconstituted with two highly purified proteins. The assay is very easy to establish and execute.

In general, the two motility assays give the same values for velocity and have similar characteristics. In each case the movement is assumed to be "unloaded" since the velocity does not depend on the concentration of myosin bound to the bead or to the glass surface above a certain threshold level (Collins *et al.*, 1990; Sellers *et al.*, 1985; Sheetz *et al.*, 1984). The velocity is also not dependent on the length of actin filaments in the sliding actin assay (Collins *et al.*, 1990; Warshaw *et al.*, 1990). In both systems the direction of movement is determined by the polarity of actin (Sellers and Kachar, 1990; Saito *et al.*, 1994). Each myosin exhibits a characteristic velocity determined by its own rate constants, which is, in turn, a function of the assay conditions. There is a wide range of velocities from different myosins (Sellers *et al.*, 1993), ranging from 0.04 μm/sec for phosphorylated human platelet myosin (Umemoto and Sellers, 1990) to greater than 4 μm/sec for rabbit skeletal muscle myosin (Toyoshima *et al.*, 1987). The velocity of movement of phosphorylated smooth muscle myosin lies between these extremes and, in general, ranges from 0.2 to 1.4 μm/sec depending on conditions and tissue source of the myosin (Umemoto and Sellers, 1990; Warshaw *et al.*, 1990; Trybus *et al.*, 1994; Kelley *et al.*, 1993). The assays have been widely used to study smooth muscle myosin and its regulation by phosphorylation and thin filament proteins.

Together, these two assays opened up new avenues of research into actin and myosin interaction. Two different adaptations of the assay allow for the measurement of force by small numbers of myosin molecules (Kishino and Yanagida, 1988; Finer *et al.*, 1994). The first of these uses a flexible, calibrated microneedle to which an actin filament is attached. The deflection of the needle as myosin moves the actin filament can be measured to determine the force (Kishino and Yanagida, 1988). In the second adaptation, an optical trap (Block, 1990) is used to measure both the force and step size of single myosin molecules (Finer *et al.*, 1994). Beads are attached to each end of an actin filament. Dual optical traps position the beads (and thus the actin filament) in the vicinity of a single or a few myosin molecules. A sensitive quadrant detector is used to detect the position of the bead as myosin moves the actin filament.

In this chapter, I will discuss the principle and design of *in vitro* motility assays and their use for myosin in general and smooth muscle myosin in particular.

Since the assay has become so routine and widely used, I will not dwell on the particular experiments that have recently made use of them, since these details will undoubtedly be discussed in the relevant chapters.

II. DESCRIPTION OF EQUIPMENT

There are several detailed accounts of the equipment and methods required to perform *in vitro* motility assays (Kron *et al.*, 1991; Sheetz *et al.*, 1986; Warrick *et al.*, 1993; Sellers *et al.*, 1993; Higashi-Fujime, 1991). Because of its ease of use and more general applicability, I will describe only the sliding filament assay in this chapter. A diagram of the equipment needed to measure *in vitro* motility is shown in Fig. 1.

To image actin filaments, it is necessary to have a microscope and a suitable light intensification system. The microscope can be in either the inverted or upright configuration and must be equipped with an epifluorescence illuminator and filter sets for measuring rhodamine fluorescence. In general, oil immersion objectives of 40–100× power with numerical apertures (N.A.) of 1.3 to 1.4 are used. It is useful to have a heat filter (to remove IR radiation) and neutral density filters on hand to attenuate the illuminating light, which is usually provided by a 100-W mercury source. The video system is coupled via a C-mount to the microscope. The temperature must be regulated, which can be accomplished in several ways. The simplest is to use an air curtain, which is a hair dryer-type fan unit with an adjustable output. The temperature at the sample position can be measured using a thermistor embedded underneath a coverslip. The temperature output of the air curtain or the distance of the air curtain from the microscope is varied to achieve the desired sample

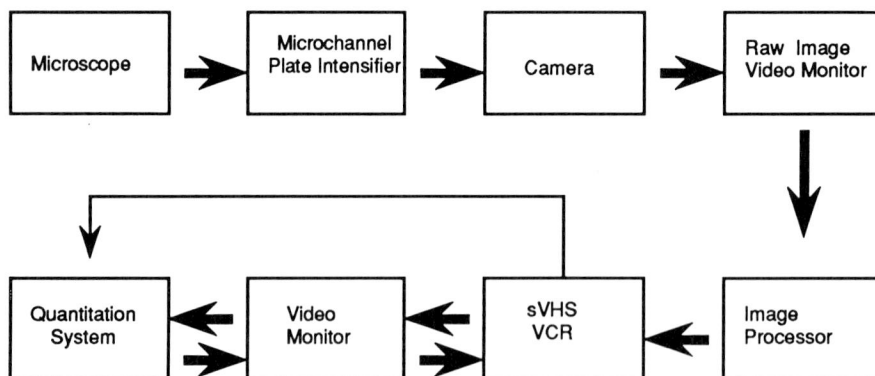

FIGURE 1   Schematic of the equipment required for the sliding actin *in vitro* motility assay. The minimal equipment would be the microscope, an imaging system, and a video monitor.

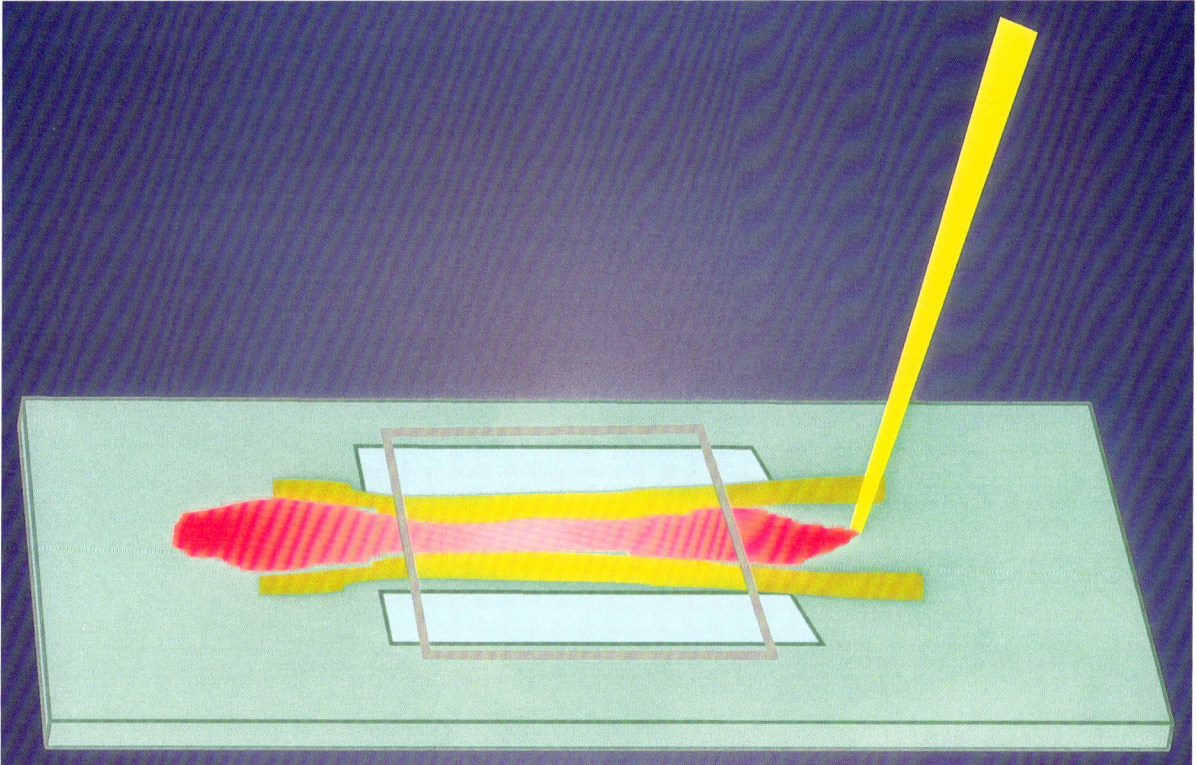

FIGURE 3 Construction of the flow cell for the sliding actin *in vitro* motility assay. The paired irregular yellow strips represent the grease that adheres the coverslip (on top) to the slide. The coverslip slivers that are used as spacers are narrower than the top coverslip. The pink liquid represents a solution of rhodamine phalloidin-labeled actin that is being applied to the flow cell.

temperature. A more sophisticated design using a water-jacketed system for the objective and the microscope stage was described by Kron *et al.* (1991). Another method involves wrapping a thermal foil around the objective. It should be noted that when oil immersion is used, the objective becomes a significant heat sink and it is usually not sufficient to merely regulate the temperature of the stage.

The low-light video imaging system is a key component of the setup. In general, two types of systems can be employed. The first type uses a silicon intensifier target (SIT) camera or an intensified silicon intensifier target (ISIT) camera. SIT cameras offer sufficient sensitivity for the imaging of the rhodamine phalloidin-labeled actin filaments provided that higher illuminating light levels are used. The ISIT cameras are more sensitive, but are more expensive and considerably more delicate to use. In both cases, the temporal resolution is poor at low light levels and one can see persistence or "comet tails" for rapidly moving actin filaments. The second choice of imaging systems is to use a microchannel plate intensifier coupled with either a CCD or Newvicon camera. These systems usually perform better at low light levels than the SIT cameras, have a longer shelf life, and are more forgiving of biochemists who are not well trained in microscopy. In particular, the microchannel plate-intensified CCD systems offer better light sensitivity and a high temporal resolution with little or no geometric distortion of the image. The latter aspect is important if quantitative measurements are to be made, and the user is advised to image a square grid to check for lack of geometric distortion by the imaging system.

The quality of the image can be dramatically improved by an image processor. Simple commercially available systems can perform frame averaging, background subtraction, and contrast enhancement to improve the signal-to-noise ratio. In addition, these systems usually offer functions such as a distance scale bar, distance measurements between points, and a time–date function. More sophisticated systems can measure the light intensity of objects, overlay images, create collages, detect motion, and display pseudocolor images. Image processing is particularly useful for situations where the actin filaments are moving slowly and longer imaging time is required. The processor can average 32–64 video frames, which allows lower illumination levels to be used in order to diminish photobleaching of the rhodamine-labeled actin.

The standard recording system is currently sVHS video recorders. This system offers considerably better resolution than the older VHS format and is not nearly as expensive as U-matic systems or optical memory disk recorders (OMDR). It is convenient to

have a machine that can play in a fast forward or fast reverse mode in order to quickly scan video tapes or to observe the movement of slowly moving actin filaments.

Usually, the most tedious aspect of motility assays is the quantitation of the data. This can be accomplished in several ways, including placing clear plastic wrap over the monitor and marking the positions of actin filaments as a function of time, although there are certainly more precise and less laborious means available. Various investigators have written tracking programs that can be used in conjunction with commercially available frame grabbers (Work and Warshaw, 1992). The degree of user input and the speed of action vary with the individual systems. One commercially available unit is the Cell Trak System from Motion Analysis (Santa Rosa, CA). This system automatically performs a gray-level threshold of the on-screen images at adjustable sampling rates of up to 60 frames/sec, determines the centroid position of all the actin filaments in each frame, connects the centroids in time and space to determine paths, calculates the instantaneous velocity between each individual pair of points in a given path, and, finally, provides a measurement of the mean velocity of each filament path. When run by a 486 DX2 66-MHz processor, this system can quantify the velocity of 30–50 actin filaments in less than one minute.

III. ASSAY PROCEDURE

A. Preparation of Rhodamine Phalloidin-Labeled Actin Filaments

Labeling actin with rhodamine phalloidin serves two critical functions. Primarily, this allows actin filaments, which are below the resolution of the light microscope, to be visualized by virtue of the light emanated by the fluorescent rhodamine group. Secondarily, the phalloidin moiety stabilizes actin filaments by decreasing the critical concentration for polymerization of actin. This keeps actin filaments polymerized under conditions where they would normally depolymerize and allows the investigator to work at actin concentrations where the movement of individual actin filaments can be discerned. There are several variations on the labeling of actin filaments. The procedure used in our laboratory is to place 60 μl of 3.3 μM (in methanol) rhodamine phalloidin (Molecular Probes, Eugene, OR) into an Eppendorf tube and dry using a Speed Vac concentrator (Savant, Hicksville, NY). The rhodamine phalloidin powder is redissolved in 3–5 μl of methanol, taking great care to dissolve all the dried powder, followed by addition of 85 μl of 20 mM KCl, 20 mM MOPS (pH 7.4), 5 mM MgCl₂, 0.1 mM EGTA, 10 mM DTT (buffer A). A freshly diluted 20 μM F-actin solution is added to a final concentration of 2 μM. This is incubated for 2 hr or, more often, overnight. The sample is centrifuged for 10 min in a Beckman TLA Ultracentrifuge at 100,000 rpm. The supernatant is removed and the pink actin pellet is gently resuspended in 100 μl of buffer A, taking care not to overly shear actin filaments. These rhodamine phalloidin-labeled filaments can be stored on ice and used for several weeks. The actin is diluted to 20 nM with buffer A immediately prior to use in the motility assay. The movement of the rhodamine phalloidin-labeled actin filaments over a myosin-coated surface is illustrated in Fig. 2.

B. Preparation of Coverslip Surfaces

Smooth muscle myosin filaments can be directly bound to a glass surface for the motility assay, but it is preferable that monomeric myosin or HMM be attached to a surface that is coated with either nitrocellulose or silicon. Nitrocellulose-coated coverslips can be prepared by placing one drop of 1% nitrocellulose (Superclean grade, EF Fullam, Schenectady, NY) in amyl acetate from a Pasteur pipet onto the surface of water in a 10-cm round glass dish. The amyl acetate evaporates in about 1 min, leaving a film of nitrocellulose on the surface of the water. Coverslips are carefully placed onto the film using forceps. The film in between coverslips is torn away using forceps and a free nitrocellulose-coated coverslip is retrieved by holding it with forceps and pushing it down into the water, where it is inverted and lifted out. The coverslips are air-dried with the nitrocellulose-coated surface side up, which usually takes about 30 min, and are used within a few hours.

Silicon-coated coverslips can be prepared by diluting dichlorodimethlysilane to 2% in chloroform. Coverslips are immersed into this solution and allowed to air-dry.

C. Construction of Flow Cells

A simple flow cell can be constructed using a microscope slide and a coverslip with slivers of a No. 1 coverslip cut with a diamond scribe as spacers (about 2–3 mm in width) (Fig. 3). Two tracks of Apiezon M grease (Biddle Instruments, Blue Bell, PA) are applied to the microscope slide using a syringe with a large-gauge needle. Two coverslip spacers are laid about 8–12 mm apart on the outside of each track of grease and an 18-mm² coverslip is placed upon the tracks and pressed lightly with forceps to form a seal. The coverslips can also be attached to the slide using clear fingernail pol-

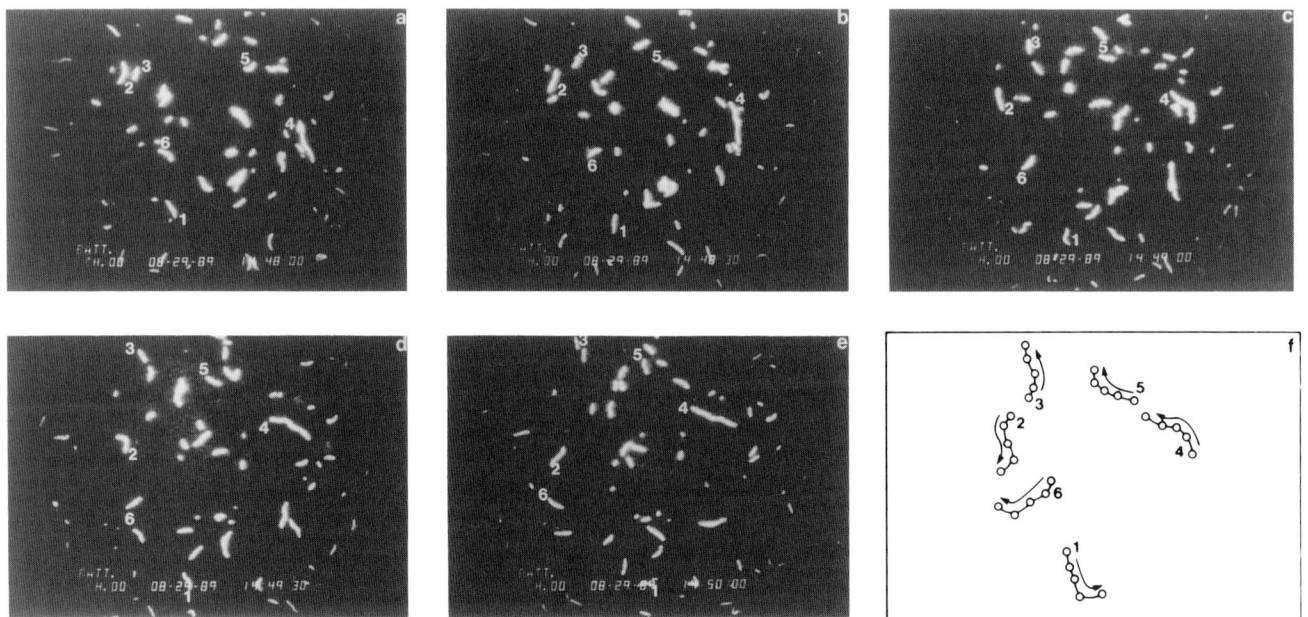

FIGURE 2   Collage demonstrating the movement of rhodamine phalloidin-labeled actin filaments over a myosin-coated surface. Panels a–e are direct photographs of the video monitor at 30-sec intervals. Panel f shows a plot of the centroid position of selected actin filaments in the field, with arrows showing the direction of movement. The figure is taken from Collins et al. (1990).

ish as a glue. If the coverslip is coated with nitro-cellulose or silicon, the coated surface should be placed film side down. The flow cell typically has a volume of 30–40 μl, but this can be reduced by narrowing the distance between spacers, by cutting the coverslip in half (9 × 18), or by using a thinner coverslip sliver (#0) as the spacer. Flow cells with a sample volume of 10 μl or less can be easily constructed.

D. Preparation of the Slide

The following is an example of the steps required for measuring the *in vitro* motility of actin filaments over a surface of monomeric smooth muscle myosin.

1. Myosin at concentrations between 30 and 200 μg/ml in 0.5 M NaCl, 10 mM MOPS (pH 7.0), 0.1 mM EGTA, 1 mM DTT (buffer B) is introduced into the flow cell, which is inclined at an angle of about 30°.
2. After about 60 sec, the flow cell is washed with 2 to 3 vol of buffer B containing 1 mg/ml bovine serum albumin to remove unbound myosin and to block the surface in order to prevent nonspecific binding of actin.
3. After about 60 sec, the flow cell is washed with 2 to 3 vol of 20 mM KCl, 5 mM MgCl$_2$, 20 mM MOPS (pH 7.2), 0.1 mM EGTA, 10 mM DTT (buffer A).
4. Two vol of 20 nM rhodamine phalloidin labeled actin in buffer B is used to wash the flow cell.
5. Following 30–60 sec of incubation, the flow cell is washed with 2 vol of the buffer to be used for the observation of *in vitro* motility [i.e., 80 mM KCl, 20 mM MOPS (pH 7.2), 5 mM MgCl$_2$, 0.1 mM EGTA, 1 mM ATP, 50 mM DTT, 0.7% methylcellulose (4000 cps), 2.5 mg/ml glucose, 0.1 mg/ml glucose oxidase, 0.02 mg/ml catalase. The last four reagents can be purchased from Sigma Chemical Company, St. Louis, MO].
6. Any solution that has spilled onto the top of the coverslip surface is blotted off and a drop of immersion oil is placed onto the coverslip. The slide is placed onto the microscope stage.
7. The fluorescently labeled actin filaments on the underside of the coverslip are brought into focus. Actin filaments should be readily apparent and moving. Ideally, the field of view should contain 15–30 actin filaments. If too many or too few actin filaments are present, the concentration of actin added to the flow cell or the time of incubation can be adjusted in step 4. Sometimes it is necessary to increase the concentration of myosin applied to the flow cell in order to increase the number of actin filaments that are attached.

If movement over myosin filaments is to be observed, it is necessary to apply myosin filaments in a low-ionic-strength buffer to the flow cell and wash with bovine serum albumin in a low-ionic-strength buffer. It has been shown that myosin filaments remain attached to the coverslip surface during the course of the experiment (Warshaw *et al.*, 1990). Smooth muscle HMM does not support movement of actin filaments under conditions where monomeric smooth muscle myosin does. There is no explanation for this phenomenon. Trybus *et al.* (1994) reported that smooth muscle HMM could be tethered to the coverslip via a monoclonal antibody against the tip of the S-2 region. HMM bound to the surface in this manner supported movement at velocities comparable to that of myosin monomers.

The composition of the *in vitro* motility buffer requires comments. Glucose, glucose oxidase, and catalase are used to scavenge oxygen and thus to reduce the amount of photobleaching of rhodamine phalloidin-labeled actin and photodamage to the proteins. High concentrations of DTT (10–100 mM) also help in this regard, as does degassing the buffers prior to use. The presence of methylcellulose is essential to observe movement of actin filaments by monomeric myosin, for it dramatically increases the viscosity and suppresses lateral diffusion of actin filaments, but does not affect the rate of *in vitro* motility (Uyeda *et al.*, 1991). This effectively prevents actin filaments from diffusing away from the surface of the coverslip if they become transiently unattached for a brief period of time. The ionic strength can be adjusted by the addition of KCl and usually ranges from 20 to 120 mM.

E. Quality Control

Ideally, greater than 80–90% of the actin filaments should be moving in a constant manner at any time. Poor quality of movement is characterized by filaments that remain attached and do not move at all and by filaments that move erratically in a stop-and-go manner. In addition, shearing of the actin filaments is sometimes a problem. The probable factor underlying poor-quality movement is the presence of damaged or rigorlike myosin heads that bind to actin in an ATP-independent manner. The use of old nitrocellulose-coated surfaces may also contribute to this phenomenon.

The most important determinant of the quality of movement is the use of fresh, pure myosin. Freshly purified smooth muscle myosin, stored in 1–5 mM DTT, typically exhibits good-quality movement for up to 10 days. Myosin can also be quick frozen in small aliquots (0.1–0.5 ml) and stored for extended periods of time in liquid nitrogen with good results. The myosin is dialyzed into buffer B at 1 mg/ml prior to freezing. Others report storing of the myosin in 50% glycerol at −20°C (Warshaw *et al.*, 1990).

Several tricks can be used to improve the quality of movement in the assay. One is to apply 2–3 vol of buffer A containing 5 μM of unlabeled actin and 1 mM ATP prior to step 4. After 1–2 min of incubation, the flow cell is washed with buffer A and then rhodamine phalloidin actin is applied as in step 4. This treatment appears to tie up the rigor heads with unlabeled (and therefore invisible) actin. The other treatment is to mix the myosin in buffer B prior to the start of an assay with a stoichiometric amount of actin in the presence of 5 mM MgCl₂ and 5 mM ATP, followed by centrifugation in a TL100 ultracentrigue for 10 min at 100,000g, which will pellet the actin and attached rigorlike heads, but leave active myosin in the supernatant. Finally, the inclusion of tropomyosin (40–100 nM) in the motility buffer increases not only the rate of movement but also the quality of movement (Umemoto and Sellers, 1990).

IV. QUANTITATION AND PRESENTATION OF RESULTS

A. Velocity Determination

As described earlier, there are several ways to quantitate the velocity of the moving actin filaments. A path plot is usually generated for each actin filament using either the leading edge or the centroid of the actin filament as a tracking point over some time interval. The mean velocity of the moving actin filament is determined from the instantaneous velocities between each of two points in the path plot. This is done for many actin filaments and then the "mean of the mean" is calculated to give a mean velocity ± standard deviation for the movement of all the actin filaments in a data set. It is important to select an appropriate sampling rate when tracking moving actin filaments. The first instinct is to sample as fast as the instrumentation allows. However, such a strategy can result in a larger error in the velocity owing to errors in determining the position of the actin filament. Vibration in the system, nonuniform light intensity, photobleaching, and inherent jitter in videotape are all sources of small errors in determining the position of the actin filament. If the sampling rate is too high, the amount of displacement per time is small while the error in determining actin filament position is fairly constant. This will result in larger errors than if the sampling rate was slower and the displacement per time was larger. A sampling rate should be chosen that gives a displacement of 2–4 pixels of the leading edge or centroid per time point. If the sampling rate is too slow, then truncation of the path can occur if the actin filament is not moving in a straight line.

The calculated movement of any given actin filament is not always uniform over the course of observation. For example, the filament may experience some intermittent motion, or abruptly change directions, which will cause it to loop back upon itself, resulting in little or no change in centroid position for a few time points. Also, actin filaments may cross each other or move into or out of the field (which affects the centroid calculation). At times, it is necessary to select only a portion of the filament path for quantitation. This selection must be performed carefully in an unbiased manner. We use the following approach. Quantitation of the movement of many actin filaments has demonstrated that those filaments that were perceived to be moving constantly typically had a standard deviation that was less than one-third of the mean velocity, whereas filaments observed to have intermittent-type movement had a larger standard deviation compared to the mean. We calculate the mean velocity along with standard error for the movement of each actin filament within the field over a given period of time and then apply a statistical filter that selects the filaments that are moving with a velocity that is three times greater than the standard deviation. Experience has shown that the resulting mean velocity of the population of actin filaments is the same as when we carefully analyze each path by eye selecting only those paths where the actin filament moved constantly throughout the duration of tracking or by selecting the portion of a given path where the movement was deemed to be constant. The statistical filtering is clearly much quicker, requires less eye strain, and introduces less user bias into the system. It does not select for rate of filament movement, but only for consistent filament movement. For a detailed discussion of this quantitation strategy, see Sellers *et al.* (1993).

B. Presentation of Results

The most common method for presentation of the results of *in vitro* motility assays is by expressing the mean velocity of the population of moving actin filaments along with the standard deviation. This method works well to describe systems where the movement is uniform. If the actin filaments are not moving entirely uniformly, then one can select those filaments that are moving uniformly as discussed earlier.

Two other methods work better when the velocity of a given filament is not constant or under conditions where some of the filaments are stationary while others are still moving. A histogram of the distribution of velocities from a given experiment will clearly show if there is a moving population and an unmoving population (Fig. 4). In this case, it is useful to image static actin filaments (i.e., in the absence of ATP) as a control and subject these to the same quantitation routine, since even unmoving filaments give apparent velocities owing to the error sources in determining actin filament position, as discussed earlier. This exercise will identify that velocity that is indistinguishable from the noise in the system.

Another graphic method is to present path plots of the movement of actin filaments within a given field. In this case, the position of the centroid or the leading edge of an actin filament is plotted at defined time intervals (Fig. 4). This type of plot is useful for demonstrating that the filaments are moving uniformly and not in an intermittent manner. It can also be used to graphically demonstrate that the average velocity under one condition is lower or higher than another condition, if the same time intervals are chosen. In general, fewer data can be presented in this method and it is clearly important that a representative field be chosen for display.

V. EFFECTS OF PHOSPHORYLATION

It is well known that phosphorylation activates the MgATPase activity of smooth muscle myosin (see

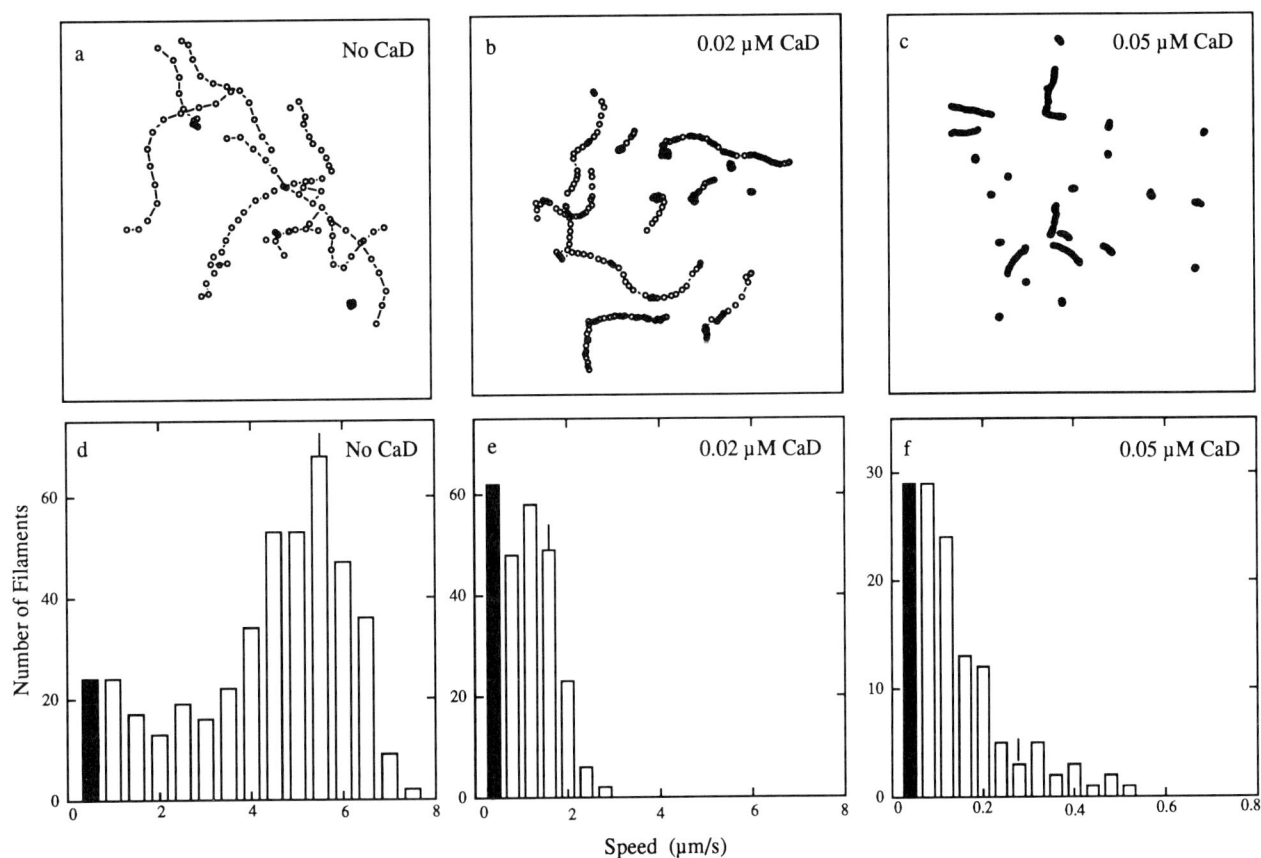

FIGURE 4 Two methods for displaying *in vitro* motility data. Shown is an experiment in which the movement of actin filaments by skeletal muscle myosin is inhibited by caldesmon (CaD). Panels a–c show path plots in which the centroid position of the actin filaments is fixed at 0.5-sec intervals. Panels d–f show histograms of the mean velocity of each individual filament from 5 to 10 video records. Note the different abcissa in panel f. The vertical lines in panels d–f show the position of the calculated mean velocity of the population of actin filaments using the statistical filtering method described in the text. The solid bars represent velocities that are indistinguishable from that of unmoving actin filaments (or the noise level in the system, see text for explanation). The data in panels d and e were analyzed at 2 frames per second sampling rate, whereas the data in panel f were analyzed at 0.2 frames per second since the actin filaments were clearly moving more slowly (compare panel c to a). This accounts for the lower noise level seen in this case.

Chapter 1, this volume). Phosphorylation is also essential for movement of actin filaments in both the *Nitella* and the sliding actin *in vitro* motility assays (Umemoto and Sellers, 1990; Warshaw *et al.*, 1990; Sellers *et al.*, 1985). The site of the regulatory phosphorylation is Ser-19 on the LC20 (Pearson *et al.*, 1984). Myosin light chain kinase (MLCK) also phosphorylates Thr-18, albeit at a much lower rate (Ikebe *et al.*, 1986). Thr-18 phosphorylation increases the actin-activated MgATPase activity (Ikebe *et al.*, 1988), but does not increase the rate of actin filament sliding in either of the two motility assays (Sellers *et al.*, 1985; Okagaki *et al.*, 1991).

Protein kinase C phosphorylates Ser-1 or Ser-2 and Thr-9 on the regulatory light chain (Bengur *et al.*, 1987; Ikebe *et al.*, 1987). This phosphorylation decreases the actin-activated MgATPase activity of smooth muscle myosin that was already phosphorylated by MLCK at Ser-19, by decreasing the apparent affinity for actin with no effect on the V_{max} (Nishikawa *et al.*, 1984). Protein kinase C phosphorylation of Ser-19 phosphorylated smooth muscle myosin does not affect the rate of movement in either of the two motility assays (Umemoto *et al.*, 1989; Okagaki *et al.*, 1991). Phosphorylation by protein kinase C in the absence of MLCK phosphorylation neither increases the actin-activated MgATPase activity nor supports *in vitro* motility (Sellers *et al.*, 1985; Nishikawa *et al.*, 1984).

The fact that these additional phosphorylations can affect the actin-activated MgATPase assay and not the *in vitro* motility assays is not problematic, as the two processes are probably limited by different steps in the kinetic cycle. The kinetic step that limits the actin-activated MgATPase assay is thought to be phosphate release or some step preceding phosphate release (Sellers, 1985). The rate of *in vitro* motility is thought to be regulated by ADP release (see Chapter 1, this volume).

Tropomyosin increases both the rate of *in vitro* motility and the actin-activated MgATPase activity (Umemoto and Sellers, 1990; Chacko and Eisenberg, 1990). The effect on the MgATPase activity is to increase the V_{max} by a factor of two with no effect on the apparent affinity for actin (Umemoto and Sellers, 1990; Chacko and Eisenberg, 1990). The mechanism by which this occurs is not known, but the presence of tropomyosin on the thin filament must affect at least two different rate constants as discussed earlier, since it accelerates both the MgATPase activity and the rate of *in vitro* motility.

Other factors can modulate the rate of *in vitro* motility. Like any enzymatic process, it is affected by temperature and the ionic conditions of the assay. Varying the ATP concentration reveals that the apparent K_d for

ATP is on the order of 25–50 μ*M* (Warshaw *et al.*, 1991; Sellers *et al.*, 1985). This does not reflect the actual affinity of ATP for myosin (which is about 5 μ*M*) since, as the ATP is lowered in this assay, myosin heads that have no ATP bound will act as a load to the continued movement of an actin filament by myosin heads that do have ATP bound. MgADP inhibits the velocity of movement with a K_i of 0.24 m*M* (Warshaw *et al.*, 1991). Increasing the ionic strength of the assay results in faster movement of actin filaments by myosin. The maximal rate of movement is observed at about 80–100 m*M* in ionic strength (Warshaw *et al.*, 1990; Umemoto and Sellers, 1990). Increasing the ionic strength beyond this usually results in slower movement and, finally, dissociation of actin filaments. Increasing the temperature of the assay increases the rate of movement with most myosins, although there are no published values for smooth muscle myosin (Collins *et al.*, 1990; Anson, 1992; Sheetz *et al.*, 1984). The rate of movement of actin filaments by phosphorylated smooth muscle myosin is not dependent on the calcium concentration in the presence of 5 m*M* MgCl$_2$ (Umemoto *et al.*, 1989).

VI. MECHANICAL EXPERIMENTS

Smooth muscle fibers generate as much isometric force per cross-sectional area as skeletal muscle fibers with only 20% as much myosin (Murphy *et al.*, 1974). There have been various possible explanations for this, including different mechanical properties of the myosin itself. Experiments with the *in vitro* motility assay provide insight into this possibility.

The duty cycle of myosin is defined as the fraction of time a myosin head is attached to actin during the ATP hydrolysis cycle (Uyeda *et al.*, 1990). Harris and Warshaw (1993) estimated the duty cycle for smooth muscle by examining the movement of actin filaments at very low concentrations of myosin bound to the surface. The number of myosin heads interacting per unit length of actin filament was estimated by measuring the amount of myosin bound to the surface (from its ATPase activity) and using the assumptions from the studies of Uyeda *et al.* (1990) about the ability of myosin bound to a fixed location to interact with an actin filament. This calculation assumes that all biochemically active cross-bridges can interact with actin and support motility at comparable rates regardless of cross-bridge orientation. By fitting the data to equation (1), values for the duty cycle (f) can be determined, where V = actin filament velocity, $a \times V_{max}$ = filament velocity at which saturation occurs, and N =

number of cross-bridges capable of interacting with actin:

$$V = (a \times V_{max}) \times [(1-f)^N] \qquad (1)$$

It was found that the duty cycles for smooth muscle myosin and skeletal muscle myosin were the same (about 4%), which meant that this could not be the determining factor behind smooth muscle myosin's apparently larger force per cross-bridge (Harris and Warshaw, 1993). These experiments were conducted under conditions of no load. It is possible that smooth and skeletal muscle myosins have different duty cycles under loaded conditions.

One way to place an apparent load into the system is to mix a myosin that has a high rate of actin filament translation with a myosin that moves actin more slowly (Sellers *et al.*, 1985; Warshaw *et al.*, 1990; Harris *et al.*, 1994). These mixing curves can be analyzed by a cross-bridge model to predict the relative force-producing capability of the two myosins (Harris *et al.*, 1994). The results suggest that smooth muscle myosin exerts 2.1 times the force per myosin cross-bridge compared to skeletal muscle myosin (Harris *et al.*, 1994).

Unphosphorylated smooth muscle myosin does not move actin filaments, but does bind actin filaments and keep them associated with the coverslip surface (Umemoto and Sellers, 1990; Warshaw *et al.*, 1990). Occasionally a few filaments can be observed to move over unphosphorylated myosin. As discussed in Chapter 1, this is consistent with the finding that phosphorylated and unphosphorylated smooth muscle HMM have fairly similar binding constants for actin in the presence of ATP (Sellers, 1985). Thus, the weakly bound unphosphorylated myosin heads are able to interact with actin sufficiently well to keep it bound to the surface. If unphosphorylated smooth muscle myosin is mixed in varying ratios with phosphorylated smooth muscle myosin, the rate of *in vitro* motility is decreased if the fraction of phosphorylated myosin is less than 50% (Warshaw *et al.*, 1990). Unphosphorylated smooth muscle myosin, when mixed with rabbit skeletal muscle myosin, exerts an even more potent inhibitory effect on the rate of movement (Warshaw *et al.*, 1990). Similar findings were also observed if mixtures of myosins are made on beads in the *Nitella*-based motility assay (Sellers *et al.*, 1985).

Using the flexible needle *in vitro* force measurement system described earlier, VanBuren *et al.* (1994) measured the force per cross-bridge of smooth muscle versus skeletal muscle myosin *in vitro*. In this elegant assay, the number of myosin heads per square micron was determined from the ATPase activity and the number of myosins capable of interacting with a fixed length of actin was calculated by assuming that all myo-

sins within 10 nm of the actin filament can interact, regardless of their orientation. The calculated force per cross-bridge head was 0.6 pN for smooth muscle myosin and 0.2 pN for skeletal muscle myosin (VanBuren *et al.*, 1994). This is believed to be an underestimate of the actual force, since studies with skeletal muscle systems suggest that only properly oriented myosin molecules generate significant force (Ishijima *et al.*, 1994). The basis for this enhanced force production from smooth muscle myosin cannot be determined from this study. It could be explained by a higher unitary force, an increased duty cycle, or a combination of the two. Studies discussed earlier suggested that the duty cycle of the two myosins is the same under conditions of zero load, but it is possible that the duty cycle does vary under loaded conditions (Harris and Warshaw, 1993).

VII. EFFECT OF LIGHT CHAIN REMOVAL

It has been shown that phosphorylation of LC20 by MLCK is essential for *in vitro* motility and actin-activated MgATPase activity. Trybus *et al.* (1994) used trifluoperazine to remove the LC20 from smooth muscle myosin. The LC20-deficient myosin did not effectively move actin filaments. Its velocity of movement was less than 10% of that of control, untreated myosin, and was not quantitatively distinguishable from that of myosin that was not moving. Motility could be restored by the readdition of LC20 either in solutions or on the nitrocellulose-coated surface. In contrast, the MgATPase activity of the light chain-deficient myosin was markedly activated even in the absence of actin. Thus, removal of LC20 resulted in loss of regulation of the MgATPase activity and decoupled the MgATPase activity from motility.

In a separate study, Trybus (1994) expressed an HMM-like fragment of smooth muscle myosin heavy chain along with both LC20 and LC17 in a baculovirus expression system. The expressed HMM-like fragment was purified with associated light chains. It moved actin filaments at rates comparable to proteolytically prepared HMM and was regulated by phosphorylation. If the expression system did not contain the LC17, the HMM moved actin filaments at about one-fourth the rate of the control HMM. Readdition of LC17 restored the rate of *in vitro* motility to about 60% of control values.

In each of these cases, the light chain deficient myosins did not function well in moving actin filaments. The myosin heavy chain sequence in this area is in the form of an extended α-helix that is stabilized by the presence of the light chains (Rayment *et al.*, 1993). Without the light chain the helix is likely to collapse.

An explanation for the poor movement in the absence of LC17 may be that the neck region of myosin is stiffened by the presence of the light chain and that a "stiff neck" is required for the generation of force and motion.

VIII. SUMMARY AND PERSPECTIVES

Additional studies using *in vitro* motility assays have not been discussed here, but will be discussed in other chapters. These include studies on the effect of caldesmon and calponin on actin filament sliding (Chapters 6 and 7, this volume), the effect of mutant light chains on myosin's regulation (Chapters 2 and 3, this volume), and the activity of various isoforms of smooth muscle myosin (Chapter 1, this volume). The assays have become commonplace and are now considered to be an essential component of smooth muscle research. An important component of smooth muscle contraction, the latch state, has not been tested *in vitro*. The ability to measure *in vitro* force may allow for this important physiological state to be tested. Other important questions concern the mechanism behind the increased velocity of the isoform of smooth muscle myosin containing an insert at loop 1 in the head region and the enhanced force production by smooth muscle versus skeletal muscle myosins. The answers to these questions will require a combination of approaches, including *in vitro* force measurements and transient kinetic analyses of the MgATPase activity. With the ability to express active mutant myosins and myosin fragments, the understanding of mechanochemical coupling and its regulation is likely to advance rapidly.

References

Anson, M. (1992). *J. Mol. Biol.* **224**, 1029–1038.

Bengur, A. R., Robinson, E. A., Appella, E., and Sellers, J. R. (1987). *J. Biol. Chem.* **262**, 7613–7617.

Block, S. M. (1990). "Noninvasive Techniques in Cell Biology," pp. 375–401. Wiley-Liss, New York.

Chacko, S., and Eisenberg, E. (1990). *J. Biol. Chem.* **265**, 2105–2110.

Collins, K., Sellers, J. R., and Matsudaira, P. (1990). *J. Cell Biol.* **110**, 1137–1447.

Finer, J. T., Simmons, R. M., and Spudich, J. A. (1994). *Nature (London)* **368**, 113–119.

Harris, D. E., and Warshaw, D. M. (1993). *J. Biol. Chem.* **268**, 14764–14768.

Harris, D. E., Work, S. S., Wright, R. K., Alpert, N. R., and Warshaw, D. M. (1994). *J. Muscle Res. Cell Motil.* **15**, 11–19.

Higashi-Fujime, S. (1991). *Int. Rev. Cytol.* **125**, 95–138.

Hynes, T. R., Block, S. M., White, B. T., and Spudich, J. A. (1987). *Cell (Cambridge, Mass.)* **48**, 953–963.

Ikebe, M., Harshorne, D. J., and Elzinga, M. (1986). *J. Biol. Chem.* **261**, 36–39.

Ikebe, M., Harshorne, D. J., and Elzinga, M. (1987). *J. Biol. Chem.* **262**, 9569–9573.

Ikebe, M., Koretz, J., and Hartshorne, D. J. (1988). *J. Biol. Chem.* **263**, 6432–6437.

Ishijima, A., Harada, Y., Kojima, H., Funatsu, T., Higuchi, H., and Yanagida, T. (1994). *Biochem. Biophys. Res. Commun.* **199**, 1057–1063.

Kachar, B., Bridgman, P. C., and Reese, T. S. (1987). *J. Cell Biol.* **105**, 1267–1271.

Kelley, C. A., Takahashi, M., Yu, J. H., and Adelstein, R. S. (1993). *J. Biol. Chem.* **268**, 12848–12854.

Kishino, A., and Yanagida, T. (1988). *Nature (London)* **334**, 74–76.

Kron, S. J., and Spudich, J. A. (1986). *Proc. Natl. Acad. Sci. U.S.A.* **83**, 6272–6276.

Kron, S. J., Toyoshima, Y. Y., Uyeda, T. Q. P., and Spudich, J. A. (1991). *In* "Methods in Enzymology" (R. B. Vallee, ed.), Vol. 196, pp. 399–416. Academic Press, San Diego, CA.

Murphy, R. A., Herlihy, J. T., and Mergerman, J. (1974). *J. Gen. Physiol.* **64**, 691–705.

Nishikawa, M., Sellers, J. R., Adelstein, R. S., and Hidaka, H. (1984). *J. Biol. Chem.* **259**, 8808–8814.

Okagaki, T., Higashi-Fujime, S., Ishikawa, R., Takano-Ohmuro, H., and Kohama, K. (1991). *J. Biochem. (Tokyo)* **109**, 858–866.

Pearson, R. B., Jakes, R., John, M., Kendrick-Jones, J., and Kemp, B. E. (1984). *FEBS Lett.* **168**, 108–112.

Rayment, I., Rypniewski, W. R., Schmidt-Bäse, K., Smith, R., Tomchick, D. R., Benning, M. M., Winkelmann, D. A., Wesenberg, G., and Holden, H. M. (1993). *Science* **261**, 50–58.

Saito, K., Aoki, T., and Yanagida, T. (1994). *Biophys. J.* **66**, 769–777.

Sellers, J. R. (1985). *J. Biol. Chem.* **260**, 15815–15819.

Sellers, J. R., and Kachar, B. (1990). *Science* **249**, 406–408.

Sellers, J. R., Spudich, J. A., and Sheetz, M. P. (1985). *J. Cell Biol.* **101**, 1897–1902.

Sellers, J. R., Cuda, G., Wang, F., and Homsher, E. (1993). *In* "Motility Assays for Motor Proteins" (J. M. Scholey, ed.), pp. 23–49. Academic Press, San Diego, CA.

Sheetz, M. P., and Spudich, J. A. (1983). *Nature (London)* **303**, 31–35.

Sheetz, M. P., Chasan, R., and Spudich, J. A. (1984). *J. Cell Biol.* **99**, 1867–1871.

Sheetz, M. P., Block, S. M., and Spudich, J. A. (1986). *In* "Methods in Enzymology" 134, (R. B. Vallee, ed.), Vol. 134, pp. 531–544. Academic Press, Orlando, FL.

Toyoshima, Y. Y., Kron, S. J., McNally, E. M., Niebling, K. R., Toyoshima, C., and Spudich, J. A. (1987). *Nature (London)* **328**, 536–539.

Trybus, K. M. (1994). *J. Biol. Chem.* **269**, 20819–20822.

Trybus, K. M., Waller, G. S., and Chatman, T. A. (1994). *J. Cell Biol.* **124**, 963–969.

Umemoto, S., and Sellers, J. R. (1990). *J. Biol. Chem.* **265**, 14864–14869.

Umemoto, S., Bengur, A. R., and Sellers, J. R. (1989). *J. Biol. Chem.* **264**, 1431–1436.

Uyeda, T. Q. P., Kron, S. J., and Spudich, J. A. (1990). *J. Mol. Biol.* **214**, 699–710.

Uyeda, T. Q. P., Warrick, H. M., Kron, S. J., and Spudich, J. A. (1991). *Nature (London)* **352**, 307–311.

VanBuren, P., Work, S. S., and Warshaw, D. M. (1994). *Proc. Natl. Acad. Sci. U.S.A.* **91**, 202–205.

Warrick, H. M., Simmons, R. M., Finer, J. T., Uyeda, T. Q. P., Chu, S., and Spudich, J. A. (1993). *In* "Motility Assays for Motor Proteins" (J. M. Scholey, ed.) pp. 1–21. Academic Press, San Diego, CA.

Warshaw, D. M., Desrosiers, J. M., Work, S. S., and Trybus, K. M. (1990). *J. Cell Biol.* **111**, 453–463.

Warshaw, D. M., Desrosiers, J. M., Work, S. S., and Trybus, K. M. (1991). *J. Biol. Chem.* **266**, 24339–24343.

Work, S. S., and Warshaw, D. M. (1992). *Anal. Biochem.* **202**, 275–285.

15

Permeabilized Smooth Muscle

GABRIELE PFITZER

Institut für Physiologie
Medizinische Fakultät der Humboldt Universität zu Berlin
Berlin, Germany

I. INTRODUCTION

Ever since Filo *et al.* (1965) showed that Ca^{2+} ions activate the contractile machinery in glycerinated smooth muscle, permeabilizing or skinning smooth muscle has been a valuable approach to study the complex regulation of smooth muscle tone. Studies in intact smooth muscle usually give only indirect information on the regulation of the contractile machinery. On the other hand, studies with purified proteins are missing the structural integrity that may be of crucial importance. For example, using a biochemically reconstituted system, 50% of the regulatory light chain of myosin (MLC) has to be phosphorylated to increase actomyosin ATPase over basal values (Persechini and Hartshorne, 1981; Merkel *et al.*, 1984). In contrast, intact and skinned smooth muscle may be fully active with only 20% of phosphorylated MLC (Cassidy *et al.*, 1981). Skinned fibers may be viewed as an intermediate between studies in solution and the intact system. In skinned or permeabilized fibers the concentration of Ca^{2+} and other ions can be strictly controlled. Intracellular signaling cascades can be investigated without the influence of transmembrane ion currents. Valuable information on the regulation of smooth muscle has been obtained by loading the preparations with inhibitory peptides (e.g., Rüegg *et al.*, 1989), or proteins such as recombinant Ras (Satoh *et al.*, 1993), which may be regarded as an experimental approach comparable to the overexpression of proteins. Contractile parameters such as force, ATPase, stiffness, and shortening velocity can be measured in permeabilized preparations. Furthermore in preparations in which the sar-

coplasmic reticulum (SR) is still intact, Ca^{2+} release from the SR can be measured. In this way, the regulatory mechanisms underlying the chemomechanical energy transduction, Ca^{2+} sensitivity modulation of the myofilaments as well as Ca^{2+} release from intracellular stores, can be studied.

II. PERMEABILIZATION PROTOCOLS

A number of protocols (Meisheri *et al.*, 1985; Pfitzer and Boels, 1991) have been used to render the cell membrane highly permeable. Initially, a glycerol extraction procedure was applied by Hasselbach and Ledermair (1958), who showed that ATP was the only energy source for contraction of uterine smooth muscle. Glycerol extraction was also used in an initial attempt to delineate the rate-limiting steps of contraction (Peterson, 1982) and the force length relation (Pfitzer *et al.*, 1982), which were comparable to those of the intact preparation. Another widely used approach is the extraction of the cell membrane with the detergent Triton X-100 (Meisheri *et al.*, 1985). Both glycerol treatment and triton skinning destroy plasmalemmal and intracellular membranes, thereby functionally isolating the contractile machinery. Using the mild detergent saponin, intracellular membrane systems (e.g., the Ca^{2+} stores) remain functional (J. Endo *et al.*, 1977; M. Endo *et al.*, 1982). A major breakthrough was the development of permeabilized smooth muscle preparations in which the coupling between surface-membrane receptors and their effectors remains functional, while at the same time the ionic composition

BIOCHEMISTRY OF SMOOTH MUSCLE CONTRACTION

can be strictly controlled and manipulated (Nishimura *et al.*, 1988; Kobayashi *et al.*, 1989). Two compounds for permeabilization are available: α-toxin from *Staphylococcus aureus*, which produces small pores (1–2 nm) allowing molecules up to a molecular mass of 1 kDa to permeate (Ahnert-Hilger *et al.*, 1989), and β-escin, which allows proteins up to even the size of antibodies to enter the cells (Iizuka *et al.*, 1994).

A. Triton Skinning

Triton skinning has been applied to a number of different smooth muscles. In general, fiber bundles are immersed in a buffer containing 1% Triton X-100, the incubation period varies from 0.5 to 16 hr, and the temperature is generally 4°C (cf. Table 1 in Meisheri *et al.*, 1985). Fibers may then be stored in a glycerol-containing buffer at −20°C, which is often referred to as freeze-glycerination. It should be noted that the properties of the triton skinned fibers may change when they are stored for more than 1 week (Wagner and Rüegg, 1986; Schmidt *et al.*, 1995). The criteria for acceptable skinning have been reviewed by Meisheri *et al.* (1985). Triton skinning destroys all membranes, and thus ATP as an energy source has to be added to the incubation buffer. Usually an ATP-regenerating system consisting of phosphocreatine kinase and creatine phosphate is also introduced to avoid accumulation of ADP and formation of rigor bridges in the core of the fiber bundles. Ca^{2+} is buffered with EGTA, and the free Ca^{2+} concentration is calculated by computer solution of the multiple ionic equilibrium equations (e.g., Fabiato and Fabiato, 1979).

Triton skinned fibers contract in response to micromolar concentrations of Ca^{2+} and are relaxed in a MgATP-containing buffer with a pCa > 8. The threshold concentration of Ca^{2+} required to induce a contraction depends on the presence of calmodulin, added to the incubation medium, and varies between 0.1 and 3 μM (Sparrow *et al.*, 1981; Arner, 1982). Maximal force obtained in triton skinned and freeze-glycerinated preparations generally amounts to 60–80% of the tension obtained, with a maximal stimulus in the intact preparation (Peterson, 1982; Arner, 1982). However, maximal unloaded shortening velocity was unaltered in the skinned preparation compared to the intact preparation (Arner, 1982). Ca^{2+} sensitivity in the absence of exogenous calmodulin is low (Sparrow *et al.*, 1981; Arner, 1982), and tension tends to decline with repeated contraction–relaxation cycles. Addition of calmodulin increases Ca^{2+} sensitivity and partially reverses the decline in force, suggesting that it may be due to loss of protein components (Kossmann *et al.*, 1987; Tansey *et al.*, 1994; Schmidt *et al.*, 1995). This is not surprising since treatment with Triton X-100 re-

moves the plasma membrane (Spedding, 1983). Loss of force may also be due to the action of Ca^{2+}-dependent proteinases, which degrade membrane plaques, cytoplasmic dense bodies and intermediate filaments (Haeberle *et al.*, 1985a). Proteolytic degradation may be prevented by addition of inhibitors such as leupeptin. The reproducibility of the contractions may also depend on the pH of the incubation medium (Pfitzer *et al.*, 1984). Also, raising the temperature to above 27°C impairs the stability of the fibers and this was associated with an increase in the resting force (Sparrow *et al.*, 1984).

Triton skinned and glycerinated fibers have been very valuable in demonstrating that phosphorylation and dephosphorylation of MLC is sufficient to induce contraction and relaxation (see Section III.B). These preparations have also been used to study the influence of ionic strength (Arheden *et al.*, 1988; Gagelmann and Güth, 1985), free Mg^{2+} (Arner, 1983; Barsotti *et al.*, 1987), pH (Mrwa *et al.*, 1974), inorganic phosphate (Schneider *et al.*, 1981), nucleotides such as ATP and ADP (Arner and Hellstrand, 1985) on isometric force development, shortening velocity, and ATP turnover. Some of these experiments have also been carried out in smooth muscle fiber bundles and single smooth muscle cells permeabilized with saponin, β-escin, or α-toxin (Saida and Nonomura, 1978; Iino, 1981; Warshaw *et al.*, 1987; Crichton *et al.*, 1993).

B. Permeabilization with Saponin

The use of saponin, a plant glycoside, for permeabilization was introduced by Endo and coworkers (1977). Saponin removes the surface membrane without impairment of the functions of SR. The plasma membrane, unlike that of triton skinned smooth muscle (Spedding, 1983), appears fairly intact in electron microscopic pictures of cross sections of permeabilized cells. However, when patches of isolated membranes were viewed face on in homogenates of permeabilized cells, numerous 70- to 80-Å holes were visible (Kargacin and Fay, 1987).

Generally, thin fiber bundles (ideally less than 150 μm thick to guarantee adequate ATP supply to the inside of the fiber) are immersed in relaxing solution containing 50 μg/ml saponin for 20 min (Endo *et al.*, 1977; Saida and Nonomura, 1978; Itoh *et al.*, 1982b). In some studies, the concentration varies between 40 and 60 μg/ml and the incubation period is increased up to 30 min (Saida, 1982). The relaxing solution usually contains EGTA (2 to 4 mM), MgATP (5mM), a buffer such as Tris maleate, and 130 mM K^+. In the earlier studies, the anion was Cl^- (Itoh *et al.*, 1982a,b) or proprionate (Saida, 1982). Some studies also included dithioerythritol to prevent oxidation of -SH groups and mitochondrial blockers. The stability of skinned skeletal fibers is much better if chloride is replaced by a different anion and the best appears to be methanesulfonate (Andrews *et al.*, 1991). The same may be true for skinned smooth muscle. In fact, in other studies methanesulfonate is frequently used as an anion (e.g., Iino, 1991; Kobayashi *et al.*, 1989; Satoh *et al.*, 1994).

Fibers treated in the described way respond rapidly to increasing Ca^{2+} concentrations. Threshold concentrations are around 2×10^{-7} M Ca^{2+} and maximum contraction was observed at 10 μM Ca^{2+}. Maximal force is comparable to or even larger than that observed before saponin treatment. This was taken as a criterion for complete skinning (Endo *et al.*, 1982). As in triton skinned fibers, tension development depends on the presence of ATP (Endo *et al.*, 1982).

Except for one study (Haeusler *et al.*, 1981), no response to agonists was observed after saponin permeabilization. Haeusler modified the saponin permeabilization using high concentrations of saponin (500 μg/ml) for a very brief period (5–6 min). Mesenteric arteries skinned in this way responded to noradrenaline with a contraction that was attributed to release of Ca^{2+} from internal stores. Saponin-treated smooth muscle, however, responds to GTPγS, a poorly hydrolyzable GTP analog, which permanently activates G proteins (Fujiwara *et al.*, 1989; Hirata *et al.*, 1992). This suggests that the disruption of pharmacomechanical coupling in saponin-treated smooth muscle is due to disruption of the coupling between surface-membrane receptors and their G proteins while the G protein effector cascade appears to be functional.

C. Permeabilization with β-Escin and α-Toxin

β-Escin and α-toxin have attracted a lot of interest since permeabilization with either one of them appears not to interfere with the receptor effector coupling. This allows the study of the intracellular signaling cascades involved in pharmacomechanical coupling. β-Escin is a saponin ester and α-toxin is a bacterial toxin from *Staphylococcus aureus*. They differ by the size of the holes they produce in the plasma membrane.

The permeabilization protocol with β-escin is similar to that applied for saponin skinning (Kobayashi *et al.*, 1989). The pores are quite large, allowing even antibodies to permeate (Iizuka *et al.*, 1994). In these preparations, the pharmacomechanical coupling is functionally intact as judged from the agonist-induced release of Ca^{2+} from intracellular stores and increase in Ca^{2+} sensitivity of the myofilaments (Kobayashi *et al.*, 1989). One criterion for complete permeabilization again is the amplitude of maximal Ca^{2+}-activated

force, which should be as large as the force elicited in the intact preparation. Incomplete permeabilization results in low steady-state force and a slow rate of tension rise. In most cases, fully permeabilized preparations are obtained by treatment with 50 μg/ml β-escin at room temperature for 30 min. However, species differences may exist (e.g., in mesenteric arteries from Sprague Dawley rats, the treatment has to be extended up to 1.5 hr; J. Beichert and G. Pfitzer, unpublished observations). Most authors include calmodulin in the incubation medium as calmodulin may be lost from the preparations. Whether or not calmodulin diffuses out of β-escin-permeabilized fibers is a matter of debate (Pfitzer *et al.*, 1991; Tansey *et al.*, 1994). Low-molecular-mass G proteins, which are candidates for mediating Ca^{2+} sensitization (Hirata *et al.*, 1992; Satoh *et al.*, 1993), are retained in the preparations as shown by GTP ligand binding blots (Satoh *et al.*, 1992). In triton skinned preparations, low-molecular-mass G proteins cannot be detected by this procedure.

Permeabilization with staphylococcal α-toxin was first applied by Cassidy *et al.* (1979). α-Toxin produces small pores of 1–2 nm effective size (Ahnert-Hilger *et al.*, 1989), which allows only molecules of a relative molecular mass of about 1 kDa to enter the cells. These preparations are closest to intact preparations. The disadvantage is that signaling cannot be probed with high-molecular-weight compounds such as heparin (Kobayashi *et al.*, 1989) or recombinant proteins such as Ras and Rho (Hirata *et al.*, 1992; Satoh *et al.*, 1993). Small molecules that may be important for the resting properties of smooth muscle or may modulate the cellular response to an agonist, however, may diffuse out of the α-toxin-permeabilized fibers. For instance, the Ca^{2+} leak from the SR is greater in the intact than in the permeabilized preparation, suggesting the loss of factors that modulate the basal properties of the SR Ca^{2+} channels (Missiaen *et al.*, 1993). Another problem is the high resting ATPase, which has been attributed to ecto-ATPases (Trinkle-Mulcahy *et al.*, 1994). The high resting ATPase compromises the maintenance of nucleotide concentrations in the tissue even in the presence of an ATP-regenerating system. It may be inhibited by 4,4'-diisothiocyanatostilbene-2,2'-disulfonic acid (DIDS) and sodium azide (Trinkle-Mulcahy *et al.*, 1994).

III. PROBING THE PHOSPHORYLATION THEORY IN TRITON SKINNED SMOOTH MUSCLE

A. Calmodulin Is Essential for Contraction

The fact that calmodulin diffuses out of triton skinned fibers could be used to demonstrate in reconstitution experiments with exogenous calmodulin that calmodulin is essential for smooth muscle contraction (Sparrow et al., 1981; Cassidy et al., 1981). The Ca^{2+}–calmodulin complex was further implicated in the regulation of smooth muscle contraction because compounds that inhibit calmodulin-dependent enzymes, such as trifluoperazine, also inhibit contraction of skinned smooth muscle (Hidaka et al., 1979; Cassidy et al., 1980; Sparrow et al., 1981; reviewed in Asano and Stull, 1985). The primary mechanism by which the Ca^{2+}–calmodulin complex induces smooth muscle contraction is by activation of myosin light chain kinase (MLCK) (Kamm and Stull, 1985), and it was suggested that the calmodulin antagonists inhibit contraction through inhibition of MLCK. However, the calmodulin antagonists are probably not specific inhibitors of calmodulin action (Zimmer and Hofmann, 1984). Moreover, calmodulin also binds to other proteins of the myofilament such as caldesmon (Sobue et al., 1982). It was therefore desirable to find an inhibitor that specifically inhibits the binding of calmodulin to MLCK. A peptide of 20 amino acid residues (RS 20) derived from the calmodulin binding domain binds calmodulin with high affinity and potently inhibits MLCK activity (Lukas et al., 1986). This peptide induces relaxation of Ca^{2+}-activated force in skinned taenia coli (Rüegg et al., 1989) and inhibits shortening of skinned isolated smooth muscle cells (Kargacin et al., 1990). However, relaxation was associated only with partial dephosphorylation of MLC suggesting that MLC phosphorylation may not be the sole determinant of contractile force (Rüegg et al., 1989).

Using freeze-dried fibers, Rüegg and coworkers (1984) estimated the fraction of calmodulin available for activation of MLCK to be in the range of 0.3 to 4 μM, which is about 10% of the total cellular calmodulin. This pool is readily exchangeable and therefore rapidly equilibrates with the incubation medium (Rüegg et al., 1984). Similar results were obtained in triton skinned tracheal smooth muscle (Tansey et al., 1994). In contrast, calmodulin appears to be retained in β-escin-permeabilized tracheal smooth muscle (Tansey et al., 1994).

B. Activation of Contraction by Phosphorylation of Myosin Light Chain

Phosphorylation of MLC is generally accepted to be the primary event involved in the initiation of smooth muscle contraction (Kamm and Stull, 1985; Somlyo and Somlyo, 1994). In triton skinned taenia coli, phosphorylation and dephosphorylation of MLC preceded contraction and relaxation, respectively (Kühn et al., 1990). Although this is in line with the hypothesis that

MLC phosphorylation regulates smooth muscle contraction, it does not establish a causal relationship, in particular since maximal force was obtained at phosphorylation values ranging from 20% to about 60% of the MLC being phosphorylated (Cassidy et al., 1981). It was therefore of prime importance to demonstrate that phosphorylation of MLC in the absence of Ca^{2+} is sufficient to induce contraction. This was achieved by two independent experiments with skinned fibers. (1) Using the ATP analog ATPγS as substrate for MLCK, MLC may be permanently phosphorylated since thiophosphorylated MLC is a poor substrate for MLC phosphatase (MLCP) (Sherry et al., 1978). To induce thiophosphorylation of MLC, skinned fibers were first incubated with ATPγS and Ca^{2+} and in the absence of ATP. After removal of ATPγS and Ca^{2+}, addition of ATP induced an irreversible, Ca^{2+}-independent contraction (Hoar et al., 1979). (2) Limited proteolysis of MLCK with chymotrypsin produces a constitutively active fragment of MLCK (Walsh et al., 1982a). Loading skinned fibers with this enzyme induced a contraction in the absence of Ca^{2+} (Walsh et al., 1982b; Gagelmann et al., 1984). These experiments in skinned fibers showed that phosphorylation of MLC is sufficient to induce smooth muscle contraction.

The phosphorylation hypothesis also requires that dephosphorylation of MLC induces relaxation. This is the case as shown by Haeberle and coworkers (1985b), who incubated glycerinated uterine smooth muscle with a purified catalytic subunit of a type-2 phosphoprotein phosphatase. This induced dephosphorylation of MLC and relaxation in the presence of saturating concentrations of Ca^{2+} and calmodulin. A very slow relaxation was found in skinned gizzard in which the endogenous MLCP activity was low. Relaxation can be normalized by addition of exogenous phosphatase (Bialojan et al., 1985).

C. The "Latch" State in Skinned Fibers

In intact smooth muscle, phosphorylation often declines while force is maintained, the so-called "latch" state (Dillon et al., 1981). Regulation of the latch state is still poorly understood. It has been very difficult to induce a latch state in skinned smooth muscle. The closest may be triton skinned chicken gizzard, which contracts independent of MLC phosphorylation (Wagner and Rüegg, 1986). In skinned chicken gizzard, but also in other types of smooth muscle (Bialojan et al., 1987; Siegman et al., 1989; Kenney et al., 1990; Schmidt et al., 1995), a steep and nonlinear relation between force and MLC phosphorylation was observed, which was postulated by the "latch model" of Hai and Murphy (1989). This model proposes that the relation between force and MLC phosphorylation depends on the activity of MLCP. However, inhibiting MLCP in skinned smooth muscle using the phosphatase inhibitor okadaic acid had no (Siegman et al., 1989) or very little effect on this relation (Schmidt et al., 1995). In contrast, evidence for the cooperative attachment of dephosphorylated cross-bridges contributing to tension maintenance at low levels of MLC phosphorylation was obtained in glycerinated smooth muscle. This was achieved (1) by analyzing the kinetics of relaxation from rigor by photolytic release of ATP from caged ATP (Arner et al., 1987; Somlyo et al., 1988; see also Section IV) and (2) by single turnover experiments on the nucleotide bound to myosin (Vyas et al., 1992).

The relation between force, Ca^{2+}, and MLC phosphorylation may also be modulated by the thin filament-linked protein, caldesmon. Loading β-escin-permeabilized arteries with a small peptide from the actin binding region of caldesmon increased force at low levels of Ca^{2+} (Katsuyama et al., 1992). It was proposed that this was due to competition of binding of endogenous caldesmon with actin, thereby relieving the inhibitory action of endogenous caldesmon. Loading triton skinned chicken gizzard fibers with exogenous caldesmon resulted in inhibition of force at low but not at high levels of MLC phosphorylation (Pfitzer et al., 1993). Loading of skinned fibers with a protein is similar to overexpression, an approach currently widely used to study intracellular signaling. Both approaches have the same caveat, namely, that the protein is present in higher than physiological concentrations. Thus the effects may be unrelated to the effects at physiological concentrations.

IV. KINETIC INVESTIGATIONS IN PERMEABILIZED SMOOTH MUSCLE USING CAGED COMPOUNDS

To fully understand the regulation of smooth muscle contraction, the rate-limiting steps of excitation–contraction coupling and of the cross-bridge cycle have to be known. Kinetic investigations in multicellular preparations and even in single cells are, however, complicated by the long diffusional delays. This was overcome by the introduction of caged compounds. In caged compounds the biologically active moiety is protected by a photolabile chemical modification that can be cleaved by light to release the active compound. The first caged compound was caged ATP (Kaplan et al., 1978). Since then, a number of caged compounds, including caged Ca^{2+}, IP_3, and GTPγS, became available (McCray and Trentham, 1989). The caged com-

pound can be diffused into permeabilized smooth muscle and equilibrated at its site of action. Then the biologically active molecule is released by photolysis with a light flash and becomes available for binding to its ligand within milliseconds.

By using caged compounds in combination with permeabilized smooth muscle preparations, valuable information regarding the rate-limiting steps between excitation and contraction (see Somlyo and Somlyo, 1990, for review; Somlyo et al., 1992), regulation of Ca^{2+} release from the SR (Walker et al., 1987; Iino and Endo, 1992), and the kinetics of the cross-bridge cycle (Arner et al., 1987; Somlyo et al., 1988; Fuglsang et al., 1993) was obtained. Differences between tonic and phasic smooth muscle could be ascribed to the different kinetics of the phosphorylation reaction and cross-bridge cycle (Fuglsang et al., 1993).

V. MODULATION OF Ca^{2+} SENSITIVITY OF CONTRACTION

A. In Triton Skinned Preparations

The first evidence that Ca^{2+} sensitivity of smooth muscle myofilaments may be modulated by second messengers was obtained in triton skinned fibers. Thus, it was shown by a number of authors that cAMP, as well as the catalytic subunit of the cAMP-dependent protein kinase, decreased Ca^{2+} sensitivity of tension development (Mrwa et al., 1979; Kerrick and Hoar, 1981; Rüegg and Paul, 1982; Meisheri and Rüegg, 1983; Pfitzer et al., 1985). As a mechanism it was suggested that cAMP-dependent protein kinase phosphorylates MLCK, thereby reducing the affinity of MLCK for calmodulin (Conti and Adelstein, 1981). The physiological relevance of this finding has been questioned for a number of reasons (1) The desensitizing effect of cAMP and the catalytic subunit of the cAMP-dependent protein kinase was overcome when the calmodulin concentration in the incubation medium was increased to 0.5 μM (Meisheri and Rüegg, 1983) or to 5 μM (Pfitzer et al., 1985), respectively. These are concentrations of calmodulin that are thought to be available for activation of MLCK under physiological conditions (Rüegg et al., 1984). (2) In the intact smooth muscle, the activity ratio of MLCK was not affected by interventions that increase the cytosolic levels of cAMP (Miller et al., 1983). On the other hand, simultaneous measurements of force and cytosolic [Ca^{2+}] in intact smooth muscle showed that β-adrenergic relaxation may be induced with only minor decreases in the cytosolic [Ca^{2+}] (Morgan and Morgan, 1984; Ozaki et al., 1990). In a study using α-toxin-permeabilized mesenteric arteries, the increase in Ca^{2+} sensitivity induced by endothelin was partially reversed by cAMP (Nishimura et al., 1992). In conclusion, although a desensitizing action of cAMP is well documented, the mechanism by which this occurs is still not clear.

In triton skinned and in α-toxin-permeabilized smooth muscle preparations, the Ca^{2+} sensitivity of force production is also decreased by cGMP (Pfitzer et al., 1984, 1986; Nishimura et al., 1992). This may be due to an up-regulation of MLCP (Pfitzer et al., 1986). Clear evidence that modulation of the activity of MLCP would affect Ca^{2+} sensitivity of tension development was in fact obtained in triton skinned smooth muscle when it was shown that inhibition of MLCP by the black sponge toxin, okadaic acid, increased Ca^{2+} sensitivity (Takai et al., 1987; Bialojan et al., 1988). On the other hand, incubation of triton skinned chicken gizzard fibers with a purified phosphatase decreased Ca^{2+} sensitivity (Bialojan et al., 1987).

B. In β-Escin- and α-Toxin-Permeabilized Smooth Muscle

The development of permeabilized smooth muscle, in which the coupling between membrane receptors and its effectors is still functional, greatly facilitated our understanding of the mechanisms underlying Ca^{2+}-sensitivity modulation that has been observed in the intact tissue (Morgan and Morgan, 1984; Bradley and Morgan, 1987; Rembold and Murphy, 1988; Himpens et al., 1990; Ozaki et al., 1990). These preparations respond to stimulation with an agonist with an increase in force at constant Ca^{2+} (Nishimura et al., 1988, 1992; Kobayashi et al., 1989; Kitazawa et al., 1989; Himpens et al., 1990; Satoh et al., 1994). Force can also decline at constant Ca^{2+}, indicating a decrease in Ca^{2+} sensitivity (Kitazawa and Somlyo, 1990). As already mentioned, the Ca^{2+} stores are functional in permeabilized smooth muscle. Thus precautions, such as using high concentrations of Ca^{2+} buffers (10 mM) and functionally eliminating the SR by treatment with A 23187, have to be taken to exclude confounding effects due to intracellular Ca^{2+} compartmentalization.

The changes in Ca^{2+} sensitivity are generally paralleled by changes in MLC phosphorylation, indicating that they result in corresponding changes of the activities of MLCK and MLCP (Kitazawa et al., 1991b; Gong et al., 1992; Kubota et al., 1992). The second messenger mediating Ca^{2+} sensitization may be arachidonic acid (Gong et al., 1992). However, diacylglycerol and activation of protein kinase C were not excluded (Collins et al., 1992; Katsuyama and Morgan, 1993; see also Somlyo and Somlyo, 1994). Moreover, the increased

Ca^{2+} sensitivity of force production may also be associated with a shift in the relation between force and MLC phosphorylation (Jiang and Morgan, 1989), indicating that additional regulatory mechanisms may contribute to Ca^{2+}-sensitivity modulation.

The agonist-induced Ca^{2+} sensitization can be mimicked by GTPγS and inhibited by GTPβS (Fujiwara et al., 1989; Kitazawa et al., 1991a,b; Satoh et al., 1994), but the G protein(s) involved have yet to be identified. β-Escin-permeabilized smooth muscle has been loaded with monomeric G proteins of the Ras superfamily (Hirata et al., 1992; Satoh et al., 1993). The constitutively active mutant of Ras (G12VRas·GTP) and native RhoA activated with GTPγS increased force at constant submaximal Ca^{2+} (Satoh et al., 1993; Hirata et al., 1992). However, as pointed out by Satoh et al. (1993), there remains the possibility that GTP or GTPγS may be released from Ras or Rho in the smooth muscle so that the Ca^{2+}-sensitizing effect would be due to activation of endogenous G proteins rather than the action of the added monomeric G proteins. This problem is less severe in the case of G12VRas·GTP, because GTP, unlike GTPγS, has only a minor Ca^{2+}-sensitization effect (Satoh et al., 1993). Otto et al. (1995) have tested the effect of the GTP-bound form of Gly14ValRhoA, which is the constitutively active mutant of RhoA, and found no effect on Ca^{2+} sensitivity. However, Val14RhoA·GTP could restore the rundown of the Ca^{2+}-sensitizing effect of the muscarinic agonist carbachol. Taken together with the finding that incubation of β-escin-permeabilized fibers with botulinum C3 exoenzyme, which inactivates RhoA, completely abolished the Ca^{2+}-sensitizing effect of carbachol, the authors proposed that Rho may be a necessary but not sufficient element in the signaling cascade leading to Ca^{2+} sensitization. The downstream effectors of the monomeric G proteins in smooth muscle are not known, but protein tyrosine kinase cascades may be involved since tyrosine kinase inhibitors inhibit the agonist-induced increase in Ca^{2+} sensitivity (Steusloff et al., 1995). Moreover, the tyrosine kinase inhibitor, tyrphostin, partially inhibited the Ras-induced increase in Ca^{2+}-activated force (Satoh et al., 1993).

VI. STUDIES ON THE Ca^{2+} RELEASE FROM INTRACELLULAR STORES IN PERMEABILIZED SMOOTH MUSCLE

Smooth muscle preparations permeabilized with saponin, β-escin, or α-toxin have significantly contributed to the understanding of Ca^{2+} release from intracellular stores. Thus, it was demonstrated in this preparation that IP$_3$ is the second messenger that links the stimulation of a membrane-bound receptor to release of Ca^{2+} from the SR (Suematsu et al., 1984; Somlyo et al., 1985; Hashimoto et al., 1986; Walker et al., 1987).

In general, the stores are loaded with Ca^{2+} buffered with millimolar concentrations of EGTA (Endo et al., 1977; Saida and Nonomura, 1978; Itoh et al., 1982a,b; Saida, 1982). The stores are fully loaded within 3 to 5 min at 1 μM Ca^{2+} (Saida, 1982), whereby the Ca^{2+} uptake depends on the Ca^{2+} concentration (Saida, 1982; Yamamoto and van Breemen, 1986). At Ca^{2+} concentrations greater than 1 μM, a Ca^{2+}-induced Ca^{2+} release was observed (Itoh et al., 1981; Saida, 1982). The deposition of Ca^{2+} in the SR of saponin-permeabilized smooth muscle was demonstrated by electron probe X-ray microanalysis (Kowasaki et al., 1985). The stored Ca^{2+} can be released by either caffeine or IP$_3$. Many studies (e.g., Endo et al., 1977; Saida and Nonomura, 1978; Itoh et al., 1982a,b; Saida, 1982) used the transient tension response elicited by the released Ca^{2+} as a measure of Ca^{2+} uptake into the SR. In order to observe a contractile effect, a solution with a weak Ca^{2+} buffering capacity (0.1 mM EGTA) has to be used so that Ca^{2+} released from the store, instead of being bound by the buffer system, is made available to the contractile elements. The limitations of this index of Ca^{2+} uptake have been discussed previously (Meisheri et al., 1985; Yamamoto and van Breemen, 1986). The most serious limitation probably is that the experimental procedures may not only affect Ca^{2+} release but also the Ca^{2+} sensitivity of the myofilaments. Therefore, some authors investigated the properties of the Ca^{2+} stores with ^{45}Ca^{2+} flux measurements (Stout and Diecke, 1983; Yamamoto and van Breemen, 1986; Missiaen et al., 1991, 1993).

A powerful new approach is the direct measurement of released Ca^{2+} with fluorescent Ca^{2+} indicators such as fura-2 or fluo-3, whereby the permeabilized smooth muscle has to be placed into microcuvettes (Iino, 1991). Since force is no longer required as an indicator, Ca^{2+} release may be measured in the absence of MgATP, which allowed the study of adenine nucleotides on the IP$_3$-induced Ca^{2+} release without the interference by Ca^{2+} uptake (Iino, 1991; Hirose and Iino, 1994). Inclusion of ryanodine allows a separate study of the IP$_3$- and caffeine-sensitive stores (Hirose et al., 1993). Experimental protocols have been designed that allow the use of high concentrations of EGTA, thereby preventing Ca^{2+}-mediated feedback regulation of the Ca^{2+} release (Hirose and Iino, 1994). Direct measurement of Ca^{2+} release with indicators becomes even more powerful in combination with caged compounds (Walker et al., 1987; Iino and Endo, 1992).

VII. CONCLUSIONS

Permeabilized smooth muscle tissues have been valuable tools to investigate the complex regulation of smooth muscle contraction. The development of permeabilized smooth muscle preparations, in which the coupling between membrane receptors and the intracellular effectors is still functionally intact allows investigation of not only the properties of the contractile machinery (which is also possible in triton skinned fibers) but also of the signaling pathways involved in pharmacomechanical coupling. A major step forward is the combination of the permeabilized fibers with the use of caged compounds. It appears that some of the signaling steps leading to Ca^{2+} sensitization of the myofilaments also mediate growth. Thus, regulation of contractility and growth or proliferation of smooth muscle may be intimately connected. Permeabilized smooth muscle may help to analyze how these pathways are connected, which may be of importance in understanding the biochemistry of vascular diseases.

References

Ahnert-Hilger, G., Mach, W., Föhr, K. J., and Gratzl, M. (1989) Methods Cell. Biol. 31, 63–90.

Andrews, M. A. W., Maughan, D. W., Nosek, T. M., and Godt, R. E. (1991). J. Gen. Physiol. 98, 1105–1125.

Arheden, H., Arner, A., and Hellstrand, P. (1988). J. Physiol. (London) 403, 539–558.

Arner, A. (1982). Pflügers Arch. 395, 277–284.

Arner, A. (1983). Pflügers Arch. 397, 6–12.

Arner, A., and Hellstrand, P. (1985). J. Physiol. (London) 360, 347–365.

Arner, A., Goody, R. S., Rapp, G., and Rüegg, J. C. (1987). J. Muscle Res. Cell Motil. 8, 377–385.

Asano, M., and Stull, J. T. (1985). In "Calmodulin Antagonists and Cellular Physiology" (H. Hidaka and D. J. Hartshorne, eds.), pp. 225–260. Academic Press, Orlando, FL.

Barsotti, R. J., Ikebe, M., and Hartshorne, D. J. (1987). Am. J. Physiol. 252, C543–C554.

Bialojan, C., Merkel, L., Rüegg, J. C., Gifford, D., and Di Salvo, J. (1985). Proc. Soc. Exp. Biol. Med. 178, 648–652.

Bialojan, C., Rüegg, J. C., and Di Salvo, J. (1987). Pflügers Arch. 410, 304–312.

Bialojan, C., Rüegg, J. C., and Takai, A. (1988). J. Physiol. (London) 398, 81–95.

Bradley, A. B., and Morgan, K. G. (1987). J. Physiol. (London) 385, 437–448.

Cassidy, P. S., Hoar, P. E., and Kerrick, W. G. L. (1979). J. Biol. Chem. 254, 11148–11153.

Cassidy, P. S., Hoar, P. E., and Kerrick, W. G. L. (1980). Pflügers Arch. 387, 115–120.

Cassidy, P. S., Kerrick, W. G. L., Hoar, P. E., and Malencik, D. A. (1981). *Pflügers Arch.* **392**, 115–120.

Collins, E. M., Walsh, M. P., and Morgan, K. G. (1992). *Am. J. Physiol.* **262**, H754–H762.

Conti, M. A., and Adelstein, R. S. (1981). *J. Biol. Chem.* **256**, 3178–3181.

Crichton, C. A., Taggart, M. J., Wray, S., and Smith G. L. (1993). *J. Physiol. (London)* **465**, 629–645.

Dillon, P. F., Aksoy, M. O., Driska, S. P., and Murphy, R. A. (1981). *Science* **211**, 495–497.

Endo, J., Kitazawa, T., Yagi, S., Iino, M., and Kabuta, Y. (1977). *In* "Excitation–Contraction Coupling in Smooth Muscle" (R. Casteels, T. Godfraind, and J. C. Rüegg, eds.), pp. 199–210. Elsevier-North Holland Biomed. Press, Amsterdam.

Endo, M., Yagi, S., and Iino, M. (1982). *Fed. Proc., Fed. Am. Soc. Exp. Biol.* **41**, 2245–2250.

Fabiato, A., and Fabiato, F. (1979). *J. Physiol. (Paris)* **75**, 463–505.

Filo, R. S., Bohr, D. F., and Rüegg, J. C. (1965). *Science* **147**, 1581–1583.

Fuglsang, A., Khromov, A., Török, K., Somlyo, A. V., and Somlyo, A. P. (1993). *J. Muscle Res. Cell Motil.* **14**, 666–673.

Fujiwara, T., Itoh, T., Kubota, Y., and Kuriyama, H. (1989). *J. Physiol. (London)* **408**, 535–547.

Gagelmann, M., and Güth, K. (1985). *Pflügers Arch.* **403**, 210–214.

Gagelmann, M., Mrwa, U., Boström, S., Rüegg, J. C., and Hartshorne, D. (1984). *Pflügers Arch.* **401**, 107–109.

Gong, M. C., Fuglsang, A., Alessi, D., Kobayashi, S., Cohen, P., Somlyo, A. V., and Somlyo, A. P. (1992). *J. Biol. Chem.* **267**, 21492–21498.

Haeberle, J. R., Coolican, S. A., Evan, A., and Hathaway, D. R. (1985a). *J. Muscle Res. Cell Motil.* **6**, 347–363.

Haeberle, J. R., Hathaway, D. R., and DePaoli-Roach, A. A. (1985b). *J. Biol. Chem.* **260**, 9965–9968.

Haeusler, G., Richards, J. G., and Thorens, S. (1981). *J. Physiol. (London)* **321**, 537–556.

Hai, C.-M., and Murphy, R. A. (1989). *Annu. Rev. Physiol.* **51**, 285–298.

Hashimoto, T., Hirata, M., Itoh, T., Kanmura, Y., and Kuriyama, H. (1986). *J. Physiol. (London)* **370**, 605–618.

Hasselbach, W., and Ledermair, O. (1958). *Pflügers Arch. Gesamte Physiol. Menschen Tiere* **267**, 532–542.

Hidaka, H., Yamak, T., Totsuka, T., and Asano, M. (1979). *Mol. Pharmacol.* **15**, 49–59.

Himpens, B., Kitazawa, T., and Somlyo, A. P. (1990). *Pflügers Arch.* **417**, 21–28.

Hirata, K., Kikuchi, A., Sasaki, T., Kuroda, S., Kaibuchi, K., Matsuura, Y., Seki, H., Saida, K., and Takai, Y. (1992). *J. Biol. Chem.* **267**, 8719–8722.

Hirose, K., and Iino, M. (1994). *Nature (London)* **372**, 791–793.

Hirose, K., Iino, M., and Endo, M. (1993). *Biochem. Biophys. Res. Commun.* **194**, 726–732.

Hoar, P. E., Kerrick, W. G. L., and Cassidy, P. S. (1979). *Science* **204**, 503–506.

Iino, M. (1981). *J. Physiol. (London)* **320**, 449–467.

Iino, M. (1991). *J. Gen. Physiol.* **98**, 681–698.

Iino, M., and Endo, M. (1992). *Nature (London)* **360**, 76–78.

Iizuka, K., Ikebe, M., Somlyo, A. V., and Somlyo, A. P. (1994). *Cell Calcium* **16**, 431–445.

Itoh, T., Izumi, H., and Kuriyama, H. (1981). *J. Physiol. (London)* **321**, 513–515.

Itoh, T., Kajiwara, M., Kitamura, K., and Kuriyama, H. (1982a). *J. Physiol. (London)* **322**, 197–125.

Itoh, T., Izumi, H., and Kuriyama, H. (1982b). *J. Physiol. (London)* **326**, 475–493.

Jiang, M. T., and Morgan, K. G. (1989). *Pflügers Arch.* **413**, 637–643.

Kamm, K. E., and Stull, J. T. (1985). *Annu. Rev. Pharmacol. Toxikol.* **25**, 593–620.

Kaplan, J. H., Forbush, B., III, and Hoffman, J. F. (1978). *Biochemistry* **17**, 1929–1935.

Kargacin, G. J., and Fay, F. S. (1987). *J. Gen. Physiol.* **90**, 49–73.

Kargacin, G. J., Ikebe, M., and Fay, F. S. (1990). *Am. J. Physiol.* **259**, C315–C324.

Katsuyama, H., and Morgan, K. G. (1993). *Circ. Res.* **72**, 651–657.

Katsuyama, H., Wang, C.-L. A., and Morgan, K. G. (1992). *J. Biol. Chem.* **267**, 14555–14558.

Kenney, R. E., Hoar, P. E., and Kerrick, W. G. L. (1990). *J. Biol. Chem.* **265**, 8642–8649.

Kerrick, W. G. L., and Hoar, P. E. (1981). *Nature (London)* **292**, 253–255.

Kitazawa, T., and Somlyo, A. P. (1990). *Biochem. Biophys. Res. Commun.* **172**, 1291–1297.

Kitazawa, T., Kobayashi, S., Horiuti, K., Somlyo, A. V., and Somlyo, A. P. (1989). *J. Biol. Chem.* **264**, 5339–5342.

Kitazawa, T., Gaylinn, B. D., Denney, G. H., and Somlyo, A. P. (1991a). *J. Biol. Chem.* **266**, 1708–1715.

Kitazawa, T., Masuo, M., and Somlyo, A. P. (1991b). *Proc. Natl. Acad. Sci. U.S.A.* **88**, 9307–9310.

Kobayashi, S., Kitazawa, T., Somlyo, A. V., and Somlyo, A. P. (1989). *J. Biol. Chem.* **264**, 17997–18004.

Kossmann, T., Fürst, D., and Small, J. V. (1987). *J. Muscle Res. Cell Motil.* **8**, 135–144.

Kowasaki, D., Shuman, H., Somlyo, A. P., and Somlyo, A. V. (1985). *J. Physiol. (London)* **366**, 153–175.

Kubota, Y. M., Nomura, M., Kamm, K. E., Mumby, M. C., and Stull, J. T. (1992). *Am. J. Physiol.* **262**, C405–C410.

Kühn, H., Tewes, A., Gagelmann, M., Güth, K., Arner, A., and Rüegg, J. C. (1990). *Pflügers Arch.* **416**, 512–518.

Lukas, T. J., Burgess, W. H., Prendergast, F. G., Lau, W., and Watterson, D. M. (1986). *Biochemistry* **25**, 1458–1464.

McCray, J. A., and Trentham, D. R. (1989). *Annu. Rev. Biophys. Chem.* **18**, 239–270.

Meisheri, K. D., and Rüegg, J. C. (1983). *Pflügers Arch.* **399**, 315–320.

Meisheri, K. D., Rüegg, J. C., and Paul, R. J. (1985). *In* "Calcium and Contractility" (A. K. Grover, and E. E. Daniel, eds.), pp. 191–224. Humana Press, Clifton, NJ.

Merkel, L., Meisheri, K. D., and Pfitzer, G. (1984). *Eur. J. Biochem.* **138**, 429–433.

Miller, J. R., Silver, P. J., and Stull, J. T. (1983). *Mol. Pharmacol.* **24**, 241–263.

Missiaen, L., De Smedt, H., Droogmans, G., Declerck, I., Plessers, L., and Casteels, R. (1991). *Biochem. Biophys. Res. Commun.* **174**, 1183–1188.

Missiaen, L., Parys, J. B., De Smedt, H., and Casteels, R. (1993). *Biochem. Biophys. Res. Commun.* **193**, 6–12.

Morgan, J. P., and Morgan, K. G. (1984). *J. Physiol. (London)* **357**, 539–551.

Mrwa, U., Achtig, I., and Rüegg, J. C. (1974). *Blood Vessels* **11**, 277–286.

Mrwa, U., Troschka, M., and Rüegg, J. C. (1979). *FEBS Lett.* **197**, 371–374.

Nishimura, J., Kolber, M., and van Breemen, C. (1988). *Biochem. Biophys. Res. Commun.* **157**, 677–683.

Nishimura, J., Moreland, S., Ahn, H. Y., Kawase, T., Moreland, R. S., and van Breemen, C. (1992). *Circ. Res.* **71**, 951–959.

Otto, B., Steusloff, A., Just, I., Aktories, K., and Pfitzer, G. (1995). *J. Muscle Res. Cell Motil.* (in press).

Ozaki, H., Kwon, S.-C., Tajimi, M., and Karaki, H. (1990). *Pflügers Arch.* **416**, 351–359.

Persechini, A., and Hartshorne, D. J. (1981). *Science* **213**, 1383–1385.

Peterson, J. W. (1982). *J. Gen. Physiol.* **79**, 437–452.

Pfitzer, G., and Boels, P. J. (1991). *Blood Vessels* **28**, 262–267.

Pfitzer, G., Peterson, J. W., and Rüegg, J. C. (1982). *Pflügers Arch.* **394**, 174–181.

Pfitzer, G., Hofmann, F., Di Salvo, J., and Rüegg, J. C. (1984). *Pflügers Arch.* **401**, 277–280.

Pfitzer, G., Rüegg, J. C., Zimmer, M., and Hofmann, F. (1985). *Pflügers Arch.* **405**, 70–76.

Pfitzer, G., Merkel, L., Rüegg, J. C., and Hofmann, F. (1986). *Pflügers Arch.* **407**, 87–91.

Pfitzer, G., Satoh, S., Steusloff, A., Andrews, M. A., and Boels, P. J. (1991). *Pflügers Arch.* **418**, R50.

Pfitzer, G., Zeugner, C., Troschka, M., and Chalovich, J. M. (1993). *Proc. Natl. Acad. Sci. U.S.A.* **90**, 5904–5908.

Rembold, C. M., and Murphy, R. A. (1988). *Circ. Res.* **63**, 593–603.

Rüegg, J. C., and Paul, R. J. (1982). *Circ. Res.* **50**, 394–399.

Rüegg, J. C., Pfitzer, G., Zimmer, M., and Hofmann, F. (1984). *FEBS Lett.* **170**, 383–386.

Rüegg, J. C., Zeugner, C., Strauss, J. D., Paul, R. J., Kemp, B., Chem, M., Li, A.-Y., and Hartshorne, D. J. (1989). *Pflügers Arch.* **414**, 282–285.

Saida, K. (1982). *J. Gen. Physiol.* **80**, 191–202.

Saida, K., and Nonomura, Y. (1978). *J. Gen. Physiol.* **72**, 1–14.

Satoh, S., Steusloff, A., and Pfitzer, G. (1992). *J. Muscle Res. Cell. Motil.* **13**, 253.

Satoh, S., Rensland, H., and Pfitzer, G. (1993). *FEBS Lett.* **324**, 211–215.

Satoh, S., Kreutz, R., Wilm, C., Ganten, D., and Pfitzer, G. (1994). *J. Clin. Invest.* **94**, 1397–1403.

Schmidt, U. S., Troschka, M., and Pfitzer, G. (1995). *Pflügers Arch.* **429**, 708–715.

Schneider, M., Sparrow, J. C., and Rüegg, J. C. (1981). *Experientia* **37**, 980–982.

Sherry, J. M. F., Gorecka, A., Aksoy, M. O., Dabrowska, R., and Hartshorne, D. J. (1978). *Biochemistry* **17**, 4411–4418.

Siegman, M. J., Butler, T. M., and Mooers, S. U. (1989). *Biochem. Biophys. Res. Commun.* **161**, 838–842.

Sobue, K. M., Moromoto, K., Inui, M., Kanda, K., and Kakiuchi, S. (1982). *Biomed. Res.* **3**, 188–196.

Somlyo, A. P., and Somlyo, A. V. (1990). *Annu. Rev. Physiol.* **52**, 857–874.

Somlyo, A. P., and Somlyo, A. V. (1994). *Nature (London)* **372**, 231–236.

Somlyo, A. V., Bond, M., Somlyo, A. P., and Scarpa, A. (1985). *Proc. Natl. Acad. Sci. U.S.A.* **82**, 5231–5235.

Somlyo, A. V., Goldman, Y. E., Fujimori, T., Bond, M., Trentham, D. R., and Somlyo, A. P. (1988). *J. Gen. Physiol.* **91**, 165–192.

Somlyo, A. V., Horiuti, K., Trentham, D. R., Kitazawa, T., and Somlyo, A. P. (1992). *J. Biol. Chem.* **267**, 22316–22322.

Sparrow, M. P., Mrwa, U., Hofmann, F., and Rüegg, J. C. (1981). *FEBS Lett.* **125**, 141–145.

Sparrow, M. P., Pfitzer, G., Gagelmann, M., and Rüegg, J. C. (1984). *Am. J. Physiol.* **246**, C308–C314.

Spedding, M. (1983). *Br. J. Pharmacol.* **79**, 225–231.

Steusloff, A., Paul, E., Semenchuk, L., Di Salvo, J., and Pfitzer, G. (1995). *Arch. Biochem. Biophys.* **304**, 386–391.

Stout, M. A., and Diecke, F. P. J. (1983). *J. Pharmacol. Exp. Ther.* **225**, 102–111.

Suematsu, E., Hirata, M., Hashimoto, T., and Kuriyama, H. (1984). *Biochem. Biophys. Res. Commun.* **120**, 481–485.

Takai, A., Bialojan, C., Troschka, M., and Rüegg, J. C. (1987). *FEBS Lett.* **217**, 81–84.

Tansy, M. G., Luby-Phelps, K., Kamm, K. E., and Stull, J. T. (1994). *J. Biol. Chem.* **269**, 9912–9920.

Trinkle-Mulcahy, L., Siegman, M. J., and Butler, T. M. (1994). *Am. J. Physiol.* **266**, C1673–C1683.

Vyas, T. B., Mooers, S. U., Narayan, S. R., Witherell, J. C., Siegman, M. J., and Butler, T. M. (1992). *Am. J. Physiol.* **263**, C210–C219.

Wagner, J., and Rüegg, J. C. (1986) *Pflügers Arch.* **407**, 569–571.

Walker, J. W., Somlyo, A. V., Goldman, Y. E., Somlyo, A. P., and Trentham, D. R. (1987). *Nature (London)* **327**, 249–252.

Walsh, M. P., Dabrowska, R., Hinkins, S., and Hartshorne, D. J. (1982a). *Biochemistry* **21**, 1919–1925.

Walsh, M. P., Bridenbaugh, R., Hartshorne, D. J., and Kerrick, W. G. L. (1982b). *J. Biol. Chem.* **257**, 5987–5990.

Warshaw, D. M., McBride, W. J., and Hubard, M. S. (1987). *Am. J. Physiol.* **252**, C418–C427.

Yamamoto, H., and van Breemen, C. (1986). *J. Gen. Physiol.* **87**, 369–389.

Zimmer, M., and Hofmann, F. (1984). *Eur. J. Biochem.* **142**, 393–397.

CALCIUM
MOVEMENTS

16

Calcium Channels and Potassium Channels

HARM J. KNOT, JOSEPH E. BRAYDEN, and MARK T. NELSON

Department of Pharmacology, Ion Channel Group
University of Vermont, College of Medicine
Medical Research Facility
Colchester, Vermont

I. INTRODUCTION

Smooth muscle contractility is regulated by intracellular Ca^{2+}, which is controlled by entry of Ca^{2+} through voltage-dependent calcium channels. Changes in the membrane potential of the smooth muscle cell and its subsequent effects on the level of intracellular free calcium $[Ca^{2+}]_i$ in the cytoplasm are the basis for this electromechanical coupling. Smooth muscle usually operates in a contracted state (often referred to as "tone") from which it can contract further or relax depending on its physiological function. A significant portion of this tone is caused by intrinsic mechanisms, but it can also be induced and modulated through mechanical and excitatory or inhibitory neurohormonal stimuli. In general, one can distinguish two types of smooth muscle: (1) Smooth muscle that has intrinsic tone responding to stimuli with graded changes in membrane potential ("tonic" smooth muscle). Examples of this type of smooth muscle are the smooth muscle in the lung (airway smooth muscle) and that of most blood vessels (vascular smooth muscle). For example, small arteries exist in a partially contracted state from which they can constrict further or dilate depending on demand for blood. A significant portion of this intrinsic arterial tone is caused by transmural pressure and has been called myogenic tone (Bayliss, 1902). Smooth muscle tone in small arteries and arterioles of the microcirculation (resistance vessels) is an important determinant of peripheral vascular resistance and blood pressure. (2) Other types of smooth muscle that exhibit slow-wave activity or generate action potentials can be called "phasic" smooth muscle. Examples of such types of smooth muscle are the smooth muscle of the gastrointestinal tract and the smooth muscle of the uterus, but also the vascular smooth muscle of the portal vein.

The plasma membrane of smooth muscle cells contains a variety of ion channels, including channels that are selective for calcium, potassium, and chloride. The distribution and properties of these channels vary in different types of smooth muscle (i.e., intestinal, vascular, tracheal) and even within the same type of smooth muscle (i.e., large arteries versus small arteries or veins). Ion channels may respond directly to changes in membrane potential (voltage dependent) or through action of hormones and neurotransmitters, either directly (ligand gated) or indirectly through second-messenger pathways as a result of agonist–receptor binding.

The rise in intracellular $[Ca^{2+}]$ that initiates and maintains contraction in smooth muscle in response to depolarization of the membrane is largely due to the influx of extracellular Ca^{2+} through voltage-dependent Ca^{2+} channels. Voltage-dependent Ca^{2+} channels respond to depolarization with an increase in their open-state probability (P_o). The influx of extracellular Ca^{2+} through these channels may lead directly to contraction, or in the case of influx as a result of action potentials may release additional Ca^{2+} through calcium-induced calcium release (CICR) from the sarcoplasmic reticulum (SR). Nonetheless, the membrane potential plays a critical role in the contractile response. The smooth muscle membrane potential depends on a balance between hyperpolarizing K+ conductances and depolarizing conductances. Opening of potassium channels hyperpolarizes the plasma membrane of smooth muscle cells, thereby reducing the open-probability of Ca^{2+} channels and, hence, leads to relaxation (see Figs. 2C–2E for an illustration). Conversely, the inhibition of potassium channels will lead to depolarization and contraction (see Figs. 1C and 1D for an illustration). Therefore, potassium channels, through membrane potential regulation, play a critical role in the control of smooth muscle contractility.

This chapter examines the properties and roles of calcium and potassium channels that have been described in smooth muscle, with the major emphasis on arterial smooth muscle. Two types of calcium channels have been identified in arterial smooth muscle. (1) Voltage-dependent ("L-type") Ca^{2+} channels increase their activity with membrane depolarization, and thereby control calcium influx into the smooth muscle cell. Therefore, these channels play a key role in excitation–contraction coupling in virtually all smooth muscle. (2) Direct ligand-gated nonselective channels have been described in some types of smooth muscle, including arterial and airway smooth muscle. The physiological role of these channels in arterial smooth muscle is at present unclear. Activation of these channels leads to membrane depolarization as result of an increase in Na^+ permeability. However, since these channels have not been identified on the molecular or single-channel level in most types of arterial smooth muscle, they will not be discussed. Four distinct types of potassium channels have been identified in the cell membrane of arterial smooth muscle cells. (1) Voltage-dependent K+ (K_V) channels increase their activity with membrane depolarization and are important regulators of smooth muscle membrane potential under physiological conditions, as well as in response to depolarizing stimuli. This type of potassium channel has been found in every type of smooth muscle that has been studied. (2) Ca^{2+}-activated K+ (K_{Ca}) channels respond to changes in intracellular Ca^{2+} to regulate membrane potential and play an important role in the control of myogenic tone in small arteries. These channels may also play a role in the repolarization after an action potential in other types of smooth muscle. (3) ATP-sensitive K+ (K_{ATP}) channels respond to changes in cellular metabolism and are targets of a variety of vasodilating and relaxing stimuli in many types of smooth muscle. (4) Inward rectifier K+ (K_{IR}) channels regulate membrane potential in smooth muscle cells from several types of resistance arteries and may be responsible for elevated external K+-induced vasodilations in these arteries.

In vascular smooth muscle, membrane potential is regulated by potassium channels and is an important determinant of arterial tone, and hence arterial diameter. A number of endogenous vasodilators [e.g., calcitonin gene-related peptide (CGRP) and adenosine] act in part through activation of K+ channels. Inhibition of K+ channels, which causes membrane potential depolarization, leads to vasoconstriction. A common feature of vasoconstrictors is that they cause membrane potential depolarization. Inhibition of K+ channels may therefore contribute to vasoconstrictor-induced membrane depolarization and the subsequent vasoconstriction. Defects in potassium channel function may lead to vasoconstriction or vasospasm as well as compromise the ability of an artery to dilate. Alteration in vascular smooth muscle K+ channel function may therefore be involved in pathological conditions of the vasculature such as vasospasm, hypertension, ischemia, hypotension during endotoxic shock, and changed vascular reactivity during diabetes.

This chapter will examine the properties (structure, distribution), pharmacology (openers and blockers as tools to study K+ channels) of calcium and potassium channels. Emphasis will be on integration of the information to develop a picture of the physiological roles of the different types of ion channels. Since potassium channels regulate arterial smooth muscle function by controlling membrane potential, the membrane potential and its relationship to smooth muscle function will be discussed first, as well as the role that K+ channels play in controlling membrane potential.

In each section on the different ion channels, some unresolved issues and future directions will be addressed. In general, little is known about the precise molecular structures of the ion channels (e.g., K+ channels) in smooth muscle and our knowledge of endogenous agents as well as key signal transduction pathways that may modulate smooth muscle ion channels is far from complete. Further, as indicated previously, the modulation and expression of ion channels vary with the type of, and even within (e.g., large versus small arteries), smooth muscle. Studies on K+ channels in nonvascular types of smooth muscle will be discussed if similar material from arterial smooth muscle is limited.

A. Measuring Ion Channels in Smooth Muscle

Our knowledge of ion channels in smooth muscle has increased significantly over the last 10 years, largely owing to two technical advances: (1) improved techniques to isolate single smooth muscle from native smooth muscle tissue (i.e., from small arteries, see

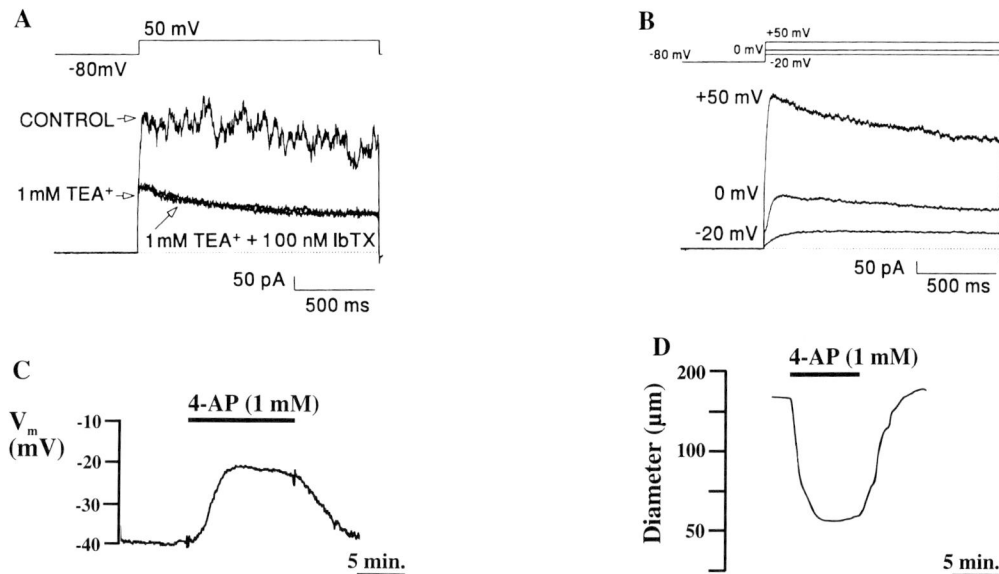

FIGURE 1 Illustration of electrophysiological properties, pharmacology, and physiology of voltage-dependent K+ channels [voltage-dependent K+ (K_V) channel and calcium-activated K+ (K_{Ca}) channel]. (A) Inhibition of K+ currents by tetraethylammonium (TEA+, 1 mM) and iberiotoxin (IbTX, 100 nM). Traces of whole-cell currents were elicited by stepping from a holding potential of −80 to +50 mV for 1.5 s in presence and absence of TEA+ or TEA+ + IbTX. The combination of TEA+ and IbTX inhibits the K_{Ca} component of the outward current. The residual currents are due to other voltage-dependent K+ (K_V) currents. Dotted line in this and all other current traces indicates zero current. (B) Three outward current traces elicited by depolarization steps from a holding potential of −80 to −20, 0, and +50 mV for 1.5 s, in presence of external TEA+ (1 mM) and nimodipine (1 μM). Bath (external) and pipette (internal) K+ were 140 and 5 mM, respectively. (Isolated arterial smooth muscle myocytes from rabbit basilar cerebral artery: panels A and B taken from Robertson and Nelson, 1994, with permission.) (C) Depolarization of a pressurized (80 mm Hg) rabbit middle cerebral artery through inhibition of voltage-dependent K+ channels by 4-aminopyridine (4-AP, 1 mM) bathed in physiological salt solution (PSS). (D) Vasoconstriction in the same intact pressurized artery by 4-AP, 1 mM [in PSS with TEA+ (1 mM) to diminish the contribution of K_{Ca} channels]. (Panel C adapted from Knot and Nelson, 1995.)

Quayle et al., 1993a,b) and (2) the patch-clamp technique, which permitted the measurement of channel currents in small cells such as arterial smooth muscle cells (see examples of whole-cell currents from single myocytes in Figs. 1A, 1B, 2A, and 2B). Using these techniques, investigators are now able to measure currents through voltage-dependent Ca^{2+} and K+ channels, K_{Ca} channels, K_{IR} channels, and K_{ATP} channels in single smooth muscle cells. On the intact tissue level, the functional manifestations of K+ channel activation or inhibition can be monitored as changes in membrane potential (measured with conventional microelectrodes; see Figs. 1C, 2C, and 2D), as changes in isometric force or isotonic displacement of strips or ring segments of smooth muscle, or in a more physiological sense as changes in the diameter of pressurized arteries (see Figs. 1D and 2E) (Duling et al., 1981). K+ channel activity has also been assessed by measuring isotopic K+ or rubidium efflux from intact segments of arteries.

B. Regulation of Vascular Arterial Tone by Membrane Potential

The membrane potential (V_m) of smooth muscle cells in the arterial wall appears to be an important regulator of vascular tone. The relationship between smooth muscle membrane potential and arterial tone is very steep, so that even membrane potential changes of a few millivolts cause significant changes in blood vessel diameter (Brayden and Nelson, 1992; Nelson et al., 1988, 1990a). Membrane potential changes would then act in concert with other mechanisms (e.g., changes in Ca^{2+} sensitivity of the contractile process, or Ca^{2+} release from the SR) to alter blood vessel diameter. Membrane potential of smooth muscle primarily regulates muscle contractility through alterations in calcium influx through voltage-dependent calcium channels (Nelson et al., 1990b). Membrane potential could also regulate Ca^{2+} entry through Na^+–Ca^{2+} exchange (Nelson et al., 1990b), if this transport

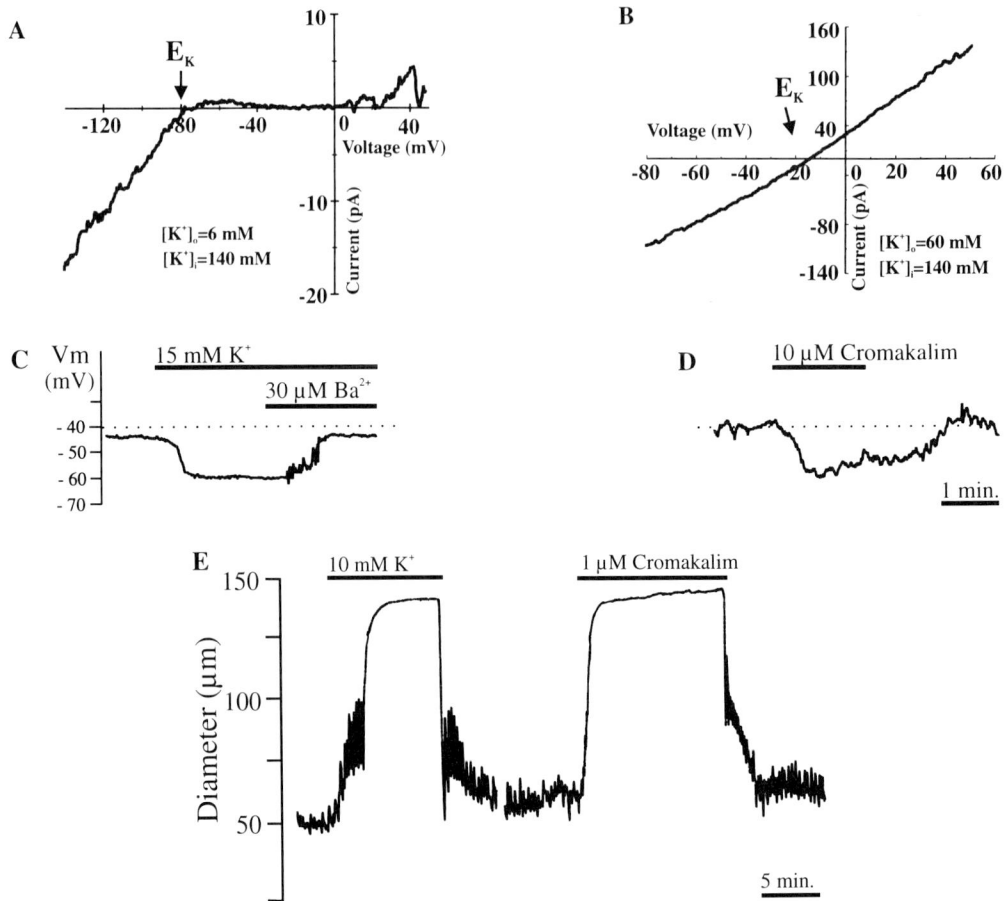

FIGURE 2 Illustration of electrophysiological properties, pharmacology, and physiology of the inward rectifier family of K^+ channels [ATP-sensitive K^+ (K_{ATP}) channel and inward rectifier K^+ (K_{IR}) channel]. (A) Current–voltage relationship of 0.5 mM barium-sensitive current in a human coronary artery (<40 μm diameter) myocyte under physiological K^+ conditions (E_K about -80 mV). Voltage ramps of 180 ms duration were applied from a holding potential of -140 to $+50$ mV without and with 0.5 mM Ba^{2+}. Currents before and after Ba^{2+} were subtracted to obtain a Ba^{2+}-sensitive current (A. D. Bonev, P. A. Zimmermann, L. A. Mulieri, H. J. Knot, and M. T. Nelson, unpublished results). This trace illustrates the typical voltage-dependent properties of an inward rectifier K^+ current. (B) Current–voltage relationship showing ionic selectivity and voltage dependence of the glibenclamide-sensitive current in a guinea pig urinary bladder smooth muscle cell. Voltage ramps of 250 ms duration were applied in presence of lemakalim (30 μM; a stereoisomer of cromakalim) without and with glibenclamide (10 μM). Currents before and after glibenclamide were subtracted to obtain a glibenclamide-sensitive current. Currents from three cells were averaged here. E_K is about -21 mV under these conditions of K^+ concentrations (taken from Bonev and Nelson, 1993a, with permission). (C) Elevated extracellular K^+ (15 mM) activates K_{IR} channels in pressurized (to 60 mm Hg) rat posterior cerebral arteries (ca. 200 μm diameter), leading to hyperpolarization close to E_K (E_K is about -61 mV under these conditions). The effect is reversed by low concentrations of extracellular barium (30 μM) (H. J. Knot and M. T. Nelson, unpublished results). (D) A pressurized (to 75 mm Hg) rabbit middle cerebral artery hyperpolarizes in response to the synthetic K_{ATP} channel opener cromakalim (10 μM). (E) Elevated extracellular potassium (10 mM) and cromakalim (1 μM) dilate a small (150 μm) pressurized (to 60 mm Hg) dog coronary artery, indicating that in this vessel both K_{IR} and K_{ATP} channels may be present (H. J. Knot, P. A. Zimmermann, and M. T. Nelson, unpublished results).

system is present in a given type of smooth muscle, as well as affect intracellular Ca^{2+} release through the voltage dependence of IP_3 production (Ganitkevich and Isenberg, 1993; Itoh *et al.*, 1992). The relationship between calcium influx through voltage-dependent calcium channels and membrane potential can be very steep, with 3 mV depolarization or hyperpolarization increasing or decreasing calcium influx as much as twofold (Nelson *et al.*, 1990b). Any physiological or pharmacological agent that alters membrane potential should cause a significant change in blood vessel diameter (see Figs. 1C, 2C, and 2D). Thus, it is not surprising that membrane hyperpolarization through activation of potassium channels is a powerful mechanism to lower blood pressure through vasodilation. For example, the synthetic K_{ATP} channel openers such as cromakalim and minoxidil hyperpolarize and dilate arteries (see Figs. 2D and 2E) (Standen et al., 1989). Similarly, elevation of external potassium from 5 to 15 mM activates K_{IR} channels in rat posterior cerebral arteries (ca. 200 μm in size) and hyperpolarizes these small arteries by almost 13 mV, causing a 40–50% increase in diameter (see also Figs. 2C and 2E) (Knot et al., 1994a). Further, inhibition of K^+ channels can cause vasoconstriction through membrane depolarization (see Figs. 1C and 1D). The K_{Ca} channel blocker, charybdotoxin, depolarized a pressurized (75 mm Hg) middle cerebral artery by 7 mV and constricted the artery by 45 μm (see Brayden and Nelson, 1992).

C. Regulation of Smooth Muscle Membrane Potential by Potassium Channels

Smooth muscle cells in arteries and arterioles, in vitro, have stable membrane potentials between −40 and −60 mV when subjected to normal levels of intravascular pressure (observe V_m in Figs. 1C, 2C, and 2D) (Brayden and Nelson, 1992; Harder, 1984; Knot and Nelson, 1995). Membrane potentials measured in vivo are in the range of −40 to −55 mV (Hirst and Edwards, 1989; Neild and Keef, 1985). Potassium permeability does not completely dominate the membrane conductance, since the membrane potential is considerably more positive than the potassium equilibrium potential (E_K is about −85 mV; see Fig. 2A). The membrane potential of arterial smooth muscle may be positive to E_K because the chloride conductance is relatively high (Hirst and Edwards, 1989). The estimated chloride equilibrium potential (E_{Cl}) is about −31 mV. As mentioned previously, small arteries have steady or slowly changing membrane potentials around −60 to −40 mV. With physiological extracellular K^+ (about 5 mM), the potassium equilibrium potential is approximately −85 mV and passive K^+ movement through any open K^+ channel will be out of the cell. The patch-clamp technique permits one to measure this movement (flux) of ions as whole-cell K^+ currents as well as single-channel current properties (see Figs. 1A, 1B, 2A, and 2B). In arterial smooth muscle cells that contain K_V, K_{Ca}, K_{IR}, and K_{ATP} channels, the estimates of channel number per cell range from about 100–500 per cell for K_{ATP} and K_{IR} channels to 1000–10,000 K_V and K_{Ca} channels per cell. K^+ channels in smooth muscle cells in intact arteries at physiological membrane po-

tentials have very low levels of activity (low P_o). However, very few K^+ channels need to be open to contribute to a cell's membrane potential because the resting input resistance of arterial smooth muscle cells is very high, on the order of 5–15 Gohm (see, e.g., Gelband and Hume, 1992; Quayle et al., 1993b; Smirnov and Aaronson, 1992), so that small changes in this activity can have a profound effect on membrane potential and ultimately on arterial diameter.

II. PROPERTIES OF CALCIUM AND POTASSIUM CHANNELS

A. Voltage-Dependent ("L-Type") Calcium Channels

Voltage-dependent Ca^{2+} channels exist in the plasma membrane of a wide variety of excitable cells, including neurons and heart, skeletal, and smooth muscle, and are among the best and most extensively studied ion channels (for reviews, see Tsien, 1983; Hess, 1990). They are responsible for the influx of calcium ions into the muscle cells, thereby regulating muscle contraction and secretion of neurotransmitters, as well as the electrical properties of many other excitable cells. Based on their pharmacology and different properties, a variety of classes of Ca^{2+} channels can be distinguished. However, the most important Ca^{2+} channel present in smooth muscle cells studied to date is the voltage-dependent ("L-type") Ca^{2+} channel. The structure of this channel is very similar to that of other voltage-dependent ion channels but it is highly selective for permeating calcium ions over all other ions, except in the absence of all divalent cations, in which case monovalent ions such as Na^+ can permeate the channel (Tsien et al., 1987). The voltage-dependent Ca^{2+} channel has been studied in cerebral arteries (Gollasch et al., 1992; Quayle et al., 1993a). The voltage-dependence of the L-type Ca^{2+} channel is very steep. The open-state probability (P_o) of this channel increases steeply with membrane depolarization (e-fold per 5 mV depolarization) (Nelson et al., 1990b). The structure of the smooth muscle "L-type" calcium channel is the topic of Chapter 17 in this volume and the interested reader is referred to that chapter. Calcium ions and the divalent barium and strontium ions can readily permeate the calcium channel. It has been proposed that the calcium channel has two divalent cation binding sites in the permeation pathway and that calcium binds with high affinity to these sites. The high flux of calcium ions through the pore of the channel is then accomplished by electrostatic repulsion between divalent cations in these two sites (Tsien et al., 1987).

B. Voltage-Dependent Potassium (K$_V$) Channels

Voltage-dependent potassium channels open when the membrane potential of the cell is depolarized (Hille, 1992). Potassium efflux through K$_V$ channels increases with depolarization through this voltage-dependent activation or increase in P_o (see Fig. 1B) (see also Nelson and Quayle, 1995, for more background information), and also because depolarization increases the electrochemical driving force for potassium efflux from the cell. Many K$_V$ channels also undergo inactivation following membrane depolarization. Because the inactivation process is slower than the activation process, current through K$_V$ channels in response to a depolarizing voltage step initially increases to a peak over time, due to voltage-dependent activation, and then decays, due to voltage-dependent inactivation. In the steady state, such as in tonic smooth muscle, current through K$_V$ channels will depend on the balance between channel activation and inactivation processes. In general, modulation of voltage-dependent channels by drugs may be accomplished by changing either the voltage-dependent activation or inactivation, both of which will affect the steady-state current.

Several families of voltage-dependent potassium channels have been cloned, and many are expressed in smooth muscle (see, e.g., Jan and Jan, 1992; Pongs, 1992; Roberds et al., 1993; Roberds and Tamkun, 1991; Overturf et al., 1994). The basic structure of K$_V$ channels appears to be well conserved between different families. Significant progress has been made in relating the primary structure of the molecule, as deduced from molecular cloning, to channel properties (for reviews, see Jan and Jan, 1992; Pongs, 1992). Each channel is probably composed of four subunits. Each subunit has six regions of hydrophobic amino acids (S1–S6), which are thought to form membrane-spanning domains, and these hydrophobic regions are linked by sequences of hydrophilic amino acids, which are exposed to the intracellular or extracellular space. Each subunit also has a cytoplasmic amino- and carboxy-terminal domain. One of the membrane-spanning domains in each subunit (the S4 region) is charged, having a basic amino acid (lysine or arginine) every third residue, and this S4 region is thought to be an important component of the channel voltage sensor. The permeation pathway is formed in part by the region linking the S5 and S6 transmembrane sequences, and this is called the H5 or pore region. Voltage-gated sodium and calcium channels are thought to have a basic architecture similar to that of K$_V$ channels, comprising four groups of six transmembrane domains, each group containing a region equivalent to S4 and H5. Unlike K$_V$ channels, which are thought to be tetramers of four subunits, each calcium or sodium channel is a single molecule rather than an aggregation of four subunits (Jan and Jan, 1992; Pongs, 1992).

All smooth muscle cells that have been investigated have at least one voltage-dependent potassium current activated by membrane depolarization. K$_V$ channels have been identified in single smooth muscle cells from coronary, cerebral (see Fig. 1 for a representative K$_V$ current), renal, mesenteric, and pulmonary artery, as well as tracheal smooth muscle (e.g., Beech and Bolton, 1989a,b; Boyle et al., 1992; Ishikawa et al., 1993; Okabe et al., 1987; Smirnov and Aaronson, 1992; Volk et al., 1991; Volk and Shibata, 1993). K$_V$ channels in vascular smooth muscle have been subclassified on the basis of voltage dependence and pharmacology. Different members of the K$_V$ channel family have also been called "delayed rectifier" and "transient outward currents" based on the kinetics of activation and inactivation (see earlier discussion). Different single-channel conductances have been reported for K$_V$ channels in vascular smooth muscle. Conductances in physiological potassium gradients fall broadly into two groups (4 to 6 mM [K$^+$]$_o$, using the cell-attached configuration of the patch-clamp technique). Small conductance channels are present in rabbit basilar artery (5.5 pS: Robertson and Nelson, 1994), rabbit coronary artery (7.3 pS: Volk and Shibata, 1993), and rabbit portal vein (5 and 8 pS: Beech and Bolton, 1989b). Larger single-channel conductances have been reported in rabbit coronary artery (70 pS with 140 mM [K$^+$]$_o$ and 140 mM [K$^+$]$_i$: Ishikawa et al., 1993) and in canine renal artery (57 pS with 5.4 mM [K$^+$]$_o$, cell attached) (Gelband and Hume, 1992). The estimated number of channels per cell varies, ranging from 750 channels in rabbit cerebral artery (Robertson and Nelson, 1994; Volk and Shibata, 1993) to 5000 channels in porcine coronary artery. The sensitivity of K$_V$ channels to membrane potential is central to their physiological role. As the membrane potential is depolarized, K$_V$ channels open and this may act to limit further membrane depolarization (see Fig. 1B for an illustration of the increase in current at more positive potentials) (Hirst and Edwards, 1989). Steady-state current through K$_V$ channels should be measurable in the physiological range of membrane potentials, and indeed steady-state, 4-aminopyridine-sensitive K$^+$ currents were measured (2.5 pA at -40 mV) in myocytes from rabbit cerebral arteries (Robertson and Nelson, 1994). Several lines of evidence suggest that there is more than one type of K$_V$ channel expressed in vascular smooth muscle cells based on differences in the voltage dependence of channel activation and inactivation, differences in sensitivity to inhibitors, and differences in single-channel conductances (Beech and Bolton, 1989a; Okabe *et al.*, 1987; Robertson and Nelson, 1994; Smirnov and Aaronson, 1992). One member of the K_V channel family, K_V 1.5, has been shown to be expressed in vascular smooth muscle cells (Overturf *et al.*, 1994; Roberds *et al.*, 1993; Roberds and Tamkun, 1991). The current was inhibited by 4-aminopyridine, with half-inhibition at a concentration of 211 μM. Single K_V 1.5 channels, expressed in *Xenopus* oocytes, had a slope conductance of 8.3 pS in symmetrical 140 mM [K^+] (Overturf *et al.*, 1994). Although the corresponding native channel is not known, these properties of the K_V 1.5 channels are similar to that of K_V channels described in cerebral and coronary arteries (see Figs. 1A and 1B taken from Robertson and Nelson, 1994; Volk *et al.*, 1991).

C. Calcium-Activated Potassium (K_{Ca}) Channels

Large-conductance potassium channels (with single-channel conductances of about 250 pS in symmetrical K^+) that are activated by intracellular calcium and membrane depolarization have been found in virtually every type of smooth muscle (see Nelson, 1993; Nelson and Quayle, 1995). These potassium channels have often been called "Big" K_{Ca} (BK) channels or "Maxi" K_{Ca} channels in contrast to the low-conductance "small" K_{Ca} (SK) channels that have also been described in smooth muscle. The open-state probability (P_o) of most K_{Ca} channels increases with membrane depolarization (2.7-fold increase per 12–14 mV depolarization) (Benham *et al.*, 1986; Langton *et al.*, 1991) and elevations of intracellular calcium through the physiological range (see Fig. 1A). Some small-conductance K_{Ca} channels, however, seem to be voltage independent. Based on these properties, K_{Ca} channels should be activated by the membrane depolarization and elevation in intracellular calcium that is associated with tone in smooth muscle, suggesting a role for these channels in regulating smooth muscle tone. A mammalian homologue (m*Slo*) of the *Drosophila* Slowpoke locus, which encodes a K_{Ca} channel, has been identified in mouse brain and skeletal muscle (Adelman *et al.*, 1992; Atkinson *et al.*, 1991; Butler *et al.*, 1993). K_{Ca} channels also share the basic structure of six membrane-spanning hydrophobic segments (S1–S6), which form the pore and contain the voltage sensor (S4), with voltage-dependent K^+ channels. However, in addition, K_{Ca} channels have four conserved segments (S7–S10) that may be involved in intracellular Ca^{2+} sensing. The single mammalian gene generates many variant K_{Ca} channel peptides, which may ex-

plain the large variation of Ca^{2+} sensitivity and conductances among K_{Ca} channels.

D. ATP-Sensitive Potassium (K_{ATP}) Channels

Potassium channels closed by intracellular adenosine-5'-triphosphate (called ATP-sensitive or ATP-dependent K^+ channels, often abbreviated as K_{ATP} channels) were first identified in cardiac muscle (Noma, 1983). Since then, they have been found in skeletal muscle, pancreatic β-cells, and certain types of neurons. K_{ATP} channels have been identified in smooth muscle (Nelson, 1993; Bonev and Nelson, 1993a; Standen *et al.*, 1989). K_{ATP} channels in arterial and urinary bladder smooth muscle are not voltage dependent (illustrated by the almost linear current–voltage relationship in Fig. 2B). The whole-cell current–voltage relationships for K_{ATP} currents activated by lemakalim or calcitonin gene-related peptide were essentially linear in symmetrical potassium (Bonev and Nelson, 1993a). Because of their ATP sensitivity, the K_{ATP} channels may respond to changes in the cellular metabolic state as well as to a number of endogenous and synthetic vasodilators. Whole-cell K_{ATP} channel currents have since then been measured in single smooth muscle cells from pulmonary arteries, renal artery, rat tail artery, canine coronary arteries (Kovacs and Nelson, 1991), and rat and rabbit mesenteric arteries (Standen *et al.*, 1989). Whole-cell K_{ATP} channel currents have also been measured in the following types of nonarterial smooth muscle: portal vein, urinary bladder (e.g., Fig. 2B taken from Bonev and Nelson, 1993a), and gallbladder (Zhang *et al.*, 1994a). There has been considerable variation in the single-channel conductances reported for K_{ATP} channels in smooth muscle, with the values falling into two groups. Small- to immediate-conductance (15–50 pS in symmetric high K^+) K_{ATP} channels have been identified in the cell membranes of smooth muscle cells from portal vein, cultured coronary artery cells, and urinary bladder (see Nelson and Quayle, 1995). Large-conductance K_{ATP} channels (130 pS in high K^+) have been found in smooth muscle cells from mesenteric arteries, rat tail artery, and canine aorta (see Nelson and Quayle, 1995). The significance of this heterogeneity of single-channel conductances remains unclear. The density of K_{ATP} channels in a single arterial smooth muscle cell appears to be low, in the range of 300–500 channels/cell or 1 channel per 10 square microns (assuming a constant single-channel conductance). Analysis of K_{ATP} currents in smooth muscle cells from pulmonary artery and urinary bladder suggested that these cells have approximately 300 and 425 channels, respectively, corresponding to about 1 channel per 10 square microns. The open-state probability of K_{ATP} channels in arterial smooth muscle is presumably quite low in the absence of activators of the channel and in the presence of physiological (millimolar) internal ATP. An estimate of activity of K_{ATP} channels in an intact cell from mesenteric arteries suggests that K_{ATP} channels in smooth muscle operate at low open-state probabilities to regulate smooth muscle membrane potential (<0.02 under control conditions and <0.10 when activated by 10 nM CGRP or 10 μM pinacidil). Therefore, the primary role of ATP may be to keep P_o low, against which other factors (e.g., vasodilators, protein kinases A and C, ADP) can regulate channel activity.

The structure of the smooth muscle K_{ATP} channel is not known, but this channel type has been cloned from heart muscle and appears to be a member of the inward rectifier K$^+$ channel family (Ashford et al., 1994).

E. Inward Rectifier Potassium (K_{IR}) Channels

Inward rectifier potassium (K_{IR}) channels are present in a variety of excitable and nonexcitable cells, including some arterial smooth muscle cells (e.g., human coronary artery, see Fig. 2A) (Edwards and Hirst, 1988; Edwards et al., 1988; Quayle et al., 1993b). The name of this channel comes from the observation that when membrane potential is controlled, for example, by voltage clamp of the cell, inward currents through the K_{IR} channel (movement of potassium ions from the extracellular solution into the cell) are larger than outward currents (see Fig. 2A). This is because the K_{IR} channel is activated by membrane hyperpolarization, in contrast to K_V and K_{Ca} channels, which are activated by membrane depolarization (see the foregoing). Although outward currents through the K_{IR} channel are small, in most physiological situations the cell membrane potential is positive to the potassium equilibrium potential (E_K). The inward rectifier K$^+$ channel will therefore normally conduct an outward, hyperpolarizing membrane current. From a physiological standpoint, these small outward currents are therefore of considerable interest. An inward rectifier potassium channel has been cloned from mouse macrophage (referred to as IRK1) (Kubo et al., 1993a). A G-protein-coupled muscarinic inward rectifier K$^+$ (GIRK1) channel has also been cloned from heart (Kubo et al., 1993b). The proposed structure of the K_{IR} (IRK1) channel has two membrane-spanning regions. In effect, K_{IR} channels lack the transmembrane domains equivalent to S1 to S4, which are present in voltage-dependent potassium channels. The linking region between the two transmembrane domains in the K_{IR} channel probably forms part of the channel pore, having a high degree of sequence homology to the proposed pore region which is well conserved across other potassium channel families (Jan and Jan, 1992; Kubo et al., 1993a; Pongs, 1992). It is not currently known how the gene products are organized into channels, for example, whether the functional channel is a tetramer, as is the case with K_V and K_{Ca} channels (see earlier discussion).

Smooth muscle inward rectifier potassium currents have only been identified in cerebral and mesenteric arterioles, and small coronary (see Fig. 2A) and cerebral arteries (200 μm in diameter) (Bonev et al., 1994; Quayle et al., 1993b). Earlier studies were conducted with single-microelectrode voltage clamp of intact segments of arteriole (Edwards and Hirst, 1988; Edwards et al., 1988). The K_{IR} channel in arterial smooth muscle, as in other cells, conducts inward current at membrane potentials negative to the potassium equilibrium potential, and smaller outward currents at membrane potentials positive to E_K. The activity of K_{IR} channels is a function of both membrane potential and of the extracellular potassium concentration (see Knot et al., 1994b). As $[K^+]_o$ is changed, the channel conducts inward current at membrane potentials negative to the new E_K (compare E_K in Figs. 2A and 2B under different conditions of $[K^+]_o$ as an illustration of how E_K shifts with $[K^+]_o$), while outward current remains small (about 1 pA). The control of channel activity by $[K^+]_o$ is a distinguishing feature of the K_{IR} channel and contrasts with the situation for other potassium channels, for example, K_{Ca} or K_V channels, where a change in $[K^+]_o$ will change the reversal potential of the current, but not the voltage range over which channels are active (see Knot et al., 1994b).

III. PHARMACOLOGY OF CALCIUM AND POTASSIUM CHANNELS IN SMOOTH MUSCLE

Inhibitors and openers are invaluable tools for studying the properties and physiological role of ion channels. Unfortunately, many of the drugs, with some exceptions, that we can use to block K$^+$ channels in order to study their function are nonselective. Table I provides a summary of inhibitors of arterial smooth muscle K$^+$ channels and their relative selectivity and potency.

A. Pharmacology of "L-Type" Calcium Channels

Several di- and trivalent cations such as cadmium, cobalt, nickel, and lanthanum block calcium channels

TABLE I Pharmacology of Inhibitors of Arterial Smooth Muscle Potassium Channels[a]

Type of ion channel	4-Amino-pyridine	Charybdotoxin or iberiotoxin	Tetraethyl-ammonium ions (TEA$^+$)	Glibenclamide	Barium ions
K_V	**K_I-200 μM–1.1 mM**	No effect	K_I-10mM	No effect at 10 μM	$K_I > 1$ mM
K_{Ca}	$K_I > 5$ mM	**K_I-1–10 nM**	**K_I-200 μM**	No effect at 10 μM	Small effect at 10 mM
K_{ATP}	K_I-200 μM	No effect	K_I-7 mM	**K_I-20–100 nM**	K_I-100 μM at −80 mV
K_{IR}	10% inhibition at 1 mM	No effect	No effect at 1 mM	No effect at 10 μM	**K_I-2 μM at −60 mV**

[a]Note. Barium block of K$^+$ channels is very voltage dependent. Block by charybdotoxin and iberiotoxin is also voltage dependent. TEA$^+$ block is voltage-dependent but not very steep. In all aforementioned cases the block increases with membrane hyperpolarization. The bold values indicate the relative selective concentration for the inhibitor of that type of potassium channel. K_I = half-inhibition constant. For references, see text.

by binding in the permeation pathway of the pore (see Tsien *et al.*, 1987). "Calcium channel antagonists" are a widely used class of drugs that act through inhibition of L-type Ca^{2+} channels. There are three classes of inhibitors: dihydropyridines (e.g., nifedipine, nisoldipine, and isradipine), phenylalkylamines (e.g., verapamil), and benzothiazepines (e.g., diltiazem). Dihydropyridine and verapamil inhibition of the L-type Ca^{2+} channel increases with membrane depolarization, which is thought to be the result of the preferential binding of these drugs to the inactivated and open states, respectively, of the channel (Nelson and Worley, 1989). Some dihydropyridines [e.g., Bay K8644, SDZ(+)202-791)] are potent activators of the L-type Ca^{2+} channel and are therefore referred to as "calcium channel agonists" (e.g., see Quayle *et al.*, 1993a; Knot *et al.*, 1991; Hof *et al.*, 1985).

B. Pharmacology of K_V Channels

4-Aminopyridine (4-AP) may be the most selective inhibitor of K_V channels in vascular smooth muscle (Table I). For this reason, 4-AP can be used to separate K_V currents from K_{Ca} currents, which are also activated by membrane depolarization (see Fig. 1A) (Beech and Bolton, 1989b; Gelband and Hume, 1992; Ishikawa *et al.*, 1993; Okabe *et al.*, 1987; Robertson and Nelson, 1994; Smirnov and Aaronson, 1992; Volk and Shibata, 1993). Measured half-inhibition constants (K_I) are 300 μM in rabbit pulmonary artery at +10 mV (Okabe *et al.*, 1987), about 1.1 mM in human mesenteric artery at +60 mV (Smirnov and Aaronson, 1992), and 700 μM in canine renal artery at +10 mV (Gelband and Hume, 1992). 4-AP does not affect K_{Ca} or K_{IR} channels in this concentration range, but may inhibit K_{ATP} currents to some extent (Quayle *et al.*, 1993b) (Table I).

Tetraethylammonium (TEA$^+$) ions inhibit K_V channels at higher concentrations than needed to inhibit K_{Ca} channels ($K_I > 5$ mM) (e.g., see Beech and Bolton, 1989b) (Table I).

C. Pharmacology of K_{Ca} Channels

Large conductance K_{Ca} channels are blocked by external tetraethylammonium ions (concentration for half-block K_I is about 200 μM), charybdotoxin (K_I = 10 nM), and iberiotoxin ($K_I < 10$ nM) (see also Fig. 1A and Table I) (e.g., Brayden and Nelson, 1992; Giangiacomo *et al.*, 1992; Langton *et al.*, 1991; Miller *et al.*, 1985). Charybdotoxin, a peptide produced by the scorpion *Leirus quinquestriatus*, and iberiotoxin from the venom of the scorpion *Buthus tamulus* block K_{Ca} channels in arterial smooth muscle. These peptides are thought to be highly selective for K_{Ca} channels. The large-conductance K_{Ca} channels in arterial smooth muscle are not inhibited by apamin, a peptide from bee venom, which blocks low-conductance calcium-activated potassium channels in other tissues (Blatz and Magleby, 1987) (Table I).

Certain synthetic potassium channel openers have been shown to activate K_{Ca} channels, such as NS-004 (Sargent *et al.*, 1993) and NS-1619. However, iberiotoxin was unable to reverse the functional effects of NS-004, and therefore its mechanism of action may also involve inhibition of calcium current rather than activation of K_{Ca} channels. Some drugs (e.g., cromakalim) that activate K_{ATP} channels have also been shown to activate K_{Ca} channels in aortic smooth muscle cells but not in other types of arterial smooth muscle. Since blockers of K_{Ca} channels (charybdotoxin and <1 mM TEA$^+$) had no effect on membrane potential hyperpolarizations or dilations to any of these drugs, K_{Ca} channels are unlikely to participate in the functional effects of these synthetic K^+ channel openers.

D. Pharmacology of K_{ATP} Channels

K_{ATP} channels in smooth muscle are inhibited by the antidiabetic sulfonylurea drugs glibenclamide and tolbutamide and by external barium (Table I; see also Fig. 2B). Glibenclamide inhibits whole-cell K_{ATP} currents in coronary and mesenteric arteries and in portal vein, with a half-inhibition constant between 20 and 200 nM. Arterial relaxations to synthetic K_{ATP} channel openers such as cromakalim, pinacidil, minoxidil sulfate, and RP49356 are half-inhibited by 72 to 148 nM glibenclamide (Meisheri et al., 1993). K_{ATP} channels are not blocked by inhibitors of K_{Ca} channels such as charybdotoxin and iberiotoxin, and are relatively insensitive to external TEA^+ (half-block constant is 7 mM) (see Nelson, 1993; Standen et al., 1989) (Table I). External barium also blocks K_{ATP} channels in smooth muscle, in a voltage-dependent manner, with 100 μM causing half-block at −80 mV (Quayle et al., 1988). Glibenclamide appears to be selective for K_{ATP} channels and does not block a variety of other ion channels, including Ca^{2+} channels, K_{IR}, K_V, and K_{Ca} channels (Table I). Therefore, glibenclamide has become an important tool in the investigation of the physiological role of K_{ATP} in intact preparations.

A number of antihypertensive drugs act through potassium channel activation and have collectively been called potassium channel openers. This class of antihypertensive drugs includes compounds that have been in use for some time as antihypertensives (e.g. minoxidil sulfate and diazoxide) and newer drugs such a pinacidil, nicorandil, cromakalim (see Figs. 2D and 2E), and RP49356. Vasodilation to all of these compounds is blocked by glibenclamide (Meisheri et al., 1993). The actions of these vasodilators are not affected by blockers of K_{Ca} channels (charybdotoxin, iberiotoxin, and low concentrations of TEA^+) (see Nelson and Quayle, 1995) (Table I). Cromakalim and pinacidil have been shown to directly activate K_{ATP} channels in vascular smooth muscle (see Nelson and Quayle, 1995) and cardiac muscle (Ashcroft and Ashcroft, 1990). These studies led to the proposal that K_{ATP} channels and not K_{Ca} channels are the targets of K^+ channels openers such as diazoxide, cromakalim, and pinacidil.

E. Pharmacology of K_{IR} Channels

The K_{IR} channel in arterial smooth muscle is very sensitive to inhibition by extracellular barium ions,

with a dissociation constant in the micromolar range (Bonev et al., 1994; Quayle et al., 1993b). Inhibition by barium ions is greater at more negative membrane potentials (see also Fig. 2C). The dissociation constant for barium block is an exponential function of membrane potential, decreasing e-fold for a 24-mV hyperpolarization, from 8 μM at −40 mV to 0.6 μM at −100 mV (Quayle et al., 1993b). Extracellular barium ions block other potassium channels, but at much higher concentrations than required to inhibit the K_{IR} channel (Table I), and barium is therefore a useful tool for investigating the physiological role of this channel. Inhibitors of other potassium channels present in arterial smooth muscle are not effective inhibitors of the K_{IR} potassium current. 4-Aminopyridine (1 mM) and glibenclamide (10 μM), which block voltage-activated and ATP-sensitive potassium channels, respectively, have little effect on the K_{IR} current in cerebral arteries. TEA^+ ions (1 mM) and charybdotoxin (100 nM), which inhibit calcium-activated potassium channels, have no effect on the K_{IR} current (Quayle et al., 1993b) (Table I).

IV. PHYSIOLOGY OF CALCIUM AND POTASSIUM CHANNELS IN SMOOTH MUSCLE

A. Physiological Role of "L-Type" Calcium Channels

Influx of extracellular Ca^{2+} through voltage-dependent calcium channels regulates contraction of smooth muscle. Under physiological conditions the single-channel current through the calcium channel has been studied in small cerebral arteries from the rat (Rubart et al., 1995). Under these conditions the single-channel current at −40 mV membrane potential is about −0.2 pA (1 pA corresponds to about 3 million Ca^{2+} ions per second) and the single-channel conductance was calculated to be 3.5 pS over the membrane potential range of −70 to −20 mV (Rubart et al., 1995). In the same study, these investigators showed that, on average, only one to five channels (from an estimated 3000 or more present in these cells) are open simultaneously in a single smooth muscle cell at physiological membrane potentials (Rubart et al., 1995). Other investigators have provided evidence that 0.5 pA of steady Ca^{2+} current across the membrane of a single cell is sufficient to elevate intracellular Ca^{2+} (Fleischmann and Kotlikoff, 1995). These data together suggest that only a few open Ca^{2+} channels may be sufficient to maintain the contractile state of smooth muscle under tonic (steady-state) conditions. The best established mechanism of modulation of the L-type Ca^{2+} channel is that by cAMP-dependent protein kinase in cardiac myocytes (Sperelakis and Schneider,

1976). It is not clear whether L-type Ca^{2+} channels in other tissues such as smooth muscle are also affected by conditions in which cAMP is elevated. Other modulatory pathways that have been proposed are (1) both activation and inhibition of the L-type Ca^{2+} channel in cardiac and smooth muscle cells by protein kinase C (PKC) and (2) direct activation of the skeletal muscle and cardiac L-type Ca^{2+} channel by a GTP binding protein (G_s-α). The smooth muscle L-type Ca^{2+} channel can be modulated by agonists such as in bronchial and vascular smooth muscle (Nelson et al., 1988; Worley et al., 1991; Kamishima et al., 1992). Modulation of the smooth muscle L-type Ca^{2+} channel remains an important subject of future research.

B. Physiological Roles of K_V Channels

Despite the wide distribution of K_V channels, few studies have been conducted on the physiological role of the channel in arterial smooth muscle. Because the channel is activated by depolarization, it may be involved in action potential repolarization in electrically excitable smooth muscle preparations such as the portal vein, and this is a principal function of the channel in other excitable cells, including neurons and cardiac muscle (Beech and Bolton, 1989b; Hille, 1992). However, many smooth muscle cells such as arterial and tracheal smooth muscle cells do not generate action potentials but respond to stimulation with graded membrane potential changes. Under those conditions, voltage-dependent potassium channels provide an important potassium conductance in the physiological membrane potential range (Fleischmann et al., 1993). As expected under these conditions, inhibition of K_V channels by 4-aminopyridine depolarizes and constricts many arteries (see Figs. 1C and 1D) (Knot and Nelson, 1995). Therefore, K_V channels play an important role in the regulation of membrane potential and may be directly modulated by vasoconstrictors and vasodilators. In accordance with this is the observation that a 4-AP-sensitive potassium current is inhibited by a vasoconstricting histamine H_1–receptor agonist in coronary arteries (Ishikawa et al., 1993).

Pulmonary arteries constrict in hypoxia, which minimizes blood perfusion in poorly ventilated areas of the lung. This hypoxic vasoconstriction contrasts with the hypoxic vasodilation seen in many small systemic arteries, which have been linked to activation of K_{ATP} channels (see the following). During hypoxia, pulmonary arteries depolarize and may even generate action potentials. The resulting pulmonary vasoconstriction is abolished by removal of extracellular calcium and by calcium channel antagonists such as verapamil, illustrating that calcium entry through

voltage-dependent calcium channels is important in the hypoxic response (McMurtry et al., 1976). Some studies suggest a role for potassium channels in hypoxia-induced membrane depolarization and constriction (e.g., Post et al., 1992; Smirnov et al., 1994). Consistent with a mechanism involving inhibition of potassium channels is the finding that K^+ channel inhibitors such as TEA^+ ions and 4-aminopyridine increase tone in isolated pulmonary vessels, and increase perfusion pressure in the isolated perfused lung. Thus, potassium channels contribute to the membrane potential in pulmonary arteries as they do in systemic arteries. Hypoxia may thus depolarize by inhibiting potassium channels. Hypoxia inhibited a voltage-activated potassium current in these arteries and the voltage dependence of the hypoxia-sensitive channel points toward a member of the K_V or K_{Ca} families (Post et al., 1992).

1. Future Directions

K_V currents in vascular smooth muscle cells have diverse properties. The existence of multiple K_V currents within a given artery, or of the differential distribution between different vascular beds, is unknown. More understanding of the properties and distribution of individual K_V channels may shed some light on their physiological role, and remains an important goal for future research. Inhibition of K_V channels may be involved in hypoxia and vasoconstrictor-induced membrane depolarization. The predicted modulation of K_V channels by vasodilators and vasoconstrictors will also be an important area of future investigation.

C. Physiological Roles of K_{Ca} Channels

Elevation of intravascular pressure depolarizes smooth muscle cells in resistance arteries, raises intracellular Ca^{2+}, and causes vasoconstriction. This partially constricted state has been referred to as "myogenic tone" and is a major contributor to peripheral resistance. Based on the properties of K_{Ca} channels, pressure-induced membrane depolarization and the resulting increase in intracellular calcium should activate K_{Ca} channels (Brayden and Nelson, 1992; Nelson, 1993). Activation of K_{Ca} channels, under these conditions, increases potassium efflux, which would counteract the depolarization and constriction caused by pressure and vasoconstrictors. Consistent with this mechanism is the observation that lowering intravascular pressure or blocking calcium channels, which would decrease intracellular calcium and thus decrease active tone, greatly attenuated the effects of K_{Ca} channel blockers on membrane potential and arterial tone (Brayden and Nelson, 1992; see also Knot and

Nelson, 1995). K_{Ca} channels may therefore serve a dynamic role in the control of arterial smooth muscle membrane potential by serving as a negative feedback pathway to regulate the degree of membrane depolarization and hence vasoconstriction caused by pressure as well as other vasoconstrictors (Brayden and Nelson, 1992; Ishikawa et al., 1993). This "feedback" mechanism, aimed at limiting the depolarization and constriction, would predict that under steady-state conditions in a vessel with tone, blockers of K_{Ca} channels should depolarize and constrict arteries with tone. This is, indeed, the case, for TEA^+, charybdotoxin, and iberiotoxin depolarize and constrict myogenic cerebral and coronary arteries (Brayden and Nelson, 1992; Knot and Nelson, 1995). The important role that K_{Ca} channels play in the control of myogenic tone makes them an ideal target for modulation through vasodilator and vasoconstrictor substances. This hypothesis is strengthened by recent evidence that a number of vasoactive substances may act in part through modulation of K_{Ca} channels (see the following).

1. Regulation of K_{Ca} Channels by Endogenous Vasoactive Substances

Most vasoconstrictors (e.g., norepinephrine, angiotensin II, endothelin, and serotonin) depolarize vascular smooth muscle (Nelson et al., 1990b). It is conceivable that inhibition of K_{Ca} channels contributes to this membrane depolarization. Angiotensin II and a thromboxane A2 agonist (U46619) have been shown to inhibit K_{Ca} channels from coronary artery smooth muscle (Scornik and Toro, 1992; Toro et al., 1990). Similarly, muscarinic receptor stimulation has been shown to inhibit K_{Ca} channels in airway and colonic smooth muscle (Cole et al., 1989). Activation of K_{Ca} channels would hyperpolarize smooth muscle and lead to muscle relaxation. Consistent with this is the finding that stimulation of β_2-adrenergic receptors activates K_{Ca} channels in airway smooth muscle cells. Thus, K_{Ca} channels may contribute to β-adrenergic bronchodilation. The activation of K_{Ca} channels in airway and coronary artery smooth muscle cells appears to be mediated by cAMP-dependent protein kinase phosphorylation as well as a direct G-protein pathway (Scornik et al., 1993). Vasorelaxation of some vascular beds (e.g., mesenteric) in response to nitric oxide appears to involve activation of K_{Ca} channels (Khan et al., 1993). New evidence indicates that cyclic guanosine monophosphate (cGMP)-dependent protein kinase can activate K_{Ca} channels in smooth muscle cells isolated from cerebral and coronary arteries (Robertson et al., 1993). This may be an important mechanism by which the potent endothelium-derived vasodilator nitric oxide

(NO), which activates cGMP-dependent protein kinase through stimulation of guanylyl cyclase and elevation of cGMP, relaxes arteries. Nitric oxide has also been reported to activate K_{Ca} channels directly in aortic smooth muscle (Bolotina et al., 1994), as well as activate K_{ATP} channels in rabbit mesenteric arteries (Murphy and Brayden, 1995).

Membrane fatty acids and membrane stretch influence the activity of single K_{Ca} channels in pulmonary smooth muscle cells and pulmonary arteries. Both effects appear to be direct (i.e., independent of changes in intracellular Ca^{2+} or of channel phosphorylation) (Kirber et al., 1992).

2. Future Directions

The molecular structure of the K_{Ca} channel in smooth muscle is not known, although based on functional similarities, it would be expected to be a member of the mSlo family. K_{Ca} channels seem to integrate many vasoactive signals that activate many important signal transduction pathways. The relative importance and regulation of K_{Ca} channels through cGMP-protein kinase, cAMP-protein kinase, or direct G-protein activation pathways will be the subject of future research.

D. Physiological Roles of K_{ATP} Channels

The K_{ATP} channel is activated by a number of endogenous vasodilators [CGRP and vasointestinal peptide (VIP)], and the associated membrane hyperpolarization causes part of the resulting vasodilation to these compounds (Standen et al., 1989). The K_{ATP} channel may also be inhibited by vasoconstrictors, which would tend to cause depolarization and constriction. The channel is involved in the metabolic regulation of blood flow; it is activated in conditions of increased blood demand, for example, in hypoxia, either by release of vasodilators from the surrounding tissue or as a direct result of hypoxia on the vascular smooth muscle cells. Finally, the channel may be active in the resting state, as inhibition of K_{ATP} channels can lead to increased resistance to blood flow in some vascular beds. In many cases the presence of K_{ATP} channels has been implicated through the ability of the glibenclamide to reverse a particular functional effect.

ATP inhibits and ADP increases K_{ATP} currents (Ashcroft and Ashcroft, 1990). Information about intracellular ATP inhibition of K_{ATP} channels in arterial smooth muscle is limited. ATP inhibited single K_{ATP} channels from aortic smooth muscle and portal vein with half-inhibition constants of about 30–40 μM (Kovacs and Nelson, 1991). Whole-cell K_{ATP} currents in canine coronary artery are inhibited by ATP with a K_I

of about 350 μM. ADP or other nucleotide diphosphates can activate the K_{ATP} channel. Based on these findings, it has been suggested that nucleotide diphosphates have a primary role in regulating the level of K_{ATP} channel activity in portal vein (see Beech and Bolton, 1989a; Nelson and Quayle, 1995). However, the physiological roles of changes in intracellular nucleotide diphosphate concentrations, as of other possible coregulators of K_{ATP} channel function such as pH, are unknown.

Hypoxia decreases resistance to blood flow in many vascular beds. This could be a compensatory mechanism to increase blood flow to the hypoxic region. In the coronary, cerebral, renal, and skeletal muscle circulations this hypoxic vasodilation is attenuated by glibenclamide, suggesting a role for K_{ATP} channels in the response (von Beckerath et al., 1991; Daut et al., 1990). Hypoxia also causes the release of adenosine from cardiac myocytes. Adenosine in a potent dilator of coronary arteries. It has been shown that adenosine activates K_{ATP} currents in single coronary artery smooth muscle cells (Dart and Standen, 1993, 1994). Activation of the K_{ATP} channel is also involved in the hypoxic vasodilation in the cerebral, renal (Loutzenhiser and Parker, 1994), skeletal muscle, and cremaster muscle and cheek pouch circulations (see Nelson and Quayle, 1995). Activation of K_{ATP} channels in these vascular beds may be a direct consequence of hypoxia on a smooth muscle oxygen sensor, an effect of hypoxia on smooth muscle cell metabolism, or through the release of vasodilator metabolites like adenosine from surrounding tissue, similar to hypoxic coronary vasodilation.

Recent evidence supports the idea that K_{ATP} channels may play a role in the maintenance of tone in certain vascular beds, such as the coronary circulation, mesenteric arteries, and arterioles in the hamster cheek pouch and cremaster muscles (when equilibrated with 0% O_2), based on the observations that glibenclamide depolarizes and constricts these arteries, or increases resistance to blood flow in perfused vascular beds [e.g., glibenclamide causes a significant membrane depolarization (ca. 5–9 mV) (Nelson et al., 1990a) of mesenteric arteries]. In contrast to the coronary and mesenteric circulations, glibenclamide (or tolbutamide) does not affect basal tone in the renal, cerebral (Brayden, 1990), and pulmonary arteries. However, in these vascular beds, K_{ATP} channels may still regulate tone during hypoxic conditions (see the foregoing). For example, glibenclamide has no effect on pulmonary perfusion pressure in either normoxia or moderate hypoxia. However, in sustained and severe pulmonary hypoxia, this initial hypoxic pulmonary vasoconstriction is followed by a decrease in

pulmonary arterial tone, which is prevented by glibenclamide (Wiener et al., 1991). Thus, activation of the K_{ATP} channels may underlie pulmonary vasorelaxation in severe hypoxia (see Nelson and Quayle, 1995).

Reactive hyperemia is the increased blood flow over normal values that is seen following a period of interruption of flow. Reactive hyperemia is attenuated by glibenclamide in both coronary and skeletal muscle vascular beds, suggesting a role of K_{ATP} channels in this phenomenon (Clayton et al., 1992). Activation of K_{ATP} channels under these conditions may occur through a buildup of vasodilator metabolites (e.g., adenosine) during the period of occlusion.

Autoregulation of blood flow is the ability of a vascular bed to maintain flow in response to changes in perfusion pressure. Inhibition of K_{ATP} channels by glibenclamide disrupts coronary and cerebral autoregulation (see Nelson and Quayle, 1995). In the cerebral circulation, this may be a consequence of the release of the vasodilator calcitonin gene-related peptide from sensory neurons in the vessel wall during a fall in perfusion pressure, since CGRP dilates cerebral arteries by activating K_{ATP} channels.

Recent evidence suggests that excessive activation of K_{ATP} channels may contribute to the precipitous life-threatening drop in blood pressure observed during endotoxic shock. This result can be reconciled with the observations that elevated levels of calcitonin gene-related peptide and elevated nitric oxide production are associated with the hypotension observed during sepsis, if both CGRP and NO act in part through activation of K_{ATP} channels. Lactic acidosis also leads to a glibenclamide-sensitive fall in blood pressure, suggesting a possible role of metabolic compromise in the endotoxic shock-induced hypotension. This interesting hypothesis remains to be examined.

1. K_{ATP} Channels as Targets of Endogenous Vasodilators and Vasoconstrictors

A number of endogenous vasodilators act at least in part through membrane hyperpolarization caused by K^+ channel activation (Nelson, 1993). Evidence suggests that many of these vasodilators hyperpolarize by activating K_{ATP} channels in arterial smooth muscle (Standen et al., 1989). However, it should be emphasized that these vasodilators can also relax arteries through mechanisms independent of K^+ channel activation. Calcitonin gene-related peptide has been shown to activate K_{ATP} channels (mesenteric arteries and gallbladder, see Nelson and Quayle, 1995), causing glibenclamide-sensitive membrane potential hyperpolarizations and dilations. CGRP may be involved in glibenclamide-sensitive cerebral autoregulation. Adenosine has also been widely demonstrated to activate K_{ATP} channels and to cause glibenclamide-sensitive hyperpolarizations of intact arteries (see Nelson and Quayle, 1995, for references).

The activation of K_{ATP} channels to many vasodilators seems to be mediated through stimulation of protein kinase A. Consistent with this is the fact that most vasodilators that cause glibenclamide-sensitive relaxation (e.g., CGRP, VIP, adenosine) increase cAMP levels in smooth muscle cells through activation of adenylate cyclase. Also, forskolin, a direct activator of adenylate cyclase, causes glibenclamide-sensitive membrane hyperpolarizations in smooth muscle cells from mesenteric arteries and gallbladder (Nelson et al., 1990a; Zhang et al., 1994b).

It has also been shown that atrial natriuretic factor (ANP), nitric oxide, and isosorbide dinitrite, which dilate arteries and increase intracellular cGMP levels, activate K_{ATP} channels suggesting the possibility that either elevation of cGMP or stimulation of cGMP-dependent protein kinase could activate K_{ATP} channels in smooth muscle (e.g., see Murphy and Brayden, 1995).

Angiotensin II, vasopressin, and endothelin inhibit K_{ATP} channels in cultured coronary artery smooth muscle cells (e.g., Miyoshi et al., 1992). Muscarinic receptor stimulation inhibited K_{ATP} channels in smooth muscle cells in urinary bladder through stimulation of protein kinase C (Bonev and Nelson, 1993b). Serotonin, phenylephrine, histamine, and neuropeptide Y also inhibit K_{ATP} currents in smooth muscle cells from mesenteric arteries through stimulation of protein kinase C (Bonev and Nelson, 1995).

2. Future Directions

An important feature of K_{ATP} channels in smooth muscle is regulation by a number of signal transduction pathways and by metabolism. Future work will be directed toward the mechanisms of and relationships between these modulatory pathways and the possible role of this channel in disease (e.g., ischemia and diabetes).

E. Physiological Roles of K_{IR} Channels

The physiological roles of the inward rectifier K^+ channel in cells other than smooth muscle include regulating the resting membrane potential, preventing membrane hyperpolarization to values more negative than the potassium equilibrium potential by the electrogenic Na^+/K^+-ATPase, and minimizing cellular potassium loss, and therefore energy expenditure, during sustained membrane depolarization. The roles of the K_{IR} channel in arterial smooth muscle are incompletely understood, but may include some of the fol-

lowing functions. The K_{IR} channel is more active at negative membrane potentials and is therefore a candidate for regulating the membrane potential of smooth muscle cells in arteries in the absence of extrinsic influences that depolarize (e.g., pressure or vasoconstrictors; see also Fig. 2A). The membrane potential of arterial smooth muscle cells at low transmural pressure (<20 mm Hg) measured in vitro lies in the range of -75 to -60 mV (Edwards and Hirst, 1988; Nelson et al., 1990b). In nonpressurized rat cerebral arterioles ($[K^+]_o = 10$ mM), barium ions (0.1 to 0.5 mM) depolarize and inhibit K_{IR} currents (see Figs. 2C and 2E) (Edwards et al., 1988; see also Fig. 3 in Edwards and Hirst, 1988). However, 50 μM barium has little effect on membrane potential in rabbit middle cerebral artery (Brayden, 1990), or on diameter in rat posterior cerebral arteries pressurized to half the mean systolic blood pressure (McCarron and Halpern, 1990a). Clearly, factors such as membrane potential, transmural pressure, vasoconstrictors, and vasodilators, as well as possible differences in channel distribution and density, are likely to influence the contribution of the K_{IR} channel to the membrane potential.

Vasodilation in response to elevated extracellular potassium is a prominent response in small cerebral and coronary vascular beds under physiological and pathophysiological conditions. However, the mechanism by which extracellular potassium dilates arteries has been unknown. Blood flow within many organs is linked to metabolic demand. This is thought to result in part from release of vasodilator metabolites from surrounding tissue. Several candidate vasodilators have been proposed, including protons, adenosine, and potassium ions. In the cerebral circulation, an increase in the extracellular potassium concentration in the range of 1 to 15 mM causes small arteries to dilate (see Fig. 2E) (e.g., Kuchinsky et al., 1972; McCarron and Halpern, 1990a). Potassium ions are released as a consequence of neuronal activity, and may be one of the factors that increase local cerebral blood flow during increased cerebral activity (Kuchinsky et al., 1972; see also Newman, 1986; Paulson and Newman, 1987). Extracellular potassium concentration increases during cerebral hypoxia, ischemia, or hypoglycemia from around 3 mM to values in excess of 10 mM (e.g., Sieber et al., 1993). Smaller increases in $[K^+]_o$ also occur during physiological changes in neuronal activity (Somjen, 1979).

Extracellular potassium dilates arteries by at least two mechanisms: activation of the electrogenic Na^+/K^+ pump and activation of inward rectifier potassium channels (Edwards et al., 1988; McCarron and Halpern, 1990a). The role of the K_{IR} channel in potassium-induced dilations was first proposed in the cerebral circulation, where several lines of evidence support this hypothesis (Edwards *et al.*, 1988). An increase in extracellular potassium concentration from 5 to 10 mM hyperpolarizes rat middle cerebral and posterior arterioles (see Fig. 2C) (Edwards *et al.*, 1988; Knot *et al.*, 1994a). This hyperpolarization is prevented by barium ions, which inhibit the K_{IR} channel in this artery (Edwards *et al.*, 1988; Knot *et al.*, 1994a) (see Fig. 2C). An increase in $[K^+]_o$ in this range also dilates pressurized rat cerebral arteries (McCarron and Halpern, 1990a; Knot *et al.*, 1994a) (see Fig. 2E). These dilations are prevented by <10 μM barium, but not by endothelium removal, by inhibitors of other potassium channels, or by a number of receptor antagonists (Knot *et al.*, 1994a). Barium ions inhibit K_{IR} currents and K^+-induced hyperpolarizations in rat cerebral arteries at a concentration of <10 μM (Edwards *et al.*, 1988; Quayle *et al.*, 1993b; Knot *et al.*, 1994a), negative to −40 mV (Table I). These observations are consistent with the hypothesis that potassium ions dilate cerebral arteries by activating the K_{IR} channel. A similar mechanism for K-induced dilation has been identified in rat and canine coronary arteries (Knot *et al.*, 1994a,b) (see Fig. 2E). An increase in the extracellular potassium ion concentration in the 1 to 5 mM range causes a transient dilation of small cerebral arteries (McCarron and Halpern, 1990a). This dilation is abolished by ouabain but not by barium ions, suggesting that over this concentration range, potassium ions dilate arteries by activating the Na^+/K^+ pump (McCarron and Halpern, 1990a).

1. Future Directions

The molecular structure of the smooth muscle K_{IR} channel is unknown, although, based on its properties, it is likely to be a member of the IRK family. Inward rectifier potassium currents have been identified in small cerebral, coronary (see Fig. 2A), and mesenteric arterioles (<200 μm diameter). The presence of K_{IR} channels may be a common feature in small arteries that determine in large part peripheral vascular resistance. It is not known if K_{IR} channels are exclusively present in small arteries, although they have not yet been reported in larger vessels. Innervation of this size artery (<100–200 μm) is usually sparse and therefore small arteries may be more prone to respond to metabolic demand from the tissue, as reflected by potassium efflux. Thus, the appearance of K_{IR} channels may reflect a transition of blood flow regulation to local (tissue) control. This is an intriguing hypothesis that remains to be tested. It will be important to study systematically the distribution of the K_{IR} channel within a vascular bed, as this may have physiological consequences. K_{IR} channels, like K_{Ca} and K_{ATP} channels,

may also be modulated by important vasoactive substances; this remains to be determined. K_{IR} channel properties may also be altered in disease, since small cerebral arteries from hypertensive rats lose their ability to dilate to external K^+ (McCarron and Halpern, 1990b). The role of this channel type in disease (e.g., cerebral and cardiac ischemia) will undoubtedly be an important area of future investigation.

V. CONCLUSIONS

The main conclusions of this chapter are: (1) Voltage-dependent Ca^{2+} channels play an important role in the excitation–contraction coupling of smooth muscle by responding to changes in membrane potential. (2) Regulation of smooth muscle membrane potential through activation or inhibition of K^+ channel activity and subsequent changes in the activity of voltage-dependent Ca^{2+} channels provide an important mechanism to relax (dilate) or contract (constrict) smooth muscle (arteries). (3) Voltage-dependent K^+ channels, K_{Ca} channels, K_{IR} channels, and K_{ATP} channels serve unique functions in the regulation of (arterial) smooth muscle membrane potential and tone. (4) K^+ channels integrate a variety of constricting and relaxing signals through regulation of the membrane potential. (5) Many of the functional and molecular aspects of the regulation of these ion channels in smooth muscle remain to be investigated.

Acknowledgments

Sponsored by the NIH (HL44455 and HL51728 to M.T.N. HL35911 to J.E.B.), NSF (DCB-9019563 to M.T.N.), and American Heart Association Vermont Affiliate (to H.J.K).

References

Adelman, J. P., Shen, K.-Z., Kavanaugh, M. P., Warren, R. A., Wu, Y.-N., Lagrutta, A., Bond, C. T., and North, R. A. (1992). *Neuron* **9**, 209–216.

Arden, W. A., Fiscus, R. R., Wang, X., Yang, L., Maley, R., Nielsen, M., Lanzo, Sh., and Gross, D. R. (1994). *Circ. Shock* **42**, 147–153.

Ashcroft, S. J. H., and Ashcroft, F. M. (1990). *Cell Signal.* **2**, 197–214.

Ashford, M. L. J., Bond, C. T., Blair, T. A., and Adelman, J. P. (1994). *Nature (London)* **370**, 456–459.

Atkinson, N. S., Robertson, G. A., and Ganetzky, B. (1991). *Science* **253**, 551–555.

Bayliss, W. M. (1902). *J. Physiol. (London)* **28**, 220–231.

Beech, D. J., and Bolton, T. B. (1989a). *J. Physiol. (London)* **412**, 397–414.

Beech, D. J., and Bolton, T. B. (1989b). *J. Physiol. (London)* **418**, 293–309.

Benham, C. D., Bolton, T. B., Lang, R. J., and Takewaki, T. (1985). *Pflügers Arch.* **403**, 120–127.

Benham, C. D., Bolton, T. B., Lang, R. J., and Takewaki, T. (1986). *J. Physiol. (London)* **371**, 45–67.

Blatz, A. L., and Magleby, K. L. (1987). *Trends Neurosci.* **10**, 463–467.

Bolotina, V. M., Najibi, S., Palacino, J. J., Pagano, P. J., and Cohen, R. A. (1994). *Nature (London)* **368**, 850–853.

Bonev, A. D., and Nelson, M. T. (1993a). *Am. J. Physiol.* **264**, C1190–C1200.

Bonev, A. D., and Nelson, M. T. (1993b). *Am. J. Physiol.* **265**, C1723–C1728.

Bonev, A. D., and Nelson, M. T. (1995). *Biophys. J.* **68**, A46.

Bonev, A. D., Robertson, B. E., and Nelson, M. T. (1994). *Biophys. J.* **66**, A327.

Boyle, J. P., Tomasic, M., and Kotlikoff, M. I. (1992). *J. Physiol. (London)* **447**, 329–350.

Brayden, J. E. (1990). *Am. J. Physiol.* **259**, H668–H673.

Brayden, J. E., and Nelson, M. T. (1992). *Science* **256**, 532–535.

Butler, A., Tsunoda, S., McCobb, D. P., Wei, A., and Salkoff, L. (1993). *Science* **261**, 221–224.

Clapp, L. H., Gurney, A. M., Standen, N. B., and Langton, P. D. (1994). *J. Membr. Biol.* **140**, 205–213.

Clayton, F. C., Hess, T. A., Smith, M. A., and Grover, G. J. (1992). *Pharmacology* **44**, 92–100.

Cole, W. C., Carl, A., and Sanders, K. M. (1989). *Am. J. Physiol.* **257**, C481–C487.

Conway, M., Nelson, M. T., and Brayden, J. E. (1994). *Am. J. Physiol.* **266**, H1322–H1326.

Dart, C., and Standen, N. B. (1993). *J. Physiol. (London)* **471**, 767–786.

Dart, C., and Standen, N. B. (1994). *J. Physiol. (London)* **483**, 29–39.

Daut, J., Maier-Rudolph, W., von Becherath, N., Mehrke, G., Günther, K., and Goedel-Meinen, L. (1990). *Science* **247**, 1341–1344.

Daut, J., Standen, N. B., and Nelson, M. T. (1994). *J. Cardiovasc. Electrophysiol.* **5**, 154–181.

Du, C., Carl, A., Smith, T. K., Khoyi, M. A., Sanders, K. M., and Keef, K. D. (1994). *J. Pharmacol. Exp. Ther.* **268**, 208–215.

Duling, B. R., Gore, R. W., Dacey, R. G., and Damon, D. N. (1981). *Am. J. Physiol.* **241**, H108–H116.

Edwards, F. R., and Hirst, G. D. S. (1988). *J. Physiol. (London)* **404**, 437–454.

Edwards, F. R., Hirst, G. D. S., and Silverberg, G. D. (1988). *J. Physiol. (London)* **404**, 455–466.

Fleischmann, B. K., and Kotlikoff, M. I. (1995). *Proc. Natl. Acad. Sci. U.S.A.* (in press).

Fleischmann, B. K., Washabau, R. J., and Kotlikoff, M. I. (1993). *J. Physiol. (London)* **469**, 625–638.

Ganitkevich, V. Y., and Isenberg, G. (1993). *J. Physiol. (London)* **470**, 35–44.

Gelband, C. H., and Hume, J. R. (1992). *Circ. Res.* **71**, 745–758.

Gelband, C. H., Ishikawa, T., Post, J. M., Keef, K. D., and Hume, J. R. (1993). *Circ. Res.* **73**, 24–34.

Giangiacomo, K. M., Garcia, M. L., and McManus, O. B. (1992). *Biochemistry* **31**, 6719–6727.

Gollasch, M., Hescheler, J., Quayle, J. M., Patlak, J. B., and Nelson, M. T. (1992). *Am. J. Physiol.* **262**, C948–C952.

Harder, D. R. (1984). *Circ. Res.* **55**, 197–202.

Hess, P. (1990). *Annu. Rev. Neurosci.* **13**, 337–356.

Hille, B. (1992). "Ionic Channels of Excitable Membranes." Sinauer Assoc., Sunderland, MA.

Hirst, G. D. S., and Edwards, F. R. (1989). *Physiol. Rev.* **69**, 546–604.

Hof, R. P., Rüegg, U. T., Hof, A., and Vogel, A. (1985). *J. Cardiovasc. Pharmacol.* **7**, 689–693.

Huang, Y., Quayle, J. M., Worley, J. F., Standen, N. B., and Nelson, M. T. (1989). *Biophys. J.* **56**, 1023–1028.

Ishikawa, T., Hume, J. R., and Keef, K. D. (1993). *J. Physiol. (London)* **468**, 379–400.

Itoh, T., Seki, N., Suzuki, S., Ito, S., Kajikura, J., and Kuriyama, H. (1992). *J. Physiol. (London)* **451**, 307–328.

Jan, L. Y., and Jan, Y. N. (1992). *Annu. Rev. Physiol.* **54**, 537–555.

Jones, T. R., Charette, L., Garcia, M. L., and Kaczorowski, G. J. (1990). *J. Pharmacol. Exp. Ther.* **255**, 697–705.

Kamishima, T., Nelson, M. T., and Patlak, J. B. (1992). *Am. J. Physiol.* **263**, C69–C77.

Khan, S. A., Mathews, W. R., and Meisheri, K. D. (1993). *J. Pharmacol. Exp. Ther.* **267**, 1327–1335.

Kirber, M. T., Ordway, R. W., Clapp, L. H., Walsh, J. V., and Singer, S. S. (1992). *FEBS Lett.* **297**, 24–28.

Knot, H. J., de Ree, M. M., Gähwiler, B. H., and Rüegg, U. T. (1991). *J. Cardiovasc. Pharmacol.* **18** (suppl. 10), S7–S14.

Knot, H. J., and Nelson, M. T. (1995). *Am. J. Physiol.* **269**, H348–H355.

Knot, H. J., Zimmermann, P. A., and Nelson, M. T. (1994a). *Biophys. J.* **66**, A144.

Knot, H. J., Quayle, J. M., Zimmermann, P. A., McCarron, J. G., Brayden, J. E., and Nelson, M. T. (1994b). *In* "The Resistance Arteries: Integration of the Regulatory Pathways" (W. Halpern, ed.), pp. 93–101. Humana Press, Totowa, NJ.

Kovacs, R., and Nelson, M. T. (1991). *Am. J. Physiol.* **261**, H604–H609.

Kubo, Y., Baldwin, T. J., Jan, Y. N., and Jan, L. Y. (1993a). *Nature (London)* **362**, 127–133.

Kubo, Y., Reuveny, E., Slesinger, P. A., Jan, Y. N., and Jan, L. Y. (1993b). *Nature (London)* **364**, 802–806.

Kuchinsky, W., Wahl, M., Bosse, O., and Thurau, K. (1972). *Circ. Res.* **31**, 240–247.

Langton, P. D., Nelson, M. T., Huang, Y., and Standen, N. B. (1991). *Am. J. Physiol.* **260**, H927–H934.

Loutzenhiser, R. D., and Parker, M. J. (1994). *J. Vasc. Res.* **31**, 30.

Matsuda, H., Saigusa, A., and Irisawa, H. (1987). *Nature (London)* **325**, 156–159.

McCarron, J. G., and Halpern, W. (1990a). *Am. J. Physiol.* **259**, H902–H908.

McCarron, J. G., and Halpern, W. (1990b). *Circ. Res.* **67**, 1035–1039.

McCarron, J. G., Quayle, J. M., Halpern, W., and Nelson, M. T. (1991). *Am. J. Physiol.* **261**, H287–H291.

McMurtry, I. F., Davidson, A. B., Reeves, J. T., and Grover, R. F. (1976). *Circ. Res.* **38**, 99–104.

Meisheri, K. D., Kahn, S. A., and Martin, J. L. (1993). *J. Vasc. Res.* **30**, 2–12.

Miller, C., Moczydlowski, E., Latorre, R., and Phillips, M. (1985). *Nature (London)* **313**, 316–318.

Miyoshi, Y., Nakaya, Y., Wakatsuki, T., Kakaya, S., Fujino, K., Saito, K., and Inoue, I. (1992). *Circ. Res.* **70**, 612–616.

Murphy, M. E., and Brayden, J. E. (1995). *J. Physiol. (London)* **486**, 47–58.

Neild, T. O., and Keef, K. (1985). *Microvasc. Res.* **30**, 19–28.

Nelson, M. T. (1993). *Trends Cardiovasc. Med.* **3**, 54–60.

Nelson, M. T., and Quayle, J. M. (1995). *Am. J. Physiol.* **268**, C799–C822.

Nelson, M. T., and Worley, J. F. (1989). *J. Physiol. (London)* **412**, 65–91.

Nelson, M. T., Standen, N. B., Brayden, J. E., and Worley, J. F. (1988). *Nature (London)* **336**, 382–385.

Nelson, M. T., Huang, Y., Brayden, J. E., Hescheler, J., and Standen, N. B. (1990a). *Nature (London)* **344**, 770–773.

Nelson, M. T., Patlak, J. B., Worley, J. F., and Standen, N. B. (1990b). *Am. J. Physiol.* **259**, C3–C18.

Newman, E. A. (1986). *Science* **233**, 453–454.

Nichols, C. G., and Lederer, W. J. (1991). *Am. J. Physiol.* **261**, H1675–H1686.

Noma, A. (1983). *Nature (London)* **305**, 147–148.

Okabe, K., Kitamura, K., and Kuriyama, H. (1987). *Pflügers Arch.* **409**, 561–568.

Overturf, K. E., Russell, S. N., Carl, A., Hart, P. J., Hume, J. R., Sanders, K. M., and Horowitz, B. (1994). *Am. J. Physiol.* **267**, C1231–C1238.

Pardo, L. A., Heinemann, S. H., Terlau, H., Ludewig, U., Lorra, C., Pongs, O., and Stuhmer, W. (1992). *Proc. Natl. Acad. Sci. U.S.A.* **89,** 2466–2470.

Paulson, O. B., and Newman, E. A. (1987). *Science* **237,** 896–898.

Pongs, O. (1992). *Trends Pharmacol. Sci.* **13,** 359–365.

Post, J., Hume, J., Archer, S. L., and Weir, E. K. (1992). *Am. J. Physiol.* **262,** C882–C890.

Quayle, J. M., Standen, N. B., and Stanfield, P. R. (1988). *J. Physiol. (London)* **405,** 677–697.

Quayle, J. M., McCarron, J. G., Asbury, J. R., and Nelson, M. T. (1993a). *Am. J. Physiol.* **264,** H470–H478.

Quayle, J. M., McCarron, J. G., Brayden, J. E., and Nelson, M. T. (1993b). *Am. J. Physiol.* **265,** C1363–C1370.

Roberds, S. L., and Tamkun, M. M. (1991). *Proc. Natl. Acad. Sci. U.S.A.* **88,** 1798–1802.

Roberds, S. L., Knoth, K. M., Po, S., Blair, T. A., Bennet, P. B., Hartshorne, R. P., Snyders, D. J., and Tamkun, M. M. (1993). *J. Cardiovasc. Electrophysiol.* **4,** 68–80.

Robertson, B. E., and Nelson, M. T. (1994). *Am. J. Physiol.* **267,** C1589–C1597.

Robertson, B. E., Schubert, R., Hescheler, J., and Nelson, M. T. (1993). *Am. J. Physiol.* **265,** C299–C303.

Rubart, M., Patlak, J., and Nelson, M. T. (1995). *Biophys. J.* **68,** A348.

Sargent, C. A., Grover, G. J., Antonaccio, M. J., and McCullough, J. R. (1993). *J. Pharmacol. Exp. Ther.* **266,** 1422–1429.

Scornik, F. S., and Toro, L. (1992). *Am. J. Physiol.* **262,** C708–C713.

Scornik, F. S., Codina, J., Birnbaumer, L., and Toro, L. (1993). *Am. J. Physiol.* **265,** H1460–H1465.

Sieber, F. E., Wilson, D. A., Hanley, D. F., and Trystman, R. J. (1993). *Am. J. Physiol.* **264,** H1774–H1780.

Smirnov, S. V., and Aaronson, P. I. (1992). *J. Physiol. (London)* **457,** 431–454.

Smirnov, S. V., Robertson, T. P., Ward, J. P. T., and Aaronson, P. I. (1994). *Am. J. Physiol.* **266,** H365–H370.

Somjen, G. G. (1979). *Annu. Rev. Physiol.* **41,** 159–177.

Sperelakis, N., and Schneider, J. (1976). *Am. J. Cardiol.* **37,** 1079–1085.

Standen, N. B., Quayle, J. M., Davies, N. W., Brayden, J. E., Huang, Y., and Nelson, M. T. (1989). *Science* **245,** 177–180.

Stanfield, P. R., Davies, N. W., Shelton, P. A., Khan, I. A., Brammar, W. J., Standen, N. B., and Conley, E. C. (1994). *J. Physiol. (London)* **475,** 1–7.

Toro, L., Amador, M., and Stefani, E. (1990). *Am. J. Physiol.* **258,** H912–H915.

Toro, L., Vaca, L., and Stefani, E. (1991). *Am. J. Physiol.* **260,** H1779–H1789.

Tsien, R. W. (1983). *Annu. Rev. Physiol.* **61,** 687–708.

Tsien, R. W., Hess, P., McCleskey, E. W., and Rosenberg, R. L. (1987). *Annu. Rev. Biophys. Biophys. Chem.* **16,** 265–290.

Volk, K. A., and Shibata, E. F. (1993). *Am. J. Physiol.* **264,** H1146–H1153.

Volk, K. A., Matsuda, J. J., and Shibata, E. F. (1991). *J. Physiol. (London)* **439,** 751–768.

von Beckerath, N., Cyrys, S., Dischner, A., and Daut, J. (1991). *J. Physiol. (London)* **442,** 297–319.

Wiener, C. M., Dunn, A., and Sylvester, J. T. (1991). *J. Clin. Invest.* **88,** 500–504.

Worley, J. F., Quayle, J. M., Standen, N. B., and Nelson, M. T. (1991). *Am. J. Physiol.* **261,** H1951–H1960.

Zhang, L., Bonev, A., Nelson, M. T., and Mawe, G. (1994a). *J. Physiol. (London)* **478,** 483–491.

Zhang, L., Bonev, A., Mawe, G., and Nelson, M. T. (1994b). *Am. J. Physiol.* **267,** G494–G499.

17

Molecular Biology and Expression of Smooth Muscle L-Type Calcium Channels

FRANZ HOFMANN and NORBERT KLUGBAUER

*Institut für Pharmakologie und Toxikologie der Technischen Universität München,
München, Germany*

I. INTRODUCTION

The rise and fall in cytosolic calcium are the principal mechanisms that initiate contraction and relaxation of vascular and other smooth muscles. The cytosolic calcium concentration is regulated by the release and reuptake of calcium from/into intracellular stores and by its flux across the plasma membrane. A large part of the calcium influx is controlled by voltage-dependent opening and closing of calcium channels. Smooth muscle cells contain two types of voltage-dependent calcium channels, the T- and L-type channels, which are the products of different genes (Hofmann *et al.*, 1994).

The T-type or low-voltage-activated calcium channels have been identified in a variety of smooth muscle cells. They activate at low membrane potentials (around -50 mV) with a maximum around -20 mV, have a small conductance (7–8 pS with 110 mM Ba^{2+} as charge carrier), inactivate rapidly, and are blocked by 10–100 μM Ni^{2+} and a variety of compounds, including some of the organic calcium channel blockers. The biological role of the T-type channels remains to be established owing to the lack of specific blockers.

L-type or high-voltage-activated calcium channels carry the majority of the calcium inward current in smooth muscle cells. They start to activate at a high membrane potential (around -30 mV) with a maximum at slightly positive membrane potentials (around $+10$ mV), have a large conductance (20–25 pS with 110 mM Ba^{2+} as charge carrier), inactivate slowly, and are readily and specifically blocked by the classic organic calcium channel blockers nifedipine (a 1,4-dihydropyridine), verapamil (a phenylalkylamine), and diltiazem (a benzothiazepine) (see Hofmann *et al.*, 1994).

II. SUBUNIT STRUCTURE AND GENES OF THE HIGH-VOLTAGE-ACTIVATED CALCIUM CHANNEL

A. Subunit Composition

The high-voltage-activated calcium channels are a complex of several proteins, which have been purified from skeletal muscle, heart, and brain but not from other tissues. The purified complex is composed of at least four proteins (Fig. 1): the α_1 subunit, which contains the binding sites for all known calcium channel blockers and the ion-conducting pore, and the intracellularly located β subunit and the α_2/δ subunit, a disulfide-linked dimer. The primary structure of these proteins has been deduced from their cloned cDNAs. The skeletal muscle calcium channel complex contains another protein, the transmembrane γ subunit, which has not been identified in other tissues so far. The α_2/δ subunit is highly conserved in most tissues, including brain, skeletal, cardiac, and smooth muscle, indicating that the high-voltage-activated calcium channels of these tissues are hetero-oligomers formed from a common α_2/δ but from different α_1 and β subunits.

B. The α_1 Subunit Genes

The α_1 subunit is the main subunit of the calcium channel and contains the ion-conducting pore, the selectivity filter of the pore, and the sites for the inorganic and organic calcium channel blockers. Six different genes (classes A, B, C, D, E, and S) have been identified for the α_1 subunit (see also Hofmann *et al.*, 1994). They encode polypeptides of predicted molecular masses of 212 to 273 kDa, which are 41 to 70%

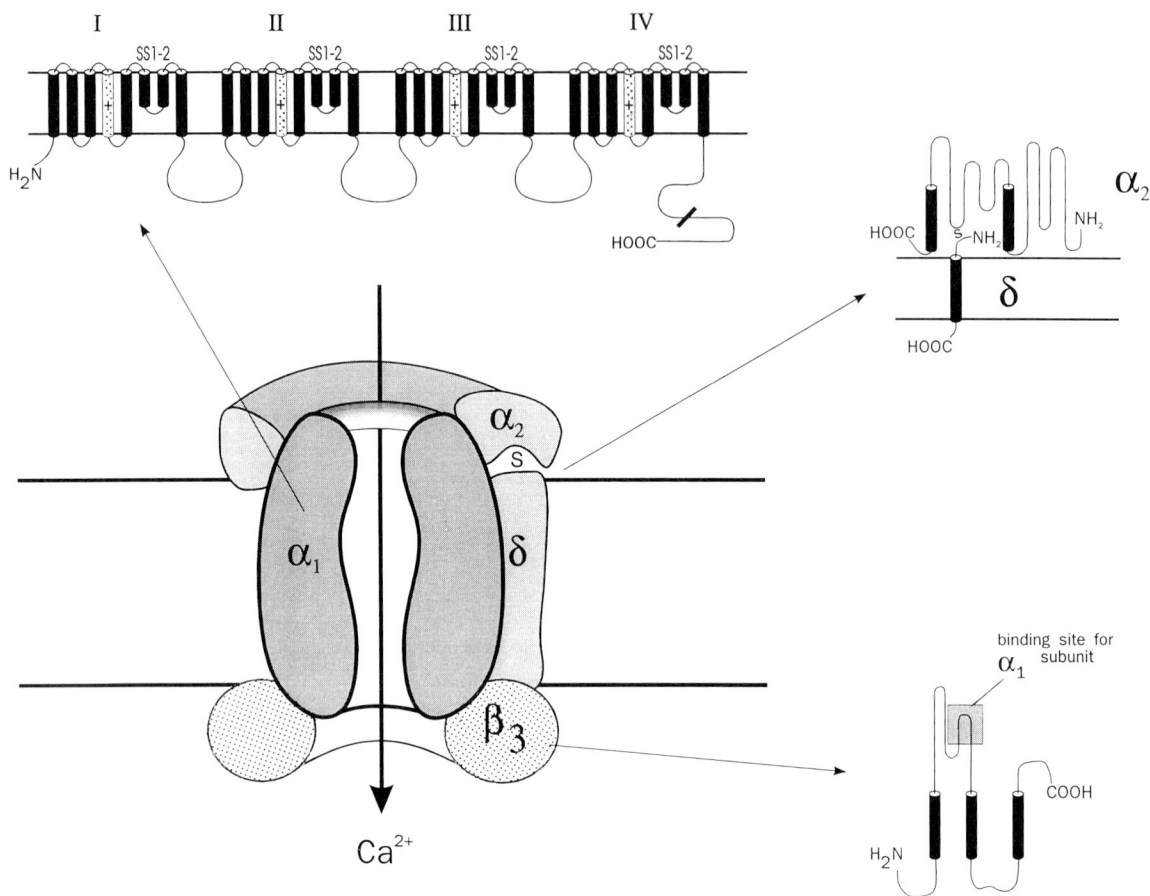

FIGURE 1   Proposed subunit composition of the smooth muscle L-type calcium channel. This channel is a three-subunit oligomer (α_1, α_2/δ, β). The putative membrane configuration of individual subunits is taken from the hydropathicity analysis of the primary sequences. I, II, III, IV, repeats of the α_1 subunit; +, proposed transmembrane amphiphilic segment (S) 4; SS1–2, pore-forming region. In some tissues, the α_1 subunit is processed after translation at the COOH terminus. The approximate location of the proteolytic cleavage is indicated by "S," the disulfide bridge between the transmembrane δ and the extracellularly located α_2 protein. The extracellular space is above the horizontal lines.

homologous to each other and are structurally similar to voltage-activated sodium channels. Complete cDNA clones of the α_1 subunits have been isolated from a variety of tissues, including heart (Mikami *et al.*, 1989; Schultz *et al.*, 1993) and smooth muscle (Biel *et al.*, 1990; Koch *et al.*, 1990).

The α_1 subunit of the L-type calcium channel is encoded by the S, C, and D genes. Skeletal muscle expresses the S gene, and heart, smooth muscle, and brain the C gene. The human C gene has been located to the distal region of chromosome 12p13 (Sun *et al.*, 1992; Schultz *et al.*, 1993). The gene contains several alternative used exons (Soldatov, 1994). The translated α_{1S} and α_{1C} subunits are processed proteolytically at the carboxy terminus by about 150 amino acids. The cDNA of a third gene (D) was isolated from neural and endocrine tissues and represents a neuroendocrine-specific L-type calcium channel. The gene products of

the A, B, and E genes have been found predominantly in brain, but they may be expressed in other tissues. The α_1 subunit cDNA of the A, B, and E genes induce high-voltage-activated calcium currents that are insensitive to nifedipine. The current induced by the A gene product is similar to that of a P-type channel (Mori *et al.*, 1991), whereas the B gene encodes an N-type channel (Dubel *et al.*, 1992; Williams *et al.*, 1992). The cDNA of the E gene induces currents of a high-voltage-activated calcium channel, which shares some similarities with a low-voltage-activated calcium channel (Soong *et al.*, 1993).

C. The α_2/δ Subunit Gene

The α_2/δ subunit is a glycosylated membrane protein of 125,018 Da, which is processed post-translationally by proteolysis, resulting in an α_2 protein containing amino acids 1 through 934 and a δ protein containing amino acids 935 through 1080 (De Jongh *et al.*, 1990). The transmembrane δ subunit anchors the extracellularly located α_2 protein by disulfide bridges to the plasma membrane (Jay *et al.*, 1991). The α_2/δ subunit is apparently a highly conserved protein. Brain also expresses a splice variant that encodes an identical δ but a slightly different α_2 protein.

D. The β Subunit Genes

Four β subunit genes (β_1–β_4) have been identified (Ruth *et al.*, 1989; Hullin *et al.*, 1992; Castellano *et al.*, 1993), each of which gives rise to several splice variants. The skeletal muscle β_1 subunit is the prototype protein. It is an intracellularly located membrane protein consisting of 524 amino acids (Ruth *et al.*, 1989), which contains some features of a cytoskeletal protein and is phosphorylated by cAMP kinase *in vitro*. The β_1 subunit is present mainly in skeletal muscle and brain. The cDNAs for the β subunit genes 2 and 3 have been isolated from a rabbit heart library (Hullin *et al.*, 1992). Like the β_1 gene, the β_2 and β_3 genes are tissue specifically expressed with transcripts of β_2 existing abundantly in heart and to a lower degree in aorta, trachea, and lung, whereas transcripts of the β_3 gene are expressed in brain and smooth muscle containing tissues such as aorta, trachea, and lung. This suggests that the β_3 gene product may be expressed predominantly in neuronal and smooth muscle cells. Western blots indicate that different splice variants of the β_3 subunit are expressed in brain and smooth muscle. The β_2 subunit is present abundantly in heart from various species. An additional β subunit (β_4) has been cloned from rat

brain (Castellano *et al.*, 1993). The β_4 mRNA has been found in brain and kidney.

III. THE SMOOTH MUSCLE L-TYPE CALCIUM CHANNEL α_1 SUBUNIT

Northern and immunological data indicate that the smooth muscle L-type calcium channel is a complex of the α_{1C}, α_2/δ, and β_3 subunits. The α_1 subunit is derived from the class C gene and has a predicted molecular mass of 242,516 Da. The smooth (Cb) and the cardiac (Ca) muscle α_1 subunits are splice products of the same gene. One major difference between the two sequences is an alternative used exon for the IVS3 segment. Some evidence suggests that the IVS3 segment of the Cb splice variant predominates in adult animals (Diebold *et al.*, 1992; Feron *et al.*, 1994). The transiently or stably expressed Ca and Cb α_1 subunits (Bosse *et al.*, 1992; Welling *et al.*, 1993a) yield no major differences in their basic electrophysiological properties, including the amplitude of inward current, steady-state activation, and inactivation. However, the Cb isoform is more sensitive to blockade by the dihydropyridine nisoldipine (Welling *et al.*, 1993a). Northern blots and polymerase chain reaction (PCR) analysis show that both splice variants are differentially expressed in heart and smooth muscle (Biel *et al.*, 1991).

A. The Voltage-Sensing Parts

Hydropathicity analysis of all known α_1 subunits predicts four repeats, which are 45 to 60% homologous (Fig. 2). Each repeat is composed of five trans-

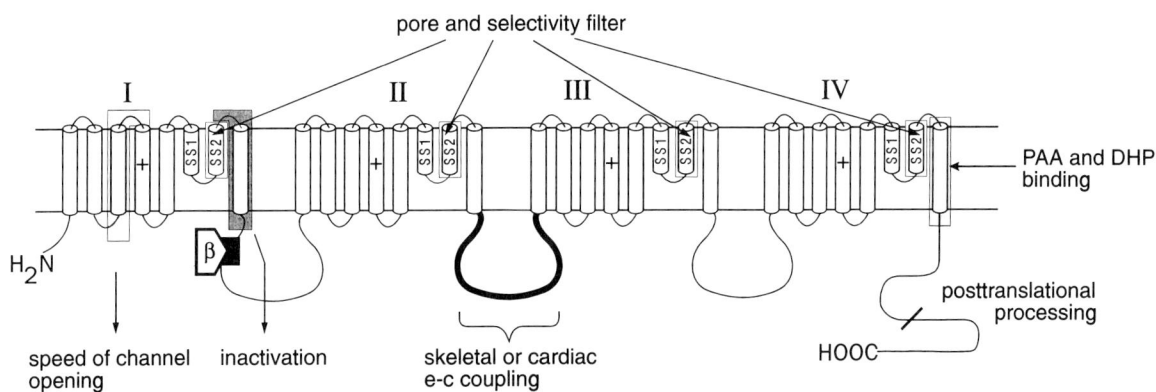

FIGURE 2   Identified structural features of the α_{1C} subunit of the L-type calcium channel. The structure of the α_1 subunit has been derived from the hydropathicity analysis of the primary sequence and includes four homologous repeats (I, II, III, IV) containing six membrane-spanning segments. +, proposed transmembrane amphiphilic segment (S) 4, which is the voltage-sensing device of the channel. The SS1–SS2 region is part of the channel pore and contains the selectivity filter; DHP and PAA, dihydropyridine and phenylalkylamine binding sites; e–c coupling, excitation–contraction coupling. The shaded areas indicate parts of the protein that are shown to be responsible for certain properties of the channel. β, binding site for the β subunits. The extracellular space is above the horizontal lines.

membrane α-helices and one amphiphilic segment, S4, which contains a positively charged residue at every third position and usually hydrophobic residues at the remaining positions. The positively charged S4 segment has been postulated to act as the voltage-sensing device responsible for channel gating. It is thought that S4 responds to a change in the membrane potential with a slight shift of its positive charges and thereby induces a conformational change in the protein that leads to channel opening. This hypothesis is supported by several mutagenesis experiments in sodium and potassium channels where a change of positive-charged residues to noncharged residues alters the voltage dependency and/or shifts the voltage range of channel activation. Repeat 1 determines the speed of channel activation. Replacement of the segment from IS3 to the beginning of IS4 with the segment from the skeletal muscle α_{1S} subunit slows the activation of the channel (Nakai *et al.*, 1994). The change in activation time is voltage independent, suggesting that this region is involved in the speed of pore opening and not in the gating of the channel. Additional chimeras indicate that the region from IS5 to IS6 and the adjacent intracellular amino acids are crucially involved in voltage-dependent inactivation of the channel (Zhang *et al.*, 1994).

B. The Selectivity Filter and Pore Region

The two short amino acid sequences located between S5 and S6, referred to as SS1 and SS2, span part of the membrane and form part of the channel pore (Guy and Conti, 1990). The glutamic acid residues that occur at equivalent positions in SS2 of the four repeats of all calcium channels (Fig. 2), but not in sodium and potassium channels, are critical in determining the ion selectivity of the calcium channel and constitute part of the channel's selectivity filter. Mutation of these glutamates alters the monovalent/divalent selectivity of the channel, as well as the affinity of blockade by divalent ions (Yang *et al.*, 1993). These mutations increase the concentration of cadmium or calcium needed to block lithium currents through the channel. Each glutamate contributes differently to the calcium selectivity of the channel. This result has led to the interesting hypothesis that the broken symmetry of the glutamates may be functionally important for calcium permeation. By misaligning the four glutamates, the calcium binding site may become more spread out in space and more fuzzy in its logic, being able to bind one calcium ion with high affinity or two calcium ions with lower affinity. The high affinity binding of one calcium ion is required to block the permeation of monovalent ions, whereas low affinity binding of two calcium ions allows permeation of calcium itself.

C. Coupling to Intracellular Proteins

The β subunit of the calcium channel binds to the calcium channel α_1 subunit at the intracellular loop between repeats I and II (Pragnell *et al.*, 1994). The smooth muscle α_1 subunit contains an insert in this region that is not present in the cardiac α_1 subunit (Biel *et al.*, 1990). This insert may be necessary to interact efficiently with the β_3 subunit of smooth muscle. The region of the β subunit that interacts with the loop I–II of the α_1 subunit has been localized to a 30-amino acid domain between the amino acids 163 and 366 of the β_3 subunit (De Waard *et al.*, 1994).

Functional expression of chimeras of the skeletal and cardiac muscle α_1 subunit in skeletal muscle cells from mdg mice showed that the cytoplasmic loop between repeats II and III determines the type of excitation–contraction coupling (Fig. 2). The loop from the skeletal muscle calcium channel α_{1S} supports contraction in the absence of calcium influx, whereas the loop from the cardiac calcium channel α_{1C} subunit induces contraction only in the presence of calcium influx (Tanabe *et al.*, 1990). These results suggest that the loop between repeats II and III interacts with the ryanodine receptor in a tissue-specific manner.

D. The Receptor Sites for Calcium Channel Blockers

Binding studies with radiolabeled dihydropyridines demonstrated that the stably expressed α_{1S} and α_{1C} subunits alone contain the allosterically coupled binding sites for the known calcium channel blockers (Kim *et al.*, 1990; Bosse *et al.*, 1992). Photoaffinity labeling of the skeletal muscle α_{1S} subunit with ludopamil, a phenylalkylamine, indicated that the IVS6 region contains part of the phenylalkylamine binding site (Catterall and Striessnig, 1992). Antipeptide-specific antibodies localized portions of the DHP binding site to the pore region and the S6 segment of repeats III and IV. Sequencing of radioactive-labeled peptides identified intracellular residues adjacent to the IVS6 segment as a part of the dihydropyridine binding site (Regulla *et al.*, 1991). Site-directed mutagenesis of the cardiac α_{1C} subunit showed that only the exchange of the region between IVS3 and IVS6 is necessary to abolish the dihydropyridine block (Tang *et al.*, 1993). An exchange of the same region of repeat III does not affect the dihydropyridine block. A further refinement of the repeat IV region suggests that part of the high-affinity binding site is on the IVS6 transmembrane segment. In agreement with the localization of the dihydropyridine binding site to repeat IV is the observation that the concentration of dihydropyridines required for a channel block is affected by the used IVS3 exon (Welling *et al.*, 1993a).

IV. FUNCTIONAL INTERACTION OF THE CALCIUM CHANNEL SUBUNITS

Although the calcium channel has an oligomeric structure, the α_1 subunit alone functions as an L-type calcium channel when expressed in *Xenopus* oocytes and CHO or HEK cells. Coexpression of the α_2/δ subunit with the α_1 subunit does not affect significantly the channel properties (Singer *et al.*, 1991). In contrast, the β subunit affects significantly the properties of the expressed α_1 subunit by binding to a conserved motif in the I–II cytoplasmic linker of the α_1 subunit (Pragnell *et al.*, 1994) (Fig. 2). Binding of the β subunit increases the number of dihydropyridine binding sites and the density of functional channels (Welling *et al.*, 1993b). The β subunit shifts the voltage dependence of channel opening and channel inactivation to negative membrane potentials and decreases the channel activation time twofold (Neely *et al.*, 1993). Depending on the used α_1 subunit gene, the four β subunits affect the time constant for channel inactivation differently. Overall, these results show decisively that efficient expression of calcium channels with normal physiological properties is greatly enhanced by coexpression of the α_1 subunit with the α_2/δ and β subunits. Each of these subunits interacts directly with the α_1 subunit to increase the expression of a functional channel or to restore some aspects of normal channel function. These findings imply that a three-subunit oligomer (α_1, α_2/δ, β) is the physiologically functional calcium channel in many cells.

V. HORMONAL REGULATION OF THE SMOOTH MUSCLE CALCIUM CHANNEL

The opening and closing of L-type calcium channels is controlled by the membrane potential. In addition, the β-adrenergic receptor controls the availability of the channel to open upon membrane depolarization. Isoproterenol increases the cardiac calcium current three- to sevenfold either by cAMP-dependent phosphorylation of the channel (Kameyama *et al.*, 1985; Hartzell and Fischmeister, 1992) or by the activated α subunits of the trimeric GTP binding protein G_S (Yatani and Brown, 1989) or by a combination of the activated α subunits of the trimeric GTP binding protein G_S and the active cAMP kinase (Cavalié *et al.*, 1991). The L-type calcium current of isolated tracheal smooth muscle cells is also stimulated by the activation of the β-adrenergic receptor (Welling *et al.*, 1992). This β-adrenergic receptor effect is mediated directly by a G-protein and not by cAMP kinase activation. Isoproterenol also increased the L-type calcium current in isolated smooth muscle cells from rabbit ear

arteries (Benham and Tsien, 1988), taenia coli (Muraki *et al.*, 1993), rat aortic myocytes (Neveu *et al.*, 1994), and rabbit portal vein (Xiong *et al.*, 1994). Stimulation of the current did not involve activation of cAMP kinase and was mediated in some cells by an atypical adrenergic receptor. These results suggested that the α_{1C} channel of smooth muscle may be regulated *in vivo* by adrenergic receptors through the activation of a G-protein. In contrast, the α_{1C} channel of cardiac muscle is regulated mainly by cAMP-dependent phosphorylation.

The primary sequences of cardiac and smooth muscle α_{1C} subunits are almost identical and contain identical phosphorylation sites. Several reports (Yoshida *et al.*, 1992; Sculptoreanu *et al.*, 1993; Perez-Reyes *et al.*, 1994) suggested that the expressed α_{1C} subunit is phosphorylated *in vivo* by cAMP kinase. The phosphorylation increased the current density and facilitated channel opening. Other groups (Kleppisch *et al.*, 1994; Singer-Lahat *et al.*, 1994) were unable to repeat these results. These latter observations are in agreement with the *in vivo* findings from smooth muscle, that is, cAMP kinase does not increase L-type calcium channel current in various smooth muscles. It is therefore conceivable that the activity of the α_{1C} channel is not regulated by cAMP kinase directly but requires a further protein such as the β subunit. The deduced amino acid sequence of the skeletal muscle β_1 subunit contains several phosphorylation sites. Two of these sites, Ser[182] and Thr[205], are phosphorylated *in vitro* by cAMP-dependent protein kinase. The equivalent to Thr[205] is conserved in the "cardiac" β_2 subunit (Thr[165] in β_{2a} and Thr[191] in β_{2b} subunit) but is not present in the "smooth muscle" β_3 subunit. The sequence following this potential phosphorylation site is highly variable and determines several splice variants (Hullin *et al.*, 1992). This variable region within the β subunits may be responsible for the tissue-specific regulation of the L-type calcium currents by hormones and neurotransmitters. Other work seemed to confirm this hypothesis (Haase *et al.*, 1993). However, coexpression of the "cardiac" calcium channel consisting of the α_{1Ca}, α_2/δ, and β_{2a} subunits in HEK cells has not led to cAMP-dependent regulation of the channel (Hofmann and Zong, 1995). This negative result is in line with the known regulation of the smooth muscle L-type calcium current. It also indicates that the basis for the different hormonal regulation of the smooth and the cardiac L-type channels has not been elucidated.

Acknowledgments

We thank Mrs. I. Schatz for typing the manuscript. The results obtained in the authors' laboratory were supported by grants from Deutsche Forschungsgemeinschaft and Fond der Chemie.

References

Benham, C. D., and Tsien, R. W. (1988). *J. Physiol. (London)* **404,** 767–784.

Biel, M., Ruth, P., Bosse, E., Hullin, R., Stühmer, P., Flockerzi, V., and Hofmann, F. (1990). *FEBS Lett.* **269,** 409–412.

Biel, M., Hullin, R., Freundner, S., Singer, D., Dascal, N., Flockerzi, V., and Hofmann, F. (1991). *Eur. J. Biochem.* **200,** 81–88.

Bosse, E., Bottlender, R., Kleppisch, T., Hescheler, J., Welling, A., Hofmann, F., and Flockerzi, V. (1992). *EMBO J.* **11,** 2033–2038.

Castellano, A., Wei, X., Birnbaumer, L., and Perez-Reyes, E. (1993). *J. Biol. Chem.* **268,** 12359–12366.

Catterall, W. A., and Striessnig, J. (1992). *Trends Pharmacol. Sci.* **13,** 256–262.

Cavalié, A., Allen, T. J. A., and Trautwein, W. (1991). *Pflügers Arch.* **419,** 433–443.

De Jongh, K. S., Warner, C., and Catterall, W. A. (1990). *J. Biol. Chem.* **265,** 14738–14741.

De Waard, M., Pragnell, M., and Campbell, K. (1994). *Neuron* **13,** 495–503.

Diebold, R. J., Koch, W. J., Ellinor, P. T., Wang, J.-J., Muthuchamy, M., Wieczorek, D. F., and Schwartz, A. (1992). *Proc. Natl. Acad. Sci. U.S.A.* **89,** 1497–1501.

Dubel, S. J., Starr, T. V. B., Hell, J., Ahlijanian, M. K., Enyeart, J. J., Catterall, W. A., and Snutch, T. P. (1992). *Proc. Natl. Acad. Sci. U.S.A.* **89,** 5058–5062.

Feron, O., Octave, J.-N., Christen, M.-O., and Godfraind, T. (1994). *Eur. J. Biochem.* **222,** 195–202.

Guy, H. R., and Conti, F. (1990). *Trends Neurosci.* **13,** 201–206.

Haase, H., Karczewski, P., Beckert, R., and Krause, E. G. (1993), *FEBS Lett.* **335,** 217–222.

Hartzell, H. C., and Fischmeister, R. (1992). *Trends Pharmacol. Sci.* **13,** 380–385.

Hofmann, F., and Zong, X.-G. (1995). *In* "Beta Adrenoceptor Agonists and the Airways" (R. Small, ed.), pp. 27–33. Royal Society of Medicine Press, London.

Hofmann, F., Biel, M., and Flockerzi, V. (1994). *Annu. Rev. Neurosci.* **17,** 399–418.

Hullin, R., Singer-Lahat, D., Freichel, M., Biel, M., Dascal, N., Hofmann, F., and Flockerzi, V. (1992). *EMBO J.* **11,** 885–890.

Jay, S. D., Sharp, A. H., Kahl, St. D., Vedvick, T. S., Harpold, M. M., and Campbell, K. (1991). *J. Biol. Chem.* **266,** 3287–3293.

Kameyama, M., Hofmann, F., and Trautwein, W. (1985). *Pflügers Arch.* **405,** 285–293.

Kim, H., Wei, X., Ruth, P., Perez-Reyes, E., Flockerzi, V., Hofmann, F., and Birnbaumer, L. (1990). *J. Biol. Chem.* **265,** 11858–11863.

Kleppisch, T., Pedersen, K., Bosse, E., Flockerzi, V., Hofmann, F., and Hescheler, J. (1994). *EMBO J.* **13,** 2502–2507.

Koch, W. J., Ellinor, P. T., and Schwartz, A. (1990). *J. Biol. Chem.* **265,** 17786–17791.

Mikami, A., Imoto, K., Tanabe, T., Niidome, T., Mori, Y., Takeshima, H., Narumiya, S., and Numa, S. (1989). *Nature (London)* **340,** 230–233.

Mori, Y., Friedrich, T., Kim, M. S., Mikami, A., Nakai, J., Ruth, P., Bosse, E., Hofmann, F., Flockerzi, V., Furuichi, T., Mikoshiba, K., Imoto, K., Tanabe, T., and Numa, S. (1991). *Nature (London)* **350,** 398–402.

Muraki, K., Bolton, T. B., Imaizumi, Y., and Watanabe, M. (1993). *J. Physiol. (London)* **471,** 563–582.

Nakai, J., Adams, B. A., Imoto, K., and Beam, K. G. (1994). *Proc. Natl. Acad. Sci. U.S.A.* **91,** 1014–1018.

Neely, A., Wei, X., Olcese, R., Birnbaumer, L., and Stefani, E. (1993). *Science* **262,** 575–578.

Neveu, D., Quignard, J. F., Fernandez, A., Richard, S., and Nargeot, J. (1994). *J. Physiol. (London)* **479**(2), 171–182.

Perez-Reyes, E., Yuan, W., Wei, X., and Bers, D. M. (1994). *FEBS Lett.* **342,** 119–123.

Pragnell, M., De Waard, M., Mori, Y., Tanabe, T., Snutch, T. P., and Campbell, K. P. (1994). *Nature (London)* **368,** 67–70.

Regulla, S., Schneider, T., Nastainczyk, W., Meyer, H. E., and Hofmann, F. (1991). *EMBO J.* **10,** 45–49.

Ruth, P., Röhrkasten, A., Biel, M., Bosse, E., Regulla, S., Meyer, H. E., Flockerzi, V., and Hofmann, F. (1989). *Science* **245,** 1115–1118.

Schultz, D., Mikala, G., Yatani, A., Engle, D. B., Iles, D. E., Segers, B., Sinke, R. J., Weghuis, D. O., Klöckner, U., Wakamori, M., Wang, J.-J., Melvin, D., Varadi, G., and Schwartz, A. (1993). *Proc. Natl. Acad. Sci. U.S.A.* **90,** 6228–6232.

Sculptoreanu, A., Rotman, E., Takahashi, M., Scheuer, T., and Catterall, W. A. (1993). *Proc. Natl. Acad. Sci. U.S.A.* **90,** 10135–10139.

Singer, D., Biel, M., Lotan, I., Flockerzi, V., Hofmann, F., and Dascal, N. (1991). *Science* **253,** 1553–1557.

Singer-Lahat, D., Lotan, I., Biel, M., Flockerzi, V., Hofmann, F., and Dascal, N. (1994). *Recept. Channels* **2,** 215–226.

Soldatov, N. M. (1994). *Genomics* **22,** 77–87.

Soong, T. W., Stea, A., Hodson, C. D., Dubel, S. J., Vincent, S. R., and Snutch, T. P. (1993). *Science* **260,** 1133–1136.

Sun, W., McPherson, J. D., Hoang, D. Q., Wasmuth, J. J., Evans, G. A., and Montal, M. (1992). *Genomics* **14,** 1092–1094.

Tanabe, T., Mikami, A., Numa, S., and Beam, K. G. (1990). *Nature (London)* **344,** 451–453.

Tang, S., Yatani, A., Bahinski, A., Mori, Y., and Schwartz, A. (1993). *Neuron* **11,** 1013–1021.

Welling, A., Felbel, J., Peper, K., and Hofmann, F. (1992). *Am. J. Physiol.* **262,** L351–L359.

Welling, A., Kwan, Y. W., Bosse, E., Flockerzi, V., Hofmann, F., and Kass, R. S. (1993a). *Circ. Res.* **73,** 974–980.

Welling, A., Bosse, E., Cavalié, A., Bottlender, R., Ludwig, A., Nastainczyk, W., Flockerzi, V., and Hofmann, F. (1993b). *J. Physiol. (London)* **471,** 749–765.

Williams, M. E., Brust, P. F., Feldman, D. H., Patthi, S., Simerson, S., Maroufi, A., McCue, A. F., Velicelebi, G., Ellis, S. B., and Harpold, M. M. (1992). *Science* **257,** 389–395.

Xiong, Z., Sperelakis, N., and Fenoglio-Preiser, C. (1994). *Pflügers Arch.* **428,** 105–113.

Yang, J., Ellinor, P. T., Sather, W. A., Zhang, J. F., and Tsien, R. W. (1993). *Nature (London)* **366,** 158–161.

Yatani, A., and Brown, A. M. (1989). *Science* **245,** 71–74.

Yoshida, A., Takahashi, M., Nishimura, S., Takeshima, H., and Kokubun, S. (1992). *FEBS Lett.* **309,** 343–349.

Zhang, J. F., Ellinor, P. T., Aldrich, R. W., and Tsien, R. W. (1994). *Nature (London)* **372,** 97–100.

18

Electromechanical and Pharmacomechanical Coupling

CHRISTOPHER M. REMBOLD

Cardiovascular Division
Departments of Internal Medicine and Physiology
University of Virginia Health Science Center
Charlottesville, Virginia

I. INTRODUCTION

Stimuli contract arterial smooth muscle by at least four mechanisms (Fig. 1): (1) Both contractile agonists and high $[K^+]_o$ depolarize smooth muscle and increase Ca^{2+} influx through voltage-dependent Ca^{2+} channels. (2) Agonists release Ca^{2+} from intracellular stores. (3) Agonists activate both voltage-dependent and voltage-independent Ca^{2+} channels, which will increase Ca^{2+} influx beyond that expected from the degree of depolarization. (4) Agonists also increase $[Ca^{2+}]_i$ sensitivity, that is, agonists increase force at a given $[Ca^{2+}]_i$. The first mechanism is electromechanical coupling because the contraction involves a change in membrane potential (E_m). The latter three mechanisms are forms of pharmacomechanical coupling because the contraction is larger than would be expected from the change in E_m. The first three mechanisms produce contraction by changing $[Ca^{2+}]_i$. The last mechanism ($[Ca^{2+}]_i$ sensitivity) produces contraction without changing $[Ca^{2+}]_i$. High $[K^+]_o$ directly contracts most smooth muscles only by the first mechanism (changing E_m). Agonists can contract smooth muscle by some or all of these four mechanisms. Agents that induce smooth muscle relaxation do so by interfering with one of these four mechanisms or by additional mechanisms (e.g., stimulation of Ca^{2+} efflux or sequestration; see Fig. 1). The goal of this chapter is to review the mechanisms involved in and compare the relative physiological importance of these electromechanical and pharmacomechanical mechanisms in smooth muscle contraction and relaxation.

II. HISTORICAL BACKGROUND

Over 100 years ago, electrical stimulation was noted to induce a twitch of frog skeletal muscle. This led to the hypothesis that muscles are electrically regulated, that is, that there was electromechanical coupling. This paradigm is still valid in skeletal and cardiac muscle. However, in smooth muscle this paradigm is incomplete. Some stimuli induce smooth muscle contraction with less depolarization than others. This phenomenon was first studied by Bulbring (1957) and was called pharmacomechanical coupling (Somlyo and Somlyo, 1968). After the distinction between electromechanical coupling and pharmacomechanical coupling was made, study of these two contractile mechanisms diverged into two disciplines.

Over the last 20 years, electromechanical coupling was primarily studied by measuring E_m in intact muscle or ionic currents in isolated cells. Several groups found that depolarization (either by high $[K^+]_o$ or contractile agonists) activated several types of Ca^{2+} channels and induced contractions (Hermsmeyer *et al.*, 1988; Nelson *et al.*, 1988). The relationship between changes in E_m and force production depended on the stimulus: contractile agonists induced contraction with less depolarization than that required by high $[K^+]_o$ (Mulvany *et al.*, 1982; Nelson *et al.*, 1988; Neild and Kotecha, 1987). Several investigators found that agonists increased L-type Ca^{2+} channel current at a constant E_m in dissociated smooth muscle cells (Nelson *et al.*, 1988). These data suggest that pharmacomechanical coupling may represent agonist-

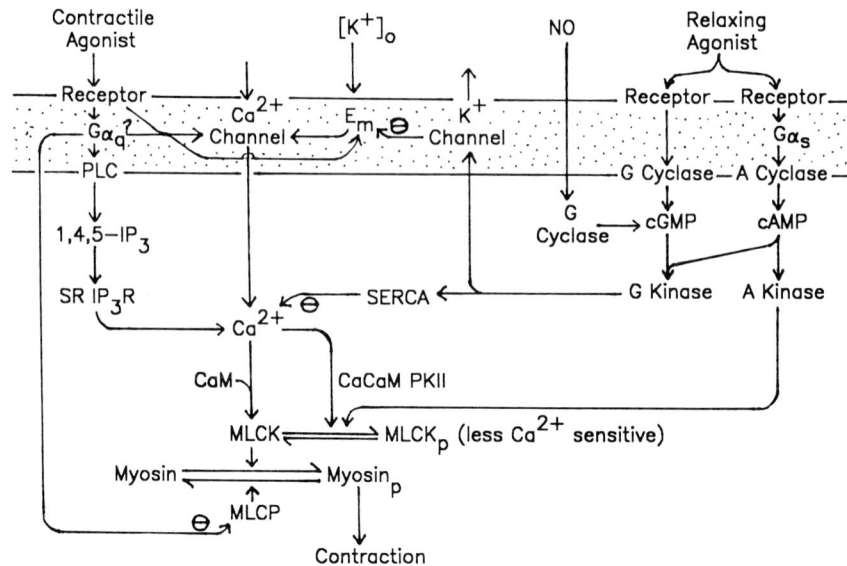

FIGURE 1   Schema of some of the hypothesized contraction and relaxation mechanisms in arterial smooth muscle. The electromechanical mechanism for contraction is depolarization (change in E_m) induced directly by increases in extracellular [K$^+$] ([K$^+$]$_o$) or indirectly by contractile agonists binding to their receptor. The depolarization increases Ca^{2+} influx through L-type Ca^{2+} channels. The increased intracellular [Ca^{2+}] ([Ca^{2+}]$_i$) binds to calmodulin (CaM) and activates myosin light chain kinase (MLCK). Myosin light chain kinase then phosphorylates the 20-kDa light chain of myosin (myosin$_p$), which induces contraction. Increases in [Ca^{2+}]$_i$ also inhibit myosin light chain kinase activity by activating Ca^{2+}-calmodulin protein kinase II (CaCaM PKII), which phosphorylates myosin light chain kinase, a process that decreases the [Ca^{2+}]$_i$ sensitivity of myosin light chain kinase.

Pharmacomechanical mechanisms for contraction include: (1) Increases in 1,4,5-inositol trisphosphate (1,4,5-IP$_3$) by receptor-dependent activation of Gα_q and phospholipase C (PLC). The 1,4,5-IP$_3$ binds to its receptor on the sarcoplasmic (endoplasmic) reticulum (SR IP$_3$R), permitting release of Ca^{2+} from the intracellular store. (2) Agonist binding to its receptor can also increase Ca^{2+} influx through L-type Ca^{2+} channels by an unknown mechanism not involving change in membrane potential (E_m). Both of these mechanisms increase [Ca^{2+}]$_i$, which then contracts the smooth muscle as described in the preceding. (3) Agonist binding to receptors also appears to decrease the activity of myosin light chain phosphatase (MLCP) through the mediation of a G protein. The decreased phosphatase activity increases myosin phosphorylation and contraction at lower levels of [Ca^{2+}]$_i$ and myosin light chain kinase activity than the preceding three mechanisms.

The electromechanical mechanism for relaxation is hyperpolarization. Endothelial-derived relaxing factor (NO) directly activates soluble guanylyl cyclase (G cyclase). Binding of relaxing agonists to their receptors directly activates particulate guanylyl cyclase or activates adenyl cyclase (A cyclase) through the mediation of Gα_s. These actions result in the formation of cyclic GMP (cGMP) and cyclic AMP (cAMP), respectively. Increases in either [cAMP] or [cGMP] activate cGMP-dependent protein kinase (G kinase), which results in activation of plasma membrane K$^+$ channels and hyperpolarization. The hyperpolarization decreases Ca^{2+} influx through L-type Ca^{2+} channels and thereby decreases [Ca^{2+}]$_i$, myosin phosphorylation, and contraction.

Pharmacomechanical mechanisms for relaxation include: (1) G kinase-dependent increases in the activity of sarcoplasmic reticulum Ca^{2+} pumps (SERCA). Ca^{2+} pumps on the plasma membrane may also be stimulated (not shown). This increase in Ca^{2+} sequestration and extrusion decreases [Ca^{2+}]$_i$ and induces relaxation as shown in the foregoing. (2) Some agents appear to decrease 1,4,5-IP$_3$ formation and may relax smooth muscle by decrease Ca^{2+} release (not shown) (3) Finally, increases in [cAMP] activate cAMP-dependent protein kinase (A kinase), which could phosphorylate myosin light chain kinase and decrease its Ca^{2+} sensitivity (this mechanism has not been demonstrated in intact smooth muscle).

induced increases in [Ca^{2+}]$_i$ beyond that expected by the level of depolarization.

Pharmacomechanical coupling was primarily studied by measuring [Ca^{2+}]$_i$ and myosin regulatory light chain phosphorylation in intact and skinned smooth muscle. Several groups found that stimuli increased

[Ca^{2+}]$_i$, which increased phosphorylation of the regulatory light chain of myosin and induced a contraction (Morgan and Morgan, 1984a; Rembold and Murphy, 1988b). The relationship between changes in [Ca^{2+}]$_i$ and myosin phosphorylation depended on the stimulus: agonists increased myosin phosphorylation and force with smaller increases in $[Ca^{2+}]_i$ than that observed with high $[K^+]_o$ (i.e., agonists induced a higher $[Ca^{2+}]_i$ sensitivity of phosphorylation than that observed with high $[K^+]_o$). These data suggest that pharmacomechanical coupling may represent agonist-induced increases in the $[Ca^{2+}]_i$ sensitivity of phosphorylation rather than increases in $[Ca^{2+}]_i$.

As will be detailed in the following, both disciplines were correct. Pharmacomechanical coupling represents increases in both $[Ca^{2+}]_i$ and the $[Ca^{2+}]_i$ sensitivity of phosphorylation beyond that expected given the level of depolarization.

III. ELECTROMECHANICAL COUPLING

Regulation of contractile force by changes in E_m is called "electromechanical coupling." This type of regulation primarily involves changes in $[Ca^{2+}]_i$.

A. Potential-Dependent Ca²⁺ Influx

Depolarization (e.g., by action potentials or change in $[K^+]_o$) increases smooth muscle $[Ca^{2+}]_i$ primarily by activation of L-type Ca^{2+} channels (Hermsmeyer et al., 1988). In most but not all smooth muscles, high $[K^+]_o$ contractions are completely dependent on extracellular Ca^{2+} and L-type Ca^{2+} channel blockers inhibit contraction. These data suggest that high $[K^+]_o$ contracts smooth muscle primarily by increasing Ca^{2+} influx through L channels. In some smooth muscle tissues, high $[K^+]_o$ can also release Ca^{2+} from the intracellular store (a form of pharmacomechanical coupling, see the following) both directly, by an unknown mechanism, and indirectly, by activation of nerve terminals and release of neurotransmitters.

B. Agonist-Dependent Depolarization

Norepinephrine (Nelson et al., 1988; Haeusler and De Peyer, 1989; Neild and Kotecha, 1987), histamine (Droogmans et al., 1977; Casteels and Suzuki, 1980; Keef and Bowen, 1989; Keef and Ross, 1986), 5-hydroxytryptamine (Neild and Kotecha, 1987), and endothelin (McPherson and Angus, 1991) depolarize intact vascular smooth muscle tissue. This depolarization increases Ca^{2+} influx by activating L-type and potentially other voltage-dependent Ca^{2+} channels.

1. Mechanisms for Agonist-Dependent Depolarization

There are several potential mechanisms for agonist-induced depolarization: (1) ATP (via the P_{2x} receptor) and carbachol reportedly open nonspecific cation channels (Benham and Tsien, 1987; Den Hertog et al., 1990; Ganitkevich and Isenberg, 1990; Sims, 1992); the resulting Na^+ influx could induce depolarization and activation of L channels. (2) Angiotensin II, U46119 (a thromboxane agonist), and acetylcholine inhibit K^+ channel activity in smooth muscle cells (Miyoshi and Nakaya, 1991; Scornik and Toro, 1992; Bonev and Nelson, 1993). If some K^+ channels are open in unstimulated cells, inhibition of these K^+ channels could depolarize the cell and increase Ca^{2+} influx. (3) Agonist-dependent increases in Ca^{2+} (e.g., by intracellular Ca^{2+} release) can activate Cl^- channels, which depolarizes the cell (Hogg et al., 1994). Agonist-dependent depolarization was not inhibited by verapamil, nifedipine, H7 (a protein kinase C inhibitor), or 2-nitro-4-carboxyphenyl-N, N-diphenylcarbamate (NCDC, a phospholipase C inhibitor) in rabbit basilar artery (Clark and Garland, 1993). These results suggest that agonist-dependent depolarization does not depend on Ca^{2+} influx, activation of protein kinase C, or activation of phospholipase C. There is also a feedback mechanism for depolarization: increases in $[Ca^{2+}]_i$, per se, appear to inhibit depolarization by $[Ca^{2+}]_i$-dependent activation of K_{Ca} (BK) channels (Brayden and Nelson, 1992).

2. Agonists Induce Higher Force Than Expected from the Degree of Depolarization

The relationship between E_m and contraction depends on the stimulus (Nelson et al., 1988; Neild and Kotecha, 1987; Mulvany et al., 1982). In rat tail artery depolarized with high $[K^+]_o$, contraction was associated with substantial depolarization. At any given E_m, agonists induced a larger contraction than observed with high $[K^+]_o$. This change in the relationship between E_m and force results from agonist-dependent activation of pharmacomechanical mechanisms in addition to the depolarization. These mechanisms include (1) an agonist-dependent increase in Ca^{2+} influx resulting in higher $[Ca^{2+}]_i$ (Nelson et al., 1988), (2) an agonist-dependent increase in the Ca^{2+} sensitivity of myosin phosphorylation (Morgan and Morgan, 1984a; Rembold and Murphy, 1988b), and potentially (3) other unknown processes.

C. cGMP-Induced Reductions in [Ca²⁺]ᵢ

Endothelial-derived relaxing factor (EDRF) (e.g., nitric oxide) and the atrial natriuretic factor(s) increase [cGMP] in smooth muscle. They relax arterial smooth muscle by three mechanisms: decreasing $[Ca^{2+}]_i$, decreasing the $[Ca^{2+}]_i$ sensitivity of phosphorylation, and uncoupling force from myosin phosphorylation (the latter two are pharmacomechanical mechanisms;

see the following). Nitrovasodilators, nitric oxide, and 8-bromo-cGMP attenuate stimulus-induced increases in $[Ca^{2+}]_i$ in isolated smooth muscle cells (Word *et al.*, 1991; Rashatwar *et al.*, 1987; Kobayashi *et al.*, 1985) and intact tissues (Karaki *et al.*, 1988; McDaniel *et al.*, 1992). There are several proposed mechanisms for cGMP-mediated reduction in $[Ca^{2+}]_i$, including (1) hyperpolarization, (2) decreased Ca^{2+} influx without changing E_n, and (3) increased Ca^{2+} efflux and/or sequestration (the latter two are also pharmacomechanical mechanisms; see the following).

1. cGMP-Dependent Hyperpolarization Resulting in Decreased Ca^{2+} Influx

cGMP-dependent protein kinase may activate K^+ channels, which will hyperpolarize the cell and decrease Ca^{2+} influx through L-type channels (Williams *et al.*, 1988; Thornbury *et al.*, 1991; Robertson *et al.*, 1993). A study in pituitary cells suggests that cGMP-dependent protein kinase phosphorylates and activates a protein phosphatase 2A, which dephosphorylates and activates a K^+ channel, thus inducing hyperpolarization (White *et al.*, 1993). These studies suggest that elevations in [cGMP] could hyperpolarize some cells by opening K^+ channels.

2. cAMP-Dependent Hyperpolarization Resulting in Decreased Ca^{2+} Influx

A mixture of high concentrations of cAMP and cAMP-dependent protein kinase increased the open probability of K_{Ca} channels reconstituted in a lipid bilayer (Savaria *et al.*, 1992). β-Adrenergic stimulation increased K_{Ca} channel activity in whole-cell clamped rat myometrium (Anwer *et al.*, 1992). In rabbit mesenteric arteries, calcitonin gene-related peptide (CGRP) appears to activate K_{ATP} channels through the mediation of protein kinase A (Quayle *et al.*, 1994). These studies suggest that elevations in [cAMP] could hyperpolarize smooth muscle by opening K_{Ca} or K_{ATP} channels.

3. The Role of Hyperpolarization in Endothelial-Dependent Relaxation

The literature on the role of hyperpolarization and endothelial-dependent relaxation is complex because there are at least two relaxing compounds released from the endothelium: nitric oxide and endothelial-derived hyperpolarizing factor (EDHF). There are conflicting reports on the mechanism of action of each agent. ADP (Brayden, 1991) and acetylcholine (Feletau and Vanhoutte, 1988; Keef and Bowen, 1989) induce relaxation via the endothelium, releasing nitric oxide and/or EDHF (Brayden and Wellman, 1989; Nagao

and Vanhoutte, 1992). In some tissues, ADP-induced relaxation is not associated with hyperpolarization (Brayden, 1991) or with changes in [cGMP] (Brayden and Wellman, 1989). The endothelium-dependent relaxation induced by acetylcholine (McPherson and Angus, 1991) or ATP (via P_2 receptor stimulation) (Keef *et al.*, 1992) can occur without inducing hyperpolarization. Nitric oxide induces hyperpolarization in some (Tare *et al.*, 1990), but not all smooth muscles (Komori *et al.*, 1988; Brayden, 1990). Furthermore, nitric oxide-dependent hyperpolarization does not appear to be necessary for nitric oxide-dependent relaxation (Plane and Garland, 1993). Finally, some vasodilating drugs hyperpolarize cells without changing [cGMP] by directly opening K^+ channels (e.g., pinacidil and cromakalim activate K^+ channels—it is controversial whether these open K_{Ca} and/or K_{ATP} channels) (Vidbaek *et al.*, 1988; Doggrell *et al.*, 1989). Some of these K^+ channel activators can also relax smooth muscle by decreasing $[Ca^{2+}]_i$ sensitivity (Okada *et al.*, 1993) and/or other pharmacomechanical mechanisms (Cox, 1991).

IV. PHARMACOMECHANICAL COUPLING

Regulation of force independent of changes in E_m is called "pharmacomechanical coupling." Pharmacomechanical coupling involves either changes in $[Ca^{2+}]_i$ or changes in the cellular response to $[Ca^{2+}]_i$ independent of changes in E_m.

A. Pharmacomechanical Coupling by Changing $[Ca^{2+}]_i$

1. Receptor-Mediated Intracellular Ca^{2+} Release

a. Receptors and G Protein Coupling   Receptor G protein coupling in smooth muscle is detailed by Bárány and Bárány in Chapter 21 (this volume). Not all agonists activate G proteins or release of Ca^{2+} from the intracellular store in smooth muscle. It appears that those receptors belonging to the so-called "seven membrane spanning family" activate phospholipase $C_{\beta 1}$ through the mediation of $G_{\alpha q}$ (Majerus, 1992; Exton, 1994). Such receptors include the α_1-adrenergic receptor (norepinephrine), the AT_1 receptor (angiotensin II), the H1 receptor (histamine), the M3 muscarinic receptor (acetylcholine), and the thrombin receptor. Some investigators suggest an alternate pathway where receptors activate the β_2 and β_3 isoforms of phospholipase C via the $\beta\gamma$ subunit of a G protein (Exton, 1994). There are some receptors [α_2-adrenergic (clonidine) and P_{2x} (ATP)] that contract smooth muscle without apparent release of Ca^{2+} from the intracellular store (Rembold *et al.*, 1991; Abe *et al.*, 1987).

b. Inositol Trisphosphate Activated phospholipase C is proposed to hydrolyze a minor membrane lipid, phosphatidylinositol 4,5-bisphosphate (PIP_2), resulting in the production of 1,4,5-IP_3 and 1,2-diacylglycerol (DAG). 1,4,5-IP_3 diffuses from the inner surface of the plasma membrane to a specific receptor on the intracellular store. The 1,4,5-IP_3 receptor is a homotetramer that forms a cloverleaf like structure with a Ca^{2+} channel in the center and one 1,4,5-IP_3 binding site on each of the four subunits (Taylor and Richardson, 1991). Opening of the Ca^{2+} channel in the 1,4,5-IP_3 receptor is highly cooperative and is also dependent on cytoplasmic [Ca^{2+}]; the channel opens optimally with [Ca^{2+}]$_i$ between 200 and 500 nM. Both higher and lower [Ca^{2+}]$_i$ inhibits Ca^{2+} release. Although 1,4,5-IP_3 appears to be the predominant polyphosphoinositol involved in Ca^{2+} release from nonmitochondrial intracellular stores (Berridge, 1987), other less well described second messengers can also release Ca^{2+} from intracellular stores. Cyclic ADP ribose binds to and opens the ryanodine receptor in smooth muscle (Lee and Aarhus, 1993).

The 1,4,5-IP_3 hypothesis has been extensively tested in smooth muscle. In skinned smooth muscle, physiological concentrations of 1,4,5-IP_3 have been shown to release sufficient Ca^{2+} rapidly enough to account for transient contractions (Somlyo *et al.*, 1985; Suematsu *et al.*, 1984; Walker *et al.*, 1987; Baron *et al.*, 1989; Abdel-Latif, 1989). This response can be blocked by intracellular heparin (Kobayashi *et al.*, 1988b), a competitive blocker of cerebellar 1,4,5-IP_3 receptors (Worley *et al.*, 1987). Agonist stimulation increased the apparent concentration of 1,4,5-IP_3 when measured as either [³H]IP_3 in cells (Griendling *et al.*, 1986) or tissues (Howe *et al.*, 1986; Long and Stone, 1987; Takuwa *et al.*, 1986; Miller-Hance *et al.*, 1988; Salmon and Bolton, 1988) or as 1,4,5-IP_3 mass with a receptor binding assay (Chilvers *et al.*, 1989). 1,4,5-IP_3 does not typically release the entire intracellular Ca^{2+} store (as measured by caffeine-induced intracellular release). In skinned smooth muscle, GTP can release Ca^{2+} from a non-1,4,5-IP_3-sensitive store (Chuch *et al.*, 1987; Kobayashi *et al.*, 1988a). This has been hypothesized to represent G protein-mediated transfer of Ca^{2+} from a non-1,4,5-IP_3-dependent store to a 1,4,5-IP_3-responsive store (Gill *et al.*, 1989; Burgoyne *et al.*, 1989).

There are some important discrepancies in the 1,4,5-IP_3 hypothesis that center about the two methods employed to measure changes in 1,4,5-IP_3 concen-

tration. In an elegant set of experiments, Baron *et al.* (1989) showed that agonist stimulation increased not just [³H]1,4,5-IP_3, but also incorporation of [³H]inositol into [³H]PIP_2. Therefore agonist-dependent increases in [³H]1,4,5-IP_3 could represent increases in [1,4,5-IP_3] and/or increases in the specific activity of [³H]1,4,5-IP_3. Baron *et al.* showed that both increased in the carbachol-stimulated canine trachealis. None of the other studies in smooth muscle measured the specific activity of PIP_2, therefore, none of these other studies proved that agonists increased [1,4,5-IP_3]. The IP_3 receptor binding assay measures 1,4,5-IP_3 mass directly without the need to radiolabel inositol. Carbachol increased 1,4,5-IP_3 mass in canine trachealis (Chilvers *et al.*, 1989). However, this assay reported very high levels of [1,4,5-IP_3]: specifically resting [1,4,5-IP_3] estimates were 12 pmol/mg protein, which corresponds to a concentration of 60–120 μM if protein were 5–10% of cell weight. This estimate of [1,4,5-IP_3] is much higher than the 0.1–1.0 μM concentrations employed in biochemical and skinned studies (Somlyo *et al.*, 1985; Suematsu *et al.*, 1984). Potentially, smooth muscle may contain a compound that interferes with the IP_3 receptor binding assay.

2. Receptor-Mediated Ca^{2+} Influx

Sustained smooth muscle contraction is dependent on extracellular Ca^{2+} (Deth and van Breemen, 1977; Ratz and Murphy, 1987). Without extracellular Ca^{2+}, agonist-induced contractions quickly relax as the intracellular Ca^{2+} pool empties (Bozler, 1969). Sustained agonist-induced contractions are associated with sustained increases in Ca^{2+} influx [as measured with $^{45}Ca^{2+}$ (van Breemen *et al.*, 1986; Forder *et al.*, 1985) or Mn^{2+} influx (Chen and Rembold, 1992)]. This Ca^{2+} influx results in sustained substantial elevations of myoplasmic [Ca^{2+}] (Rembold and Murphy, 1988b; Gunst and Bandyopadhyay, 1989; Himpens and Somlyo, 1988), although not all investigators find sustained large increases in [Ca^{2+}]$_i$ (Morgan and Morgan, 1984a; Rasmussen *et al.*, 1987). Some, but not all, of the increased Ca^{2+} influx depends on agonist-dependent depolarization. To account for the sustained increase in [Ca^{2+}]$_i$ and tonic contraction, some investigators hypothesized the existence of receptor-operated, Ca^{2+}-permeable channels (Bolton, 1979). ATP and carbachol appear to activate nonspecific ion channel in some smooth muscle (Benham and Tsien, 1987; Ganitkevich and Isenberg, 1990). Activation of these channels could increase Ca^{2+} influx directly. Contractile agonists can also directly increase conductance through L-type Ca^{2+} channels without changes in E_m by changing the current–voltage characteristics of the

L channel (Nelson *et al.*, 1988; Ohya and Sperelakis, 1991).

3. cGMP-Dependent Inhibition of Intracellular Ca^{2+}L Release

Preincubation of rat aorta with nitroprusside attenuated agonist-induced increases in inositol monophosphate (Rapoport, 1986) and 1,4,5-inositol trisphosphate mass (Langlands and Diamond, 1990).

4. cGMP-Dependent Decreases in Ca^{2+} Influx Independent of Changes in E_m

Nitroprusside (Clapp and Gurney, 1991) and 8-bromo-cGMP (Ishikawa *et al.*, 1993) decreased L channel conductance in patch-clamped vascular smooth muscle cells, suggesting that cGMP can decrease Ca^{2+} influx by inactivating L channels independent of changes in E_m. Nitrovasodilators reduced Mn^{2+} influx (a surrogate for Ca^{2+} influx) (Chen and Rembold, 1992) and [Ca^{2+}]$_i$ (McDaniel *et al.*, 1992; Karaki *et al.*, 1988) in intact arterial smooth muscle. These data suggest that elevations in [cGMP] reduce [Ca^{2+}]$_i$ at least partially by decreasing Ca^{2+} influx, potentially by either hyperpolarization or direct action on L channels.

5. cGMP-Dependent Enhancement of Ca^{2+} Extrusion and Sequestration

There are several proposed mechanisms for cGMP-dependent increases in Ca^{2+} extrusion and sequestration: (1) cGMP-dependent protein kinase can phosphorylate phospholamban, a process that activates the sarcoplasmic reticulum Ca^{2+}-ATPase by removing the inhibition caused by unphosphorylated phospholamban (Lincoln and Cornwell, 1991; Twort and van Breemen, 1988; Kimura *et al.*, 1982); (2) cGMP-dependent protein kinase may also activate a plasma membrane Ca^{2+}-ATPase (Furakawa *et al.*, 1988; Popescu *et al.*, 1985); and (3) 8-bromo-cGMP increased Na$^+$/Ca^{2+} exchange in isolated rat aortic smooth muscle cells (Furukawa *et al.*, 1991). Supporting these mechanisms was the finding that nitroglycerin enhanced ^{45}Ca^{2+} efflux in acetylcholine-stimulated coronary artery cells (Itoh *et al.*, 1983).

6. cAMP-Induced Reduction in [Ca^{2+}]$_i$

Physiological stimuli that increase [cAMP] include those agonists that activate the β_2-adrenergic, P$_1$ purinergic (adenosine A2), or calcitonin gene-related peptide receptors. cAMP relaxes swine carotid artery by at least two mechanisms: decreasing [Ca^{2+}]$_i$ and decreasing the [Ca^{2+}]$_i$ sensitivity of phosphorylation (see the following). Elevations in [cAMP] decreased [Ca^{2+}]$_i$ in some (Kimura *et al.*, 1982; Parker *et al.*, 1987;

Gunst and Bandyopadhyay, 1989; McDaniel *et al.*, 1991) but not all smooth muscles (Takuwa *et al.*, 1988; Morgan and Morgan, 1984b). Adenosine decreased ^{45}Ca^{2+} influx in coronary artery (Motulsky and Michel, 1988), suggesting that increases in [cAMP] may decrease [Ca^{2+}]$_i$ by decreasing Ca^{2+} influx. Low-dose forskolin, a specific activator of adenyl cyclase, increased [cAMP], reduced Mn^{2+} influx (a surrogate for Ca^{2+} influx), reduced [Ca^{2+}]$_i$, and did not change the [Ca^{2+}]$_i$ sensitivity of phosphorylation in intact swine carotid artery (McDaniel *et al.*, 1991; Chen and Rembold, 1992). These data suggest that cAMP reduces [Ca^{2+}]$_i$ at least partially by decreasing Ca^{2+} influx, potentially by either hyperpolarization or direct action on L channels. Forskolin and 8-bromo-cAMP had biphasic effects on L-type Ca^{2+} channel conductance in patch-clamped vascular smooth muscle cells; low doses increased conductance and high doses decreased conductance (Ishikawa *et al.*, 1993).

7. Activation of Protein Kinase G by cAMP

The primary mechanisms for the cAMP-mediated reduction in [Ca^{2+}]$_i$ may be by activation of cGMP-dependent protein kinase (G kinase). In a smooth muscle cell line deficient in G kinase, elevations in [cAMP] increased [Ca^{2+}]$_i$ (Lincoln *et al.*, 1990). If G kinase is reintroduced by electroporation, elevations in [cAMP] decreased [Ca^{2+}]$_i$. These data suggest that G kinase was responsible for the cAMP-induced decrease in [Ca^{2+}]$_i$. A biochemical explanation for this result is that both cGMP and cAMP activate G kinase, although it requires approximately 10 times higher [cAMP] than [cGMP] to activate G kinase (Francis *et al.*, 1988). In the swine carotid, resting [cAMP] is typically 10 times greater than [cGMP] (Rembold, 1989). Therefore, elevations in [cAMP] should activate both G kinase and cAMP-dependent protein kinase. Elevations in [cGMP] should primarily activate G kinase unless the increase in [cGMP] is large.

B. Pharmacomechanical Coupling by Regulation of the [Ca^{2+}]$_i$ Sensitivity of Myosin Light Chain Phosphorylation

Modulation of myosin phosphorylation independent of changes in [Ca^{2+}]$_i$ is another form of pharmacomechanical coupling.

1. [Ca^{2+}]$_i$-Dependent and Independent Phosphorylation of Myosin by Myosin Light Chain Kinase

In vitro, myosin light chain kinase activity depends on both [Ca^{2+}] and [calmodulin] (Kamm and Stull, 1989). Activated myosin light chain kinase primarily phosphorylates the 20-kDa light chain of myosin on serine 19 (Ikebe et al., 1988), and this phosphorylation increases the actin-activated ATPase activity of myosin. Based on these biochemical studies, myosin light chain kinase activity and therefore myosin phosphorylation levels should be proportional to myoplasmic $[Ca^{2+}]$. Varying levels of $[K^+]_o$ induced a single relation between changes in aequorin-estimated myoplasmic $[Ca^{2+}]$ and myosin phosphorylation (Rembold and Murphy, 1986). Contractile agonists also induced a single relation between changes in aequorin-estimated myoplasmic $[Ca^{2+}]$ and myosin phosphorylation (Rembold and Murphy, 1988b; Rembold, 1990). However, the relationships induced by depolarization and contractile agonists differed: contractile agonists sensitized the myosin phosphorylation system. Contractile agonists induced a higher $[Ca^{2+}]_i$ sensitivity of phosphorylation than that observed with high $[K^+]_o$ depolarization (Rembold and Murphy, 1988b; Rembold, 1990; Taylor et al., 1989; Kitazawa and Somlyo, 1990). These data suggest that factors beyond $[Ca^{2+}]_i$ regulate myosin phosphorylation levels. The Ca^{2+} sensitivity of phosphorylation in the swine carotid artery is regulated over a wide range by various stimuli as shown in Table I. This system may be even more complex in some smooth muscles (Jiang and Morgan, 1989; Aburto et al., 1993), suggesting that these other smooth muscles may have additional regulatory systems.

Initially, a dissociation between $^{45}Ca^{2+}$ influx and force (van Breemen et al., 1986) and a dissociation between myoplasmic $[Ca^{2+}]$ and force (Morgan and Morgan, 1984a) were the first clues that the $[Ca^{2+}]_i$ sensitivity of phosphorylation varied with the form of stimulation. Subsequently, many investigators measured $[Ca^{2+}]_i$ and force simultaneously and suggested that the $[Ca^{2+}]_i$ sensitivity of force depends on the stimulus (Jensen et al., 1992; Bruschi et al., 1988; Morgan and Morgan, 1984a; Abe and Karaki, 1989; Karaki et al., 1988; Himpens and Somlyo, 1988). However,

interpretation of $[Ca^{2+}]_i$–force relationships can be difficult because the relation between phosphorylation and force is time dependent and not necessarily linear. For example, if the agonist-induced $[Ca^{2+}]_i$ and phosphorylation transient are attenuated, force develops much more slowly than that observed if the agonist had induced a large $[Ca^{2+}]_i$ and phosphorylation transient (Ratz and Murphy, 1987; Rembold and Murphy, 1989). Apparent alterations in the "$[Ca^{2+}]_i$ sensitivity of force" may be caused by (1) changes in the time course of $[Ca^{2+}]_i$ and myosin phosphorylation, (2) changes in the $[Ca^{2+}]_i$ sensitivity of phosphorylation, and/or (3) changes in the relationship between myosin phosphorylation and force. Also see Chapter 27 by Kamm and Grange (this volume) for a more detailed discussion of Ca^{2+} sensitivity.

2. Regulation of the Ca^{2+} Sensitivity of Myosin Light Chain Kinase by Its Phosphorylation

Myosin light chain kinase can be phosphorylated on several residues (Adelstein et al., 1978). Myosin light chain kinase phosphorylation at "site A" decreases the Ca^{2+} sensitivity of myosin light chain kinase (Stull et al., 1990). Both cAMP-dependent protein kinase and Ca^{2+}-calmodulin-dependent protein kinase II phosphorylate myosin light chain kinase on site A in vitro. Myosin light chain kinase phosphorylation on site A depends primarily on $[Ca^{2+}]_i$ regardless of the stimuli (Stull et al., 1990); Van Riper et al., 1995), probably via activation of Ca^{2+}-calmodulin-dependent protein kinase II (Tansey et al., 1992). High $[Ca^{2+}]_i$ increases myosin light chain kinase phosphorylation and therefore decreases its sensitivity to $[Ca^{2+}]_i$. These findings suggest that myosin light chain kinase phosphorylation cannot explain the variable $[Ca^{2+}]_i$ sensitivity of phosphorylation observed with agonists and high $[K^+]_o$. However, myosin light chain kinase phosphorylation may be responsible for the phenomenon of Ca^{2+}-induced desensitization, which may act to prevent any adverse effects resulting

TABLE I Relative $[Ca^{2+}]$ Sensitivity of Phosphorylation in the Swine Carotid Stimulated by Various Contractile or Relaxing Agents[a]

High Sensitivity ←							→ Low Sensitivity
1 μM PDBu	>	Histamine Phenylephrine Histamine & NaF 0.01 μM PDBu Histamine & FSK	>	Histamine & $[K^+]_o$ NaF & $[K^+]_o$	>	$[K^+]_o$ Bay K 8644 ATP PDBu & $[K^+]_o$	<	Stretch $[K^+]_o$ & FSK

[a]The $[Ca^{2+}]_i$ sensitivity of phosphorylation was determined from the relation between aequorin estimated $[Ca^{2+}]_i$ and myosin phosphorylation levels (Rembold, 1990, 1992). PDBu, phorbol dibutyrate; FSK, forskolin; $[K^+]_o$, depolarization.

from prolonged increases in $[Ca^{2+}]_i$ (Stull *et al.*, 1990; Himpens *et al.*, 1989; Ozaki *et al.*, 1991).

3. Myosin Phosphatase Could Be Regulated

Exogenous application of GTP analogues or contractile agonists increased the $[Ca^{2+}]_i$ sensitivity of phosphorylation in smooth muscle permeabilized with staphylococcal α-toxin (Nishimura *et al.*, 1988; Kitazawa *et al.*, 1991a). Exogenous application of either histamine or AlF_4^- (a nonspecific activator of G proteins) to depolarized intact tissues also increased the $[Ca^{2+}]_i$ sensitivity of phosphorylation (Rembold, 1990). These data suggest that agonist-dependent activation of a G protein may sensitize the myosin phosphorylation system to $[Ca^{2+}]_i$. Agonist-dependent increases in the Ca^{2+} sensitivity of phosphorylation has been called agonist-induced sensitization and may result from inhibition of myosin phosphatase. High concentrations of phenylephrine (100 μM) and 3 μM GTP-γ-S slowed dephosphorylation rates by 50% in α-toxin skinned rabbit portal vein (Kitazawa *et al.*, 1991b). Dephosphorylation rates were also slower in GTP-γ-S-treated smooth muscle homogenates (Kubota *et al.*, 1992). These results suggest that myosin phosphatase may be regulated. The mechanism for phosphatase regulation is unknown. Very high concentrations of arachidonic acid (300 μM) decrease the activity of smooth muscle myosin phosphatase (Gong *et al.*, 1992). However, such levels of arachidonic acid have yet to be demonstrated in intact smooth muscle.

4. Role of Protein Kinase C in the $[Ca^{2+}]_i$ Sensitivity of Phosphorylation

Phorbol diesters, activators of protein kinase C, contract smooth muscle Chatterjee and Tejada, 1986; Jiang and Morgan, 1987; Nishimura *et al.*, 1990). Phorbol diesters increase the $[Ca^{2+}]_i$ sensitivity of force (Chatterjee and Tejada, 1986; Jiang and Morgan, 1987; Nishimura *et al.*, 1990) as well as the $[Ca^{2+}]_i$ sensitivity of phosphorylation (Rembold and Murphy, 1988a). This large increase in $[Ca^{2+}]_i$ sensitivity has been called a $[Ca^{2+}]_i$-independent contraction by some investigators because the response required no increase in $[Ca^{2+}]_i$. Down-regulation of protein kinase C by prolonged incubation in phorbol myristic acid inhibited phorbol dibutyrate-induced increase in $[Ca^{2+}]_i$ sensitivity and contraction (Hori *et al.*, 1993). These data are consistent with phorbol diesters contracting smooth muscle by a form of pharmacomechanical coupling.

However, these data to not demonstrate that protein kinase C is definitively involved in agonist-induced increases in the $[Ca^{2+}]_i$ sensitivity of phosphorylation. Many of the early data on the role of

protein kinase C in smooth muscle contraction depended on the use of "specific" protein kinase C inhibitors. It is now clear that many of these inhibitors are only partially specific or nonspecific for protein kinase C, because the catalytic site appears to be similar in most serine/threonine protein kinases (Wilkinson and Hallam, 1994). Some of these inhibitors and even the phorbol diesters have other nonspecific actions, especially at the high concentration employed in some studies. Furthermore, down-regulation of protein kinase C with phorbol myristic acid had no effect on agonist-induced increases in the $[Ca^{2+}]_i$ sensitivity of phosphorylation in one smooth muscle (Hori *et al.*, 1993). These data suggest that phorbol diesters and agonists may increase the $[Ca^{2+}]_i$ sensitivity of phosphorylation by parallel mechanisms. Currently, the role of protein kinase C in agonist-induced contraction remains unclear.

5. Tyrosine Phosphorylation May Be Important in the Regulation of $[Ca^{2+}]_i$ Sensitivity

Smooth muscle tissues contain large amounts of tyrosine kinases such as pp60[c-src] (Di Salvo *et al.*, 1989). Putative tyrosine kinase inhibitors (geldanomycin, tyrphostin, and genistein) inhibited agonist-induced contraction in intact smooth muscle (Di Salvo *et al.*, 1993). Genistein inhibited Ca^{2+}-induced contractions in α-toxin skinned guinea pig ileal longitudinal smooth muscle (Pfitzer *et al.*, 1993). Tyrphostin did not have measurable effects on high $[K^+]_o$-induced contractions in intact vessels (Di Salvo *et al.*, 1993). These data suggest that inhibition of tyrosine kinases may specifically interfere with contractile agonist-induced increases in the Ca^{2+} sensitivity of force. However, tyrosine kinase inhibitors could also induce relaxation in intact smooth muscle by interfering with agonist-induced increases in $[Ca^{2+}]_i$. Genistein reportedly decreased whole-cell Ca^{2+} currents in rabbit ear artery with an IC_{50} of 36 μM (Wijetunge *et al.*, 1992). There are reports that contractile stimuli induce tyrosine phosphorylation of a number of proteins in smooth muscle (Yang, *et al.*, 1993; Hisada *et al.*, 1993). We found that histamine induced tyrosine phosphorylation of four proteins, however, the time course of tyrosine phosphorylation was slower than the time course of contraction and relaxation (C. M. Rembold and B. A. Weaver, unpublished data). These data suggest that phosphorylation of these four proteins studies was not primarily involved in either the initial or sustained phase of histamine-induced contraction. Histamine-dependent increases in phosphorylation of these four proteins on tyrosine residues may be either (1) permissive in the regulation of sustained contraction or (2) involved in other smooth muscle function such as growth, differentiation, and/or secretion. Also see Chapter 22 by Di Salvo *et al.* (this volume) for a more detailed discussion of the role of tyrosine phosphorylation in smooth muscle contraction.

6. $[Ca^{2+}]_i$ Sensitivity and the Accuracy of $[Ca^{2+}]_i$ Estimates

A potential cause for differences in the $[Ca^{2+}]_i$ sensitivity of phosphorylation is that the methodology for measuring $[Ca^{2+}]_i$ could be misleading. For example, $[Mg^{2+}]_i$ affects aequorin calibration (Blinks *et al.*, 1982) and aequorin overestimates $[Ca^{2+}]_i$ if $[Ca^{2+}]_i$ is inhomogeneous (Rembold, 1989; Kargacin and Fay, 1991). Protein binding, extracellular leakage, and loading of Fura 2 into organelles affects Fura 2 calibration (Uto *et al.*, 1991; Shuttleworth and Thompson, 1991). We investigated this possibility by measuring myoplasmic $[Ca^{2+}]$ with both Fura 2 and aequorin in intact swine carotid media. Both indicators revealed that histamine induced a smaller increase in $[Ca^{2+}]_i$ than that observed with high $[K^+]_o$ at similar levels of force. Histamine induced a higher $[Ca^{2+}]_i$ sensitivity of phosphorylation than that observed with high $[K^+]_o$ with both $[Ca^{2+}]_i$ indicators (Gilbert *et al.*, 1991). These data, obtained with two mechanistically different $[Ca^{2+}]_i$ indicators, suggest that the differences observed in the $[Ca^{2+}]_i$ sensitivity of phosphorylation are real. We have observed some differences in aequorin and Fura 2 $[Ca^{2+}]_i$ signals with other stimuli, for example, Ca^{2+} repletion after Ca^{2+} depletion (C. M. Rembold, D. A. Van Riper, and X.-L. Chen, 1995). These differences may represent focal increases in $[Ca^{2+}]_i$ (Rembold, 1989).

7. cAMP-Dependent Decreases in the $[Ca^{2+}]_i$ Sensitivity of Phosphorylation

cAMP-dependent protein kinase phosphorylates myosin light chain kinase on site A *in vitro* and therefore can decrease its Ca^{2+} sensitivity. Ten μM forskolin treatment increased total myosin light chain kinase phosphorylation (site unknown) in intact tissues (de Lanerolle *et al.*, 1984). However, myosin light chain kinase phosphorylation did not significantly change during isoproterenol-induced relaxation of bovine trachealis (Stull *et al.*, 1990; Miller *et al.*, 1983). Forskolin induced swine carotid relaxation primarily by decreasing $[Ca^{2+}]_i$ without increasing myosin light chain kinase phosphorylation (McDaniel *et al.*, 1991; Van Riper *et al.*, 1995).

8. cGMP-Dependent Decreases in the $[Ca^{2+}]_i$ Sensitivity of Phosphorylation

Several studies have shown a decrease in the $[Ca^{2+}]_i$ sensitivity of force in nitroprusside-relaxed in-

tact tissues (Karaki *et al.*, 1988) and cGMP-relaxed α-toxin skinned tissues (Nishimura and van Breemen, 1989). In the swine carotid, nitrovasodilators decrease the $[Ca^{2+}]_i$ sensitivity of phosphorylation and uncoupled force from myosin phosphorylation (McDaniel *et al.*, 1992). In Triton X-100 skinned guinea pig mesenteric arteriole, addition of cGMP-preactivated cGMP-dependent protein kinase reduced myosin phosphorylation and contractile force proportionally despite a constant $[Ca^{2+}]_i$, suggesting that cGMP primarily decreased the $[Ca^{2+}]_i$ sensitivity of phosphorylation in these skinned tissues (Pfitzer and Boels, 1991). The mechanism for cGMP-dependent decreases in the $[Ca^{2+}]_i$ sensitivity of phosphorylation is unknown. cGMP-dependent protein kinase cannot directly phosphorylate myosin light chain kinase on site A *in vitro*. If [cGMP] increases to high levels, cGMP could activate cAMP-dependent protein kinase, which could phosphorylate myosin light chain kinase. Several substrates for cGMP-dependent protein kinase have been found (Baltensperger *et al.*, 1990), however, with the exception of the protein phosphatase 2A data in parotid (White *et al.*, 1993), the link to relaxation by decreases in the $[Ca^{2+}]_i$ sensitivity of phosphorylation remains elusive.

V. REGULATION OF CONTRACTILE FORCE

Activated myosin light chain kinase phosphorylates the 20-kDa light chain of myosin on serine 19. This phosphorylation is associated with an increase in the actin-activated myosin ATPase activity (Ikebe *et al.*, 1988). With most stimuli, contractile force depends on increases in myosin phosphorylation (Hai and Murphy, 1989; Rembold, 1990). There are several examples of uncoupling force from myosin phosphorylation (i.e., decreasing force without proportional decreases in myosin phosphorylation): Ca^{2+} repletion protocols (Gerthoffer, 1987), high extracellular $[Mg^{2+}]$ (D'Angelo *et al.*, 1992; Bárány and Bárány, 1993), nitrovasodilators (McDaniel *et al.*, 1992), okadaic acid (Tansey *et al.*, 1990), and Ca^{2+} depletion (Aburto *et al.*, 1993). These exceptions to the dependence of force on myosin phosphorylation suggest that regulatory systems other than myosin light chain phosphorylation may also modulate contractile force in some smooth muscles. The relaxation induced by high extracellular $[Mg^{2+}]$ (D'Angelo *et al.*, 1992) or nitrovasodilators (McDaniel *et al.*, 1992) was not caused by phosphorylation of myosin on sites other than serine 19. Therefore, these relaxations may represent uncovering of thin filament regulatory mechanisms. These possibilities are discussed in chapters 6, 13, and 24–27.

VI. RELATIVE IMPORTANCE OF THE ELECTROMECHANICAL AND PHARMACOMECHANICAL CONTRACTILE MECHANISMS

We evaluated the relative importance of these electromechanical and pharmacomechanical mechanisms in the rat tail artery by measuring E_m, $[Ca^{2+}]_i$, and force. Both 0.1 and 0.3 μM phenylephrine depolarized the smooth muscle, increased Fura 2 estimated $[Ca^{2+}]$, and induced a contraction (Fig. 2). This increase in $[Ca^{2+}]_i$ was partially induced by depolarization increasing Ca^{2+} influx through L-type Ca^{2+} channels, because the Ca^{2+} channel blocker diltiazem decreased both Fura 2 estimated $[Ca^{2+}]_i$ and force to near resting

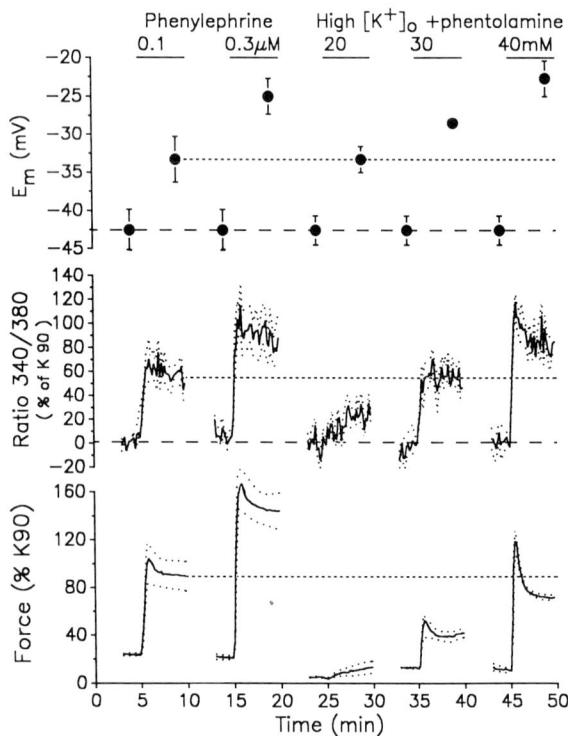

FIGURE 2   Comparison of E_m, Fura 2 fluorescence, and isometric force in de-endothelialized rat tail artery segments stimulated with phenylephrine or high $[K^+]_o$ (in the presence of 3 μM phentolamine to block the effects of endogenous norepinephrine release). Tissues were stimulated for 5 min with the following stimuli: 0.1 μM phenylephrine at 5 min, 0.3 μM phenylephrine at 15 min, 20 μM $[K^+]_o$ at 25 min, 30 mM $[K^+]_o$ at 35 min, and 40 mM $[K^+]_o$ at 45 min. The top panel shows mean E_m ± 1 SEM (symbols without error bars imply that the SEM is smaller than the symbol size). The middle panel shows the normalized mean Fura 340/380 ratio (solid line) ± 1 SEM (dotted lines). The bottom panel is normalized force in grams from the Fura 2 experiments (solid and dotted lines). Force from the E_m experiments was similar to the force measured in the Fura 2 experiments. The Fura 340/380 ratio was normalized to the measured response 5 min after stimulation with 90 mM $[K^+]_o$. Data are replotted from Chen and Rembold (1995).

values without significantly changing E_m (Chen and Rembold, 1995). This demonstrates that phenylephrine partially induced contraction by depolarization, an electromechanical mechanism. We then added 3 μM phentolamine to block the effects of high $[K^+]_o$ induced release of norepinephrine on α_1 receptors. Twenty mM $[K^+]_o$ depolarized the smooth muscle to a similar extent as that observed with 0.1 μM phenylephrine (in Fig. 2, the horizontal, finely dashed lines show the level of the 0.1 μM phenylephrine response). However, 20 mM $[K^+]_o$ did not increase $[Ca^{2+}]_i$ or induce as large a contraction as was observed with 0.1 μM phenylephrine. This demonstrates that phenylephrine also induced contraction by increasing $[Ca^{2+}]_i$ beyond that expected given the depolarization. This is a pharmacomechanical mechanism probably representing phenylephrine-dependent increased Ca^{2+} influx (see Chen and Rembold, 1995, for a full discussion). Thirty mM $[K^+]_o$ increased $[Ca^{2+}]_i$ to a level similar to that observed with 0.1 μM phenylephrine. However, 30 mM $[K^+]_o$ did not induce as large a contraction as was observed with 0.1 μM phenylephrine. This is a second pharmacomechanical mechanism representing phenylephrine-dependent increased $[Ca^{2+}]_i$ sensitivity of phosphorylation. These data show that a relatively low concentration of phenylephrine (0.1 μM) invoked one electromechanical and two pharmacomechanical mechanisms. Additional experiments (not shown) demonstrated that the third pharmacomechanical mechanism, Ca^{2+} release from the intracellular store, occurred only with high concentrations of phenylephrine (≤10 μM) (Chen and Rembold, 1995).

VII. SUMMARY AND PERSPECTIVES

In contrast with skeletal or cardiac muscle, smooth muscle evolved with a diversity of mechanisms for regulating contraction or relaxation. These mechanisms involve many aspects of Ca^{2+} regulation and also some aspects of the cellular response to changes in $[Ca^{2+}]_i$. This diversity may allow smooth muscle to retain a semblance of normal function despite elimination of one mechanism by a genetic abnormality, a pathophysiology state, or pharmacological therapy. This diversity also presents an opportunity for development of pharmacological agents that have relative specificities for certain types of smooth muscle.

Although it is convenient for physiologists to categorize smooth muscle excitation–contraction coupling mechanisms, it is important to understand that electromechanical and pharmacomechanical coupling mechanisms are complexly intertwined in intact smooth muscle. For example, any of the three systems that increase $[Ca^{2+}]_i$ will activate K_{Ca} (BK) channels, which will hyperpolarize the cell and reduce Ca^{2+} influx through L channels (Ganitkevich and Isenberg, 1990; Brayden and Nelson, 1992). Conversely, hyperpolarization inhibits IP_3 synthesis in some smooth muscles (Itoh *et al.*, 1992).

Finally, many of the foregoing discoveries resulted from the application of scientific reductionism to smooth muscle. Scientific reductionism has its limitations. These limitations are elegantly described in the chapter entitled "Ant Fugue" in "Godel, Escher, Bach" (Hofstadter, 1979). New methodologies reveal a diversity of smooth muscle contractile and relaxing mechanisms when smooth muscle is studied *ex vivo*, skinned, and homogenized. However, we reductionists should not lose sight of our primary goal to investigate how smooth muscles function *in vivo*. It is crucial that smooth muscle physiologists study not only new mechanisms, but also the relative importance of these mechanisms and how these mechanisms interact.

Acknowledgments

The author would like to thank Xiao-Liang Chen, Elizabeth Gilbert D'Angelo, Nancy McDaniel, Richard A. Murphy, Dee A. Van Riper, and Barbara Weaver for their involvement with my research over the years. C. M. Rembold is a Lucille P. Markey Scholar and grants from the Lucille P. Markey Charitable Trust, the PHS (1R01 HL38918), and the Virginia Affiliate of the American Heart Association supported this research.

References

Abdel-Latif, A. A. (1989). *Life Sci.* **45**, 757–786.

Abe, A., and Karaki, H. (1989). *J. Pharmacol. Exp. Ther.* **249**, 895–900.

Abe, K., Matsuki, N., and Kasuya, Y. (1987). *Jpn. J. Pharmacol.* **45**, 249–261.

Aburto, T. K., Lajoie, C., and Morgan, K. G. (1993). *Circ. Res.* **72**, 778–785.

Adelstein, R. S., Conti, M. A., and Hathaway, D. R. (1978). *J. Biol. Chem.* **253**, 8347–8350.

Anwer, K., Toro, L., Oberti, C., Stefani, E., and Sanborn, B. M. (1992). *Am. J. Physiol.* **263**, C1049–C1056.

Baltensperger, K., Chiesi, M., and Carafoli, E. (1990). *Biochemistry* **29**, 9753–9760.

Baron, C. B., Pring, M., and Coburn, R. F. (1989). *Am. J. Physiol.* **256**, C375–C383.

Bárány, M., and Bárány, K. (1993). *Arch. Biochem. Biophys.* **305**, 202–204.

Benham, C. D., and Tsien, R. W. (1987). *Nature (London)* **328**, 275–278.

Berridge, M. J. (1987). *Annu. Rev. Biochem.* **56**, 159–193.

Blinks, J. R., Wier, W. G., Hess, P., and Prendergast, F. G. (1982). *Prog. Biophys. Mol. Biol.* **4026**, 1–114.

Bolton, T. B. (1979). *Physiol. Rev.* **59**, 606–718.

Bonev, A. D., and Nelson, M. T. (1993). *Am. J. Physiol.* **265**, C1723–C1728.

Bozler, E. (1969). *Am. J. Physiol.* **216**, 671–673.

Brayden, J. E. (1990). *Am. J. Physiol.* **259**, H668–H673.

Brayden, J. E. (1991). *Circ. Res.* **69**, 1415–1420.

Brayden, J. E., and Nelson, M. T. (1992). *Science* **256**, 532–535.

Brayden, J. E., and Wellman, G. C. (1989). *J. Cereb. Blood Flow Metab.* **9**, 256–263.

Bruschi, G., Bruschi, M. E., Regolisti, G., and Borghetti, A. (1988). *Am. J. Physiol.* **254**, H840–H854.

Bulbring, E. (1957). *J. Physiol. (London)* **135**, 412–425.

Burgoyne, R. D., Cheek, T. R., Morgan, A., O'Sullivan, A. J., Moreton, R. B., Berridge, M. J., Mata, A. M., Coyler, J., Lee, A. G., and East, J. M. (1989). *Nature (London)* **342**, 72–74.

Casteels, R., and Suzuki, H. (1980). *Pflügers Arch.* **387**, 17–25.

Chatterjee, M., and Tejada, M. (1986). *Am. J. Physiol.* **251**, C356–C361.

Chen, X.-L., and Rembold, C. M. (1992). *Am. J. Physiol.* **263**, C468–C473.

Chen, X.-L., and Rembold, C. M. (1995). *Am. J. Physiol.* **268**, H74–H81.

Chilvers, E. R., Challiss, R. A. J., Barnes, P. J., and Nahorski, S. R. (1989). *Eur. J. Pharmacol.* **164**, 587–590.

Chuch, S. H., Mullaney, J. M., Ghosh, T. K., Zachary, A. L., and Gill, D. L. (1987). *J. Biol. Chem.* **262**, 13857–13864.

Clapp, L. H., and Gurney, A. M. (1991). *Pflügers Arch.* **418**, 462–470.

Clark, A. H., and Garland, C. J. (1993). *Eur. J. Pharmacol.* **235**, 113–116.

Cox, R. H. (1991). *Adv. Exp. Med. Biol.* **308**, 27–43.

D'Angelo, E. K. G., Singer, H. A., and Rembold, C. M. (1992). *J. Clin. Invest.* **89**, 1988–1994.

de Lanerolle, P., Nishikawa, M., Yost, D. A., and Adelstein, R. S. (1984). *Science* **223**, 1415–1417.

Den Hertog, A., Nelemans, S. A., Molleman, A., Hoiting, B. H., Van den Akker, J., and Duin, M. (1990). In "Frontiers in Smooth Muscle Research" (N. Sperilakis and J. D. Wood, eds.), pp. 183–192. Alan R. Liss, New York.

Deth, R. C., and van Breemen, C. (1977). *J. Membr. Biol.* **30**, 363–380.

Di Salvo, J., Gifford, D., and Kokkinakis, A. (1989). *J. Biol. Chem.* **264**, 10773–10778.

Di Salvo, J., Steusloff, A., Semenchuk, L., Satoh, S., Kolquist, K., and Pfitzer, G. (1993). *Biochem. Biophys. Res. Commun.* **190**, 968–974.

Doggrell, S. A., Smith, J. W., Downing, O. A., and Wilson, K. A. (1989). *Eur. J. Pharmacol.* **174**, 131–133.

Droogmans, G., Raeymaekers, L., and Casteels, R. (1977). *J. Gen. Physiol.* **70**, 129–148.

Exton, J. H. (1994). *Annu. Rev. Physiol.* **56**, 349–369.

Feletau, E., and Vanhoutte, P. M. (1988). *Br. J. Pharmacol.* **93**, 515–524.

Forder, J., Scriabine, A., and Rasmussen, H. (1985). *J. Pharmacol. Exp. Ther.* **235**, 267–273.

Francis, S. H., Noblett, B. D., Todd, B. W., Wells, J. N., and Corbin, J. D. (1988). *Mol. Pharmacol.* **34**, 506–517.

Furukawa, K., Tawada, Y., and Shigekawa, M. (1988). *J. Biol. Chem.* **263**, 8058–8065.

Furukawa, K.-I., Ohshima, N., Tawada-Iwata, Y., and Shigekawa, M. (1991). *J. Biol. Chem.* **266**, 12337–12341.

Ganitkevich, V., and Isenberg, G. (1990). *Circ. Res.* **67**, 525–528.

Gerthoffer, W. T. (1987). *J. Pharmacol. Exp. Ther.* **240**, 8–15.

Gilbert, E. K., Weaver, B. A., and Rembold, C. M. (1991). *FASEB J.* **5**, 2593–2599.

Gill, D. L., Ghosh, T. K., and Mullaney, J. M. (1989). *Cell Calcium* **10**, 363–374.

Gong, M. C., Fuglsang, A., Alessi, D., *et al.* (1992). *J. Biol. Chem.* **267**, 21492–21498.

Griendling, K. K., Rittenhouse, S. E., Brock, T. A., Ekstein, L. S., Jr.,

Gimbrone, M. A., and Alexander, R. W. (1986). *J. Biol. Chem.* **261,** 5901–5906.

Gunst, S. J., and Bandyopadhyay, S. (1989). *Am. J. Physiol.* **257,** C355–C364.

Haeusler, G., and De Peyer, J. (1989). *Eur. J. Pharmacol.* **166,** 175–182.

Hai, C.-M., and Murphy, R. A. (1989). *Annu. Rev. Physiol.* **51,** 285–298.

Hermsmeyer, K., Sturek, M., and Rusch, N. J. (1988). *Ann. N.Y. Acad. Sci.* **522,** 25–31.

Himpens, B., and Somlyo, A. P. (1988). *J. Physiol. (London)* **395,** 507–530.

Himpens, B., Matthijs, G., and Somlyo, A. P. (1989). *J. Physiol. (London)* **413,** 489–503.

Hisada, T., Walsh, J. V., Jr., and Singer, J. J. (1993). *Pflügers Arch.* **422,** 393–396.

Hofstadter, D. R. (1979). "Godel, Escher, Bach," pp. 311–336. Vintage Books, New York.

Hogg, R. C., Wang, Q., and Large, W. A. (1994). *Br. J. Pharmacol.* **111,** 1333–1341.

Hori, M., Sato, K., Miyamoto, S., Ozaki, H., and Karaki, H. (1993). *Br. J. Pharmacol.* **110,** 1527–1531.

Howe, P. H., Akhtar, R. A., Naderi, S., and Abdel-Latif, A. A. (1986). *J. Pharmacol. Exp. Ther.* **239,** 574–583.

Ikebe, M., Koretz, J., and Hartshorne, D. J. (1988). *J. Biol. Chem.* **263,** 6432–6437.

Ishikawa, T., Hume, J. R., and Keef, K. D. (1993). *Circ. Res.* **73,** 1128–1137.

Itoh, T., Kuriyama, H., and Ueno, H. (1983). *J. Physiol. (London)* **343,** 233–252.

Itoh, T., Seki, N., Suzuki, S., Ito, S., Kajikuri, J., and Kuriyama, H. (1992). *J. Physiol. (London)* **451,** 307–328.

Jensen, P. E., Mulvany, M. J., and Aalkjær, C. (1992). *Pflügers Arch.* **420,** 536–543.

Jiang, M. J., and Morgan, K. G. (1987). *Am. J. Physiol.* **253,** H1365-H1371.

Jiang, M. J., and Morgan, K. G. (1989). *Pflügers Arch.* **413,** 637–643.

Kamm, K. E., and Stull, J. T. (1989). *Annu. Rev. Physiol.* **51,** 299–313.

Karaki, H., Sato, K., Ozaki, H., and Murakami, K. (1988). *Eur. J. Pharmacol.* **156,** 259–266.

Kargacin, G., and Fay, F. S. (1991). *Biophys. J.* **60,** 1088–1100.

Keef, K. D., and Bowen, S. M. (1989). *Am. J. Physiol.* **257,** H1096–H1103.

Keef, K. D., and Ross, G. (1986). *Am. J. Physiol.* **250,** H524–H529.

Keef, K. D., Pasco, J. S., and Eckman, D. M. (1992). *J. Pharmacol. Exp. Ther.* **260,** 592–600.

Kimura, M., Kimura, I., and Kobayashi, S. (1982). *Biochem. Pharmacol.* **31**(19), 3077–3083.

Kitazawa, T., and Somlyo, A. P. (1990). *Biochem. Biophys. Res. Commun.* **172,** 1291–1297.

Kitazawa, T., Gaylinn, B. D., Denney, G. H., and Somlyo, A. P. (1991a). *J. Biol. Chem.* **266,** 1708–1715.

Kitazawa, T., Masuo, M., and Somlyo, A. P. (1991b). *Proc. Natl. Acad. Sci. U.S.A.* **88,** 9307–9310.

Kobayashi, S., Kanaide, H., and Nakamura, M. (1985). *Science* **229,** 553–555.

Kobayashi, S., Somlyo, A. P., and Somlyo, A. V. (1988a). *J. Physiol. (London)* **403,** 601–619.

Kobayashi, S., Somlyo, A. V., and Somlyo, A. P. (1988b). *Biochem. Biophys. Res. Commun.* **153,** 625–631.

Komori, K., Lorenz, R. R., and Vanhoutte, P. M. (1988). *Am. J. Physiol.* **255,** H207–H212.

Kubota, Y., Nomura, M., Kamm, K. E., Mumby, M. C., and Stull, J. T. (1992). *Am. J. Physiol.* **262,** C405–C410.

Langlands, J. M., and Diamond, J. (1990). *Biochem. Biophys. Res. Commun.* **173,** 1258–1265.

Lee, H. C., and Aarhus, R. (1993). *Biochim. Biophys. Acta Protein Struct. Mol. Enzymol.* **1164,** 68–74.

Lincoln, T. M., and Cornwell, T. L. (1991). *Blood Vessels,* **28,** 129–137.

Lincoln, T. M., Cornwell, T. L., and Taylor, A. E. (1990). *Am. J. Physiol.* **258,** C399–C407.

Long, C. J., and Stone, I. W. (1987). *J. Pharm. Pharmacol.* **39,** 1010–1014.

Majerus, P. W. (1992). *Annu. Rev. Biochem.* **61,** 225–250.

McDaniel, N. L., Rembold, C. M., Richard, H. L., and Murphy, R. A. (1991). *J. Physiol. (London)* **439,** 147–160.

McDaniel, N. L., Chen, X.-L., Singer, H. A., Murphy, R. A., and Rembold, C. M. (1992). *Am. J. Physiol.* **263,** C461–C467.

McPherson, G. A., and Angus, J. A. (1991). *Br. J. Pharmacol.* **103,** 1184–1190.

Miller, J. R., Silver, P. J., and Stull, J. T. (1983). *Mol. Pharmacol.* **24,** 235–242.

Miller-Hance, W. C., Miller, J. R., Wells, J. N., Stull, J. T., and Kamm, K. E. (1988). *J. Biol. Chem.* **263,** 13979–13982.

Miyoshi, Y., and Nakaya, Y. (1991). *Biochem. Biophys. Res. Commun.* **181,** 700–706.

Morgan, J. P., and Morgan, K. G. (1984a). *J. Physiol. (London)* **351,** 155–167.

Morgan, J. P., and Morgan, K. G. (1984b). *J. Physiol. (London)* **357,** 539–551.

Motulsky, H. J., and Michel, M. C. (1988). *Am. J. Physiol.* **255,** E880–E885.

Mulvany, M. J., Nilsson, H., and Flatman, J. A. (1982) *J. Physiol. (London)* **332,** 363–373.

Nagao, T., and Vanhoutte, P. M. (1992). *Br. J. Pharmacol.* **107,** 1102–1107.

Neild, T. O., and Kotecha, N. (1987). *Circ. Res.* **60,** 791–795.

Nelson, M. T., Standen, N. B., Brayden, J. E., and Worley, J. F., III (1988). *Nature (London)* **336,** 382–385.

Nishimura, J., and van Breemen, C. (1989). *Biochem. Biophys. Res. Commun.* **163,** 929–935.

Nishimura, J., Kolber, M., and van Breemen, C. (1988). *Biochem. Biophys. Res. Commun.* **157,** 677–683.

Nishimura, J., Khalil, R. A., Drenth, J. P., and van Breemen, C. (1990). *Am. J. Physiol.* **259,** H2–H8.

Ohya, Y., and Sperelakis, N. (1991). *Circ. Res.* **68,** 763–771.

Okada, Y., Yanagisawa, T., and Taira, N. (1993). *Naunyn-Schmiedeberg's Arch. Pharmacol.* **347,** 438–444.

Ozaki, H., Gerthoffer, W. T., Publicover, N. G., Fusetani, N., and Sanders, K. M. (1991). *J. Physiol. (London)* **440,** 207–224.

Parker, I., Ito, Y., Kuriyama, H., and Miledi, R. (1987). *Proc. R. Soc. London* **230,** 207–214.

Pfitzer, G., and Boels, P. J. (1991). *Blood Vessels* **28,** 262–267.

Pfitzer, G., Steusloff, A., Kolquist, K., and Di Salvo, J. (1993). *Biophys. J.* **64,** A257.

Plane, F., and Garland, C. J. (1993). *Br. J. Pharmacol.* **110,** 651–656.

Popescu, L. M., Foril, C. P., Hinescu, M., Panoiu, C., Cinteza, M., and Gherasim, L. (1985). *Biochem. Pharmacol.* **34,** 1857–1860.

Quayle, J. M., Bonev, A. D., Brayden, J. E., and Nelson, M. T. (1994). *J. Physiol. (London)* **475,** 9–13.

Rapoport, R. M. (1986). *Circ. Res.* **58,** 407–410.

Rashatwar, S. S., Cornwell, T. L., and Lincoln, T. M. (1987). *Proc. Natl. Acad. Sci. U.S.A.* **84,** 5685–5689.

Rasmussen, H., Takuwa, Y., and Park, S. (1987). *FASEB J.* **1,** 177–185.

Ratz, P. H., and Murphy, R. A. (1987). *Circ. Res.* **60,** 410–421.

Rembold, C. M. (1989). *J. Physiol. (London)* **416,** 273–290.

Rembold, C. M. (1990). *J. Physiol. (London)* **429,** 77–94.

Rembold, C. M. (1992). *Hypertension* **20,** 129–137.

Rembold, C. M., and Murphy, R. A. (1986). *Circ. Res.* **58,** 803–815.

Rembold, C. M., and Murphy, R. A. (1988a). *Am. J. Physiol.* **255**, C719–C723.

Rembold, C. M., and Murphy, R. A. (1988b). *Circ. Res.* **63**, 593–603.

Rembold, C. M., Van Riper, D. A., and Chen, X.-L., *J. Physiol. (London)* in press (1995).

Rembold, C. M., and Murphy, R. A. (1989). *Am. J. Physiol.* **257**, C122–C128.

Rembold, C. M., Weaver, B. A., and Linden, J. (1991). *J. Biol. Chem.* **266**, 5407–5411.

Robertson, B. E., Schubert, R., Hescheler, J., and Nelson, M. T. (1993). *Am. J. Physiol.* **265**, C299–C303.

Salmon, D. M. W., and Bolton, T. B. (1988). *Biochem. J.* **254**, 553–557.

Savaria, D., Lanoue, C., Cadieux, A., and Rousseau, E. (1992). *Am. J. Physiol.* **262**, L327–L336.

Scornik, F. S., and Toro, L. (1992). *Am. J. Physiol.* **262**, C708–C713.

Shuttleworth, T. J., and Thompson, J. L. (1991). *J. Biol. Chem.* **266**, 1410–1414.

Sims, S. M. (1992). *J. Physiol. (London)* **449**, 377–398.

Somlyo, A. V., and Somlyo, A. P. (1968). *J. Pharmacol. Exp. Ther.* **159**, 129–145.

Somlyo, A. V., Bond, M., Somlyo, A. P., and Scarpa, A. (1985). *Proc. Natl. Acad. Sci. U.S.A.* **82**, 5231–5235.

Stull, J. T., Hsu, L.-C., Tansey, M. G., and Kamm, K. E. (1990). *J. Biol. Chem.* **265**, 16683–16690.

Suematsu, E., Hirata, M., Hashimoto, T., and Kuriyama, H. (1984). *Biochem. Biophys. Res. Commun.* **120**, 481–485.

Takuwa, Y., Takuwa, N., and Rasmussen, H. (1986). *J. Biol. Chem.* **261**, 14670–14675.

Takuwa, Y., Takuwa, N., and Rasmussen, H. (1988). *J. Biol. Chem.* **263**, 762–768.

Tansey, M. G., Hori, M., Karaki, H., Kamm, K. E., and Stull, J. T. (1990). *FEBS Lett.* **270**, 219–221.

Tansey, M. G., Word, R. A., Hidaka, H., Singer, H. A., Schworer, C. M., Kamm, K. E., and Stull, J. T. (1992). *J. Biol. Chem.* **267**, 12511–12516.

Tare, M., Parkington, H. C., Coleman, H. A., Neild, T. O., and Dusting, G. J. (1990). *Nature (London)* **346**, 69–71.

Taylor, C. W., and Richardson, A. (1991). *Pharmacol. Ther.* **51**, 97–137.

Taylor, D. A., Bowman, B. F., and Stull, J. T. (1989). *J. Biol. Chem.* **264**, 6207–6213.

Thornbury, K. D., Ward, S. M., Dalziel, H. H., Carl, A., Westfall, D. P., and Sanders, K. M. (1991). *Am. J. Physiol.* **261**, G553–G557.

Twort, C. H., and van Breemen, C. (1988). *Circ. Res.* **62**, 961–964.

Uto, A., Arai, H., and Ogawa, Y. (1991). *Cell Calcium* **12**, 29–37.

van Breemen, C., Lukeman, S., Leijten, P., Yamamoto, H., and Loutzenhiser, R. (1986). *J. Cardiovasc. Pharmacol.* **8**, S111–S116.

Van Riper, D. A., Weaver, B. A., Stull, J. T., and Rembold, C. M. (1995). *Am. J. Physiol.* **268**, H2466–H2475.

Vidbaek, L. M., Aalkjær, C., and Mulvany, M. J. (1988). *Br. J. Pharmacol.* **95**, 103–108.

Walker, J. W., Somlyo, A. V., Goldman, Y. E., Somlyo, A. P., and Trentham, D. R. (1987). *Nature (London)* **327**, 249–252.

White, R. E., Lee, A. B., Shcherbatko, A. D., Lincoln, T. M., Schonbrunn, A., and Armstrong, D. L. (1993). *Nature (London)* **361**, 263–266.

Wijetunge, S., Aalkjær, C., Schachter, M., and Hughes, A. D. (1992). *Biochem. Biophys. Res. Commun.* **189**, 1620–1623.

Wilkinson, S. E., and Hallam, T. J. (1994). *Trends Pharmacol. Sci.* **15**, 53–57.

Williams, D. L., Jr., Katz, G. M., Roy-Contancin, L., and Reuben, J. P. (1988). *Proc. Natl. Acad. Sci. U.S.A.* **85**, 9360–9364.

Word, R. A., Casey, M. L., Kamm, K. E., and Stull, J. T. (1991). *Am. J. Physiol.* **260**, C861–C867.

Worley, P. F., Baraban, J. M., Supattapone, S., Wilson, V. S., and Snyder, S. H. (1987). *J. Biol. Chem.* **262**, 12132–12136.

Yang, S.-G., Saifeddine, M., Laniyonu, A., and Hollenberg, M. D. (1993). *J. Pharmacol. Exp. Ther.* **264**, 958–966.

19

Calcium Pumps

LUC RAEYMAEKERS and FRANK WUYTACK

Laboratorium voor Fysiologie
K. U. Leuven, Campus Gasthuisberg O/N
Leuven, Belgium

I. INTRODUCTION

Ca^{2+} pumps in the plasma membrane (PM) compensate the influx of Ca^{2+} along a steep electrochemical gradient, thereby allowing the cells to maintain a steady state with respect to total Ca^{2+} content. In most cell types, two different transporters operate in parallel: the Ca^{2+}-transport ATPase and the Na^+–Ca^{2+} exchanger. These transporters also participate in returning the cell to the resting state at the end of a stimulus by decreasing the cytosolic Ca^{2+} concentration ($[Ca^{2+}]_i$) below 100 nM. In the latter function they are assisted by Ca^{2+} pumps of the intracellular Ca^{2+} stores, presumably the endoplasmic reticulum (ER) or a subcompartment of it. These pumps transport Ca^{2+} from the cytoplasm into the lumen of the ER, thereby accumulating Ca^{2+} up to a concentration in the millimolar range. This high luminal Ca^{2+} content represents a store of Ca^{2+} that can be released upon excitation of the cell, and also plays a role in the regulation of the synthesis, folding, and sorting of proteins in the ER (Broström and Broström, 1990; Sambrook, 1990; Suzuki *et al.*, 1991; Wileman *et al.*, 1991). Besides these basic functions that are necessary for the survival of the cell, Ca^{2+} pumps contribute to the fine-tuning of cellular function by acting as targets for the action of second messengers and through the developmental- or tissue-dependent expression of functionally different isoforms.

The Ca^{2+}-transport ATPases belong to the group of P-type transport ATPases, which are characterized by the transfer of the γ-phosphate of ATP to the enzyme to form a transient phosphorylated intermediate. The Ca^{2+}-transport ATPases of the PM and of the ER are sufficiently different to categorize them into distinct gene families, designated PMCA (plasma membrane calcium-transporting ATPase) and SERCA [sarco(endo)plasmic reticulum calcium-transporting ATPase], respectively. They differ in molecular weight, immunological reactivity, and the differential interaction with the regulator proteins calmodulin and phospholamban. Both gene families are encoded by multiple genes. The isoform diversity is further increased by alternative processing of the primary transcripts. The study of the properties and the differential expression of these different isoforms has only just started. The first part of this review will deal with a short summary and an updating of what is currently known about the structure and diversity of Ca^{2+}-transport ATPases. More background information can be found in reviews on the PM Ca^{2+} pump (including Garrahan and Rega, 1990; Strehler, 1991; Carafoli, 1992, 1994; Wuytack and Raeymaekers, 1992; Penniston and Enyedi, 1994) or the SR-ER-type Ca^{2+} pump (MacLennan, 1990; Inesi *et al.*, 1990; Inesi and Kirtley, 1990) or both (Carafoli and Chiesi, 1992; Grover and Khan, 1992). The main part of this review will focus specifically on Ca^{2+} pumps in smooth muscle cells, particularly on the differential expression of isoforms and the role of Ca^{2+} pumps as mediators of the effects of second messengers. Related topics have been reviewed by Raeymaekers and Wuytack (1993), by O'Donnell and Owen (1994), and by Lompré *et al.* (1994).

II. ROLE OF Ca²⁺ PUMPS IN THE REGULATION OF CYTOSOLIC Ca²⁺

In smooth muscle cells, the PM transporters are not outnumbered to such an extent by the intracellular Ca^{2+} pumps as in skeletal muscle. In some smooth muscles, the number of PM Ca^{2+}-transport sites may even be higher than that in the ER (Raeymaekers *et al.*, 1985). The smaller number of ER Ca^{2+} pumps in smooth muscle cannot be explained by a smaller volume of intracellular Ca^{2+} stores in smooth as compared to striated muscle (Somlyo, 1985; Ashida *et al.*, 1988). Rather, it is due to a lower density of Ca^{2+} pumps in the ER membranes (see Section IV.A). This explains why the rate of the decrease of $[Ca^{2+}]_i$ at the end of a stimulus, and hence the rate of relaxation, is much slower in smooth muscle than in striated muscle. Moreover, the surface-to-volume ratio of smooth-muscle cells is much higher than that of skeletal muscle cells, creating a constellation in which the PM could have a relatively more important contribution to Ca^{2+} fluxes. Furthermore, in smooth muscle cells an important fraction of the ER is located peripherally just beneath the PM, often in association with caveolae (Gabella, 1989; Villa *et al.*, 1993). This organization may reflect a specialized direct interaction between the ER and the PM via protein–protein interactions analogous to that between the SR terminal cisternae and the T-tubuli in skeletal muscle cells (Irvine, 1990), or an indirect interaction by the creation of a restricted Ca^{2+} compartment separate from the bulk cytosol if the buffering capacity of the superficial ER is sufficiently high. The latter model is also known as the buffer barrier hypothesis (reviewed in van Breemen and Saida, 1989; Chen and van Breemen, 1992). This complex morphological arrangement makes it difficult to obtain sufficient spatial and temporal resolution of Ca^{2+} signals to elucidate the operation of the different Ca^{2+} regulatory processes and their relative contribution to the generation of the Ca^{2+} transient. Despite these difficulties, some quantitative estimates of Ca^{2+}-pump-mediated fluxes have been made.

It has been estimated that the caffeine-releasable content of the ER of rat mesenteric artery smooth muscle cells constitutes about 80% of total cellular Ca^{2+} (Baró *et al.*, 1993). The Ca^{2+} content of the ER represents about 10 times the amount of Ca^{2+} needed to trigger a single contraction. The rate of uptake into the ER would correspond to a rate of decrease of $[Ca^{2+}]_i$ of 45 nM/sec, which is sufficiently fast for ER Ca^{2+} pumping to be a major mechanism of Ca^{2+} removal during relaxation (Kargacin and Kargacin, 1994).

Kargacin and Fay (1991) concluded from computer simulations on one- and two-dimensional diffusion models of a smooth muscle cell that the rate of decline of Ca^{2+} transients is mainly determined by Ca^{2+} accumulation in the ER, whereas the Ca^{2+} efflux across the PM appeared more likely to be involved in the long-term regulation of Ca^{2+}.

An estimate of the relative importance of Ca^{2+}-ATPase and Na^+–Ca^{2+} exchange has been made in cultured rat aortic smooth muscle cells by Furukawa *et al.* 1988). At $[Ca^{2+}]_i$ below 1 μM, the contribution of the former was significantly greater than that of the latter.

One should be cautious, however, to extrapolate such quantitative data to all smooth muscle types or even to similar ones of other species because of the heterogeneity of smooth muscle. For instance, Stehno-Bittel and Sturek (1992) found evidence for preferential release of Ca^{2+} from the ER toward the sarcolemma in bovine but not in porcine coronary artery. The ER also differs between different smooth muscles in the expression of different types of calcium-buffering proteins (Raeymaekers *et al.*, 1993) and of the regulator protein phospholamban (see Section V). Furthermore, the data of Eggermont *et al.* (1988) indicate that the ratio of the number of ER and PM pump proteins also varies considerably between different smooth muscle types.

The use of the SERCA-selective inhibitors thapsigargin, cyclopiazonic acid (CPA), and 2,5-di(*tert*-butyl)-1,4-benzohydroquinone (tBu-BHQ) has confirmed the important contribution of intracellular Ca^{2+} uptake to the regulation of $[Ca^{2+}]_i$ in smooth muscle cells. This topic has been reviewed by Darby *et al.* (1993).

III. GENERAL PROPERTIES OF Ca²⁺ PUMPS AND THEIR REGULATORS

A. General Properties of Ca²⁺ Pumps

The reaction mechanism of the SERCA- and PMCA-type Ca^{2+} pumps is essentially identical, involving the transient formation and hydrolysis of an acylphosphate bond to an aspartate residue. For more information on the elementary steps of the reaction cycle the reader is referred to the reviews cited in the introduction and references therein.

Our knowledge on the structure of Ca^{2+}-ATPases is mainly based on studies of the Ca^{2+} pump of the SR of fast skeletal muscle (a SERCA1 gene product). However, the general structural model that emerged from these studies is probably also valid for the other SERCAs as well as for the PM Ca^{2+} pump. The predicted structure of the Ca^{2+}-transport ATPase incorporates 10 membrane-spanning helices, most of them connected outside the membrane by short peptide stretches, except for two extended hydrophilic loops at the cytoplasmic side, a shorter one between membrane-spanning domains M2 and M3, and a long loop between M4 and M5 (Fig. 1). The longer loop contains the nucleotide-binding domain and the phosphate-accepting aspartate residue (Green and Stokes, 1992). PMCA, but not SERCA pumps, possess a third cytoplasmic domain formed by the C terminus. Binding of the transported Ca^{2+} probably occurs in the membrane domain at several sites distributed over four transmembrane helices (Clarke *et al.*, 1989).

The alignment of the SERCA and PMCA amino acid sequences not only shows a similar homologous conservation of the number of putative transmembrane regions but also indicates several regions of high sequence conservation, particularly surrounding the phosphorylation site and in the ATP-binding domain. The alignment also shows several insertions or deletions in PMCA and SERCA, as shown schematically in Fig. 1. PMCA proteins have both an extended N terminus and a markedly longer C-terminal tail that together largely explain the higher molecular mass of about 130 kDa, as compared to about 105 kDa for SERCA.

The C-terminal domain of PMCA pumps is crucial for the pump's regulation by calmodulin, acidic phospholipids, and protein kinases. Activation of the PM Ca^{2+}-transport ATPase by calmodulin results in an increase of the V_{max} and in an increase of the affinity for Ca^{2+} (Niggli *et al.*, 1981). The binding of calmodulin allowed the purification of the PM Ca^{2+} pump by affinity chromatography from erythrocytes (Niggli *et al.*, 1979), and from pig stomach (Wuytack *et al.*, 1981) and bovine aorta (Furukawa and Nakamura, 1984) smooth muscle. In the absence of calmodulin, the calmodulin-binding domain is able to interact with two regions far upstream in the pump sequence (Fig. 1), resulting in inhibition of the activity. On binding of calmodulin, this self-inhibition is removed (Falchetto *et al.*, 1991, 1992).

The alignment of the SERCA and PMCA sequences also shows in PMCA pumps an insert just upstream of the third transmembrane segment that is thought to bind negatively charged phospholipids and thereby to contribute to the stimulatory effect of these compounds (Zvaritch *et al.*, 1990). In the SERCA sequences there is an insert downstream of the phosphorylation site that is the binding site for the unphosphorylated form of the regulatory protein phospholamban (James *et al.*, 1989) (Fig. 1). Unphosphorylated phospholam-

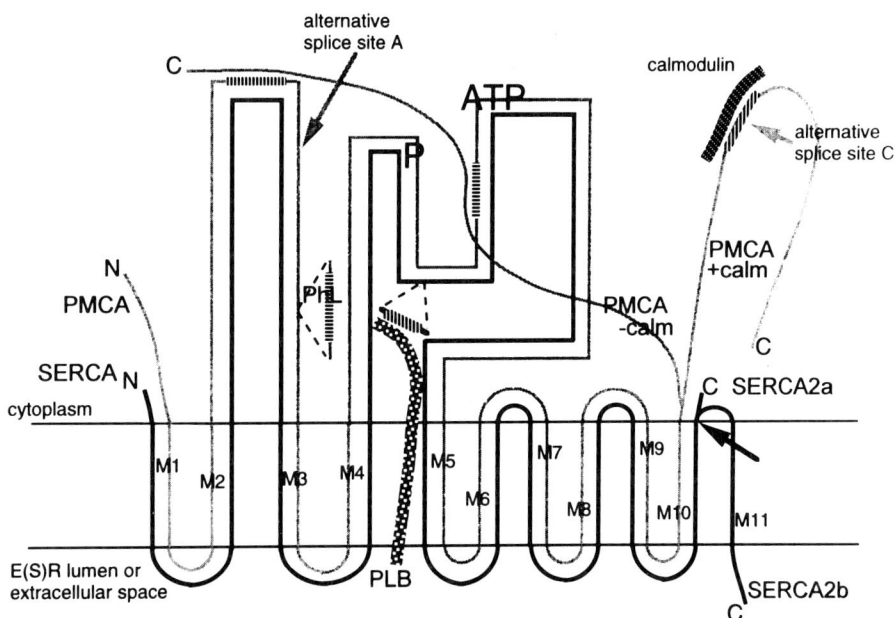

FIGURE 1 Schematic alignment of the major domains in the SERCA (black line) and PMCA (gray line) primary structure, taking into account the approximate positions of the transmembrane domains (M), the large cytosolic regions, insertions, binding sites (striped regions), and the positions corresponding to sites of alternative splicing (arrows). PLB, phospholamban, represented as a monomer; PhL, phospholipid-binding site; ATP, ATP-binding loops; P, phosphorylation site.

ban inhibits the pump, mainly by decreasing the apparent affinity of the Ca^{2+} pump for Ca^{2+}. Relief of the inhibition is brought about by phosphorylation by cyclic AMP-dependent protein kinase (cAK), cyclic GMP-dependent protein kinase (cGK), or calmodulin-dependent protein kinase II. The role of unphosphorylated phospholamban bound to the SERCA pump can be considered analogous to that of the calmodulin-binding domain of PMCA, which is autoinhibitory. Thus the role of the phosphorylation of SERCA pump-associated phospholamban corresponds to that of binding of calmodulin to the PMCA pump. This view is further strengthened by sequence homology between phospholamban and the calmodulin-binding domain of the PMCA-type Ca^{2+} pump (Chiesi et al., 1991).

PMCA and SERCA primary transcripts are alternatively spliced, generating a large number of isoforms, particularly in PMCA.

B. Ca^{2+}-Pump Isoforms

1. Isoforms of the Endoplasmic Reticulum-Type Ca^{2+} Pumps (SERCA Pumps)

The SERCA family consists of three genes, SERCA1, SERCA2, and SERCA3. The primary transcripts of SERCA1 and SERCA2 are subject to alternative processing, each transcript giving rise to two protein isoforms. SERCA1 and SERCA3 show a restricted expression pattern, whereas SERCA2 is ubiquitously expressed. In the following, attention will be focused on the SERCA2 isoform since the ER Ca^{2+} pump of smooth muscle is transcribed exclusively from this gene, as shown by immunological studies (Wuytack et al., 1989a) and mRNA analysis (Lytton et al., 1989; Eggermont et al., 1989). In SERCA2b the last 4 amino acids of SERCA2a are replaced by a stretch of 49 amino acids. Since this stretch contains a hydrophobic segment it could form an additional transmembrane domain. In this case the C termini of SERCA2a and SERCA2b would be located on opposite sites of the membrane. Campbell et al. (1992) provided experimental evidence for such an arrangement. An interesting finding is that SERCA2b presents a higher affinity for Ca^{2+} but a lower turnover rate than SERCA2a (Verboomen et al., 1992). The stretch of the last 12 residues of SERCA2b is of critical importance for this difference (Verboomen et al., 1994).

SERCA2b is probably ubiquitously expressed and can be considered as a "housekeeping" isoform. Alternative splicing of SERCA2 to the 2a isoform only occurs in cardiac muscle, slow skeletal muscle, and to some extent also in smooth muscle (Gunteski-Hamblin et al., 1988; Eggermont et al., 1990b). The nucleotide sequences governing this muscle-specific splicing are being investigated by Van Den Bosch et al. (1994).

2. Isoforms of the Plasmalemmal Ca^{2+} Pumps (PMCA Pumps)

It is beyond the scope of this chapter to overview the complex isoform diversity generated by alternative splicing of the primary transcripts of the four recognized PMCA genes. The isoform diversity has been updated and reviewed by Carafoli (1994). The following is intended to represent the most essential information.

Alternative splicing of PMCA transcripts occurs at two sites. The site designated "C" affects the presence and the position of a consensus recognition sequence for cAK, and in some isoforms also the sequence of the calmodulin-binding domain. This domain may occur in two forms, named "a" and "b." The splice variant containing the b domain is the most common, whereas the a form is expressed in a limited range of tissues. The PMCA1 isoform containing domain a does not possess the cAK phosphorylation site. In addition, this isoform has a lower affinity for calmodulin, resulting in a lower apparent affinity for Ca^{2+} (Enyedi et al., 1994).

A second site of alternative splicing (designated site "A") occurs in PMCA2 and is located upstream of the phospholipid-binding domain and downstream of a region involved in the activation of the pump by calmodulin (Fig. 1). Although it is conceivable that these isoforms would differ in their regulation by phospholipids, the expression of the corresponding proteins in Sf9 insect cells failed to reveal different functional responses to activating phospholipids (Hilfiker et al., 1994).

Because of the limited availability of pure protein preparations of specific isoforms, little is known of the functional properties of the different isoforms and their regulation.

C. Structure and Function of Phospholamban

Phospholamban is a homopentamer, each subunit consisting of 52 amino acids. The subunits consist of a 30-residue N-terminal hydrophilic region containing the phosphorylation sites, and a C-terminal hydrophobic domain responsible for oligomerization and for anchoring phospholamban in the membrane. The N-terminal domain is a basic amphiphilic α-helix, which resembles the calmodulin-binding domain of many proteins (Chiesi et al., 1991). Phospholamban is phosphorylated at serine-16 by cAK and cGK and at threonine-17 by Ca^{2+}/calmodulin-dependent protein kinase II. It has been proposed that unphosphorylated phospholamban inhibits the pump by inducing the formation of oligomeric complexes of the Ca^{2+}-pump protein and that phosphorylation of phospholamban releases the pump from this kinetically unfavorable associated state (Voss *et al.*, 1994).

IV. EXPRESSION OF Ca^{2+} PUMPS AND PHOSPHOLAMBAN IN SMOOTH MUSCLE CELLS

A. Expression of the Endoplasmic Reticulum Ca^{2+} Pump

Ca^{2+} uptake in ER vesicles is usually measured in the presence of oxalate. Oxalate very quickly permeates ER membranes and coprecipitates with calcium that is transported into the lumen. This does not occur in PM vesicles (Wibo *et al.*, 1981; Raeymaekers *et al.*, 1983). ER vesicles can be separated from PM by increasing the density of the ER vesicles by calcium-oxalate loading (Raeymaekers *et al.*, 1980; Raeymaekers and Hasselbach, 1981) or by selectively increasing the density of the PM vesicles with digitonin (Wibo *et al.*, 1981; Raeymaekers *et al.*, 1985). The highest rate of Ca^{2+} uptake reported for purified ER from smooth muscle, about 100 nmol/mg/min (Raeymaekers *et al.*, 1985), is still at least one order of magnitude lower than that in cardiac SR. Correspondingly lower values are also seen when determining the Ca^{2+}-stimulated ATPase activity, the steady-state level of phosphoprotein intermediate, immunoreactivity, and the number of Ca^{2+}-pump particles visible after freeze-fracturing (Raeymaekers *et al.*, 1980; Raeymaekers and Hasselbach, 1981; Wuytack *et al.*, 1984, 1989b; Khan and Grover, 1993). In the smooth muscle of the rabbit stomach the difference in expression level as compared to cardiac muscle was 70-fold at the protein level, 3.5-fold at the mRNA level, and only 1.2-fold at the level of nuclear mRNA precursors. This observation suggests that differences in expression level are caused by post-transcriptional steps (Khan *et al.*, 1990; Khan and Grover, 1993).

Based on immunological evidence and on the analysis of the tryptic fragments of the phosphorylated transport intermediate, it was demonstrated that the smooth muscle ER Ca^{2+} pump was more similar to that of cardiac and slow skeletal muscle than to that of fast skeletal muscle (Wuytack *et al.*, 1989a). This finding was confirmed by cDNA sequencing, showing that the ER Ca^{2+} pump of smooth muscle is transcribed from the same gene (SERCA2) as in slow skeletal and cardiac muscle (Lytton *et al.*, 1989; Eggermont

et al., 1989). These new tools were then applied to investigate the tissue-specific expression of the two isoforms generated by alternative splicing of the SERCA2 gene and to quantitate their expression relative to that in cardiac muscle. Knowledge on the differential expression of SERCA2a and SERCA2b is important because of the functional differences between these splice variants (Section IV.A).

Antibodies that discriminate between SERCA2a and SERCA2b were used in combination with non-discriminating antibodies for the relative quantification of these isoforms in different smooth muscles of the pig (Wuytack *et al.*, 1989b; Eggermont *et al.*, 1990b). SERCA2b was the predominant isoform in all smooth muscle tissues examined (Fig. 2). This result was confirmed for several smooth muscles of the rabbit by discrimination of SERCA2a and SERCA2b isoforms based on a small difference in electrophoretic migration (Spencer *et al.*, 1991).

The analysis of mRNA levels in pig tissues revealed four classes of SERCA2-derived mRNAs. Class 1 (4.4 kb) encodes SERCA2a, whereas classes 2 (4.4 kb), 3 (8.0 kb), and 4 (5.6 kb) encode the SERCA2b isoform. These four mRNA classes present a tissue-specific dis-

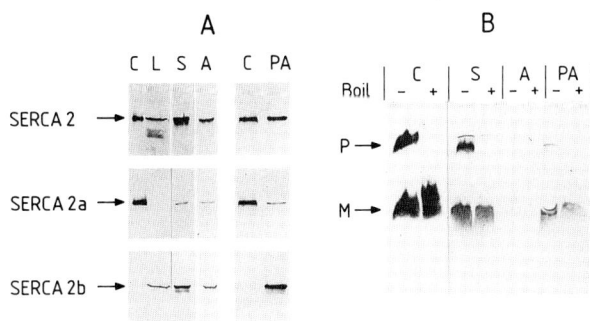

FIGURE 2 Western blots showing the relative expression of the SERCA2 isoforms and of phospholamban in pig smooth muscle. (A) SR- or ER-enriched membrane fractions of cardiac muscle (lane C), liver (L), stomach smooth muscle (S), aorta (A), and pulmonary artery (PA). SERCA2, nondiscriminating monoclonal antibody; SERCA2a, polyclonal SERCA2a-specific antiserum; SERCA2b, polyclonal SERCA2b-specific antiserum. Approximately equal amounts of SERCA2 protein were applied per lane. Note that the SERCA2a band in smooth muscle is much weaker than that in cardiac muscle, whereas the SERCA2b signal in smooth muscle is comparable to that in liver. Lanes 1–4 and 5–6 were taken from separate experiments. (B) Western blots of similar fractions using a monoclonal antibody against phospholamban. As in A, approximately equal amounts of SERCA2-immunoreactive protein were applied. Samples were (+) or were not (−) boiled in sample buffer before application. P denotes the phospholamban pentamer, which decomposes into the monomer (M) upon boiling. Note that the phospholamban signal varied between the different smooth muscle tissues, in spite of the fact that an approximately equal amount of SERCA2 protein was applied. Reproduced with permission from Eggermont *et al.* (1990b).

tribution. Class 4 is confined to neuronal cells. In pig smooth muscle, all three remaining mRNA types are expressed (Eggermont *et al.*, 1990a). The significance of the presence of three mRNA types that are translated into one single type of SERCA protein (e.g., differential stability of the messenger) remains speculative.

Using an RNAse protection assay, Eggermont *et al.* (1990b) estimated that the SERCA2b messages (class 2 + class 3) accounted for 72–81% of the total SERCA2 mRNA in four different smooth muscles of the pig, the remaining encoding SERCA2a (Fig. 3). Lytton *et al.* (1989) obtained a value of 85% mRNA encoding SERCA2b in rabbit aorta and urinary bladder, and 95% in stomach and small and large intestine. Also Khan *et al.* (1990) found almost exclusively the SERCA2b isoform mRNA in rabbit stomach smooth muscle. Also rat uterus SERCA mRNA is predominantly SERCA2b. Its expression increases slightly from Day 15 of pregnancy to delivery (Khan *et al.*, 1993). Anger *et al.* (1994) failed to observe any developmental regulation of SERCA2b during ontogeny in the rat.

A major unresolved question concerns the subcellular localization of the SERCA2a and SERCA2b isozymes. If the tails of these isoforms would contain different targeting signals, their compartmentalized expression would create discrete regions of cytosolic and ER-luminal Ca^{2+} handling. This question is particularly relevant with respect to the presence of both muscle- and non-muscle-type Ca^{2+}-binding proteins in the ER of smooth muscle. Besides the typical non-muscle ER proteins Bip (immunoglobulin-binding protein) and calreticulin, smooth muscle ER contains several muscle-type Ca^{2+}-binding proteins: the cardiac form of calsequestrin (Wuytack *et al.*, 1987; Raeymaekers *et al.*, 1993; Villa *et al.*, 1993), sarcalumenin, and the histidine-rich Ca^{2+}-binding protein (Raeymaekers *et al.*, 1993). In vas deferens smooth muscle, calsequestrin was found preferentially clustered within smooth-surfaced structures at the periphery of the cell, whereas nonmuscle proteins were evenly distributed throughout the endomembrane system (Villa *et al.*, 1993). The localization of the muscle-type isoform SERCA2a and of phospholamban with respect to these Ca^{2+}-binding proteins has not yet been investigated.

B. Expression of Phospholamban

In cardiac muscle, the increased activity of the Ca^{2+} pump of the SR as a consequence of the phosphorylation of phospholamban plays a key role in the positive chronotropic and inotropic action of β-receptor agonists (Tada and Kadoma, 1989; Luo *et al.*, 1994). The highest levels of expression of phospholamban are found in cardiac and in slow skeletal muscle, the two muscle types that express SERCA2a as the major Ca^{2+}-pump isoform. Fast skeletal muscle, which expresses SERCA1, does not contain phospholamban. Coexpression studies have shown, however, that phospholamban is able to interact with the SERCA1 pump (Fujii *et al.*, 1990).

Phospholamban was initially detected in smooth muscle in purified ER preparations from pig stomach and rabbit aorta as electrophoretic bands comigrating with the 25-kDa pentameric and 5-kDa monomeric forms of cardiac phospholamban. The dissociation of the pentamer depends on the conditions of pretreatment of the sample with the denaturing electrophoresis buffer. Phospholamban protein has also been detected in many other vascular and gastrointestinal smooth muscles, including rat aortic cells (Sarcevic *et al.*, 1989; Karczewski *et al.*, 1992), canine ileum and iliac artery (Ferguson *et al.*, 1988), and bovine aorta (L. Raeymaekers, unpublished observations; Watras, 1988). However, phospholamban protein was undetectable in pig aorta (Raeymaekers and Jones, 1986; Eggermont *et al.*, 1990b). To our knowledge, published data on myometrium are lacking. The amino acid sequence of phospholamban of pig stomach smooth muscle determined from cDNA sequencing is identi-

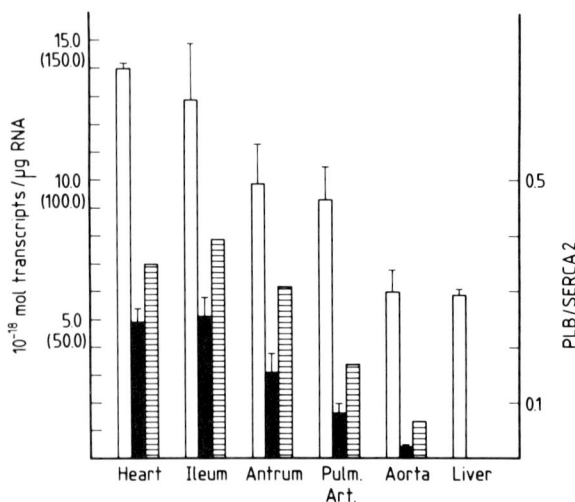

FIGURE 3 mRNA levels of SERCA2 and phospholamban and the ratio of phospholamban to SERCA2 determined with an RNAse protection assay in pig cardiac muscle, stomach, pulmonary artery, aorta, and pig liver. Open bars represent total SERCA2 mRNA levels determined with a nondiscriminating anti-sense RNA probe; solid bars represent phospholamban mRNA levels; striped bars represent ratio of phospholamban mRNA to SERCA2 mRNA. For cardiac mRNA levels the scale has been compressed 10-fold. Reproduced with permission from Eggermont *et al.* (1990b).

cal to that of dog cardiac muscle (Verboomen *et al.*, 1989).

The level of phospholamban protein relative to that of SERCA Ca^{2+}-ATPase in isolated ER vesicles from smooth muscle has been compared in bovine aorta (Watras, 1988), in bovine main pulmonary artery, and in several smooth muscles of the pig (Eggermont *et al.*, 1990b). Watras (1988) and Raeymaekers *et al.* (1990) found a ratio of approximately 1. A possible drawback of these experiments is that they depended on the preservation of the pump's activity during isolation of the membranes. In the porcine smooth muscles investigated by Eggermont *et al.* (1990b), the level of phospholamban protein relative to that of SERCA2 protein was smaller than in cardiac muscle (Fig. 2). It is not clear whether this different result is due to the tissue difference or to the use of a more reliable method.

The expression of phospholamban at the mRNA level was examined by Eggermont *et al.* (1990b) on total RNA extracted from several smooth muscle tissues of the pig using RNAse protection assay. The highest phospholamban mRNA level was observed in the pig ileum, followed by the pig stomach. In these tissues, the amount of both mRNAs was about 10- to 15-fold lower than in cardiac muscle, implying a similar value of the ratio phospholamban mRNA/SERCA2 mRNA in both tissues (Fig. 3). The discrepancy with the estimates of both gene products at the protein level can possibly be explained by the higher value of protein/mRNA in cardiac muscle expressing mainly SERCA2a compared to smooth muscle expressing mainly SERCA2b (Grover and Khan, 1992; see Section IV.A).

In rat aortic cells, phospholamban is expressed in differentiated freshly dispersed cells and is absent from dedifferentiated late-passage cells as shown by differential cDNA screening (Shanahan *et al.*, 1993).

C. Expression of the Plasma Membrane Ca^{2+} Pump

In pig stomach smooth muscle, PMCA1b was the only isoform detected either by screening of a cDNA library or by analysis of mRNA by Northern blotting and by the polymerase chain reaction (PCR) (De Jaegere *et al.*, 1990). This finding contrasts with the presence of at least three different PMCA-related polypeptide bands on Western blots. Since there was no evidence for artefactual postmortem proteolysis, De Jaegere *et al.* (1990) concluded that the PMCA1b protein is post-translationally modified in smooth muscle. In rabbit stomach and aorta, Khan and Grover (1991) found by PCR that 83 to 86% of the PMCA1 message is accounted for by PMCA1b, the remainder represent-

ing PMCA1a. Both isoforms differ in their calmodulin-binding domain and by the presence (PMCA1b) or absence (PMCA1a) of a phosphorylation site for cAMP-dependent protein kinase. In rat myometrium, Adamo and Penniston (1992) detected several PMCA splice variants by using PCR: fragments were amplified corresponding to PMCA1b, the only transcript of the PMCA1 gene, to a PMCA2 variant containing the longest insert (135 base pairs) at splice site A (designated PMCA2w), and a fragment corresponding to PMCA2a. The relative amounts of these messengers have not been determined quantitatively.

Immunocytochemistry at the ultrastructural level on mouse ileum smooth muscle showed that the localization of the PM Ca^{2+} pump was restricted to caveolae, although many of these invaginations were unlabeled (Fugimoto, 1993). The functional relevance of this observation remains to be determined. Since a protein resembling the inositol 1,4,5-trisphosphate receptor has been localized to caveolae (Fujimoto *et al.*, 1992), these structures may play an important role in Ca^{2+} handling.

In cultured smooth muscle cells of the rat, the expression of the PM Ca^{2+} pump is rapidly and stongly up-regulated by cyclic AMP and protein kinase C (PKC). Both pathways are synergistic (Kuo *et al.*, 1993).

V. REGULATION OF THE SERCA-TYPE Ca^{2+} PUMPS IN SMOOTH MUSCLE CELLS

A. Regulation by Cyclic Nucleotide-Dependent Protein Kinases

1. Phospholamban Phosphorylation in Isolated Membrane Vesicles

When protein kinase substrates of isolated cardiac SR are analyzed by denaturing gel electrophoresis, protein phosphorylation is almost exclusively observed in bands corresponding to phospholamban. In contrast, multiple phosphorylated bands are observed in smooth muscle ER. Phospholamban in ER vesicles from pig stomach smooth muscle was phosphorylated by exogenously added cAK or cGK, and by the activation of an endogenous Ca^{2+}/calmodulin-dependent protein kinase, endogenous cAK (Raeymaekers and Jones, 1986; Raeymaekers *et al.*, 1990), and by an endogenous cGK (Cornwell *et al.*, 1991). cGMP was much more effective in activating endogenous protein kinase activity than cAMP, indicating an association of cGK with the vesicles that contain phospholamban. The colocalization of cGK with phospholamban was confirmed by immunocytochemistry on whole cells. The localization was mainly perinuclear (Cornwell *et al.*, 1991). In the presence of exogenous cGK there was a more preferential phosphorylation of phospholamban relative to the other substrates than in the presence of cAK (Raeymaekers *et al.*, 1988; Huggins *et al.*, 1989).

2. Study of Ca^{2+}-Transport Regulation in Isolated Membrane Vesicles

In SR of cardiac muscle, phosphorylation of phospholamban increases the affinity of the pump for Ca^{2+}. At low Ca^{2+} concentrations ($<10^{-6} M$), a several fold stimulation of the rate of Ca^{2+} uptake is observed. However, the stimulation by cAK of the rate of Ca^{2+} uptake in isolated ER from smooth muscle is much smaller than in cardiac SR (Watras, 1988; Raeymaekers *et al.*, 1990). The most important possibility to be considered to explain this difference is a lower ratio of the number of phospholamban molecules to ATPase molecules in smooth muscle ER as compared to cardiac SR. As discussed in Section IV.B, this possibility cannot be excluded at present. A further difference between smooth muscle ER and cardiac SR is the predominance of the SERCA2b isoform in the first and the predominance of the SERCA2a isoform in the latter. However, Verboomen *et al.*, (1992) have shown by cotransfection experiments that phospholamban is about equally effective in inhibiting both splice variants.

3. Evidence for Regulated Ca^{2+} Uptake in the ER of Smooth Muscle Cells *in Vivo*

One of the newer insights concerning the regulation of $[Ca^{2+}]_i$ is that cGMP would be more effective in decreasing $[Ca^{2+}]_i$ than cAMP (Felbel *et al.*, 1988; Cornwell and Lincoln, 1989; Lincoln *et al.*, 1990; Lermioglu *et al.*, 1991; Ushio-Fukai *et al.*, 1993). Despite the difficulties involved in assigning the causes of modulation of $[Ca^{2+}]_i$ in smooth muscle cells to specific flux components, a great number of studies on intact smooth muscle cells have presented evidence for a modulation by second messengers of Ca^{2+} uptake into the ER. These data till 1992 have been discussed in a previous review (Raeymaekers and Wuytack, 1993). Additional evidence was obtained from experiments using thapsigargin, CPA, and tBu-BHQ as inhibitors of the ER Ca^{2+} pump. These inhibitors reduced the relaxation by cGMP (Luo *et al.*, 1993) or cAMP (Low *et al.*, 1993a).

Especially relevant are studies on the phosphorylation of phospholamban in smooth muscle cells *in vivo*. Cornwell *et al.* (1991) and Karczewski *et al.* (1992) convincingly demonstrated phosphorylation of phospholamban in rat aorta incubated with agents that increase cGMP.

It can be expected that the availability of phospholamban-deficient mice (Luo *et al.*, 1994) will be very useful in future studies of the role of phospholamban phosphorylation in smooth muscle cells.

B. Regulation of the SERCA Pump by Ca^{2+}/Calmodulin-Dependent Protein Kinase

The SERCA2 pump of cardiac muscle is regulated in a dual way by Ca^{2+}/calmodulin-dependent protein kinase II, via an indirect route by the previously mentioned phosphorylation of phospholamban, and via a direct phosphorylation of the SERCA2 pump on serine-38, as described by Toyofuku *et al.* (1994). Since the phosphorylation site is located near the N terminus, this phosphorylation probably occurs in both SERCA2a and SERCA2b. Whereas phospholamban phosphorylation increases the affinity for Ca^{2+}, the direct phosphorylation increases the V_{max}. Therefore, both forms of phosphorylation may be synergistic in stimulating the Ca^{2+} pump.

In isolated ER vesicles from smooth muscle, the addition of calmodulin induces a submaximal phosphorylation of phospholamban via an endogenous protein kinase (Raeymaekers and Jones, 1986). The addition of calmodulin does not, however, significantly stimulate the Ca^{2+} uptake in isolated ER vesicles (L. Raeymaekers, unpublished observations) or in skinned smooth muscle cells (Stout and Silver, 1992). Since this lack of effect could be due to washout of the endogenous kinase, further experiments are needed using purified Ca^{2+}/calmodulin-dependent protein kinase.

VI. REGULATION OF THE PMCA-TYPE Ca^{2+} PUMPS IN SMOOTH MUSCLE CELLS

A. Regulation of the Plasmalemmal Ca^{2+} Pump by Calmodulin

One of the most typical properties of the PM Ca^{2+} pumps is their regulation by the direct interaction with calmodulin in its Ca^{2+}-bound state. The observed effects of calmodulin cannot be explained by an indirect action via Ca^{2+}/calmodulin-dependent protein kinase (Zhang *et al.*, 1992). Alternative splicing at splice site C generates two main classes of calmodulin-binding domains, designated **a** (with insert) and **b** (without insert). The **a** form has about a 10-fold lower affinity for Ca^{2+}/calmodulin (Enyedi *et al.*, 1991). Interestingly, these differences resulted in an apparent reduced Ca^{2+} affinity of the splice variants expressing the **a** domain (Enyedi *et al.*, 1994). Smooth muscle cells express mainly PMCA1b (presenting the **b** domain) and to a lesser extent also isoforms processing the **a** domain (Section IV.C). However, at present the functional implications of the splice variants for Ca^{2+} metabolism are not clear. A quantitative study at the protein level in more types of smooth muscles will be needed in order to make correlations with function.

B. Regulation of the Plasmalemmal Ca^{2+} Pump by Phospholipids

The PM Ca^{2+}-transport ATPase can be purified in the presence of detergent and added phospholipids, or it can be purified in the absence of phospholipids but in the presence of 20% glycerol. The ATPase is not active in the latter condition but can be reactivated by adding phospholipids (Kosk-Kosicka and Inesi, 1985), allowing the systematic study of the lipid species on the ATPase activity. The activation curve of the PM Ca^{2+}-ATPases purified from stomach smooth muscle and from erythrocytes was bell-shaped, with higher concentrations of acidic phospholipids inhibiting the ATPase activity. The relative efficiency of different acidic phospholipids was $PIP_2 > PIP > PI \approx PS \approx PA$ (phosphatidyinositol 4,5-bisphosphate > phosphatidylinositol 4-phosphate > phosphatidylinositol $\approx$ phosphatidylserine $\approx$ phosphatidic acid, respectively) (Missiaen et al., 1989b). The same sequence was found if the lipid–protein interaction was assayed directly by fluorescence energy transfer using modified phosphoinositides (Verbist et al., 1991). Acidic phospholipids increased both the V_{max} and the affinity for Ca^{2+}. The activation of the PM Ca^{2+} pump by the phosphoinositides has been implicated in its regulation by protein kinases and by hormones, as discussed in the following.

C. Regulation of the Plasmalemmal Ca^{2+} Pump by Protein Kinases

In smooth muscle cells a phosphorylation of the PM Ca^{2+}-pump protein has been clearly demonstrated only for PKC. The regulation of the PMCA ATPase by cGK does not require concomitant phosphorylation of the pump. Although a modulation by cAK of the PM Ca^{2+} pump of erythrocytes and cardiac cells has been reported, such an effect could not be demonstrated for the ATPase from smooth muscle (see the following).

The PM Ca^{2+} pump of smooth muscle may be preferentially fueled by ATP produced by a membrane-associated glycolytic system (Hardin et al., 1992).

When endogenous glycolysis was the sole source of ATP, phosphorylation of the membranes by endogenous protein kinase activity induced a stimulation of the Ca^{2+} uptake that was more pronounced (up to three-fold stimulation) than in the presence of exogenously added ATP (Hardin et al., 1993). The characterization of these endogenous kinases is required for further evaluation of these results.

1. Regulation of the Plasmalemmal Ca^{2+} Pump by Cyclic Nucleotide-Dependent Protein Kinases

Measurements of Ca^{2+} fluxes and of the changes of $[Ca^{2+}]_i$ in smooth muscle cells strongly suggest a stimulation of the PM Ca^{2+} pump by cGK. Kobayashi et al. (1985) studied the Ca^{2+} release induced by caffeine in Ca^{2+}-free solution. The cGMP-elevating agent nitroglycerin did not affect the size of the first response to caffeine, but reduced the size of the subsequent responses, indicating that the extrusion of Ca^{2+} from the cell had been accelerated during exposure to the drug. Itoh et al. (1985) suggested that cGMP mainly acts on cellular Ca^{2+} by activation of Ca^{2+} extrusion, in contrast to cAMP, which would increase the amount of Ca^{2+} stored in the cell. Hassid and Yu (1989) observed that in the presence of atriopeptin the rate of Ca^{2+} efflux did not change despite the decrease of the cytoplasmic Ca^{2+} concentration, indicating that the affinity of the Ca^{2+}-extrusion system for Ca^{2+} was increased. Furukawa et al. (1988) discriminated Ca^{2+}-pump-dependent and Na^+–Ca^{2+}-exchange-dependent components of the Ca^{2+} efflux. An increase of cGMP specifically stimulated the component of the Ca^{2+} efflux, which was mediated by the Ca^{2+} pump, especially at a low cytoplasmic Ca^{2+} concentration. cAMP had no effect. Likewise, Magliola and Jones (1990) observed a stimulation of the Ca^{2+} extrusion in rat aorta by the cGMP-elevating agent sodium nitroprusside.

In agreement with the effects observed on smooth muscle cells *in vivo*, Furukawa and Nakamura (1987), Vrolix et al. (1988), and Imai et al. (1990) observed that cGMP-dependent phosphorylation increased the affinity for Ca^{2+} and the maximum Ca^{2+}-uptake activity of the PM Ca^{2+}-ATPase purified from aortic microsomes. However, Vrolix et al. (1988) and Baltensperger et al. (1988) failed to observe a phosphorylation of the Ca^{2+}-transport enzyme concomitantly with the stimulation of the ATPase activity. Instead, in several preparations of purified PM Ca^{2+}-transport ATPases, traces of proteins that are substrates for cGK have been detected. A phosphorylated protein migrating in denaturing gels at the position of the Ca^{2+}-ATPase was identified as a cytoskeletal component (Baltensberger *et al.*, 1990). All of these data indicate that the PM Ca^{2+}-transport ATPase is indirectly stimulated by cGK via the phosphorylation of another protein that remains associated with it during purification. This protein has not yet been characterized. It has been proposed by Vrolix *et al.* (1988) that this protein is a cGK-stimulated PI kinase and that the enhanced formation of PIP would stimulate the Ca^{2+} pump, PIP being more effective than PI (see Section VI.B). However, this hypothesis could not be confirmed (Yoshida *et al.*, 1991). One of the cGK substrates copurifying with the ATPase from porcine aorta has been identified as a 240-kDa inositol 1,4,5-trisphosphate (IP_3) receptor-like protein (Koga *et al.*, 1994). A similar protein has been colocalized with the PM Ca^{2+}-ATPase in caveolae (see Section IV.C). As yet, the mechanism of the regulation of the PM Ca^{2+} pump by cGK remains unresolved.

2. Regulation of the Plasmalemmal Ca^{2+} Pump by Protein Kinase C

Activation of PKC in intact cultured vascular smooth muscle cells augments the maximum rate of the Ca^{2+}-ATPase-mediated component of the Ca^{2+} efflux without a significant change in the affinity for Ca^{2+} (Furukawa *et al.*, 1989a). In contrast, treatment of the PM Ca^{2+} pump in isolated membrane vesicles or in the purified state with PKC increased the affinity of the Ca^{2+}-ATPase for Ca^{2+} (Fukuda *et al.*, 1990). Variable functional effects were also observed on the Ca^{2+}-ATPase from human erythrocytes by Wang *et al.* (1991). The functional effect of the phosphorylation by PKC on the Ca^{2+}-ATPase depended on the isozyme of the kinase used and on the lipid associated with the ATPase. Depending on these conditions, PKC may stimulate the purified erythrocyte ATPase, or antagonize its activation by calmodulin. To clarify the effect of PKC on the PM Ca^{2+} pump, more structural information will be needed on the phosphorylation site(s) in the different Ca^{2+} pump isoforms. Up till now it has been found that the Ca^{2+}-ATPase from human erythrocytes is phosphorylated by rat brain type III PKC at threonine and serine residues located in the C-terminal domain (Wang *et al.*, 1991). A synthetic peptide corresponding to the calmodulin-binding domain and containing the target threonine in the phosphorylated state was less efficient in inhibiting the truncated pump, and was also less efficient in binding calmodulin (Hofmann *et al.*, 1994). It is interesting to note that this type of regulation is functionally similar to the effect of phosphorylation of phospholamban on the SERCA pump.

In addition to its effect on the activity of the PM Ca^{2+} pump, PKC is also able to rapidly influence the expression level of this ATPase, as has been mentioned in Section IV.C.

D. Regulation of the Plasmalemmal Ca^{2+} Pump by Hormones and Other Signals

The Ca^{2+}-mobilizing hormones carbachol and oxytocin have been found to modulate the activity of the PM Ca^{2+} pump. Missiaen *et al.* (1988) treated muscle strips of pig gastric smooth muscle with carbachol. Subsequently the muscle was homogenized and membrane vesicles were isolated. Pretreatment with the muscarinic agonist reduced the activity of the Ca^{2+} pump. Equivalent results were obtained on the myometrium of rats primed with diethylstilbestrol and stimulated with oxytocin (Popescu *et al.*, 1985; Enyedi *et al.*, 1989). Oxytocin was also found to inhibit the activity of the PM Ca^{2+} pump (as well as of the ER Ca^{2+} pump) *in uteri* from pregnant rats close to term (Magosci and Penniston, 1991). The inhibition of the Ca^{2+} pump by oxytocin may contribute to the $[Ca^{2+}]_i$-elevating and contractile effects of this hormone during parturition.

Since in the experiments just described the agonists were added to the cells *in vivo*, the complexity of the pathways that are simultaneously activated does not allow one to delineate the mechanism(s) involved in the modulation of the activity of the Ca^{2+} pump. The duration of treatment with the hormones (15 min) is very probably too short for the recruitment of different pump isoforms. One possible explanation is a change in the concentration of the stimulatory phosphoinositides (see Section VI.B) induced by the activation of phospholipase C. Another possibility could be a change in the level of phosphorylation of the Ca^{2+}-ATPase by PKC (see Section VI., C.2) or the modulation of other proteins, for example, the Na^+–H^+ exchanger. Activation of PKC is known to activate the Na^+–H^+ exchanger, leading to alkalinization of the cell (see Rothstein, 1989, for review). Alkalinization will activate the PM Ca^{2+}-ATPase by increasing its affinity for Ca^{2+} (Missiaen *et al.*, 1989a).

It should also be mentioned that several groups have reported on the inhibition by IP_3 of the PM Ca^{2+} pump in coronary artery smooth muscle (Popescu *et al.*, 1986), cardiac muscle (Kuo and Tsang, 1988), and erythrocytes (Davis *et al.*, 1991). In search for a possible mechanism of this effect, it was found that IP_3 inhibits calmodulin binding to the red cell membranes (Davis *et al.*, 1991).

Ca^{2+}-transport ATPases [PMCA (Niggli *et al.*, 1982) as well as SERCA (Martonosi *et al.*, 1985)] operate as obligatory Ca^{2+}–H^+ exchangers. Since the number of protons exchanged for Ca^{2+} has not been firmly established, it is not known whether the PM Ca^{2+} pump is electrogenic or not. Electrogenic behavior would be in line with the sensitivity of Ca^{2+} extrusion in smooth muscle cells to the membrane potential (Furukawa *et al.*, 1989b).

VII. Ca^{2+} PUMPS IN PATHOLOGICAL CONDITIONS

In hypertensive animals, altered Ca^{2+}-pump activity in vascular smooth muscle cells may contribute to the increase of blood pressure. In a previous review on altered Ca^{2+} metabolism in spontaneously hypertensive rats, we concluded that most of the results pointed to a diminished ER Ca^{2+} transport (Raeymaekers and Wuytack, 1993). Low *et al.* (1993b) came to the same conclusion using the inhibitor CPA as a tool, whereas new results from Levitsky *et al.* (1993) favor the opposite view. These authors found an increased ER function in the abdominal aorta of SHR rats and an elevated level of SERCA2 mRNA.

The data on alterations of PM Ca^{2+} transport in hypertensive animals are conflicting, with some authors reporting a decrease and others an increase. An overview of these conflicting results is given by Monteith *et al.* (1994).

Reactive oxygens, especially hydrogen peroxide, superoxide, and perhydroxyl radicals, have been implicated in a variety of pathophysiological conditions. In cardiac muscle, the Ca^{2+}-transport ATPase of the SR in particular is highly susceptible (Kukreja *et al.*, 1988; Scherer and Deamer, 1986). The PM and especially the ER Ca^{2+} pumps of pig coronary artery are also susceptible to superoxide radicals (Grover and Samson, 1988, 1989) and to hydrogen peroxide (Grover *et al.*, 1992). The effect of superoxide is probably due to the irreversible modification of sulfhydryl groups (Scherer and Deamer, 1986; Suzuki and Ford, 1991). Also the vasoconstrictive side effect of alloxan has been ascribed to the formation of hydrogen peroxide and the subsequent inhibition of Ca^{2+} transport (Kwan and Beazley, 1988).

References

Adamo, H. P., and Penniston, J. T. (1992). *Biochem. J.* **283,** 355–359.

Anger, M., Samuel, J.-L., Marotte, F., Wuytack, F., Rappaport, L., and Lompré, A.-M. (1994). *J. Mol. Cell. Cardiol.* **26,** 539–550.

Ashida, T., Schaeffer, J., Goldman, W. F., Wade, J. B., and Blaustein, M. P. (1988). *Circ. Res.* **62,** 854–863.

Baltensperger, K., Carafoli, E., and Chiesi, M. (1988). *Eur. J. Biochem.* **172,** 7–16.

Baltensperger, K., Chiesi, M., and Carafoli, E. (1990). *Biochemistry* **29,** 9753–9760.

Baró, I. O'Neill, S. C., and Eisner, D. A. (1993). *J. Physiol. (London)* **465,** 21–41.

Broström, C. O., and Broström, M. A. (1990). *Annu. Rev. Physiol.* **52,** 577–590.

Campbell, A. M., Kessler, P. D., and Fambrough, D. M. (1992). *J. Biol. Chem.* **267,** 9321–9325.

Carafoli, E. (1992). *J. Biol. Chem.* **267,** 2115–2118.

Carafoli, E. (1994). *FASEB J.* **8,** 993–1002.

Carafoli, E., and Chiesi, M. (1992). *Curr. Top. Cell. Regul.* **32,** 209–241.

Chen, Q., and van Breemen, C. (1992). *Advances in Second Messenger and Phosphoprotein Res.* **26,** 335–350.

Chiesi, M., Vorherr, T., Falchetto, R., Waelchli, C., and Carafoli, E. (1991). *Biochemistry* **30,** 7978–7983.

Clarke, D. M., Loo, T. W., Inesi, G., and MacLennan, D. H. (1989). *Nature (London)* **339,** 476–478.

Cornwell, T. L., and Lincoln, T. M. (1989). *J. Biol. Chem.* **264,** 1146–1155.

Cornwell, T. L., Pryzwansky, K. B., Wyatt, T. A., and Lincoln, T. M. (1991). *Mol. Pharmacol.* **40,** 923–931.

Darby, P. J., Kwan, C. Y., and Daniel, E. E. (1993). *Biol. Signals* **2,** 293–304.

Davis, F. B., Davis, P. J., Lawrence, W. D., and Blas, S. (1991). *FASEB J.* **5,** 2992–2995.

De Jaegere, S., Wuytack, F., Eggermont, J. A., Verboomen, H., and Casteels, R. (1990). *Biochem. J.* **271,** 655–660.

Eggermont, J. A., Vrolix, M., Raeymaekers, L., Wuytack, F., and Casteels, R. (1988). *Circ. Res.* **62,** 266–278.

Eggermont, J. A., Wuytack, F., De Jaegere, S., Nelles, L., and Casteels, R. (1989). *Biochem. J.* **260,** 757–761.

Eggermont, J. A., Wuytack, F., and Casteels, R. (1990a). *Biochem. J.* **266,** 901–907.

Eggermont, J. A., Wuytack, F., Verbist, J., and Casteels, R. (1990b). *Biochem. J.* **271,** 649–653.

Enyedi, A., Brandt, J., Minami, J., and Penniston, J. T. (1989). *Biochem. J.* **261,** 23–28.

Enyedi, A., Filoteo, A. G., Gardos, G., and Penniston, J. T. (1991). *J. Biol. Chem.* **266,** 8952–8956.

Enyedi, A., Verma, A. K., Heim, R., Adamo, H. P., Filoteo, A. G., Strehler, E. E., and Penniston, J. T. (1994). *J. Biol. Chem.* **269,** 41–43.

Falchetto, R., Vorherr, T., Brunner, J., and Carafoli, E. (1991). *J. Biol. Chem.* **266,** 2930–2936.

Falchetto, R., Vorherr, T., and Carafoli, E. (1992). *Protein Sci.* **1,** 1613–1621.

Felbel, J., Trockur, B., Ecker, T., Landgraf, W., and Hofmann, F. (1988). *J. Biol. Chem.* **263,** 16764–16771.

Ferguson, D. G., Young, E. F., Raeymaekers, L., and Kranias, E. G. (1988). *J. Cell Biol.* **107,** 555–562.

Fujii, J., Maruyama, K., Tada, M., and MacLennan, D. H. (1990). *FEBS Lett.* **273,** 232–234.

Fujimoto, T. (1993). *J. Cell. Biol.* **120,** 1147–1157.

Fujimoto, T., Nakade, S., Miyawaki, A., Mikoshiba, K., and Ogawa, K. (1992). *J. Cell Biol.* **119,** 1507–1513.

Fukuda, T., Ogurusu, T., Furukawa, K.-I., and Shigekawa, M. (1990). *J. Biochem. (Tokyo)* **108,** 629–634.

Furukawa, K.-I., and Nakamura, H. (1984). *J. Biochem. (Tokyo)* **96,** 1343–1350.

Furukawa, K.-I., and Nakamura, H. (1987). *J. Biochem. (Tokyo)* **101,** 287–290.

Furukawa, K.-I., Tawada, Y., and Shigekawa, M. (1988). *J. Biol. Chem.* **263,** 8058–8065.

Furukawa, K.-I., Tawada, Y., and Shigekawa, M. (1989a). *J. Biol. Chem.* **264**, 4844–4849.

Furukawa, K.-I., Tawada-Iwata, Y., and Shigekawa, M. (1989b). *J. Biochem. (Tokyo)* **106**, 1068–1073.

Gabella, G. (1989). In "Handbook of Physiology" (S. G. Schultz, J. D. Wood, and B. B. Rauner, eds.), Sect. 6, Vol. I, Part I, pp. 103–139. Am. Physiol. Soc., Bethesda, MD.

Garrahan, P. J., and Rega, A. F. (1990). In "Intracellular Calcium Regulation" (F. Bronner, ed.), pp. 271–303. Alan R. Liss, New York.

Green, N. M., and Stokes, D. L. (1992). *Acta Physiol. Scand.* **146**, 59–68.

Grover, A. K., and Khan, I. (1992). *Cell Calcium* **13**, 9–17.

Grover, A. K., and Samson, S. E. (1988). *Am. J. Physiol.* **255**, C297–C303.

Grover, A. K., and Samson, S. E. (1989). *Am. J. Physiol.* **256**, C666–C673.

Grover, A. K., Samson, S. E., and Fomin, V. P. (1992). *Am. J. Physiol.* **263**, H537–H543.

Gunteski-Hamblin, A.-M., Greeb, J., and Shull, G. E. (1988). *J. Biol. Chem.* **263**, 15032–15040.

Hardin, C. D., Raeymaekers, L., and Paul, R. J. (1992). *J. Gen. Physiol.* **99**, 21–40.

Hardin, C. D., Zhang, C., Kranias, E. G., Steenaart, N. A. E., Raeymaekers, L., and Paul, R. J. (1993). *Am. J. Physiol.* **265**, H1326–H1333.

Hassid, A., and Yu, Y.-M. (1989). *J. Cardiovasc. Pharmacol.* **14**, 534–538.

Hilfiker, H., Guerini, D., and Carafoli, E. (1994). *J. Biol. Chem.* **269**, 26178–26183.

Hofmann, F., Anagli, J., Carafoli, E., and Vorherr, T. (1994). *J. Biol. Chem.* **269**, 24298–24303.

Huggins, J. P., Cook, E. A., Piggott, J. R., Mattinsley, T. J., and England, P. J. (1989). *Biochem. J.* **260**, 829–835.

Imai, S., Yoshida, Y., and Sun, H.-T. (1990). *J. Biochem. (Tokyo)* **107**, 755–761.

Inesi, G., and Kirtley, M. E. (1990). *J. Membr. Biol.* **116**, 1–8.

Inesi, G., Sumbilla, C., and Kirtley, M. E. (1990). *Physiol. Rev.* **70**, 749–760.

Irvine, R. F. (1990). *FEBS Lett.* **623**, 5–9.

Itoh, T., Kanmura, Y., Kuriyama, H., and Sasagiori, T. (1985). *Br. J. Pharmacol.* **84**, 393–406.

James, P., Inui, M., Tada, M., Chiesi, M., and Carafoli, E. (1989). *Nature (London)* **342**, 90–92.

Karczewski, P., Kelm, M., Hartmann, M., and Schrader, J. (1992). *Life Sciences* **51**, 1205–1210.

Kargacin, G., and Fay, F. S. (1991). *Biophys. J.* **60**, 1088–1100.

Kargacin, M. E., and Kargacin, G. J. (1994). *Biophys. J.* **66**, A98.

Khan, I., and Grover, A. K. (1991). *Biochem. J.* **277**, 345–349.

Khan, I., and Grover, A. K. (1993). *Cell Calcium* **14**, 17–23.

Khan, I., Spencer, G. G., Samson, S. E., Crine, P., Bioleau, G., and Grover, A. K. (1990). *Biochem. J.* **268**, 415–419.

Khan, I., Tabb, T., Garfield, R. E., Jones, L. R., Fomin, V. P., Samson, S. E., and Grover, A. K. (1993). *Cell Calcium* **14**, 111–117.

Kobayashi, S., Kanaide, H., and Nakamura, M. (1985). *Science* **229**, 553–556.

Koga, T., Yoshida, Y., Cai, J.-Q., Islam, M. O., and Imai, S. (1994). *J. Biol. Chem.* **269**, 11640–11647.

Kosk-Kosicka, D., and Inesi, G. (1985). *FEBS Lett.* **189**, 67–71.

Kukreja, R., Okabe, E., Schrier, G. M., and Hess, M. L. (1988). *Arch. Biochem. Biophys.* **261**, 447–457.

Kuo, T. H., and Tsang, W. (1988). *Biochem. Biophys. Res. Commun.* **152**, 1111–1116.

Kuo, T. H., Liu, B. F., Diglio, C., and Tsang, W. (1993). *Arch Biochem. Biophys.* **305**, 428–433.

Kwan, C.-Y., and Beazley, J. S. (1988). *J. Bioenerg. Biomembr.* **20**, 517–531.

Lermioglu, F., Goyal, J., and Hassid, A. (1991). *Biochem. J.* **274**, 323–328.

Levitsky, D. O., Clergue, M., Lambert, F., Souponitskaya, M. V., Le, J. T., Lecarpentier, Y., and Lompre, A. M. (1993). *J. Biol. Chem.* **268**, 8325–8331.

Lincoln, T. M., Cornwell, T. L., and Taylor, A. E. (1990). *Am. J. Physiol.* **258**, C399–C407.

Lompré, A.-M., Anger, M., and Levitsky, D. (1994). *J. Mol. Cell. Cardiol.* **26**, 1109–1121.

Low, A. M., Darby, P. J., Kwan, C. Y., and Daniel, E. E. (1993a). *Eur. J. Pharmacol.* **230**, 53–62.

Low, A. M., Kwan, C. Y., and Daniel, E. E. (1993b). *Pharmacology* **47**, 50–60.

Luo, D. L., Nakazawa, M., Ishibashi, T., Kato, K., and Imai, S. (1993). *J. Pharmacol. Exp. Ther.* **265**, 1187–1192.

Luo, W., Grupp, I. l., Harrer, J., Ponniah, S., Grupp, G., Duffy, J. J., Doetschman, T., and Kranias, E. G. (1994). *Circ. Res.* **75**, 401–409.

Lytton, J., Zarain-Herzberg, A., Periasamy, M., and MacLennan, D. H. (1989). *J. Biol. Chem.* **264**, 7059–7065.

MacLennan, D. H. (1990). *Biophys. J.* **58**, 1355–1365.

Magliola, L., and Jones, A. W. (1990). *J. Physiol. (London)* **421**, 411–424.

Magocsi, M., and Penniston, J. T. (1991). *Biochim. Biophys. Acta* **1063**, 7–14.

Martonosi, A., Kracke, G., Taylor, K. A., Dux, L., and Peracchia, C. (1985). *Soc. Gen. Physiol. Ser.* **39**, 57–85.

Missiaen, L., Kanmura, Y., Wuytack, F., and Casteels, R. (1988). *Biochem. Biophys. Res. Commun.* **150**, 681–686.

Missiaen, L., Droogmans, G., De Smedt, H., Wuytack, F., Raeymaekers, L., and Casteels, R. (1989a). *Biochem. J.* **262**, 361–364.

Missiaen, L., Raeymaekers, L., Wuytack, F., Vrolix, M., De Smedt, H., and Casteels, R. (1989b). *Biochem J.* **263**, 687–694.

Monteith, G. R., Chen, S., and Roufogalis, B. D. (1994). *J. Pharmacol. Toxicol. Methods* **31**, 117–124.

Niggli, V., Penniston, J. T., and Carafoli, E. (1979). *J. Biol. Chem.* **254**, 9955–9958.

Niggli, V., Adunyah, E. S., Penniston, J. T., and Carafoli, E. (1981). *J. Biol. Chem.* **256**, 395–401.

Niggli, V., Sigel, E., and Carafoli, E. (1982). *J. Biol. Chem.* **257**, 2350–2356.

O'Donnell, M. E., and Owen, N. E. (1994). *Physiol. Rev.* **74**, 683–721.

Penniston, J. T., and Enyedi, A. (1994). *Cell. Physiol. Biochem.* **4**, 148–159.

Popescu, L. M., Nutu, O., and Panoiu, C. (1985). *Biosci. Rep.* **5**, 21–28.

Popescu, L. M., Hinescu, M. E., Musat, S., Ionescu, M., and Pistritzu, F. (1986). *Eur. J. Pharmacol.* **123**, 167–169.

Raeymaekers, L., and Hasselbach, W. (1981). *Eur. J. Biochem.* **116**, 373–378.

Raeymaekers, L., and Jones, L. R. (1986). *Biochim. Biophys. Acta* **882**, 258–265.

Raeymaekers, L., and Wuytack, F. (1993). *J. Muscle Res. Cell Motil.* **14**, 141–157.

Raeymaekers, L., Agostini, B., and Hasselbach, W. (1980). *Histochemistry* **65**, 121–129.

Raeymaekers, L., Wuytack, F., Eggermont, J., De Schutter, G., and Casteels, R. (1983). *Biochem. J.* **210**, 315–322.

Raeymaekers, L., Wuytack, F., and Casteels, R. (1985). *Biochim. Biophys. Acta* **815**, 441–454.

Raeymaekers, L., Hofmann, F., and Casteels, R. (1988). *Biochem. J.* **252**, 269–273.

Raeymaekers, L., Eggermont, J. A., Wuytack, F., and Casteels, R. (1990). *Cell Calcium* **11**, 261–268.

Raeymaekers, L., Verbist, J., Wuytack, F., Plessers, L., and Casteels, R. (1993). *Cell. Calcium* **14**, 581–589.

Rothstein, A. (1989). *Rev. Physiol. Biochem. Pharmacol.* **112**, 235–257.

Sambrook, J. F. (1990). *Cell (Cambridge, Mass.)* **61**, 197–199.

Sarcevic, B., Brookes, V., Martin, T. J., Kemp, B. E., and Robinson, P. J. (1989). *J. Biol. Chem.* **264**, 20648–20654.

Scherer, N. M., and Deamer, D. W. (1986). *Arch. Biochem. Biophys.* **246**, 589–601.

Shanahan, C. M., Weissberg, P. L., and Metcalfe, J. C. (1993). *Circ. Res.* **73**, 193–204.

Somlyo, A. P. (1985). *Circ. Res.* **57**, 497–507.

Spencer, G. G., Yu, X., Khan, I., and Grover, A. K. (1991). *Biochim. Biophys. Acta* **1063**, 15–20.

Stehno-Bittel, L., and Sturek, M. (1992). *J. Physiol. (London)* **451**, 49–78.

Stout, M. A., and Silver, P. J. (1992). *J. Cell. Physiol.* **153**, 169–175.

Strehler, E. E. (1991). *J. Membr. Biol.* **120**, 1–15.

Suzuki, C. K., Bonifacino, J. G., Lin, A. Y., Davis, M. M., and Klausner, R. D. (1991). *J. Cell Biol.* **114**, 189–205.

Suzuki, Y. J., and Ford, G. D. (1991). *Am. J. Physiol* **261**, H568–H574.

Tada, M., and Kadoma, M. (1989). *BioEssays* **10**, 157–163.

Toyofuku, T., Kurzydlowski, K., Narayanan, N., and MacLennan, D. H. (1994). *J. Biol. Chem.* **269**, 26492–26496.

Ushio-Fukai, M., Abe, S., Kobayashi, S., Nishimura, J., and Kanaide, H. (1993). *J. Physiol. (London)* **462**, 679–696.

van Breemen, C., and Saida, K. (1989). *Annu. Rev. Physiol.* **51**, 315–329.

Van Den Bosch, L., Eggermont, J., De Smedt, H., Mertens, L., Wuytack, F., and Casteels, R. (1994). *Biochem. J.* **302**, 559–566.

Verbist, J., Gadella, T. W. J., Jr., Raeymaekers, L., Wuytack, F., Wirtz, K. W. A., and Casteels, P. (1991). *Biochim. Biophys. Acta* **1063**, 1–6.

Verboomen, H., Wuytack, F., Eggermont, J. A., De Jaegere, S., Mis-

siaen, L., Raeymaekers, L., and Casteels, R. (1989). *Biochem. J.* **262**, 353–356.

Verboomen, H., Wuytack, F., De Smedt, H., Himpens, B., and Casteels, R. (1992). *Biochem. J.* **286**, 591–596.

Verboomen, H., Wuytack, F., Van den Bosch, L., Mertens, L., and Casteels, R. (1994). *Biochem. J.* **303**, 979–984.

Villa, A., Podini, P., Panzeri, M. C., Soling, H. D., Volpe, P., and Meldolesi, J. (1993). *J. Cell Biol.* **121**, 1041–1051.

Voss, J., Jones, L. R., and Thomas, D. D. (1994). *Biophys. J.* **67**, 190–196.

Vrolix, M., Raeymaekers, L., Wuytack, F., Hofmann, E., and Casteels, R. (1988). *Biochem. J.* **255**, 855–863.

Wang, K. K. W., Wright, L. C., Machan, C. L., Allen, B. G., Conigrave, A. D., and Roufogalis, B. D. (1991). *J. Biol. Chem.* **266**, 9078–9085.

Watras, J. (1988). *J. Mol. Cell. Cardiol.* **20**, 711–723.

Wibo, M., Morel, N., and Godfraind, T. (1981). *Biochim. Biophys. Acta* **649**, 651–660.

Wileman, T., Kane, L. P., Carson, G. R., and Terhorst, C. (1991). *J. Biol. Chem.* **266**, 4500–4507.

Wuytack, F., and Raeymaekers, L. (1992). *J. Bioenerg. Biomembr.* **24**, 285–300.

Wuytack, F., De Schutter, G., and Casteels, R. (1981). *FEBS Lett.* **129**, 297–300.

Wuytack, F., Raeymaekers, L., Verbist, J., De Smedt, H., and Casteels, R. (1984). *Biochem. J.* **224**, 445–451.

Wuytack, F., Raeymaekers, L., Verbist, J., Jones, L. R., and Casteels, R. (1987). *Biochim. Biophys. Acta* **899**, 151–158.

Wuytack, F., Kanmura, Y., Eggermont, J. A., Raeymaekers, L., Verbist, J., Hartweg, D., Gietzen, K., and Casteels, R. (1989a). *Biochem. J.* **257**, 117–123.

Wuytack, F., Eggermont, J. A., Raeymaekers, L., Plessers, L., and Casteels, R. (1989b). *Biochem. J.* **264**, 765–769.

Yoshida, Y., Sun, H.-T., Cai, J.-Q., and Imai, S. (1991). *J. Biol. Chem.* **266**, 19819–19825.

Zhang, C., Paul, R. J., and Kranias, E. G. (1992). *Am. J. Physiol.* **263**, H366–H371.

Zvaritch, R., James, P., Vorherr, T., Falchetto, R., Madyanov, N., and Carafoli, E. (1990). *Biochemistry* **29**, 8070–8076.

SIGNAL TRANSDUCTION

20

The Nitric Oxide–Cyclic GMP Signaling System

THOMAS M. LINCOLN, TRUDY L. CORNWELL, PADMINI
KOMALAVILAS, LEE ANN MACMILLAN-CROW, and NANCY BOERTH

The Department of Pathology
Division of Molecular and Cellular Pathology
The University of Alabama at Birmingham
Birmingham, Alabama

I. MECHANISMS OF NITRIC OXIDE SIGNALING IN VASCULAR CELLS

With the discovery by Furchgott and Zawadzki (1980) and Furchgott *et al.* (1981) that vascular endothelium controls the tone of the vessel, a new concept in the regulation of vascular function emerged. It is now recognized that endothelial-derived messenger molecules regulate the flow of blood through the vessel by controlling the contractile activity of vascular smooth muscle cells. Indeed, the endothelium would seem to be the ideal "sensor" for regulating vascular smooth muscle tone inasmuch as changes in flow, the presence of humoral factors, and the adherence of the formed elements in the blood would interact initially with endothelial surfaces. Within the last decade, a great deal of progress has been made in identifying various endothelial-derived messenger molecules that regulate contractile activity of vascular smooth muscle. These messengers include eicosanoids, peptides, such as endothelin, and endothelial-derived relaxing factors. This chapter will focus on endothelial-derived nitric oxide (NO) as a biological signaling molecule for vascular smooth muscle cells (VSMC).

In reality, the potential biological role for NO in regulating vascular smooth muscle relaxation is a comparatively dated concept. Before Furchgott's pioneering studies, several laboratories had identified the NO free radical as the active moiety produced by nitrogen oxide-containing vasodilators such as nitroprusside and nitroglycerine (Gruetter *et al.*, 1980; Arnold *et al.*, 1977; DeRubertis and Craven, 1977). Studies by Murad

(Katsuki *et al.*, 1977) and Ignarro (Gruetter *et al.*, 1979) in the 1970s subsequently defined NO as the activator of soluble guanylate cyclase in vascular smooth muscle, which led to increases in intracellular guanosine 3′,5′-monophosphate (cGMP) and to eventual relaxation of contracted arterial and bronchiolar smooth muscle strips. What was not so widely appreciated in these earlier studies, however, was the fact that NO was *a biologically produced* messenger, as opposed to a pharmacologically produced mediator. It was not until the mid-1980s that several investigators, including Furchgott (1988) Ignarro *et al.* (1987), and Moncada (Palmer *et al.*, 1987), discovered that the pharmacological properties of authentic NO were identical with those of the endothelial-derived relaxing factor (EDRF) (Table I). These findings, together with the studies reported by Hibbs *et al.* (1987) and Iyenga *et al.*, (1987), which indicated that L-arginine was the precursor for the production of nitrogen oxides by activated macrophages, led Moncada and co-workers to the concept that NO is biologically produced from L-arginine in vascular endothelial cells (Palmer *et al.*, 1987).

The biological significance of NO as a signaling molecule for vascular cells has been extended to other cell types. The chemical properties of NO make it an ideal regulator for local tissue regulation; rather than interacting with specific receptors on cell surfaces, NO is a highly diffusible and permeant molecule that enters groups of cells within a defined radius. Thus, small groups of cells may be signaled separately from the larger tissue. Because many neurons release NO as a

Property	EDFR	NO
Half-life in solution	$\approx$ 5 sec	$\approx$ 5 sec
Effects on cyclic GMP	Elevates	Elevates
Effects of pyrogallol	Inhibits	Inhibits
Effects of hemoglobin	Inhibits	Inhibits
Effects of SOD	Prolongs	Prolongs
A_{max} after reacting with hemoglobin	406 nm	

neurotransmitter, the breadth of signaling by NO is dependent on the level of recruitment for neural firing, for instance. In vascular beds, the short half-life of NO makes it an ideal mediator for regulating blood flow to tissues in order to exactly meet the minute-to-minute or even second-to-second tissue demand.

The production of NO in vascular endothelium as well as in other cell types is catalyzed by the enzyme nitric oxide synthase (NOS). Three different classes of NOS have been described through cloning techniques and other approaches: (1) an endothelial form of NOS whose activity is stimulated by the binding of the Ca^{2+}–calmodulin complex upon cellular activation of Ca^{2+} mobilization; (2): a neural form of NOS whose activation is through a similar mechanism (i.e., intracellular Ca^{2+} mobilization) but whose properties are distinct from the endothelial enzyme; and (3) an in-

ducible form of NOS whose activity is independent of changes in intracellular Ca^{2+} but whose mRNA is rapidly induced by a variety of biological modifier molecules such as cytokines. This latter form of NOS has different properties from either the endothelial or neural forms, including a high V_{max} for NO production from L-arginine. For more information on the properties of NOS, the reader is referred to several excellent review articles (Marletta, 1993; Nathan, 1992; Moncada and Higgs, 1993).

From the foregoing discussion, it is clear that NO might have a wide variety of signaling roles in biological systems. As illustrated in Fig. 1, NO produced from L-arginine by generator cells interacts with a variety of enzymes in responding cells to produce biological effects. At low concentrations of NO, such as those produced by vascular endothelial cells upon activation by Ca^{2+}-mobilizing hormones (e.g., acetylcholine and bradykinin), the major cellular response is the activation of soluble guanylate cyclase and the elevation in cGMP levels. In fact, soluble guanylate cyclase is a heme-containing protein that has a high affinity for NO. The binding of submicromolar concentrations of NO to the heme moiety of guanylate cyclase evokes a large increase in the V_{max} of the enzyme (over 100-fold in most cases) for GTP conversion to cGMP. Thus, small "puffs" of NO produced by vascular endothelial cells can result in significant increases in cGMP levels in smooth muscle cells leading to relaxation. VSMC normally do not produce their own NO since they

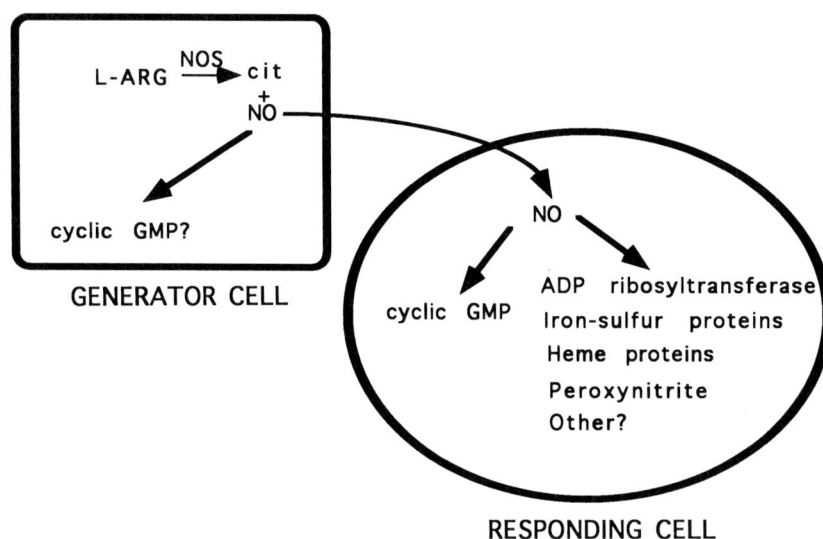

FIGURE 1   NO way for cell signaling. The model demonstrates that NO acting in responding cells affects cellular activities through both cGMP-dependent and non-cGMP-dependent mechanisms (cit = citrulline). Reprinted with permission from R. G. Landes, Co., Austin, Texas, from "Cyclic GMP: Biochemistry Physiology and Pathophysiology," by T. M. Lincoln, Copyright 1994.

rarely express NOS, and thus are signaled to relax by the endothelium. In pathophysiological instances, however, cytokines derived from immunoactivated cells cause the induction of iNOS in VSMC, resulting in large increases in NO and cGMP production. This mechanism appears to underlie the hypotension created in endotoxic shock (Nathan, 1992).

The comparatively high levels of NO such as those produced by stimulated macrophages may produce a wide variety of other cellular responses. As depicted in Fig. 2, high concentrations of NO have been reported to produce a number of cGMP-independent effects in tissues, such as the inhibition of enzymes containing ferrous-sulfhydryl groups, the production of peroxynitrite as a result of its reaction with superoxide anions, and the ADP-ribosylation of proteins under certain conditions (Lancaster and Hibbs, 1990; Kwom et al., 1991; Beckman et al., 1990; Salvemini et al., 1993; Ischiropoulos, et al., 1992). Some of these actions of NO may be responsible for the cellular killing activity of immunostimulated macrophages. Therefore, although the physiological significance of these actions is not always clear, the pathophysiological effects of these alternate NO signaling pathways are receiving a great deal of attention. Peroxynitrite formation has been associated with irreversible nitration of tyrosine residues in proteins—an event that may have pathophysiological significance (Ischiropoulos et al., 1992). On the other hand, the actions of very high concentrations of NO may mask its effects related to cGMP signaling. Perhaps even more importantly, the effects of high concentrations of NO may be erroneously attributed to cGMP signaling. As illustrated in Fig. 2, cGMP-dependent signaling actions of NO should be observed at comparatively low concentrations of NO-generating agents. It is well known, for instance, that complete vascular relaxation is achieved using submicromolar concentrations of NO-generating drugs such as nitroprusside and S-nitroso-N-acetylpenicillamine (SNAP). Higher concentrations of NO-generating drugs will more than likely produce cGMP-*independent* actions, including nonspecific effects of high concentrations of intracellular cGMP. For example, Cornwell et al., (1994a) have found that extraordinarily high concentrations of NO produce such profound increases in intracellular cGMP that this second messenger is capable of binding to and activating the cyclic AMP-dependent protein kinase (PKA) in VSMC. Activation of PKA has been associated with the inhibition of proliferation of several cultured cell models including that of VSMC. Thus, although these effects of NO are mediated by cGMP, the downstream effects of cGMP elevation in cultured VSMC may not be of physiological significance since they are produced by PKA activation. Thus, to understand the role of cGMP in mediating the effects of NO signaling in vascular cells or any other cell type for that matter, it is important to realize that the cGMP-specific effects may *only be associated with low (i.e., < 1 μM) concentrations of NO or NO-generating drugs.*

II. MECHANISM OF cGMP-EVOKED RELAXATION OF VASCULAR SMOOTH MUSCLE

Early studies in the laboratories of Ignarro (Napoli et al., 1980), Schultz (Schultz et al., 1977), and Lincoln (1983) indicated that cGMP was capable of relaxing various smooth muscle preparations contracted with either receptor-occupying agonists or with depolariz-

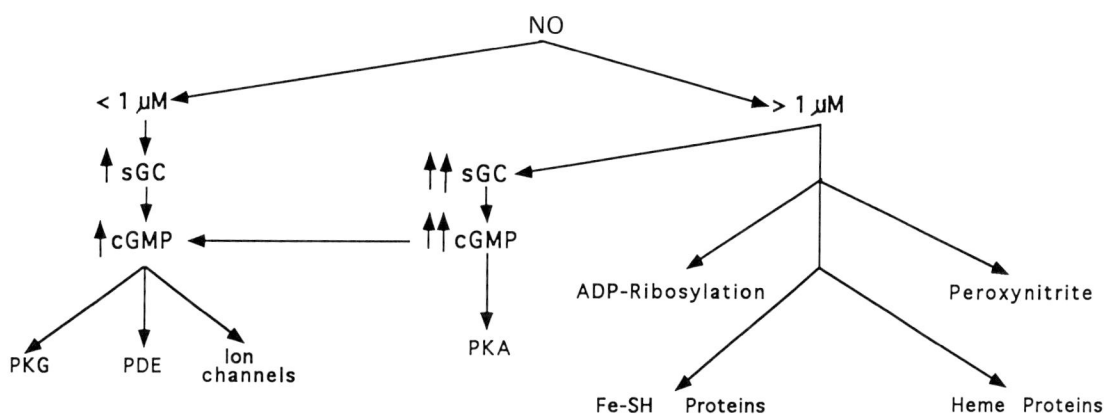

FIGURE 2 Mechanisms of NO signaling. NO, generated endogenously by cells or produced by NO-generator drugs, may have effects on cells related to its concentration. At micromolar concentrations, NO selectively activates soluble guanylate cyclase (sGC). At higher concentrations, NO may produce a variety of effects in addition to the activation of sGC. Robust stimulation of sGC by supra-micromolar concentrations of NO may produce inordinately high concentrations of cGMP that may "cross over" and activate PKA.

ing concentrations of potassium. Inasmuch as each contractile mechanism is associated with increases in intracellular concentrations of free cytosolic Ca^{2+} in the smooth muscle cell ($[Ca^{2+}]_i$), our laboratory proposed that cGMP leads to a lowering of $[Ca^{2+}]_i$ to produce relaxation (Lincoln, 1983; Johnson and Lincoln, 1985). These findings were subsequently confirmed using Ca^{2+}-sensitive enzyme activation and intracellular fluorescent dyes as indicators of the regulation of $[Ca^{2+}]_i$ by cGMP (Johnson and Lincoln, 1985; Cornwell and Lincoln, 1988; Rashatwar *et al.*, 1987). As shown in Fig. 3, the reduction in $[Ca^{2+}]_i$ produced by 8-bromo-cGMP in primary cultures of rat aortic VSMC correlated with the inhibition of Ca^{2+}-dependent activation of phosphorylase *a* conversion in the isolated rat aortic strips, suggesting that cGMP reduces the level of *physiologically* meaningful $[Ca^{2+}]_i$ in vascular smooth muscle. Because depolarization elevates $[Ca^{2+}]_i$ primarily through the opening of voltage-gated channels, whereas G-protein-coupled agonists such as vasopressin and angiotensin elevate $[Ca^{2+}]_i$ through the activation of phospholipase C and the generation of inositol 1,4,5-trisphosphate (IP_3), it ap-

pears likely that there are multiple sites of action of cGMP to reduce $[Ca^{2+}]_i$ in the VSMC. Some of these actions will be discussed in the following.

A. Ca^{2+} Sequestration

Early studies suggested that cGMP could activate Ca^{2+}-pumping mechanisms in rat aortic VSMC to reduce $[Ca^{2+}]_i$ (Rashatwar *et al.*, 1987; Furukawa and Nakamura, 1987; Vrolix *et al.*, 1988; Raeymaekers *et al.*, 1988). One potential mechanism by which this could occur is through the activation of sarcoplasmic reticulum Ca^{2+}-ATPase via the phosphorylation of the Ca^{2+}-ATPase regulatory protein, phospholamban (PLB). PLB has been shown to be a good substrate for cGMP-dependent protein kinase (PKG) both *in vitro* and in the intact smooth muscle cell (Raeymaekers *et al.*, 1988; Sarcevic *et al.*, 1989; Karczewski *et al.*, 1992; Cornwell *et al.*, 1991; Huggins *et al.*, 1989), suggesting that active Ca^{2+} sequestration resulting from PKG-dependent phosphorylation of PLB contributes to the relaxing actions of cGMP. Elevations in $[Ca^{2+}]_i$ by both depolarization and Ca^{2+} mobilization evoked by IP_3 could be attenuated by this mechanism.

B. Activation of Ca^{2+}-Activated K^+ Channels (BK Channels)

The role of NO and cGMP in the hyperpolarization of the smooth muscle cell membrane has been the subject of much scrutiny and controversy in the last few years. On the one hand, there is a wealth of evidence that supports the notion that in *some* smooth muscle cell preparations, NO or cGMP activates BK channels. However, in a few of these instances, NO has been found to activate BK channels *independent* of cGMP elevations. And then there are some smooth muscle cell types where BK channels appear to play no role in NO- or cGMP-mediated relaxation. These findings have led to the realization that there appears to be cell-specific effects for NO and cGMP for the activation of BK channels. These findings are summarized here.

Early studies using both gastrointestinal smooth muscle (Thornbury *et al.*, 1991; Ward *et al.*, 1992) and vascular smooth muscle (Krippeit-Drews *et al.*, 1992; Chen and Rembold, 1992; Tare *et al.*, 1990) demonstrated that analogs of cGMP increase BK channel activity leading to hyperpolarization and relaxation. The functional significance of these effects was supported by the finding that the BK channel inhibitor, charybdotoxin, attenuated NO- and cGMP-mediated relaxation in several preparations, including coronary arterial strips and tracheal smooth muscle strips (Hamaguchi *et al.*, 1991). However, other findings

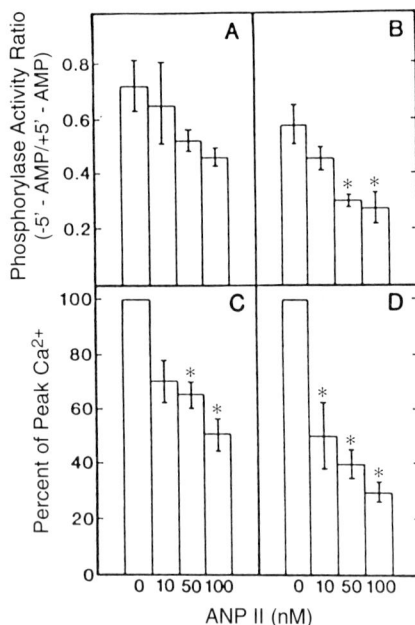

FIGURE 3 Effects on the guanylate cyclase activator, atrial natriuretic peptide II (ANP II), on Ca^{2+}-dependent activation of phosphorylase *a* formation and intracellular Ca^{2+} levels in potassium-stimulated (A and C) or vasopressin-stimulated (B and D) primary cultures of rat aortic VSMC. Cultures were incubated for one minute with the Ca^{2+}-activating stimulus in the absence or presence of different concentrations of ANP II. Asterisk denotes significance from 0 ANP II at $P < 0.05$. For details on the methodology, consult Corwell and Lincoln (1988). Reprinted with permission from the *Journal of Pharmacol. Exp. Therap.* and Williams and Wilkins, Co., Copyright 1988.

demonstrated that (1) charybdotoxin did not inhibit relaxation of contracted rat aortic smooth muscle strips (Hamaguchi *et al.*, 1991) and (2) NO gas itself stimulated BK channels in rabbit aortic smooth muscle cells in the presence of guanylate cyclase inhibitors that blocked rises in intracellular cGMP (Bolotina *et al.*, 1994; Lei *et al.*, 1992). Therefore, as an exclusive mechanism of cGMP-evoked relaxation, activation of BK channels is not supported by the literature. Rather, this mechanism is but one of several known effects of cGMP and PKG for the reduction of $[Ca^{2+}]_i$ and smooth muscle relaxation.

In those smooth muscle preparations where cGMP has been shown to stimulate BK channels, there appears to be an important role for PKG. Robertson *et al.* (1993) demonstrated that PKG activated BK single-channel currents in both cell-attached and inside-out configurations. The major effect of PKG was to increase opening probability, an effect that was most apparent at depolarized potentials. Likewise, Archer and coworkers (1994) demonstrated that PKG activators [i.e., 8-Br-cGMP and (Sp)-cGMP] were potent activators in stimulating BK channels in rat pulmonary arterial smooth muscle. These results are important in that PKA is a well-known activator of BK channels in nonvascular and vascular smooth muscle cells (Jun-Ichi *et al.*, 1988; Kume *et al.*, 1989). A different situation was suggested by White *et al.* (1993) in the GH_4C_3 pituitary cell line. In these cells, natriuretic peptides inhibited hormone secretion by stimulating BK channels. Cyclic GMP analogs and purified PKG mimicked the effects of atrial natriuretic peptides (ANP), but the effects of PKG were blocked using the protein phosphatase inhibitor okadaic acid. These authors suggested that the phosphorylated channel (probably by PKA) is relatively inactive, and the PKG-stimulated *dephosphorylation* of the channel resulted in its activation. The results implied that PKG activated an okadaic acid-sensitive protein phosphatase in this preparation. Given that there are no reports demonstrating PKG-dependent BK channel phosphorylation, these results add an interesting wrinkle to the role of phosphorylation in the regulation of BK channel function.

C. Inhibition of Phospholipase C

Several studies suggest that cGMP and PKG inhibit agonist-evoked phospholipase C formation (Rapoport, 1986; Takai *et al.*, 1981; Hirata *et al.*, 1990). Ruth *et al.* (1993) reported that CHO cells overexpressing PKG demonstrated an attenuated thrombin-stimulated IP_3 response in the presence of cGMP. Control cells not expressing PKG were insensitive to the effects of cGMP analogs. These results supported early work in platelets and rat aorta, which indicated that cGMP was capable of blocking phospholipase C activation. To date the mechanism by which cGMP inhibits IP_3 formation is not known. Several heterotrimeric G-proteins (e.g., G_i, G_o, and G_z) that have the potential to stimulate phospholipase C upon agonist activation have been examined as substrates for PKG, but none appears to be phosphorylated directly by this kinase (Lincoln, 1991). Others (G_q, G_{11}) have not been reported to be substrates for PKG. Likewise, most of the phospholipase C forms identified to date have not been shown to be phosphorylated by PKG although they are substrates for PKA and protein tyrosine kinases (Rhee *et al.*, 1993). Therefore, agonist-evoked decreases in phospholipase C activation may be inhibited by cGMP but the mechanism is obscure.

It is also possible that PKG may inhibit the actions of IP_3 to mobilize Ca^{2+} as opposed to the inhibition of IP_3 formation. Studies by Komalavilas and Lincoln (1994) and those by Imai and coworkers (Koga *et al.*, 1994) have shown that the IP_3 receptor protein from rat cerebellum and aorta is phosphorylated by PKG. These results are provocative in that both PKA and PKG catalyze the phosphorylation of this protein *in vitro* on the same seryl residue; in the intact cell, however, only PKG activation leads to its phosphorylation (Komalavilas and Lincoln, 1995). The functional significance of PKG-catalyzed IP_3 receptor phosphorylation is unknown at this time. There is also the question of the physiological importance of the effects of PKG to regulate either IP_3 formation or action in VSMC. Given the findings that cGMP and PKG stimulate BK channels and hyperpolarize the smooth muscle membrane in some preparations at least, the role of cGMP in the inhibition of agonist-stimulated phosphoinositide turnover or IP_3 action would seem secondary. On the other hand, it is important to keep in mind that cGMP may regulate several steps in the mobilization and removal of $[Ca^{2+}]_i$ simultaneously in the cell. Alternatively, one preparation (e.g., coronary arterial smooth muscle) may utilize primarily one pathway for Ca^{2+} removal (i.e., BK channel activation) whereas another preparation (e.g., platelet) may utilize another (i.e., phospholipase C activation). Still other smooth muscle tissues such as the rat aorta may rely on Ca^{2+} removal mechanisms such as activation of sarcoplasmic reticulum Ca^{2+} pumps through PLB phosphorylation to at least supplement other Ca^{2+} regulatory processes. We would propose that the effects of cGMP are varied and complex, and that this second messenger regulates a number of important events controlling intracellular Ca^{2+} homeostasis, as illustrated in Fig. 4.

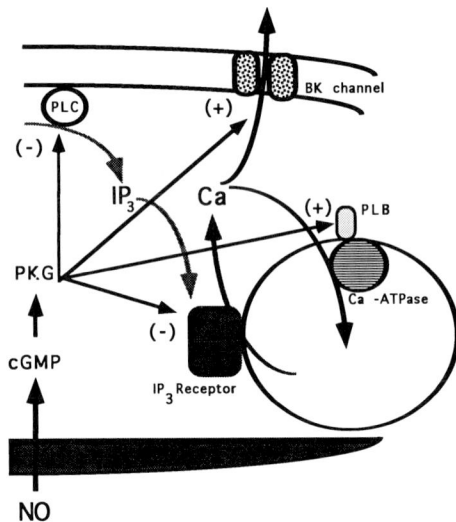

FIGURE 4   Complexity of actions of PKG in various smooth muscle cells. PKG may inhibit phospholipase C (PLC) activation or IP_3 receptor gating to inhibit the mobilization of $[Ca^{2+}]_i$ within the cell. PKG may also stimulate Ca^{2+} removal from the cytoplasm by activating BK channels or stimulating Ca^{2+}-ATPase activity through the phosphorylation of phospholamban (PLB). Other modes of action not depicted in the model include inhibition of contractile protein function, inhibition of L-type Ca^{2+} channels, and regulation of cytoskeletal events.

III. ROLE OF cGMP-DEPENDENT PROTEIN KINASE IN THE NITRIC OXIDE–cGMP SIGNALING PATHWAYS IN SMOOTH MUSCLE CELLS

From the previous discussions, it seems obvious that PKG mediates many of the effects of NO and cGMP signaling in VSMC. In fact, early studies indicated that smooth muscle was one of the richest sources of PKG in mammalian tissues. It is important to realize, however, that PKG is but one of several types of receptor proteins for cGMP in tissues. Other proteins have been shown to mediate the effects of cGMP elevations, including ion channels that are directly regulated by the nucleotide, cyclic nucleotide phosphodiesterases that are either activated or inhibited by cGMP, and, finally, PKA itself can be directly activated by high concentrations of cGMP. The role of cGMP-regulated ion channels and cGMP-regulated phosphodiesterases in the regulation of cell function has been the subject of previous review articles and will not be addressed further (Lincoln, 1994; Lincoln and Cornwell, 1993). The major receptor protein for cGMP in smooth muscle is PKG, and therefore signaling through this kinase will be the focus of this section.

PKG is actually a family of enzymes found in eukaryotic cells. Two PKG genes have been described

in *Drosophila* that encode at least six separate cDNAs (Kalderon and Rubin, 1989). In mammalian tissues, two major PKG forms have been cloned (Wernet *et al.*, 1989; Sandberg *et al.*, 1989). A type I PKG exists as two isoforms whose mRNAs are alternately spliced within the first two exons yielding a Iα and a Iβ isoform (Lincoln *et al.*, 1988; Francis *et al.*, 1989). The type II PKG is slightly larger than the type I PKG and appears to have a more limited tissue distribution (DeJong, 1981; Uhler, 1993). The structures of the various PKGs cloned from eukaryotic sources are shown in Fig. 5 and others are described in Table II. The tissue-specific expression of PKG is also highly variable and appears to be under physiological control. Factors such as cell density (Cornwell *et al.*, 1994b) and gonadal steroid hormones (T. L. Cornwell, R. A. Word, and T. M. Lincoln, unpublished observations) have been shown to regulate the type I PKG expression in smooth muscle cells, for instance. In addition to smooth muscle, platelets and cerebellar Purkinje cells also express high levels of the enzyme. Lower but measurable levels of PKG are found in cardiac myocytes, leukocytes (e.g., neutrophils and monocytes), endothelial cells, and certain secretory cells. It has been difficult to detect PKG in skeletal muscle myocytes, most central nervous system neurons, and hepatocytes, for example. On the other hand, the type II PKG is expressed at comparatively high levels in intestinal epithelial cells and mouse brain.

The reasons underlying the differences in expression are unclear at this time. It is known, however, that PKG Iα and Iβ have distinct regulatory features. The Iβ isoform has a greater K_{act} for cGMP than does the Iα isoform (approximately 5-10-fold greater), making the Iβ isoform less susceptible to activation by cGMP (Wolfe *et al.*, 1989; Ruth *et al.*, 1991). The difference in K_{act} between the two type I isoforms is also apparent using cyclic nucleotide analogs to activate the kinases. In general, 8-substituted cyclic nucleotides (e.g., 8-bromo or 8-*para*-chlorophenylthio-substituted cAMP and cGMP) have higher affinities for type Iα than for type Iβ. This property allowed Sekhar and coworkers (1992) to examine the potency of a series of cyclic nucleotide analogs to relax contracted tracheal and coronary artery smooth muscle. The results of this study suggested that the type Iα isoform, but not the type Iβ isoform, of PKG is directly responsible for mediating relaxation of the tissues. These results would seem to fit with other findings, which indicate that the type Iα PKG is expressed at higher levels in vascular smooth muscle than the type Iβ (Lincoln *et al.*, 1988). Nevertheless, the precise role for the two isoforms of PKG is unknown at this time.

The type II PKG appears to be expressed at high

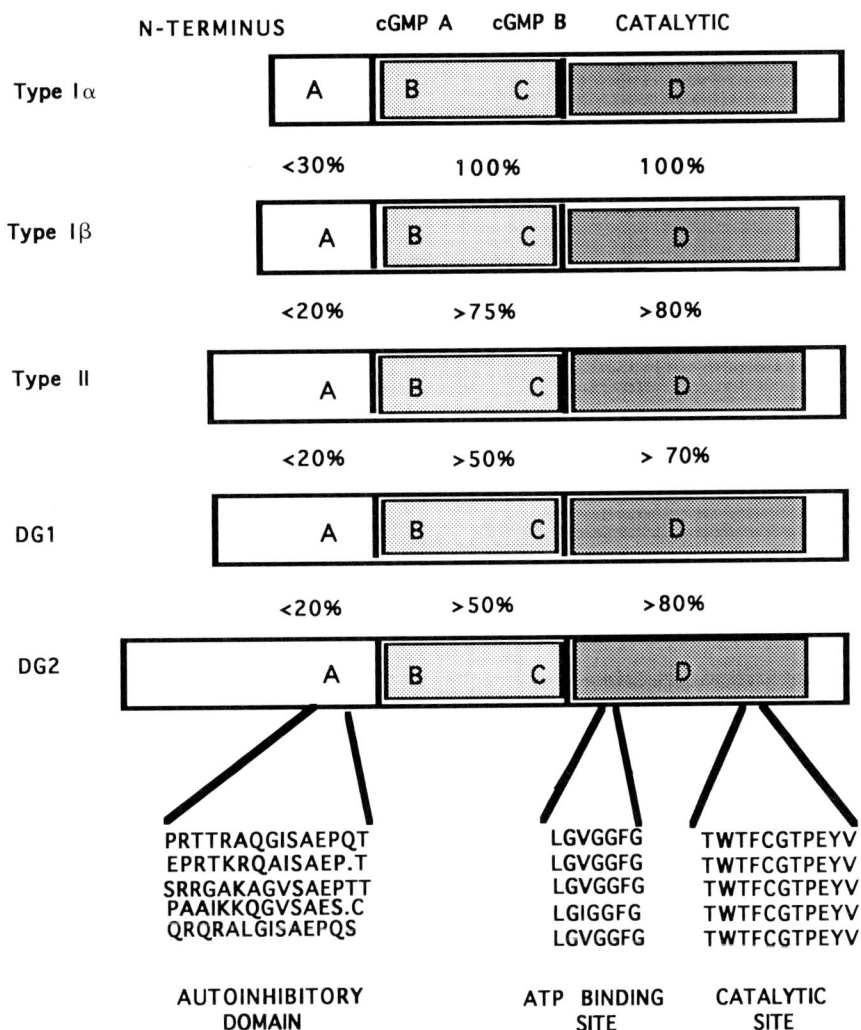

FIGURE 5 PKG forms in eukaryotic cells. The enzymes isolated from mammalian sources (type Iα and Iβ and type II) and the *Drosophila* enzymes DG1 and DG2 have their domains aligned and homology indicated. The positions of the autoinhibitory domains, ATP binding sites, and catalytic sites are shown. Reprinted with permission from the R. G. Landes Company, Austin, Texas, from "Cyclic GMP: Biochemistry, Physiology and Pathophysiology," by T. M. Lincoln, Copyright 1994.

levels in intestinal epithelium (DeJong, 1981), where it may mediate water and solute transport by endogenous guanylate cyclase activators such as natriuretic peptides and guanylin. Unlike the type I PKG, which is predominantly soluble in vascular smooth muscle, the type II appears to be anchored to the brush-border membrane. A type II form of the PKG has been cloned from mouse brain and demonstrates significant homology with the type I isoforms throughout the catalytic domain and the cGMP binding domains (Uhler, 1993). The amino terminus, however, demonstrates the least similarity with the type I isoforms. As alluded to earlier, an interesting feature of PKG appears to be the variability in sequence at the amino-terminal domain consisting of approximately 100 residues. This region of the enzyme contains the autoinhibitory site, the autophosphorylation sites, and a "leucine zipper" domain responsible for the dimerization of the subunits. Inasmuch as the PKG appears to be capable of functioning as a monomer (e.g., like protein kinase C), the physiological significance or role of the dimerization domain has been studied. In human neutrophils (Wyatt *et al.*, 1991; Pryzwansky *et al.*, 1995), VSMC (Cornwell *et al.*, 1991; MacMillan-Crow and Lincoln, 1994), and endothelial cells (MacMillan-Crow *et al.*, 1994), a significant amount of type I PKG is bound to the cytoskeleton. Confocal laser scanning microscopy and immunofluorescence imaging have shown that

TABLE II   Properties of Cyclic GMP-Dependent
Protein Kinases

Species	Forms	Subunit M_r	Subunit No.
Bovine	Iα	76,418	Dimer
	Iβ	77,803	Dimer
Human	Iβ	77,803	Dimer
Mouse	II	≈87,000	Dimer(?)
Drosophila	DG1	>76,000	—
	DG2T1	>76,000	—
	DG2T2	>76,000	—
Paramecium		88,000	Monomer
Tetrahymena		100,000	Monomer
D. discoideum		82,000	Monomer
Silkworm	Egg	72,000	Monomer
	Pupae	70,000(?)	Dimer
Ascaris suum		81,000(?)	Dimer

PKG is localized with the intermediate filament protein, vimentin, at least transiently in the cell (Cornwell et al., 1991; MacMillan-Crow et al., 1994; Pryzwansky et al., 1990). Studies using purified proteins suggest that PKG binds to vimentin with high affinity ($K_D = 50$ nM) and specificity with a stoichiometry of 1 mol PKG monomer: 1 mol vimentin monomer (MacMillan-Crow and Lincoln, 1994). Furthermore, immunoprecipitation of PKG also precipitates vimentin from smooth muscle and endothelial cells, suggesting that the two proteins interact with each other in the intact cell (MacMillan-Crow and Lincoln, 1994; MacMillan-Crow et al., 1994). The region of PKG that binds vimentin has been localized to the amino-terminal portion of the enzyme and particularly the dimerization domain (MacMillan-Crow and Lincoln, 1994). Proteolytically generated monomers of PKG Iα, for example, do not bind vimentin.

One function of PKG localization to the cytoskeleton in the intact cell may be to situate the kinase near substrates. This may be particularly important for PKG in that PKG catalyzes the phosphorylation of several proteins in the cell that are also substrates for other kinases such as PKA and protein kinase C (i.e., PLB and the IP$_3$ receptor). Thus, by virtue of localization with certain substrate proteins, PKG has ready access to them upon activation. Studies by Cornwell et al. (1991) have suggested that the localization of PKG to the sarcoplasmic reticulum in rat aortic VSMC enables the kinase to catalyze the phosphorylation of the substrates localized to the sarcoplasmic reticulum. A similar situation is known to exist for the regulatory subunit IIβ (RIIβ) of PKA in brain and other tissues. In

this instance, RIIβ anchors to microtubule-associated protein 2 (MAP 2) in brain, thus facilitating phosphorylation of this protein (Luo et al., 1990; Scott et al., 1990; Carr et al., 1992). The anchoring domain of RIIβ is also located in the dimerization (i.e., amino terminus) domain of the protein. Scott and coworkers have subsequently isolated a family of RIIβ-anchoring proteins from several sources and have defined the binding motif in these proteins (Hausken et al., 1994). Of particular interest for PKG is the fact that RIIβ-anchoring proteins bind to the dimerization domain of RIIβ (Hausken et al., 1994). Anchoring of other kinases, such as protein kinase C and protein tyrosine kinases as well as protein phosphatases, is now recognized as an important mechanism for the interaction between enzyme and substrate (Hubbard and Cohen, 1993). Given that kinases, phosphatases, and the substrates upon which they act are present in the cell in only small amounts, anchoring and targeting of kinases and phosphatases would appear to ensure rapid and efficient control of phosphorylation.

IV. ACTIVATION OF cGMP-DEPENDENT PROTEIN KINASE BY cAMP IN SMOOTH MUSCLE

One of the confusing aspects in the regulation of smooth muscle function has been the apparent redundant role of cAMP and cGMP in mediating relaxation. It has been known since the mid-1970s that agents that elevate cAMP (β-adrenergic agonists, forskolin, prostaglandins) or cGMP (nitric oxide-containing vasodilators, natriuretic peptides) mediate relaxation of contracted smooth muscle. Indeed, the pharmaceutical industry has invested heavily in the generation of cyclic nucleotide phosphodiesterase inhibitors as potential antiasthmatic and antihypertensive drugs. Still, the signaling pathways involved in the mediation of cyclic nucleotide effects remained unknown until the last few years. The original hypothesis for cAMP-mediated relaxation, that is, the phosphorylation of myosin light chain kinase by PKA and subsequent inhibition of calmodulin binding (Conti and Adelstein, 1981), has been more or less replaced by the newer findings that cAMP mediates smooth muscle relaxation by activation of PKG (Francis et al., 1988; Lincoln et al., 1990). Francis and coworkers (1988) used a series of cyclic nucleotide analogs as activators of both PKG and PKA and demonstrated that a positive correlation exists between their potency to relax contracted coronary artery and tracheal smooth muscle and their capacity to activate PKG—not PKA. Lincoln and coworkers (1990) used a different approach to demonstrate the critical role for PKG in mediating cAMP-dependent effects in VSMC. This laboratory determined that PKG expression was lost upon passaging of rat aortic VSMC in culture (Cornwell and Lincoln, 1989). Loss of PKG expression resulted in a loss of the capacity of cGMP to lower $[Ca^{2+}]_i$ in the cultured cells; restoration of PKG to the cells by introducing the enzyme into the cells via an osmotic procedure restored the sensitivity of the cells to cGMP (Cornwell and Lincoln, 1989). On the other hand, PKA expression was not reduced upon repetitive passaging of the cells (Lincoln et al., 1990). Agents that elevated cAMP in passaged VSMC (isoproterenol, forskolin) did not reduce $[Ca^{2+}]_i$ even though PKA was present. When PKG was restored to the passaged VSMC, however, elevations in cAMP resulted in the reduction in $[Ca^{2+}]_i$.

These results seemed to underscore the importance of PKG in mediating reductions in $[Ca^{2+}]_i$ to both cAMP and cGMP, and support the concepts developed by Francis et al. (1988) and as illustrated in Fig. 6 that PKG may be the actual focal point for smooth muscle relaxation. The concept of cyclic nucleotide "cross-over," that is, the capacity for cAMP to activate endogenous PKG and for cGMP to activate endogenous PKA, should be kept in mind when constructing protocols designed to elucidate cyclic nucleotide-dependent effects in cells. As will be discussed in the following sections, interpretations involving the role of NO and cGMP in the regulation of VSMC proliferation need to take into account the potential role for cGMP to "cross-over" and activate PKA in passaged VSMC.

V. PROTEIN SUBSTRATES FOR cGMP-DEPENDENT PROTEIN KINASE

There has already been some discussion of the role of PKG in protein phosphorylation in the intact cell. Proteins regulating Ca^{2+} transport, phospholipase C, and Ca^{2+}-activated K^+ channels are all potential substrates for PKG even if the precise proteins that are phosphorylated are not known at this time. Perhaps the best-characterized protein substrate for PKG in terms of kinetics, stoichiometry, and sites of phosphorylation is an actin binding protein originally isolated from platelets and known as VASP (vasodilator-stimulated phosphoprotein). VASP is a 46-kDa protein whose phosphorylation by PKA and PKG in the intact cell correlates with platelet shape changes and cytoskeletal rearrangements (Waldmann et al., 1987; Halbrugge and Walter, 1989; Reinhard et al., 1992). Walter and coworkers have studied the phosphorylation of VASP in response to NO-dependent elevations in cGMP in intact platelets and have identified two PKG phosphorylation sites in the protein—site 1 is RKVSKQE and site 2 is RRVSNAG (Halbrugge et al., 1990; Butt et al., 1994). Site 1 is particularly interesting in that the site contains an RK sequence instead of an RR sequence before the serine. Studies in Corbin's laboratory on the phosphorylation site in the type V cGMP phosphodiesterase indicate that PKG tolerates the lysine residue in the dibasic sequence better than does PKA (Colbran et al., 1992). Hence, this could be one specificity determinant for PKG-dependent phosphorylation. The role of VASP in mediating the effects of PKG to reduce $[Ca^{2+}]_i$ in platelets was also studied by this group. VASP phosphorylation by either eleva-

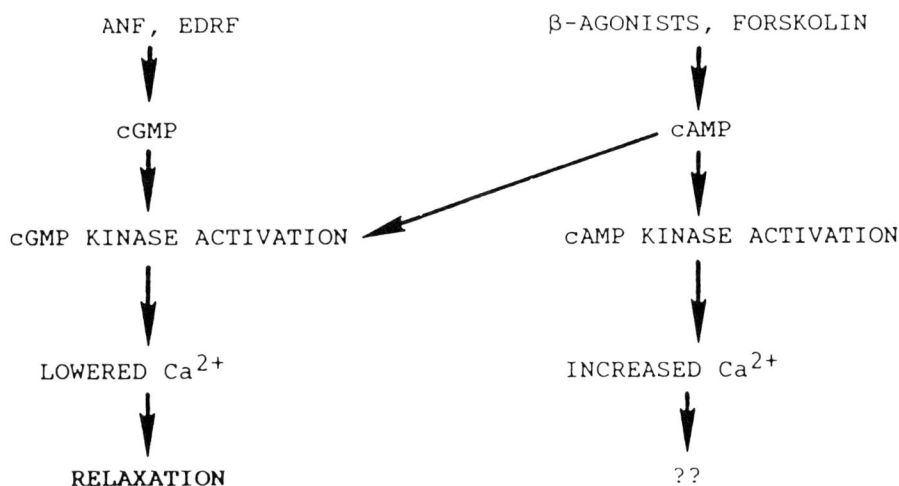

FIGURE 6 Activation of PKG by cAMP and cGMP in vascular smooth muscle cells.

tions in cGMP or cAMP did not correlate with the inhibition of Ca²⁺ release in thrombin-activated platelets (Meinecke *et al.*, 1994). Therefore other phosphorylated proteins may be involved in the regulation of Ca²⁺ mobilization by PKG, and other roles for VASP must exist in the platelet and other cells.

Cloning of VASP indicates that it is a structurally distinct cytoskeletal protein with little homology with other known actin binding proteins (Haffner *et al.*, 1995). Inasmuch as VASP appears to be localized to focal adhesions in intact platelets, the interesting possibility exists that PKG regulates focal adhesion assembly. In a report by Murphy-Ullrich and coworkers (1994), selective inhibitors of PKG, but not PKA or protein kinase C, inhibited thrombospondin-evoked focal adhesion disassembly, suggesting that PKG mediates thrombospondin action. This would be a new role for PKG in regulating cellular shape and motility. The role of VASP in this process is currently unknown, but the availability of cDNA probes for controlling expression of this unique protein may provide answers.

VI. ROLE OF NITRIC OXIDE AND cGMP IN VASCULAR SMOOTH MUSCLE CELL PROLIFERATION AND DIFFERENTIATION

Several important studies appeared in the late 1980s suggesting that NO inhibits VSMC proliferation *in vitro* (Garg and Hassid, 1989a,b; Abell *et al.*, 1989; Kariya *et al.*, 1989). These studies had immediate impact in that VSMC proliferation *in vivo* is associated with various vascular disorders such as atherosclerosis and restenosis following vascular injury. The major findings from these studies were that high concentrations of NO-generating drugs (Garg and Hassid, 1989a,b), natriuretic peptides (Abell *et al.*, 1989), or analogs of cGMP (Garg and Hassid, 1989b) blocked DNA synthesis and proliferation of repetitively passaged VSMC. Studies by Cornwell *et al.* (1994a) confirmed these findings and demonstrated that the high levels of cGMP generated in response to NO production (either by NO-generating vasodilators or by iNOS induction via interleukin 1β in VSMC) "cross over" and activate the PKA. It has long been known that cAMP inhibits the growth of cultured cells including VSMC (Morisaki *et al.*, 1988; Fukumoto *et al.*, 1988), and activation of PKA by high concentrations of cGMP appears to mimic the actions of cAMP. Whether this is a pathophysiological role for NO and cGMP—that is, cytokine-evoked iNOS induction with subsequent cGMP generation—or an artefact of the manipulation of the culture conditions must await further results from *in vivo* studies. Nevertheless, it is conceivable that PKG plays a role in VSMC phenotypic modulation

(Campbell and Campbell, 1987). This process, which is the dedifferentiation of VSMC from a "contractile" state to a "secretory" state, occurs in culture as well as *in vivo* in response to injury. The secretion of extracellular matrix proteins as well as the loss of contractile proteins during phenotypic modulation is thought to underlie the changes in contractile and proliferative behavior of cultured VSMC compared with their counterparts in the medial layer of the intact vessel. Because PKG expression is also reduced during culture, it seemed reasonable that PKG may regulate one or more properties of the phenotype of VSMC. Studies by Boerth and Lincoln (1994) demonstrated that transfection of cultured, dedifferentiated VSMC with a cDNA encoding an active catalytic domain of type I PKG promoted morphological changes of VSMC that resembled, on the surface at least, the differentiated

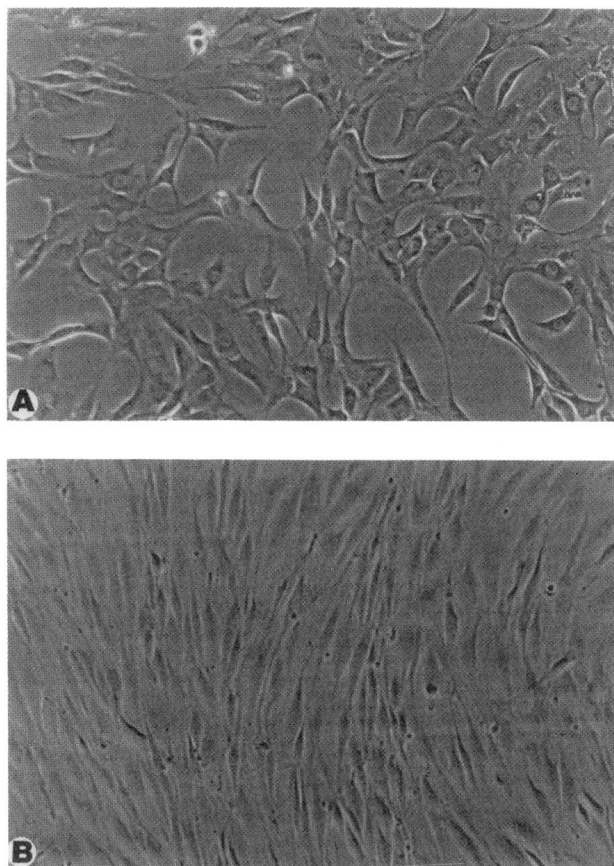

FIGURE 7   Effects of the expression of the constitutively active catalytic domain of PKG on the morphology of rat aortic SMC. Rat aortic SMC were stably transfected with pMEP4 vector containing the cDNA encoding PKG 366–671 as described in Boerth and Lincoln (1994) or control pMEP4 vector (A). Following induction of expression of the cDNA with 10% FBS + 100 μM ZnSO₄ for 2 days, cells were photographed through a phase-contrast Nikon microscope. Magnification is × 800. Data taken from Boerth and Lincoln (1994), with permission of the authors and Elsevier Press.

phenotype (Fig. 7). The expression of the constitutively active catalytic domain of PKG obviates many of the problems addressed earlier with NO signaling pathways, namely, the problems associated with using high concentrations of NO generators to elevate cGMP for long periods of time in culture and the multiplicity of effects of cGMP itself to activate ion channels, phosphodiesterases, and PKA. Thus, the catalytic domain of PKG may represent a useful tool with which to study *specific* effects of PKG, such as the regulation of VMSC phenotype, that may be unrelated to other pathways in the NO–cGMP signaling system. Clearly, more work needs to be done in understanding the potential role of PKG in regulating VSMC growth and differentiation.

VII. SUMMARY

In this chapter we have addressed a number of effects of NO and cGMP signaling in smooth muscle and other cells. The widespread interest in NO has brought renewed interest in this intracellular signaling pathway, but there are cautions that must be exercised in dissecting the role of cGMP in the NO signaling pathway. In particular, NO is a highly reactive species that at high concentrations has a number of actions unrelated to cGMP-mediated pathways in cells. Non-specific effects of both NO and cGMP have been found in the published literature and it is important that results be interpreted in light of the numerous non-specific actions of this signaling pathway. Of all the intracellular effectors of the NO–cGMP signaling pathway, perhaps the most important is PKG. This enzyme appears to mediate the vasorelaxation associated with elevations in cGMP and cAMP, as well as the inhibition of platelet activation. Specific mechanisms of action of PKG are only now being identified; such effects as the inhibition of Ca^{2+} mobilization, activation of Ca^{2+}-activated K^+ channels, and reorganization of the cytoskeleton all appear to be important roles for PKG in the cell. Other roles for PKG that may include regulation of contractile protein sensitivity to Ca^{2+}, regulation of cell attachment and motility, and even gene regulation may be discovered in the near future. Clearly this enzyme occupies a central role in the regulation of many important cellular activities.

References

Abell, T. J., Richards, A. M., Ikram, H., Espiner, E. A., and Yandle, T. (1989). *Biochem. Biophys. Res. Commun.* **160**, 1392–1396.

Archer, S. L., Huang, J. M. C., Hampl, V., Nelson, D. P., Shultz, P. J., and Weir, E. K. (1994). *Proc. Natl. Acad. Sci. U.S.A.* **91**, 7583–7587.

Arnold, W. P., Mittal, C. K., Katsuki, S., and Murad, F. (1977). *Proc. Natl. Acad. Sci. U.S.A.* **74**, 3203–3207.

Beckman, J. S., Beckman, T. W., Chen, J., Marshall, P. A., and Freeman, B. A. (1990). *Proc. Natl. Acad. Sci. U.S.A.* **87**, 1620–1624.

Boerth, N. J., and Lincoln, T. M. (1994). *FEBS Lett.* **342**, 255–260.

Bolotina, V. M., Najibi, S., Palacino, J. J., Pagano, P. J., and Cohen, R. A. (1994). *Nature (London)* **368**, 850–853.

Butt E., Abel, K., Krieger, M., Palm, D., Hoppe, V., Hoppe, J., and Walter, U. (1994). *J. Biol. Chem.* **269**, 14509–14517.

Campbell, G. R., and Campbell, J. H. (1987). *In* "Vascular Smooth Muscle in Culture," J. H. Campbell and G. R. Campbell, eds.), Vol. 1, pp. 39–56. CRC Press, Boca Raton, FL.

Carr, D. W., Hausken, Z. E., Fraser, I. D. C., Stofko-Hahn, R. E., and Scott, J. D. (1992). *J. Biol. Chem.* **267**, 13376–13382.

Chen, X.-L., and Rembold, C. M. (1992). *Am. J. Physiol.* **263**, C468–C473.

Colbran, J. L., Francis, S. H., Leach, A. B., Thomas, M. K., Jiang, H., McAllister, L. M., and Corbin, J. D. (1992). *J. Biol. Chem.* **267**, 9589–9594.

Conti, M. A., and Adelstein, R. S. (1981). *J. Biol. Chem.* **256**, 3178–3181.

Cornwell, T. L., and Lincoln, T. M. (1988). *J. Pharmacol. Exp. Ther.* **247**, 524–530.

Cornwell, T. L., and Lincoln, T. M. (1989). *J. Biol. Chem.* **264**, 1146–1155.

Cornwell, T. L., Pryzwansky, K. B., Wyatt, T. A., and Lincoln, T. M. (1991). *Mol. Pharmacol.* **40**, 923–931.

Cornwell, T. L., Arnold, E., Boerth, N. J., and Lincoln, T. M. (1994a). *Am. J. Physiol.* **36**, C1405–C1413.

Cornwell, T. L., Soff, G. A., Traynor, A. E., and Lincoln, T. M. (1994b). *J. Vasc. Res.* **31**, 330–337.

DeJong, H. R. (1981). *Adv. Cyclic Nucleotide Res.* **14**, 315–333.

DeRubertis, F. R., and Craven, P. A. (1977). *Biochim. Biophys. Acta* **499**, 337–351.

Francis, S. H., Noblett, B. D., Todd, B. W., Wells, J. N., and Corbin, J. D. (1988). *Mol. Pharmacol.* **504**, 506–517.

Francis, S. H., Woodford, T. A., Wolfe, L., and Corbin, J D. (1989). *Second Messengers Phosphoproteins* **12**, 301–310.

Fukumoto, Y., Kawahara, Y., Kariya, K., Araki, S., Fukuzaki, H., and Takai, Y. (1988). *Biochem. Biophys. Res. Commun.* **157**, 337–345.

Furchgott, R. F. (1988). *In* "Mechanisms of Vasodilation" (P. Vanhoutte, ed.), Vol. 4, pp. 401–414. Raven Press, New York.

Furchgott, R. F., and Zawadzki, J. V. (1980). *Nature (London)* **288**, 373–376.

Furchgott, R. F., Zawadzki, J. V., and Cherry, P. D. (1981). *In* "Vasodilation" (P. Vanhoutte and I. Leusen, eds.), pp. 49–66. Raven Press, New York.

Furukawa, K.-I., and Nakamura, H. (1987). *J. Biochem. (Tokyo)* **101**, 287–290.

Garg, U. C., and Hassid, A. (1989a). *Am. J. Physiol.* **257**, F60–F66.

Garg, U.C., and Hassid, A. (1989b). *J. Clin. Invest.* **83**, 1774–1777.

Gruetter, C. A., Barry, B. K., McNamara, D. B., Gruetter, D. Y., Kadowitz, P. J., and Ignarro, L. J. (1979). *J. Cyclic Nucleotide Res.* **5**, 211–224.

Gruetter, C. A., Barry, B. K., McNamara, D. B., Kadowitz, P. J., and Ignarro, L. J. (1980). *J. Pharmacol. Exp. Ther.* **214**, 9–15.

Haffner, C., Jarchau, T., Reinhard, M., Hoppe, J., Lohmann, S. M., and Walter, U. (1995). *EMBO J.* **14**, 19–27.

Halbrugge, M., and Walter, U. (1989). *Eur. J. Biochem.* **185**, 41–50.

Halbrugge, M., Friedreich, C., Eigenthaler, M., Schanzenbächer, P., and Walter, U. (1990). *J. Biol. Chem.* **265**, 3088–3093.

Hamaguchi, M., Ishibashi, T., and Imai, S. (1991). *J. Pharmacol. Exp. Ther.* **262**, 263–270.

Hausken, Z. E., Coghlan, V. M., Schafer-Hastings, C. A., Reimann, E. M., and Scott, J. D. (1994). *J. Biol. Chem.* **269**, 24245–24251.

Hibbs, J. B., Vavrin, Z., and Taintor, R. R. (1987). *J. Immunol.* **138**, 550–565.

Hirata, M., Kohse, K. P., Chang, C.-H., Ikebe, T., and Murad, F. (1990). *J. Biol. Chem.* **265**, 1268–1273.

Hubbard, M. J., and Cohen, P. (1993). *Trends Biochem. Sci.* **18**, 172–177.

Huggins, J. P., Cook, E. A., Piggott, J. R., Mattinsley, T. J., and England, P. J. (1989). *Biochem. J.* **260**, 829–835.

Ignarro, L. J., Byrns, R. E., Buga, G. M., and Wood, K. S. (1987). *Circ. Res.* **61**, 866–879.

Ischiropoulos, H., Zhu, L., Chen, J., Tsai, M., Martin, J. C., Smith, C. D., and Beckman, J. S. (1992). *Arch. Biochem. Biophys.* **298**, 431–437.

Iyenga, R., Stuehr, D. J., and Marletta, M. A. (1987). *Proc. Natl. Acad. Sci. U.S.A.* **84**, 6369–6373.

Johnson, R. M., and Lincoln, T. M. (1985). *Mol. Pharmacol.* **27**, 333–342.

Jun-Ichi, S., Akaike, N., Kanaide, H., and Nakamura, M. (1988). *Am. J. Physiol.* **255**, H754–H759.

Kalderon, D., and Rubin, G. M. (1989). *J. Biol. Chem.* **264**, 10738–10748.

Karczewski, P., Kelm, M., Hartmann, M., and Schrader, J. (1992). *Life Sci.* **51**, 1205–1210.

Kariya, K., Kawahara, Y., Araki, S., Fukuzaki, H., and Takai, Y. (1989). *Atherosclerosis (Shannon, Irel.)* **80**, 143–147.

Katsuki, S., Arnold, W. P., and Murad, F. (1977). *J. Cyclic Nucleotide Res.* **3**, 239–247.

Koga, T., Yoshida, Y., Cai, J. Q., and Imai, S. (1994). *J. Biol. Chem.* **269**, 11640–11647.

Komalavilas, P., and Lincoln, T. M. (1994). *J. Biol. Chem.* **269**, 8701–8707.

Komalavilas, P., and Lincoln, T. M. (1995). *FASEB J.* **9**, A609.

Krippeit-Drews, P., Norel, N., and Godfraind, T. (1992). *J. Cardiovasc. Pharmacol.* **20**, S72–S75.

Kume, H., Takai, A., Tokuno, H., and Tomita, T. (1989). *Nature (London)* **341**, 152–153.

Kwon, N. S., Stuehr, D. J., and Nathan, C. F. (1991). *J. Exp. Med.* **174**, 761–767.

Lancaster, J. R., and Hibbs, J. B. (1990). *Proc. Natl. Acad. Sci. U.S.A.* **87**, 1223–1227.

Lei, S. Z., Pan, Z.-H., Aggarwal, S. K., Chen, H.-S. V., Hartman, J., Sucher, N. J., and Lipton, S. A. (1992). *Neuron* **8**, 1087–1089.

Lincoln, T. M. (1983). *J. Pharmacol. Exp. Ther.* **224**, 100–107.

Lincoln, T. M. (1991). *Second Messengers Phosphoproteins* **13**, 99–109.

Lincoln, T. M. (1994). "Cyclic GMP: Biochemistry, Physiology, and Pathophysiology." R. G. Landes Co., Austin TX (distributed by CRC Press, Boca Raton, FL).

Lincoln, T. M., and Cornwell, T. L. (1993). *FASEB J.* **7**, 328–338.

Lincoln, T. M., Thompson, M., and Cornwell, T. L. (1988). *J. Biol. Chem.* **163**, 17632–17637.

Lincoln, T. M., Cornwell, T. L., and Taylor, A. E. (1990). *Am. J. Physiol.* **258**, C399–C407.

Luo, Z., Shafit-Zagardo, B., and Erlichman, J. (1990). *J. Biol. Chem.* **265**, 21804–21810.

MacMillan-Crow, L. A., and Lincoln, T. M. (1994). *Biochemistry* **33**, 8035–8043.

MacMillan-Crow, L. A., Murphy-Ullrich, J. E., and Lincoln, T. M. (1994). *Biochem. Biophys. Res. Commun.* **201**, 531–537.

Marletta, M. (1993). *J. Biol. Chem.* **268**, 12231–12234.

Meinecke, M., Geiger, J., Butt, E., Sandberg, M., Jahnsen, T., Chakraborty, T., Walter, U., Jarchau, T., and Lohmann, S. M. (1994). *Mol. Pharmacol.* **46**, 283–290.

Moncada, S., and Higgs, A. (1993). *N. Engl. J. Med.* **329**, 2002–2012.

Morisaki, N., Kanzaki, T., Motoyama, N., Saito, Y., and Yoshida, S. (1988). *Atherosclerosis (Shannon, Irel.)* **71**, 165–171.

Murphy-Ullrich, J. E., Pallero, M. A., Lincoln, T. M., Cornwell, T. L., Erickson, H. P., and Sagel, E. H. (1994). *Mol. Biol. Cell.* **5**, 378a.

Napoli, S. A., Gruetter, C. A., Ignarro, L. J., and Kadowitz, P. J. (1980). *J. Pharmacol. Exp. Ther.* **212**, 469–473.

Nathan, C. F. (1992). *FASEB J.* **6**, 3051–3064.

Palmer, R. M. J., Ferrige, A. G., and Moncada, S. (1987). *Nature (London)* **327**, 524–526.

Pryzwansky, K. B. Wyatt, T. A., Nichols, H., and Lincoln, T. M. (1990). *Blood* **76**, 612–618.

Pryzwansky, K. B., Wyatt, T. A., and Lincoln, T. M. (1995). *Blood* **85**, 222–230.

Raeymaekers, L., Hofmann, F., and Casteels, R. (1988). *Biochem. J.* **252**, 269–273.

Rapoport, R. M. (1986). *Circ. Res.* **58**, 407–410.

Rashatwar, S. S., Cornwell, T. L., and Lincoln, T. M. (1987). *Proc. Natl. Acad. Sci. U.S.A.* **84**, 5685–5689.

Reinhard, M., Halbrugge, M., Scheer, U., Wiegand, C., Jakusch, B. M., and Walter, U. (1992). *EMBO J.* **11**, 2063–2070.

Rhee, S. G., Lee, C. W., and Jhon, D. Y. (1993). *Adv. Second Messenger Phosphoprotein Res.* **28**, 57–64.

Robertson, B. E., Schubert, R., Hescheler, J., and Nelson, M. T. (1993). *Am. J. Physiol.* **265**, C299–C303.

Ruth, P., Landgraf, W., Keilbach, A., May, B., Engleme, C., and Hofman, F. (1991). *Eur. J. Biochem.* **202**, 1339–1344.

Ruth, P., Wang, G.-X., Boekhoff, I., May, B., Pfeifer, A., Penner, R., Korth, M., Breer, H., and Hofmann, F. (1993). *Proc. Natl. Acad. Sci. U.S.A.* **90**, 2623–2627.

Salvemini, D., Misko, T. P., Masferrer, J. L., Seibert, K., Currie, M. G., and Needleman, P. (1993). *Proc. Natl. Acad. Sci. U.S.A.* **90**, 7240–7244.

Sandberg, M., Natarajan, V., Ronander, I., Kalderon, D., Walter, U., Lohmann, S. M., and Jahnsen, T. (1989). *FEBS Lett.* **255**, 321–329.

Sarcevic, B., Brookes, V., Martin, T. J., Kemp, B. E., and Robinson, P. J. (1989). *J. Biol. Chem.* **264**, 20648–20654.

Schultz, K. D., Schultz, K., and Schultz, G. (1977). *Nature (London)* **265**, 750–751.

Scott, J. D., Stofko, R. E., McDonald, J. R., Comer, J. D. Vitalis, E. A., and Mangeli, J. (1990). *J. Biol. Chem.* **265**, 21561–21566.

Sekhar, K. R., Hatchett, R. J., Shabb, J. B., Francis, S. H., Wells, J. N., and Corbin, J. D. (1992). *Mol. Pharmacol.* **42**, 103–108.

Takai, Y., Kaibuchi, K., Matsubara, T., and Nishizuka, Y. (1981). *Biochem. Biophys. Res. Commun.* **101**, 61–67.

Tare, M., Parkington, H. C., Coleman, H. A., Neild, T. O., and Dusting, G. L. (1990). *Nature (London)* **346**, 69–71.

Thornbury, K. D., Ward, S. M., Dalziel, H. H., Carl, A., Westfall, D. P., and Sanders, K. M. (1991). *Am. J. Physiol.* **261**, G553–G557.

Uhler, M. (1993). *J. Biol. Chem.* **268**, 13586–13591.

Vrolix, M., Raeymaekers, L., Wuytack, F., Hofmann, F., and Casteels, R. (1988). *Biochem. J.* **255**, 855–863.

Waldmann, R., Nieberding, M., and Walter, U. (1987). *Eur. J. Biochem.* **167**, 441–448.

Ward, S. M., Dalziel, H. H., Bradley, M. E., Buxton, I. L. O., Keef, K., Westfall, D. P., and Sanders, K. M. (1992). *B. J. Pharmacol.* **107**, 1075–1082.

Wernet, W., Flockerzi, V., and Hofmann, F. (1989). *FEBS Lett.* **251**, 191–196.

White, R. E., Lee, A. B., Shcherbatko, A. D., Lincoln, T. M., Schonbrunn, A., and Armstrong, D. L. (1993). *Nature (London)* **361**, 263–266.

Wolfe, L., Corbin, J. D., and Francis, S. H. (1989). *J. Biol. Chem.* **264**, 7734–7741.

Wyatt, T. A., Lincoln, T. M., and Pryzwansky, K. B. (1991). *J. Biol. Chem.* **266**, 21274–21280.

21

Inositol 1,4,5-Trisphosphate Production

MICHAEL BÁRÁNY

Department of Biochemistry
College of Medicine, University of Illinois at Chicago
Chicago, Illinois

KATE BÁRÁNY

Department of Physiology and Biophysics
College of Medicine, University of Illinois at Chicago
Chicago, Illinois

I. INTRODUCTION

Inositol-containing phospholipids are minor components of plasma membranes of eukaryotic cells and constitute 3–5% of the total phospholipids (Abdel-Latif, 1986). Approximately 10–20% of phosphatidylinositol (PI) is phosphorylated at positions 4 and 5 of the inositol moiety to yield phosphatidylinositol 4,5,-biphosphate (PIP_2). Stimulation of cells with hormones, neurotransmitters, or growth factors activates isozymes of phospholipase C, enzymes specific for the hydrolysis of phosphoinositides (PI-PLC), to produce messenger molecules from PIP_2, that is, D-*myo*-inositol-1,4,5,-trisphosphate (IP_3) and diacylglycerol (DAG). IP_3 mobilizes Ca^{2+} from intracellular stores (Berridge and Irvine, 1989) and thereby plays a central role in smooth muscle contraction.

Historically, Hokin and Hokin had discovered, in 1953, that acetylcholine (ACh) stimulated the turnover of PIs in pancreas. Labeling the tissue with ^{32}P and measuring the incorporation of the label into phospholipids indicated a cycle of degradation and synthesis of PIs (Agranoff, 1986). Similar experiments on rabbit iris smooth muscle showed that ACh increased the breakdown of PIP_2, prelabeled with [^{32}P]phosphate (Abdel-Latif *et al.*, 1977). For a while it was thought that the interaction of ACh with the surface receptor results in the breakdown of PIP_2 rather than producing an active hydrolytic product. This view changed rapidly after the discovery that the breakdown of PIP_2 generated IP_3 (Berridge, 1983), and a large number of papers appeared demonstrating that the increase of IP_3 paralleled smooth muscle contraction (Abdel-Latif,

1986; Carsten and Miller, 1990; Somlyo and Somlyo, 1992).

In this review we discuss the factors that control the production of IP_3, with a special emphasis on the involvement of G-proteins in the regulation of PI-PLC activity. The role of IP_3 in the receptor-mediated intracellular Ca^{2+} release is discussed in Chapters 18–20.

II. ANALYSIS OF INOSITOL PHOSPHATES AND PHOSPHOLIPIDS

The inositol ring contains six hydroxyl residues, with positions 1,4, and 5 phosphorylated in IP_3 (Fig. 1). A scheme for the metabolism of phosphoinositides is shown on Fig. 2 (Dean and Beaven, 1989).

For studying changes in concentrations of inositol phosphates during smooth muscle contraction, the muscles are usually labeled with D-*myo*-[^3H]inositol, or occasionally with [^{32}P]orthophosphate, in the resting state. Labeling requires several hours to reach an isotopic equilibrium, when the specific activities of all inositol-containing compounds, phosphatidylinositols and inositol phosphates (IPs), are identical. In case of ^{32}P labeling, at equilibrium, the specific activities of the phosphate groups of PIP_2 and the [γ-^{32}P]ATP are identical. Thus, if the specific activity of [γ-^{32}P]ATP is determined, the mass of the inositol-containing compounds can be calculated. The equilibrated muscles are stimulated for various times and frozen in liquid nitrogen.

In case of smooth muscle tissue cultures, LiCl (1–10 mM) is added to the medium just prior to the addition

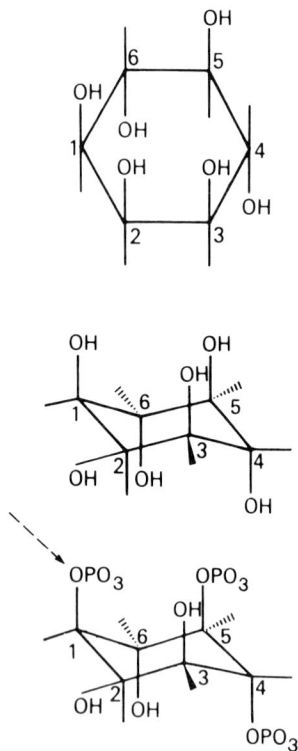

FIGURE 1   D-*myo*-Inositol (planar and chair configuration) and D-*myo*-inositol 1,4,5-trisphosphate [I(1,4,5)P$_3$]. The arrow indicates the site of the ester link with diacylglycerol in phosphatidylinositol. The negative charge of the phosphate groups is not indicated. From Dean and Beaven (1989, Fig. 1, p. 200).

of the stimulant (Dean and Beaven, 1989). For intact muscle, rat aorta rings, the LiCl concentration is increased to 25 mM (Pijuan *et al.*, 1993). Li$^+$ inhibits the inositol phosphate phosphomonoesterases and, thereby, increases the accumulation of phosphorylated inositol phosphates.

The frozen muscle is pulverized, and the powder is extracted with a trichloroacetic acid (TCA) or perchloric acid (PCA) solution. After removal of the precipitated proteins, the supernatant extracts are freed of excess acid. TCA is removed by extraction with diethyl ether, and PCA is removed by precipitation with KOH at neutral pH. The inositol phosphates in the watery extract are separated by chromatography.

Alternatively the muscle powder can be extracted with a mixture of chloroform: methanol: HCl (Pijuan *et al.*, 1993). The organic phase contains the PIs (and DAGs). The watery phase contains inositol, glycerophosphorylinositol (GPI), and all of the inositol phosphates. Acidified methanol–chloroform extraction releases more water-soluble [^{3}H]inositol label to the aqueous phase than does TCA extraction.

In Table II of the review of Palmer and Wakelam (1989), various techniques for the mass measurement of inositol phosphates, and the sensitivity of these techniques, are described. In the following, only the most common procedures will be discussed.

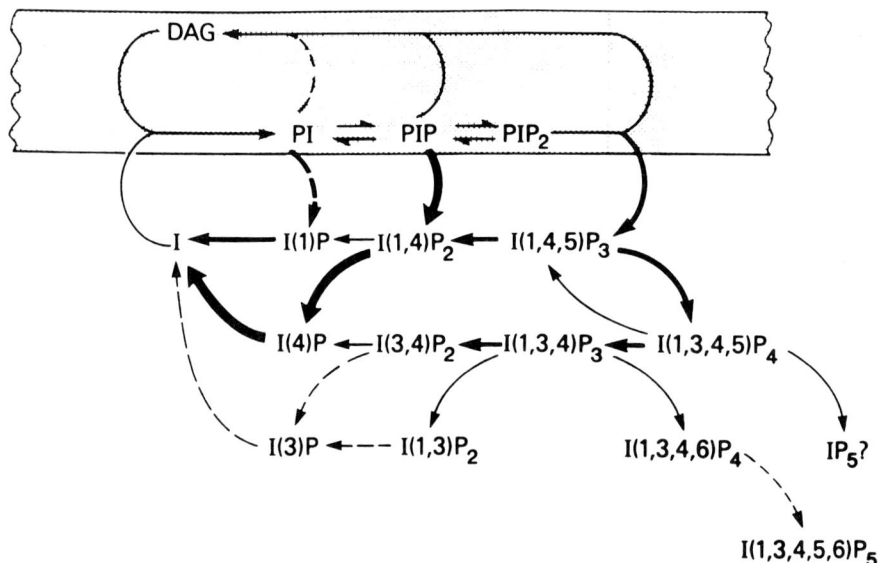

FIGURE 2   Pathways of metabolism of the inositol phosphates. The pathways shown by solid arrows have been established from studies with RBL-2H3 cells. The pathways shown by dashed arrows have been inferred from studies with other types of cells. The numbering system is for D-*myo*-inositol. From Dean and Beaven (1989, Fig. 2, p. 200).

A. Separation and Detection of Inositol Phosphates

Ion-exchange chromatography is used most frequently, because of its simplicity and low cost. Extracts usually containing [³H]inositol-labeled inositol phosphates are placed on a Dowex formate column from which GPI, IP, IP$_2$ (inositol bisphosphate), and IP$_3$ are eluted sequentially with buffers of ammonium formate of increasing molarity.

High-performance liquid chromatography (HPLC) has much better resolution than the Dowex column, that is, it also resolves the individual isomers of the inositol phosphates (Button et al., 1994; Foster et al., 1994). Most HPLC procedures employ anion-exchange columns and an aqueous mobile phase of ammonium phosphate to elute the water-soluble inositols. Double-labeled (³H and ³²P) inositol compounds can be used to follow by HPLC the fate of the inositol and phosphate residues during smooth muscle contraction.

A metal-dye detection system has been developed that permits picomolar-range HPLC analysis of inositol phosphates from nonradioactively labeled tissues (Mayr, 1990), and was applied for the determination of masses of inositol phosphates in resting and stimulated skeletal muscles (Mayr and Thieleczek, 1991).

B. Quantification of Inositol and Phosphate

Inositol derived by dephosphorylation of purified inositol phosphates may be quantified. Phosphates are removed from the inositol ring, either by acid hydrolysis or by treatment with alkaline phosphatase, and the free inositol is determined by enzymatic conversion to myo-inose with myo-inositol dehydrogenase and NAD⁺. The NADH thus formed is oxidized back to NAD⁺ with oxoloacetate to form malate, which is measured fluorometrically. This coupled reaction is stoichiometric and is sufficiently sensitive to assay 0.2 to 8 nmol of inositol per sample (Dean and Beaven, 1989; Palmer and Wakelam, 1989).

The P$_i$ liberated by acid or enzymatic hydrolysis of the inositol phosphates may also be assayed. An anion-exchange HPLC column, which separates the inositol phosphates, is coupled to a second column that contains immobilized alkaline phosphatase. The P$_i$ that is released by the enzyme is measured colorimetrically. The system can detect 1 nmol of inositol phosphates in a single sample to indicate levels of IP$_3$ of 13 to 40 nmol/g of tissue. The sensitivity of the phosphate assay is increased to the picomolar level with malachite green as a reagent (Dean and Beaven, 1989; Palmer and Wakelam, 1989).

C. Determination of Inositol Phospholipids

The [³H]inositol-labeled phospholipids are separated from other phospholipids on thin-layer chromatography (TLC) plates. After elution, the radioactive inositol phospholipids are cochromatographed with standards and visualized by exposure to iodine vapors, or by spraying with 10% $CuSO_4$ in 8% H_3PO_4, then scanned for radioactivity (Garnier et al., 1994; Baron et al., 1989).

III. ENZYMES INVOLVED IN THE TURNOVER OF PHOSPHATE IN PHOSPHOINOSITIDES

A. Kinases

1. Phosphoinositide Kinases

PI is synthesized from myo-inositol and DAG. First DAG is phosphorylated to phosphatidic acid (PA), then PA and CTP react to form CDP-diacylglycerol. The synthesis of PI (from CDP-diacylglycerol and inositol) occurs in the endoplasmic reticulum (ER) and it is transported from there to other membranes by a PI-specific transfer protein. Once at the plasma membrane, PI can be sequentially phosphorylated by PI 4-kinase and PI(4)P 5-kinase to generate PI(4)P and PI(4,5)P$_2$ (Majerus, 1992). PI 4-kinase is membrane associated in most tissues, and has been purified to homogeneity from bovine uterus; it has a molecular mass of 55 kDa and is specific for the 4 position (Carsten and Miller, 1990). PI(4)P 5-kinase has been found in both the soluble and particulate fractions of cell homogenates. A 53-kDa form of this enzyme has been purified from human erythrocyte membranes. The activities of PI 4- and PI(4)P 5-kinases are stimulated by Mg^{2+} and inhibited by Ca^{2+} (Abdel-Latif, 1986).

2. IP$_3$ 3-Kinase

IP$_3$ 3-kinase is a widely distributed soluble enzyme that converts I(1,4,5)P$_3$ to I(1,3,4,5)P$_4$ in the presence of Mg^{2+} and ATP (Majerus, 1992). The 3-kinase has a low K_m for IP$_3$ (0.2 to 1.5 μM), thus the kinase can compete effectively with the inositol polyphosphate 5-phosphatases for IP$_3$, since the K_m for the latter enzymes are higher (7–25 μM). The native enzyme is composed of two catalytic subunits, a 53-kDa protein and calmodulin (CaM). Physiological concentrations of Ca^{2+} stimulate IP$_3$ 3-kinase activity via the calmodulin subunit. The enzyme has been purified from porcine aortic smooth muscle (Yamaguchi et al., 1987), the cytosolic fraction of which contains enough CaM so that the enzyme is fully active without the addition of exogenous CaM. The high CaM concentration of smooth muscle could allow simultaneous activation of myosin light chain kinase (MLCK) and IP$_3$ 3-kinase in the initial phase of contraction.

B. Phosphatases

1. Inositol Polyphosphate Phosphatases

In rabbit iris sphincter smooth muscle microsomal fraction, there are phosphomonoesterases that degrade IP$_3$ to IP$_2$, IP$_2$ to IP, and IP to free *myo*-inositol and P$_i$ (Abdel-Latif, 1986). The IP$_3$ phosphatase has been shown to specifically remove the 5-phosphate from IP$_3$ and from cyclic IP$_3$ to produce IP$_2$ and cyclic IP$_2$, respectively. The polyphosphoinositide phosphatase has also been reported to dephosphorylate inositol tetrakisphosphate (IP$_4$) to IP$_3$. These enzyme activities are both cytosolic and membranous, dependent on Mg^{2+}, and not inhibited by Li$^+$. The IP$_3$ 5-phosphatase was studied in the microsomal fraction of bovine iris sphincter muscle (Wang *et al.*, 1994). It hydrolyzed IP$_3$ to I(1,4)P$_2$ with an apparent K_m of 28 μM; Mg^{2+} was required for its activity, Ca^{2+} ($>0.5 \mu M$) was inhibitory, and Li$^+$ or phosphorylation of the microsomal fraction with cAMP-dependent protein kinase or protein kinase C (PKC) had no effect on the activity of the enzyme.

New IP$_3$ 5-phosphatase inhibitors have been prepared: L-*myo*-inositol 1,4,5-trisphosphorothioate and *myo*-inositol 1,3,5-trisphosphorothioate, which inhibited IP$_3$ metabolism with concomitant elevation of the heparin-sensitive IP$_3$-induced release of ^{45}Ca^{2+} in T cells (Ward *et al.*, 1994); and L-*chiro*-inositol 1,4,6-triphosphate and the corresponding trisphosphorothioate compound L-*chiro*-I(1,4,6)PS$_3$, which potentially and selectively inhibited inositol polyphosphate metabolism without affecting Ca^{2+} stores in electrically permeabilized neuroblastoma cells (Hansbro *et al.*, 1994). These novel 5-phosphatase inhibitors may provide a starting point for development of cell-permeable analogues for further studies of the functions of IP$_3$ and IP$_4$ in the regulation of [Ca^{2+}]$_i$ in smooth muscle.

2. Inositol Polyphosphate 4-Phosphatase

Inositol polyphosphate 4-phosphatase converts I(1,3,4)P$_3$ to I(1,3)P$_2$ and I(3,4)P$_2$ to I(3)P. It does not require metal ions, is not inhibited by Li$^+$, and does not hydrolyze the 4-phosphate from other inositol polyphosphates. The purified enzyme has an apparent molecular mass of 110 kDa, and its apparent K_m values are 40 and 25 μM for I(1,3,4)P$_3$ and I(3,4)P$_2$, respectively. The ratio of inositol polyphosphate 1-phosphatase to inositol polyphosphate 4-phosphatase activity in hydrolyzing their common substrate I(1,3,4)P$_3$ varies among tissues; only 5–20% of I(1,3,4)P$_3$ is utilized by the 4-phosphatase (Majerus, 1992).

3. Inositol Polyphosphate 3-Phosphatase

Inositol polyphosphate 3-phosphatase catalyzes the hydrolysis of the 3-position phosphate bond of I(1,3)P$_2$ to form inositol 1-phosphate and P$_i$. Two isoforms of this enzyme, designated types I and II, have been isolated from rat brain; in the native form these enzymes are dimers with apparent molecular masses of 110 and 147 kDa, respectively. Both isoforms of the 3-phosphatase hydrolyze PI 3-phosphate to form PI and P$_i$. Thus far this is the only example in inositol phosphate metabolism of an enzyme that hydrolyzes both a lipid- and a corresponding water-soluble substrate (Majerus, 1992).

4. Inositol Polyphosphate 1-Phosphatase

Inositol polyphosphate 1-phosphatase, discovered in brain homogenates, hydrolyzes I(1,3,4)P$_3$ and I(1,4)P$_2$ to I(3,4)P$_2$ and I(4)P, respectively. The enzyme is a monomeric protein of 44 kDa, and the K_m for I(1,4)P$_2$ is 4 to 5 μM and that for I(1,3,4)P$_3$ is 20 μM. The enzyme requires Mg^{2+} and is inhibited at physiological Ca^{2+} concentrations. Li$^+$ inhibits both I(1,3,4)P$_3$ and I(1,4)P$_2$ hydrolysis; this inhibitory potency is the basis of using Li$^+$ for treatment of psychiatric disorders (Majerus, 1992). The enzyme was also found in the soluble fraction of bovine iris smooth muscle (Wang *et al.*, 1994). It hydrolyzed inositol phosphate to free inositol and P$_i$, with an apparent K_m of 89 mM. The enzyme was Mg^{2+} dependent, and Ca^{2+} ($>100 \mu M$) and Li$^+$ were inhibitory.

The crystal structure of recombinant bovine inositol polyphosphate 1-phosphatase (1-ptase) was determined in the presence of Mg^{2+} at 2.3-Å resolution (York *et al.*, 1994). The fold of 1-ptase is similar to that of two other metal-dependent/Li$^+$-sensitive phosphatases, inositol monophosphate phosphatase and fructose 1,6-biphosphatase. Comparison of the active-site pockets of these proteins will likely provide insight into substrate binding, the mechanisms of metal-dependent catalysis, and Li$^+$ inhibition.

5. Inositol Monophosphatase

Inositol monophosphatase hydrolyzes phosphate groups of all inositol monophosphates with the exception of inositol 2-phosphate; it requires Mg^{2+} for activity and is inhibited by Li$^+$. Its apparent molecular mass is 55 kDa.

6. Conclusions

From these data it appears that through the action of specific inositol phosphatases both IP_3 and IP_4 are sequentially dephosphorylated to free inositol (cf. Fig. 2). The dephosphorylation of IP_3 requires Mg^{2+} and physiological concentrations of Ca^{2+}. The inositol phosphate phosphomonoesterase is inhibited by Li^+, but the IP_3 5'-phosphomonoesterase is not inhibited (Carsten and Miller, 1990). Interestingly, soluble and particulate extracts from porcine skeletal muscle also metabolize IP_3 and IP_4 to inositol in a stepwise fashion (Foster *et al.*, 1994). Apparently, smooth and skeletal muscles have the same set of inositol polyphosphate phosphatases, although their functional role in skeletal muscle is not known.

IV. PHOSPHOINOSITIDE-SPECIFIC PHOSPHOLIPASE C

A. Isozymes

PI-PLC catalyzes the hydrolysis of PI, PI(4)P, and $PI(4,5)P_2$ to yield the 1-phosphates of IP, IP_2, and IP_3 and cyclic inositol phosphates. The other product of the PI hydrolysis is DAG. Three types of PI-PLCs (β,γ,δ) have been described based on the sequence homology and immunological cross-reactivity, each of which is a discrete gene product (Rhee and Choi, 1992a). The β- and δ-isozymes show the greatest specificity for PIP_2 relative to PI. All three forms of PI-PLC are single-polypeptide enzymes, with molecular masses of 150–154 kDa for PI-PLC-β, 145–148 kDa for PI-PLC-γ, and 85–88 kDa for PI-PLC-δ. The sequences of the three types of PI-PLC enzymes exhibit significant similarity (Fig. 3). This is apparent in two domains, one of about 170 amino acids and the other of about 260 amino acids. The two domains, designated X and Y in Fig. 3, are about 60 and 40% identical, respectively, between the three PI-PLCs. Each of the three enzymes contains an amino-terminal 300-amino acid region that precedes the X domain. There is no sequence similarity in this region (Rhee and Choi, 1992a). The amino-terminal domain of PI-PLC-δ_1 appears to be involved in the binding of both IP_3 and PIP_2 (Yagisawa *et al.*, 1994).

Based on the conserved domain X or Y, cDNAs were isolated from various tissues that revealed that each PI-PLC type (β, γ, and δ) contains several distinct members. Greek letters have been used to designate the types of PI-PLC with different primary structures and Arabic subscript numerals to designate the members of each type. In the 400-amino acid insert between X and Y, PI-PLC-γ_1 contains three regions that are re-

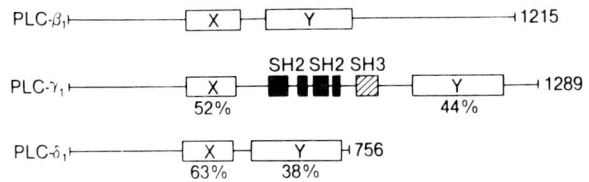

FIGURE 3 Linear display of three types of mammalian PI-PLCs (β, γ, δ types) represented by PLC-β_1, PLC-γ_1, and PLC-δ_1. The open boxes X and Y denote the regions of approximately 170 and 260 amino acids, respectively, of similar sequences found in the three types of mammalian PI-PLCs. The numbers at the right of the representations are the numbers of amino acid residues in each molecule. The percentage values under the X and Y domains of PLC-γ_1 and PLC-δ_1 refer to the percentage identity with the corresponding domains in PLC-β_1. Reproduced with permission from S. G. Rhee and K. D. Choi, "Multiple forms of phospholipase C isozymes and their activation mechanisms," *Advances in Second Messenger and Phosphoprotein Research*, 1992, Vol. 26, pp. 35–61, Fig. 1.

lated in sequence to limited portions of the *src* product. The three regions correspond to duplicates of SH_2 and to SH_3 (*src* homology 2 and 3), which were first recognized as highly conserved sequences in the regulatory domains of a number of nonreceptor tyrosine kinases (Rhee and Choi, 1992a).

As pointed out by Majerus (1992), none of these three PI-PLC isoforms contains a membrane-spanning sequence. Since the PI substrates for these PI-PLC's are in the membrane bilayers, PI-PLC must bind to membranes before hydrolyzing PIs. The translocation of PI-PLC-γ_1 (induced by its tyrosine phosphorylation) from the cytosol to the cellular membrane (Rhee and Choi, 1992a) may be an example of such a binding.

B. Activation

There are two basic mechanisms by which agonists activate PIP_2 hydrolysis. In the case of hormones, neurotransmitters, and certain other agonists, the signal is transduced by G-proteins from receptors with seven membrane-spanning segments to β-isozymes of PI-PLC. In the case of growth factors, activation of their receptors results in enhanced tyrosine kinase activity (Rhee and Choi, 1992b), which leads to phosphorylation of specific tyrosine residues in the cytoplasmic domains of the receptors, to which γ-isozymes of PI-PLC become associated (Rhee and Choi, 1992a; Exton, 1994; Lee and Severson, 1994).

1. Activation by G-Proteins

The guanine nucleotide binding proteins (G-proteins) that mediate regulation of several effector molecules by various agonists are heterotrimers consisting of α-, β- and γ-subunits. The α-subunits appear to be most diverse and are believed to be responsible for the specificity of the interaction of different G-proteins with their effectors.

Numerous experiments support a simple model for the activation of G-proteins (Fig. 4). In the basal state, the α-subunit contains bound GDP, and association of α- and $\beta\gamma$-subunits is highly favored, keeping the G-protein in the inactive form. Stimulation of the G-protein results when it binds GTP rather than GDP. Receptors interact most efficiently with the heterotrimeric form of the G-protein and accelerate activation by increasing the rate of dissociation of GDP and enhancing association of GTP. Activation of G-protein-coupled receptors results in the dissociation of heterotrimeric G-proteins into α-subunits and $\beta\gamma$-dimers. Finally, the G-protein α-subunit has an intrinsic hydrolytic activity that slowly converts GTP to GDP and returns the G-protein to its inactive form (Sternweis and Smrcka, 1992).

It has been known for some time that G-proteins mediate the effects of agonists on the breakdown of PIP_2 in various mammalian cells (Cockcroft and Gomperts, 1985; Gonzales and Crews, 1985; Litosch et al., 1985). Numerous studies have demonstrated that GTP and its nonhydrolyzable analogue, GTPγS, and aluminum fluoride (AlF_4^-), a universal activator of G-proteins (see the following), stimulate PI turnover in permeabilized cells, crude membrane fractions, or partially purified enzyme preparations (Majerus,

1992; Exton, 1994). This has led to the conclusion that a G-protein is involved in coupling between receptors and PI-PLC. Two examples will be described.

G-protein control of PI hydrolysis was demonstrated in permeabilized segments of rat tail artery prelabeled with [^{3}H]inositol (LaBelle and Murray, 1990). Norepinephrine (NE) and GTPγS were both able to increase levels of IP, IP_2, and IP_3 in the segments. The effects of both NE and GTPγS on the segments were not additive, suggesting that GTPγS interacts with the same G-protein that NE must interact with in order to stimulate PIP_2 breakdown in the muscle cells. The NE-stimulated increases in IP, IP_2, and IP_3 were insensitive to pertussis toxin, indicating that a pertussis toxin-insensitive G-protein is required to mediate the effect of NE on PI-PLC in the arterial segments. Amiot et al. (1993) showed that guanine nucleotides decreased the binding of bombesin to rat myometrial membranes and accelerated the rate of ligand dissociation, reflecting the coupling of the receptors to G-proteins. The results also revealed that the bombesin peptide activates the PI-PLC pathway in the myometrium in a pertussis toxin-insensitive manner.

2. Gα_q Family

The G-proteins involved in the regulation of PI-PLC-mediated hydrolysis of PIP_2 have been identified (Pang and Sternweis, 1990; Strathmann and Simon, 1990). They belong to α-subunits of G-proteins, called Gα_q, are insensitive to pertussis toxin, and contain two polypeptides, Gα_q (molecular mass 42 kDa) and Gα_{11} (43 kDa), which show 88% sequence identity and are approximately equal in their ability to activate PI-PLC-β_1 specifically (Majerus, 1992; Exton, 1994). Both of these G-proteins are widely present in mammalian tissues.

Pertussis toxin-resistant G-proteins are known to lack the cysteine residue in the C terminus and this is also characteristic for the G-proteins of the Gα_q family (Exton, 1994). Antisera prepared against the C-terminal peptide sequences, common to Gα_q and Gα_{11}, were found to block GTPase and PI-PLC activity, in cell membrane preparations, stimulated by specific ligands such as tromboxane A_2, bradykinin, angiotensin II, vasopressin, and histamine. The data indicate that the Gα_q proteins are responsible for regulation of PI-PLC by G-protein-linked receptors (Rhee and Choi, 1992a; Sternweis and Smrcka, 1992).

Activation of G_q/G_{11} by carbachol in phospholipid vesicles containing M_1 muscarinic receptors has also been demonstrated. The time course of binding of ^{35}S-labeled GTPγS closely followed the activation of purified PI-PLC-β_1 as measured by the hydrolysis of ^{3}H-labeled PIP_2 to ^{3}H-labeled IP_3. The half-maximal

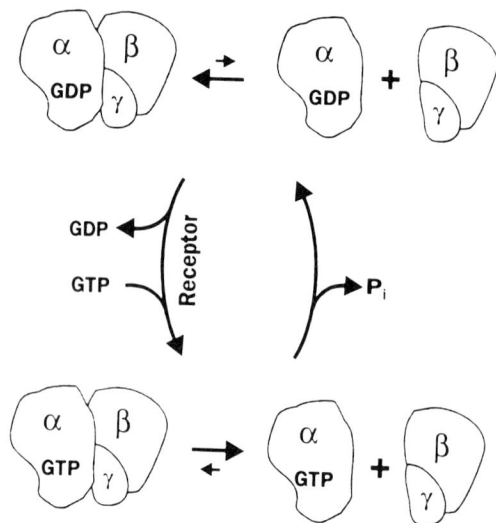

FIGURE 4 Model for activation of G-proteins. The functional state of a G-protein is determined by its bound nucleotide. With GDP, the G-protein is inactive and subunit association is favored. With bound GTP, the G-protein is activated and the affinity between its α- and $\beta\gamma$-subunits is markedly reduced. Receptors stimulate G-proteins by catalyzing exchange of GTP for GDP. Reproduced with permission from Sternweis and Smrcka (1992), Fig. 1, p. 502.

concentrations of carbachol for stimulation of GTPγS binding and of PI-PLC activity were also similar. Co-reconstitution of M_1 receptor, G_q/G_{11}, and PI-PLC-β_1 allowed the demonstration of GTPγS-dependent, carbachol-stimulated hydrolysis of PIP$_2$. From these data it was concluded "that the three components are sufficient to allow *in vitro* reconstitution of agonist-stimulated IP$_3$ formation" (Exton, 1994).

It was discovered that PI-PLC-β_1 greatly stimulated the GTPase activity of G_q/G_{11}. The ability of the phospholipase to act as a GTPase-activating protein (GAP) was specific in that it was not exerted against G_o or G_s. It was also blocked by monoclonal antibodies against PI-PLC-β_1. The physiological function of GTPase activation by PI-PLC-β_1 may be related to the rapid turnoff of the enzyme that is observed when an agonist is removed or an antagonist is introduced (Exton, 1994).

3. Activation by Growth Factors

The PI-PLC-γ isozyme exists in two forms (γ_1 and γ_2), which participate in signal transductions induced by polypeptide growth factors such as platelet-derived growth factor (PDGF), epidermal growth factor (EGF), or fibroblast growth factor (FGF) (Rhee and Choi, 1992a). Binding of growth factors to their receptors in many cell lines leads to dimerization of the receptor and activation of its tyrosine kinase activity, which leads to phosphorylation of several tyrosine residues in its C terminus (cytoplasmic receptor tail). Phosphorylation of the receptor tail then provides new binding sites for SH2-containing proteins, such as PI-PLC-γ_1, initiating tyrosine phosphorylation (cf. Fig. 5). The major sites of tyrosine phosphorylation induced by growth factors are residues 771, 783, and 1254 in PI-PLC-γ_1 (Exton, 1994).

4. Activation by Aluminum Fluoride

Classic experiments have demonstrated that aluminum fluoride (generated by a combination of aluminum chloride and sodium fluoride) bypasses the need for membrane receptor occupancy by directly activating G-proteins, leading to a persistent stimulus (Sternweis and Gilman, 1982; Blackmore and Exton, 1986; O'Shea *et al.*, 1987). It has been shown that AlF$_4^-$ activates the transducin G-protein family by binding to the inactive G-protein–GDP complex with a geometry resembling a pentavalent intermediate for GTP hydrolysis (Sondek *et al.*, 1994). AlF$_4^-$ stimulated the breakdown of PIP$_2$ in various cells (Blackmore *et al.*, 1985; Claro *et al.*, 1989), including cells from myometrium (Marc *et al.*, 1988) and ileum (Watson *et al.*, 1988). AlF$_4^-$ also increased the concentration of inositol phosphates in rat tail arterial segments (LaBelle and Murray, 1990).

Using intact smooth muscle, Phillippe (1994a) reported that AlF$_4^-$ increased the phasic contractions of pregnant rat myometrial strips. The AlF$_4^-$-stimulated contractions were suppressed by 2-nitro-4-carboxyphenyl-N,N-diphenylcarbamate (NCDC), an alleged inhibitor of PI-PLC. NCDC also suppressed the oxytocin-stimulated contractions of nonpregnant rat myometrial strips (Phillippe, 1994b). In contrast, NCDC had only a minimal effect on KCl-stimulated tonic myometrial contractions. Thus, these studies suggest the involvement of G-proteins and PI-PLC in AlF$_4^-$-stimulated myometrial contractions.

Watson *et al.* (1988) reported that nifedipine, a Ca^{2+} channel blocker, inhibited the fluoride (actually AlF$_4^-$)-induced contractions in guinea pig ileum longitudinal smooth muscle. The data were consistent with a model in which the activation of a G-protein by AlF$_4^-$ leads to the following sequential events: activation of PI-PLC, release of intracellular Ca^{2+}, opening of voltage-operated (i.e., dihydropyridine sensitive) Ca^{2+} channels, and contraction.

5. Ca^{2+} Stimulation

The activity of PI-PLC (determined in membrane fragments from ^{32}P-labeled iris sphincter smooth muscle by measuring the release of ^{32}P-labeled IP$_3$ from the endogenous ^{32}P-labeled PIP$_2$) was found to be sensitive to Ca^{2+}, with half-maximal stimulation at about 1.1 μM Ca^{2+} (Howe *et al.*, 1986).

Stimulation of PI-PLC by Gα_q was measured over a large range of Ca^{2+} concentrations. In addition to increasing the maximal activity of PI-PLC, Gα_q also changed the apparent affinity of the enzyme for Ca^{2+} from about 1 to 0.1 μM. Thus, PI-PLC activation by G-proteins is achieved by increasing both the intrinsic activity of the enzyme and its sensitivity to the concentration of free Ca^{2+} (Rhee and Choi, 1992a).

C. Pathways for Activation

Figure 5 illustrates the three different mechanisms for the activation of PI-PLC isoforms (Lee and Severson, 1994). The most common one involves the Gα_q proteins. In the example shown, NE interacts with the α_1-adrenergic receptor and the signal produced activates Gα_q by exchanging its bound GDP to bound GTP, followed by the activation of the β_1- and β_3-isoforms of PI-PLC. The hydrolysis of PIP$_2$, catalyzed by PI-PLC, leads to IP$_3$ and DAG formation; the former liberates Ca^{2+} from intracellular stores, for example, the sarcoplasmic reticulum (SR), to initiate smooth muscle contraction (Chapters 18 and 20, this volume), and the latter activates PKC (Section VIII.C). PI-PLC can also be regulated by pertussis toxin-

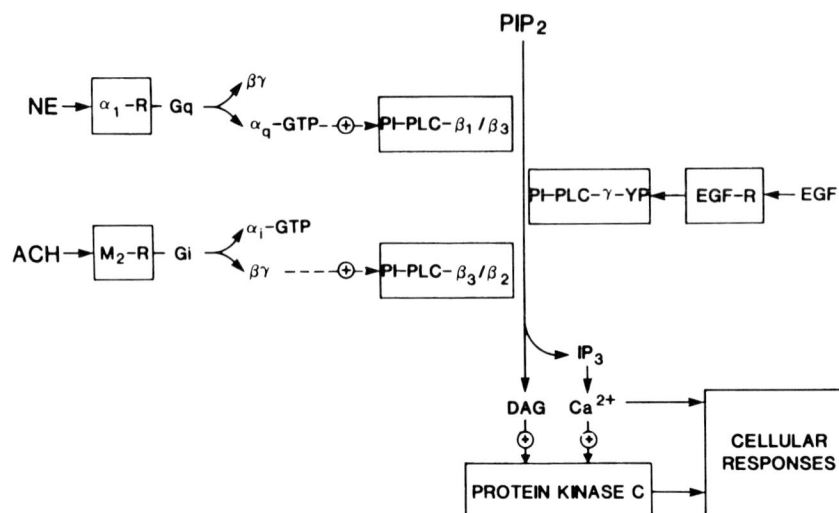

FIGURE 5 Pathways for activation of PI-PLC isoforms. The upper left row shows the most common pathway for PI-PLC-β isoform activation, initiated by stimulation of an α_1-adrenergic receptor (α_1-R) with NE, and involving Gα_q-proteins. The lower left row shows the activation of PI-PLC-β isoforms, initiated by acetylcholine (ACH) stimulation of M$_2$-muscarinic receptor (M$_2$-R), and mediated by the βγ-subunit of the pertussis toxin-sensitive G-protein (G$_i$). The right part of the figure shows the activation of PI-PLC-γ isoforms, initiated by the binding of EGF to its receptors, and executed by the tyrosine phosphorylation (YP) of PI-PLC-γ. In all three cases the activated PI-PLC hydrolyzes PIP$_2$ to form the messengers, IP$_3$ and DAG. For more details see the text. Reproduced with permission from Lee and Severson (1994), Fig. 3, p. C662.

sensitive G-proteins, which are subtypes of G$_i$ and G$_o$ (Exton, 1994). Recent evidence suggests that the βγ-subunits of these proteins may be responsible for PI-PLC activation. This is illustrated in Fig. 5: ACh stimulation of an M$_2$-muscarinic receptor activates the βγ-subunits of G$_i$, which transmit the activation to the β$_3$- and β$_2$-isoforms of PI-PLC. The messengers IP$_3$ and DAG, generated by the hydrolysis of PIP$_2$ through action of PI-PLC, elicit cellular responses and PKC activation as described before. The PI-PLC domain that interacts with α_q is the C terminal, whereas that which interacts with βγ-subunits is toward the N terminus, but the molecular details of either interaction are incomplete (Exton, 1994).

PI-PLC activation by EGF is initiated by the binding of EGF to its specific receptors containing sites autophosphorylated by their intrinsic tyrosine kinase. The autophosphorylated site of the receptor then interacts with the SH2 domain of PI-PLC-γ (Rhee and Choi, 1992a; Exton, 1994), and PI-PLC-γ will be activated as a consequence of its tyrosine phosphorylation. This results in IP$_3$ and DAG release from PIP$_2$.

There are potential roles that different isoforms of G-protein responsive PI-PLC might fill. One might be to provide a unique response in specific tissues or cells. It is possible that specific distribution of PI-PLC-β

isozymes is related to the distribution of specific regulatory G-proteins (Sternweis and Smrcka, 1992).

D. Inhibition

Two inhibitors are known. (1) Neomycin is an antibiotic that interacts with polyphosphoinositides. Neomycin inhibited the Ca^{2+}-dependent histamine secretion from GTPγS-loaded mass cells (Cockcroft and Gomperts, 1985), the GTPγS-induced contractions in skinned rabbit main pulmonary arteries (Kobayashi *et al.*, 1988), the carbachol-induced contractions in β-escin-permeabilized guinea pig ileum longitudinal muscles (Kobayashi *et al.*, 1989), and the oxytocin- and AlF$_4^-$-stimulated contractions in rat myometrium (Phillippe, 1994c). However, neomycin is not a specific PI-PLC inhibitor, because it also inhibited the caffeine-induced contractions, albeit to a lesser extent (Kobayashi *et al.*, 1989; see also Carsten and Miller, 1990). (2) NCDC probably acts as a carbamylating agent. NCDC inhibited the AlF$_4^-$- and NE-stimulated contractions in rat uterus (Phillippe, 1994a,b), and the serotonin-stimulated contraction in rabbit basilar artery (Clark and Garland, 1993). The latter authors used NCDC as a putative inhibitor for both PKC and PI-PLC, thus NCDC may not be specific for PI-PLC either.

V. INOSITOL 1,4,5-TRISPHOSPHATE AND CONTRACTION

A correlation between smooth muscle stimulation and increased metabolism of inositol phosphates has been known for some time, for example, in carbachol- or NE-stimulated rabbit iris (Abdel-Latif et al., 1977), in carbachol- or ACh-stimulated trachea (Baron et al., 1984; Hashimoto et al., 1985; Takuwa et al., 1986; Duncan et al., 1987), in angiotensin- or vasopressin-stimulated vascular muscle (Smith et al., 1984; Doyle and Rüegg, 1985), in ACh- or histamine-stimulated visceral muscle (Best et al., 1985), in oxytocin-stimulated human myometrium (Amiot et al., 1993), or in bombesin-stimulated rat myometrium (Schrey et al., 1988).

Time course studies on changes in concentrations of inositol phosphates in stimulated smooth muscles demonstrated a rapid accumulation of IP_3. Thus, carbachol-stimulated rabbit iris smooth muscle, pre-labeled with myo-[^{3}H]inositol, showed a release of ^{3}H-labeled IP_3 from ^{3}H-labeled PIP_2 in 15–30 s. The accumulation of ^{3}H-labeled IP_3 preceded that of ^{3}H-labeled IP_2, which was followed by ^{3}H-labeled IP in about 2 min (Akhtar and Abdel-Latif, 1984). These increases were dose dependent, were observed in the absence of extracellular Ca^{2+}, and were blocked by atropin, a muscarinic antagonist. Similarly, in guinea pig myometrium stimulated by carbachol or oxytocin, a release of ^{3}H-labeled IP_3 was found in 30 s, which was followed by IP_2 and IP (Marc et al., 1986). The ionophore A23187 as well as K^+ depolarization failed to increase the accumulation of inositol phosphates. Carbachol-stimulated longitudinal smooth muscle exhibited a threefold increase in ^{3}H-labeled IP_2 within 2 s, and there was also a simultaneous increase in ^{3}H-labeled IP_2 (Salmon and Bolton, 1988). ^{3}H-labeling of IP_4 was not significantly increased until 60 s after carbachol stimulation, and the accumulation of $I(1,3,4)P_3$ was relatively small. Furthermore, in vascular smooth muscle cells, endothelin induced formation of IP_3 in 15 s (Marsden et al., 1989). These results, along with others, for example, NE-stimulated rat tail arteries (Gu et al., 1991) or NE-stimulated rat aorta (Pijuan et al., 1993), established that receptor activation in smooth muscle is coupled to an enhanced inositol phosphate metabolism. The accumulation of inositol phosphates was not controlled by either the extracellular or intracellular Ca^{2+} concentration, in agreement with the pharmacomechanical coupling mechanism (Somlyo and Somlyo, 1968).

The kinetics of IP_3 production showed that, after the initial increase, IP_3 levels returned to basal levels in 30 s in carbachol-stimulated bovine tracheal muscle

(Chilvers et al., 1991), or in 3 min in NE-stimulated rat aorta (Pijuan et al., 1993). These investigators have also noted that, in contrast to the transient increase in IP_3 concentration, the accumulation of its isomer $I(1,3,4)P_3$ and its degradation product $I(1)P$ was maintained. The metabolism of [^{3}H]inositol-labeled IP_3 was studied in permeabilized rat aortic smooth muscle cells (Rossier et al., 1987). Several labeled metabolites were detected: IP_2 and IP_4 (inositol 1,3,4,5-tetrakisphosphate) reached a maximum in 2 min, whereas the production of $I(1,3,4)P_3$ was delayed. A correlation between the formation of IP_4 and that of $I(1,3,4)P_3$ was observed, suggesting that the former is the precursor of the latter; the formation of both IP_4 and $I(1,3,4)P_3$ was Ca^{2+} sensitive. The authors have suggested two distinct pathways for the metabolism of IP_3 in vascular smooth muscle cells: (1) dephosphorylation leading to IP_2 and IP formation, and (2) a Ca^{2+}-sensitive phosphorylation/dephosphorylation pathway involving formation of IP_4 and leading to formation of $I(1,3,4)P_3$.

Howe et al. (1986) included myosin light chain (MLC) phosphorylation into the correlation studies. Using different concentrations of carbachol, time, and temperature, a good correlation was found between hydrolysis of PIP_2, MLC phosphorylation, and contraction in rabbit iris sphincter smooth muscle. This supports the concept that the carbachol-induced signal transduction in iris muscle occurs through enhanced PIP_2 turnover, appearance of IP_3, the elevation of $[Ca^{2+}]_i$, and MLC phosphorylation.

The described correlations between increased $[IP_3]_i$ and smooth muscle contraction are based on labeling the muscle with [^{3}H]inositol, but as pointed out by Rembold (Chapter 18, this volume), there is no absolute evidence that agonist stimulation increases $[IP_3]_i$. This is based on the experiments of Baron et al. (1989), which demonstrated that stimulation increases the incorporation of [^{3}H]inositol not only into IP_3 but also into PIP_2. Baron et al. (1989) have estimated the turnover time of [^{3}H]inositol in PIP_2 as 4.0 min; with such a short turnover time, the initial increases in ^{3}H-labeled IP_3 concentrations (measured as increase in radioactivity) may be due to increased initial [^{3}H]inositol incorporation into PIP_2. Conflicting reports on the agonist-stimulated increase in IP_3 in rat aorta are also discussed by Pijuan et al. (1993). It is known that resting smooth muscle contains a surprisingly high IP_3 concentration, 1–4 μM (Shears, 1992); measurements of a small increment in IP_3 with such a high background may not be easy.

As discussed in Section II, in ^{32}P-labeled tissues the specific activity of [γ-^{32}P]ATP can be used as a reference for quantification of the concentrations of inositol phospholipids and inositol phosphates. Indeed, such a procedure showed a 20-fold increase in IP_3 concentration upon thrombin stimulation in platelets (Palmer and Wakelam, 1989, and references therein). Several other examples for the increase of IP_3 mass in stimulated cells were described (Palmer and Wakelam, 1989). Furthermore, by the HPLC/metal-dye detection, a nonradioactive method, a significant increase in IP_3 concentration was found upon tetanic stimulation of vertebrate skeletal muscles (Mayr and Thieleczek, 1991). Thus, there is also a possibility that in smooth muscle, agonist stimulation results in an elevated level of $[IP_3]_i$.

A. Compartmentalization

Baron et al. (1992) have shown compartmentalization of IP_3 in tracheal smooth muscle. There is a pool of IP_3 in unstimulated muscle that contains about 15 times more IP_3 than is necessary to maximally release Ca^{2+} from the SR. This pool has limited access to IP_3 5-phosphatase and 3-kinase, and to IP_3-sensitive SR. During carbachol-induced contraction, IP_3 can be released from this sequestered compartment (without increase in the total IP_3 content), and now IP_3 has access to the degrading enzymes and to the IP_3-sensitive SR. The second compartment, in which IP_3 is nonsequestered, would control the Ca^{2+} release by SR. This hypothesis explains the discrepancies about IP_3 increase in stimulated muscles, described before. Metabolically distinct pools of inositol phospholipids and inositol phosphates have been proposed to exist in other cells as well (Palmer and Wakelam, 1989; Shears, 1992). Furthermore, in a review by Mikoshiba et al. (1994), "at least" three types of IP_3 receptors have been described. More information about the nature of the structures involved in the compartmentalization would be of importance.

VI. RELATIONSHIP BETWEEN SMOOTH MUSCLE STIMULATION, INOSITOL 1,4,5-TRISPHOSPHATE RELEASE, AND RISE IN INTRACELLULAR Ca^{2+}

Following the hypothesis of Berridge and Irvine (1984) that IP_3 is the "missing link" between plasma membrane receptors and internal Ca^{2+} stores, several papers demonstrated that IP_3 acts by mobilizing Ca^{2+} from intracellular pools (Carsten and Miller, 1990; Coburn and Baron, 1990; Somlyo and Somlyo, 1992, 1994). Since sodium azide or oligomycin, inhibitors of mitochondrial oxidative phosphorylation, did not affect the Ca^{2+} release, it became generally accepted that the intracellular pool corresponds to the SR.

IP_3 caused Ca^{2+} release and tension development in rabbit pulmonary arterial smooth muscle, permeabilized with saponin or digitonin, in a dose-dependent manner (Somlyo et al., 1985). The amount of Ca^{2+} released by IP_3 was estimated to be sufficient to cause contraction of the muscle. The messenger role of IP_3 has been proven by the use of the photosensitive precursor of IP_3, the P-1(2-nitrophenyl)ethyl ester of IP_3 (caged IP_3), a compound resistant to IP_3 5-phosphatase (Walker et al., 1987). The kinetics of Ca^{2+} release and contraction, studied by photolytic release of IP_3 from caged IP_3, showed an increase of Ca-Fluo-3 signal in approximately 10 ms, consistent with the messenger role of IP_3 (Somlyo et al., 1992). The time scale of biochemical events in pharmacomechanical coupling of smooth muscle was also described by Somlyo and Somlyo (1992). At 22°C, it takes about 1.2 s for the muscle to develop force after the receptor has been stimulated. Out of the 1.2 s, 0.5–1 s is needed for IP_3 production and 0.2–0.3 s to produce force after MLCK has been activated by Ca^{2+}-calmodulin, whereas the IP_3-mediated Ca^{2+} release and the phosphorylated MLC-initiated tension require only 20–30 ms. Previous measurements of IP_3 production, MLC phosphorylation, and isometric force in neurally stimulated bovine tracheal muscle at 37°C showed an increase in IP_3 by 500 ms, closely followed by activation of MLCK, whereas MLC phosphorylation increased after 500 ms preceding the development of maximal isometric force (Miller-Hance et al., 1988). Thus, there is a reasonable agreement in the biochemical kinetics between permeabilized and intact smooth muscle.

VII. INOSITOL TETRAKISPHOSPHATE

IP_4 has been suggested to be involved in regulating $[Ca^{2+}]_i$ in smooth muscle (Fukuda et al., 1994, and references therein). This is based on two criteria: (1) IP_3 kinase, the enzyme that forms IP_4, is present in virtually every animal cell, suggesting physiological importance; and (2) there are specific IP_4 receptor proteins that bind IP_4, though having low affinity to IP_3, suggesting that IP_4 has a distinct function that is different from that of IP_3 (Irvine, 1992; Fukuda et al., 1994). According to the model (Irvine, 1992), IP_4 and IP_3 synergistically activate their receptors, that is, open their Ca^{2+} channels, and so both Ca^{2+} mobilization from the intracellular stores and Ca^{2+} entry can occur. The agonist-induced initial $[Ca]_i$ signal in smooth muscle cells is derived from the Ca^{2+}-releasing action of IP_3 on SR. It is known that in virtually all cells that utilize the IP_3-mediated mechanism, the initial release of $[Ca^{2+}]_i$ is followed or accompanied by an accelerated entry of

Ca^{2+} into the cytoplasm across the cell membrane. It is postulated that IP$_4$ is involved in the Ca^{2+}-entry process. This is supported by experiments of Hashii et al. (1994), who applied IP$_4$ from patch pipette into the cytoplasm of mouse fibroblast cells and found a transient increase in cytoplasmic free Ca^{2+} concentration. Stimulation of the cells with bradykinin increased the level of IP$_4$. The data suggest that mouse fibroblast cells have a Ca^{2+} influx pathway gated with IP$_4$. In contrast, the data of Verjans et al. (1994) suggest that in Xenopus oocytes, IP$_3$ alone plays the crucial role in the activation of capacitative Ca^{2+} entry.

VIII. DIACYLGLYCEROL

As described before, PIP$_2$ hydrolysis by PI-PLC results in production of IP$_3$ and DAG; the latter messenger activates PKC-catalyzed serine phosphorylation of cellular proteins. Since DAG can be generated from other phospholipids, notably phosphatidylcholine (PC), the PKC pathway can be stimulated independently of changes in [Ca^{2+}]$_i$ (Lee and Severson, 1994). Because the concentration of PI is much greater than that of PIP$_2$, PI-PLC generates DAG mainly from the hydrolysis of PI. However, PC is by far the most abundant of the phospholipids, therefore it is the major source for DAG.

PKC activity in intact cells can be stimulated by phorbol esters, for example, phorbol 12,13-dibutyrate (PDBu), which are not metabolized by cells and, therefore, can produce a prolonged stimulation of PKC. In addition, cell-permeant DAG analogues such as dioctanoglycerol (DiC8) and 1-oleoyl-2-acetylglycerol (OAG) can activate PKC.

A. Metabolism of Diacylglycerol

Metabolism of DAG attenuates PKC activation, resulting in signal termination. DAG can be phosphorylated to PA by a diacylglycerol kinase or degraded by specific lipases. Sequential hydrolysis of DAG by diacyl- and monoacylglycerol lipases to free fatty acids and glycerol appears to be the predominant route for removal of the DAG in vascular smooth muscle (Lee and Severson, 1994).

B. Smooth Muscle Stimulation and Diacylglycerol Accumulation

In many tissues, the generation of DAG is biphasic (Lee and Severson, 1994). The hydrolysis of PIP$_2$ leads to an immediate and transient production of DAG, and this is followed by sustained DAG production from the

hydrolysis of PC and other phospholipids. Biphasic accumulation of DAG in intact vascular smooth muscle has been demonstrated by measuring changes in DAG and choline concentrations in response to angiotensin II, vasopressin, and NE. The formation of DAG was also biphasic in cultured rat arterial smooth muscle cells stimulated by bradykinin, vasopressin, and angiotensin II, with a transient peak at 5 s followed by a sustained increase from 60 to 600 s (Dixon et al., 1994).

The sustained accumulation of DAG in angiotensin II-stimulated cultured vascular smooth muscle cells was correlated with receptor sequestration; when the internalization of angiotensin II receptor was inhibited by phenylarsine oxide, the sustained DAG accumulation was also inhibited (Griendling et al., 1987). This suggests that agonist–receptor processing is required for production of the sustained DAG signal.

C. Role of Protein Kinase C in Smooth Muscle

Both Ca^{2+}-dependent (α,β) and Ca^{2+}-independent (ε,ζ) forms of PKC have been identified in smooth muscle (Chapters 12 and 24, this volume), and "at least" 20 endogenous substrates have been identified (Carsten and Miller, 1990). The Ca^{2+}-independent PKC isoforms were shown to have roles in vascular smooth muscle: ε-PKC underwent translocation in ferret aorta contracted in Ca^{2+}-free bathing solutions (Chapter 24, this volume). Furthermore, the Ca^{2+}-independent PKC isoforms were functioning in the contractility of Ca^{2+}-deficient porcine carotid arteries (Bárány et al., 1992).

Activation of PKC with PDBu inhibited neurotransmitter-stimulated phasic contractions in guinea pig ileum and rat uterus (Baraban et al., 1985), electrical or oxytocin-stimulated contractions in rat uterus (Savineau and Mironneau, 1990), and the AlF$_4^-$- or oxytocin-stimulated contractions in pregnant and nonpregnant rat myometrium (Phillippe, 1994a,d). These results suggest that PKC plays a role in the PI signaling pathway in myometrium. The inhibitory effect of PDBu on the production of IP$_3$, IP$_2$, and IP was actually found in porcine trachealis muscle stimulated by carbachol (Baba et al., 1989). Since this muscle is tonic, it appears that PKC can modify metabolism of inositol phosphates in both phasic and tonic smooth muscles.

In cultured canine tracheal smooth muscle cells, phorbol 12-myristate 13-acetate (PMA) inhibited the histamine-induced IP$_3$ formation (Murray et al., 1989), in cultured rat aortic smooth muscle cells phorbol esters inhibited agonist-induced PIP$_2$ hydrolysis, and in cultured rabbit aortic smooth muscle the serotonin-stimulated PI breakdown was blocked after acute exposure to phorbol esters (Lee and Severson, 1994).

Phorbol esters reduced hormone responses in cells as well as stimulation in membranes by guanine nucleotides, suggesting that PKC acts as a feedback regulator (Sternweis and Smrcka, 1992). The site of PKC action appears at a postreceptor site either on the interacting sites between the G-protein subunits and PI-PLC or on PI-PLC itself.

A role for PKC in the maintenance of Ca^{2+} homeostasis has been proposed. Phorbol esters, DiC8, and OAG have been shown to modulate the activities of several Ca^{2+} influx and efflux pathways (Lee and Severson, 1994). In basilar artery smooth muscle, PKC exerted its effect, at least in part, on membrane Ca^{2+} channels (Clark and Garland, 1993). The capacitative Ca^{2+} entry in *Xenopus* oocytes was shown to be regulated by PKC (Petersen and Berridge, 1994). For further discussions on the role of PKC in smooth muscle, we refer to Chapters 12 and 24–26 (this volume).

IX. SUMMARY AND PERSPECTIVES

The agonist-activated phosphoinositide signaling system is well recognized as a primary event following stimulation of smooth muscle. The interaction between agonists and specific receptors results in the activation of PI-PLC, which leads to the production of two second messengers: IP_3 and DAG. Diacylglycerol is also generated from PC and other phospholipids, but little is known about the phospholipases that catalyze DAG production from sources other than PIP_2. It is possible that two separate biochemical pathways exist for IP_3 and DAG production, controlled by separate signal transduction pathways emanating from the same or separate receptor(s).

IP_3 is the intracellular Ca^{2+}-mobilizing molecule in smooth muscle and its action is terminated by rapid metabolism through two routes: phosphorylation by a 3-kinase and dephosphorylation by a 5-phosphatase. Inositol polyphosphate 5-phosphatases play a control role in the regulation of various agonist-evoked responses. DAG is the physiological activator of PKC, although the role of PKC in smooth muscle is not well defined. DAG, an apolar molecule, must exert its activity in the membrane compartments of smooth muscle. Chemical reactions in lipid environment are much less understood than those in aqueous media and, therefore, future research has to elucidate the mechanism of DAG-dependent activation of PKC in membranes.

There is common agreement that the α-subunits of the G_q family of G-proteins stimulate β-isozymes of PI-PLC, and that those receptors that elicit PI breakdown couple the $G\alpha_q$-proteins in a pertussis toxin-insensitive manner. There is also a possibility that the

specificity of interaction of different G-proteins with their effectors is determined not only by the nature of their α subunits, but also by their β- and γ-subunit composition. However, currently there is no explanation of why the concentrations of βγ required for activation are two to three orders of magnitude higher than those of α_q (Exton, 1994).

The determination of the specificity of interaction of G-proteins with receptor subtypes, as well as with different effectors, remains to be investigated. Other important questions are how in molecular terms the interactions result in structural changes, and how these in turn cause activity changes. Progress in analyzing the crystal structure of transducin complexed with its activators (Noel *et al.*, 1993; Lambright *et al.*, 1994; Sondek *et al.*, 1994) raises hopes that answers to these questions will also be obtained for the smooth muscle G-proteins.

PI-PLC is the key enzyme of IP_3 production. Some information already exists on its domains that interact with G-proteins and participate in the phosphodiesteratic cleavage of PIP_2. A further understanding of the structure–function relationship of PI-PLC is awaiting the determination of its three-dimensional structure. An important question to be addressed is the molecular mechanism of the G-protein-mediated activation of PI-PLC in its cellular environment, because the PI-PLC isozymes are water soluble, whereas the G-proteins are mainly lipid soluble.

Analysis of polyphosphatidylinositol (PPI) and inositol phosphate metabolism may also contribute to the understanding of smooth muscle diseases. Button *et al.* (1994) showed receptor type-selective patterns in the formation of inositol phosphates and 3-hydroxyphosphorylated PPIs in pulmonary artery smooth muscle cells. Continuation of these studies may help in the elucidation of processes related to smooth muscle cell growth and the pathogenesis of hypertension. The anticipated new knowledge about details of the IP_3 production pathway(s) will open new avenues for gaining insight into the mechanism of several diseases, including those of smooth muscle.

Acknowledgments

We thank Anna M. Pravdik for her enthusiastic assistance in the preparation of this chapter and Janice Gentry for careful typing of this manuscript. This work was supported by the N. H. Pierce gift to the College of Medicine of the University of Illinois at Chicago and by a grant from the Campus Research Board.

References

Abdel-Latif, A. A. (1986). *Pharmacol. Rev.* **38**, 227–272.
Abdel-Latif, A. A., Akhtar, R. A., and Hawthorne, J. N. (1977). *Biochem. J.* **162**, 61–73.

Agranoff, B. W. (1986). *Fed. Proc., Fed. Am. Soc. Exp. Biol.* **45**, 2627–2628.

Akhtar, R. A., and Abdel-Latif, A. A. (1984). *Biochem. J.* **224**, 291–300.

Amiot, F., Leibner, D., Marc, S., and Harbon, S. (1993). *Am. J. Physiol.* **265**, C1579–C1587.

Baba, K., Baron, C. B., and Coburn, R. F. (1989). *J. Physiol. (London)* **412**, 23–42.

Baraban, J. M., Gould, J. R., Peroutka, S. J., and Snyder, S. H. (1985). *Proc. Natl. Acad. Sci. U.S.A.* **82**, 604–607.

Bárány, K., Polyák, E., and Bárány, M. (1992). *Biochim. Biophys. Acta* **1134**, 233–241.

Baron, C. B., Cunningham, M., Strauss, J. F., III, and Coburn, R. F. (1984). *Proc. Natl. Acad. Sci. U.S.A.* **81**, 6899–6903.

Baron, C. B., Pring, M., and Coburn, R. F. (1989). *Am. J. Physiol.* **256**, C375–C383.

Baron, C. B., Pompeo, J. N., and Azim, S. (1992). *Arch. Biochem. Biophys.* **292**, 382–387.

Berridge, M. J. (1983). *Biochem. J.* **212**, 849–858.

Berridge, M. J., and Irvine, R. F. (1984). *Nature (London)* **312**, 315–321.

Berridge, M. J., and Irvine, R. F. (1989). *Nature (London)* **341**, 197–205.

Best, L., Brooks, K. J., and Bolton, T. B. (1985). *Biochem. Pharmacol.* **34**, 2297–2301.

Blackmore, P. F., and Exton, J. H. (1986). *J. Biol. Chem.* **261**, 11056–11063.

Blackmore, P. F., Bocckino, S. B., Waynick, L. E., and Exton, J. H. (1985). *J. Biol. Chem.* **260**, 14477–14483.

Button, D., Rotham, A., Bongiorno, C., Kupperman, E., Wolner, B., and Taylor, P. (1994). *J. Biol. Chem.* **269**, 6390–6398.

Carsten, M. E., and Miller, J. D. (1990). In "Uterine Function, Molecular and Cellular Aspects" (M. E. Carsten and J. D. Miller, eds.), pp. 121–167. Plenum, New York.

Chilvers, E. R., Batty, I. H., Challiss, R. A. J., Barnes, P. J., and Nahorski, S. R. (1991). *Biochem. J.* **275**, 373–379.

Clark, A. H., and Garland, C. J. (1993). *Eur. J. Pharmacol.* **235**, 113–116.

Claro, E., Garcia, A., and Picatoste, F. (1989). *Biochem. J.* **261**, 29–35.

Coburn, R. F., and Baron, C. B. (1990). *Am. J. Physiol.* **258**, L119–L133.

Cockcroft, S., and Gomperts, B. D. (1985). *Nature (London)* **314**, 534–536.

Dean, N. M., and Beaven, M. A. (1989). *Anal. Biochem.* **183**, 199–209.

Dixon, B. S., Sharma, R. V., Dickerson, T., and Fortune, J. (1994). *Am. J. Physiol.* **266**, C1406–C1420.

Doyle, V. M., and Rüegg, U. T. (1985). *Biochem. Biophys. Res. Commun.* **131**, 469–476.

Duncan, R. A., Krzanowski, J. J., Davis, J. S., Polson, J. B., Coffey, R. G., Shimoda, T., and Szentiványi, A. (1987). *Biochem. Pharmacol.* **36**, 307–310.

Exton, J. H. (1994). *Annu. Rev. Physiol.* **56**, 349–369.

Foster, P. S., Hogan, S. P., Hansbro, P. M., O'Brien, R., Potter, B. V. L., Ozaki, S., and Denborough, M. A. (1994). *Eur. J. Biochem.* **222**, 955–964.

Fukuda, M., Aruga, J., Niinobe, M., Aimoto, S., and Mikoshiba, K. (1994). *J. Biol. Chem.* **269**, 29206–29211.

Garnier, M., Lamacz, M., Tonon, M. C., and Vaudry, H. (1994). *Proc. Natl. Acad. Sci. U.S.A.* **91**, 11743–11747.

Gonzales, R. A., and Crews, F. T. (1985). *Biochem. J.* **232**, 799–804.

Griendling, K. K., Delafontaine, P., Rittenhouse, S. E., Gimbrone, M. A., Jr., and Alexander, R. W. (1987). *J. Biol. Chem.* **262**, 14555–14562.

Gu, H., Martin, H., Barsotti, R. J., and Labelle, E. F. (1991). *Am. J. Physiol.* **261**, C17–C22.

Hansbro, P. M., Foster, P. S., Liu, C., Potter, B. V. L., and Denborough, M. A. (1994). *Biochem. Biophys. Res. Commun.* **200**, 8–15.

Hashii, M., Hirata, M., Ozaki, S., Nozawa, Y., and Higashida, H. (1994). *Biochem. Biophys. Res. Commun.* **200**, 1300–1306.

Hashimoto, T., Hirata, M., and Ito, Y. (1985). *J. Pharmacol.* **86**, 191–199.

Hokin, M. R., and Hokin, L. E. (1953). *J. Biol. Chem.* **209**, 549–558.

Howe, P. H., Akhtar, R. A., Naderi, S., and Abdel-Latif, A. A. (1986). *J. Pharmacol. Exp. Ther.* **239**, 574–583.

Irvine, R. F. (1992). *Adv. Second Messenger Phosphoprotein Res.* **26**, 161–185.

Kobayashi, S., Somlyo, A. P., and Somlyo, A. V. (1988). *J. Physiol. (London)* **403**, 601–619.

Kobayashi, S., Kitazawa, T., Somlyo, A. V., and Somlyo, A. P. (1989). *J. Biol. Chem.* **264**, 17997–18004.

LaBelle, E. F., and Murray, B. M. (1990). *FEBS Lett.* **268**, 91–94.

Lambright, D. G., Noel, J. P., Hamm, H. E., and Sigler, P. B. (1994). *Nature (London)* **369**, 621–628.

Lee, M. W., and Severson, D. L. (1994). *Am. J. Physiol.* **267**, C659–C678.

Litosch, I., Wallis, C., and Fain, J. N. (1985). *J. Biol. Chem.* **260**, 5464–5471.

Majerus, P. W. (1992). *Annu. Rev. Biochem.* **61**, 225–250.

Marc, S., Leiber, D., and Harbon, S. (1986). *FEBS Lett.* **201**, 9–14.

Marc, S., Leiber, D., and Harbon, S. (1988). *Biochem. J.* **255**, 705–713.

Marsden, P. A., Danthuluri, N. R., Brenner, B. M., Ballermann, B. J., and Brock, T. A. (1989). *Biochem. Biophys. Res. Commun.* **158**, 86–93.

Mayr, G. W. (1990). In "Methods in Inositide Research" (R. F. Irvine, ed.), pp. 83–108. Raven Press, New York.

Mayr, G. W., and Thieleczek, R. (1991). *Biochem. J.* **280**, 631–640.

Mikoshiba, K., Furuichi, T., and Miyawaki, A. (1994). *Semin. Cell Biol.* **5**, 273–281.

Miller-Hance, W. C., Miller, J. R., Wells, J. N., Stull, J. T., and Kamm, K. E. (1988). *J. Biol. Chem.* **263**, 13979–13982.

Murray, R. K., Bennett, C. F., Fluharty, S. J., and Kotlikoff, M. I. (1989). *Am. J. Physiol.* **257**, L209 L216.

Noel, J. P., Hamm, H. E., and Sigler, P. B. (1993). *Nature (London)* **366**, 654–663.

O'Shea, J. J., Urdahl, K. B., Luong, H. T., Chused, T. M., Samelson, L. E., and Klausner, R. D. (1987). *J. Immunol.* **139**, 3463–3469.

Palmer, S., and Wakelam, M. J. O. (1989). *Biochim. Biophys. Acta* **1014**, 239–246.

Pang, I. H., and Sternweis, P. C. (1990). *J. Biol. Chem.* **265**, 18707–18712.

Petersen, C. C. H., and Berridge, M. J. (1994). *J. Biol. Chem.* **269**, 32246–32253.

Phillippe, M. (1994a). *Am. J. Obstet. Gynecol.* **170**, 981–990.

Phillippe, M. (1994b). *J. Soc. Gynecol. Invest.* **1**, 49–54.

Phillippe, M. (1994c). *Biochem. Biophys. Res. Commun.* **205**, 245–250.

Phillippe, M. (1994d). *Biol. Reprod.* **50**, 855–859.

Pijuan, V., Sukholutskaya, I., Kerrick, W. G., Lam, M., van Breemen, C., and Litosch, I. (1993). *Am. J. Physiol.* **264**, H126–H132.

Rhee, S. G., and Choi, K. D. (1992a). *Adv. Second Messenger Phosphoprotein Res.* **26**, 35–61.

Rhee, S. G., and Choi, K. D. (1992b). *J. Biol. Chem.* **267**, 12393–12396.

Rossier, M. F., Capponi, A. M., and Vallotton, M. B. (1987). *Biochem. J.* **245**, 305–307.

Salmon, D. M. W., and Bolton, T. B. (1988). *Biochem. J.* **254**, 553–557.

Savineau, J. P., and Mironneau, J. (1990). *J. Pharmacol. Exp. Ther.* **255**, 133–139.

Schrey, M. P., Conford, P. A., Read, A. M., and Steer, P. J. (1988). *Am. J. Obstet. Gynecol.* **159**, 964–970.

Shears, S. B. (1992). *Adv. Second Messenger Phosphoprotein Res.* **26**, 63–92.

Smith, J. B., Smith, L., Brown, E. R., Barnes, D., Sabir, M. A.,

Davies, J. S., and Farese, R. V. (1984). *Proc. Natl. Acad. Sci. U.S.A.* **81,** 7812–7816.

Somlyo, A. P., and Somlyo, A. V. (1992). *In* "The Heart and Cardiovascular System" (H. A. Fozzard, R. B. Jennings, E. Haber, A. M. Katz, and H. E. Morgan, eds.), 2nd ed., pp. 1294–1324. Raven Press, New York.

Somlyo, A. P., and Somlyo, A. V. (1994). *Nature (London)* **372,** 231–236.

Somlyo, A. V., and Somlyo, A. P. (1968). *J. Pharmacol. Exp. Ther.* **159,** 129–145.

Somlyo, A. V., Bond, M., Somlyo, A. P., and Scarpa, A. (1985). *Proc. Natl. Acad. Sci. U.S.A.* **82,** 5231–5235.

Somlyo, A. V., Horiuti, K., Trentham, D. R., Kitazawa, T., and Somlyo, A. P. (1992). *J. Biol. Chem.* **267,** 22316–22322.

Sondek, J., Lambright, D. G., Noel, J. P., Hamm, H. E., and Sigler, P. B. (1994). *Nature (London)* **372,** 276–279.

Sternweis, P. C., and Gilman, A. G. (1982). *Proc. Natl. Acad. Sci. U.S.A.* **79,** 4888–4891.

Sternweis, P. C,. and Smrcka, A. V. (1992). *Trends Biochem. Sci.* **17,** 502–506.

Strathmann, M., and Simon, M. I. (1990). *Proc. Natl. Acad. Sci. U.S.A.* **87,** 9113–9117.

Takuwa, Y., Takuwa, N., and Rasmussen, H. (1986). *J. Biol. Chem.* **261,** 14670–14675.

Verjans, B., Petersen, C. C. H., and Berridge, M. J. (1994). *Biochem. J.* **304,** 679–682.

Walker, J. W., Somlyo, A. V., Goldman, Y. E., Somlyo, A. P., and Trentham, D. R. (1987). *Nature (London)* **327,** 249–252.

Wang, X. L., Akhtar, R. A., and Abdel-Latif, A. A. (1994). *Biochim. Biophys. Acta* **1222,** 27–36.

Ward, S. G., Lampe, D., Liu, C., Potter, B. V. L., and Westwick, J. (1994). *Eur. J. Biochem.* **222,** 515–523.

Watson, S. P., Stanley, A. F., and Sasaguri, T. (1988). *Biochem. Biophys. Res. Commun.* **153,** 14–20.

Yagisawa, H., Hirata, M., Kanematsu, T., Watanabe, Y., Ozaki, S., Sakuma, K., Tanaka, H., Yabuta, N., Kamata, H., Hirata, H., and Nojima, H. (1994). *J. Biol. Chem.* **269,** 20179–20188.

Yamaguchi, K., Hirata, M., and Kuriyama, H. (1987). *Biochem. J.* **244,** 787–791.

York, J. D., Ponder, J. W., Chen, Z. W., Mathews, F. S., and Majerus, P. W. (1994). *Biochemistry* **33,** 13164–13171.

22

Protein Tyrosine Phosphorylation and Regulation of Intracellular Calcium in Smooth Muscle Cells

JOSEPH DI SALVO, NIHAL KAPLAN, and LORI A. SEMENCHUK

Department of Medical and Molecular Physiology
School of Medicine, University of Minnesota-Duluth
Duluth, Minnesota

I. INTRODUCTION

A. Underlying Hypothesis

This chapter focuses on functional relationships between receptor activation of cultured vascular smooth muscle cells (VSMC), increases in protein tyrosine phosphorylation, and increases in the intracellular concentration of Ca^{2+} ($[Ca^{2+}]_i$). Our studies have been guided by the hypothesis that phosphorylation of tyrosine residues in proteins is an important mechanism for regulating receptor-activated contraction of smooth muscle (Di Salvo *et al.*, 1988, 1989, 1993a,b, 1994; Semenchuk and Di Salvo, 1994). We suggest that at least two sites in the signaling pathway between receptor activation and contraction may be regulated by protein tyrosine phosphorylation (Fig. 1). The first site probably involves mechanisms that couple receptor activation to the resulting increase in $[Ca^{2+}]_i$. Our data indicate that tyrosine phosphorylation participates in signaling pathways that regulate release of intracellular Ca^{2+} from the sarcoplasmic reticulum (SR), and in pathways that regulate influx of Ca^{2+} from the extracellular space. The second site of tyrosine phosphorylation appears to be triggered by the increase in $[Ca^{2+}]_i$ and may be coupled to regulation of Ca^{2+} sensitivity of actin–myosin interaction. Collaborative studies in G. Pfitzer's laboratory (Heidelberg), which were performed with permeabilized smooth muscle, indicate that GTP-dependent Ca^{2+} sensitization of the contractile apparatus may be partly regulated by tyrosine phosphorylation of the myosin heavy chains (Steusloff *et al.*, 1994).

We do not suggest that enhanced protein tyrosine phosphorylation is the only regulatory mechanism for receptor-activated increases in $[Ca^{2+}]_i$ or for GTP-dependent Ca^{2+} sensitization. Instead, we suggest that tyrosine phosphorylation is a previously unrecognized regulatory mechanism that functions in concert with other mechanisms that regulate $[Ca^{2+}]_i$ and contraction of smooth muscle.

II. CELLULAR FUNCTIONS OF PROTEIN TYROSINE PHOSPHORYLATION

In this section we summarize studies with multiple cell types which suggest that functional links exist between enhanced protein kinase activity and diverse cellular mechanisms. We regret that not all relevant studies could be cited. When and where appropriate, we refer to the reader to pertinent reviews.

A. Overview

Interest in elucidating functional roles of protein tyrosine phosphorylation increased rapidly following the demonstration that pp60[v-src], the protein encoded by the viral oncogene *v-src*, is a protein tyrosine kinase (Collet and Erickson, 1978; Hunter and Sefton, 1980; Parsons and Weber, 1990). pp60[v-src] is the causative agent for the Rous chicken sarcoma. Further interest in tyrosine phosphorylation was stimulated by the discoveries that the receptors for certain mitogens, including platelet-derived growth factor (PDGF), epidermal growth factor (EGF), and nerve growth factor (NGF), were ligand-activated tyrosine kinases (Ullrich

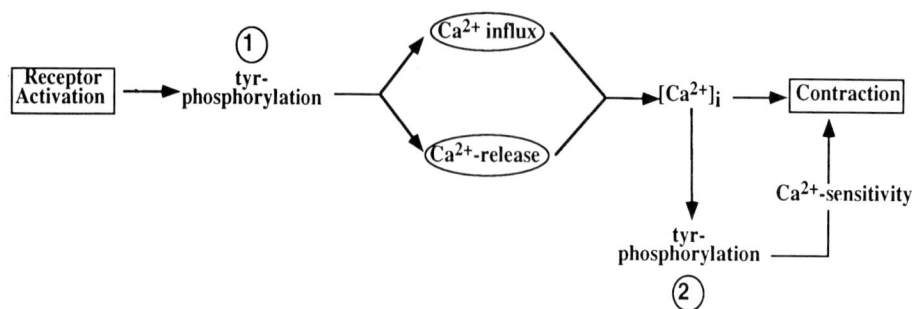

FIGURE 1   Tentative working hypothesis. On the basis of data obtained with intact smooth muscle, permeabilized smooth muscle, and cultured smooth muscle, we suggest that enhanced protein tyrosine phosphorylation participates in at least two signaling steps between receptor activation and contraction. First, we suggest that protein tyrosine phosphorylation resulting from receptor activation participates in mechanisms that regulate cellular Ca^{2+}. Second, we suggest that the rise in cellular Ca^{2+} resulting from either receptor activation (step 1) or by direct addition of Ca^{2+} to permeabilized preparations induces a second round of tyrosine phosphorylation (step 2) that is linked to control of Ca^{2+} sensitivity of the actin–myosin interaction. From Di Salvo et al. (1994), Fig. 6, p. 1439.

and Schlessinger, 1990; Fantl et al., 1993). Several gene families have now been identified that encode multiple types of protein tyrosine kinases. These include kinases that are bound to the cytoplasmic face of the plasma membrane such as pp60[c-src] (the cellular counterpart of pp60[v-src]), receptor tyrosine kinases that span the plasma membrane such as the PDGF receptor, and cytosolic tyrosine kinases such as c-abl. The functions of cellular and or oncogenic variants of these molecules are dependent on their tyrosine kinase activity. For example, mutant forms of the PDGF receptor that do not express ligand-dependent tyrosine kinase activity fail to evoke PDGF-mediated DNA synthesis. Similarly, mutants of pp60[v-src] that do not express tyrosine kinase activity are no longer able to induce neoplastic transformations.

Not surprisingly, much effort has been devoted to unraveling growth-related signaling pathways that are regulated by protein tyrosine phosphorylation and the mechanisms underlying such regulation. However, it is increasingly clear that protein tyrosine kinase activity also participates in diverse signaling pathways. Such pathways include antigen-mediated lymphocytic activation (Niklinska et al., 1992), long-term potentiation in hippocampal tissue slices (O'Dell et al., 1991), cell adhesion and cell morphology (Turner, 1994), down-regulation of neuronal receptors (Huganir and Greengard, 1990), modulation of β-adrenergic signaling pathways in cultured fibroblasts (Bushman et al., 1990), secretion of catecholamines from adrenal medullary cells (Ely et al., 1990), platelet activation (Ozaki et al., 1993), and probably others as well.

In the late 1980s we reported that pp60[c-src] tyrosine kinase activity was surprisingly and unusually high in

extracts from smooth muscle (Di Salvo et al., 1989). In extracts from coronary and aortic vascular smooth muscle, the kinase activity attributable to pp60[c-src] (20–30 fmol ^{32}P/min/mg) was 400- to 600-fold greater than it was in extracts from either cardiac or skeletal muscle. We suggested that the high tyrosine kinase activity might be coupled to mechanisms that regulate smooth muscle contraction.

Consistent with our suggestion, earlier studies had shown that contraction of smooth muscle occurred in response to mitogens known to activate receptor tyrosine kinases. For example, both EGF and PDGF contracted rat aortic strips in vitro (Berk et al., 1985, 1986). The response to PDGF was unaltered by α-adrenergic blockade with phentolamine, or by inhibition of prostaglandin synthesis with indomethacin. Moreover, PDGF-induced contraction was both delayed in onset and associated with a delayed increase in $[Ca^{2+}]_i$ detected by fura-2 fluorescence measurements. The $[Ca^{2+}]_i$ measurements were performed in a suspension of rat aortic smooth muscle cells and revealed a slow rise in $[Ca^{2+}]_i$ that peaked at about 320 nM and then declined to a lower level of about 150 nM. However, studies performed with single aortic smooth muscle cells loaded with aequorin showed that PDGF caused a slow increase in $[Ca^{2+}]_i$ that was sustained during continued exposure to PDGF (Morgan et al., 1985). Whether these kinetic differences reflect differences in Ca^{2+} probes, or differences between a mixed population of cells versus single cells, is unknown (see Section IV).

In contrast to results obtained in rat aortic preparations, constriction of ileocolic artery produced in response to EGF was blocked by indomethacin, suggesting the response was due to EGF-induced synthesis of prostaglandins (Muramatsu *et al.*, 1985). Further complexities were revealed by observations showing that both EGF and PDGF inhibited contractile responses in helical strips of mesenteric vascular smooth muscle, and that they relaxed cerebral and coronary vascular smooth muscle (Bassett *et al.*, 1988). Unfortunately, none of these studies established that the effects of the compounds were related to or dependent on tyrosine kinase activity. Nevertheless, studies reported from Hollenberg's laboratory (Yang *et al.*, 1992; Hollenberg, 1994) and from our laboratory (Di Salvo *et al.*, 1993a,b) showed that genistein and tyrphostin, two structurally different inhibitors of tyrosine kinase activity (Akiyama *et al.*, 1987; Lyall *et al.*, 1989), markedly and reversibly inhibited receptor-activated contractions in a variety of smooth muscle preparations.

B. Cellular Ca²⁺

Ample evidence suggests that functional links exist between tyrosine kinase activity and $[Ca^{2+}]_i$ (Williams, 1989; Fantl *et al.*, 1993). Activation of EGF and PDGF receptor tyrosine kinases in several different cell types is associated with a 2- to 10-fold increase in $[Ca^{2+}]_i$. The growth factor responses are characterized by an early transient increase in intracellular Ca^{2+} that appears to be due to release of intracellular Ca^{2+}, followed by a lower sustained increase in $[Ca^{2+}]_i$ that appears to be due to influx of extracellular Ca^{2+} (Tsien and Tsien, 1990). The molecular mechanisms underlying these changes in $[Ca^{2+}]_i$ have not been completely elucidated.

Ca^{2+} transients also occur in signaling pathways that do not involve receptor-tyrosine kinases. Available evidence strongly suggests that activation of antigen-specific receptors in B and T lymphocytes from various species is dependent on enhanced protein tyrosine phosphorylation even though the receptors are not tyrosine kinases. Moreover, the antigen-stimulated increase in nonreceptor tyrosine kinase activity is linked to increases in $[Ca^{2+}]_i$ (Alexander and Cantrell, 1989; Atluru *et al.*, 1991; Lane *et al.*, 1991). Herbimycin A, a benzoquinoid ansamycin antibiotic that inhibits the activity and synthesis of tyrosine kinases (Uehara *et al.*, 1986), (1) reverses transformation of fibroblasts induced by pp60ᵛ⁻ˢʳᶜ, (2) inhibits antigen- or lectin-induced activation of cultured and freshly isolated peripheral T cells, (3) suppresses protein tyrosine phosphorylation, and (4) suppresses increases in $[Ca^{2+}]_i$ that are usually associated with T-cell activation. Similarly, genistein, a bacterial flavinoid that selectively inhibits tyrosine kinase activity (Akiyama *et al.*, 1987), also suppressed activation of T cells.

Other studies also raise the possibility that tyrosine kinase activity may directly regulate both resting levels of $[Ca^{2+}]_i$ and increases in $[Ca^{2+}]_i$ that occur during T-cell receptor activation (Niklinska *et al.*, 1992),. High-level expression of pp60ᵛ⁻ˢʳᶜ, induced by retroviral gene transfer in T-cell hybridomas, increased resting levels of intracellular Ca^{2+}. This increase was not due to increased basal levels of inositol 1,4,5-triphosphate (IP_3). Moreover, the elevated $[Ca^{2+}]_i$ was reduced to control values in the presence of Herbimycin A, suggesting that it was ascribable to elevated kinase activity. Importantly, the cells containing pp60ᵛ⁻ˢʳᶜ also exhibited a twofold augmented increase in $[Ca^{2+}]_i$ when the T-cell receptor was activated. This augmented response was due to an increase in influx of extracellular Ca^{2+} that was independent of IP_3.

III. PROTEIN TYROSINE PHOSPHORYLATION AND $[Ca^{2+}]_i$ IN SMOOTH MUSCLE

In this section we review studies performed in our laboratory that strongly suggest that tyrosine kinase activity and protein tyrosine phosphorylation are important mechanisms for regulating receptor-activated increases in $[Ca^{2+}]_i$ in smooth muscle.

A. Vanadate-Induced Contraction in Intact Taenia coli

To our knowledge, the first direct evidence suggesting that enhanced protein tyrosine phosphorylation might be a regulatory mechanism for Ca^{2+}-dependent contraction of smooth muscle was obtained in studies on vanadate-induced contraction of guinea pig taenia coli (Di Salvo *et al.*, 1993b). Vanadate exhibits multiple biological activities, including inhibition of diverse ATPases and inhibition of tyrosine phosphatase activity (Gresser *et al.*, 1987; Tonks *et al.*, 1988). It is also a potent contractile agent for visceral and vascular smooth muscle (Shimada *et al.*, 1986; Sanchez-Ferrer *et al.*, 1988). Because earlier studies in this and other laboratories showed that inhibitors of tyrosine kinase activity suppressed contraction of smooth muscle (Section II), we suspected that vanadate-induced contraction might be due to its efficacy as an inhibitor of tyrosine phosphatase activity. Our experiments showed that treatment of taenia coli with 1.5 mM vanadate induced a tonic contraction that was associated with enhanced tyrosine phosphorylation of at least three substrates $(M_r, 85,000, 116,000, and 205,000; Figs. 2A and 2B)$. Maximal force attained in response to vanadate was about one-fourth to one-third of the maximal force attained during the phasic contraction elicited by stimu-

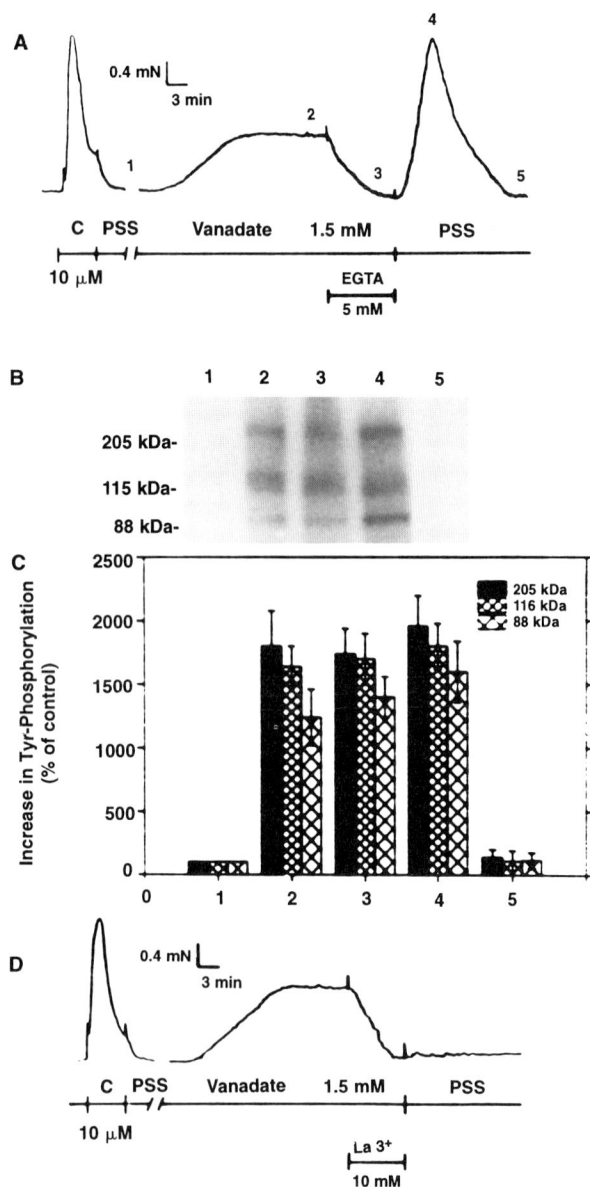

lation of muscarinic receptors with 10 μM carbachol. Removal of the vanadate resulted in relaxation and dephosphorylation of the substrates (Di Salvo et al., 1993b). The steady-state vanadate-induced contraction also relaxed when extracellular Ca^{2+} was chelated with EGTA (Fig. 2A), however, tyrosine phosphorylation persisted (Fig. 2B). Replacing the vanadate–EGTA medium with physiological salt solution (PSS) containing Ca^{2+} (1.6 mM) resulted in a rapid transient contraction (Fig. 2A). Force attained during this *spontaneous* contraction was comparable to the maximal force attained with carbachol. Interestingly, relaxation of strips contracted with vanadate also occurred when La^{3+} (10 mM $LaCl_3$) was added to the medium, an agent known to irreversibly block diverse types of Ca^{2+} channels. Like the relaxation produced by chelation of extracellular Ca^{2+}, tyrosine phosphorylation persisted during the La^{3+}-induced relaxation. However, replacement of the La^{3+}–vanadate medium with fresh medium did not result in a spontaneous transient contraction (Fig. 2D).

Accordingly, the maintenance of tyrosine phosphorylation during relaxation induced by chelation of extracellular Ca^{2+} with EGTA probably reflects persistent inhibition of tyrosine phosphatase activity by the continued presence of vanadate. The tyrosine-phosphorylated substrates may be poised to promote Ca^{2+} entry so that replacement of the medium with fresh Ca^{2+}–PSS allows for rapid entry of Ca^{2+} and a large spontaneous transient contraction. In contrast, no spontaneous contraction occurred in the experiment with La^{3+} and vanadate because La^{3+} blockade of the Ca^{2+} channels persists when the medium is replaced with fresh Ca^{2+}–PSS (Fig. 2D). Based on this interpretation of the data, further studies were performed to more directly test the hypothesis that protein tyrosine phosphorylation is an important mechanism for regulating $[Ca^{2+}]_i$ in smooth muscle cells.

FIGURE 2   Vanadate-induced contraction of taenia coli is dependent on extracellular Ca^{2+}. (A) A segment of taenia coli was contracted with 10 μM carbachol (left), washed in normal PSS, and contracted with 1.5 mM vanadate. Addition of 5 mM EGTA to chelate Ca^{2+} resulted in complete relaxation. Replacement of the medium with fresh PSS (vanadate and EGTA free) elicited a spontaneous and rapid transient contraction. (B) Western blot analysis with monoclonal antiphosphotyrosine antibodies and ^{125}I-labeled protein A revealed that vanadate-induced contraction was associated with tyrosine phosphorylation of at least three substrates exhibiting apparent masses of 80–85, 116, and 205 kDa (lane 2, point 2). This high level of phosphorylation persisted even after relaxation occurred by halting Ca^{2+} influx with EGTA (lane 3, point 3), and it was still present at the peak of the spontaneous contraction that occurred in fresh PSS (i.e., vanadate and EGTA free; lane 4, point 4). Dephosphorylation occurred when force returned to baseline (lane 5, point 5). (C) Summary of results obtained in six experiments. Control level of phosphorylation (basal level) was determined by densitometric analysis of the autoradiograms (from B) and taken to be

100%. The level of phosphorylation at the other points was expressed as percentage of control. (D) A segment of taenia coli from the same animal in A was contracted with 10 μM carbachol (left), washed with normal PSS, and contracted with 1.5 mM vanadate (middle). Addition of 10 mM $LaCl_3$ resulted in relaxation. Unlike results obtained with EGTA (A), replacement of the medium with fresh PSS did not elicit a spontaneous transient contraction. However, in accordance with results obtained in the presence of EGTA, high levels of phosphorylation were maintained after relaxation with $LaCl_3$ and dephosphorylation occurred within 15 min after relaxation (not shown). Similar results were obtained in four experiments. From Di Salvo et al. (1993b), Fig. 3, p. 390.

B. Receptor Activation, Tyrosine Kinase Activity, and $[Ca^{2+}]_i$ in Cultured Vascular Smooth Muscle Cells

Single-cell fura-2 imaging analysis (Fig. 3) was used to study functional relationships between (1) receptor activation of smooth muscle cells, (2) changes in $[Ca^{2+}]_i$, (3) participation of tyrosine kinase activity in the Ca^{2+} response, and (4) changes in protein tyrosine phosphorylation. Experiments were performed with VSMC cultured from canine femoral arterial smooth muscle and with cell lines derived from rat embryonic aorta (A10 and A7r5 cells). Depending on the cell type studied, receptor activation was evoked by stimulation of α_1-adrenergic receptors with phenylephrine, stimulation, stimulation of 5HT receptors with serotonin, or stimulation of receptors with vasoactive peptides including [Arg[8]]-vasopressin (AVP) or endothelin-1 (Et).

1. Canine Femoral Vascular Smooth Muscle Cells

In canine VSMC, pronounced heterogeneity of the cellular Ca^{2+} response evoked by receptor activation was apparent (Fig. 4). Such heterogeneity was evident with respect to (1) time elapsed between addition of agonist to the incubation medium and onset of the rise in $[Ca^{2+}]_i$, (2) rate of rise in $[Ca^{2+}]_i$, (3) magnitude of the increase in $[Ca^{2+}]_i$, (4) duration of the early transient increase in $[Ca^{2+}]_i$, (5) rate and extent of decline of the transient Ca^{2+} response, (6) magnitude of the delayed sustained increase in $[Ca^{2+}]_i$ following the transient response, and (7) the presence or absence of Ca^{2+} oscillations. Although heterogeneity of this kind has been reported previously for a variety of cell types, its underlying mechanism(s) and its physiological significance are unknown (Tsien and Tsien, 1990).

Preincubation of the cells with genistein invariably blocked the Ca^{2+} response evoked by either phenylephrine or serotonin (Figs. 4C and 4D). Though not shown, the inhibitory effects of genistein were reversible as replacement of the medium with genistein-free PSS resulted in recovery of responsiveness to both phenylephrine and serotonin. Previous studies showed that the concentration of genistein used in these experiments virtually abolished the tyrosine kinase activity of pp60[c-src] in extracts from smooth muscle and other

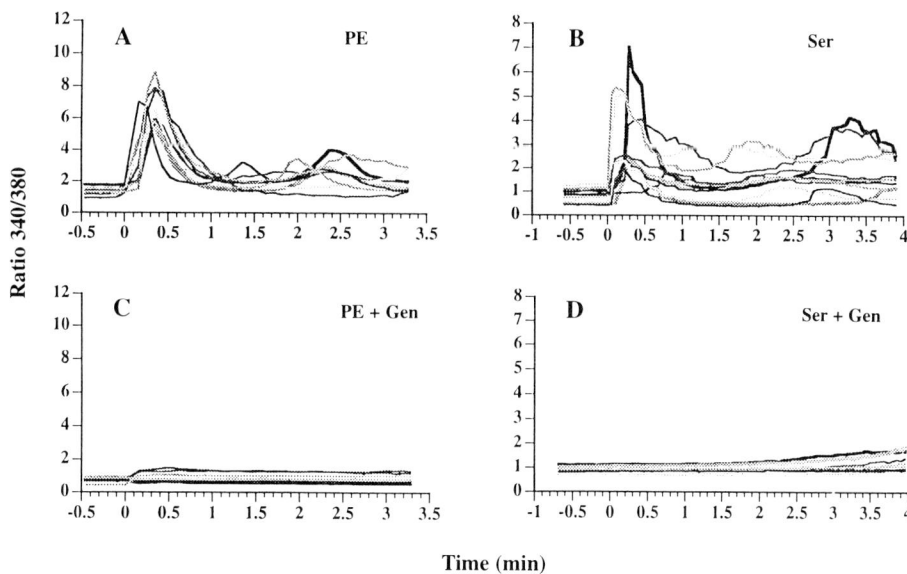

Time (min)

FIGURE 4   Genistein markedly inhibits the increase in $[Ca^{2+}]_i$ evoked by stimulation of α_1-adrenergic and serotonin receptors in cultured canine femoral VSMC. Cells were grown on coverslips and loaded with fura-2 AM (Grynkiewicz et al., 1985). At zero time (abscissa), they were stimulated with either 100 μM phenylephrine (PE; panel A) or 100 nM serotonin (Ser; panel B). Changes in $[Ca^{2+}]_i$ were monitored in real time by recording changes in the 340/380 fluorescence ratio of fura-2 ratio (ordinate) using single-cell imaging technology (Universal Imaging). Each tracing shows the receptor-activated changes in the 340/380 ratio in a single cell. In all cells, receptor activation induced a pronounced transient increase in $[Ca^{2+}]_i$ that was followed by either (1) one or more Ca^{2+} oscillations of varying magnitude and duration or (2) a lower sustained increase in Ca^{2+}. Preincubation of the cells with 110 μM genistein (Gen; 45 min) virtually abolished the Ca^{2+} response induced by both phenylephrine (C) and serotonin (D). In situ calibration of fluorescence in 16 cells studied showed that a ratio of 1 represents 140 nM Ca^{2+} and a ratio of 3 represents 965 nM Ca^{2+}. See Fig. 3 for photographs of imaging analysis.

tissues (O'Dell *et al.*, 1991; Di Salvo *et al.*, 1993a). In contrast, this concentration of genistein had little or no effect on the activity of several serine–threonine kinases such as protein kinase A (PKA) or myosin light chain kinase (MLCK). Therefore, it is likely that the inhibitory effect of genistein on Ca^{2+} responses in VSMC is ascribable to inhibition of tyrosine kinase activity.

2. A7r5 and A10 Vascular Smooth Muscle Cells

Cellular heterogeneity was also apparent during the Ca^{2+} response of A7r5 cells to AVP (Fig. 3 and Fig. 5A) and in the response of A10 cells to Et (Fig. 5B). However, these cell lines did not exhibit Ca^{2+} oscillations during stimulation with either AVP (A7r5 cells) or Et (A10 cells). Receptor activation with AVP or Et evoked a rapid transient increase in $[Ca^{2+}]_i$ followed by a smaller sustained increase in cellular Ca^{2+}, which was maintained throughout the period of observation. Preincubation of the cells with 110 μM genistein always inhibited both the early transient increase in $[Ca^{2+}]_i$ and the later sustained plateau component of the Ca^{2+} response (Figs. 5B and 5D). These results strongly suggest that tyrosine kinase activity may be required for increases in $[Ca^{2+}]_i$ resulting from activation of different types of receptors in different types of VSMC. Boldly stated, tyrosine kinase activity may be obligatory for mediating receptor-activated increases in $[Ca^{2+}]_i$.

The two major sources that contribute to a rise in $[Ca^{2+}]_i$ are the release of intracellular Ca^{2+} from the SR and the influx of extracellular Ca^{2+} (van Breemen and Saida, 1989; Tsien and Tsien, 1990; Missiaen *et al.*, 1992; Pozzan *et al.*, 1994). Therefore, two important questions are (1) does genistein-sensitive tyrosine kinase activity participate in regulatory mechanisms that couple receptor activation and release of intracellular Ca^{2+}, and (2) does tyrosine kinase activity participate in mechanisms that couple receptor activation to influx of extracellular Ca^{2+}? To address these questions, we studied the effects of genistein on the Ca^{2+} response evoked by AVP or Et in the presence or absence of extracellular Ca^{2+}. Ca^{2+} responses elicited in the absence of extracellular Ca^{2+} (0.5 mM EGTA and no added Ca^{2+}) purportedly reflect release of intracellular Ca^{2+} from vesicular storage sites such as the SR. Therefore, as amply documented in many studies, comparison of Ca^{2+} response elicited in the presence and absence of extracellular Ca^{2+} allows for assessment of the relative contribution of influx and release pathways to the evoked change in $[Ca^{2+}]_i$.

In about 70% of the cells studied in the absence of extracellular Ca^{2+}, the early transient component of the Ca^{2+} response to AVP was depressed by about 60%

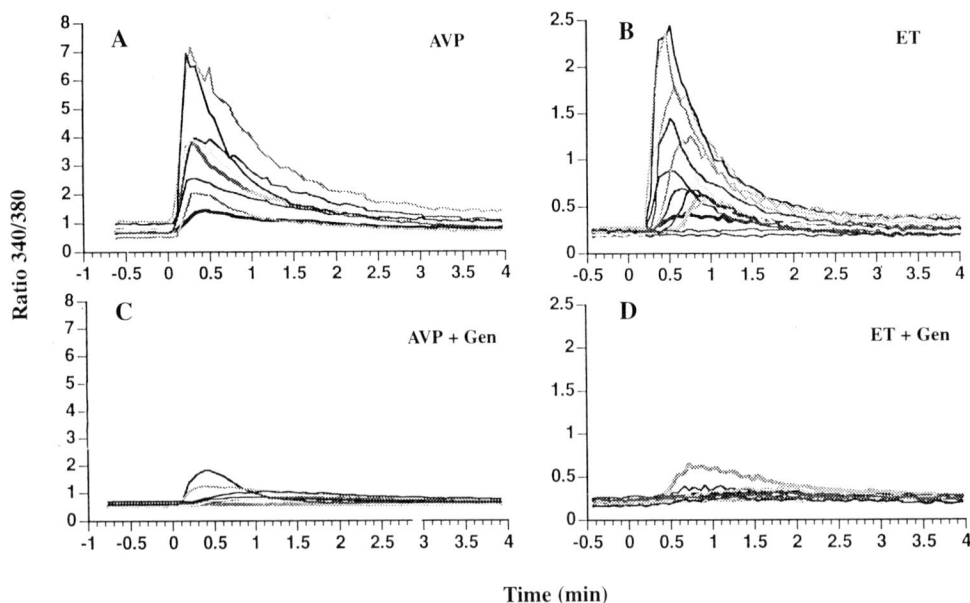

FIGURE 5 Genistein markedly inhibits the increases in $[Ca^{2+}]_i$ evoked by stimulation of A7r5 cells with 20 nM AVP (A, C) or stimulation of A10 VSMC with 400 nM Et (B, D). Procedures were as in Fig. 3. Note that apparent heterogeneity among cells stimulated with AVP (A) and among cells stimulated with Et (B) as shown by cell-to-cell differences in time elapsed between time of stimulation and the rise in $[Ca^{2+}]_i$, magnitude and duration of the early transient response, and magnitude of the delayed sustained response. Also see Fig. 3. for additional details.

relative to the response obtained in cells from the same population studied in the presence of extracellular Ca^{2+} (Fig. 6A). This suggests that 40% of the initial Ca^{2+} transient in these cells probably was due to release of intracellular Ca^{2+}, whereas 60% of the transient was probably due to influx of extracellular Ca^{2+}.

However, the Ca^{2+} transient was independent of extracellular Ca^{2+} in about 30% of the cells studied (Fig. 6B). That is, the magnitude of the Ca^{2+} transient evoked by AVP was essentially the same in the presence or absence of extracellular Ca^{2+}. Accordingly, virtually all of the increase in $[Ca^{2+}]_i$ occurring during the transient response in these cells was probably due to release of intracellular Ca^{2+}. The basis for this pronounced division in the relative contribution of stored intracellular Ca^{2+} to the total Ca^{2+} in the transient is unknown. A division of this kind would not be recognized in analytical procedures that only sense the average change in fluorescence intensity of a population of cells viewed simultaneously (e.g., a population of cells examined in a fluorometric cuvette). In contrast, single-cell imaging analysis permits viewing and evaluation of receptor-activated Ca^{2+} responses in individual cells.

In all cells, preincubation with 110 μM genistein virtually abolished the early transient Ca^{2+} response to AVP in either the presence or absence of extracellu-lar Ca^{2+} (Figs. 5 and 6). Therefore, it is likely that tyrosine kinase activity is involved in both the release and influx components of the early transient Ca^{2+} response evoked by AVP in A7r5 cells. The delayed lower sustained increase in $[Ca^{2+}]_i$ was always eliminated in the absence of extracellular Ca^{2+} (Fig. 6). In accordance with other studies (Tsien and Tsien, 1990), the delayed Ca^{2+} response is due entirely to influx of extracellular Ca^{2+}. As noted earlier in this section, genistein invariably inhibited the delayed Ca^{2+} response (Fig. 5).

C. Smooth Muscle Activation and Enhanced Protein Tyrosine Phosphorylation: ras*GAP* Is a Substrate

In this section we review data showing that activation of smooth muscle preparations induces tyrosine phosphorylation of several substrates. As shown in Fig. 7, a similar set of substrates (M_r 42,000–205,000) are tyrosine phosphorylated during (1) activation of muscarinic receptors in intact taenia coli, (2) vanadate-induced contraction of the same preparation, (3) activation of α_1-adrenergic receptors in VSMC cultured from canine femoral artery, and (4) Ca^{2+}-activated contraction of permeabilized ileal longitudinal smooth muscle. It is likely that one or more of these substrates

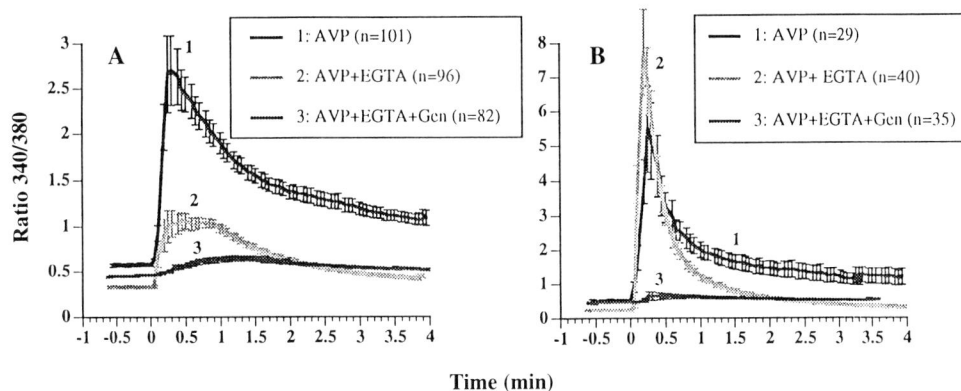

Time (min)

FIGURE 6 Genistein suppresses AVP receptor-activated increases in influx of extracellular Ca^{2+} and release of intracellular Ca^{2+} from the SR of A7r5 cells. To simplify data analysis and minimize problems associated with cell-to-cell heterogeneity, Ca^{2+} responses were expressed as mean values + 1 SE. Stimulation with 20 nM AVP was performed in the presence (1.6 mM) and absence of extracellular Ca^{2+} (PSS containing 0.5 mM EGTA and no added Ca^{2+}): this allowed for assessment of the relative contribution of Ca^{2+} influx and Ca^{2+} release to the receptor-activated increase in $[Ca^{2+}]_i$ (see text for further discussion). (A) In 70% of the cells studied in the absence of extracellular Ca^{2+}, the early transient increase in $[Ca^{2+}]_i$ was markedly suppressed and the delayed sustained component of the Ca^{2+} response was eliminated. Further inhibition of the transient response occurred when the cells were preincubated with 110 μM genistein for 1 hr. (B) In contrast, in 30% of the cells, the early transient increase in $[Ca^{2+}]_i$ was essentially unaltered in the absence of extracellular Ca^{2+}. However, in accordance with results shown in A, the delayed response was eliminated, and preincubation with genistein virtually abolished the early transient rise in $[Ca^{2+}]_i$. Results similar to those shown in A and B were obtained in A10 VSMC stimulated with 400 nM Et.

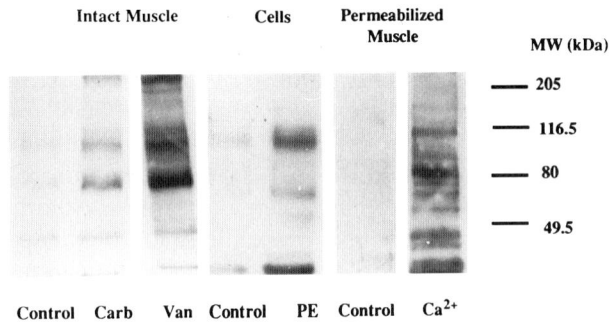

FIGURE 7 A similar set of substrates is tyrosine phosphorylated during activation of either intact taenia coli, cultured VSMC, or staphylococcal α-toxin-permeabilized ileal longitudinal smooth muscle. In these experiments, tyrosine-phosphorylated substrates were detected by immunoblotting with antiphosphotyrosine antibodies and enhanced chemiluminescence technology rather than the less sensitive ^{125}I-labeled protein A technology used in Fig. 2. Stimulation of guinea pig taenia coli with either 10 μM carbachol (Carb) or 1.5 mM vanadate (Van) resulted in pronounced tyrosine phosphorylation of at least nine substrates with apparent masses of 42–45, 50, 70, 80–85, 95, 100, 110, 116, and 205 kDa. In like fashion, stimulation of canine femoral VSMC with 100 μM phenylephrine (PE) resulted in enhanced tyrosine phosphorylation of a similar set of substrates (however, note that qualitative differences were evident with respect to some substrates, such as the one of 205 kDa). Similarly, the same substrates appeared to be tyrosine phosphorylated when permeabilized ileal smooth muscle was contracted with Ca^{2+} (pCa 4.5). From Di Salvo et al. (1994), Fig. 5, p. 1438.

of genistein that does not inhibit tyrosine kinase activity (Nakashima et al., 1991; Lee et al., 1993), did not alter receptor-activated increases in either protein tyrosine phosphorylation or the rise in $[Ca^{2+}]_i$ (Semenchuk and Di Salvo, 1994). These results are consistent with receptor-activated increases in tyrosine phosphorylation being functionally coupled to receptor-activated increases in cellular Ca^{2+} (Fig. 1).

Because rasGAP (ras GTPase-activating protein), a protein of 116,000–120,000 Da, participates in multiple signaling pathways involving ras, a monomeric G-protein (Bourne et al., 1991) present in smooth muscle (Adam and Hathaway, 1993), we reasoned that the tyrosine-phosphorylated substrate of 116,000 Da might be rasGAP. This hypothesis was confirmed by Western blot analysis of proteins that were electrophoretically separated from extracts of VSMC that had been activated with serotonin. That is, stripping electrophoretic nitrocellulose membranes that had been immunoblotted for phosphotyrosine proteins and reprobing the same membrane with antibodies for rasGAP revealed that the tyrosine-phosphorylated substrate of 116,000 Da was specifically recognized (Fig. 8B). The only substrate that reacted with the rasGAP antibody was the polypeptide of 116,000 Da. Moreover, immunoprecipitation of the cell extracts with rasGAP anti-

is coupled to regulatory mechanisms for receptor-activated increases in $[Ca^{2+}]_i$ and Ca^{2+} sensitization (Fig. 1). The electrophoretic pattern of tyrosine phosphorylation that we observed is strikingly similar to the pattern of tyrosine-phosphorylated substrates found in extracts from rat aortic smooth muscle cells that had been stimulated with different vasoactive agents (Tsuda et al., 1991; Molloy et al., 1993). These data make the point that stimulation of protein tyrosine phosphorylation can occur in smooth muscle preparations by mechanisms that do not involve activation of receptor tyrosine kinases.

Activation of serotonin receptors in canine femoral VSMC evoked tyrosine phosphorylation of a group of substrates similar to those that were tyrosine phosphorylated during stimulation of $α_1$-adrenergic receptors with phenylephrine (Fig. 8A). Preincubation of the cells with 110 μM genistein suppressed tyrosine phosphorylation that was evoked by stimulation of either serotonin or adrenergic receptors (Fig. 8A, lanes 4 and 5). The same concentration of genistein that inhibited receptor-activated increases in $[Ca^{2+}]_i$ (Fig. 5) also inhibited receptor-activated increases in protein tyrosine phosphorylation. In contrast, preincubation of the VSMC with 110 μM diadzein, a structural analog

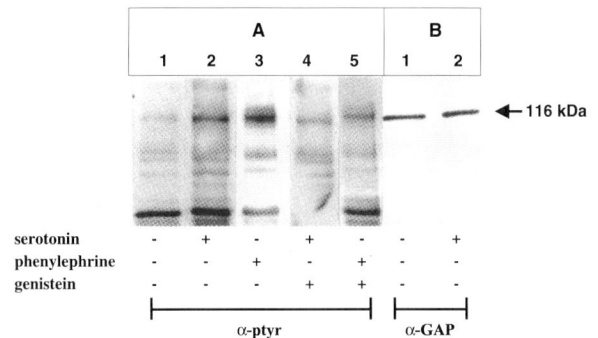

FIGURE 8 Genistein inhibits protein tyrosine phosphorylation of rasGAP and other substrates that is evoked by phenylephrine or serotonin in canine femoral VSMC. Treatment conditions are listed below the figure. (A) Cells were exposed to 100 μM phenylephrine or 100 nM serotonin for 18 sec in the presence or absence of 100 μM genistein. They were processed for electrophoretic separation of proteins, transferred to nitrocellulose membranes, and immunoblotted with a monoclonal phosphotyrosine antibody (α-ptyr) for analysis of phosphotyrosine proteins (Semenchuk and Di Salvo, 1994). (B) The immunoblotted nitrocellulose membranes were stripped of phosphotyrosine antibody and lanes from control and serotonin treated cells were reprobed with a polyclonal antibody (α-GAP) specific for rasGAP. Only the substrate of 116 kDa was recognized by the rasGAP antibody. Note that the intensity of the rasGAP band was the same for both samples, confirming that the protein loads (10 μg) were also the same.

body before electrophoresis selectively removed the 116,000-Da substrate (data not shown). Interestingly, Molloy *et al.* (1993) noted that they were unable to detect significant increases in tyrosine phosphorylation of *ras*GAP during receptor activation of rat aortic smooth muscle cells. This disparity may reflect procedural differences. They studied rat cells stimulated with angiotensin and employed immunoprecipitation for monitoring substrate phosphorylation, whereas we studied canine cells stimulated with serotonin or phenylephrine and used single-label immunoblotting of whole-cell extracts for assessing changes in protein tyrosine phosphorylation.

Comparison of the time course for tyrosine phosphorylation of *ras*GAP and the time course for the increase in $[Ca^{2+}]_i$ during receptor activation with phenylephrine showed that (1) the onset of phosphorylation precedes initiation of the transient rise in $[Ca^{2+}]_i$, (2) attains maximal level immediately before the maximal rise in $[Ca^{2+}]_i$, and (3) declines before the Ca^{2+} transient declines to a sustained suprabasal level (Fig. 9). These data suggest that temporal changes in tyrosine phosphorylation of *ras*GAP may be coupled to temporal changes in $[Ca^{2+}]_i$ evoked by receptor activation of femoral VSMC. Indeed, the data in Fig. 9 indicate that the transient rise in Ca^{2+} during the early transient increase in $[Ca^{2+}]_i$ is directly proportional to the rise in tyrosine phosphorylation of *ras*GAP.

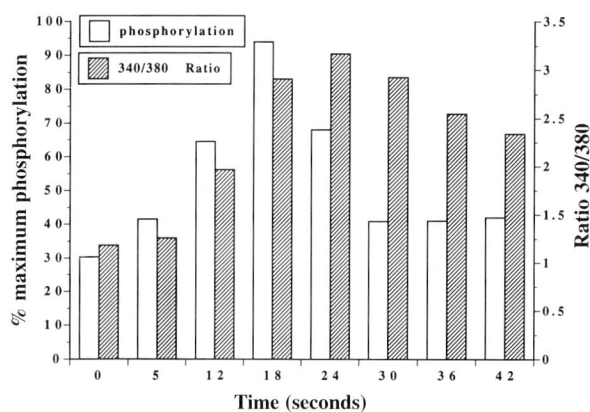

FIGURE 9 Tyrosine phosphorylation of *ras*GAP occurs before the transient increase in $[Ca^{2+}]_i$ and declines before Ca^{2+} declines during activation of α_1-adrenergic receptors with phenylephrine. All experiments were performed with canine femoral VSMC. Each hatched bar represents the mean values for Ca^{2+} determination in 64 cells, whereas the open bar represents mean values for tyrosine phosphorylation of *ras*GAP in four experiments. Maximal phosphorylation as determined by densitometric analysis was taken to be 100% and all other points were expressed as percentage of maximum. Similar results were obtained during the transient increase in $[Ca^{2+}]_i$ evoked by stimulation of serotonin receptors (not shown).

IV. SUMMARY, WORKING MODEL, AND FUTURE DIRECTIONS

$[Ca^{2+}]_i$ is a key regulator for contraction of smooth muscle (van Breemen and Saida, 1989; Somlyo and Somlyo, 1994). The major point of this chapter is that tyrosine kinase activity is a previously unrecognized mechanism for coupling receptor activation of smooth muscle and cellular Ca^{2+} (Figs. 1–9). Several lines of evidence suggest that receptor-activated increases in protein tyrosine phosphorylation function in mechanisms that regulate influx of extracellular Ca^{2+} and in mechanisms that regulate release of intracellular Ca^{2+} from the SR. First, vanadate-induced contraction appears to be functionally linked to enhanced protein tyrosine phosphorylation, which is probably coupled to influx of extracellular Ca^{2+} (Fig. 2). Second, inhibition of tyrosine kinase activity with genistein suppressed vanadate-induced contraction and tyrosine phosphorylation in intact taenia coli, and it also suppressed muscarinic receptor-activated contraction and tyrosine phosphorylation in the same preparation (Di Salvo *et al.*, 1994). Third, genistein virtually abolished receptor-activated increases in $[Ca^{2+}]_i$ that were evoked by stimulation of receptors for phenylephrine, serotonin, AVP, or Et in different types of cultured VSMC (Figs. 4 and 5). Fourth, genistein inhibited both influx of extracellular Ca^{2+} and release of intracellular Ca^{2+} (Fig. 6 and Section III.B). Fifth, activation of diverse types of receptors markedly enhanced tyrosine phosphorylation of a similar set of substrates including *ras*GAP in intact smooth muscle and cultured smooth muscle cells (Figs. 7 and 8). Moreover, the increase in tyrosine phosphorylation of *ras*GAP occurred before the increase in cellular Ca^{2+}, and then declined before cellular Ca^{2+} declined (Fig. 9). Sixth, genistein inhibited both the increase in tyrosine phosphorylation and the increase in cellular Ca^{2+}, suggesting that these two responses were functionally coupled to receptor activation.

Two major challenges for future studies are to (1) elucidate molecular mechanisms whereby activation of structurally diverse receptors that do not exhibit ligand-dependent tyrosine kinase activity (e.g., α_1-adrenergic receptor) promotes enhanced protein tyrosine phosphorylation of specific substrates in smooth muscle cells, and (2) determine how phosphorylation of one or more of these substrates is coupled to regulation of Ca^{2+} influx and Ca^{2+} release. The temporal correlation between tyrosine phosphorylation of *ras*GAP and the transient increase in $[Ca^{2+}]_i$ (Fig. 9) suggests that *ras*GAP may participate in regulating increases in cellular Ca^{2+} evoked by receptor activation of VSMC. Activated *ras* (i.e., *ras* bound to GTP) reportedly stimulates Ca^{2+} channels in neuronal cells (Heschler *et al.*, 1991). However, *ras*GAP promotes conversion of the biologically active *ras*GTP to inactive *ras*GDP (Bourne *et al.*, 1991). *ras*GAP is also known to form complexes with several tyrosine kinases, including pp60[c-src] (Ellis *et al.*, 1990; Gibbs *et al.*, 1990; Brott *et al.*, 1991; Moran *et al.*, 1991; Liu and Pawson, 1991). Moreover, tyrosine phosphorylation of *ras*GAP downregulates its ability to inactivate *ras*GAP.

Based on these considerations, a useful working model is that receptor activation of smooth muscle may promote complex formation between *ras*GAP and the nonreceptor tyrosine kinase pp60[c-src], which we have previously shown to be unusually high in smooth muscle (Di Salvo *et al.*, 1988, 1989). This *ras*GAP–pp60[c-src] complex might enhance tyrosine phosphorylation of *ras*GAP and down-regulate its activity. In this setting, *ras*GTP would be more available to promote Ca^{2+} influx and contribute to receptor-activated increases in $[Ca^{2+}]_i$. Conceivably, activated *ras* may also directly promote release of Ca^{2+} from the SR, or indirectly promote release by activation of a specific isoform of phospholipase c (PLC) and formation of IP_3. Alternatively, receptor activation of cytoplasmic tyrosine kinase activity (e.g., pp60[c-src]) could tyrosine phosphorylate and directly activate $PLC\gamma_1$. Such a mechanism reportedly occurs in cultured rat aortic smooth muscle cells stimulated with angiotensin II (Marrero *et al.*, 1994). In either case, treatment of smooth muscle with tyrosine kinase inhibitors, such as genistein, would be expected to reduce receptor-activated tyrosine phosphorylation of *ras*GAP so that it exhibits high GAP activity. Accordingly, *ras*GTP would be rapidly inactivated to *ras*GDP and result in (1) suppression of receptor-activated increases in $[Ca^{2+}]_i$ (Figs. 4–6) and (2) a decrease in contractile force (Di Salvo *et al.*, 1993a). In contrast, inhibition of protein tyrosine phosphatase activity (e.g., vanadate) would be expected to enhance tyrosine phosphorylation and down-regulation of *ras*GAP. This would effectively increase *ras*GTP and result in (1) higher $[Ca^{2+}]_i$ and (2) an increase in contractile force (Fig. 2). An attractive feature of this working model is that it is testable.

Other significant questions that require resolution include: (1) How is activation of receptors that are not ligand-dependent tyrosine kinases functionally coupled to enhanced protein tyrosine kinase activity? (2) Is pp60[c-src] the tyrosine kinase that is activated? (3) Are the Ca^{2+} influx channels that appear to be regulated by tyrosine phosphorylation during the early transient rise in $[Ca^{2+}]_i$ the same as, or different from, the influx channels that are regulated during the delayed lower sustained Ca^{2+} response? (4) Does regulation of Ca^{2+} influx and release channels by tyrosine kinase activity involve direct tyrosine phosphorylation of the channels, phosphorylation of associated regulatory proteins, or generation of second messengers that modulate channel activity? (5) What is the nature of the protein tyrosine phosphatase(s) that reverse the effects of receptor-activated increases in tyrosine phosphorylation? (6) What are the identities and functions of the other tyrosine-phosphorylated substrates that are detected during receptor activation of smooth muscle cells? Addressing these and other questions implicit in the tentative working model promises to provide new insights into molecular mechanisms that couple receptor activation of smooth muscle, enhanced protein phosphorylation of specific substrates, and receptor-activated increases in $[Ca^{2+}]_i$.

Acknowledgments

Studies performed in our laboratory were supported by a grant from the NIH (HL 49536) and the Edwin Eddy Foundation.

References

Adam, L. P., and Hathaway, D. R. (1993). *FEBS Lett.* **32,** 56–60.

Akiyama, T., Ishida, J., Nakagawa, S., Ogawara, H., Watanabe, S., Itoh, N., Shibuya, M., and Fukami, Y. (1987). *J. Biol. Chem.* **262,** 5592–5595.

Alexander, D. R., and Cantrell, D. A. (1989). *Immunol. Today* **10,** 200–205.

Atluru, D., Jackson, T. M., and Atluru, S. (1991). *Clin. Immunol. Immunopathol.* **3,** 379–387.

Bassett, J. E., Bowen-Pope, D. F., Takayusi, M., and Dacey, R. G. (1988). *Microvasc. Res.* **35,** 368–373.

Berk, B. C., Brock, T. A., Webb, R. C., Taubman, M. B., Atkinson, W. J., Gimbrone, M. A., and Alexander, R. W. (1985). *J. Clin. Invest.* **75,** 1083–1086.

Berk, B. C., Alexander, R. W., Brock, T. A., Gimbrone, M. A., and Webb, R. C. (1986). *Science* **232,** 87–90.

Bourne, H. R., Sanders, D. A., and McCormick, F. (1991). *Nature (London)* **349,** 117–127.

Brott, B. K., Decker, S., Shafer, J., Gibbs, J. B., and Jove, R. (1991). *Proc. Natl. Acad. Sci. U.S.A.* **88,** 755–775.

Bushman, W. A., Wilson, L. K., Luttrell, D. K., Moyers, J. S., and Parsons, S. J. (1990). *Proc. Natl. Acad. Sci. U.S.A.* **87,** 7462–7466.

Collett, M. S., and Erickson, R. C. (1978). *Proc. Natl. Acad. Sci. U.S.A.* **75,** 2021–2024.

Di Salvo, J., Gifford, D., and Kokkinakis, A. (1988). *Biochem. Biophys. Res. Commun.* **153,** 388–394.

Di Salvo, J., Gifford, D., and Kokkinakis, A. (1989). *J. Biol. Chem.* **264,** 10773–10778.

Di Salvo, J., Steusloff, A., Semenchuk, L., Satoh, S., Kolquist, K., and Pfitzer, G. (1993a). *Biochem. Biophys. Res. Commun.* **190,** 968–974.

Di Salvo, J., Semenchuk, L. A., and Lauer, J. (1993b). *Arch. Biochem. Biophys.* **304,** 386–391.

Di Salvo, J., Pfitzer, G., and Semenchuk, L. A. (1994). *Can. J. Physiol. Pharmacol.* **72,** 1434–1439.

Ellis, C., Moran, M., McCormick, F., and Pawson, T. (1990). *Nature (London)* **343**, 377–381.

Ely, C. M., Oddie, K., Litz, J. S., Rossomando, A. J., Kanner, S. B., Sturgill, T. W., and Parsons, S. J. (1990). *J. Cell Biol.* **110**, 731–742.

Fantl, W. J., Johnson, D. E., and Williams, L. T. (1993). *Annu. Rev. Biochem.* **62**, 453–481.

Gibbs, J. B., Marshall, M. S., Scolnick, E. M., Dixon, R. A. F., and Vogel, U. S. (1990). *J. Biol. Chem.* **265**, 20437–20442.

Gresser, M. J., Tracey, A. S., and Stankiewicz, P. J. (1987). *Adv. Protein Phosphatases* **4**, 35–57.

Grynkiewicz, G., Poenie, M., and Tsien, R. Y. (1985). *J. Biol. Chem.* **260**, 3440–3450.

Heschler, J., Klinz, F. J., Schultz, G., and Wittinghofer, A. (1991). *Cell. Signal.* **3**, 127–133.

Hollenberg, M. D. (1994). *Trends Pharmacol. Sci.* **15**, 108–114.

Huganir, R. L., and Greengard, P. (1990). *Neuron* **5**, 555–567.

Hunter, T., and Sefton, B. M. (1980). *Proc. Natl. Acad. Sci. U.S.A.* **77**, 1311–1315.

Lane, P. J. L., Ledbetter, J. A., McConnell, F. M., Draves, K., Deans, J., Schievers, G. L., and Clark, E. A. (1991). *J. Immunol.* **146**, 715–722.

Lee, K.-M., Toscas, K., and Villareal, M. L. (1993). *J. Biol. Chem.* **268**, 9945–9948.

Lui, X., and Pawson, T. (1991). *Mol. Cell Biol.* **11**, 2511–2516.

Lyall, R. M., Zilberstein, A., Gazit, A., Gilon, C., Levitzki, A., and Schlessinger, J. (1989). *J. Biol. Chem.* **264**, 14503–14509.

Marrero, M. B., Paxton, W. G., Duff, J. L., Berk, B. C., and Bernstein, K. E. (1994). *J. Biol. Chem.* **269**, 10935–10939.

Missiaen, L., De Smedt, H., Droogmans, G., Himpens, B., and Casteels, R. (1992). *Pharmacol. Ther.* **56**, 191–231.

Molloy, C. J., Taylor, D. S., and Weber, H. (1993). *J. Biol. Chem.* **268**, 7338–7345.

Moran, M. F., Polakis, P., McCormick, F., Pawson, T., and Ellis, C. (1991). *Mol. Cell. Biol.* **11**, 1804–1812.

Morgan, K. G., DeFeo, T. T., Wenc, K., and Weinstein, R. (1985). *Pflügers Arch.* **405**, 77–79.

Muramatsu, I., Hollenberg, M. D., and Lederis, K. (1985). *Physiol. Pharmacol.* **63**, 994–999.

Nakashima, S., Koike, T., and Nozawa, Y. (1991). *Mol. Pharmacol.* **39**, 475–480.

Niklinska, B. B., Yamada, H., O'Shea, J. J., June, C. H., and Ashwell, J. D. (1992). *J. Biol. Chem.* **267**, 7154–7159.

O'Dell, T. J., Kandel, E. R., and Grant, G. N. (1991). *Nature (London)* **353**, 558–560.

Ozaki, Y., Yatomi, Y., Jinnai, Y., and Kume, S. (1993). *Biochem. Pharmacol.* **46**, 395–403.

Parsons, J. T., and Weber, M. J. (1990). *Curr. Top. Microbiol. Virol.* **147**, 80–127.

Pozzan, T., Rizzuto, R., Volpe, P., and Meldolesi, J. (1994). *Physiol. Rev.* **74**, 595–636.

Sanchez-Ferrer, C. F., Marin, J., Lluch, M., Valverde, A., and Salaices, M. (1988). *Br. J. Pharmacol.* **93**, 53–60.

Semenchuk, L. A., and Di Salvo, J. (1994). *Biophys. J.* **66**, A131.

Shimada, T., Shimamura, K., and Sunano, S. (1986). *Blood Vessels* **23**, 113–124.

Somlyo, A. P., and Somlyo, A. V. (1994). *Nature (London)* **372**, 231–236.

Steusloff, A., Paul, E., Semenchuk, L., Di Salvo, J., and Pfitzer, G. (1994). *Biophys. J.* **66**, A242.

Tonks, N. K., Diltz, C. D., and Fischer, E. H. (1988). *J. Biol. Chem.* **263**, 6722–6730.

Tsien, R. W., and Tsien, R. Y. (1990). *Annu. Rev. Cell Biol.* **6**, 715–760.

Tsuda, T., Kawahara, Y., Shii, K., Koide, M., Ishida, Y., and Yokoyama, M. (1991). *FEBS Lett.* **285**, 44–48.

Turner, C. E. (1994). *BioEssays* **16**, 47–52.

Uehara, Y., Mari, M., Takeuchi, T., and Umezama, M. (1986). *Mol. Cell. Biol.* **6**, 2198–2203.

Ullrich, A., and Schlessinger, J. (1990). *Cell (Cambridge, Mass)* **61**, 203–212.

van Breemen, C., and Saida, K. (1989). *Annu. Rev. Physiol.* **51**, 129–145.

Williams, L. T. (1989). *Science* **243**, 1564–1570.

Yang, S.-G., Saifeddine, M., and Hollenberg, M. D. (1992). *Can. J. Physiol. Pharmacol.* **70**, 85–93.

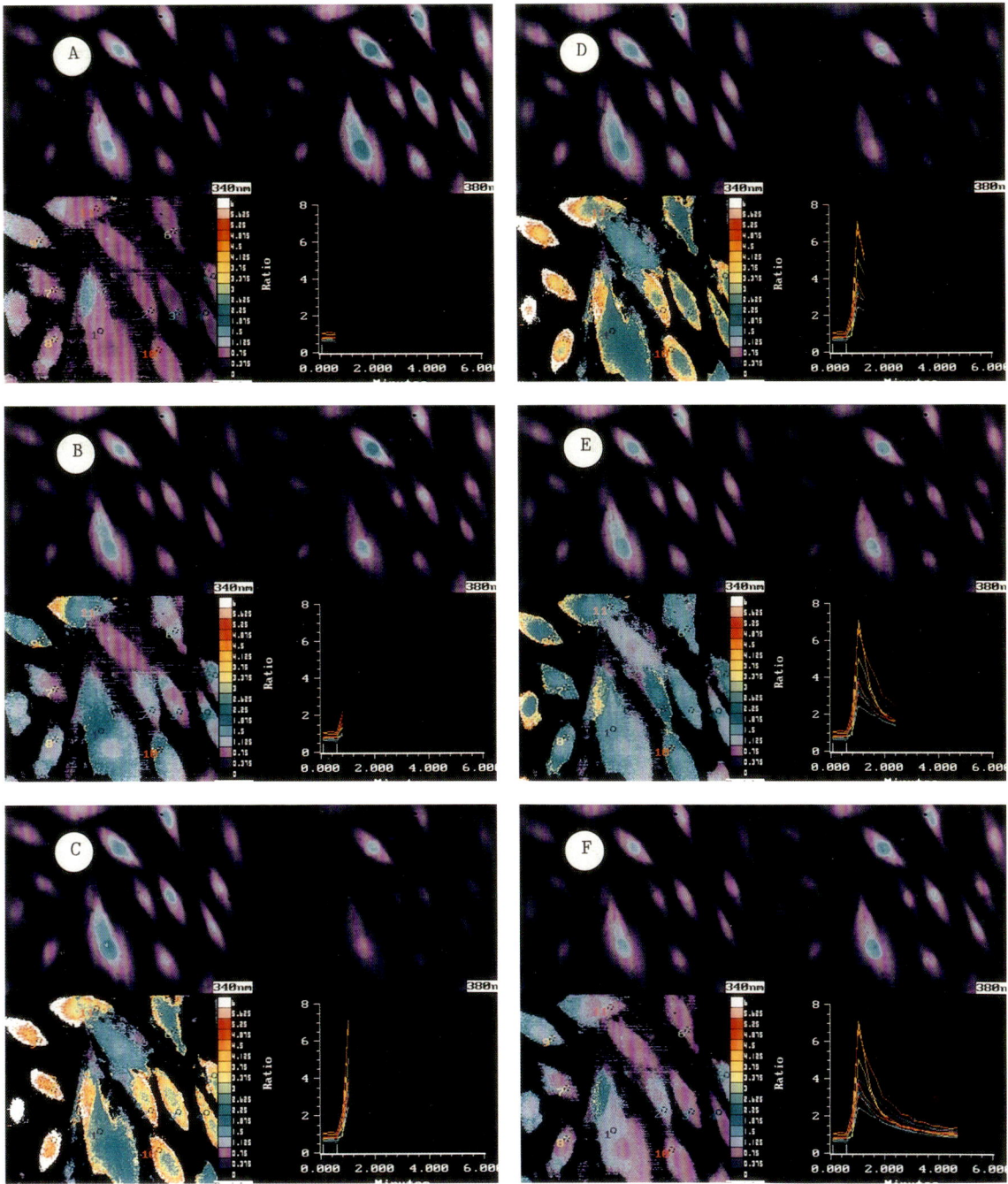

FIGURE 3 Fura-2 image analysis of multiple individual cells reveals that stimulation of AVP receptors in A7r5 cells evokes an increase in cellular Ca^{2+}. A7r5 cells were grown on glass coverslips and processed for fura-2 single-cell imaging analysis of changes in $[Ca^{2+}]_i$ evoked by 20 nM AVP as in Fig. 4. Regions of identical size were selected for analysis in 11 different cells in the same field. Each panel (A–F) contains four quadrants showing results obtained by on-line electronic photography *in real time*. The upper left quadrant shows the intensity of fluorescence during excitation at 340 nm, whereas the upper right quadrant shows the intensity at 380 nm. The bottom left quadrant is the computer-generated image of the 340/380 ratio depicted in pseudocolor. The color scale for the ratio is given immediately to the right. The bottom right quadrant shows progressive changes in the 340/380 ratio (ordinate) during the period of observation (abscissa, min) recorded in real time. Each line in the graph traces the time course for the calcium response in each individual cell. The first small vertical bar in the graph identifies the start of the experiment, whereas the second bar denotes the addition of AVP to a final concentration of 20 nM. Panels A–F show time-dependent changes in fluorescence intensity at 380 and 340 nm and the changes in 340/380 ratio that occur in response to stimulation of AVP receptors. Note that stable baselines are obtained for each cell and that the intensity of fluorescence is greater at 380 nm than at 340 nm (A). After the addition of AVP, the 340/380 ratio rapidly rises (B, C), reaches a maximum (i.e., transient increase in $[Ca^{2+}]_i$), and then declines to a lower sustained value that is higher than basal level (i.e., delayed sustained increase in $[Ca^{2+}]_i$; D–F). As $[Ca^{2+}]_i$ increases (B, C), the fluorescence intensity at 340 nm increases, whereas the intensity at 380 nm decreases. In contrast, during the declining phase of the transient Ca^{2+} response, the intensity at 340 nm decreases whereas the intensity at 380 nm increases (D, E). Also note that the pseudocolor image quadrants (lower left in A and B) show that the early rise in $[Ca^{2+}]_i$ is more pronounced in the periphery of the cells than in the interior of the cells.

23

Cyclic ADP-Ribose and Calcium Signaling

ANTONY GALIONE and JASWINDER SETHI

University Department of Pharmacology
Oxford
United Kingdom

I. INTRODUCTION

Stimulus-evoked Ca^{2+} release from intracellular stores represents an important component of Ca^{2+} signals (Pozzan *et al.*, 1994). Ca^{2+} signals regulate cellular responses in both excitable and nonexcitable cells, such as muscle contraction, secretion, fertilization, or changes in membrane excitability. Ca^{2+} mobilization from internal stores is effected by two families of closely related Ca^{2+} release channels, inositol 1,4,5-trisphosphate receptors (IP_3Rs) and ryanodine receptors (RyRs) (Furuichi *et al.*, 1994). Although the signal transduction pathways regulating Ca^{2+} release via IP_3Rs have been well documented, where it appears to function as a ubiquitous pathway for generating intracellular Ca^{2+} signals (Berridge, 1993a), RyR-mediated Ca^{2+} signaling is less well understood. However, an awareness of the importance of this second pathway for Ca^{2+} mobilization in an equally diverse set of cellular systems has increased (Coronado *et al.*, 1994; Meissner, 1994; Ogawa, 1994; Sorrentino and Volpe, 1993).

RyRs were first identified as the major pathway for Ca^{2+} release from the sarcoplasmic reticulum during excitation–contraction coupling in skeletal and cardiac muscles (for review, see Fleischer and Inui, 1989). In mammalian cells, three distinct isoforms (RyR1, RyR2, and RyR3) have been characterized and shown to be present in a large range of both excitable and nonexcitable cells (Sorrentino and Volpe, 1993). Molecular characterization of RyRs has shown that they have extensive structural homology with IP_3Rs (Furuichi *et al.*, 1994), and these are mirrored by several functional similarities (Berridge and Dupont, 1994). An important feature of both RyRs and IP_3Rs is that they can both be activated by increases in intracellular free calcium concentrations ($[Ca^{2+}]_i$) (Galione *et al.*, 1993a). This phenomenon, called Ca^{2+}-induced Ca^{2+} release (CICR), has been proposed as a mechanism for cardiac muscle excitation–contraction coupling (Fleischer and Inui, 1989) and also for generating the spatiotemporal complexities of Ca^{2+} signals widely observed (Berridge and Dupont, 1994), where many stimuli often evoke Ca^{2+} spiking and where local Ca^{2+} transients may propagate as regenerative intra- and intercellular Ca^{2+} waves.

With an increasing number of cells shown to express RyRs, much attention is now being focused on the functions of RyRs in many different systems. In this chapter, we will review the current understanding of the role of the novel NAD^+ metabolite, cyclic ADP-ribose (cADPR), as a possible physiological regulator of IP_3-insensitive Ca^{2+} release via RyR-like channels in cells including smooth muscle.

II. IDENTIFICATION OF CYCLIC ADP-RIBOSE AS A Ca^{2+} MOBILIZING AGENT

A. The Discovery of cADPR as a Ca^{2+} Mobilizing Molecule

Much of what we know about cADPR as a Ca^{2+} mobilizing agent has come from studies of sea urchin eggs. They have been extensively used as a model system to investigate the molecular mechanisms of fertilization (Whitaker and Swann, 1993) and also for understanding the basic mechanisms of intracellular Ca^{2+} signaling. The reasons for this are that they are large and robust cells that display a large and stereotypic calcium wave upon insemination *in vitro* (Fig. 1). They are amenable for microinjection and available in large numbers, and microsomal fractions with multiple Ca^{2+} mobilizing mechanisms still intact can be readily prepared.

Although sea urchin eggs were among the first intact cells in which IP_3 was shown to mobilize calcium from intracellular stores (Whitaker and Irvine, 1984), subsequent studies in egg homogenates have shown that additional Ca^{2+} mobilizing mechanisms were present (Clapper *et al.*, 1987). Addition of NAD^+, whose levels dramatically change during fertilization (Epel, 1964), to egg homogenates supplemented with an ATP regenerating system caused a large release of Ca^{2+} from nonmitochondrial stores. In contrast to the rapid calcium release induced by IP_3 from this preparation, there was a significant latency before calcium

release was observed in response to NAD^+ (Clapper *et al.*, 1987). Furthermore, if NAD^+ was preincubated with egg extracts, then the mixture could cause Ca^{2+} release without an apparent delay. These data suggested that NAD^+ was not a Ca^{2+} mobilizing agent itself, but rather was converted to the active species perhaps by an enzyme in the egg extract. Later analyses of the "enzyme-activated"-NAD^+, as it was called, led to the identification of a novel cyclized ADP-ribose, cADPR (Lee *et al.*, 1989).

B. Structural Determination of cADPR

Since NAD^+ seemed to be converted to a very potent Ca^{2+} mobilizing agent, it was important for Lee and his colleagues to identify the active metabolite. The metabolite was purified by high-performance liquid chromatography (HPLC), and then subjected to structural analyses. A number of different approaches were used: nuclear magnetic resonance (NMR) and mass spectroscopy suggested that the metabolite was

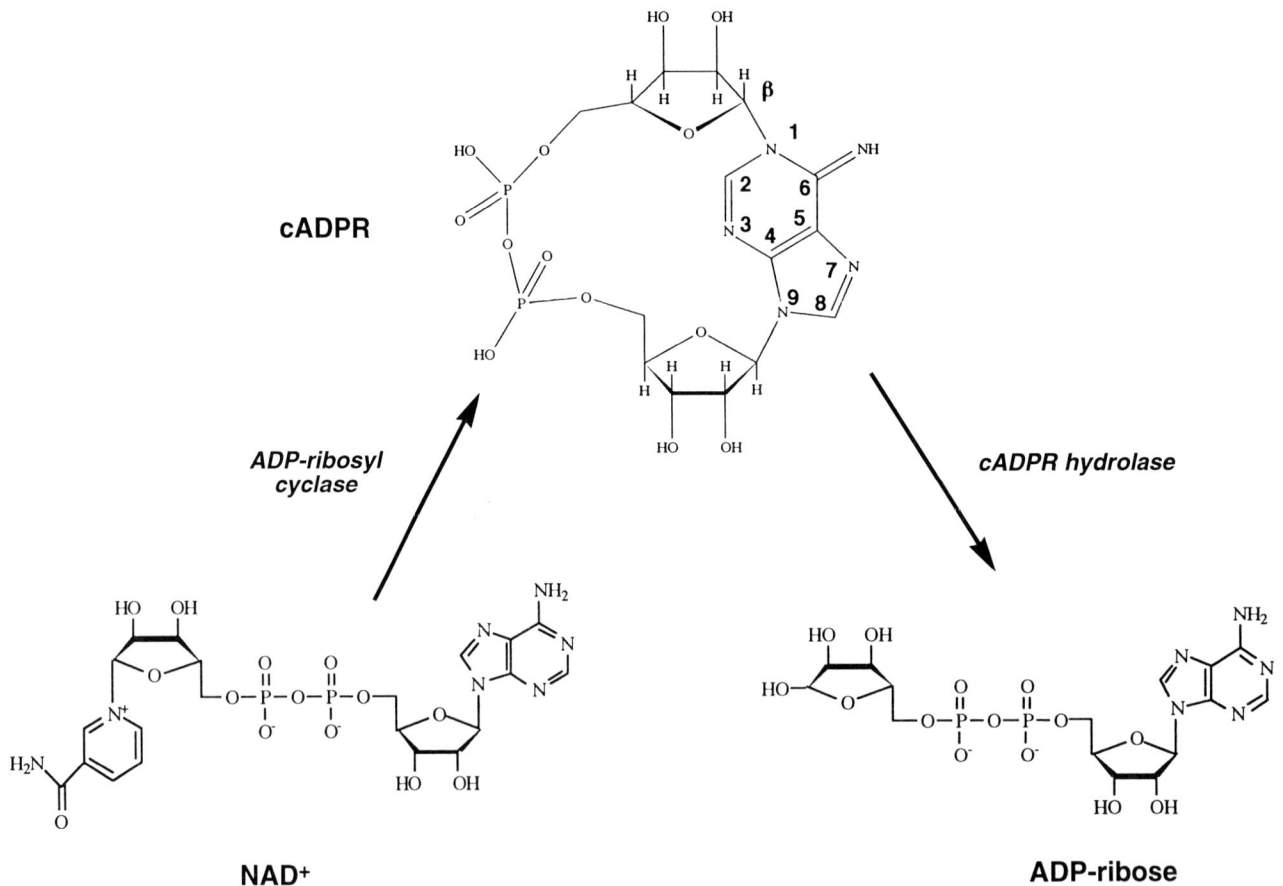

FIGURE 2   Metabolism of cyclic ADP-ribose. cADPR is synthesized from β-NAD^+ by ADP-ribosyl cyclases and hydrolyzed to its inactive metabolite, ADP-ribose, by cADPR hydrolases. In mammalian cells, cyclase and hydrolase activities have been shown to be expressed on one bifunctional protein.

a cyclized ADP-ribose with an N-glycosyl bond between the anomeric carbon of the terminal ribose and the N^6-amino group of the adenine (Lee *et al.*, 1989). However, comparison of results from ultraviolet absorption spectroscopy of a number of adenine nucleotides with cADPR suggested that a N^1 rather than the N^6 linkage was likely (Kim *et al.*, 1993a). This structure (Fig. 2) has been confirmed by both X-ray crystallography (Lee *et al.*, 1994b) and a stereoselective chemical synthesis of the compound from β-NAD$^+$ (Yamada *et al.*, 1994).

III. CYCLIC ADP-RIBOSE METABOLISM

A. Measurements of cADPR in Tissues

Much of the current interest in cADPR as a Ca^{2+} mobilizing agent has come from reports that it is an endogenous molecule, whose resting levels have been measured in a number of animal tissues (Walseth *et al.*, 1991). However, the number of reports are few due to the heavy reliance for its quantitation on a bioassay involving Ca^{2+} release from sea urchin egg and other microsomes. The procedure involves several steps. The first is to freeze-clamp tissues to prevent metabolism of cADPR by cADPR hydrolases (see Section III.C). Tissues are then ground and subjected to acid extraction and the metabolites are purified by HPLC. The cADPR fraction, identified by its coelution with radiolabeled tracer, is then collected, concentrated, and added to sea urchin egg homogenates or other cADPR-sensitive microsomes. A fluorimetric assay for Ca^{2+} release then forms the basis of the mass assay. The authenticity of cADPR in samples can then be validated by inhibition of Ca^{2+} release with the specific cADPR antagonist, 8-amino-cADPR (see Section IV.A), or by desensitization of the microsomes by prior release of microsomal Ca^{2+} by a maximal dose of cADPR (Dargie *et al.*, 1990). This assay has been used not only to quantify cADPR (Walseth *et al.*, 1991) in tissues but also to identify ADP-ribosyl cyclase activities (Howard *et al.*, 1993; Summerhill *et al.*, 1993; Lee *et al.*, 1993b).

In rat tissues, this technique showed that levels in heart were ~1 pmol/mg protein, whereas those in liver were ~3 pmol/mg protein (Walseth *et al.*, 1991). Using rat cerebellar microsomes as a cADPR assay, Takasawa *et al.* (1993a) found that the Ca^{2+} mobilizing activities of extracts of rat islets of Langerhans, whose active factor resembled cADPR, increase fourfold upon glucose stimulation. This raises the interesting possibility that cADPR may play an important role in stimulus-secretion coupling in pancreatic β cells (see Sections III.C and V.C).

The developments of further sensitive cADPR assays would greatly increase our knowledge concerning possible mechanisms of regulation of cADPR metabolism in cells. So far, there has been little success in raising an antibody to cADPR, and since little is known about cADPR binding proteins (Walseth *et al.*, 1993), measurements based on a radioreceptor binding assay must wait.

B. ADP-Ribosyl Cyclases

Enzymes involved in the metabolism of cADPR have been reported in many tissues. A scheme for the synthesis and a degradation of cADPR is depicted in Fig. 2.

1. Characterization of ADP-Ribosyl Cyclases

The demonstration that a component of sea urchin egg extracts could convert NAD$^+$ into an active Ca^{2+} mobilizing metabolite (Clapper *et al.*, 1987) suggested the presence of an activity responsible for catalyzing the conversion. This activity was shown to be widespread in a large number of mammalian cell extracts (Rusinko and Lee, 1989). A metabolite from liver extracts was compared with that from sea urchin egg extracts and found to be identical both in functional Ca^{2+} mobilizing activity and in structural terms from NMR and mass spectroscopic studies (Rusinko and Lee, 1989). The activity was sensitive to heat denaturation and protease treatment, suggesting an enzyme. A soluble form of ADP-ribosyl cyclase, as the NAD$^+$ cyclizing enzyme was called, was first purified and sequenced from the ovotestis of the marine mollusc *Aplysia californica* (Glick *et al.*, 1991; Hellmich and Strumwasser, 1991; Lee and Aarhus, 1991). It is a 29-kDa enzyme that has been immunolocalized to granules in the ovotestis. Its function is unknown. In most mammalian tissues as well as in the sea urchin egg, ADP-ribosyl cyclase activities are membrane bound (Lee, 1991; Lee and Aarhus, 1993). Another unusual feature of the *Aplysia* isoform is that it is not associated with cADPR hydrolase activity, and thus is not a bifunctional protein (see Section III.D). This has led to incubation of NAD$^+$ with the *Aplysia* enzyme becoming the most widely used route for commercial cADPR synthesis. Several membrane-associated cyclases have been characterized. CD38, a 40-kDa transmembrane plasma membrane protein from lymphocytes (Malavasi *et al.*, 1994), and various homologous proteins such as CD38H (Koguma *et al.*, 1994) from rat islets of Langerhans, a 39-kDa protein from canine spleen (Kim *et al.*, 1993b), and a surface molecule on erythrocytes (Lee *et al.*, 1993b) have been shown to be ADP-ribosyl cyclases (see Section III.D). These proteins are bifunctional, in that they also express cADPR hydrolase activity.

2. Measurement of ADP-Ribosyl Cyclase Activities

The synthesis and degradation of cADPR are catalyzed by ADP-ribosyl cyclase and cADPR hydrolase, respectively. Problems in detecting the formation of cADPR have led to frequent identification of CD38 as a NAD+ glycohydrolase (Kim *et al.*, 1993b). ADP-ribosyl cyclase and CD38 can also cyclize nicotinamide guanine dinucleotide (NGD+), producing a novel nucleotide whose analyses by high-performance liquid chromatography and mass spectroscopy indicate that it is cyclic GDP-ribose (cGDPR) (Graeff *et al.*, 1994b). cGDPR has a structure similar to that of cADPR except with a guanine replacing an adenine. cGDPR is also a more stable compound, showing greater resistance to hydrolysis by cADPR hydrolases (Graeff *et al.*, 1994a). Spectroscopic analyses have indicated that cGDPR is fluorescent and has an absorption spectrum different from that of both NGD+ and GDPR, providing a method for monitoring its enzymatic formation. The use of NGD+ as substrate for assaying the cyclization reaction may make it a useful way for distinguishing CD38-like enzymes from degradative NADases (Graeff *et al.*, 1994a) (see Section III.D).

C. cADPR Hydrolases

The discovery of the Ca^{2+} mobilizing actions of cADPR, its presence in cells, and enzymatic activities to synthesize it are all consistent with a possible role as a Ca^{2+} mobilizing messenger. If this is the case, then the message also needs to be rapidly turned off or removed after the signaling process is complete. A membrane-bound enzyme that can do just this has been demonstrated and found to be as equally widespread as ADP-ribosyl cyclase (Lee and Aarhus, 1993). It has been called cADPR hydrolase, as it cleaves the N-glycosidic bond in cADPR to ADP-ribose. The latter is inactive as a Ca^{2+} mobilizing agent (Dargie *et al.*, 1990). cADPR hydrolase activities are often found associated with ADP-ribosyl cyclases, and both activities have been found on the same polypeptide (Kim *et al.*, 1993b) (see Section III.B.1), although an important exception is the *Aplysia* ADP-ribosyl cyclase, which makes it useful for synthesizing large quantities of cADPR (Lee and Aarhus, 1991). cADPR hydrolase may be inhibited by millimolar ATP levels (Takasawa *et al.*, 1993b). It has been proposed that this may be one way in which glucose might increase cADPR levels in pancreatic β cells (Takasawa *et al.*, 1993a) (see Section V.C), via an increase in intracellular ATP, which would serve to inhibit the breakdown of cADPR (Takasawa *et al.*, 1993b).

D. CD38 and Homologous Proteins

CD38 is a multifunctional protein expressed primarily on lymphoid cells but also in other tissues (Malavasi *et al.*, 1994). It has been shown to be a bifunctional enzyme expressing both ADP-ribosyl cyclase and cADPR hydrolase activities (Howard *et al.*, 1993; Summerhill *et al.*, 1993), and has sequence homology with the *Aplysia* ADP-ribosyl cyclase enzyme (States *et al.*, 1992). Activation of the CD38-mediated signaling processes with agonistic antibodies, which bind to CD38, can regulate lymphocyte proliferation and other cellular responses. Interestingly, in the human T-cell Jurkat cell line, it leads to Ca^{2+} mobilization (F. Malavasi, personal communication). Since the presence of RyR3-like protein has been reported in Jurkat cells that may play a role in the proliferation of these cells (Hakamata *et al.*, 1994), it is possible that CD38-catalyzed production of cADPR leads to Ca^{2+} mobilization via RyRs. This hypothesis is further given credence by the recent demonstration that cADPR can mobilize Ca^{2+} from permeabilized Jurkat cells (A. Guse *et al.*, 1995). However, a tantalizing problem with this hypothesis is that the enzymatic domains of CD38 are extracellular, and there is no evidence at the moment for either the enzymatic domains of CD38 becoming exposed to the cytoplasm by internalization or for transport of cADPR into the cell; either mechanism would allow cADPR to reach its presumed intracellular site of action (Malavasi *et al.*, 1994).

The introduction of site-directed mutations to CD38 cysteine residues 119 and 201 results in a CD38 protein that exhibited only ADP-ribosyl cyclase activity (Tohgo *et al.*, 1994). Furthermore, *Aplysia* ADP-ribosyl cyclase into which mutations K95C and E176C were introduced (these correspond to residues 119 and 201 of human CD38) exhibited not only ADP-ribosyl cyclase activity but also cADPR hydrolase. These data indicate that cysteine residues 119 and 201 in CD38 have important roles in the hydrolysis of cADPR (Tohgo *et al.*, 1994).

ADP-ribosyl cyclase and cADPR hydrolase activities have been shown to be widespread in mammalian and other tissues (see Sections III.B and III.C). However, the demonstration that these enzymes are bifunctional in that they are resident on the same protein has led to the suggestion that many common ecto-NAD+ glycohydrolases, which covert NAD+ to ADP-ribose, may do so via cADPR as an intermediate (Kim *et al.*, 1993b). This might suggest that many of these reported ADP-ribosyl cyclases in tissues (Rusinko and Lee, 1989) might play little role in intracellular cADPR signaling. However, the development of the NGD$^+$/cGDPR technique (see Section III.B.2) to measure ADP-ribosyl cyclase activity per se has shown that in many cases NAD$^+$ glycohydrolases do not exhibit any ADP-ribosyl cyclase activity (Graeff et al., 1994b).

E. Regulation of cADPR Synthesis

For cADPR to be considered as a classic intracellular messenger, intracellular levels of cADPR must be regulated by extracellular signals. There are few data on how cADPR levels might be regulated in cells, and sensitive ways of measuring cADPR in cells are widely employed (see Section III.A). One exception is in sea urchin eggs, where a role for cGMP has been proposed (Galione et al., 1993b). cGMP itself has been established as a messenger for a number of different signaling molecules (Goy, 1991). It is synthesized by intracellular guanylyl cyclases of which there are two distinct isoforms utilizing GTP as a substrate. One guanylate cyclase isoform is particulate and is the intracellular domain of a family of single transmembrane-spanning receptors (Maack, 1992). The extracellular domains represent receptors for a variety of peptides, including atrial natriuretic peptide and sea urchin egg chemoattractant peptides. The second form is cytosolic and can be activated by the gaseous transmitter molecules nitric oxide (NO) and carbon monoxide (Schmidt et al., 1993).

Microinjection of cGMP into sea urchin eggs results in a large mobilization of Ca^{2+} from internal stores, which is not blocked by the IP$_3$R antagonist heparin (Whalley et al., 1992), but is inhibited by the RyR antagonist ruthenium red (Galione et al., 1993b). The pharmacology of cGMP-induced Ca^{2+} release has been analyzed in sea urchin egg homogenates. Release of Ca^{2+} in response to cGMP requires the presence of micromolar quantities of NAD$^+$, the precursor of cADPR, and is blocked by the cADPR antagonist 8-amino-cADPR (Willmott et al., 1995). The activation of the Ca^{2+} release mechanism by cGMP appears to be indirect and has a latency of between 10 and 300 s inversely dependent on the cGMP concentration. It is blocked by inhibitors of cGMP-dependent protein kinases (cGK) (Butt et al., 1990), suggesting that a phosphorylation step is involved. Furthermore, analysis of NAD$^+$ levels and its metabolites in egg homogenates treated with cGMP suggest that cGMP stimulates the conversion of β-NAD$^+$ into cADPR and ADP-ribose (Galione et al., 1993b). However, cGMP did not affect the rate of hydrolysis of ^{3}H-labeled cADPR to ADPR, suggesting

that it was not inhibiting cADPR hydrolases (Galione et al., 1993b).

NO, which elevates cGMP levels in the sea urchin egg, probably by activating soluble guanylate cyclases, also leads to Ca^{2+} mobilization in both intact eggs (Fig. 3) and egg homogenates (Willmott et al., 1995), with a pharmacology suggesting that it activates a pathway involving both cGMP and cADPR (Lee, 1994). It will be interesting to see if such a pathway operates in mammalian cells too.

IV. MECHANISM OF CYCLIC ADP-RIBOSE-INDUCED Ca^{2+} RELEASE

cADPR has now been shown to have Ca^{2+} mobilizing activity in a wide range of cells in addition to the sea urchin egg. Although the mechanism of its action has been best characterized in the sea urchin egg, in other cell preparations where the pharmacology and characteristics of cADPR-mediated Ca^{2+} release have been investigated, there is a close similarity between its mode of action and that of Ca^{2+} release mediated via RyRs (Galione and White, 1994). This has led to the hypothesis that cADPR is an endogenous regulator of RyRs, perhaps by an analogous mechanism by which IP$_3$R activates Ca^{2+} release via IP$_3$Rs (Galione, 1992).

A. Pharmacology of cADPR-Induced Ca^{2+} Release

The finding that a ryanodine-sensitive CICR mechanism was expressed in sea urchin eggs in addition to that activated by IP$_3$ allowed the possibility that cADPR exerted its Ca^{2+} mobilizing effects through this former mechanism to be tested. Cross-desensitization studies suggested that cADPR and the pharmacological RyR activators, caffeine and ryanodine, all released Ca^{2+} from the same Ca^{2+} pool, whereas experiments with the classic RyR blockers indicated that cADPR acted via a RyR-like Ca^{2+} release channel (Galione et al., 1991). These results led to the hypothesis that cADPR acts via a RyR-like mechanism, distinct from the IP$_3$R. Furthermore, agents that can activate RyRs, including divalent cations, caffeine (Lee, 1993), and ryanodine (Buck et al., 1994), can potentiate Ca^{2+} release by cADPR in sea urchin egg homogenates.

Important tools in studying cADPR-mediated Ca^{2+} mobilization have been the use of 8-substituted analogues of cADPR (Walseth and Lee, 1993). 8-Amino-, 8-bromo-, and 8-azido-cADPR can be synthesized by incubating their respective 8-substituted NAD$^+$ analogues with the *Aplysia* ADP-ribosyl cyclase. The cADPR analogues are selective competitive antagonists of cADPR- but not IP₃-induced Ca²⁺ release, and compete with ³²P-labeled cADPR binding to sea urchin egg microsomes. In particular, the use of the most potent of these derivatives, 8-amino-cADPR, in microinjection studies, has suggested the importance of cADPR- and RyR-mediated Ca²⁺ in the fertilization Ca²⁺ wave in sea urchin eggs (Lee *et al.*, 1993a), and for cADPR as a possible regulator of the RyR in cardiac excitation–contraction coupling (Rakovic *et al.*, 1995). However, these 8-substituted cADPR analogues do not block Ca²⁺ release by caffeine in sea urchin eggs, suggesting that if cADPR and caffeine are both binding to RyRs, they do so at distinct sites (Walseth and Lee, 1993).

B. The Identity of the cADPR Receptor

1. Specific Binding of cADPR to Sea Urchin Egg Microsomes

Saturable and specific ³²P-labeled cADPR binding to sea urchin egg microsomes has been reported (Lee, 1991). This binding was not affected by high micromolar concentrations of either its precursor, NAD⁺, or its breakdown product, ADP-ribose, although it could be totally inhibited by 0.3 μM unlabeled cADPR. Scatchard analysis indicated a B_{max} of 25 fmol/mg protein and a K_d of around 17 nM. However, the presence of dialyzable endogenous cADPR in the microsomal preparations may obscure the K_d determination, since in dialyzed membranes considerably lower K_d values have been obtained (Lee *et al.*, 1994c).

2. RyRs as cADPR Receptors

Direct evidence that cADPR may function as an endogenous modulator has come from studying the effects of cADPR on the gating properties of cardiac RyRs incorporated into lipid bilayers (Meszaros *et al.*, 1993). cADPR (1 μM) increased the open probability of RyR2 but not skeletal muscle RyR1, although this remains controversial (see Section V.A.).

3. cADPR Binding Proteins

Another strategy for identifying cADPR receptors has come from photoaffinity labeling of cADPR binding sites. One of the 8-substituted derivatives of cADPR mentioned in Section IV.A, 8-azido-cADPR, can be used as a photoaffinity probe. [³²P]8-Azido-cyclic ADPR specifically labels two proteins from sea urchin egg preparations of approximate molecular masses of 100 and 140 kDa (Walseth *et al.*, 1993), and a 45-kDa protein in cardiac muscle and a 75-kDa protein from mammalian brain (T. F. Walseth, personal com-

munication). The relationship of these proteins to RyRs has yet to be determined, but interesting labeling of the 100-kDa protein is blocked by caffeine (Walseth *et al.*, 1993). These proteins may be proteolytical fragments of the much larger reported RyRs, smaller isoforms such as the small 75-kDa RyR protein from brain (Takeshima *et al.*, 1993), or perhaps a distinct family of RyR modulatory proteins that have recently been reported, one class of which are the immunophilins (Brillantes *et. al.*, 1994).

4. cADPR-Regulated Ca²⁺ Release Channel of Plant Microsomes

cADPR has been shown to release Ca²⁺ from plant microsomal vesicles via a ryanodine- and ruthenium red-sensitive Ca²⁺ release channel (Allen *et al.*, 1994). This discovery may underscore the ubiquity of the cADPR signaling pathway. Fusion of these plant vesicles with lipid bilayers has demonstrated a large conductance cADPR-regulated Ca²⁺ release channel. Whether this channel represents a plant homologue of RyRs remains to be determined.

C. Regulation of cADPR-Induced Ca²⁺ Release by Calmodulin

Calmodulin confers cADPR sensitivity upon sea urchin egg microsomes, since after the microsomes have been purified from egg homogenates on a Percoll gradient, they are rendered insensitive to cADPR with regard to Ca²⁺ release (Lee *et al.*, 1994a). cADPR sensitivity can be restored by adding back cytosolic extracts whose active constituent was identified as calmodulin. This was substantiated by the finding that the addition of calmodulin alone was sufficient to restore cADPR sensitivity or that the calmodulin antagonist, W7, inhibited cADPR-induced Ca²⁺ mobilization. It is interesting to note that calmodulin binding sites on mammalian RyRs have been characterized (Wagenknecht *et al.*, 1994). However, calmodulin has been shown to directly inhibit CICR activity of mammalian RyR at the single-channel level (Smith *et al.*, 1989). It remains to be seen if lack of calmodulin or other accessory factors can explain the apparent difficulties of some workers in seeing effects of cADPR on Ca²⁺ release from certain broken-cell mammalian systems (Berridge, 1993b).

V. cADPR-MEDIATED Ca²⁺ RELEASE IN MAMMALIAN CELLS

Although cADPR appears to mobilize intracellular calcium in sea urchin eggs by activating RyR-like cal-

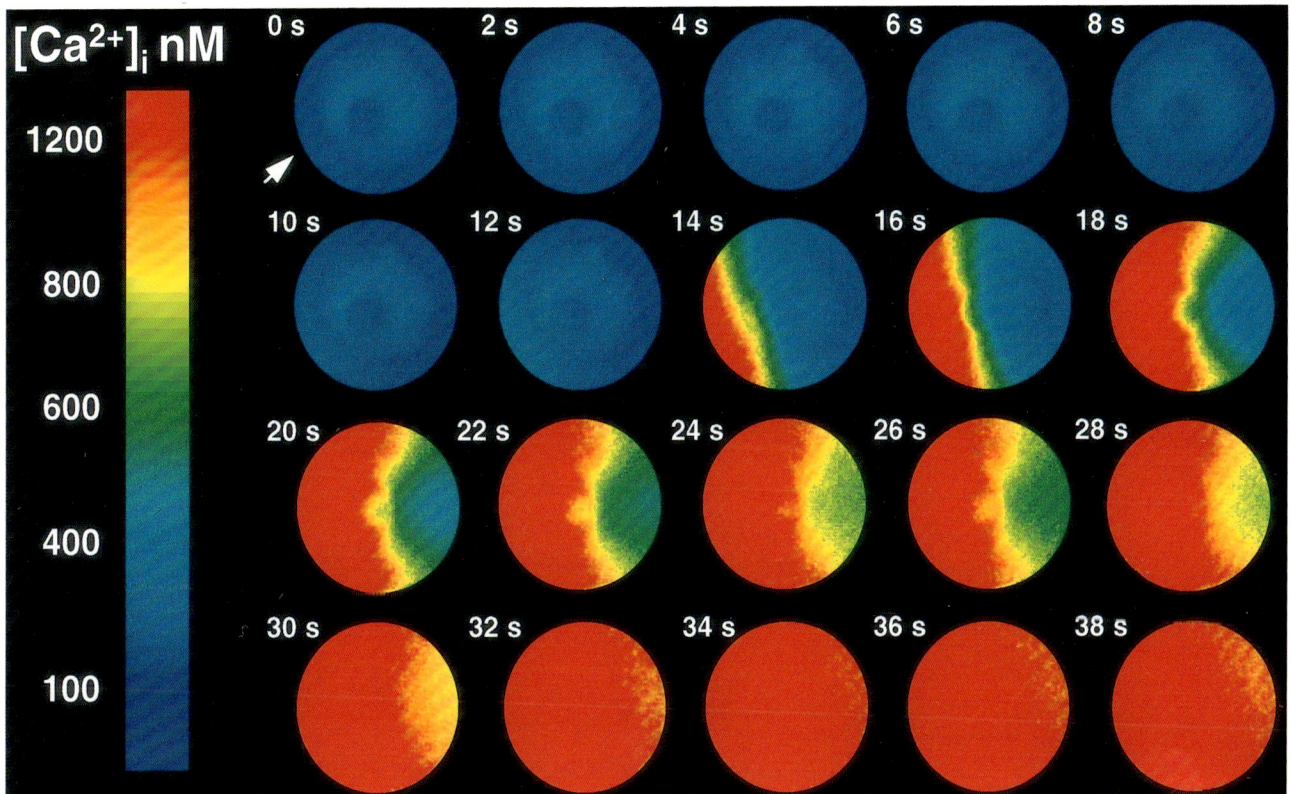

FIGURE 1 The fertilization Ca^{2+} wave in the sea urchin egg. A sperm-induced Ca^{2+} wave at fertilization in a sea urchin egg can be visualized by digital ratio-imaging techniques. Images of cytoplasmic Ca^{2+} are measured with the calcium-sensitive dye fura-2. High calcium concentrations are shown in red and resting calcium concentrations are shown in blue. The images shown were recorded 2 s apart. The egg is 100 μm in diameter. The fertilization wave started at the site of sperm entry, indicated by the arrow. The wave crosses the egg at an approximate velocity of 10 μm/s.

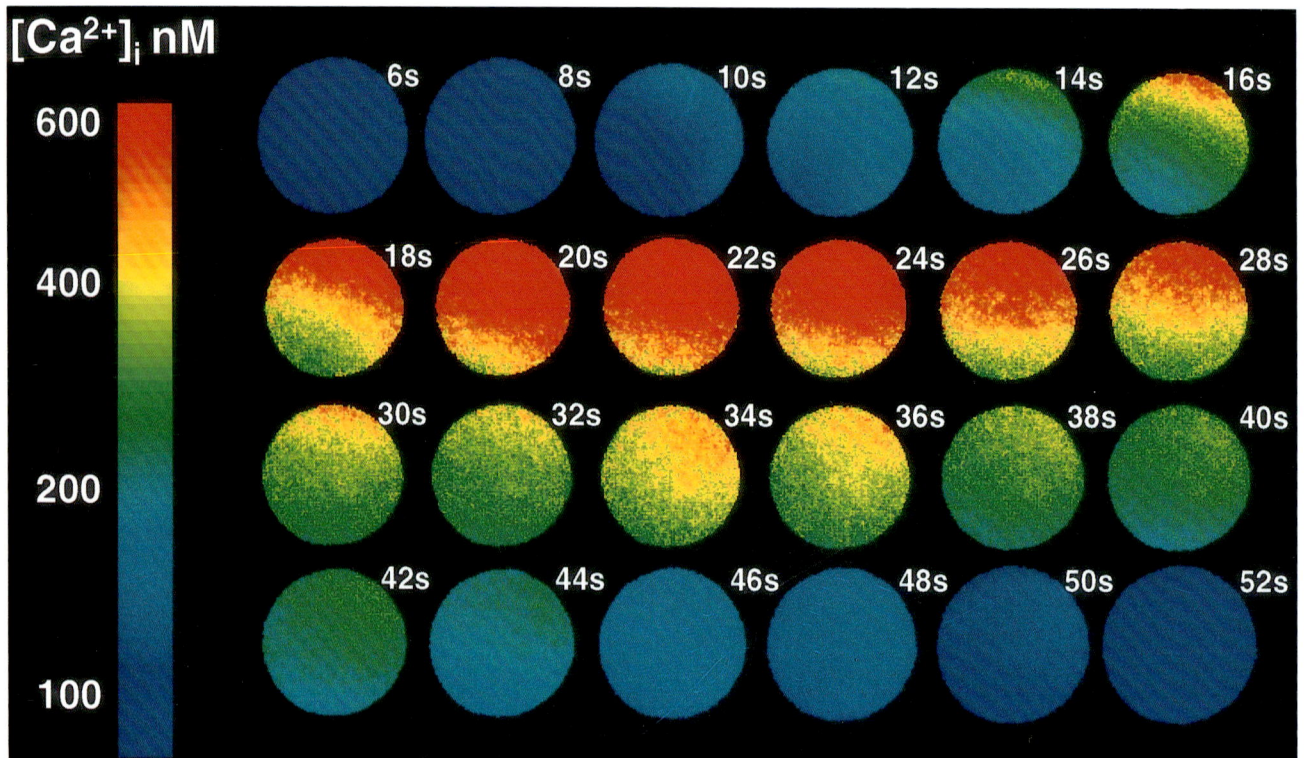

FIGURE 3 Nitric oxide-induced Ca^{2+} wave in the sea urchin egg. Images of changes in intracellular Ca^{2+} in sea urchin eggs in response to addition of seawater containing NO. Digital ratio images measured with the Ca^{2+}-sensitive dye fura-2 are displayed at different times after exposure to NO (100 μM) in the bathing sea water. The images shown were measured 2 s apart. There was a latency of at least 14 s before a calcium rise was detected, and the calcium often increased in one region of the egg before spreading across the cell. The increase in calcium was transient, with a return to baseline often occurring within 10 s of its initiation.

cium release channel, its identity is unknown (see Section IV.B). A preliminary molecular characterization of RyRs in sea urchin eggs has revealed the presence of a protein that reacts with antibodies raised against mammalian skeletal muscle RyRs with an approximate molecular mass of 380 kDa (McPherson *et al.*, 1992). This is considerably smaller than the individual subunit size of the three main RyRs found in mammalian cells. Its localization to the egg cortex contradicts evidence from confocal fluorescent microscopy, which indicates that ryanodine receptors can mediate the propagation of calcium waves deep into the egg cytoplasm (Galione *et al.*, 1993a). Whether this protein is modulated by cADPR is unknown. However, cADPR has also been shown to activate calcium release in mammalian cells, where the RyRs are better characterized (see Section IV.B.2).

A. Cardiac Muscle

Since muscle cells contain the highest densities of RyRs and cADPR levels and ADP-ribosyl cyclase activities have been measured in mammalian heart (Walseth *et al.*, 1991), cardiac muscle was an obvious choice in which to investigate the effects of cADPR in mammalian tissues.

cADPR can stimulate calcium efflux from cardiac but not skeletal muscle sarcoplasmic reticulum (SR) vesicles, and this effect is blocked by ryanodine (Meszaros *et al.*, 1993). cADPR increases the open probability of single RyR2s after cardiac SR vesicles are fused into lipid bilayers, but has little effect on RyR1 incorporated by the fusion of skeletal SR vesicles. cADPR also increases the magnitude of specific [^{3}H]ryanodine binding to cardiac SR vesicles, suggesting interaction between cADPR and ryanodine binding. Such data, though suggesting that cADPR is able to bring about conformation changes in RyR2, do not preclude the possibility that cADPR effects might be indirect, for example, by interacting with an accessory protein that may be associated with the channel. Although cADPR appears to be capable of regulating calcium release from cardiac SR in intact cardiac myocytes (Rakovic and Terrar, 1994), another study of RyR2 properties in lipid bilayers suggests that the effects of cADPR are nonspecific and unphysiological since they are not apparent in the presence of physiological levels of ATP (Sitsapesan *et al.*, 1994). A further study suggests that cADPR has no effect at all on muscle RyRs (Fruen *et al.*, 1994). If normal channel properties are preserved in lipid bilayers, then such apparently contradictory data might suggest that other factors such as calmodulin (see Section IV.C) are involved in conferring cADPR

sensitivity upon SR release channels in intact cells (Lee *et al.*, 1994a).

B. Neurons

Nervous tissue is a rich source of RyRs, although they are present in much lower densities than in muscle (Meissner, 1994). As in cardiac muscle, cADPR is present in nervous tissue (Walseth *et al.*, 1991) as are ADP-ribosyl cyclases and cADPR hydrolases (Lee and Aarhus, 1993; Rusinko and Lee, 1989). There are now several reports documenting cADPR effects on calcium movements in neuronally derived preparations (Meszaros *et al.*, 1993; Takasawa *et al.*, 1993a; White *et al.*, 1993). First, cADPR can stimulate calcium release from brain or cerebellar microsomes. This effect is restricted to cADPR and not shared by either its precursor NAD$^+$ or its metabolite ADP-ribose. In dorsal root ganglion cells, patch pipette perfusion of cells with cADPR concentrations as low as 10 nM induced oscillations in calcium-dependent membrane currents (Currie *et al.*, 1992). The effects of cADPR were blocked by depleting the caffeine-sensitive store, and were enhanced by increasing the intracellular free calcium concentration, consistent with cADPR modulating a CICR mechanism on a caffeine-sensitive calcium pool. A study in bull frog sympathetic ganglion cells, where depolarization-induced calcium influx triggers CICR and a pronounced train of calcium spikes if the CICR mechanism has been sensitized to calcium by caffeine, has indicated that cADPR can also sensitize CICR (Hua *et al.*, 1994). Although calcium influx alone does not trigger CICR, in the presence of cADPR it can now do so. Furthermore, in the presence of low concentrations of caffeine, which allows influx to elicit only a modest Ca^{2+} release, in the presence of cADPR calcium spiking was observed.

C. Pancreatic Cells

Effects of cADPR on Ca^{2+} release mechanisms have been reported in both endocrine and exocrine pancreatic cells.

cADPR has been reported to stimulate Ca^{2+} release from microsomes derived from pancreatic β cells, and to stimulate insulin release from permeabilized islets of Langerhans (Takasawa *et al.*, 1993a). Such release is not blocked by heparin but is inhibited by ryanodine. Furthermore, increasing extracellular glucose has been reported to raise cADPR levels in rat islets of Langerhans (see Section III.C). However, other reports have failed to document such an effect (Islam *et al.*, 1993; Rutter *et al.*, 1994).

Pancreatic acinar cells respond to a number of secretagogues with increases in $[Ca^{2+}]_i$, resulting in enhanced fluid secretion and exocytosis of granules containing digestive enzymes (Petersen et al., 1994). Confocal imaging of $[Ca^{2+}]_i$ has shown that both acetylcholine and cholecystokinin (CCK) produce calcium spikes in the apical region that can propagate in an apical to basal direction. Both IP_3 and ryanodine receptors have been implicated in controlling these complex calcium signals since intracellular application of IP_3 via patch pipette or extracellular application of caffeine can mimic some features of the agonist-induced calcium signals (Thorn et al., 1994). A report has shown that intracellular perfusion of cells with cADPR induces short-lived Ca^{2+} spikes in the apical region of cells (Thorn et al., 1994). The frequency of Ca^{2+} spikes increases with increasing pipette concentration of cADPR. Ryanodine abolishes cADPR-induced spikes and reduces agonist-induced responses, but does not reduce the effects of IP_3, if anything, ryanodine enhances the effects of IP_3. A complication in this study is that heparin abolishes not only the effects of IP_3 but, in contrast to its selective inhibition of cADPR-induced Ca^{2+} release in sea urchin eggs (Dargie et al., 1990), also abolishes the effects of caffeine and cADPR in acinar cells. This has been interpreted as evidence that the spatiotemporal complexity in Ca^{2+} signals in exocrine pancreatic cells results from the interplay between IP_3 and RyRs.

D. Possible Roles of cADPR in the Regulation of Smooth Muscle Contraction

1. Role of RyRs in Smooth Muscle Signal Transduction

The excitation–contraction coupling processes of smooth muscle cells employ both extra- and intracellular sources of Ca^{2+}. As in skeletal and cardiac muscle cells, the intracellular source of Ca^{2+} is the sarcoplasmic reticulum. Although RyRs are the major pathway for Ca^{2+} release in striated muscle (Fleischer and Inui, 1989), both RyR and IP_3Rs are present in smooth muscle (Wibo and Godfraind, 1994; Zhang et al., 1993), but there is a much clearer role for IP_3Rs in pharmaco-mechanical release (Somlyo and Somlyo, 1994).

All three mammalian RyR isoforms have been found to be expressed in smooth muscle. In aortic smooth muscle the RyRs are functionally similar to RyR1 and RyR2 (Hermann-Frank et al., 1991). Reconstitution into lipid bilayers of a CICR channel from vascular smooth muscle indicated a ryanodine-sensitive channel with a similar pharmacology to those from skeletal and cardiac muscles (Hermann-Frank et al., 1991). Analysis of mRNA in this tissue

indicated the presence of mRNA for RyR2 (Moschella and Marks, 1993). In uterine smooth muscle, a RyR3-related receptor has been indicated (Hakamata et al., 1992; Lynn et al., 1993). In situ hybridization indicates that RyR3 is expressed along with RyR1/RyR2 in smooth muscle from stomach, gut, and aorta (Giannini et al., 1992). In amphibian gastric muscle, a RyR-like Ca^{2+} release channel has been characterized that differs from the RyR1 and RyR2 isoforms (Xu et al., 1994). Although use of ryanodine to functionally remove ryanodine-sensitive stores indicates that a substantial portion of agonist-induced releasable stores express RyRs (Iino et al., 1988), experiments with the IP_3R antagonist heparin indicate that in many smooth muscle cell types, IP_3 is probably the major mechanism for Ca^{2+} mobilization (Somlyo and Somlyo, 1994). However, there are notable exceptions. Some smooth muscle types appear to express only one class of Ca^{2+} release channel. Whereas rabbit circular intestinal muscle seems mainly to contain IP_3Rs (Murthy et al., 1991), rabbit longitudinal muscle SR predominantly contains RyRs (Kuemmerle et al., 1994). This strongly suggests that pharmacomechanical coupling proceeds via a RyR mechanism. CICR mediated by either IP_3Rs or RyRs may give rise to Ca^{2+} waves seen in muscle cells (Mahoney et al., 1993; Wier and Blatter, 1991), which resemble those seen in the sea urchin egg (Fig. 1).

2. cADPR in Smooth Muscle

At present there are no data available reporting endogenous levels of cADPR in any smooth muscle cells. However, a preliminary report suggests that cADPR might regulate the RyR-linked Ca^{2+} release mechanism in longitudinal muscle (Kuemmerle and Makhlouf, 1994). Specific binding of ^3H-labeled cADPR was reported to longitudinal muscle SR, with an IC_{50} for displacement by unlabeled cADPR of 2 nM. This binding was displaced by ryanodine but not by IP_3. Furthermore, this binding was strongly Ca^{2+} dependent, as would be expected of a regulator of CICR. These binding studies were substantiated by functional studies (Kuemmerle and Makhlouf, 1994). cADPR mobilized Ca^{2+} (as measured by fura-2 fluorescence) and caused contractions in permeabilized longitudinal muscle with EC_{50}'s of between 2 and 4 nM. The pharmacology of cADPR- and ryanodine-induced $^{45}Ca^{2+}$ fluxes appeared to be similar: they were blocked by RyR antagonists dantrolene and ruthenium red, augmented by caffeine, but were not affected by agents acting at IP_3Rs. Furthermore, cADPR at nanomolar concentrations sensitized the CICR mechanism in this preparation. However, in circular muscle cADPR had no effect, which is consistent with the evidence that they predominantly contain IP$_3$Rs and not RyRs on their SR. Possible involvement of cADPR in regulating smooth muscle contraction is shown in Fig. 4.

However, cADPR may not be a universal regulator of RyRs in smooth muscle cells. In permeabilized strips of guinea pig vas deferens smooth muscle, cADPR at concentrations as high as 100 μM, in contrast to the effects of IP$_3$ and caffeine, failed to contract these preparations (Nixon *et al.*, 1994).

If cADPR is a mediator of pharmacomechanical coupling in some smooth muscle cells, then agonists should modulate its intracellular levels. This has not been demonstrated yet, however, two cell-surface receptor agonists that would be good candidates are cholecystokinin (CCK) in rabbit longitudinal muscle

and 5-hydroxytryptamine (5-HT) in guinea pig tracheal muscle. In longitudinal smooth muscle, the peptide CCK-8 induces contraction by increasing $[Ca^{2+}]_i$. Part of the source of Ca^{2+} is by a CCK-8-mediated Ca^{2+} influx that is blocked by nifedipine, but this Ca^{2+} influx is augmented by a CICR mechanism that it activates (Kuemmerle and Makhlouf, 1994). Pharmacological removal of Ca^{2+} stores by treatment with either thapsigargin, which blocks Ca^{2+} sequestration mechanisms, or ryanodine blocks the CICR component and allows only a transient attenuated contraction that is blocked by nifedipine. Since CCK-8 produces minimal increases in IP$_3$, then the major CICR mechanism seems to be via a RyR-like Ca^{2+} release mechanism, which may be sensitized to CICR by cADPR.

Another candidate receptor for signal transduction through the cADPR pathway is the 5-HT$_{2A}$ receptor in tracheal smooth muscle. Here 5-HT in the absence of extracellular Ca^{2+} can mobilize Ca^{2+} without increasing IP$_3$ formation, an effect that is blocked by prior incubation with ryanodine (Watts *et al.*, 1994). It will be interesting to see if cADPR plays a role in transducing the effects of these two receptor types linked to the contractile processes in smooth muscle.

Another possible way of regulating cADPR levels in cells is via the NO/cGMP signaling pathway (see Section III.E). However, since NO is a potent smooth muscle relaxing agent, it may be considered an unlikely pathway leading to Ca^{2+} mobilization in smooth muscle. However, a possible role for a NO/cGMP/

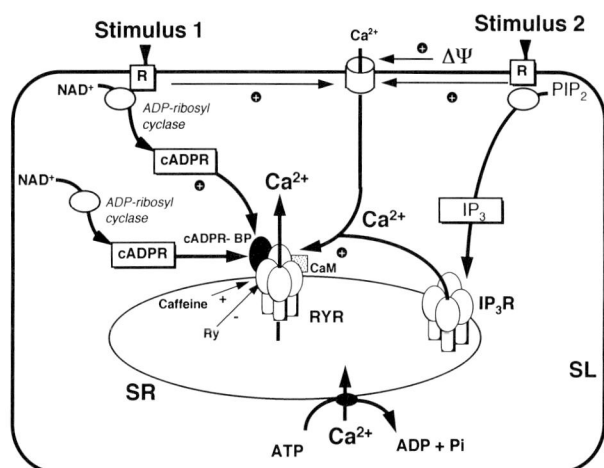

FIGURE 4 Scheme for a possible role for cyclic ADP-ribose in pharmacomechanical coupling in smooth muscle cells. Release of Ca^{2+} by cADPR is a complex process and at present not well understood. Ca^{2+} release by cADPR appears to be ultimately via a ryanodinelike receptor Ca^{2+} release channel, but requires calmodulin, and may indirectly involve the interaction of cADPR binding proteins with the Ca^{2+} release channel. cADPR may act in two distinct ways. The first mode is by acting as a classic second messenger by analogy with IP$_3$, whereby extracellular ligands (stimulus 1) activate cADPR synthesis leading a increase in $[cADPR]_i$, resulting in the activation of Ca^{2+} release channels in the SR. The second model proposes a more subtle role for cADPR as a modulator of CICR via Ca^{2+} release channels in the SR. In this model the level of cADPR in the cytoplasm sensitizes Ca^{2+} release channels to activation by Ca^{2+}, so both cADPR and an increase in $[Ca^{2+}]_i$ are required to trigger CICR. Thus the immediate trigger for mobilization is a rise in intracellular free Ca^{2+}, which may occur via activation of Ca^{2+} influx across the plasma membrane either activated by the ligand at its cell-surface receptor or by depolarization ($\Delta\Psi$) in the case of voltage-gated Ca^{2+} channels, or by release of Ca^{2+} from IP$_3$-sensitive Ca^{2+} intracellular stores activated by ligands coupled to the inositol lipid pathway (stimulus 2), which is associated with an increase in Ca^{2+} sarcolemmal permeability. These two modes are not mutually exclusive, since IP$_3$ appears to play roles in both initiating and propagating Ca^{2+} signals.

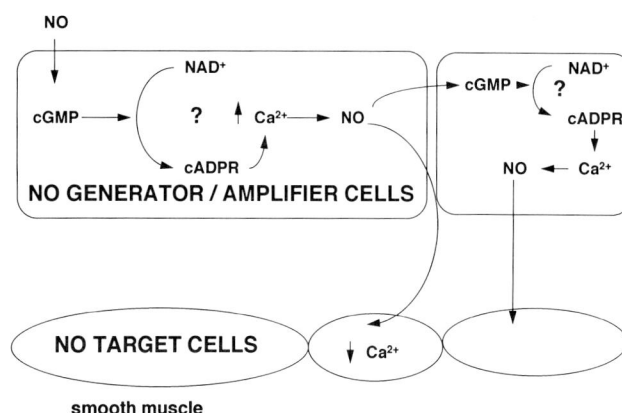

smooth muscle

FIGURE 5 Scheme for interactions between smooth muscle cells and nitric oxide-generating cells. NO causes the mobilization of intracellular Ca^{2+} in interstitial cells in the mammalian gut. This effect is blocked by ryanodine. The rise in Ca^{2+} causes more production of NO by activated NO synthase. The NO produced is released from the cells and penetrates neighboring smooth muscle cells, leading to relaxation. Since, in the sea urchin egg, NO mobilizes Ca^{2+} by increasing cGMP, which in turn activates ADP-ribosyl cyclase, it is possible that in interstitial cells, cADPR mediates this effect.

cADPR pathway in regulating smooth muscle contraction may involve interactions with other cell types. Interstitial cells of Cajal in the mammalian gut, which may be important in controlling pacemaker activity in these tissues (Thuneburg, 1982), in common with the sea urchin egg, also displays a Ca^{2+} mobilization response when treated with NO (Publicover et al., 1993). Furthermore, such release is blocked by ryanodine. Since NO synthesis by NO synthases can be controlled in these cells by intracellular Ca^{2+} (Moncada et al., 1991), this mechanism may play a role in the regenerative production of NO and Ca^{2+} in these cells. The NO produced can diffuse from the interstitial cells and into the neighboring smooth muscle where it triggers relaxation (Ignarro, 1991). Whether cADPR mediates the NO-induced Ca^{2+} mobilization in interstitial cells has not been examined. A hypothetical model for cADPR involvement in these phenomena is shown in Fig. 5.

References

Allen G. J., Muir, S. R., and Sanders, D. (1994). *Science* **268**, 735–737.

Berridge, M. J. (1993a) *Nature (London)* **361**, 315–325.

Berridge, M. J. (1993b). *Nature (London)* **365**, 388–389.

Berridge, M. J., and Dupont, G. (1994). *Curr. Opin. Cell Biol.* **6**, 267–274.

Brillantes, A.-M.B., Ondrias, K., Jayaraman, T., Scott, A., Kobrinsky, S. E., Ehrlich, B. E., and Marks, A. R. (1994). *Cell (Cambridge, Mass.)* **77**, 513–523.

Buck, W. R., Hoffmann, E. E., Rakow, T. L., and Shen, S. S. (1994). *Dev. Biol.* **163**, 1–10.

Butt, E., van Bemmelen, M., Fischer, L., Walter, U., and Jastorff, B. (1990). *FEBS Lett.* **263**, 47–50.

Clapper, D., Walseth, T., Dargie, P., and Lee, H. C. (1987). *J. Biol. Chem.* **262**, 9561–9568.

Coronado, R., Morrissette, J., Sukhareva, M., and Vaughan, D. M. (1994). *Am. J. Physiol.* **266**, C1485–C1504.

Currie, K., Swann, K., Galione, A., and Scott, R. (1992). *Mol. Biol. Cell* **3**, 1415–1422.

Dargie, P. J., Agre, M. C., and Lee, H. (1990). *Cell Regul.* **1**, 279–290.

Epel, D. (1964). *Biochem. Biophys. Res. Commun.* **17**, 69–73.

Fleischer, S., and Inui, M. (1989). *Annu. Rev. Biophys. Biophys. Chem.* **18**, 333–364.

Fruen, B. R., Mickelson, J. R., Shomer, N. H., Velez, P., and Louis, C. F. (1994). *FEBS Lett.* **352**, 123–126.

Furuichi, T., Kohda, K., Miyawaki, A., and Mikoshiba, K. (1994). *Curr. Opin. Neurobiol.* **4**, 294–303.

Galione, A. (1992). *Trends Pharmacol. Sci.* **13**, 304–306.

Galione, A., and White, A. (1994). *Trends Cell Biol.* **4**, 431–436.

Galione, A., Lee, H. C., and Busa, W. B. (1991). *Science* **253**, 1143–1146.

Galione, A., McDougall, A., Busa, W., Willmott, N., Gillot, I., and Whitaker, M. (1993a). *Science* **261**, 348–352.

Galione, A., White, A., Willmott, N., Turner, M., Potter, B. V., and Watson, S. P. (1993b). *Nature (London)* **365**, 456–459.

Giannini, G., Clementi, E., Ceci, R., Marziali, G., and Sorrentino, V. (1992). *Science* **257**, 91–93.

Glick, D., Hellmich, M., Beushausen, S., Tempst, P., Bayley, M., and Strumwasser, F. (1991). *Cell Regul.* **2**, 211–218.

Goy, M. F. (1991). *Trends Neurosci.* **14**, 293–299.

Graeff, R. M., Mehta, K., and Lee, H. C. (1994a). *Biochem. Biophys. Res. Commun.* **205**, 722–725.

Graeff, R. M., Walseth, T. F., Fryxell, K., Branton, W. D., and Lee, H. C. (1994b). *J. Biol. Chem.* **269**, 30260–30267.

Guse, A. H., da Silva, C. P., Emmrich, F., Ashamu, G. A., Potter, B. V. L. and Mayr, G. W. (1995) *J. Clin. Invest.* (in press).

Hakamata, Y., Nakai, J., Takeshima, H., and Imoto, K. (1992). *FEBS Lett.* **312**, 229–235.

Hakamata, Y., Nishimura, S., Nakai, J., Nakashima, Y., Kita, T., and Imoto, K. (1994). *FEBS Lett.* **352**, 206–210.

Hellmich, M., and Strumwasser, F. (1991). *Cell Regul.* **2**, 193–202.

Hermann-Frank, A., Darling, E., and Meissner, G. (1991). *Pflügers Arch.* **418**, 353–359.

Howard, M., Grimaldi, J. C., Bazan, J., Santos-Argumedo, L., Parkhouse, R. M. E., Walseth, T. F., and Lee, H. C. (1993). *Science* **262**, 1056–1059.

Hua, S.-Y., Tokimasa, T., Takasawa, S., Furuya, Y., Nohmi, M., Okamoto, H., and Kuba, K. (1994). *Neuron* **12**, 1073–1079.

Ignarro, L. J. (1991). *Biochem. Pharmacol.* **41**, 485–490.

Iino, M., Kobayashi, T., and Endo, M. (1988). *Biochem. Biophys. Res. Commun.* **152**, 417–422.

Islam, M. S., Larsson, O., and Berggren, P.-O. (1993). *Science* **262**, 584–585.

Kim, H., Jacobson, E. L., and Jacobson, M. K. (1993a). *Biochem. Biophys. Res. Commun.* **194**, 1143–1147.

Kim, H., Jacobson, E. L., and Jacobson, M. K. (1993b). *Science* **261**, 1330–1333.

Koguma, T., Takasawa, S., Tohgo, A., Karasawa, T., Furuya, Y., Yonekura, H., and Okamoto, H. (1994). *Biochim. Biophys. Acta* **1223**, 160–162.

Kuemmerle, J. F., and Makhlouf, G. M. (1994). *Gastroenterology* **106**, A821.

Kuemmerle, J. F., Murthy, K. S., and Makhlouf, G. M. (1994). *Am. J. Physiol.* **266**, C1421–C1431.

Lee, H. C. (1991). *J. Biol. Chem.* **266**, 2276–2281.

Lee, H. C. (1993). *J. Biol. Chem.* **268**, 293–299.

Lee, H. C. (1994). *News Physiol. Sci.* **9**, 134–137.

Lee, H. C., and Aarhus, R. (1991). *Cell Regul.* **2**, 203–209.

Lee, H. C., and Aarhus, R. (1993). *Biochim. Biophys. Acta* **1164**, 68–74.

Lee, H. C., Walseth, T. F., Bratt, G. T., Hayes, R. N., and Clapper, D. L. (1989). *J. Biol. Chem.* **264**, 1608–1615.

Lee, H. C., Aarhus, R., and Walseth, T. F. (1993a). *Science* **261**, 352–355.

Lee, H. C., Zocchi, E., Guida, L., Franco, L., Benatti, U., and De Flora, A. (1993b). *Biochem. Biophys. Res. Commun.* **191**, 639–645.

Lee, H. C., Aarhus, R., Graeff, R., Gurnack, M. E., and Walseth, T. F. (1994a). *Nature (London)* **370**, 307–309.

Lee, H. C., Aarhus, R., and Levitt, D. (1994b). *Nat. Struct. Biol.* **1**, 143–144.

Lee, H. C., Galione, A., and Walseth, T. F. (1994c). *Vitam. Horm. (N.Y.)* **48**, 199–254.

Lynn, S., Morgan, J. M., Gillespie, J. I., and Greenwell, J. R. (1993). *FEBS Lett.* **330**, 227–230.

Maack, T. (1992). *Annu. Rev. Physiol.* **54**, 11–27.

Mahoney, M. G., Slakey, L. L., Hepler, P. K., and Gross, D. J. (1993). *J. Cell Sci.* **104**, 1101–1107.

Malavasi, F., Funaro, A., Roggero, S., Horenstein, A., Calosso, L., and Mehta, K. (1994). *Immunol. Today* **15**, 95–97.

McPherson, S. M., McPherson, P. S., Mathews, L., Campbell, K. P., and Longo, F. J. (1992). *J. Cell Biol.* **116**, 1111–1121.

Meissner, G. (1994). *Annu. Rev. Physiol.* **56**, 485–508.

Meszaros, L. G., Bak, J., and Chu, A. (1993). *Nature (London)* **364**, 76–79.

Moncada, S., Palmer, R. M. J., and Higgs, E. A. (1991). *Pharmacol. Rev.* **34**, 109–142.

Moschella, M. C., and Marks, A. R. (1993). *J. Cell Biol.* **120**, 1137–1146.

Murthy, K. S., Grider, J. R., and Makhlouf, G. M. (1991). *Am. J. Physiol.* **261**, G937–G944.

Nixon, G. F., Mignery, G. A., and Somlyo, A. V. (1994). *J. Muscle Res. Cell Motil.* **15**, 682–700.

Ogawa, Y. (1994). *Crit. Rev. Biochem. Mol. Biol.* **29**, 229–274.

Petersen, O. H., Petersen, C., and Kasai, H. (1994). *Annu. Rev. Physiol.* **56**, 297–319.

Pozzan, T., Rizzuto, R., Volpe, P., and Meldolesi, J. (1994). *Physiol. Rev.* **74**, 595–636.

Publicover, N. G., Hammond, E. M., and Sanders, K. M. (1993). *Proc. Natl. Acad. Sci. U.S.A.* **90**, 2087–2091.

Rakovic, S., and Terrar, D. A. (1994). *J. Physiol. (London)* **475**, 81P.

Rakovic, S., Galione, A., Ashamu, G. A., Potter, B. V. L., and Terrar, D. A (1995). *J. Physiol.* **483**, 17P.

Rusinko, N., and Lee, H. C. (1989). *J. Biol. Chem.* **264**, 11725–11731.

Rutter, G. A., Theler, J.-M., Li, G., and Wollheim, C. B. (1994). *Cell Calcium* **16**, 71–80.

Schmidt, H. H. H. W., Lohmann, S. M., and Walter, U. (1993). *Biochim. Biophys. Acta* **1178**, 153–175.

Sitsapesan, R., Mcgarry, S. J., and Williams, A. J. (1994). *Circ. Res.* **75**, 596–600.

Smith, J. S., Rousseau, E., an Meissner, G. (1989). *Circ. Res.* **64**, 352–359.

Somlyo, A. P., and Somlyo, A. V. (1994). *Nature (London)* **372**, 231–236.

Sorrentino, V., and Volpe, P. (1993). *Trends Pharmacol. Sci.* **14**, 98–103.

States, D., Walseth, T., and Lee, H. C. (1992). *Trends Biochem. Sci.* **17**, 495.

Summerhill, R. J., Jackson, D. G., and Galione, A. (1993). *FEBS Lett.* **335**, 231–233.

Takasawa, S., Nata, K., Yonekura, H., and Okamoto, H. (1993a). *Science* **259**, 370–373.

Takasawa, S., Tohgo, A., Noguchi, N., Koguma, T., Nata, K., Sugimoto, T., Yonekura, H., and Okamoto, H. (1993b). *J. Biol. Chem.* **268**, 26052–26054.

Takeshima, H., Nishimura, S., Nishi, M., Ikeda, M., and Sugimoto, T. (1993). *FEBS Lett.* **322**, 105–111.

Thorn, P., Gerasimenko, O., and Petersen, O. H. (1994). *EMBO J.* **13**, 2038–2043.

Thuneburg, L. (1982). *Adv. Anat. Embryol. Cell Biol.* **71**, 1–130.

Tohgo, A., Takasawa, S., Noguchi, N., Koguma, T., Nata, K., Sugimoto, T., Furuya, Y., Yonekura, H., and Okamoto, H. (1994). *J. Biol. Chem.* **269**, 28555–28557.

Wagenknecht, T., Berkowitz, J., Grassucci, R., Timerman, A. P., and Fleischer, S. (1994). *Biophys. J.* **67**, 2286–2295.

Walseth, T. F., and Lee, H. C. (1993). *Biochim. Biophys. Acta* **1178**, 235–242.

Walseth, T. F., Aarhus, R., Zeleznikar, R., and Lee, H. C. (1991). *Biochim. Biophys. Acta* **1094**, 113–120.

Walseth, T. F., Aarhus, R., Kerr, J. A., and Lee, H. C. (1993). *J. Biol. Chem.* **268**, 26686–26691.

Watts, S. W., Cox, D. A., Johnson, B. G., Schoepp, D. D., and Cohen, M. L. (1994). *J. Pharmacol. Exp. Ther.* **271**, 832–844.

Whalley, T., McDougall, A., Crossley, I., Swann, K., and Whitaker, M. (1992). *Mol. Biol. Cell* **3**, 373–383.

Whitaker, M. J., and Irvine, R. F. (1984). *Nature (London)* **312**, 636–638.

Whitaker, M. J., and Swann, K. (1993). *Development (Cambridge, UK)* **117**, 1–12.

White, A. M., Watson, S. P., and Galione, A. (1993). *FEBS Lett.* **318**, 259–263.

Wibo, M., and Godfraind, T. (1994). *Biochem. J.* **297**, 415–423.

Wier, W. G., and Blatter, L. A. (1991). *Cell Calcium* **12**, 241–254.

Willmott, N., Walseth, T. F., Lee, H. C., White, A. M., Sethi, J., and Galione, A. (1995). Submitted for publication.

Xu, L., Lai, F. A., Cohn, A., Etter, E., Guerrero, A., Fay, F. S., and Meissner, G. (1994). *Proc. Natl. Acad. Sci. U.S.A.* **91**, 3294–3298.

Yamada, S., Gu, Q. M., and Sih, C. J. (1994). *J. Am. Chem. Soc.* **116**, 10787–10788.

Zhang, Z., Kwan, C,. and Daniel, E. E. (1993). *Biochem. J.* **290**, 259–266.

24

Enzyme Translocations during Smooth Muscle Activation

RAOUF A. KHALIL

Program in Smooth Muscle Research
Harvard Medical School and Cardiovascular Division
Beth Israel Hospital
Boston, Massachusetts

KATHLEEN G. MORGAN

Program in Smooth Muscle Research
Harvard Medical School and Cardiovascular Division
Beth Israel Hospital; and
Boston Biomedical Research Institute
Boston, Massachusetts

I. INTRODUCTION

Smooth muscle signal transduction has been the topic of several excellent review articles (Stull *et al.*, 1991; Andrea and Walsh, 1992; Somlyo and Somlyo, 1994; Lee and Severson, 1994) and the reader is referred to these articles as well as other relevant chapters of this book for more details on current findings and controversies in the area of signal transduction in smooth muscle. To place the topic of enzyme trafficking within the broader context of smooth muscle contractile mechanisms, an outline of the currently described mechanisms of smooth muscle contraction is shown in Fig. 1.

The more classic pathway of smooth muscle contractile activation is illustrated in the right-hand side of this panel, where activation leads to an increase in intracellular ionized calcium concentration ($[Ca^{2+}]_i$) by virtue of Ca^{2+} entry through channels or exchangers (Lyu *et al.*, 1992; van Breemen *et al.*, 1985; Khalil *et al.*, 1987) or the release of Ca^{2+} from the sarcoplasmic reticulum. $[Ca^{2+}]_i$ in combination with calmodulin activates myosin light chain (MLC) kinase to cause phosphorylation of the 20-kDa MLC, which in turn results in increased actin-activated myosin ATPase activity, increase cross-bridge cycling velocity, and, as a result, an increase in contractile force (Aksoy *et al.*, 1976; Sobieszek, 1977).

Work from several laboratories (Seumatsu *et al.*,

1991; Gong *et al.*, 1992; Tansey *et al.*, 1992; Masuo *et al.*, 1994) has clearly shown that the amount of MLC phosphorylation that occurs at any Ca^{2+} level is variable, that is, the Ca^{2+} "sensitivity" of MLC phosphorylation can be altered. A decreased Ca^{2+} sensitivity of MLC kinase appears to occur through a mechanism involving activation of Ca^{2+}/calmodulin-dependent kinase II (Tansey *et al.*, 1992, 1994). Similarly, the activity of MLC phosphatase appears to be regulated, although the details of the mechanisms involved remain controversial. Protein kinase C (PKC) and a small GTP binding protein may be involved in a pathway leading to inhibition of MLC phosphatase (Gong *et al.*, 1992).

Although there is no doubt that MLC phosphorylation plays a major role in the regulation of contractile force in smooth muscle, there is now a wealth of data indicating that under appropriate conditions, the relationship between the level of MLC phosphorylation and the magnitude of steady-state force can also be regulated (Jiang and Morgan, 1989; Gerthoffer *et al.*, 1989; Rembold, 1990; Suematsu *et al.*, 1991). Considerable evidence now indicates that PKC is involved in a parallel signal transduction pathway terminating in regulation at the level of the thin filament. Two suggested mechanisms are illustrated on the left-hand side of Fig. 1. These mechanisms will be discussed in greater detail at the end of this chapter.

It is clear from inspection of even the simplified signal transduction scheme outlined in Fig. 1 that con-

---

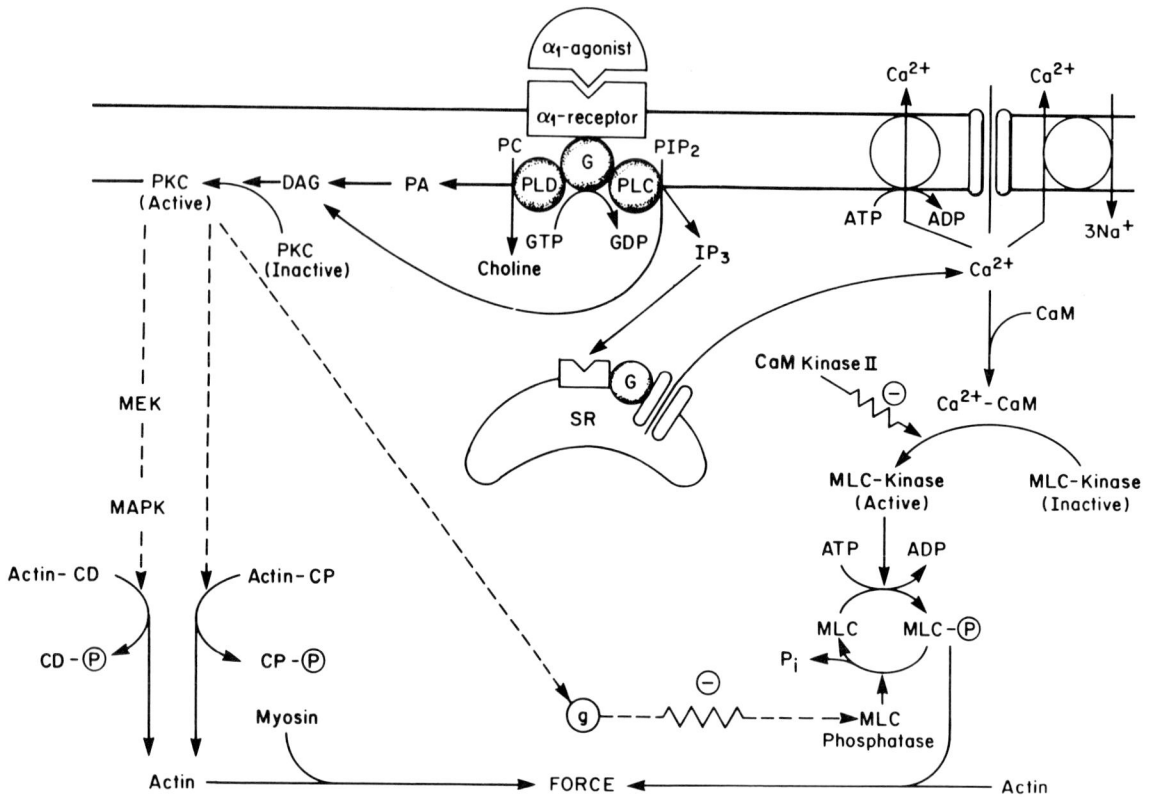

FIGURE 1   Signal transduction mechanisms in vascular smooth muscle. G, heterotrimeric GTP binding protein; PLC, phospholipase C; PIP_2, phosphatidylinositol 4,5 bisphosphate; PC, phosphatidylcholine; IP_3, inositol 1,4,5-trisphosphate; DAG, diacylglycerol; SR, sarcoplasmic reticulum; MLC, myosin light chain; PKC, protein kinase C; CD, caldesmon; CP, calponin; MAPK, mitogen-activated protein kinase; MEK, MAP kinase kinase; g, small GTP binding protein. Dashed lines indicate the involvement of more than one kinase of cofactor.

siderable trafficking of enzymes, substrates, and cofactors must occur during the transmission of a signal from the plasmalemma to the contractile proteins in the core of the cell. The targeting mechanisms by which intracellular enzyme traffic is directed will be the major topic of this chapter. Besides providing "information transfer" from extracellular signals to the interior of the cell, the targeting mechanisms allow enzymes to colocalize with physiologically relevant substrates at specific locations in the cell where important cofactors (such as ionic or lipid components) may be available in greater concentrations than in the bulk of the cytosol. Local gradients in ionic composition or concentrations of other cofactors may be an important mechanism explaining intracellular selectivity of multipotent kinases. Finally, intracellular trafficking mechanisms may allow individual kinases to play different roles in the differentiated contractile smooth muscle cell from those they play in the synthetic, proliferative smooth muscle cell. In this regard, the nucle-

ar membrane and intranuclear space provide an attractive possible mechanism for placing geographic limitations on the distribution of kinases in the differentiated cell while, on the other hand, providing targeting mechanisms that are perhaps timed with appropriate events during the cell cycle in the proliferating cell. The techniques that have been used to identify enzyme traffickings in smooth muscle, together with what has been found regarding intracellular sorting mechanisms in the smooth muscle cell, will be discussed below.

II. METHODS OF STUDY

The subcellular distribution and translocation of enzymes have been studied using three major approaches: biochemical approaches, digital imaging microscopy, and pharmacological approaches. Because most of the work in this field has not been performed in contractile smooth muscle, the application of these approaches will be discussed in general terms in regard to a number of different cell types.

A. Biochemical Approaches

Biochemical assays have been used to measure the relative distribution and/or activity of enzymes in cell fractions. However, the degree to which cell fractionation affects the distribution or activity of the enzyme is not clear. The customary use of EDTA or EGTA in cell lysis buffers to block Ca^{2+}-dependent proteases can have significant effects on the distribution of Ca^{2+}-dependent kinases. Also, the particulate fraction in these assays often contains the lumped contribution of the plasma membrane, the nuclear envelope, and other intracellular membranes. PKC is one of the enzymes that have been studied extensively in cell fractions (Takai et al., 1979, 1981). These studies have shown that translocation of the phorbol ester binding sites (PKC receptors) and PKC activity from the cytosolic fraction to the particulate fraction is associated with PKC activation. However, fractionation assays for PKC translocation have shown Ca^{2+} requirements between 10 μM and 1 mM $[Ca^{2+}]_i$ (Takai et al., 1979, 1981), concentrations greater than those expected in vivo. An in situ estimate of the $[Ca^{2+}]_i$ requirement for translocation of α-PKC (using digital imaging techniques) reported an apparently more physiological value of 200 nM (Khalil et al., 1994).

B. Digital Imaging Microscopy

1. Fluorescence Microscopy in Live Cells

Fluorescence microscopy allows the study of the subcellular distribution of enzymes in intact cells. Relatively few probes are available to directly monitor the distribution of kinases in living cells. One example is the fluorescent phorbol ester derivative Bodipy phorbol, which has been used to track PKC distributions in live cells (Khalil and Morgan, 1991). This compound is membrane permeant but has the disadvantage of retaining PKC agonist activity. A group of compounds displaying the opposite disadvantage, that is, retention of PKC antagonist activity, has become available (Chen and Poenie, 1993). The use of a ratiometric indicator to monitor cAMP-dependent protein kinase distribution and activity in live cells has been described (Bacskai et al., 1993), but at this time its use in smooth muscle has not been reported. Another related approach is to introduce labeled antibodies into permeabilized physiologically intact cells (Iizuka et al.,

1994). The difficulty of this approach is the problem of whether nonspecific binding sites can be "blocked" in the intact cell.

2. Direct or Indirect Immunofluorescence

Most immunofluorescence protocols involve fixation of cells at a specific time point of activation to preserve the distribution of intracellular proteins, followed by cell permeabilization to allow the entry of the antibody into the cell. The primary antibody can either be directly tagged with a fluorophor or, in the indirect approach, a secondary anti-IgG or anti-IgM antibody is tagged. The direct approach is simpler and less prone to background problems. In contrast, the indirect approach offers the advantage of widely available labeled antibodies and the increased sensitivity that results from amplification of the signal (for more details, see Harlow and Lane, 1988).

3. Inhibition of Shortening in Contractile Cells

A technical problem pertinent to the study of enzyme distribution in smooth muscle cells is that the changes in cell shape induced by cell shortening will interfere with accurate fluorescence measurements. To avoid shortening-induced fluorescence artifacts, methods are needed to prevent cell contraction while at the same time having minimal effect on the trafficking of the enzyme to be studied. One approach is to inhibit kinase activity specifically in the catalytic domain and thus dissociate the enzyme translocations from the phosphotransferase activity of the kinase. For example, the PKC inhibitors H-7 and staurosporine compete at the ATP binding domain, thus inhibiting the catalytic activity of the protein kinase and smooth muscle contraction without inhibiting the redistribution of the enzyme (Khalil and Morgan, 1991, 1992; Kiley et al., 1992; Nishizuka, 1992).

A second approach is to directly inhibit cross-bridge cycling. Hypertonic physiological saline solutions containing nonpenetrating solutes (such as sucrose) have been shown to decrease isometric tension in intact muscle fibers (Hodgkin and Horowicz, 1957; Cecchi and Bagni, 1994). Reports from Godt's laboratory (Godt et al., 1984) have pointed to a molecular mechanism at the myofilament level involving the lattice spacing of actin and myosin filaments. Both of these methods have limitations and necessitate that controls be carefully thought out.

4. Mathematical Deconvolution/Confocal Microscopy/Digital Image Analysis

Two challenges posed by fluorescence microscopy are the intrinsically poor resolution of the fluorescent images and the qualitative nature of the image format. An image obtained at a specific focal plane of a relatively thick cell with a conventional fluorescence microscope is contaminated by out-of-focus fluorescence from focal planes above and below. The out-of-focus fluorescent blur can be removed either physically by the use of a confocal microscope or mathematically by the use of deblurring algorithms. A variety of algorithms have been developed, ranging from simple inverse filters to complex three-dimensional constrained iterative algorithms. The more complex algorithms generally produce better resolution and accuracy but are more time-consuming and require more powerful computers. A more detailed discussion of deblurring and confocal techniques is presented in Taylor and Wang (1989).

Immunofluorescent images provide only qualitative information on the subcellular enzyme distribution. For quantitative measurement of spatial and temporal changes in fluorescence, image analysis algorithms are needed. We have used several simple algorithms for quantitation of cell fluorescence (Khalil and Morgan, 1991, 1992; Khalil et al., 1992). For example, to quantitate the expression of a particular enzyme, the total cell fluorescent intensity can be integrated and divided by the number of pixels in the cell image. The resulting number will be proportional to enzyme concentration but independent of cell size. Total cell fluorescence can also be dissected into compartments, such as surface membrane, nuclear, and cytosolic. The average surface membrane signal and the average cytosolic signal can be used to calculate the surface/cytosol fluorescence ratio and provide information on the translocation of an enzyme to the surface membrane. Similarly, the nuclear/total fluorescence ratio can be used as an indication of the localization of an enzyme to the nucleus. By using a ratio for comparison, the effects of differences in cell thickness and fluorescence labeling in different cells can be minimized.

C. Pharmacological Approaches

A third approach to studying enzyme translocations is to use specific pharmacological tools. These tools provide controls for both biochemical and immunofluorescence studies and often provide insight into possible mechanisms of translocations.

1. Inhibitors of Lipid Modification

As discussed in Section IV.B, lipid modification has been shown to play a critical role in the targeting of GTP binding proteins to cell membranes (Casey, 1994). G proteins may form complexes with, and conse-

quently cause the targeting of kinases to, membranes or, alternatively, the kinase of interest may contain a sequence that can undergo lipid modification. Specific inhibitors of these pathways are available and their effectiveness in interfering with signal transduction would suggest the involvement of such a pathway (Metz et al., 1993; Garcia et al., 1993; James et al., 1993; Graham et al., 1994).

2. Protein Kinase C Inhibitors

A wide variety of PKC inhibitors are now available. Because of sequence similarity with other kinases, inhibitors targeted at the ATP binding domain of PKC, such as staurosporine and H-7, are relatively less selective (Rüegg and Burgess, 1989). Calphostin C competes at the diacylglycerol binding site (Kobayashi et al., 1989) and is thought to be relatively more selective. The pseudosubstrate inhibitor peptides that compete at the substrate binding domain are also thought to have relatively high selectivity (House and Kemp, 1987).

3. Tyrosine Kinase Inhibitors

Tyrosine phosphorylation of plasmalemmal receptors or other proteins provides high-affinity binding sites for src homology 2 (SH2)-containing proteins and represents a key targeting mechanism for enzyme translocation. Several groups of tyrosine kinase inhibitors are currently available and have been shown to interfere with enzyme translocation.

Naturally occurring flavonoids, such as quercetin and genistein, inhibit tyrosine kinase activity by competing with ATP (Akiyama et al., 1987), and have been shown to inhibit cAMP-dependent protein kinase, PKC, and other ATP-requiring enzymes (Gazit et al., 1989) at very low concentrations. Other naturally occurring compounds, such as erbstatin, herbimycin A, lavendustin A, and aeroplysinin-1, are slightly more specific (Levitzki and Gilon, 1991) because they compete with the phosphoryl acceptor substrate, tyrosine. Tyrphostins are synthetic compounds that structurally resemble the phenolic moiety of tyrosine (Yaish et al., 1988; Gazit et al., 1989). They inhibit tyrosine kinase by competing with the substrate, tyrosine, and inhibit serine/threonine kinases such as PKC and cyclic AMP-dependent and cyclic GMP-dependent protein kinases only at very high concentrations (Yaish et al., 1988; Gazit et al., 1989; Levitzki and Gilon, 1991). Tyrphostins are relatively nontoxic and have reversible inhibitory effects.

4. Peptide Inhibitors

Pseudosubstrate inhibitor peptides are now available for a host of protein kinases in addition to PKC,

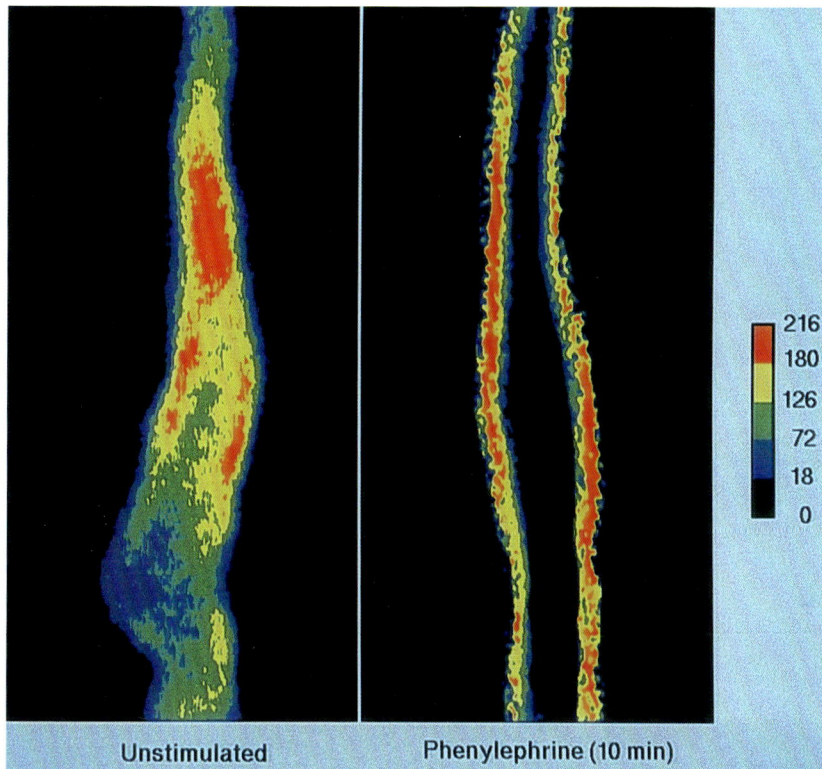

FIGURE 2 Distribution of α-PKC in resting (left) and phenylephrine-stimulated (right) ferret portal vein cell. Pseudocolors represent increasing amounts of α-PKC staining, blue < green < yellow < red.

for example, cyclic AMP-dependent kinase and Ca^{2+}/calmodulin-dependent kinase II. These peptides mimic the pseudosubstrate domain of the individual kinases and hence compete with the substrate binding domain for the endogenous substrate. For a discussion of the relative specificity of these inhibitors see Soderling (1990).

5. Phosphatases/Inhibitors/Molecular Approaches

Inhibitors of serine/threonine phosphatases, such as okadaic acid or microcystin, can be used to investigate the effects of serine/threonine phosphorylation on kinase targeting. Tyrosine phosphatase inhibitors can be used to investigate the effect of tyrosine phosphorylation on kinase targeting in a specific cell type. For example, orthovanadate has been shown to induce translocation of the immunoreactivity of phospholipase C gamma from the cytosolic to the membrane fraction of mast cells (Atkinson et al., 1993). Conversely, in permeabilized preparations, the phosphatases themselves, if known, can be added to inhibit the putative pathway.

Transfection of kinase-negative mutants with various kinases has been particularly useful in elucidating signal transduction pathways in cultured cells (Gonzalez et al., 1993; Schaap et al., 1993). However, since smooth muscle cells begin to undergo phenotypic modulation almost immediately upon the introduction of cell culture conditions, it has not yet been possible to adapt these approaches to differentiated smooth muscle cells.

III. EVIDENCE FOR TRANSLOCATION OF ENZYMES DURING SMOOTH MUSCLE ACTIVATION

A. Protein Kinase C

Cell fractionation studies in bovine carotid artery first demonstrated that phorbol esters, histamine, angiotensin, and endothelin, but not KCl, induced a translocation of PKC from the cytosol to the membrane fraction (Haller et al., 1990), although the identity of the membranes involved was not determined (Table I). Work with the non-isoform-specific fluorescent PKC probe Bodipy phorbol (Khalil and Morgan, 1991) in live ferret portal vein cells indicated the presence of PKC in the cytosol, at the surface membrane, in the nucleus, and surrounding the perinuclear organelles (Table I). No changes in the nuclear and perinuclear distribution of PKC were detected during activation by phorbol esters or physiological agonists. However, quantitative image analysis demonstrated a significant increase in PKC translocation to the surface membrane during activation. Interestingly, an inward movement of the cell permeant probe to the cytosol was visualized at a time point *before* the first detectable translocation of the label to the surface membrane (Khalil and Morgan, 1991). PKC has been identified as a family of at least 11 different isoforms with differing dependency on Ca^{2+}, phospholipids, and diacylglycerol for activation. The "conventional" PKC isoforms (cPKC) include α, βI, βII, and γ isoforms and require

TABLE I Protein Kinase C Distributions in Smooth Muscle

Enzyme	Cell type	Resting state	Activated state	Reference
Protein kinase C	Bovine carotid artery	Cytosolic fraction	Membrane fraction	Haller et al. (1990)
	Ferret portal vein	Cytosol perinuclear	Surface membrane perinuclear	Khalil and Morgan (1991)
	Rat stomach fundus	Cytosolic fraction	Membrane fraction	Secrest et al. (1991)
	Bovine trachea	Cytosolic fraction	Membrane fraction	Langlands and Diamond (1994)
α-PKC	Carotid artery	Cytosolic fraction	Particulate fraction	Singer et al. (1992)
	Bovine aorta	Cytosol	Particulate fraction	Watanabe et al. (1989)
	Ferret portal vein	Cytosol	Surface membrane	Khalil et al. (1994)
	Cultured rat aorta cells	Cytosolic fraction	Nuclear fraction	Haller et al. (1994)
β-PKC	Carotid artery	Cytosolic fraction	Particulate fraction	Singer et al. (1992)
	Cultured rat aorta cells	Cytosolic fraction	Nuclear fraction	Haller et al. (1994)
δ-PKC	Rat aorta	Cytoskeletal/organelles	Cytoskeletal/organelles	Liou and Morgan (1994)
ε-PKC	Ferret aorta	Cytosol	Surface membrane	Khalil et al. (1992)
ζ-PKC	Ferret aorta and portal vein	Perinuclear	Intranuclear	Khalil et al. (1992)
	Rat aorta	Perinuclear	Intranuclear	Liou and Morgan (1994)

Ca^{2+} for activation, whereas the "new" PKC isoforms (nPKC) δ, ε, η, and θ, as well as the "atypical" PKC isoforms (aPKC) ζ, λ, and ι, do not require a significant increase in Ca^{2+} for their activation (Nishizuka, 1992). Antibodies against specific PKC isoforms have been used to determine the location of different PKC isoforms under resting and activated conditions (Table I).

Studies on the expression of the Ca^{2+}-dependent PKC isoforms have shown that α-PKC is universally present in essentially all cells studied, whereas γ-PKC appears to be expressed exclusively in neural cells. Both α-PKC and β-PKC have been identified in vascular smooth muscle cells and have been shown to undergo a classic cytosol-to-membrane translocation upon agonist activation (Fig. 2 and Table I). Interestingly, in nonmuscle cells and cultured vascular smooth muscle cells, both the α and β isoforms have shown nuclear translocation and PKC has been suggested to be a physiologically relevant nuclear lamin kinase leading to time-dependent solubilization of lamin B, indicative of mitotic nuclear envelope breakdown (Hocevar et al., 1993; Leach et al., 1992; Haller et al., 1994).

Studies on the Ca^{2+}-independent PKC isoforms have shown relatively large quantities of ε-PKC in the ferret aorta and only small quantities in ferret portal vein. Ferret aorta cells have been shown to contract in Ca^{2+}-free bathing solutions supplemented with EGTA and in the absence of changes in $[Ca^{2+}]_i$ (Collins et al., 1992). In these cells, the kinetics of ε-PKC translocation from the cytosol to the surface membrane has been shown to precede or coincide with contraction (Khalil et al., 1992; Khalil and Morgan, 1993), suggesting a cause-and-effect relationship.

We have detected significant amounts of ζ-PKC in ferret aorta, rat aorta, and ferret portal vein. In resting cells, ζ-PKC is localized mainly in a perinuclear region. Upon stimulation with phenylephrine, ζ-PKC translocates from the perinuclear to the intranuclear compartment (Khalil et al., 1992; Liou and Morgan, 1994). That this Ca^{2+}-independent isoform is expressed in smooth muscle preparations that do not display a Ca^{2+}-independent contraction, as well as the fact that the isoform is located in the vicinity of the nucleus, suggests a role for ζ-PKC in gene expression and cell growth. Interestingly, increased nuclear translocation of ζ-PKC in hypertrophied rat aorta cells has also been reported (Liou and Morgan, 1994). These studies are consistent with reports suggesting that ζ-PKC is critical for mitogenic signal transduction (Berra et al., 1993).

In the normal rat aorta, significant amounts of

δ-PKC have been found localized in a perinuclear area similar in distribution to that of mitochondria. The functional significance of δ-PKC has not yet been identified but experiments on hypertrophied rat aorta cells have shown a subtle redistribution of δ-PKC toward the periplasmalemmal space (Liou and Morgan, 1994).

Whether other isoforms of PKC are present in smooth muscle is currently unknown. However, experiments with a non-isoform-specific fluorescent phorbol ester have shown high concentrations of perinuclear PKC in the ferret portal vein (Khalil and Morgan, 1991). Thus far only ζ-PKC has been identified near the nucleus of these cells. Since ζ-PKC is lacking phorbol ester binding sites, it is unlikely that binding ζ-PKC explains this signal. In general, the diversity of isoform-specific distribution and translocation suggests that PKC isoforms may perform specific functions in different cell types.

B. Mitogen-Activated Protein Kinase

Mitogen-activated protein (MAP) kinase is one of several kinases that play a pivotal role in the transduction of extracellular mitogenic signals to the nucleus (Rossomando et al., 1989; Boulton et al., 1990). MAP kinase is Ser/Thr protein kinase that is fully activated by dual phosphorylation at Thr and Tyr residues (Anderson et al., 1990) and has been shown to activate nuclear transcription factors and stimulate cell growth and proliferation. In cultured HeLa cells (Chen et al., 1992), COS-7 cells (Gonzalez et al., 1993), fibroblasts (Lenormand et al., 1993), and human vascular smooth muscle cells (R. A. Khalil, K. C. Kent, and K. G. Morgan, unpublished data), MAP kinase is mainly cytosolic under resting conditions, but translocates into the nucleus during activation by mitogens.

Other evidence has shown significant tyrosine kinase and MAP kinase activities in adult, terminally differentiated smooth muscle cells (Di Salvo et al., 1989; Adam et al., 1992; Childs et al., 1992) upon their stimulation with nonmitogenic agonists. MAP kinase transiently translocates to the surface membrane during early activation, but during maintained activation a second redistribution of MAP kinase from the surface membrane to the cytoskeleton is observed (Khalil and Morgan, 1993). The initial association of MAP kinase appears to require upstream PKC activity but occurs in a tyrosine phosphorylation-independent manner. In contrast, the delayed redistribution of MAP kinase occurs in a tyrosine phosphorylation-dependent manner and appears to target MAP kinase to the contractile proteins (Khalil et al., 1995).

C. Other Kinases

Many other kinases and cofactors have been shown to undergo important translocations in nonmuscle cells, but in general these observations have not yet been extended to differentiated smooth muscle.

1. Raf-1

Biochemical studies have suggested the involvement of both Ras and Raf-1 in a signal transduction pathway originating at tyrosine kinase receptors and leading to cell growth and proliferation (Pelech and Sanghera, 1992). A mechanism by which the interaction of Ras with Raf-1 may lead to activation of Raf-1 has been reported (Stokoe et al., 1994). It has been suggested that Ras functions in the activation of Raf-1 by recruiting Raf-1 to the plasma membrane, where a separate, Ras-independent activation of Raf-1 occurs. The targeting of Raf-1 from the cytosol to the plasma membrane appears to require the addition of the COOH-terminal 17 amino acids of Ras to Raf-1. The sequence has been shown to contain a polylysine domain and a CAAX motif. Interestingly, a mutant in which the polylysine domain was replaced with polyglutamine could be farnesylated, but did not localize to the plasma membrane.

Raf-1 is a serine/threonine kinase that functions by phosphorylating and activating yet other kinases such as MAP kinase kinase (Kyriakis et al., 1992). Interestingly, Raf-1 may also function as a substrate for PKC (Macdonald et al., 1993; Kolch et al., 1993), a process that may control the growth of undifferentiated cells. The fact that Raf-1 is also abundant in differentiated cell types, particularly smooth muscle (Adam and Hathaway, 1993), and that it can be activated by PKC (Kolch et al., 1993) suggest that Raf-1 may play a signaling role in contractile smooth muscle cells.

2. Diacylglycerol Kinase

Soluble diacylglycerol kinase has been reported to translocate from the cytosol to the membranes of thymocytes in a Ca^{2+}-dependent manner, and it has been suggested that under physiological conditions $[Ca^{2+}]_i$ may play a key role in the regulation of diacylglycerol kinase action by controlling the enzyme interaction with membrane phospholipids (Sakane et al., 1991). On the other hand, in a reconstituted system, Kahn and Besterman (1991) reported that the cytosol-to-membrane translocation of diacylglycerol kinase occurred in a diacylglycerol-dependent but Ca^{2+}-independent manner. Similarly, Maroney and Macara (1989) have suggested that the translocation of diacylglycerol kinase activity in fibroblasts is regulated

primarily by substrate concentration. It remains to be shown if diacylglycerol kinase translocation plays a regulatory role in smooth muscle cells.

3. Phospholipase C

Of the three phospholipase C (PLC) isoforms (β, γ, δ), enzyme translocation has been most studied for the γ isoform and is generally acknowledged to be regulated by SH2 groups (Lee and Severson, 1994). For example, in HER 14 cells, epidermal growth factor and platelet-derived growth factor promote translocation of PLC-γ from the cytosolic to the membrane fraction (Kim et al., 1990). On the other hand, in rat hepatocyte, epidermal growth factor causes significant translocation of PLC-γ from the cytosol to the cytoskeleton fraction (Yang et al., 1994). The subcellular distribution of PLC isoforms in smooth muscle is currently unknown and should be an important area for future investigation.

4. Phospholipase A2

In unstimulated macrophages, 90% of the total activity of phospholipase A2 is localized to the cytosol. 1-Oleoyl-2-acetyl-glycerol (OAG) causes translocation of phospholipase A2 to the membrane fraction (Schonhardt and Ferber, 1987). Also, an increase in Ca^{2+} concentration has been shown to induce translocation of phospholipase A2 from the cytosolic fraction to the membrane fraction of rat brain (Yoshihara and Watanabe, 1990) and rat liver (Krause et al., 1991).

5. Myosin Light Chain Kinase

Evidence in the literature suggests that MLC kinase (MLCK) is permanently targeted to the vicinity of the light chains (Adelstein and Klee, 1981; Sobieszek, 1991), and does not undergo translocations. However, this is largely based on the fact that the kinase is found bound to the insoluble fraction during purification procedures. To the best of our knowledge, studies directly assessing the subcellular location of MLCK *in situ* have not been performed.

6. Cyclic AMP-Dependent Kinase

Cyclic AMP-dependent protein kinase is a ubiquitous enzyme in many cell types including smooth muscle. In T cells, cAMP-dependent protein kinase has been shown to translocate from the cytosol to the plasmalemmal receptor (Skalhegg et al., 1994). In *Aplysia* sensory neurons, perinuclear increases in cAMP have been shown to slowly cause the translocation of the freed catalytic subunit of protein kinase A into the nucleus to an extent proportional to its dissociation from the regulatory subunit (Bacskai et al.,

1993). Whether these patterns of translocation occur in smooth muscle is currently unknown.

7. Ca²⁺/Calmodulin-Dependent Protein Kinase

Under conditions of increased synthesis of cAMP in *Aplasia* neurons, a 55-kDa subunit of Ca^{2+}/calmodulin-dependent kinase (CaMKII) has been shown to undergo a translocation from a membrane–cytoskeleton complex to the cytosol (Saitoh and Schwartz, 1985). In this system it has been suggested that the translocation is caused by phosphorylation of CaMKII and is associated with its activation, thus becoming independent of added Ca^{2+}/calmodulin.

IV. MECHANISMS OF TRANSLOCATION

The question arises as to the driving force behind the enzyme translocations and trafficking discussed in the foregoing. Two major possibilities may occur: (1) an ATP-requiring transport process could be involved, that is, some sort of subcellular motor, or (2) simple diffusion may provide the driving force, while targeting mechanisms allow high-affinity binding once the kinase diffuses into the vicinity of its target. Although the time course of motor-driven transport mechanisms is appropriate to explain the time course of the observed translocations in the smooth muscle cell, we are not aware of studies investigating this possibility in differentiated smooth muscle cells. With respect to targeting mechanisms, there are at least four different categories of targeting mechanisms possible. These mechanisms are discussed in the following sections.

A. Conformation-Induced Changes in Hydrophobicity

The classic example of this targeting mechanism is the translocation of PKC. Binding of Ca^{2+} or diacylglycerol to PKC in the presence of phosphatidylserine causes a conformational change in the molecule that results in exposure of the pseudosubstrate domain (Orr *et al.*, 1992; Bosca and Moran, 1993) and presumably increases hydrophobicity of the molecule, which facilitates binding of the enzyme to membrane lipids.

B. Lipid Modification

Lipid modification of small G proteins has been shown to change their activity as well as their subcellular distribution. For instance, replacement of a geranylgeranyl group for the native farnesyl group can make p21^ras an inhibitor of cell growth and inhibit its translocation to the surface membrane. Recent evidence has suggested a role for small G binding proteins in changing the myofilament $[Ca^{2+}]$ sensitivity. It will be of interest to determine whether similar lipid modifications of small G proteins in the smooth muscle cell would alter contractility.

Three major types of lipid modification can occur: (1) palmitoylation, by the addition of the saturated 16-carbon fatty acyl group, palmitoyl; (2) myristoylation, by the addition of the related 14-carbon myristoyl group; and (3) isoprenylation, by the addition of either a 15-carbon farnesyl or a 20-carbon geranylgeranyl moiety. Any of these modifications will increase the lipophilicity of the protein and is expected to lead to membrane targeting.

At the present time, palmitoylation is thought to be the most likely dynamic modulator of protein function, because of the greater lability of the thioester bond involved (Casey, 1994). However, it should be noted that prenylated proteins can also undergo reversible carboxymethylation of the modified cysteine after cleavage of the C-terminal AAX group, and this process may therefore play a role in targeting during signal transduction cascades (Tan and Rando, 1992). It is also possible that prenylation may mediate protein–protein interactions by having other proteins either mask or uncover a prenyl group (Marshall, 1993).

The classic example of a myristoylated protein is the MARCKS protein (myristoylated alanine-rich C kinase substrate). MARCKS is an actin filament cross-linking protein regulated by PKC and calcium-calmodulin (Aderem *et al.*, 1988; Hartwig *et al.*, 1992). Myristoylation of MARCKS is required for effective binding to the actin network at the plasma membrane (Thelen *et al.*, 1991). PKC-mediated phosphorylation of MARCKS causes its displacement from the membrane and interferes with actin cross-linking. Subsequent dephosphorylation of MARCKS is accompanied by its reassociation with the membrane through its stably attached, myristic acid, membrane-targeting moiety (Thelen *et al.*, 1991). PKC itself does not appear to contain a myristoylation consensus sequence, but when mutants containing such a sequence were constructed, no significant increase in membrane association was seen between myristoylated and nonmyristoylated mutants (James and Olson, 1992). The possible importance of MARCKS in smooth muscle has not yet been demonstrated.

C. Phosphorylation

Once a protein is in close opposition to the charges of the inner surface of the plasmalemma, the charge difference caused by phosphorylation of the protein may significantly affect the affinity of the protein for the lipid environment. For example, the phosphorylation of the MARCKS protein has been shown to have an electrostatic effect of equal importance to that of myristoylation in determining the relative affinity of the protein for the membrane (Taniguchi et al., 1993). Interestingly, phosphorylation of PKC has also been shown to be required for it to act as an effector-dependent kinase (Cazaubon and Parker, 1993). The phosphorylation sites required have been identified as located in the catalytic domain in both α-PKC (Cazaubon and Parker, 1994) and β-PKC (Zhang et al., 1994). Whether this phosphorylation is regulated in a dynamic fashion is unknown.

D. Targeting Sequences

Phosphorylated tyrosine residues and surrounding amino acids serve as high-affinity binding sites for cellular proteins that carry SH2 recognition domains. These cellular proteins include p21^{ras}, guanosine triphosphatase-activating protein (GAP), p85 PI 3-kinase binding protein, phospholipase C-γ, and the adaptor protein Grb2/Sem5.

SH2 and SH3 domains represent one type of targeting sequence that might be present either in the kinase undergoing translocation or in the protein to which the kinase is targeted. In theory, many other examples of intracellular receptors for protein domains may exist. It has been suggested that PKC contains activator-independent binding sites for arginine-rich polypeptides distal to the catalytic site (Leventhal and Bertics, 1993). These binding sites may allosterically activate PKC but they may also allow targeting of PKC to specific subcellular locations. Similarly, Mochly-Rosen et al. (1991) have suggested that cardiac myocytes and fibroblasts contain specific receptors for PKC, so-called RACKS (Ron et al., 1994), that allow the targeting of PKC to cytoskeletal elements. A peptide inhibitor derived from the sequence of two PKC binding proteins (annexin I and RACKI) has been shown to interfere with translocation of β-PKC (Ron and Mochly-Rosen, 1994).

Nuclear proteins appear to accumulate in the nucleus because they contain nuclear targeting signals that allow selective entry through the nuclear pores. The first nuclear targeting signal to be analyzed at the amino acid level was that of SV40 large T antigen. The short sequence PKKKRKV appears to be sufficient to direct proteins to the cell nucleus. One particular amino acid in this sequence may be crucial for efficient nuclear targeting. A biparticle nuclear targeting sequence has also been described consisting of two basic regions of three to four residues each separated by a spacer of approximately 10 amino acids (Robbins et al., 1991). Similarly, tethering and intracellular targeting domains for cyclic AMP-dependent protein kinase II beta have been demonstrated to be present in the "A Kinase Anchor Proteins" (AKAPs) present in neurons (Glantz et al., 1993).

V. KINASE CASCADES IN SMOOTH MUSCLE CONTRACTION

In the introduction of this chapter, we presented a simplified model for signal transduction from the plasmalemmal receptors to the contractile proteins. In recent years, more protein kinases as well as protein substrates have been suggested as regulators of smooth muscle function. Protein kinases such as PKC, MAP kinase, and c-Raf-1, as well as adaptor proteins, and guanidine nucleotide exchange factors have been implicated in regulating smooth muscle growth. Most of these kinases have also been identified in differentiated smooth muscle cells and have been suggested to activate a cascade of events leading to enhancement of smooth muscle contraction. As mentioned in the introduction, it has been proposed that PKC may inhibit MLC phosphatase, thus increasing the amount of MLC phosphorylation and as a result enhancing the magnitude of force (Gong et al., 1992). Others have suggested, however, that PKC might cause contraction by increasing the phosphorylation of the actin binding protein calponin. Phosphorylation of calponin reverses its inhibition of actin-activated myosin ATPase, and thus allows more actin to interact with myosin and increase force. Interestingly, imaging of calponin in toad gastric (Walsh et al., 1993) and ferret portal vein (Parker et al., 1994) smooth muscle has shown a filamentous distribution along the longitudinal axis of the cell under resting conditions. In the activated ferret portal cell, however, calponin appears to undergo a redistribution to a subplasmalemmal structure (Parker et al., 1994).

Significant tyrosine kinase and MAP kinase activities have been reported in several vascular preparations; however, the role of these kinases in regulating smooth muscle contraction is not clearly understood. Studies from our laboratory have shown that contraction of ferret aorta cell can be mediated by a cascade of protein kinase activations (Khalil et al., 1995) in the absence of significant increases in [Ca^{2+}]$_i$ and MLC phosphorylation (Bergh-Menice et al., 1995). On the basis of existing evidence, we propose the following enzyme cascade leading to a Ca^{2+}-independent smooth muscle contraction. As shown in Fig. 3, activation of a ferret aorta cell with an agonist stimulates the breakdown of membrane phospholipids and increases

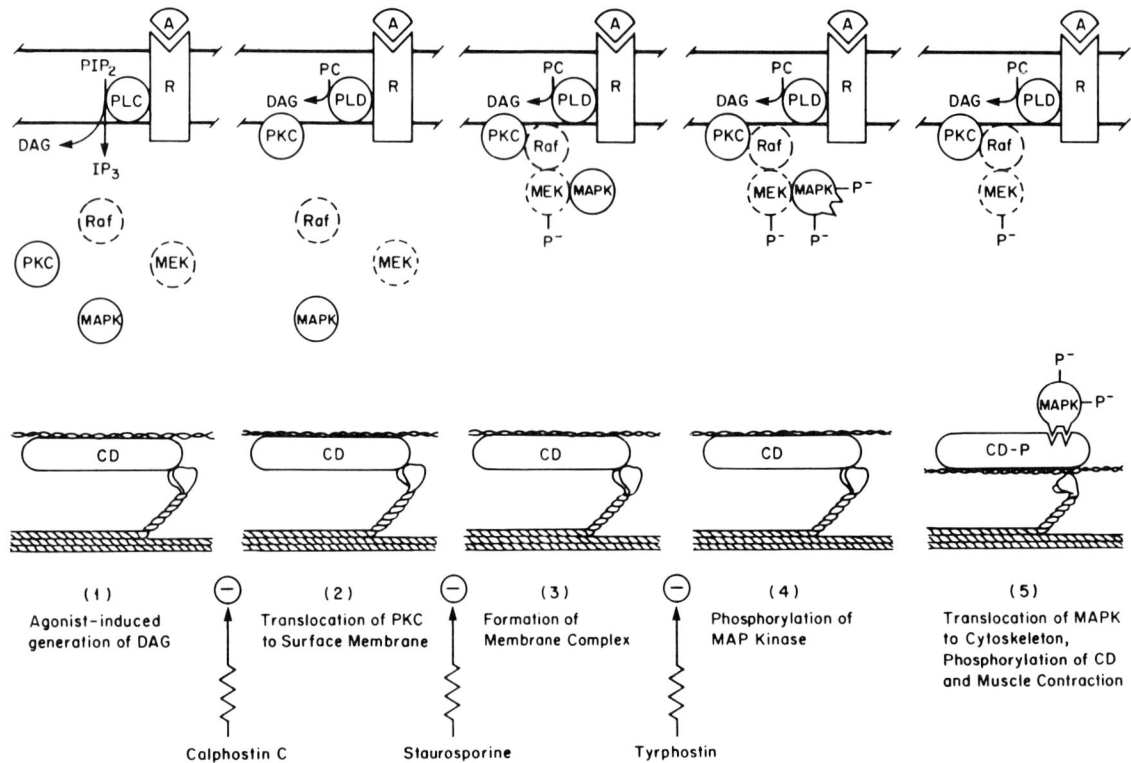

FIGURE 3 Kinase cascade leading to smooth muscle contraction. DAG, diacylglycerol; PKC, protein kinase C; MEK, MAP kinase kinase; MAPK, MAP kinase; CD, caldesmon.

the production of diacylglycerol (DAG) (Griendling *et al.*, 1986). DAG causes translocation of cystosolic ε-PKC to the surface membrane, where it is fully activated. Activated ε-PKC, by an as yet unidentified mechanism, stimulates the translocation of cytosolic MAP kinase kinase (MEK) and MAP kinase to the plasmalemma, where they form a surface membrane kinase complex (Moodie *et al.*, 1993). PKC causes phosphorylation and activation of MEK, which in turn phosphorylates MAP kinase at both threonine and tyrosine residues. Tyrosine phosphorylation targets MAP kinase to the cytoskeleton (Khalil and Morgan, 1993), where it phosphorylates caldesmon (Adam *et al.*, 1989, 1992; Childs *et al.*, 1992). Phosphorylation of caldesmon reverses its inhibition of MgATPase activity and thus increases actin–myosin interaction and smooth muscle contraction (Sobue and Sellers, 1991; Katsuyama *et al.*, 1992).

VI. SUMMARY AND PERSPECTIVES

It is clear that far more information is available on enzyme trafficking for nonmuscle and cultured smooth muscle cells than that for differentiated, contractile smooth muscle cells. However, it is expected, especially with advances in digital imaging techniques, that this situation will rapidly change. It is anticipated that information on spatial trafficking of enzymes during contractile activation of smooth muscle will be of considerable benefit in choosing the physiologically relevant biochemical pathway utilized by the smooth muscle cell from the multitude of possible pathways suggested by *in vitro* studies.

References

Adam, L. P., and Hathaway, D. R. (1993). *FEBS Lett.* **322,** 56–60.

Adam, L. P., Haeberle, J. R., and Hathaway, D. R. (1989). *J. Biol. Chem.* **264,** 7698–7703.

Adam, L. P., Gapinski, C. J., and Hathaway, D. R. (1992). *FEBS Lett.* **302,** 223–226.

Adelstein, R. S., and Klee, C. B. (1981). *J. Biol. Chem.* **256,** 7501–7509.

Aderem, A. A., Albert, K. A., Keum, M. M., Wang, J. K. T., Greengard, P., and Cohn, Z. A. (1988). *Nature (London)* **332,** 362–364.

Akiyama, T., Ishida, J., Nakagawa, S., Ogawara, H., Watanabai, S., Itoh, N., Shibuya, M., and Fukami, Y. (1987). *J. Biol. Chem.* **262,** 5592–5595.

Aksoy, M. O., Williams, D., Sharkey, E. M., and Hartshorne, D. J. (1976). *Biochem. Biophys. Res. Commun.* **69,** 35–41.

Anderson, N. G., Maller, J. L., Tonks, N. K., and Sturgill, T. W. (1990). *Nature* **343,** 651–653.

Andrea, J. E., and Walsh, M. P. (1992). *Hypertension* **20**, 585–595.

Atkinson, T. P., Lee, C. W., Rhee, S. G., and Hohman, R. J. (1993). *J. Immunol.* **151**, 1448–1455.

Bacskai, B. J., Hochner, B., Mahaut-Smith, M., Adams, S. R., Kaang, B.-K., Kandel, E. R., and Tsien, R. Y. (1993). *Science* **260**, 222–226.

Bosca, L., and Moran, F. (1993). *Biochem. J.* **290**, 827–832.

Boulton, T. G., Yancopoulos, G. D., Gregory, J. S., Slaughter, C., Moomaw, C., Hsu, Y., and Cobb, M. H. (1990). *Science* **249**, 64–67.

Casey, P. J. (1994). *Curr. Opin. Cell Biol.* **6**, 219–225.

Cazaubon, S. M., and Parker, P. J. (1993). *J. Biol. Chem.* **268**, 17559–17563.

Cazaubon, S. M., and Parker, P. J. (1994). *Biochem. J.* **301**, 443–448.

Cecchi, G., and Bagni, M. A. (1994). *News Physiol. Sci.* **9**, 3–7.

Chen, C. S., and Poenie, M. (1993). *J. Biol. Chem.* **268**, 15812–15822.

Chen, R. H., Sarnecki, C., and Blenis, J. (1992). *Mol. Cell. Biol.* **12**, 915–927.

Childs, T. J., Watson, M. H., Sanghera, J. S., Campbell, D. L., Pelech, S. L., and Mak, A. S. (1992). *J. Biol. Chem.* **267**, 22853–22859.

Collins, E. M., Walsh, M. P., and Morgan, K. G. (1992). *Am. J. Physiol.* **252**, H754–H762.

Di Salvo, J., Gifford, D., and Kokkinakis, A. (1989). *J. Biol. Chem.* **264**, 10773–10778.

Garcia, A. M., Rowell, C., Ackermann, K., Kowalczyk, J. J., and Lewis, M. D. (1993). *J. Biol. Chem.* **268**, 18415–18418.

Gazit, A., Yaish, P., Gilon, C., and Levitsky, A. (1989). *J. Med. Chem.* **32**, 2344–2352.

Gerthoffer, W. T., Murphy, K. A., and Gunst, S. J. (1989). *Am. J. Physiol.* **257**, C1062–C1068.

Glantz, S. B., Li, Y., and Rubin, C. S. (1993). *J. Biol. Chem.* **268**, 12796–12804.

Godt, R. E., Kirby, A. C., and Gordon, A. M. (1984). *Am. J. Physiol.* **246**, C148–C153.

Gong, M. C., Cohen, P., Kitazawa, T., Skebe, M., Masuo, M., Somlyo, A. P., and Somlyo, A. V. (1992). *J. Biol. Chem.* **267**, 14662–14668.

Gonzalez, F. A., Seth, A., Raden, D. L., Bowman, D. S., Fay, F. S., and Davis, R. J. (1993). *J. Cell Biol.* **122**, 1089–1101.

Graham, S. L., deSolmes, S. J., Guiliani, E. A., Kohl, N. E., Mosser, S. D., Oliff, A. I., Pompliano, D. L., Rands, E., Breslin, M. J., and Deana, A. A. (1994). *J. Med. Chem.* **37**, 725–732.

Griendling, K. K., Rittenhouse, S. E., Brock, T. A., Ekstein, L. S., Gimbrone, M. A., Jr., and Alexander, R. W. (1986). *J. Biol. Chem.* **261**, 5901–5906.

Haller, H., Smallwood, J. I., and Rasmussen, H. (1990). *Biochem. J.* **270**, 375–381.

Haller, H., Quass, P., Lindschau, C., Luft, F. C., and Distler, A. (1994). *Hypertension* **23**, 848–852.

Harlow, E. D., and Lane, D., eds. (1988). "Antibodies: A Laboratory Manual." Cold Spring Harbor Lab., Cold Spring Harbor, NY.

Hartwig, J. H., Thelen, M., Rosen, A., Jammey, P. A., Nairn, A. C., and Aderem, A. (1992). *Nature (London)* **356**, 618–622.

Hocevar, B. A., Burns, D. J., and Fields, A. P. (1993). *J. Biol. Chem.* **268**, 7545–7552.

Hodgkin, A. L., and Horowicz, P. (1957). *J. Physiol. (London)* **136**, 17P–18P.

House, C., and Kemp, B. E. (1987). *Science* **238**, 1726–1728.

Iizuka, K., Somlyo, A. P., and Somlyo, A. V. (1994). *Biophys. J.* **66**, A409.

James, G., and Olson, E. (1992). *J. Cell Biol.* **116**, 863–874.

James, G. L., Goldstein, J. L., Brown, M. S., Rawson, T. E., Somers, T. C., McDowell, R. S., Crowley, C. H., Lucas, B. K., Levinson, A. D., and Marsters, J. C. (1993). *Science* **260**, 1937–1942.

Jiang, J. J., and Morgan, K. G. (1989). *Pflügers Arch.* **413**, 637–643.

Kahn, D. W., and Besterman, J. M. (1991). *Proc. Natl. Acad. Sci. U.S.A.* **88**, 6137–6141.

Katsuyama, H., Wang, C.-L.A., and Morgan, K. G. (1992). *J. Biol. Chem.* **267**, 14555–14558.

Khalil, R., Lodge, N., Saida, K., and van Breemen, C. (1987). *J. Hypertens.* **5**, S5–S15.

Khalil, R. A., and Morgan, K. G. (1991). *Circ. Res.* **69**, 1626–1631.

Khalil, R. A., and Morgan, K. G. (1992). *J. Physiol. (London)* **455**, 585–599.

Khalil, R. A., and Morgan, K. G. (1993). *Am. J. Physiol.* **265**, C406–C411.

Khalil, R. A., Lajoie, C. A., Resnick, M. S., and Morgan, K. G. (1992). *Am. J. Physiol.* **263**, C714–C719.

Khalil, R. A., Lajoie, C., and Morgan, K. G. (1994). *Am. J. Physiol.* **266**, C1544–C1551.

Khalil, R. A., Menice, C. B., Wang, C. L. A., and Morgan, K. G. (1995). *Circ. Res.* **76**, 1101–1108.

Kiley, S. C., Parker, P. J., Fabbro, D., and Jaken, S. (1992). *Carcinogenesis (London)* **13**, 1997–2001.

Kobayashi, E., Nakano, H., Morimoto, M., and Tamaoki, T. (1989). *Biochem. Biophys. Res. Commun.* **159**, 548–553.

Kolch, W., Heldecker, G., Kochs, G., Hummel, R., Vahidi, H., Mischak, H., Finkenzeller, G., Marme, D., and Rapp, U. R. (1993). *Nature (London)* **364**, 249–252.

Krause, H., Dieter, P., Schulze-Specking, A., Ballhorn, A., and Decker, K. (1991). *Eur. J. Biochem.* **199**, 355–359.

Kyriakis, J. M., App, H., Zhang, X., Banergee, P., Brautigan, D. L., Rapp, U. R., and Avrush, J. (1992). *Nature (London)* **358**, 417–421.

Langlands, J. M., and Diamond, J. (1994). *Eur. J. Pharmacol.* **266**, 229–236.

Leach, K. L., Ruff, V. A., Jarpe, M. B., Adams, L. D., Fabbro, D., and Raben, D. M. (1992). *J. Biol. Chem.* **267**, 21816–21822.

Lee, M. W., and Severson, D. L. (1994). *Am. J. Physiol.* **267**, C659–C678.

Lenormand, P., Sardet, C., Pages, G., L'Allemain, G., Brunet, A., and Pouyssegur, J. (1993). *J. Cell Biol.* **122**, 1079–1088.

Leventhal, P. S., and Bertics, P. J. (1993). *J. Biol. Chem.* **268**, 13906–13913.

Levitzki, A., and Gilon, C. (1991). *Trends Pharmacol. Sci.* **12**, 171–174.

Liou, Y.-M., and Morgan, K. G. (1994). *Am. J. Physiol.* **267**, C980–C989.

Lyu, R.-M., Smith, L., and Smith, J. B. (1992). *Am. J. Physiol.* **263**, C628–C634.

Macdonald, S. G., Crews, C. M., Wu, L., Driller, J., Clark, R., Erickson, R. L., and McCormick, F. (1993). *Mol. Cell. Biol.* **13**, 6615–6620.

Maroney, A. C., and Macara, I. G. (1989). *J. Biol. Chem.* **264**, 2537–2544.

Marshall, C. J. (1993). *Science* **259**, 1865–1866.

Masuo, M., Reardon, S., Ikebe, M., and Kitazawa, T. (1994). *J. Gen. Physiol.* **104**, 265–286.

Menice, C. B., Lajoie, C., and Morgan, K. G. (1995). *Biophys. J.* **68**, A169.

Metz, S. A., Rabaglia, M. E., Stock, J. B., and Kowlura, A (1993). *Biochem. J.* **295**, 31–40.

Mochly-Rosen, D., Khaner, H., Lopez, J., and Smith, B. L. (1991). *J. Biol. Chem.* **266**, 14866–14868.

Moodie, S. A., Willumsen, B. M., Weber, M. J., and Wolfman, A. (1993). *Science* **260**, 1658–1661.

Nishizuka, Y. (1992). *Science* **258**, 607–614.

Orr, J. W., Keranen, L. M., and Newton, A. C. (1992). *J. Biol. Chem.* **267**, 15263–15266.

Parker, C. P., Takahashi, K., Tao, T., and Morgan, K. G. (1994). *Am. J. Physiol.* **267**, C1262–C1270.

Pelech, S. L., and Sanghera, J. S. (1992). *Science* **257**, 1355–1356.

Rembold, C. M. (1990). *J. Physiol. (London)* **429**, 77–94.

Robbins, J., Dilworth, S. M., Laskey, R. A,. and Dingwall, C. (1991). *Cell (Cambridge, Mass.)* **64**, 615–623.

Ron, D., and Mochly-Rosen, D. (1994). *J. Biol. Chem.* **269**, 21395–21398.

Ron, D., Chen, C. H., Caldwell, J., Jamieson, L., Orr, E., and Mochly-Rosen, D. (1994). *Proc. Natl. Acad. Sci. U.S.A.* **91**, 839–843.

Rossomando, A. J., Payne, D. M., Webber, M. J., and Sturgill, T. W. (1989). *Proc. Natl. Acad. Sci.* **86**, 6940–6943.

Rüegg, U. T., and Burgess, G. M. (1989). *Trends Pharmacol. Sci.* **10**, 218–220.

Saitoh, T., and Schwartz, J. H. (1985). *J. Cell Biol.* **100**, 835–842.

Sakane, F., Yamada, K., Imai, S., and Kanoh, H. (1991). *J. Biol. Chem.* **266**, 7096–7100.

Schaap, D., van der Wal, J., Howe, L. R., Marshall, C. J., and van Blitterswijk, W. J. (1993). *J. Biol. Chem.* **268**, 20232–20236.

Schonhardt, T., and Ferber, E. (1987). *Biochem. Biophys. Res. Commun.* **149**, 769–775.

Secrest, R. J., Lucaites, V. L., Mendelsohn, L. G., and Cohen, M. L. (1991). *J. Pharmacol. Exp. Ther.* **256**, 103–109.

Singer, H. A., Schworer, C. M., Sweeley, C., and Benscoter, H. (1992). *Arch. Biochem. Biophys.* **299**, 320–329.

Skalhegg, B. S., Tasken, K., Hansson, V., Huifeldt, H. S., Jahnsen, T., and Lea, T. (1994). *Science* **263**, 84–87.

Sobieszek, A. (1977). *Eur. J. Biochem.* **73**, 477–483,.

Sobieszek, A. (1991). *J. Mol. Biol.* **220**, 947–957.

Sobue, K., and Sellers, J. R. (1991). *J. Biol. Chem.* **266**, 12115–12118.

Soderling, T. R. (1990). *J. Biol. Chem.* **265**, 1823–1826.

Somlyo, A. V., and Somlyo, A. P. (1994). *Nature (London)* **372**, 231–236.

Stokoe, D., Macdonald, S. G., Cadwallader, K., Symons, M., and Hancock, J. F. (1994). *Science* **264**, 1463–1467.

Stull, J. T., Gallagher, P. J., Herring, B. P., and Kamm, K. E. (1991). *Hypertension* **17**, 723–732.

Suematsu, E., Resnick, M., and Morgan, K. G. (1991). *Am. J. Physiol.* **261**, C253–C258.

Takai, Y., Kishimoto, A., Kikkawa, U., Mori, T., and Nishizuka, Y. (1979). *Biochem. Biophys. Res. Commun.* **91**, 1218–1224.

Takai, Y., Kishimoto, A., Kawahara, Y., Minakuchi, R., Sano, K., Kikkawa, U., Mori, T., Yu, B., Kaibuchi, K., and Nishizuka, Y. (1981). *Adv. Cyclic Nucleotide Res.* **14**, 301–313.

Tan, E. W., and Rando, R. R. (1992). *Biochemistry* **31**, 5572–5578.

Taniguchi, H., and Manenti, S. (1993) *J. Biol. Chem.* **268**, 9960–9963.

Tansey, M. G., Word, R. A., Hidaka, H., Singer, H. A., Schworer, C. M., Kamm, K. E., and Stull, J. T. (1992). *J. Biol. Chem.* **267**, 12511–12516.

Tansey, M. G., Luby-Phelps, K., Kamm, K. E, and Stull, J. T. (1994). *J. Biol. Chem.* **269**, 9912–9920.

Taylor, D. L., and Wang, Y. (1989). *In* "Fluorescence Microscopy of Living Cells in Culture" (D. L. Taylor and Y. Wang, eds.), Part B, pp. 1–478. Academic Press, San Diego, CA.

Thelen, M., Rosen, A., Nairn, A. C., and Aderem, A. (1991). *Nature (London)* **351**, 320–322.

van Breemen, C., Hwang, K., Loutzenhiser, R., Lukeman, S., and Yamamoto, H. (1985). *In* "Cardiovascular Effects of Dihydropyridine-Type Calcium Antagonists and Agonists" (A. Fleckenstein, C. van Breeman, R. Gross, and F. Hoffmeister, eds.), pp. 58–71. Springer-Verlag, New York.

Walsh, M. P., Carmichael, J. D., and Kargacin, G. J. (1993). *Am. J. Physiol.* **265**, 1371–1378.

Watanabe, M., Hachiya, T., Hagiwara, M., and Hidaka, H. (1989). *Arch. Biochem. Biophys.* **273**, 165–169.

Yaish, P., Gazit, A., Gilon, C, and Levitski, A. (1988). *Science* **242**, 933–945.

Yang, L. J., Rhee, S. G., and Williamson, J. R. (1994). *J. Biol. Chem.* **269**, 7156–7162.

Yoshihara, Y., and Watanabe, Y. (1990). *Biochem. Biophys. Res. Commun.* **170**, 484–490.

Zhang, J., Wang, L., Schwurtz, J., Bond, R. W., Bishop, W. R. (1994) *J. Biol. Chem.* **269**, 19578–19584.

CONTRACTION
AND
RELAXATION

Protein Phosphorylation during Contraction and Relaxation

MICHAEL BÁRÁNY

Department of Biochemistry
College of Medicine, University of Illinois at Chicago
Chicago, Illinois

KATE BÁRÁNY

Department of Physiology and Biophysics
College of Medicine, University of Illinois at Chicago
Chicago, Illinois

I. INTRODUCTION

Studies on protein phosphorylation in smooth muscle played a major role in establishing protein phosphorylation as a key regulatory mechanism in cellular physiology. This started with the observation that the 20-kDa myosin light chain (LC20) became phosphorylated in ^{32}P-labeled porcine carotid arterial muscles, contracted by K^+ or norepinephrine (NE) (Barron *et al.*, 1979). The Ca^{2+} requirement for LC20 phosphorylation was also noted at the same time. Subsequently, reversible phosphorylation and dephosphorylation of LC20 during the contraction–relaxation cycle of the arterial smooth muscle was demonstrated (Barron *et al.*, 1980). Soon thereafter, concurrent changes in stress, shortening velocity, and LC20 phosphorylation were described at the initial phase of arterial contraction (Dillon *et al.*, 1981). Unexpectedly, LC20 phosphorylation declined while stress was maintained and this led to the postulation of the "latch" state, that is, the slowly cycling dephosphorylated cross-bridges during stress maintenance (Dillon *et al.*, 1981).

For a while, protein phosphorylation in smooth muscle meant only LC20 phosphorylation, but in 1989 Adam *et al.* described the second phosphoprotein caldesmon (CD), which was phosphorylated already in the resting arterial muscle and further phosphorylated upon muscle stimulation. The sensitivity of the ^{32}P-labeling technique also revealed the existence of other phosphoproteins in arterial muscle, such as desmin (DS) and a 28-kDa protein (Bárány *et al.*, 1992a), and research on the physiological significance of these phosphoproteins began.

This chapter reviews the extensive literature on LC20 phosphorylation and the research on the other phosphoproteins. The reader will find additional references in Chapters 26 and 27 (this volume). The *in vitro* phosphorylation of LC20 is described in Chapter 2 and the phosphorylation of myosin light chain kinase (MLCK) is outlined in Chapters 9 and 27.

II. QUANTIFICATION OF PROTEIN PHOSPHORYLATION

As discussed in this chapter, the extent of protein phosphorylation varies greatly among different laboratories, sometimes leading to contradictory results. Therefore, it is appropriate to describe in detail the methods that are used for the quantification.

A. ^{32}P Labeling

In our laboratory, the smooth muscles (porcine carotid arteries, veins, uteri, trachea, stomach, and bladder) are ^{32}P-labeled by incubation with carrier-free [^{32}P]orthophosphate, 2 mCi per 70 ml physiological salt solution (PSS) (Bárány *et al.*, 1992a), and bubbled with a gas mixture of 95% O_2–5% CO_2 at 37°C, pH 7.4, for 90–120 min. The muscle strips are then washed 15 times with PSS in 30 min to remove ^{32}P from the extracellular space of the muscles. Subsequently the physiological experiments are performed, and the muscles are frozen in liquid nitrogen in various functional states (Bárány *et al.*, 1992a; Bárány and Bárány, 1993a).

The frozen muscles are pulverized by percussion using liquid nitrogen-chilled stainless-steel mortars and pestles in a cold room of 4°C. The frozen powder is immediately homogenized with 3% perchloric acid (PCA) and centrifuged at 4°C, and the supernatants are saved for the determination of the specific radioactivity of ^{32}P-labeled phosphocreatine (PCr) of the muscle, which is equal to that of the [γ-^{32}P]ATP of the muscle (Kopp and Bárány, 1979). The PCA-treated muscle residues are washed with a solution containing 2% trichloroacetic acid (TCA) and 5 mM NaH$_2$PO$_4$ several times to remove unbound [^{32}P]phosphates, and then dissolved in 1.2% SDS–125 mM Na$_2$HPO$_4$, pH 8.5, at room temperature. After dialysis against 5000 vols of 0.02% SDS–2 mM (NH$_4$)HCO$_3$ at 25°C overnight, insoluble connective tissue is removed by centrifugation at 100,000g, and the protein content of the supernatants is determined (Bárány and Bárány, 1959). Aliquots of the supernatants are freeze-dried, and 400-μg protein samples are subjected to two-dimensional (2D) gel electrophoresis (Bárány et al., 1983). The specific protein spot is cut from the stained wet gel and digested with 30% H$_2$O$_2$, and the radioactivity is determined by liquid scintillation counting. The incorporation of [^{32}P]phosphate into the protein is quantified from: the counts in the H$_2$O$_2$-digested protein spot, the specific radioactivity of [^{32}P]PCr in the muscle, the known amount of total protein applied onto the gel, and the specific protein content of the total protein. When the content of a specific protein in a muscle is unknown, it has to be determined; electrophoresis on one-dimensional 5–7.5% polyacrylamide gels followed by measuring the percentage staining intensity of the specific protein band, relative to the 100% staining intensity of all protein bands, is an appropriate procedure. The incorporation is expressed in terms of mol [^{32}P]phosphate/mol protein and computed as

$$\frac{\text{mol } [^{32}\text{P}]\text{phosphate}}{\text{mol protein}} = \frac{\text{cpm in the protein spots}}{\text{cpm/mol } [^{32}\text{P}]\text{PCr} \times \text{mol protein}}$$

The extensive PCA and TCA treatments in this procedure denature (and thereby inactivate) the enzymes involved in protein phosphorylation and also the proteolytic enzymes, therefore, the results reflect the physiological state of the proteins in the muscle. Furthermore, virtually all proteins are extracted from the muscle, and thus the information gained refers to the entire protein phosphorylation pattern of the cell.

It should be noted that the [^{32}P]phosphate content of a phosphoprotein does not necessarily correspond to its total phosphate content. Usually, endogenous protein-bound phosphate exchanges with the [^{32}P]-phosphate in the medium during a prolonged incubation. In case of doubt, both the [^{32}]phosphate and the total phosphate content of the protein have to be determined (Homa and Bárány, 1983).

B. Densitometry and Immunoblotting

Densitometry and immunoblotting are used for quantification of the phosphorylated LC20. The proteins are extracted with a solution containing 8 M urea from an acetone-dried muscle powder, and the phosphorylated and nonphosphorylated forms of LC20 are separated by glycerol–urea polyacrylamide gel electrophoresis (PAGE) (Hathaway and Haeberle, 1985) and quantified by densitometry. Alternatively, the acetone-dried powder is extracted with a solution containing 1% SDS, and the LC20s in the extract are separated by 2D gel electrophoresis (Driska et al., 1981). Many authors do not take into account the phosphorylation of the minor LC20 isoform or the diphosphorylation of the major and minor isoforms, and therefore underestimate the actual phosphoryl content of LC20.

Moore and Stull (1984) extract the liquid nitrogen-frozen muscle with a solution containing 100 mM Na$_4$P$_2$O$_7$, 100 mM NaF, 5 mM EGTA, and 500 mM KCl. Native myosin is isolated from the extract on pyrophosphate–PAGE (Silver and Stull, 1982a) and subjected to isoelectric focusing for separation and densitometry of the LC20s.

In the radioimmunoblotting method of Hathaway and Haeberle (1985), the phosphorylated and non-phosphorylated LC20 forms are separated by glycerol–urea PAGE, then transferred by electroblotting from gels to nitrocellulose paper, and incubated with antiserum for LC20; the LC20–antibody complex is labeled with ^{125}I-labeled protein A and the radioactivity is determined. This method of Hathaway and Haeberle was modified by Persechini et al. (1986), who incubated the LC20 on the nitrocellulose paper with peroxidase-linked LC20–antibody, visualized the complex as a blue band after treatment with 4-chloro-1-naphthol, and determined the extent of LC20 phosphorylation by densitometry.

III. LC20 PHOSPHORYLATION

Figure 1 illustrates phosphorylation of LC20 in resting, stretched, and K$^+$-stimulated ^{32}P-labeled arterial muscle. Four spots of LC20 are resolved by staining on 2D gels (top), referred to as Spots 1, 2, 3, and 4 from lower to higher pH. Major changes are seen in staining intensity of Spot 4, which decreases (relative to the resting state) upon stretching, stimulation, or both stretching and stimulation, and of Spot 3, which in-

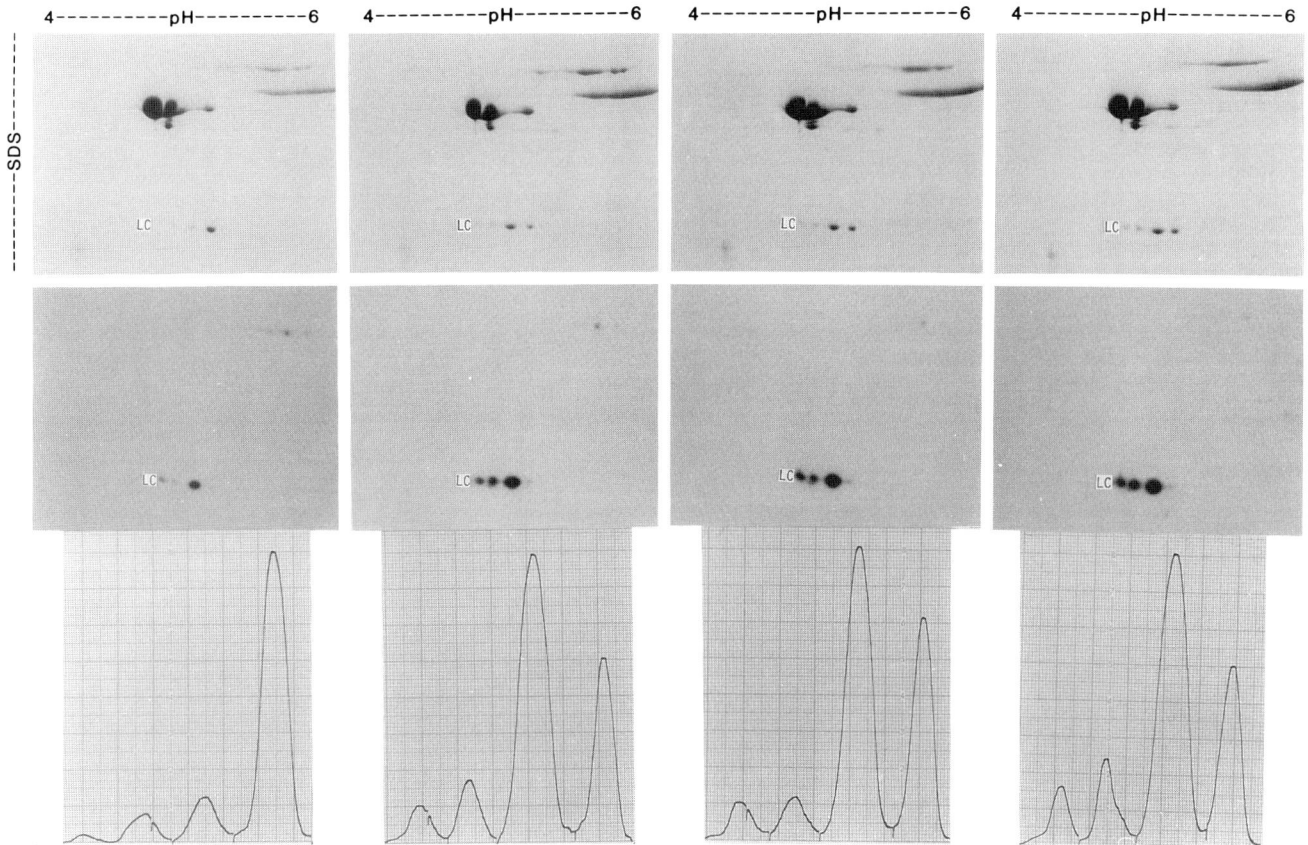

FIGURE 1 LC20 phosphorylation in porcine carotid arterial muscles. Top: staining profiles; middle: corresponding autoradiograms; bottom: densitometric tracings of LC20. First frame: muscle at rest; second frame: muscle stretched; third frame: muscle K$^+$-stimulated; fourth frame: muscle stretched and then K$^+$-stimulated. LC, LC20. From Bárány et al. (1985a, Fig. 1, p. 7127).

creases simultaneously under these conditions (bottom). Such changes in staining intensities are accepted criteria for protein phosphorylation (Aksoy and Murphy, 1983). The autoradiogram (middle) shows the corresponding changes in radioactivity. The major incorporation of [^{32}P]phosphate occurs into *Spot* 3, but the incorporation is very significant also into *Spots* 1 and 2.

LC20 phosphorylation is Ca^{2+} dependent in intact muscle (Barron et al., 1979, 1980), however, under special conditions significant phosphorylation may occur in the absence of Ca^{2+} (Bárány et al., 1992b; Table V).

A. During the Initial Phase of Contraction

It is generally accepted that phosphorylation of LC20 is the primary mechanism for initiating smooth muscle contraction (Hai and Murphy, 1989; Bárány and Bárány, 1990; de Lanerolle and Paul, 1991; Stull et al., 1991; Hartshorne and Kawamura, 1992; Somlyo and Somlyo, 1992). Table I lists the extent of LC20 phosphorylation in resting and stimulated muscles in

26 different cases. The data comprise 15 different muscles, tonic and phasic, stimulated in 10 different ways and analyzed in 12 different laboratories. Table I shows a large span in the increment of LC20 phosphorylation upon muscle stimulation: Δmol phosphate/mol LC20 varies from 0.14 to 0.81 between stimulated and resting muscles. In a previous review, Kamm and Stull (1985b) collected data from 14 different cases and found that Δmol phosphate varied from 0.22 to 0.68. It is not clear what causes this large discrepancy, inadequate phosphorylated LC20 determination or the tissue itself. Nevertheless, it is a warning sign not to develop quickly new theories for the molecular mechanism of smooth muscle contraction based on a low level of LC20 phosphorylation.

The data of Table I indicate: (1) The maximal phosphorylation in the initial phase of the contraction is close to 0.5 mol P/mol LC20 (0.54 ± 0.19, $n = 26$), suggesting that the phosphorylation of at least one myosin head is required for the contractile interaction of myosin and actin. (2) There are certain drugs that do

TABLE I   **Extent of LC20 Phosphorylation in Resting and Stimulated Smooth Muscles**

Muscle	Stimulus	Temperature (°C)	Resting	Stimulated	Reference
			(mol P/mol LC20)		
Porcine carotid artery	10 μM histamine, rhythmic	37	0.12	0.49	Driska et al. (1989)
	10 μM histamine	37	0.08	0.63	McDaniel et al. (1992)
	10 μM histamine	37	0.25[a]	0.80[a]	Bárány et al. (1992a)
	100 mM K$^+$	37	0.25[a]	0.70[a]	Bárány et al. (1992b)
	Electrical	37	0.07	0.54	Singer and Murphy (1987)
	0.3 μM endothelin-1	37	0.05	0.35	Moreland et al. (1992)
	0.1 μM endothelin-1	37	0.09[a]	0.34[a]	Bárány and Bárány (1993a)
Rabbit femoral artery	pCa 5	25	0.20	0.97	Kitazawa et al. (1991a)
Ferret aorta	10 μM phenylephrine	35	0.12	0.47	Jiang and Morgan (1989)
Bovine trachea	Electrical	37	0.05	0.64	Kamm and Stull (1985a)
	1 μM carbachol	37	0.09	0.68	Silver and Stull (1982a)
Porcine trachea	100 μM carbachol	37	0.10[a]	0.35[a]	Bárány and Bárány (1993a)
Rat uterus	Spontaneous, rhythmic	37	0.35[a]	0.80[a]	Csabina et al. (1987)
	100 μM carbachol	37	0.19[a,b]	1.00[a]	Bárány et al. (1985c)
	100 mM K$^+$	21	0.05	0.46	Haeberle et al. (1985a)
Porcine uterus	100 μM carbachol	37	0.14[a,c]	0.58[a]	Bárány and Bárány (1993b)
Canine jugular vein	10 μM norepinephrine	37	0.21	0.46	Aksoy et al. (1986)
Canine femoral vein	10 μM norepinephrine	37	0.19	0.54	Aksoy et al. (1986)
Guinea pig portal vein	pCa 5	25	0.07	0.58	Kitazawa et al. (1991a)
Canine colon	100 μM acetylcholine	37	0.16	0.30	Gerthoffer et al. (1991)
	60 mM K$^+$	37	0.11	0.29	Gerthoffer et al. (1991)
Rabbit taenia coli	Electrical	37	0.12	0.32	Butler et al. (1986)
Guinea pig gallbladder	100 μM acetylcholine	37	0.06	0.48	Washabau et al. (1991)
	80 μM K$^+$	37	0.06	0.42	Washabau et al. (1991)
Porcine urinary bladder	109 mM K$^+$	37	0.10[a]	0.51[a]	Bárány and Bárány (1993a)
Porcine stomach	100 μM carbachol	37	0.08[a]	0.38[a]	Bárány and Bárány (1993a)

Note. Unless otherwise indicated, LC20 phosphorylation was quantified from densitometric scans of stained gels, with 100% phosphorylation being equal to 1.0 mol/mol. The maximal phosphorylation values are listed for the *stimulated* muscles.

[a]LC20 phosphorylation was quantified by [^{32}P]phosphate incorporation.

[b]1 μM isoproterenol.

[c]10 μM isoproterenol.

not induce maximal LC20 phosphorylation, for example, endothelin. (3) Electrical stimulation is not superior to agonist-induced LC20 phosphorylation. (4) There is no clear difference in the extent of phosphorylation between tonic and phasic muscles. (5) There is no clear indication for a temperature dependency of the phosphorylation. (6) Resting muscles do contain phosphorylated LC20. (7) Changes in LC20 phosphate content during rhythmic contraction imitate those occurring during stimulated contraction.

Studies of various laboratories showed that the initial levels of LC20 phosphorylation correlate with the onset of force development (Obara et al., 1987; Jiang and Morgan, 1989; Kitazawa et al., 1991a; unpublished results of this laboratory).

1. In Disease

In an experimentally induced disease, intimal hyperplasia in rabbit carotid artery, LC20 phosphorylation was increased following stimulation with 30 μM prostaglandin F$_{2\alpha}$, compared with the stimulated control artery (Seto et al., 1993). This was explained by a desensitization of the dephosphorylation system in the hyperplastic tissue.

B. During the Steady State of Contraction

LC20 phosphorylation is transient in the contractile phase of mechanical activity, that is, it declines at the time when the muscle remains in sustained contraction (Dillon *et al.*, 1981). Table II compares the phosphate content of LC20 at the initial stimulation with that at the steady state. In the majority of the experiments K$^+$ was used as a stimulant. With 100–110 mM K$^+$, the initial phosphorylation ranges from 0.30 to 0.74 mol P/mol LC (average of 0.50 ± 0.13, $n = 9$) and the steady-state phosphorylation ranges from 0.10 to 0.45 (0.27 ± 0.10, $n = 9$); on the average, approximately half of the initial value remains in the steady state. With 20–80 mM K$^+$, the initial phosphorylation ranges from 0.22 to 0.43 mol P/mol LC (0.33 ± 0.07, $n = 7$) and the steady-state phosphorylation ranges from 0.07 to 0.41 (0.26 ± 0.10, $n = 7$); actually three out of the

seven cases show no difference between initial and steady states.

Table II also lists the effect of agonist, electrical, and *p*Ca stimulation on the phosphorylation: at the initial state the range is from 0.33 to 0.80 mol P/mol LC (0.55 ± 0.15, $n = 11$) and at the steady state it is from 0.18 to 0.61 (0.35 ± 0.16, $n = 11$). With certain stimuli, for example, 100 μM histamine or 10 nM cholecystokinin, the steady-state values are close or identical to the initial ones.

The data of Table II have been compiled from 27 cases using 12 different smooth muscles, in 14 different laboratories. The results are somewhat controver-

TABLE II **Extent of LC Phosphorylation during the Initial and Steady-State Phases of Smooth Muscle Contraction**

Muscle	Stimulus	Temperature (°C)	Initial	Steady State (mol P/mol LC20)	Reference
Porcine carotid artery	110 mM K$^+$	37	0.54	0.34[c]	Moreland *et al.* (1987)
	110 mM K$^+$	23	0.56	0.27[e]	Moreland *et al.* (1987)
	40 mM K$^+$	37	0.38	0.31[c]	Moreland *et al.* (1987)
	110 mM K$^+$	36	0.32	0.16[h]	Adam *et al.* (1990)
	109 mM K$^+$	37	0.65	0.23[g]	Dillon *et al.* (1981)
	100 mM K$^+$	37	0.74[i]	0.38[i,h]	Bárány *et al.* (1991a)
Rabbit pulmonary artery	109 mM K$^+$, 20 mM K^{+a}	Room	0.50	0.23[c]	Himpens *et al.* (1988)
Guinea pig ileum	109 mM K$^+$	Room	0.30	0.10[e]	Himpens *et al.* (1988)
Rat uterus	100 mM K$^+$	21	0.46	0.30[i]	Haeberle *et al.* (1985a)
Guinea pig taenia caeci	100 mM K$^+$	37	0.45	0.45[b]	Obara *et al.* (1987)
Guinea pig gallbladder	80 mM K$^+$	37	0.43	0.41[d]	Washabau *et al.* (1994)
	60 mM K$^+$	37	0.33	0.34[d]	Washabau *et al.* (1994)
Canine colon	60 mM K$^+$	37	0.29	0.22[c]	Gerthoffer *et al.* (1991)
Canine trachea	60 mM K$^+$	37	0.28	0.22[b]	Gunst *et al.* (1994)
Rabbit thoracic artery	40 mM K$^+$	37	0.38	0.07[f]	Seto *et al.* (1990)
Ferret aorta	21 mM K$^+$	35	0.22	0.22[c]	Jiang and Morgan (1989)
Porcine carotid artery	100 μM histamine	37	0.80[i]	0.60[i,h]	Bárány *et al.* (1992a)
	3 μM histamine	37	0.47	0.31[b]	McDaniel *et al.* (1992)
Rabbit thoracic artery	7 μM histamine	37	0.47	0.28[f]	Seto *et al.* (1990)
Porcine uterus	100 μM histamine	37	0.72[i]	0.61[i,c]	K. Bárány and M. Bárány, unpublished
Bovine trachea	Electrical	37	0.64	0.20[c]	Kamm and Stull (1985a)
	1 μM carbachol	36	0.75	0.26[c]	Silver and Stull (1982b)
Canine trachea	1 μM carbachol	37	0.52	0.39[c]	Hai and Karlin (1993)
Porcine carotid artery	0.3 μM endothelin-1	37	0.35	0.22[c]	Moreland *et al.* (1992)
Ferret aorta	10 μM phenylephrine	35	0.47	0.18[c]	Jiang and Morgan (1989)
Guinea pig gallbladder	10 nM cholecystokinin	37	0.55	0.58[d]	Washabau *et al.* (1994)
Guinea pig ileum	*p*Ca 6.3	Room	0.33	0.25[b]	Kitazawa and Somlyo (1990)

Note. The initial phosphorylation was determined within 0.3–3.0 min after the stimulus.

[a]Each K$^+$ stimulus lasted 15 min, in the order given. The steady-state phosphorylation was determined: [b]10, [c]15, [d]20, [e]30, [f]40, [g]50, [h]60, and [i]90 min after the stimulus.

[i]Phosphorylation was quantified by [^{32}P]phosphate incorporation.

sial and do not lead to a clear conclusion as to what is the relationship between steady-state LC20 phosphorylation and stress. However, such a relationship was found within a single laboratory (D'Angelo *et al.*, 1992) by stimulating porcine carotid arteries with various agonists, activators of contraction, or depolarization. A hyperbolic relation was found between percentage phosphorylation and steady-state stress, with the maximal stress, approximately 1.75×10^5 N/m², at about 0.35 mol phosphate per mol LC20. This fits the four-state cross-bridge model for regulation of smooth muscle contraction (Murphy, 1994) (see Chapter 26, this volume).

C. During Contraction in the Absence of Ca²⁺

There are a few reports claiming smooth muscle contraction without LC20 phosphorylation in Ca²⁺-free medium: oxytocin-stimulated rat uterus (Oishi *et al.*, 1991), 12-deoxyphorbol 13-isobutyrate-treated rat aorta (Sato *et al.*, 1992), or rabbit saphenous vein contracted with the relatively selective α₂-agonist UK 14304 (Aburto *et al.*, 1993). Two problems are apparent with these papers: (1) the low level of LC20 phosphorylation (0.25–0.35 mol P/mol LC20) in the control muscles stimulated in Ca²⁺-containing PSS (cf. with the much higher phosphorylation values in Table I), and (2) the low force output of the experimental muscles, for example, only 10% of the control value (Oishi *et al.*, 1991). Thus, these reports do not overrule the concept that LC20 phosphorylation is required for smooth muscle contraction.

D. Without Contraction

Stretching of arterial smooth muscle (Ledvora *et al.*, 1983) or rat uterine muscle (Csabina *et al.*, 1986) induced phosphorylation of LC20 to the same extent as was observed in muscles contracted by agonists or K⁺ (Fig. 1). The extent of phosphorylation was proportional to the applied stretch, reaching a plateau at 1.6 times the resting length (Bárány *et al.*, 1985a); no active force development could be registered on maximally stretched muscles upon stimulation. Thus, active force development and LC20 phosphorylation were separated by stretching, evidently by eliminating the overlap between myosin and actin filaments in the muscle. Mobilization of Ca²⁺ from intracellular sources was necessary for the stretch-induced LC20 phosphorylation; the phosphorylation decreased as a function of time, suggesting the involvement of mechanosensitive Ca²⁺ transients (Lansman and Franco, 1991; Oike *et al.*, 1994). Tryptic phosphopeptide mapping identified MLCK as the enzyme that phosphorylated LC20 in the stretched muscle (Bárány *et al.*,

1990a). Based on these findings it seems likely that the physiological stretch that regulates the arterial blood pressure works through Ca²⁺-mediated MLCK activation. On the other hand, the myogenic tone of veins is not associated with LC20 phosphorylation but possibly with protein kinase C (PKC)-induced phosphorylation of CD (Laporte *et al.*, 1994).

E. Is It Related to Cross-bridge Cycling Rate?

Since the discovery of Dillon *et al.* (1981) that the rate of LC20 phosphorylation declines during force development, the question remained as to what mechanism controls the cross-bridge cycling rate in smooth muscle. A major experimental effort in many laboratories brought controversial results: several studies (Aksoy *et al.*, 1982; Kamm and Stull, 1985a; Walker *et al.*, 1994, and references herein) have found a correlation between shortening velocity (an estimate of cross-bridge cycling rate) and LC20 phosphorylation, whereas others have reported that the correlation does not hold (Siegman *et al.*, 1984; Haeberle *et al.*, 1985a; Butler *et al.*, 1986; Moreland *et al.*, 1987; Gerthoffer *et al.*, 1991; Fulginiti *et al.*, 1993; Gunst *et al.*, 1994; Washabau *et al.*, 1994). A part of the discrepancy may be due to technical difficulties, such as measuring shortening velocity in a tissue in which "there is no control over initial lengths" (Murphy, 1994), and another part may be caused by the low-sensitivity methods used to follow the time course of changes in LC20 phosphorylation. An alternative explanation for the discrepancy is the postulation of a mechanism, other than LC20 phosphorylation, for slowing the cross-bridge cycle. A second Ca²⁺-dependent mechanism was discussed some time ago (Hartshorne, 1987). However, this turned out to be the Ca²⁺-dependent phosphorylation of MLCK resulting in rapid decrease of LC20 phosphorylation (Tansey *et al.*, 1994) (see Chapter 27, this volume). Ca²⁺ binding to the thin filament proteins, CD and calponin (CP), may be considered as an LC20-independent regulatory mechanism, though the affinities of CD and CP for Ca²⁺/calmodulin appear to be two or three orders of magnitude weaker than the affinity of Ca²⁺/calmodulin for MLCK (Stull *et al.*, 1991).

The exchange of the covalently bound phosphate of LC20 during contraction (see Section V) supports a relationship between LC20 phosphorylation and cross-bridge cycling rate.

F. How Is It Coupled to the Interaction between Myosin and Actin?

Phosphorylation of LC20 in smooth muscle precedes tension development (Driska *et al.*, 1981; Bárány *et al.*, 1985b), indicating that this phosphorylation must initiate a reaction on myosin that is the prerequisite for the subsequent contraction. The $10S \rightarrow 6S$ conformational transition of myosin would be the likely candidate, because it is known from *in vitro* studies that dephosphorylated smooth muscle myosin forms an assembly-incompetent monomer that unfolds and assembles into filaments upon phosphorylation (Trybus, 1991). However, cryosections of relaxed and contracted gizzard muscle labeled with monoclonal antibody specific for the folded monomer did not detect a significant amount of monomeric myosin in either the resting or contracted state, thus ruling out the possibility of a myosin assembly upon smooth muscle stimulation, at least in the gizzard (Horowitz *et al.*, 1994). LC20 phosphorylation evidently changes the conformation of the ATPase site of myosin, enabling the activation of its MgATPase by actin, that is, phosphorylation releases the "suppressive" effect of the unphosphorylated LC20 (Hasegawa *et al.*, 1990). There is also a hypothesis (Hartshorne and Kawamura, 1992) that phosphorylation increases flexibility in the head–neck junction of the myosin molecule, changing the constricted position of the heads, and thereby facilitating its interaction with actin. The covalently linked dianionic phosphate group on the head increases its net negative charge, which may produce an outward movement of the head from the thick filament backbone through repulsive forces (Padrón *et al.*, 1991). Furthermore, the increased negative charge on the cross-bridge may provide an attractive force toward the actin filaments so that the phosphate on the head may be involved in ionic and hydrogen bond interactions to arginine (and lysine) side chains of the actin (cf. Johnson and Barford, 1993). Such interaction would facilitate the combination of myosin and actin filaments during smooth muscle activation (cf. Bárány and Bárány, 1990).

The idea that LC20 phosphorylation induces a structural change or a proper orientation of the contractile proteins is supported by the experiments of Butler *et al.* (1986). These authors stimulated rabbit taenia coli muscles electrically for 20 s and measured force and LC20 phosphorylation, then after a 30-s waiting period restimulated the muscles and measured the same parameters. Force was returned to 90% of its previous value after the second stimulation, while LC20 was hardly rephosphorylated. Such phenomena were also observed in this laboratory with K$^+$-stimulated arterial muscles. The data suggests that the initial LC20 phosphorylation fulfills its mission on the contractile apparatus, reducing the need for a maximal phosphorylation when the muscle is challenged for the second time. Experimentation on the LC20 "memory" may provide new information about the mechanism of muscle activation.

G. Diphosphorylation

Diphosphorylation of LC20 has been observed in intact smooth muscles such as carbachol- or oxytocin-stimulated rat uterus (Bárány *et al.*, 1985c), K$^+$-stimulated porcine carotid artery (Erdödi *et al.*, 1987), carbachol-stimulated bovine trachea (Colburn *et al.*, 1988), prostaglandin F$_{2\alpha}$-stimulated rabbit thoracic artery (Seto *et al.*, 1990), phorbol dibutyrate (PDBu)-treated carotid artery (Rokolya *et al.*, 1991), or endothelin-1-stimulated lamb tracheal muscle (Katoch, 1993). Tryptic phosphopeptide mapping of the diphosphorylated LC20 revealed peptides, characteristic for MLCK-catalyzed phosphorylation (Colburn *et al.*, 1988; Rokolya *et al.*, 1991) and characteristic for both MLCK and PKC phosphorylations (Rokolya *et al.*, 1991). Upon stimulation, 10–20% of the total LC20 becomes diphosphorylated (Bárány *et al.*, 1985c; Csabina *et al.*, 1987; Colburn *et al.*, 1988). It was suggested that the rate of force generation is accelerated with increased diphosphorylation of LC20 (Seto *et al.*, 1990).

IV. LC20 DEPHOSPHORYLATION

A. Related to Relaxation

Two avenues were explored to find the relationship between LC20 dephosphorylation and smooth muscle relaxation: correlation and kinetic analysis.

In a correlation analysis, skinned rat uterine muscles contracted by saturating Ca^{2+} and calmodulin in the PSS (presumably at room temperature) were relaxed by adding the catalytic subunit of a type-2 phosphatase to the PSS (Haeberle *et al.*, 1985b). Relaxation was associated with a significant, but not a complete, LC20 dephosphorylation. Furthermore, Ca^{2+}-contracted skinned guinea pig taenia coli fibers were relaxed by a synthetic peptide, corresponding to the calmodulin recognition sequence of MLCK, whereas two-thirds of LC20 remained phosphorylated (Rüegg *et al.*, 1989). In contrast, Hoar *et al.* (1985) did not find relaxation when LC20 was dephosphorylated in Ca^{2+}-contracted skinned gizzard muscle fibers, at 23°C, with the myosin light chain phosphatase SMP-IV.

In canine tracheal muscles contracted with 1 μM methacholine, at 37°C, and relaxed with 0.4 μM atropin or 40 μM forskolin, myosin phosphorylation and force decayed simultaneously (de Lanerolle, 1988). In rabbit tracheal smooth muscle stimulated with carbachol, at 37°C, LC20 dephosphorylation occurred at about the same rate as the decline in stress upon stimulus washout (Gerthoffer and Murphy, 1983). In porcine carotid arterial muscle stimulated at 37°C with K$^+$, histamine, or NE, and completely relaxed by washings with PSS, the [^{32}P]phosphate content of LC20 approached that of the resting muscle (Bárány et al., 1992a).

The experiments on intact muscles, unlike those on skinned fibers, agree with the concept that relaxation of contracted smooth muscle is associated with LC20 dephosphorylation (Stull et al., 1991; Hartshorne and Kawamura, 1992; Somlyo and Somlyo, 1994).

In kinetic analysis, the rate constants of both the dephosphorylation and relaxation were determined in several cases: in bovine tracheal muscle after 15 s of electrical stimulation, at 37°C, LC20 was dephosphorylated with an apparent first-order rate constant (k) of 0.26 s^{-1}, whereas k for relaxation was 0.13 s^{-1} (Kamm and Stull, 1985a); in histamine-treated rhythmically contracting porcine carotid arteries, at 37°C, the k values were 0.17 and 0.02 s^{-1}, respectively (Driska et al., 1989); in permeabilized rabbit portal vein stimulated at pCa 5 for 10 min, at 15°C, the values were 0.02 and <0.01 s^{-1}, respectively (Kitazawa et al., 1991b); and in rabbit urinary bladder muscle after 10 s of electrical stimulation, at 37°C, the values were 1.15 and 0.22 s^{-1}, respectively (Kwon and Murphy, 1994). In rabbit taenia coli electrically stimulated for 25 s at 18°C, the estimated dephosphorylation rate constant was 0.07 s^{-1} (Butler et al., 1986), whereas in permeabilized rabbit portal vein stimulated with pCa 4.5, at 20°C, the rate constant was 0.37 s^{-1} (Butler et al., 1994); the relaxation rate was not measured in these experiments.

These kinetic data show that LC20 dephosphorylation is much faster than smooth muscle relaxation, in contrast to the correlation data discussed in the foregoing, which show no major difference between extent of dephosphorylation and relaxation. A possible explanation for this discrepancy: if the spontaneous dephosphorylation of LC20 is neglected in the kinetic analysis, the rate of dephosphorylation will increase artificially.

The kinetic data also show other inconsistencies. For instance, at 37°C the dephosphorylation rate varies 7-fold and the relaxation rate 11-fold. The large temperature variations in the dephosphorylation rates may not be explained by the high temperature dependence of smooth muscle myosin light chain phosphatase (MLCP) activity (Mitsui et al., 1994). Actually, at room temperature the rate varies 5-fold within the same laboratory. Apparently more work is needed to find the relationship between rates of LC20 dephosphorylation and smooth muscle relaxation.

B. Uncoupled from Relaxation

Several laboratories reported smooth muscle relaxation without LC20 dephosphorylation: in tracheal muscles contracted with carbachol and relaxed either by a stepwise reduction of extracellular Ca^{2+} concentration (Gerthoffer, 1986) or with okadiac acid (Tansey et al., 1990) or contracted with endothelin-1 and relaxed with isoproterenol (Katoch, 1992); and in arterial muscles contracted with histamine and relaxed with either nitroprusside or nitroglycerine (McDaniel et al., 1992) or with Mg^{2+} (D'Angelo et al., 1992). Thus, with two different tonic smooth muscles, conditions exist under which relaxation is uncoupled from LC20 dephosphorylation. Dissociation of relaxation and LC20 dephosphorylation was also described for a phasic smooth muscle, porcine uterus, using several contracting and relaxing agents (Bárány and Bárány, 1993b). Furthermore, this was shown for another phasic muscle, porcine bladder (Bárány et al., 1994).

Smooth muscle relaxation without LC20 dephosphorylation can be elicited when the muscle is contracted with agonists and relaxed with agents in the presence of the agonists (Table III). Under these conditions, the stimulus for Ca^{2+} release from the intracellular stores is on, thereby MLCK is kept in an activated state and LC20 remains phosphorylated. Figure 2 illustrates the tension records of uterine muscles when resting, contracted, and contracted then relaxed. The spontaneous activity of the muscle was immediately stopped by MgCl$_2$. Carbachol kept the muscle in a sustained contraction with only a slight decline in tension during 17 min. Mg^{2+} readily relaxed the carbachol-contracted muscle, and 10% of the tension remained after 17 min. Figure 3 shows the gel electrophoretograms of the proteins from these muscles. The stain distribution in the LC20 spots of the contracted then relaxed muscle is similar to that in the contracted muscle, the monophosphorylated spot being the major one. In contrast, in the resting muscle most of the stain is in the nonphosphorylated spot. These differences in staining intensities are paralleled by the differences in the intensities of LC20 spots on autoradiograms, which show very little radioactivity in the resting muscle in contrast to the dark spots in the contracting muscle. The radioactivity of LC20 in the contracted then relaxed muscle was slightly reduced relative to that in the contracted muscle.

In Table III the fractional muscle relaxation and the fractional LC20 dephosphorylation are compared. Three smooth muscles are used: arteries, uteri, and bladder contracted then relaxed by the combination of several agents. The muscles were frozen before they

TABLE III Fractional Relaxation and Fractional Myosin Light Chain Dephosphorylation in Various Porcine Smooth Muscles

Muscle	Conditions[a]	Fractional relaxation[b]	Fractional dephosphorylation[c]	n
Artery	Histamine + Mg^2	0.77 ± 0.13	0.14 ± 0.08	13
	Histamine + theophylline	0.74 ± 0.19	0.29 + 0.15	7
	Histamine + papaverine	0.84 ± 0.12	0.40 ± 0.12	6
	Serotonin + Mg^{2+}	0.85 ± 0.11	0.20 ± 0.07	6
	Serotonin + papaverine	0.93 ± 0.09	0.36 ± 0.16	4
	NE + papaverine	0.73 ± 0.16	0.30 ± 0.09	4
	NE + Mg^{2+}	0.78	0.15	2
Uterus	Carbachol + Mg^{2+}	0.91 ± 0.07	0.23 ± 0.12	14
	Carbachol + papaverine	0.96 ± 0.04	0.24 ± 0.17	10
	Carbachol + theophylline	0.92 ± 0.02	0.22 ± 0.09	4
	Carbachol + isoproterenol	0.76 ± 0.07	0.27 ± 0.06	4
	Histamine + Mg^{2+}	0.94 ± 0.12	0.19 ± 0.05	4
	Histamine + papaverine	0.83 ± 0.06	0.26 ± 0.16	4
	Oxytocin + Mg^{2+}	0.85 ± 0.10	0.26 ± 0.08	4
Bladder	Carbachol + Mg^{2+}	0.90	0.15	2
	Carbachol + sodium nitroprusside	0.83	0.39	2

[a] All experiments were carried out at 37°C. The muscles were incubated in PSS for 30 min, then the muscles were ^{32}P-labeled with carrier-free [^{32}P]orthophosphate for 90 min, followed by 30-min washout of ^{32}P from the extracellular space of the muscles. Subsequently the muscles were contracted by an agent and when peak force was reached the relaxing agent was added into the bath in the presence of the contracting agent. The course of relaxation was followed and when the force decreased to 5–35% of its peak value the muscles were frozen in liquid nitrogen. The workup of the frozen muscles and the quantification of the mol [^{32}P]phosphate per mol LC20 is described in Section II.A.1. Conditions for artery: histamine, 50–100 μM for 2–5 min; serotonin, 1 mM for 2 min; NE, 100 μM for 3 min; Mg^{2+}, 10–50 mM for 10–35 min; theophylline, 10 mM for 15–35 min; papaverine, 0.5 mM for 10–25 min. Conditions for uterus: carbachol, 50–100 μM for 1–5 min; histamine, 100 μM for 2 min; oxytocin, 1 μM for 0.5 min; Mg^{2+}, 30–50 mM for 5–30 min; papaverine, 0.05–0.5 mM for 10–30 min; theophylline, 10 mM for 8–12 min. Conditions for bladder: carbachol, 100 μM for 2 min; Mg^{2+}, 10 mM for 7–10 min; sodium nitroprusside, 1 mM for 10–12 min.

[b] Fractional relaxation $= 1 - \left[\dfrac{\text{Force of contracted then relaxing muscle}}{\text{Force of contracted muscle}} \right]$

[c] Fractional dephosphorylation $= \dfrac{(P_C - P_R) - (P_{CR} - P_R)}{P_C - P_R}$

where P_C = phosphorylation of LC20 in contracted muscle; P_R = phosphorylation of LC20 in resting muscle; P_{CR} phosphorylation of LC in contracted then relaxing muscle; phosphorylation is expressed in terms of mol [^{32}P]phosphate per mol LC20.

reached the fully relaxed state. In all experiments shown in Table III there are major differences between values of fractional relaxation and fractional dephosphorylation. The relaxation values are close to the maximum, 1.0, whereas the dephosphorylation values remain in the low fractional decimals. This means that LC20 dephosphorylation lags well behind muscle relaxation. From the data it becomes evident that the dephosphorylation is not a prerequisite for smooth muscle relaxation. One must also conclude that the contraction–relaxation cycle of smooth muscle cannot be explained by an exclusive myosin light chain phosphorylation–dephosphorylation mechanism.

It was of interest to delineate the relationship between [Ca]$_i$ and stress in the contracted then relaxed muscles under conditions when LC20 phosphoryla-

tion remained high. Christopher M. Rembold kindly carried out a few experiments with porcine carotid arterial muscles stimulated with 100 μM histamine and relaxed with either 10 mM theophylline or 0.5 mM papaverine in the presence of histamine. The intracellular Ca^{2+} concentration was measured by the aequorin technique (D'Angelo et al., 1992; McDaniel et al., 1992). The two different relaxing agents, theophylline and papaverine, behaved differently in regulating [Ca]$_i$. In the histamine–theophylline combination, [Ca]$_i$ remained above the baseline when relaxation was near complete, whereas in the histamine–papaverine combination, both [Ca]$_i$ and stress reached the resting value.

These preliminary data, along with those of D'Angelo et al. (1992) and McDaniel et al. (1992), suggest that

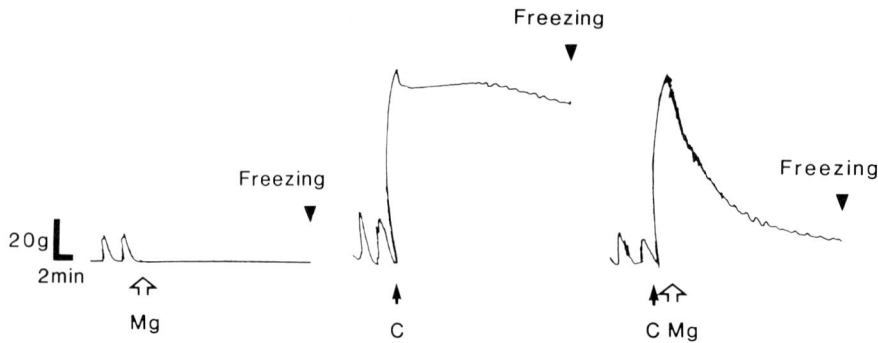

FIGURE 2   Tension records of porcine uterine longitudinal muscles frozen in resting (left), contracted (middle), and contracted then relaxed (right) states. Resting: the spontaneous activity of the uterine muscle was stopped by treatment with 50 mM $MgCl_2$ (Mg) in PSS. Contracted: the muscle was contracted by 100 μM carbachol (C) in PSS. Contracted then relaxed: the muscle was contracted with 100 μM carbachol, after 1.5 min $MgCl_2$ was added to 50 mM final concentration to the carbachol containing PSS. All three muscles were frozen in liquid nitrogen 17 min after the start of the experiment.

there may be two different mechanisms to bring about relaxation without significant LC20 dephosphorylation: (1) elevated level of $[Ca]_i$ with histamine–theophylline (current work) and histamine–nitroglycerine or histamine–nitroprusside combination (McDaniel *et al.*, 1992), and (2) reduced $[Ca]_i$, with histamine–papaverine (current work) and histamine–Mg^{2+} combination (D'Angelo *et al.*, 1992). Elucidation of the detailed mechanisms of these relaxations will be of interest.

1. Mg^{2+}-Induced Relaxation

Antagonistic effects of Ca^{2+} and Mg^{2+} have been established in vascular smooth muscle, and Mg^{2+} is believed to be a naturally occurring Ca^{2+} antagonist (Zhang *et al.*, 1992). Mg^{2+} blocks Ca^{2+} channels in ventricular muscle, thereby reducing inward Ca^{2+} movement (Ataka *et al.*, 1993). A concentration dependence of the Mg^{2+}-induced relaxation on porcine uterine muscles contracted by carbachol or histamine was found (Table IV). The force reaches an equilibrium at each Mg^{2+} concentration, the higher the $[Mg^{2+}]_o$ the less the remaining force and the less the half-time for relaxation. Since $[Mg^{2+}]_i$ did not significantly increase with $[Mg^{2+}]_o$ (D'Angelo *et al.*, 1992), the results suggest that the 16.7–50.0 mM external Mg^{2+} competes with the 2.5 mM Ca^{2+} in the PSS, and consequently prevents the entry of the external Ca^{2+} into the myo-

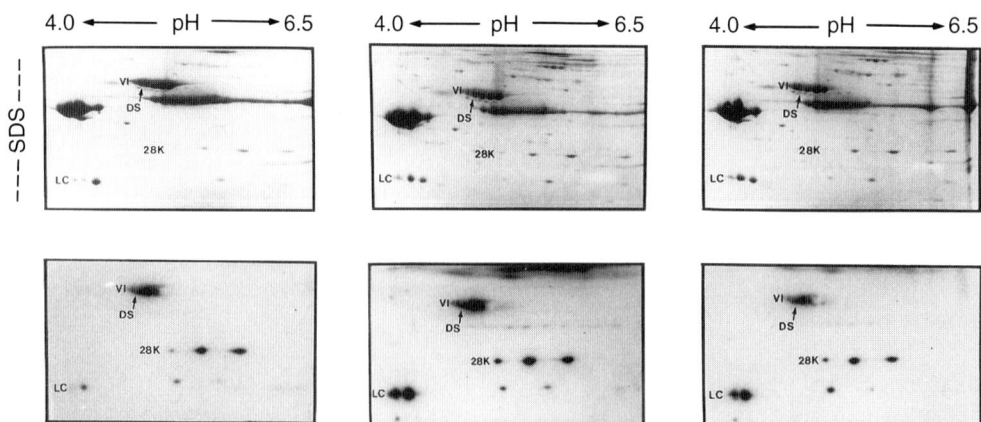

FIGURE 3   Protein phosphorylation patterns of porcine uterine proteins isolated from ^{32}P-labeled muscles. Resting (left), contracted (middle), and contracted then relaxed (right), as described in the legend of Fig. 2. Top: staining profiles (15% gels in the second dimension); bottom: corresponding autoradiograms (the exposure times were adjusted to the small differences in the specific radioactivities of $[^{32}P]Cr$ among the muscles). LC, LC20; 28K, 28-kDa protein; DS, desmin; VI, vimentin.

TABLE IV   Concentration Dependency of Mg^{2+}-Induced Relaxation on Carbachol- or Histamine-Contracted Porcine Uterine Muscles[a]

Mg^{2+} concentration (mM)	Remaining force (%)		Half-time for relaxation (min)	
	100 μM carbachol	100 μM histamine	100 μM carbachol	100 μM histamine
16.7	36 ± 5	39 ± 8	4.2 ± 0.8	3.4 ± 1.0
33.4	17 ± 2	20 ± 3	2.1 ± 1.0	1.8 ± 0.4
50.0	9 ± 2	10 ± 1	1.2 ± 0.3	0.9 ± 0.1

[a] The longitudinal strips of the uterine muscles were incubated in PSS (Csabina et al., 1986) at 37°C for 2 h. The muscles were contracted with either carbachol or histamine for 1 min, then $MgCl_2$ was added to 16.7 mM to the bath, and the muscles relaxed to a steady-state force, 36 and 39% of the maximal force, respectively. After 10 min at the equilibrium force level, the muscles were washed 15 times with PSS over a period of 45 min and the muscles fully relaxed; then the muscles were recontracted with carbachol or histamine, producing force equal to that in the first contraction. Relaxation was brought about by 33.4 mM Mg^{2+} in the presence of carbachol or histamine; after the equilibrium force was reached, the muscles were fully relaxed by washings with PSS then recontracted with carbachol or histamine, producing again the original force. Relaxation was repeated with 50 mM Mg^{2+}; after reaching the equilibrium force, the muscles were fully relaxed by washings and then recontracted with carbachol or histamine, producing again the original force. Four experiments were performed with each of carbachol or histamine.

plasm. Apparently, the external Mg^{2+} does not interfere with the Ca^{2+} efflux and, therefore, the Mg^{2+}-induced relaxation can be explained by decreasing $[Ca^{2+}]_i$ as suggested by D'Angelo et al. (1992).

LC20 phosphorylation remains 87–91% in carbachol- or histamine-contracted muscle treated with Mg^{2+} for 1 h. Even after such a prolonged exposure to Mg^{2+} the muscles regain full contractility upon extensive washings with PSS.

V. EXCHANGE OF THE COVALENTLY BOUND PHOSPHATE OF LC20

If the cross-bridge cycle of smooth muscle contraction is correlated with cyclic phosphorylation–dephosphorylation of LC20, then the covalently bound phosphoryl group of LC20 should exchange with the terminal phosphate group of ATP in the medium. Indeed, this was found: using double-labeled ATP ([γ-^{32}P]- and caged [γ-^{33}P]ATP), Butler et al. (1994) reported a rapid turnover of LC20-bound phosphate in permeabilized rabbit portal vein activated by Ca^{2+} at 20°C. The incorporation of ^{33}P from [γ-^{33}P]ATP into

LC20, initially labeled with ^{32}P from [γ-^{32}P]ATP, under nearby steady-state conditions occurred with an apparent first-order rate constant of $0.37\ s^{-1}$, corresponding to the rate of MLCP activity at 20°C. Kamm and Stull (1986) determined the rate constant of LC20 phosphorylation (MLCK activity) in electrically stimulated bovine tracheal muscles at 37°C, and found a value of $1.1\ s^{-1}$. Thus, the rate of MLCK and MLCP activity may be comparable in smooth muscle contracting at 37°C.

Bárány et al. (1991a) determined the exchange of LC20-bound phosphate during sustained contraction of porcine carotid arteries at 37°C. LC20 was phosphorylated by K^+ contraction in the arteries and the LC-bound nonradioactive phosphate was exposed to carrier-free [^{32}P]orthophosphate in the K^+-stimulating solution containing the arteries. The covalently bound LC20–phosphate in the K^+-contracted muscle was completely exchanged by [^{32}P]phosphate in 60 min (the earliest time when it could be determined).

Thus, during both force development and force maintenance a cyclic phosphorylation–dephosphorylation of LC takes place. This must occur through the reactions

$$\text{LC20-P} \longrightarrow \text{LC20} + \text{P} \qquad (1)$$
$$\text{LC20} + [\gamma\text{-*P}]\text{ATP} \longrightarrow \text{LC20-*P} + \text{ADP} \qquad (2)$$

That is, LC20-P is dephosphorylated by MLCP to yield free LC20, and subsequently MLCK transfers the terminal *P-labeled (^{32}P or ^{33}P) phosphate of ATP to LC20.

This is strong evidence for the involvement of LC20 phosphorylation–dephosphorylation in the contractile event and for the existence of slowly cycling latch-bridges (Dillon et al., 1981). It supports the idea that LC20 phosphorylation–dephosphorylation is related to the cross-bridge cycling rate (Section III.E), and it indicates that the cyclic phosphorylation contributes to the energy utilization of smooth muscle contraction (Walker et al., 1994), estimated to be about one-third of the suprabasal energy requirement (Bulter et al., 1994).

The complete exchange of LC20-bound phosphate in muscle that is in sustained contraction and, therefore, contains a maximum number of myosin filaments attached to actin filaments indicates that the bound phosphate does not participate in bond formation between myosin and actin. This refers to P-Ser-19 in the polypeptide chain of LC20. Similar rapid labeling of LC20-bound phosphate was found in PDBu-contracted arterial muscles (Bárány et al., 1992b), containing significant amounts of P-Ser-1 or P-Ser-2. Thus, the NH_2-terminal portion of the LC20 molecule is readily accessible to MLCP and MLCK in the intact muscle.

VI. CALDESMON PHOSPHORYLATION

The biochemical characteristics of CD and its phosphorylation *in vitro* are described in Chapters 6 and 13 (this volume). Adam *et al.* (1989) were the first to show CD phosphorylation in intact smooth muscle, namely, porcine carotid arteries. In resting muscles, CD was phosphorylated to 0.45 mol [^{32}P]phosphate/mol CD, and upon 60-min treatment with K$^+$, PDBu, histamine, or ouabain the incorporation increased to 0.74, 1.08, 0.93, and 1.08 mol [^{32}P]phosphate/mol CD, respectively. Endothelin-1 also stimulated CD phosphorylation from 0.35 to 0.52 mol [^{32}P]phosphate/mol CD in 60 min (Adam *et al.*, 1990). During relaxation of the K$^+$-stimulated muscle, CD was partially dephosphorylated. CD phosphorylation was confirmed in the carotid arterial muscles during short (1–5 min) and prolonged (60 min) stimulations (Bárány *et al.*, 1992a,b) (Table V) and a partial dephosphorylation of CD was also observed in the PDBu-contracted muscle. The time-dependent ^{32}P labeling of CD in smooth muscle may be explained by its endogenous phosphorylation, requiring the sequential actions of CD phosphatase and CD kinase for [^{32}P]phosphate incorporation into CD to occur. Adam *et al.* (1990) suggested that sustained arterial muscle contraction may require two reactions: (1) some phosphorylation of LC20 to form force-bearing cross-bridges and (2) CD phosphorylation to slow the detachment rate of the bridges.

The significance of CD phosphorylation is further evidenced by the work of Adam *et al.* (1995b), which demonstrated that the level of mitogen-activated protein kinase (MAPK), the enzyme presumably involved in CD phosphorylation *in vivo*, is regulated in response to both mechanical load and pharmacological stimulation. Further work will likely delineate the precise role of CD phosphorylation in smooth muscle.

VII. DESMIN PHOSPHORYLATION

Smooth muscle is rich in a three-dimensional cytoskeletal network, built of 10-nm intermediate filaments (Stromer, 1990). They contain a 53-kDa major protein DS and a 54-kDa minor protein vimentin (VI). The concentration of DS is about half that of tropomyosin. By 2D gel electrophoretic analysis, purified DS was found to contain two isoelectric variants, α-(more acidic, p*I* 5.4) and β-desmins (p*I* 5.5). *In vitro* experiments have shown that both cAMP-dependent protein kinase and PKC can phosphorylate VI (Inagaki *et al.*, 1987) and DS (Inagaki *et al.*, 1988), and the resulting increase in the phosphorylation level is correlated with their disassembly. Park and Rasmussen (1986) were the first to demonstrate phosphorylation of DS in intact smooth muscle by treating bovine trachea with carbachol for 1 h. Both the α and β isoforms of DS were phosphorylated. The 2D gels of our laboratory resolved five phosphorylated DS spots in stimulated porcine carotid arteries (Bárány *et al.*, 1992a). Upon prolonged stimulation of the arterial muscles, the molar [^{32}P]phosphate content of DS exceeded that of LC20 (Table V). Similarly, high DS phosphorylation was found in porcine uterus, trachea, stomach, and bladder. Reversible phosphorylation of DS was shown during a 60-min contraction–relaxation–contraction cycle of porcine carotid arteries (Bárány *et al.*, 1992a) and of porcine uterine muscles to a smaller extent. The reversible phosphorylation of DS gives support to the idea of Rasmussen *et al.* (1987) that phosphorylation of the "filamin–actin–desmin fibrillar domain" occurs during sustained arterial contraction. Phosphorylation of DS could partially depolymerize the DS filaments (Inagaki *et al.*, 1988), and thereby could change the configuration of the filamin–actin–desmin domain. Small *et al.* (1986) postulated two functionally

TABLE V Protein Phosphorylation in Arterial Smooth Muscle Contracted by Phorbol 12,13-Dibutyrate and Other Agents[a]

Treatment (60 min)	Mol [^{32}P]phosphate/mol protein				
	LC20	28 kDa	DS	CD	*n*
8 μM PDBu	0.51 ± 0.04	0.93 ± 0.10	2.06 ± 0.24	1.45 ± 0.21	6
8 μM PDBu + 0.5 mM EGTA	0.47 ± 0.06	0.92 ± 0.04	1.94 ± 0.15	1.44 ± 0.25	6
109 mM KCl	0.37 ± 0.05	0.45 ± 0.11	0.63 ± 0.15	0.59 ± 0.09	6
100 μM NE	0.34 ± 0.07	0.49 ± 0.07	0.89 ± 0.12	0.62 ± 0.10	6
100 μM *histamine*	0.60 ± 0.09	0.48 ± 0.05	0.94 ± 0.20	0.74 ± 0.14	6

[a]Adapted from Bárány *et al.* (1992a, Table II, p. 575) and Bárány *et al.* (1992b, Tables I and III, p. 236).

distinct systems in smooth muscle: an actomyosin system required for contraction and an intermediate filament–actin system required for muscle tone. The latter system could maintain force with low expenditure of energy in sustained smooth muscle contraction. Indeed, a study on the phosphorylation of the dense-plaque proteins talin and paxillin during tracheal smooth muscle contraction (Pavalko *et al.*, 1995) supports the hypothesis that reorganization of cytoskeletal–membrane interactions may account for functional properties of smooth muscle tissues.

A. Disease and Pregnancy

A highly phosphorylated form of DS was described in an autosomal dominant familial myopathy (Rappaport *et al.*, 1988). Modification of DS structure and/or the enzymatic balance that controls the phosphorylation–dephosphorylation of DS may be the cause of this disease.

An approximately twofold increase in DS phosphorylation was observed in porcine uterus in advanced pregnancy relative to that in the nonpregnant uterus

(M. Bárány and K. Bárány, unpublished results). These data further underline the significance of DS phosphorylation in biology.

VIII. PHOSPHORYLATION OF THE 28 kDa PROTEIN

^{32}P labeling of arterial muscle revealed a 28-kDa protein with three phosphorylated isoforms (Bárány *et al.*, 1992a), containing the same phosphopeptides in different proportions (Bárány *et al.*, 1992b). This phosphoprotein shows a similarity to the 27-kDa heat-shock protein (HSP27) isolated from smooth muscle cells of rabbit rectosigmoid by Bitar *et al.* (1991). HSP27 was labeled with [^{32}P]orthophosphate and it was resolved into four spots by 2D gel electrophoresis; three of the four spots were radioactive. Utilizing permeabilized smooth muscle cells and monoclonal antibodies to HSP27, Bitar *et al.* showed that HSP27 is a mediator of sustained smooth muscle contraction. An increased phosphorylation of the 28-kDa protein during sustained carbachol-induced contraction of porcine uterine muscle was found, compared with the

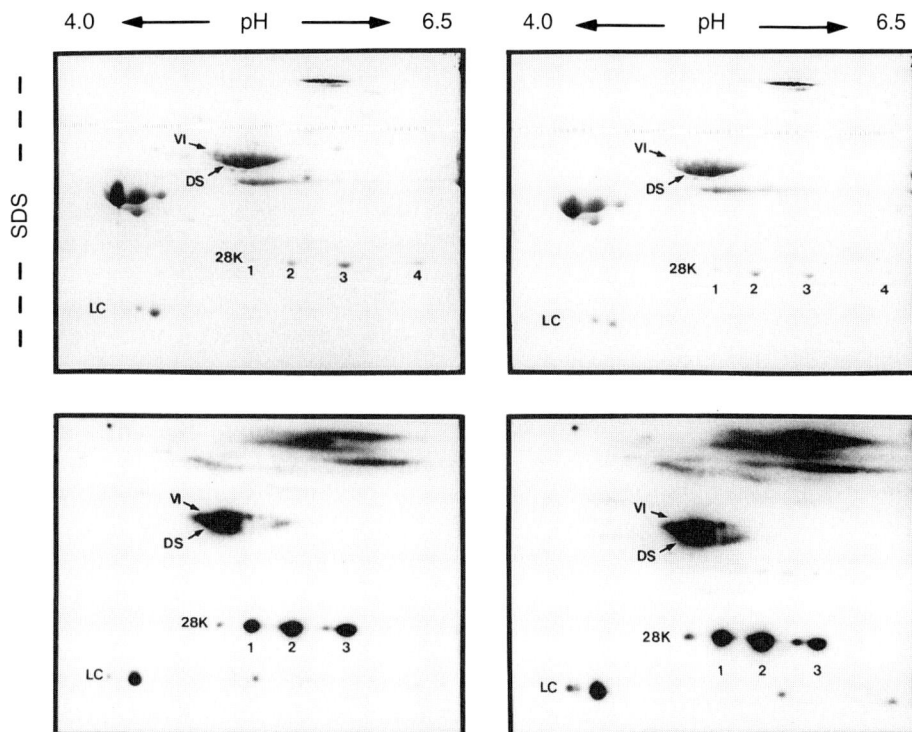

FIGURE 4   Protein phosphorylation in ^{32}P-labeled porcine uterine longitudinal muscles during prolonged carbachol contraction. Top: staining profiles; bottom: corresponding autoradiograms. Left: resting muscle treated with 10 μM isoproterenol for 30 min and then frozen; right: muscle contracted with 100 μM carbachol for 30 min and then frozen. LC, LC20; 28K, 28-kDa protein; DS, desmin; VI, vimentin. The spots of the 28-kDa protein are numbered from 1 to 4, from acidic to basic. Note that there are four stained spots but only three radioactive spots.

isoproterenol-treated resting muscle, as shown on the autoradiograms of Fig. 4. The figure also illustrates a shift in the distribution of the [^{32}P]phosphate among the isoforms: it decreases in the most basic monophosphorylated isoform *Spot* 3, whereas it increases in the more acidic isoforms, *Spots* 2 and 1. Phosphorylation of the 28-kDa protein in other smooth muscles, such as porcine trachea, stomach, and bladder, was also observed.

A heat-shock phosphoprotein of 28 kDa, with properties similar to those of HSP27, was described in mammalian cells (Arrigo and Welch, 1987). Apparently the same protein has been found in bovine aortic cells (Demolle *et al.*, 1988) and guinea pig or rabbit hearts (Edes and Kranias, 1990; Talosi and Kranias, 1992). Heat-shock proteins are known to bind and thereby stabilize an otherwise unstable conformer of another protein (Hendrick and Hartl, 1993). The amino acid sequence of HSP27 shows a striking homology with mammalian α-crystallin (Hickey *et al.*, 1986), a protein proposed to be a myofibril stabilizer (Atomi *et al.*, 1991). It is possible that phosphorylation of the 28-kDa protein during sustained smooth muscle contraction initiates its binding to a myofibrillar protein, resulting in a conformational modification of the contractile system. The high content of the 28-kDa protein in smooth muscle (comparable to that of LC20) invites studies to establish its role.

IX. ABSENCE OF CALPONIN PHOSPHORYLATION DURING SMOOTH MUSCLE CONTRACTION

The characteristic properties of calponin are described in Chapter 7. Interest in CP phosphorylation *in situ* originates from the observation of Winder and Walsh (1990) that phosphorylation of CP *in vitro* reverses its inhibition on smooth muscle actomyosin ATPase. The idea of Winder and Walsh that smooth muscle contraction may be regulated by CP phosphorylation has been tested in several laboratories (Bárány *et al.*, 1991b; Gimona *et al.*, 1992; Bárány and Bárány, 1993a; Winder *et al.*, 1993; Adam *et al.*, 1995a, and in preliminary reports referenced in these papers). *Qualitatively*, no CP phosphorylation was found in porcine carotid arteries contracted with four different agents for various times as compared with resting muscles (Bárány *et al.*, 1991b). Similarly, no CP phosphorylation was detected in the following contracting and resting smooth muscles: guinea pig taenia coli, porcine stomach, and chicken gizzard (Gimona *et al.*, 1992). On the other hand, Winder *et al.* (1993) described CP phosphorylation in carbachol-contracted

toad stomach muscle. Figure 5 shows qualitative CP phosphorylation in resting and contracting arterial smooth muscles (Bárány and Bárány, 1993a). There is no difference in the staining distribution of the six CP spots between these muscles (top panels). The same was true for resting and contracting uterine, tracheal, stomach, and bladder smooth muscles, indicating no CP phosphorylation during contraction. The bottom panels of Fig. 5 show the very weak radioactivity in the CP spots. Six days were required for autoradiography to visualize these spots. With such a long exposure time, many other spots that are not detectable by staining exhibit strong radioactivity. Furthermore, no difference is seen in CP radioactivity between contracting and resting muscle. In contrast, the radioactivity of LC20 from contracting muscle is several times higher than that in the resting muscle.

Quantitative analysis of CP phosphorylation in these five different smooth muscles revealed a very low CP phosphorylation, ranging from 0.002 to 0.010 mol [^{32}P]phosphate/mol CP (Bárány and Bárány, 1993a). There was no increase in CP phosphorylation upon contraction with any stimulus for any time. On the other hand, in all cases there was a major increase in LC20 phosphorylation upon contraction of the same muscles, the [^{32}P]phosphate content of LC20 being 50- to 250-fold higher in the contracting muscle than that of CP in the same muscle. These results were confirmed by Adam *et al.* (1995a), who have used the CD phosphorylation as a reference for CP phosphorylation. In resting porcine carotid arteries, CD was phosphorylated to a level of 0.41 mol [^{32}P]phosphate/mol CD, whereas CP was phosphorylated to levels less than 0.01 mol/mol. Stimulation of the arteries with three different agents for various times did not increase the phosphate content of CP. Therefore, the quantitative data also indicate that CP phosphorylation does not accompany contraction of smooth muscle. The observed CP phosphorylation in contracted stomach muscle by Winder *et al.* (1993) is based on the use of very high ^{32}P labeling of the muscle that produced a nonspecific CP phosphorylation as evidenced by the high phosphorylation of many proteins in the contracted muscle, as compared to the resting muscle (cf. Fig. 11 in Winder *et al.*, 1993).

The finding that CP is phosphorylated *in vitro* but not in the intact muscle is reminiscent of that of the inhibitory component of skeletal muscle troponin, which can be phosphorylated (Stull *et al.*, 1972) and dephosphorylated (England *et al.*, 1972) *in vitro* but is not phosphorylated during contraction of intact skeletal muscle (Bárány *et al.*, 1974). Apparently, protein residues that are free for phosphorylation *in vitro* may participate in bond formation or are buried *in situ* and,

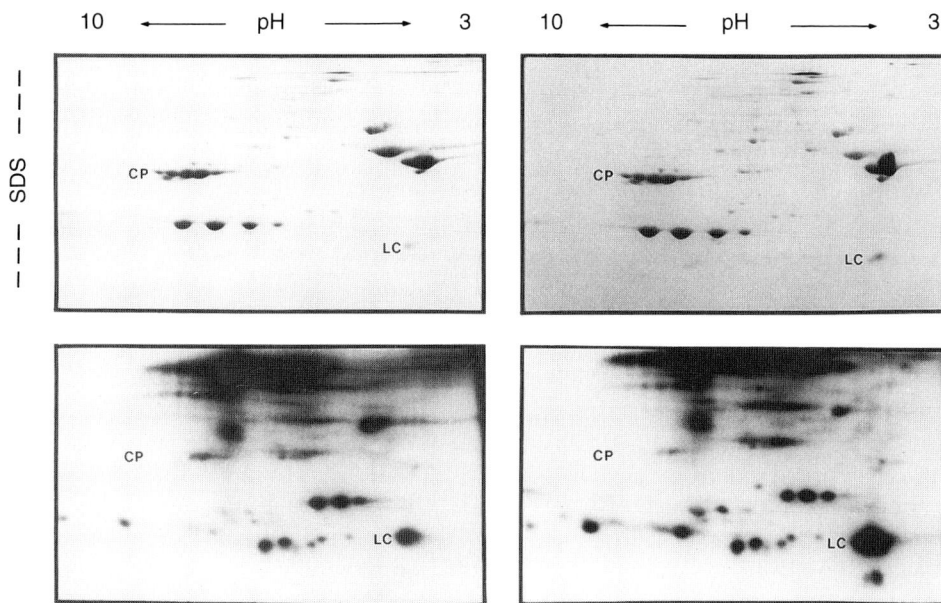

FIGURE 5 Analysis of CP phosphorylation in porcine carotid arterial smooth muscle by the 2D electrophoresis procedure of O'Farrell *et al.* (1977), nonequilibrium pH gradient electrophoresis in the first dimension and SDS–PAGE in the second dimension. Top: staining profiles; bottom: corresponding autoradiograms. Left: resting muscle; right: contracting muscle. CP, calponin; LC, LC20. From Bárány and Bárány (1993a, Fig. 1, p. 231).

therefore, are not available for enzymes involved in the turnover of protein-bound phosphate.

X. PROTEIN PHOSPHORYLATION DURING THE RESTING–CONTRACTION–RELAXATION–CONTRACTION CYCLE

Since several proteins are phosphorylated in smooth muscle, the question is which of these phosphorylations has physiological relevance. To answer this question, we carried porcine carotid arterial smooth muscles through a resting–contraction–relaxation–contraction cycle under the influence of various contracting agents for short (1–5 min) and long (60 min) contraction times and determined the [^{32}P]phosphate content of the four major phosphoproteins—LC20, the 28-kDa protein, DS, and CD—at each state of the contraction cycle (Bárány *et al.*, 1992a). In the short contraction–relaxation–contraction cycle of the arteries lasting minutes, induced by K$^+$, histamine, or NE, only LC20 underwent a phosphorylation–dephosphorylation–rephosphorylation cycle. In the contraction–relaxation–contraction cycle of long duration, induced by the same agents, cyclic phosphorylation of both LC20 and DS was observed. With 60-min PDBu stimulation, the phosphorylations of LC20, DS, and CD were cycling. Experiments were also carried out with porcine uterine

muscles in the short contraction–relaxation–contraction cycle with carbachol, oxytocin, and histamine as contracting agents. Only LC20 was phosphorylated–dephosphorylated and rephosphorylated in consort with the contraction cycle.

Since LC20 phosphorylation–dephosphorylation goes parallel with the mechanical cycle of the muscle, it has the potential for regulating smooth muscle contraction. The phosphorylation of DS may be involved in force maintenance during sustained contraction.

XI. PROTEIN PHOSPHORYLATION IN PHORBOL ESTER-TREATED SMOOTH MUSCLE

Phorbol esters are specific activators of PKC and, therefore, have been used by several investigators (referenced in Bárány *et al.*, 1992b) to assess the role of PKC-induced phosphorylation in smooth muscle contractility. PDBu is the most commonly used phorbol ester, because it is more potent than the other phorbol derivatives. At a low (0.8 μ*M*) concentration PDBu elicited a slowly developing contraction in arteries, reaching in 1 h about 50% of the K$^+$-induced force; LC20 phosphorylation was above resting level and partially due to PKC (Bárány *et al.*, 1990b). Fulginiti *et al.* (1993) have demonstrated that the stress–stiffness relationship in porcine carotid arterial strips was not different during KCl and PDBu stimulation, suggesting similar cross-bridge interactions. Interestingly, in stretched rabbit facial veins the phorbol ester-induced potentiation of myogenic tone was not associated with increases in Ca^{2+} influx, myoplasmic free Ca^{2+} concentration, or LC20 phosphorylation (Laporte *et al.*, 1994).

Arterial muscle tension in the absence of Ca^{2+}, equal to that of the K^+-induced tension, was first shown by Singer (1990) using a high concentration (10 μM) of PDBu. Following his work, very similar physiological and biochemical parameters of arterial muscles reversibly contracted with 8 μM PDBu in the presence or absence of Ca^{2+} were described (Bárány *et al.*, 1992b). Tryptic phosphopeptide maps of LC20 from the PDBu-treated muscles in the presence or absence of Ca^{2+} showed that two-thirds of LC20 phosphorylation was attributable to MLCK and one-third to PKC. Thus, 8 μM PDBu activates MLCK more than PKC. Previously, Rembold and Murphy (1988) found that PDBu stimulation of arterial smooth muscle is associated with increase in $[Ca^{2+}]_i$ and hypothesized that activation of MLCK can explain the contraction. Nevertheless, since LC20 phosphorylation also takes place in the absence of Ca^{2+}, forms of MLCK and PKC, which are independent of Ca^{2+}, must also be present in the muscle, and PDBu activates such kinases. Indeed, Ca^{2+}/calmodulin-independent MLCK and Ca^{2+}-independent PKC in smooth muscles have been noted (Ikebe *et al.*, 1987; Parente *et al.*, 1990).

Table V quantifies the incorporation of [^{32}P]phosphate into LC20, the 28-kDa protein, DS, and CD of arterial muscles contracted with PDBu and other agents for 60 min. The 8 μM PDBu-induced LC20 phosphorylation, in both the presence and absence of Ca^{2+}, exceeds that produced by KCl or NE, and it is comparable to that in histamine-stimulated muscles. Under the same conditions, PDBu, with or without Ca^{2+}, increases the [^{32}P]phosphate content of the 28-kDa protein, DS, and CD by two- or threefold. It is surprising that phosphorylations amounting from 0.5 to 2.0 mol phosphate/mol protein can proceed in arterial muscle treated with PDBu in a Ca^{2+}-independent manner.

In contrast, porcine uterine muscle did not contract with 8 μM PDBu in the absence of Ca^{2+} (K. Bárány and M. Bárány, unpublished results). The lack of force development was accompanied by lack of LC20 phosphorylation. However, incorporation of [^{32}P]phosphate into the 28-kDa protein, DS, VI, and CD did not differ. Thus, Ca^{2+} is required for the PDBu-induced LC20 phosphorylation, whereas the Ca^{2+}-independent phosphorylation works for the other proteins. Apparently, uterine muscle does not have Ca^{2+}-independent MLCK, indicating an essential difference between the phasic uterus and tonic artrial muscle that is worthy of further investigation.

Several authors have concluded that PKC plays no major role in smooth muscle contraction (Colburn *et al.*, 1988; Sutton and Haeberle, 1990; Hartshorne and Kawamura, 1992). On the other hand, two other studies suggest that activation of PKC in vascular smooth muscle increases the Ca^{2+} sensitivity of contraction (Jiang *et al.*, 1994; Masuo *et al.*, 1994). Accordingly, the significance of PKC in smooth muscle remains unsettled. The role of PKC in smooth muscle is further discussed in Chapters 12, 21, 24, and 26 (this volume).

XII. SUMMARY AND PERSPECTIVES

LC20 phosphorylation initiates smooth muscle contraction most likely by modifying the conformation of myosin. The phosphorylation reaction is independent from the contractile event per se, namely, it also occurs in stretched muscle that is unable to contract. In general, it is accepted that LC20 phosphorylation is a prerequisite for contraction and that during the contractile process LC20 undergoes a phosphorylation–dephosphorylation cycle. A priori cyclic phosphorylation of LC20 is not an absolute requirement for the contractile activity, because thiophosphorylation of LC20 (not hydrolyzed by MLCP) in skinned or permeabilized smooth muscle produces isometric force proportionally to the fraction of LC20 thiophosphorylated (Kenney *et al.*, 1990; Vyas *et al.*, 1992). Thus, cyclic phosphorylation of LC20 in contracting muscle must have a special function.

Since the determination of LC20 phosphate content varies among laboratories by two- to threefold, one cannot establish a stoichiometry of the phosphorylation in either the initial or steady state of contraction. Measurements in this laboratory with arterial and uterine muscles, which produced maximal stress and phosphorylation, suggest that both myosin heads are phosphorylated in the initial and one head in the sustained phase of contraction. The other extreme, contraction without phosphorylation, does not occur in our experience.

The relationship between relaxation and LC20 phosphorylation is not established. Various smooth muscles can be relaxed without a significant LC20 dephosphorylation (Table III), indicating that mechanism(s) other than LC20 dephosphorylation must exist for relaxation. This is illustrated by experiments with okadaic acid (OA), a potent inhibitor of MLCP (Mitsui *et al.*, 1992). Okadaic acid elicited contraction in skinned chicken gizzard fibers in the absence of Ca^{2+} via alteration of the MLCK–MLCP balance in favor of

MLCK (Hartshorne *et al.*, 1989). Nevertheless, OA completely relaxed the carbachol contracted intact tracheal smooth muscle without any LC20 dephosphorylation (Tansey *et al.*, 1990). The relaxing effect of OA was associated with a decrease in $[Ca^{2+}]_i$ by 93%, supporting the concept that OA reduces Ca^{2+} influx by increasing the phosphorylation of a Ca^{2+} channel (Sadighian *et al.*, 1993). It is possible that the decrease of $[Ca^{2+}]_i$, and not LC20 dephosphorylation, controls the relaxation process. This important topic needs exploration.

In addition to LC20 phosphorylation, a second Ca^{2+}-dependent regulatory mechanism involving CD and CP has been postulated (Stull *et al.*, 1991; Hartshorne and Kawamura, 1992). However, there is no evidence that phosphorylation of CD and CP participates in the regulation of intact muscle contraction. On the other hand, conformational changes in CD and CP were detected between histamine-contracted and resting intact porcine carotid arterial muscles (Bárány *et al.*, 1992c). *In vitro* experiments demonstrated changes in the orientation of CD bound to actin, caused by the binding of S-1 (Szczesna *et al.*, 1994). Thus, it is reasonable to assume that such changes can also occur in CD and CP of the intact muscle, because these proteins are part of the thin filament and consequently the combination of actin with the cross-bridges should have a direct effect on the orientation of CD and CP on the surface of actin. There is no evidence that protein phosphorylation is the only biochemical reaction that regulates smooth muscle contraction–relaxation; reactions other than protein phosphorylation may also be involved. Biochemically, the contraction–relaxation cycle of smooth muscle can be divided into two parts: (1) the combination of actin and myosin and (2) the dissociation of actomyosin. The biochemical cycle may not by symmetrical, that is, the actomyosin dissociation in the muscle may not be a simple reversal of actin to myosin combination. Since the LC20-bound phosphate does not participate in the actin–myosin bond (Section V), there is no compelling reason to assume that LC20 dephosphorylation should separate myosin from actin. It seems more likely that smooth muscle relaxation is achieved by perturbing the bonds between actin and myosin in actomyosin. The decrease in $[Ca^{2+}]_i$ may change the interaction between CD, CP, and actin, and the modified actin filament could dissociate from the cross-bridges, in the presence of ATP, to relax the muscle. Multiple studies are required to determine whether CD and CP are participating in smooth muscle relaxation.

Much has to be learned about the function of the 28-kDa phosphoprotein, which is present in large quantities in all smooth muscles. Upon muscle stimulation, the phosphate content of the 28-kDa protein increases but the conditions of the dephosphorylation are not known. Even the localization of the protein in the substructure of the muscle has to be investigated.

Of all the phosphoproteins in smooth muscle, DS contains most of the bound phosphate (because the DS content of the muscle is high and its phosphate content is also high, see Table V). Only a part of the phosphate is mobile (Bárány *et al.*, 1992a), whereas the rest withstands functional changes. This may be a phosphate reservoir of smooth muscle.

The work of Adam *et al.* (1989, 1990, 1995b) provided our basic and advanced knowledge about CD phosphorylation. The impressive progress of this laboratory predicts a clear understanding of the role of CD phosphorylation in smooth muscle contraction in the near future.

In principle, one can classify the phosphoproteins into two groups: (1) functional, whose phosphorylation is correlated with contraction, and (2) structural, whose phosphate content remains rather steady during the contraction cycle. Structural phosphoproteins could make contact with other proteins to form a specific protein network. Alternatively, they may bind divalent metals, Ca^{2+} or Mg^{2+}. The common experience of the slow turnover of phosphate in these proteins also suggests that the covalently bound phosphate is not free and, therefore, not readily available for protein kinases and phosphatases. Future investigation should provide information about the role of the structural phosphoproteins in smooth muscle.

Acknowledgments

We thank Christopher M. Rembold for the Ca^{2+} determinations and Janice Gentry for careful typing of the manuscript. This work was supported by Grant AM 34602 from the National Institutes of Health and the N. H. Pierce gift to the College of Medicine of the University of Illinois at Chicago.

References

Aburto, T. K., Lajoie, C., and Morgan, K. G. (1993). *Circ. Res.* **72**, 778–785.

Adam, L. P., Haeberle, J. R., and Hathaway, D. R. (1989). *J. Biol. Chem.* **264**, 7698–7703.

Adam, L. P., Milio, L., Brengle, B., and Hathaway, D. R. (1990). *J. Mol. Cell. Cardiol.* **22**, 1017–1023.

Adam, L. P., Haeberle, J. R., and Hathaway, D. R. (1995a). *Am. J. Physiol.* **268**, C903–C909.

Adam, L. P., Franklin, M. T., Raff, G. J., and Hathaway, D. R. (1995b). *Circ. Res.* **76**, 183–190.

Aksoy, M. O., and Murphy, R. A. (1983). *In* "Biochemistry of Smooth Muscle" (N. L. Stephens, ed.), Vol. I, pp. 141–166. CRC Press, Boca Raton, FL.

Aksoy, M. O., Murphy, R. A., and Kamm, K. D. (1982). *Am. J. Physiol.* **242**, C109–C116.

Aksoy, M. O., Stewart, G. J., and Harakal, C. (1986). *Biochem. Biophys. Res. Commun.* **135,** 735–741.

Arrigo, A. P., and Welch, W. J. (1987). *J. Biol. Chem.* **262,** 15359–15369.

Ataka, K., Chen, D., McCully, J., Levitsky, S., and Feinberg, H. (1993). *J. Mol. Cell. Cardiol.* **25,** 1387–1390.

Atomi, Y., Yamada, S, Strohman, R., and Nonomura, Y. (1991). *J. Biochem. (Tokyo)* **110,** 812–822.

Bárány, K., and Bárány, M. (1990). *In* "Uterine Function" (M. E. Carsten and J. D. Miller, eds.), pp. 71–98. Plenum, New York.

Bárány, K., Sayers, S. T., DiSalvo, J., and Bárány, M. (1983). *Electrophoresis* **4,** 138–142.

Bárány, K., Ledvora, R. F., Mougios, V., and Bárány, M. (1985a). *J. Biol. Chem.* **260,** 7126–7130.

Bárány, K., Ledvora, R. F., and Bárány, M. (1985b). *In* "Calmodulin Antagonists and Cellular Physiology" (H. Hidaka and D. J. Hartshorne, eds.), pp. 199–223. Academic Press, New York.

Bárány, K., Csabina, S., and Bárány, M. (1985c). *Adv. Protein Phosphatases* **2,** 37–58.

Bárány, K., Rokolya, A., and Bárány, M. (1990a). *Biochem. Biophys. Res. Commun.* **173,** 164–171.

Bárány, K., Rokolya, A., and Bárány, M. (1990b). *Biochim. Biophys. Acta* **1035,** 105–108.

Bárány, M., and Bárány, K. (1959). *Biochim. Biophys. Acta* **35,** 293–309.

Bárány, M., and Bárány, K. (1993a). *Biochim. Biophys. Acta* **1179,** 229–233.

Bárány, M., and Bárány, K. (1993b) *Arch. Biochem. Biophys.* **305,** 202–204.

Bárány, M., Bárány, K., Gaetjens, E., and Horváth, B. Z. (1974). *In* "Exploratory concepts in Muscular Dystrophy II" (A. T. Milhorat, ed.), pp. 451–462. Excerpta Medica, Amsterdam.

Bárány, M., Rokolya, A., and Bárány, K. (1991a). *Arch. Biochem. Biophys.* **287,** 199–203.

Bárány, M., Rokolya, A., and Bárány, K. (1991b). *FEBS Lett.* **279,** 65–68.

Bárány, M., Polyák, E., and Bárány, K. (1992a). *Arch. Biochem. Biophys.* **294,** 571–578.

Bárány, M., Polyák, E., and Bárány, K. (1992b). *Biochim. Biophys. Acta* **1134,** 233–241.

Bárány, K., Polyák, E., and Bárány, M. (1992c). *Biochem. Biophys. Res. Commun.* **187,** 847–852.

Bárány, M., Hegedüs, L., and Bárány, K. (1994). *Biophys. J.* **66,** A139.

Barron, J. T., Bárány, M., and Bárány, K. (1979). *J. Biol. Chem.* **254,** 4954–4956.

Barron, J. T., Bárány, M., Bárány, K., and Storti, R. V. (1980). *J. Biol. Chem.* **255,** 6238–6244.

Bitar, K. N., Kaminski, M. S., Hailat, N., Cease, K. B., and Strahler, J. R. (1991). *Biochem. Biophys. Res. Commun.* **181,** 1192–1200.

Butler, T. M., Siegman, M. J., and Mooers, S. U. (1986). *Am. J. Physiol.* **251,** C945–C950.

Butler, T. M., Narayan, S. R., Mooers, S. U., and Siegman, M. J. (1994). *Am. J. Physiol.* **267,** C1160–C1166.

Colburn, J. C., Michnoff, C. H., Hsu, L. C., Slaughter, C. A., Kamm, K. E., and Stull, J. T. (1988). *J. Biol. Chem.* **263,** 19166–19173.

Csabina, S., Bárány, M., and Bárány, K. (1986). *Arch. Biochem. Biophys.* **249,** 374–381.

Csabina, S., Bárány, M., and Bárány, K. (1987). *Comp. Biochem. Physiol. B* **87B,** 271–277.

D'Angelo, E. K. G., Singer, H. A., and Rembold, C. M. (1992). *J. Clin. Invest.* **89,** 1988–1994.

de Lanerolle, P. (1988). *J. Appl. Physiol.* **64,** 705–709.

de Lanerolle, P., and Paul, R. J. (1991). *Am. J. Physiol.* **261,** L1–L14.

Demolle, D., Lecomte, M., and Boeynaems, J. M. (1988). *J. Biol. Chem.* **263,** 18459–18465.

Dillon, P. F., Aksoy, M. O., Driska, S. P., and Murphy, R. A. (1981). *Science* **211,** 495–497.

Driska, S. P., Aksoy, M. O., and Murphy, R. A. (1981). *Am. J. Physiol.* **240,** C222–C233.

Driska, S. P., Stein, P. G., and Porter, R. (1989). *Am. J. Physiol.* **256,** C315–C321.

Edes, I., and Kranias, E. G. (1990). *Circ. Res.* **67,** 394–400.

England, P. J., Stull, J. T., and Krebs, E. G. (1972). *J. Biol. Chem.* **247,** 5275–5277.

Erdödi, F., Bárány, M., and Bárány, K. (1987). *Circ. Res.* **61,** 898–903.

Fulginiti, J., III, Singer, H. A., and Moreland, R. S. (1993). *J. Vasc. Res.* **30,** 315–322.

Gerthoffer, W. T. (1986). *Am. J. Physiol.* **250,** C597–C604.

Gerthoffer, W. T., and Murphy, R. A. (1983). *Am. J. Physiol.* **244,** C182–C187.

Gerthoffer, W. T., Murphey, K. A,. Mangini, J., Boman, S., and Lattanzio, F. A., Jr. (1991). *Am. J. Physiol.* **260,** G958–G964.

Gimona, M., Sparrow, M. P., Strasser, P., Herzog, M., and Small, J. V. (1992). *Eur. J. Biochem.* **205,** 1067–1075.

Gunst, S. J., Al-Hassani, M. H., and Adams, L. P. (1994). *Am. J. Physiol.* **266,** C684–C691.

Haeberle, J. R., Hott, J. W., and Hathaway, D. R. (1985a). *Pflügers Arch.* **403,** 215–219.

Haeberle, J. R., Hathaway, D. R., and DePaolí-Roach, A. A. (1985b). *J. Biol. Chem.* **260,** 9965–9968.

Hai, C. M., and Karlin, N. (1993). *Arch. Biochem. Biophys.* **301,** 299–304.

Hai, C. M., and Murphy, R. A. (1989). *Annu. Rev. Physiol.* **51,** 285–298.

Hartshorne, D. J. (1987). *In* "Physiology of the Gastrointestinal Tract" (L. R. Johnson, ed.), 2nd ed., pp. 423–482. Raven Press, New York.

Hartshorne, D. J., and Kawamura, T. (1992). *News Physiol. Sci.* **7,** 59–64.

Hartshorne, D. J., Ishihara, H., Karaki, H., Ozaki, H., Sato, K., Hori, M., and Watabe, S. (1989). *Adv. Protein Phosphatases* **5,** 219–231.

Hasegawa, Y., Tanahashi, K., and Morita, F. (1990). *J. Biochem. (Tokyo)* **108,** 909–913.

Hathaway, D. R., and Haeberle, J. R. (1985). *Am. J. Physiol.* **249,** C345–C351.

Hendrick, J. P., and Hartl, F. U. (1993). *Annu. Rev. Biochem.* **62,** 349–384.

Hickey, E., Brandon, S. E., Potter, R., Stein, G., Stein, J., and Weber, L. A. (1986). *Nucleic Acids Res.* **14,** 4127–4145.

Himpens, B., Matthijs, G., Somlyo, A. V., Butler, T. M., and Somlyo, A. P. (1988). *J. Gen. Physiol.* **92,** 713–729.

Hoar, P. E., Pato, M. D., and Kerrick, W. G. L. (1985). *J. Biol. Chem.* **260,** 8760–8764.

Homa, F. L., and Bárány, M. (1983). *Comp. Biochem. Physiol. B* **76B,** 801–810.

Horowitz, A., Trybus, K. M., Bowman, D., and Fay, F. S. (1994). *J. Cell Biol.* **126,** 1195–1200.

Ikebe, M., Stepinska, M., Kemp, B. E., Means, A. R., and Hartshorne, D. J. (1987). *J. Biol. Chem.* **260,** 13828–13834.

Inagaki, M., Nishi, Y., Nishizawa, K., Matsuyama, M., and Sato, C. (1987). *Nature (London)* **328,** 649–652.

Inagaki, M., Gonda, Y., Matsuyama, M., Nishizawa, K., Nishi, Y., and Sato, C. (1988). *J. Biol. Chem.* **263,** 5970–5978.

Jiang, M. J., and Morgan, K. G. (1989). *Pflügers Arch.* **413,** 637–643.

Jiang, M. J., Chan, C. F., and Chang, Y. L. (1994). *Life Sci.* **54,** 2005–2013.

Johnson, L. N., and Barford, D. (1993). *Annu. Rev. Biophys. Biomol. Struct.* **22,** 199–232.

Kamm, K. E., and Stull, J. T. (1985a). *Am. J. Physiol.* **249,** C238–C247.

Kamm, K. E., and Stull, J. T. (1985b). *Annu. Rev. Pharmacol. Toxicol.* **25,** 593–620.

Kamm, K. E., and Stull, J. T. (1986). *Science* **232,** 80–82.

Katoch, S. S. (1992). *Indian J. Exp. Biol.* **30,** 252–254.

Katoch, S. S. (1993). *Indian J. Physiol. Pharmacol.* **37,** 183–188.

Kenney, R. E., Hoar, P. E., and Kerrick, W. G. L. (1990). *J. Biol. Chem.* **265,** 8642–8649.

Kitazawa, T., and Somlyo, A. P. (1990). *Biochem. Biophys. Res. Commun.* **172,** 1291–1297.

Kitazawa, T, Gaylinn, B. D., Denney, G. H., and Somlyo, A. P. (1991a). *J. Biol. Chem.* **266,** 1708–1715.

Kitazawa, T., Masuo, M., and Somlyo, A. (1991b). *Proc. Natl. Acad. Sci. U.S.A.* **88,** 9307–9310.

Kopp, S. J., and Bárány, M. (1979). *J. Biol. Chem.* **254,** 12007–12012.

Kwon, S. C., and Murphy, R. A. (1994). *Biophys. J.* **66,** A408.

Lansman, J. B., and Franco, A., Jr. (1991). *J. Muscle Res. Cell Motil.* **12,** 409–411.

Laporte, R., Haeberle, J. R., and Laher, I. (1994). *J. Mol. Cell. Cardiol.* **26,** 297–302.

Ledvora, R. F., Bárány, K., VanderMeulen, D. L., Barron, J. T., and Bárány, M. (1983). *J. Biol. Chem.* **258,** 14080–14083.

Masuo, M., Reardon, S., Ikebe, M., and Kitazawa, T. (1994). *J. Gen. Physiol.* **104,** 265–286.

McDaniel, N. L., Chen, X. L., Singer, H. A., Murphy, R. A., and Rembold, C. M. (1992). *Am. J. Physiol.* **263,** C461–C467.

Mitsui, T., Inagaki, M., and Ikebe, M. (1992). *J. Biol. Chem.* **267,** 16727–16735.

Mitsui, T., Kitazawa, T., and Ikebe, M. (1994). *J. Biol. Chem.* **269,** 5842–5848.

Moore, R. L., and Stull, J. T. (1984). *Am. J. Physiol.* **247,** C462–C471.

Moreland, R. S., Cilea, J., and Moreland, S. (1992). *Am. J. Physiol.* **262,** C862–C869.

Moreland, S., Moreland, R. S., and Singer, H. A. (1987). *Pflügers Arch.* **408,** 139–145.

Murphy, R. A. (1994). *FASEB J.* **8,** 311–318.

Obara, K., Kunimoto, M., Ito, Y., and Yabu, H. (1987). *Comp. Biochem. Physiol. A* **87A,** 503–508.

O'Farrell, P. Z., Goodman, H. M., and O'Farrell, P. H. (1977). *Cell (Cambridge, Mass.)* **12,** 1133–1142.

Oike, M., Droogmans, G., and Nilius, B. (1994). *Proc. Natl. Acad. Sci. U.S.A.* **91,** 2940–2944.

Oishi, K., Takano-Ohmuro, H., Minakawa-Matsuo, N., Suga, O., Karibe, H., Kohama, K., and Uchida, M. K. (1991). *Biochem. Biophys. Res. Commun.* **176,** 122–128.

Padrón, R., Panté, N., Sosa, H., and Kendrick-Jones, J. (1991). *J. Muscle Res. Cell Motil.* **12,** 235–241.

Parente, J. E., Walsh, M. P., Kerrick, W. G. L., and Hoar, P. E. (1990). *In* "Frontiers in Smooth Muscle Research" (N. Sperelakis and J. D. Wood, eds.), pp. 149–157. Wiley-Liss, New York.

Park, S., and Rasmussen, H. (1986). *J. Biol. Chem.* **261,** 15734–15739.

Pavalko, F. M., Adam, L. P., Wu, M. F., Walker, T. L., and Gunst, S. J. (1995). *Am. J. Physiol.* **268,** C563–C571.

Persechini, A., Kamm, K. E., and Stull, J. T. (1986). *J. Biol. Chem.* **261,** 6293–6299.

Rappaport, L., Contard, F., Samuel, J. L., Delcayre, C., Marotte, F., Tome, F., and Fardeau, M. (1988). *FEBS Lett.* **231,** 421–425.

Rasmussen, H., Takuwa, Y., and Park, S. (1987). *FASEB J.* **1,** 177–185.

Rembold, C. M., and Murphy, R. A. (1988). *Am. J. Physiol.* **255,** C719–C723.

Rokolya, A., Bárány, M., and Bárány, K. (1991). *Biochim. Biophys. Acta* **1057,** 276–280.

Rüegg, J. C., Zeugner, C., Strauss, J. D., Paul, R. J., Kemp, B., Chem, M., Li, A. Y., and Hartshorne, D. J. (1989). *Pflügers Arch. Eur. J. Physiol.* **414,** 282–285.

Sadighian, J. J., Lattanzio, F. A., and Ratz, P. H. (1993). *Biophys. J.* **64,** A260.

Sato, K., Hori, M., Ozaki, H., Takano-Ohmuro, H., Tsuchiya, T., Sugi, H., and Karaki, H. (1992). *J. Pharmacol. Exp. Ther.* **261,** 497–505.

Seto, M., Sasaki, Y., and Sasaki, Y. (1990). *Pflügers Arch.* **415,** 484–489.

Seto, M., Yano, K., Sasaki, Y., and Azuma, H. (1993). *Exp. Mol. Pathol.* **58,** 1–13.

Siegman, M. J., Butler, T. M., Mooers, S. U., and Michale, K. A (1984). *Pflügers Arch.* **401,** 385–390.

Silver, P. J., and Stull, J. T. (1982a). *J. Biol. Chem.* **257,** 6137–6144.

Silver, P. J., and Stull, J. T. (1982b). *J. Biol. Chem.* **257,** 6145–6150.

Singer, H. A (1990). *Am. J. Physiol.* **259,** C631–C639.

Singer, H. A., and Murphy, R. A (1987). *Circ. Res.* **60,** 438–445.

Small, J. V., Furst, D. O., and DeMey, J. (1986). *J. Cell Biol.* **102,** 210–220.

Somlyo, A. P., and Somlyo, A. V. (1992). *In* "The Heart and Cardiovascular System" (H. A. Fozard, R. B. Jennings, E. Haber, A. M. Katz, and H. E. Morgan, eds.), pp. 1295–1324. Raven Press, New York.

Somlyo, A. P., and Somlyo, A. V. (1994). *Nature (London)* **372,** 231–236.

Stromer, M. H. (1990). *In* "Cellular and Molecular Biology of Intermediate Filaments" (R. D. Goldman and P. M. Steinert, eds.), pp. 19–36. Plenum, New York.

Stull, J. T., Broström, C. O., and Krebs, E. G. (1972). *J. Biol. Chem.* **247,** 5272–5274.

Stull, J. T., Gallagher, P. J., Herring, B. P., and Kamm, K. E. (1991). *Hypertension* **17,** 723–732.

Sutton, T. A., and Haeberle, J. R. (1990). *J. Biol. Chem.* **265,** 2749–2754.

Szczesna, D., Graceffa, P., Wang, C. L. A., and Lehrer, S. S. (1994). *Biochemistry* **33,** 6716–6720.

Talosi, L., and Kranias, E. G. (1992). *Circ. Res.* **70,** 670–678.

Tansey, M. G., Hori, M., Karaki, H., Kamm, K. E., and Stull, J. T. (1990). *FEBS Lett.* **270,** 219–221.

Tansey, M. G., Luby-Phelps, K., Kamm, K. E., and Stull, J. T. (1994). *J. Biol. Chem.* **269,** 9912–9920.

Trybus, K. M. (1991). *Cell Motil. Cytoskeleton* **18,** 81–85.

Vyas, T. B., Mooers, S. V., Narayan, S. R., Witherell, J. C., Siegman, M. J., and Butler, T. M. (1992). *Am. J. Physiol.* **263,** C210–C219.

Walker, J. S., Wingard, C. J., and Murphy, R. A. (1994). *Hypertension* **23,** 1106–1112.

Washabau, R. J., Wang, M. B., Dorst, C. L., and Ryan, J. P. (1991). *Am. J. Physiol.* **260,** G920–G924.

Washabau, R. J., Wang, M. B., Dorst, C., and Ryan, J. P. (1994). *Am. J. Physiol.* **266,** G469–G474.

Winder, S. J., and Walsh, M. P. (1990). *J. Biol. Chem.* **265,** 10148–10155.

Winder, S. J., Allen, B. G., Fraser, E. D., Kang, H. M., Kargacin, G. J., and Walsh, M. P. (1993). *Biochem. J.* **296,** 827–836.

Zhang, A., Cheng, T. P. O., and Altura, B. M. (1992). *Biochim. Biophys. Acta* **1134,** 25–29.

26

Regulation of Cross-bridge Cycling in Smooth Muscle

JOHN D. STRAUSS and RICHARD A. MURPHY

Department of Molecular Physiology and Biological Physics
University of Virginia
Charlottesville, Virginia

I. INTRODUCTION

This chapter focuses on the mechanisms regulating smooth muscle (SM) cross-bridges and the biological response (force production and shortening). Despite several reports of Ca^{2+}-independent activation of the contractile system (Gerthoffer, 1987; Matsuo *et al.*, 1989; Oishi *et al.*, 1991; Khalil and Morgan, 1992), changes in $[Ca^{2+}]$ are generally accepted to physiologically regulate contraction of SM. Native SM actomyosin ATPase activity is sensitive to changes in $[Ca^{2+}]$ and experiments using Ca^{2+}-sensitive dyes directly support the central role of Ca^{2+} in the control of contraction in living tissue (Filo *et al.*, 1965; Gerthoffer *et al.*, 1989; Brozovich and Morgan, 1989; Kamm and Grange, Chapter 27, this volume). However, it is also generally accepted that Ca^{2+} does not directly activate SM cross-bridges (Ito and Hartshorne, 1990). Therefore, regulation of SM cross-bridges must occur through mechanisms secondary to elevation in $[Ca^{2+}]$.

Experimental evidence to date implies that regulation in SM is complex. Many of the mechanisms are proposed on inferential grounds and attempt to explain specific observations without an analysis of the capacity to predict the overall mechanical output and energetic properties of SM. We begin by consideration of the information required to confirm a role for potential regulatory mechanisms.

II. CRITERIA FOR IDENTIFICATION OF PHYSIOLOGICAL CROSS-BRIDGE REGULATORY MECHANISMS

By generalizing the analysis of Krebs and Beavo (1979), minimum criteria can be established for identi-

fying the regulatory systems of the SM cross-bridge (Table I). The first is self-evident, although experimentally challenging. A regulated parameter exists in at least two distinct states or a continuum of values *in vivo*. Active stress (force/cell cross-sectional area) is obviously a regulated parameter in muscle. Stress is accepted as the manifestation of a population of cross-bridges in at least two states: force-producing (strong binding) and non-force-producing. In SM, cross-bridge cycling rates are also regulated and must be quantified. The second criterion (Table I) is to establish a correlation between the magnitude of changes in the regulatory element and a regulated response such as force. Specificity requires that the change in the regulatory component is always associated with the response; that is, there is a qualitative and quantitative correlation between a change in regulatory component and the activity of the cross-bridge. The third criterion testing the predictive value of a hypothesis is an appropriate temporal correlation in the proposed sequence of events leading to the biological response. Reversibility is the fourth criterion proposed here. In muscle, reversibility implies that a proposed regulatory mechanism(s) can quantitatively and temporally explain relaxation as the components of the regulatory system return to the prestimulated conditions. The last criterion is that the mechanism must be demonstrated in cells or isolated tissues that retain the characteristics of the *in vivo* organ.

III. EMPIRICAL OBSERVATIONS IN SMOOTH MUSCLE

There is a wealth of data on the behavior of SM cross-bridges in the literature. The experiments range

1. Identification of both a biological response(s) (e.g., biochemistry, mechanics, and energetics) and a potential regulatory element (LC20 phosphorylation, DAG concentration, calcium concentration, etc.)

2. Quantify the correlation between changes in the state of the proposed regulatory component and parameters associated with the biological responses (force generation, shortening, ATPase activity, etc.)

3. Demonstrate a temporal correlation between a change in the regulatory component and a change in contractile state, consistent with a cause-and-effect relationship

4. Demonstration of reversibility of the response to a prestimulated state upon return of the regulatory component to the starting conditions

5. Correlate changes in regulatory element and biological response in intact isolated cells or tissue that retain characteristics of the *in vivo* situation

from enzyme biochemistry and tissue physiology to the latest optical imaging and molecular genetic technology. These data support and set limits on potential regulatory mechanisms, and provide a framework of behavior that must be explained by any models proposed.

A. Protein Biochemistry

Phosphorylation of LC20 is the most studied regulatory mechanism (see Bárány and Bárány, Chapter 25, this volume). Unphosphorylated SM myosin, unlike skeletal or cardiac myosin, has a very low ATPase activity that is poorly enhanced by actin, actin–tropomyosin, or a number of other actin–protein combinations (Hartshorne and Gorecka, 1980). This led to early concepts of thick filament-based regulation in which SM myosin requires some additional "activating" factor that striated muscle proteins do not (Hartshorne and Gorecka, 1980). Therefore, the first challenge was to discover mechanisms by which the actomyosin interaction may be enhanced. Phosphorylation of LC20 on serine residue in the 19th position of the primary structure (Ser[19]) increases actin-activated ATPase activity by one or two orders of magnitude (Sobieszek and Small, 1977; Adelstein and Eisenberg, 1980; Hartshorne, 1987). The potential for a central regulatory role of phosphorylation independent of [Ca^{2+}] was demonstrated in SM tissues using phosphatase inhibitors such as okadaic acid. Such treatments lead to high levels of LC20 phosphorylation and force with little change in [Ca^{2+}] (Karaki *et al.*, 1989; Obara and de Lanerolle, 1989). The molecular

mechanism by which LC20 phosphorylation enhances actin–myosin interaction is not fully understood. However, LC20 phosphorylation at Ser[19] is associated with a change in the conformation of the myosin hexamer *in vitro* (Ikebe *et al.*, 1988; Trybus, 1989). Alternate phosphorylation sites have been identified (Ikebe, 1989; Ikebe *et al.*, 1986; see Bárány and Bárány, Chapter 22, this volume), but the physiological relevance of these is questionable. In fact, most of the current data suggest that Ser[19] appears to be the major site of phosphorylation, with a small amount of phosphorylation in threonine at position 18 (Kamm *et al.*, 1989; Monical *et al.*, 1993).

The phosphorylation of Ser[19] is catalyzed almost exclusively by myosin light chain kinase (MLCK) in native filaments (Walsh, 1991; Sweeney and Stull, 1990). The activity of MLCK is Ca^{2+}-calmodulin dependent and may be itself regulated by phosphorylation (de Lanerolle *et al.*, 1984; see Stull *et al.*, Chapter 9, this volume). Finally, LC20 may be readily dephosphorylated when MLCK activity is reduced, leading to a reduced ATPase activity. Though LC20 may be dephosphorylated by several different phosphatases, a specific myosin light chain phosphatase (MLCP) has been identified (Shirazi *et al.*, 1994; Somlyo and Somlyo, 1994). The evidence that LC20 phosphorylation regulates actin-activated ATPase activity of SM myosin meets criteria 1–4 of Table I.

There are a number of actin-associated proteins proposed to be involved in the regulation of SM. These are discussed in the following sections and elsewhere in this volume. Although it remains debatable whether these proteins contribute to the regulation of the SM cross-bridge *in vivo*, there is ample biochemical evidence in which the activity of actomyosin preparations and *in vitro* motility assays are affected by caldesmon (CD), calponin (CP), and tropomyosin (TM), as well as several other proteins.

B. Permeabilized Fibers

Permeabilized muscle preparations have been used extensively to study cross-bridge behavior. There are various methods for permeabilization that permit strict control of the myofilament ionic environment while retaining the ability to generate force. Almost universally, phosphorylation of LC20 precedes contraction in these preparations and is generally required for maintaining contraction (Vyas *et al.*, 1994; Moreland *et al.*, 1988; Kenney *et al.*, 1990). Since some of these preparations permit access of fairly large molecules to the contractile filaments, the concept of myosin phosphorylation regulating contraction could be directly tested. Phosphorylation by a partially proteolyzed, Ca^{2+}-independent form of MLCK (Walsh *et al.*, 1982) and thiophosphorylation (Cassidy *et al.*, 1979) provided additional evidence that the phosphorylation of LC20 via MLCK independent of $[Ca^{2+}]$ is essential and adequate for the initiation of SM contraction. Further, the degree of phosphorylation correlated with modulation of ATPase activity and velocity (Hellstrand and Arner, 1985; Sellers *et al.*, 1985). The difficulty in extrapolating the behavior of these preparations to *in vivo* conditions lies in the fact that force is typically directly proportional to LC20 phosphorylation in permeabilized preparations (Tanner *et al.*, 1988; Kenney *et al.*, 1990). This contrasts to living tissues that show classic latch behavior where the dependence of force on LC20 phosphorylation is quasi-hyperbolic (Murphy, 1994). In general, data on permeabilized SM were consistent with biochemical studies meeting criteria 1–4 (Table I) to establish LC20 phosphorylation, per se, as a requirement for cross-bridge cycling. However, criterion 5 is not fully met. More detailed information on the behavior of permeabilized preparations is reviewed by Pfitzer (Chapter 15, this volume).

C. Isolated Tissues and Latch Behavior

Isolated tissue has a characteristic behavior that has been extensively described. Stimulation of isolated tonic SM tissue induces a transient increase in $[Ca^{2+}]$ that is correlated with a number of parameters, such as phosphorylation of myosin, shortening velocity, and ATP consumption (Murphy, 1989, 1994). Force, however, is not directly dependent on phosphorylation and can be maintained after Ca^{2+} and LC20 phosphorylation fall to modest sustained values. Phasic smooth muscle behaves in the same way at 37°C (Fischer and Pfitzer, 1989; Kwon *et al.*, 1992). These characteristics of high force with slowed cross-bridge cycling rates and lowered phosphorylation form the core of what has become known as "latch" phenomena. Experimental temperature is an important variable that alters this behavior, such that at 20–25°C, latch phenomena may be minimal (Murphy, 1994). Figure 1 relates phosphorylation, force, and shortening velocity data measured in swine carotid medial preparations for a wide variety of stimuli. The dependence of both stress and velocity on LC20 phosphorylation (Fig. 1) must be explained by any model of regulation. One of the more unique properties of smooth muscle is that its energetic relationships are also functions of LC20 phosphorylation (Aksoy *et al.*, 1983; Krisanda and Paul, 1988; Paul, 1989; Wingard *et al.*, 1994). Any regulatory mechanism proposed must also account for this behavior. Although Ca^{2+}-stimulated LC20 phosphorylation is clearly associated with contraction in intact smooth

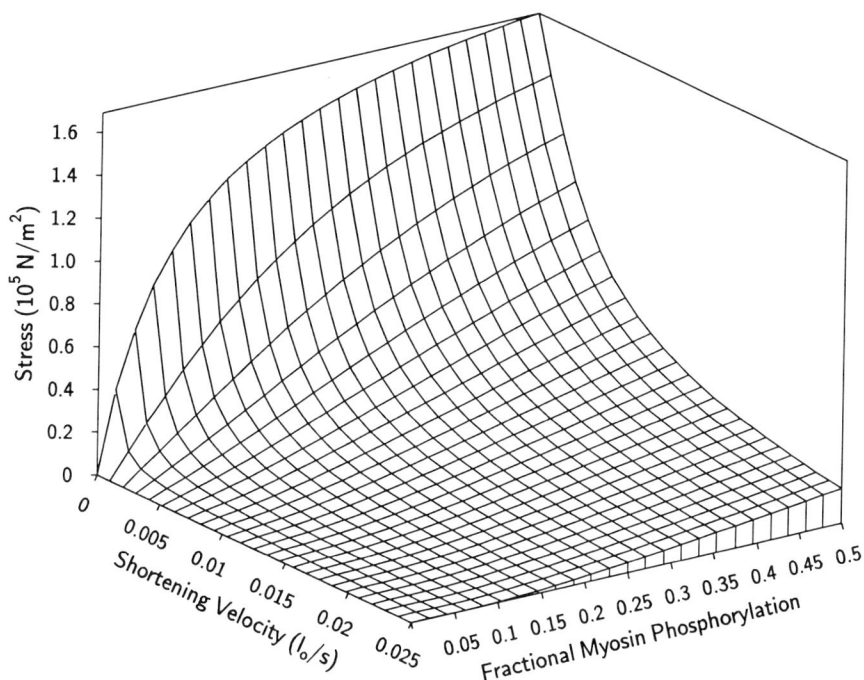

FIGURE 1 Three-dimensional plot of the relations between phosphorylation, force, and shortening velocity for the swine carotid artery at 37°C. The relation between velocity and force is not unique and depends on cross-bridge phosphorylation from Murphy (1994).

muscle, the observations differ from predictions based on studies of isolated proteins or permeabilized preparations. Force is not a simple function of LC20 phosphorylation. This has fueled suggestions that additional regulatory mechanisms are involved in latch.

D. Uncoupling of Ca²⁺, Phosphorylation, and Cross-bridge Cycling

One of the more easily demonstrable exceptions to the general rule of obligatory phosphorylation of LC20 for contraction is stimulation of protein kinase C (PKC). With long incubation times, phorbol esters or other activators of PKC elicit slow contractions without a proportional increase in either Ca²⁺ or LC20 phosphorylation (Khalil and Morgan, 1992; Rasmussen et al., 1987; Singer, Chapter 12, this volume). The consensus is that PKC does not phosphorylate the contractile proteins directly, but may act on proteins that either influence Ca²⁺ concentration or modulate cross-bridge attachment and cycling (Lee and Severson, 1994). Caldesmon was a candidate for the role of PKC intermediary based on a report that correlated CD phosphorylation and the sustained phase of contraction (Adam et al., 1989). It has since been determined that PKC phosphorylates CD on a site that is not phosphorylated in vivo, hence making CD an unlikely participant (Crist et al., 1992). Rasmussen and colleagues (1987, 1990) have provided a catalog of proteins phosphorylated by PKC. Several are cytoskeletal proteins and may mediate mechanical changes that do not directly involve cross-bridge interaction. Matsuo et al., (1989) suggest that activation of PKC decreases the rate of dephosphorylation of LC20, thereby altering phosphorylation levels at any given MLCK activity. This could explain some reports of PKC-dependent Ca²⁺ sensitization as well as apparently "Ca²⁺-independent" activation (Khalil and Morgan, 1992).

Inconsistent with any phosphorylation-based mechanisms are reports of phosphorylation-independent contractions and phosphorylation of LC20 without force production. In addition to PKC, these reports involve Ca²⁺ depletion–repletion protocols (Merkel et al., 1990), pharmacological stimuli (Gerthoffer, 1987; Oishi et al., 1991), and a component of relaxation with nitrovasodilators that depress force without a proportional reduction in LC20 phosphorylation (McDaniel et al., 1992). Such reports are the best evidence for secondary regulatory systems involved in the cross-bridge cycle. However, the circumstances that produce Ca²⁺ and phosphorylation-independent contraction do not correspond to known physiological situations. Furthermore, small changes in LC20 phosphorylation are difficult to detect given the limits of resolution for this measurement and potentially underlie significant alterations in force at low levels of activation.

IV. POTENTIAL SITES OF REGULATION IN THE CROSS-BRIDGE CYCLE

The basic mechanism for force generation is accepted to be the same in all three vertebrate muscle types: the cyclic interaction between actin and myosin (Taylor, 1987). The binding of ATP to actomyosin (AM), hydrolysis of ATP, and eventual release from AM of ADP and Pᵢ results in a large change in free energy that is translated into a force-producing conformation change and translocation of thin and thick filaments relative to one another.

A. The ATPase Cycle

The ATPase cycle underlying chemomechanical transduction is believed to be similar in striated and SM cross-bridges (Taylor, 1987). Since turnover is regulated in SM (latch phenomenon), one or more of the biochemical steps in this cycle are presumably regulated by factors other than load or myosin isoforms. The cycle is complex, with a number of reactions (r₁ through r₇ in Fig. 2) that may be potentially regulated. Work on skeletal muscle cross-bridge kinetics is more advanced, partly owing to allosteric regulation where only the number of cycling cross-bridges changes. Attachment of a cross-bridge occurs spontaneously, but results in a nonproductive attachment of extremely short duration (r₂) (Adelstein and Eisenberg, 1980; Brenner, 1987). The transition from this "weakly attached" state to a "strongly" attached, load-generating state is regulated by the conformation of the troponin–tropomyosin complex and involves either the isomerization step (r₃) or phosphate release (r₄). It is generally accepted that steps downstream from the transition from weak to strong binding states in skeletal muscle are essentially unregulated and the rates of cross-bridge cycling are functions of the specific myosin isozyme, the substrate concentrations, and loading (Brenner, 1987).

A similar kinetic cycle for SM actomyosin ATPase is generally accepted, but there is uncertainty concerning the regulated step(s). Phosphorylation of myosin seems to be permissive with respect to actin binding, that is, it affects recruitment of cross-bridges. It is likely that phosphorylation affects a step prior to phosphate release since this is generally accepted to be the step affected by actin binding, and phosphorylation of myosin has little effect on phosphate release in the absence of actin (Hartshorne and Gorecka, 1980).

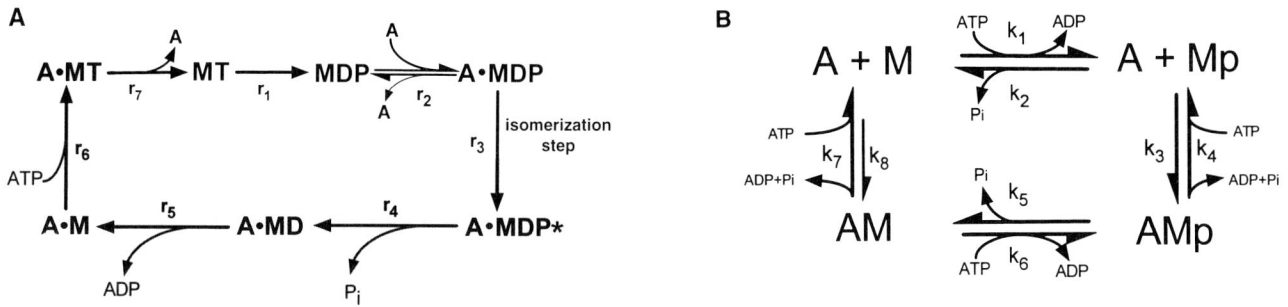

FIGURE 2   (A) Myosin ATPase cycle. MT, MD, and MDP indicate myosin species with an ATP, ADP, and ADP + P_i bound, respectively. A·M(x) denotes actin binding to myosin species. Asterisk indicates a putative conformational isomerization step that does not involve a loss of substrates. The reaction steps r_x are numbered starting with ATP hydrolysis. Shaded area indicates strong-binding states that are potentially subject to regulation. (B) Schematic representation of a general four-state model for regulation of smooth muscle cross-bridge cycling. The notion is consistent with the kinetic model for regulation proposed by Hai and Murphy (1988). this scheme depicts only free versus attached and phosphorylated versus dephosphorylated cross-bridges, where Mp and AMp are free and attached phosphorylated cross-bridges, respectively. AM, which indicates an attached, dephosphorylated cross-bridge, or "latch-bridge," is not equivalent with A·M in panel A, which is the nucleotide-free state.

Since cross-bridge cycling rates are also regulated in SM, it may be concluded that a second rate-limiting step in the SM myosin ATPase is regulated. Regulation of detachment rates in SM could potentially occur at any step subsequent to the attachment. ATP binds so readily to nucleotide-free actomyosin (r_6) that it is unlikely that regulation of this step would affect the cycle rate, or that the nucleotide-free (M·T) rigor state exists on a time frame of any consequence (Somlyo and Somlyo, 1990, 1994). However, ADP release (r_5) is a likely candidate for regulation as it is normally the rate-limiting step in the cross-bridge cycle (Somlyo and Somlyo, 1994; Butler et al., 1990; Murphy, 1994; Hai and Murphy, 1989b). An alternative mechanism, although unlikely, is stabilization of the second A·MDP state.

Perhaps the most controversial category of hypotheses to explain latch in SM proposes arresting the ATPase cycle in a force-bearing state. Such a mechanism is implied in schemes that propose a "cross-linking" protein, such as older hypotheses for the role of CD. This would lead to a state similar to the "catch" mechanism in molluscan muscle where cross-bridges are not cycling. The catch mechanism is ideally suited to the function of molluscan shell closure muscle, but it is difficult to conceptualize as a part of a SM system that is still capable of actively shortening. Furthermore, the force-generating and load-bearing characteristics of SM are similar at high phosphorylation levels and in latch as predicted if cross-bridges are the only element responsible for force generation or maintenance (Singer et al., 1986). The concept of "catch" where attached cross-bridges resist imposed loads

without ATP consumption or shortening is fundamentally different from "latch."

Progress has been made in defining the molecular structure of both actin (Holmes et al., 1990) and myosin (Rayment et al., 1993). A structural model is emerging with predicted specific conformational states that may correspond to the various biochemical states of the enzymatic cycle. Moreover, the structure of myosin (albeit skeletal myosin) includes relative locations of the LC20 as well as the essential light chain (Rayment et al., 1993). Details of how phosphorylation of LC20 or bound nucleotide may affect this structure will provide insights into the physical constraints of regulation.

B. Cross-bridge Kinetics and Regulation

1. Homogeneous versus Heterogeneous Populations of Cross-bridges

Huxley's model (1957) for muscle contraction states that the mechanical and enzymatic characteristics can be described by the overall apparent attachment and detachment rates of the cross-bridge. Contraction was described as a transition between free and attached states (Huxley, 1957). This simple yet elegant two-state model produces several specific predictions about the relationships between force production, ATPase rates, and shortening as a function of these two rate constants. Although this model is somewhat oversimplified given the actual number of biochemical states, and several of its assumptions may not strictly hold in

SM, it implies that the cross-bridge theoretically can be regulated in two basic ways: by regulating attachment (or perhaps more appropriately the transition to a force-producing state) and by regulating detachment.

There are two fundamental ways in which any parameter manifested as the result of the activity of populations of discrete entities may exist over a range of values. First, each element of the population may have a continuously variable activity and the population continuum is a function of a homogeneous population of elements, each having the same activity at a given time. This situation would require cooperativity between elements. Alternatively, the macroscopic parameter may be the mean of a heterogeneous population of elements having two or more discrete values. Thus, the variable ATPase and velocity observed in SM may be the result of a homogeneous population of cross-bridges, each with a continuously variable cycle rate and shortening velocity; or it may be the result of at least two populations of cross-bridges with discrete kinetic properties that represent the extremes of the range of ATPase rate and shortening velocities.

The organization of the filaments must allow phosphorylated cross-bridges to affect the activity of dephosphorylated cross-bridges if the mechanical properties of individual cross-bridges are homogeneous. There is some evidence for this in skeletal muscle, where cross-bridge attachment enhances both the Ca^{2+} affinity of unoccupied troponin C (TnC) at neighboring sites and potentially the binding of myosin to "unactivated" sites (Morano and Rüegg, 1991; Brandt et al., 1990). However, the effect is relatively small and similar data have not been reported for SM. The alternate concept of "averaging" the velocity of a heterogeneous population is directly supported by substitution experiments by Larsson and Moss (1993) in which skeletal muscle shortening velocity in a given muscle fiber was a function of contractile protein isoform composition, and by in vitro motility assays that show actin filament sliding velocity is a function of the myosin isoform mixture (Umemoto and Sellers, 1990). The intrinsic maximal shortening velocity of a given SM tissue generally correlates with the in vivo distribution of myosin isoforms, which have inherent and distinct ATPase activities (Malmqvist and Arner, 1991; Hewett et al., 1993).

It is not easy to explain how shortening velocity may be averaged by mixing molecular motors with different speeds within a contractile unit. It may be hypothesized that the slower cross-bridges act as an internal load on the faster cross-bridges (Aksoy et al., 1983). However, there is energetic evidence against such an internal load in SM (Butler et al., 1986). Data from an in vitro motility assay, on the other hand, indi-

cate that nonphosphorylated cross-bridges may indeed act as a load on rapidly cycling cross-bridges (Warshaw et al., 1989, 1990). It is sufficient for the present discussion to accept the empirical evidence that mixing myosins with different intrinsic cycle rates leads to mechanical properties intermediate between the properties predicted for a pure population of either isoform.

2. Kinetics and Regulation

Studies of striated muscle indicate that Ca^{2+} binding to troponin enables strong binding of cross-bridges (r_3 in Fig. 2A) and thus affects recruitment of cross-bridges into force-generating states. However, such a regulatory mechanism governing only attachment will not affect cycle rate. The latter requires that at least two steps in a cross-bridge cycle must be subject to modification. The first is attachment, allowing the transition from relaxation to contraction. Regulation of some subsequent step that influences the rate of detachment would in principle explain variable cross-bridge cycling rates manifested in shortening velocity and ATP consumption. Since SM myosin exists as a heterogeneous population, that is, some combination of phosphorylated and unphosphorylated myosin, it is reasonable to model contraction as a manifestation of a heterogeneous population of cross-bridges with discrete states. Two populations are consistent with biochemical and biophysical evidence that phosphorylation of myosin leads to a transition from one discrete state to another, whether this transition is thought of in terms of inactive to cycling, $10S$ to $6S$, unfolded to folded, or nonmotile to motile.

Two cross-bridge species allow four cross-bridge states: free and attached, phosphorylated and dephosphorylated. Covalent regulation can generate these four states if attached as well as free cross-bridges are substrates for MLCK and MLCP. Applying this concept to the cross-bridge cycle in Fig. 2A, the 7 reaction species of this scheme could exist as either phosphorylated or nonphosphorylated. Thus the smooth muscle cross-bridge may cycle through 14 reaction species rather than 7. A multiplicity of cross-bridge cycles would be possible, depending on the activity of MLCK and MLCP. The functional significance of the different cycles depends on a change in the rate-limiting step with phosphorylation state. Note that each of these potential cycles involves attachment and detachment with hydrolysis of ATP by the cross-bridge as well as additional ATP hydrolysis through phosphorylation/dephosphorylation of LC20.

The detachment rates of two cycles that occur in a SM must differ to obtain variable cross-bridge cycling rates, that is, k_4 and k_7 of Fig. 2B are unequal. Both of these lumped detachment rate constants (Fig. 2B) include reactions r_4, r_5, and r_6 in Fig. 2A, so all are potential candidates for modification. ATP binding to a nucleotide-free cross-bridge, whether phosphorylated on Ser[19] or not, is very rapid (Taylor, 1987), so r_6 is an unlikely candidate. By the same token, r_5 is the normal rate-limiting step in the cross-bridge cycle, and the one that differs among myosin isoforms, so it is the most plausible candidate for regulation. The issue to be considered in the next section concerns how detachment rates might be regulated in SM. The simplest possibility is that the LC20 phosphorylation state affects detachment rates and this is fundamental to several current hypotheses. However, in principle, mechanisms that are totally independent of cross-bridge phosphorylation could modulate detachment, and there are several candidates.

V. PROPOSED REGULATORY MECHANISMS

There are several general categories of mechanisms for the regulation of any enzyme system (Table II). All of these mechanisms have been shown to modulate cross-bridge cycling *in vitro*. Specific examples in SM will be considered in two general categories: thin filament-linked regulation and thick filament-linked regulation. The environment of a cross-bridge *in vivo* is exceptionally stable and factors such as temperature that profoundly affect cross-bridge cycling have no regulatory significance (at least in mammals); how-

ever, both allosteric and covalent regulation are possible and likely mechanisms in SM.

A. Thick Filament Regulation

1. Phosphorylation Switch Hypotheses

The concept that LC20 phosphorylation regulates contraction in SM by enabling cross-bridge attachment predicts that force would be proportional to LC20 phosphorylation and that Huxley's two-state model (Huxley, 1957) could predict the biophysical response of SM. Though this hypothesis is consistent with many observations in experiments where the flux in LC20 phosphate is abolished or greatly reduced (isolated proteins, permeabilized cells and tissues, or intact tissues at reduced temperatures), it is inconsistent with many observations under physiological conditions in intact tissues at 37°C.

These observations raise a number of recurring issues regarding the role of phosphorylation (Murphy, 1994). One issue is whether the concept that phosphorylation is necessary to initiate cross-bridge attachment is wrong. We maintain that this is unlikely. Also at issue is whether the models incorporating LC20 phosphorylation neglect other mechanisms regulating cross-bridges. Other unresolved questions are whether both free and attached cross-bridges are potential or equivalent substrates for MLCK and MLCP, and whether attached cross-bridges may cooperatively influence the cross-bridge cycle. There is now considerable evidence for mechanisms modulating the linkage between cell Ca^{2+} and phosphorylation, that is, calcium sensitivity. Such mechanisms are not always distinguished from those mechanisms that regulate the relationship between LC20 phosphorylation and cross-bridge cycling. Observations of changes in the dependence of force on Ca^{2+} do not implicate additional cross-bridge regulatory elements if the change is the Ca^{2+} sensitivity of LC20 phosphorylation.

2. Phosphorylation with Four Cross-bridge States

Hai and Murphy (1988, 1989b) propose a kinetic model in which there are two distinct populations of attached, force-bearing cross-bridges having different detachment kinetics; these are phosphorylated and dephosphorylated (Fig. 2B). Hai and Murphy postulate that LC20 phosphorylation is obligatory for cross-bridge attachment. Therefore, "latch-bridges" (AM in Fig. 2B) can be formed only by dephosphorylation of attached cross-bridges in this hypothesis (the value of k8 is assumed to be negligible). In fitting the model to experimental data, several assumptions were made. For modeling purposes, it was assumed that free and

TABLE II General Mechanisms for Regulating Enzyme Systems

Mode	Characteristics	Examples
1. Allosteric	Fast response, acts quantally to change the state of a protein, readily reversible	Ca^{2+}–troponin system, voltage-dependent ion channels
2. Covalent	Slower response, generally more complex system requiring several regulatory components; can be highly selective	All phosphorylation-dependent systems (e.g., cAMP-dependent protein kinase, PKC, MAP kinase, etc.), glycosylation, and acetylation
3. General environmental modification	Lacks selectivity	Changes in ionic strength, pH, osmolarity, and temperature

attached cross-bridges were equivalent substrates for MLCK and MLCP. Little evidence exists to either support or refute this assumption. The complex mechanical behavior of the carotid media (Fig. 1) could be quantitatively predicted if the detachment of a dephosphorylated cross-bridge is five-fold slower than its corresponding step for the phosphorylated cross-bridge. Since MLCP activity is postulated to be appreciable, substantial energy would be used for regulation by phosphorylation and dephosphorylation. The significance of this energy expenditure, as well as the relation between model predictions and experimental observations, has been addressed (Murphy, 1994; Walker et al., 1994; Wingard et al., 1994).

The Hai and Murphy model proposes that cross-bridge cycling is determined only by the ratio of MLCK to MLCP activity, and thus LC20 phosphorylation. The model predicts the behavior of swine carotid artery for which it was developed (Hai and Murphy, 1989a; Paul, 1990) and, as such, it meets the criteria listed in Table I. However, this model is not incompatible with the coexistence of additional regulatory elements.

If one accepts the evidence that force development and maintenance in smooth muscle are due to cross-bridge cycling, then dephosphorylated cross-bridges must participate (Hai and Murphy, 1989b; Butler et al., 1994). As a result, a large potential exists for regulation of and by the phosphatase (Butler et al., 1994; Driska, 1987; Gong et al., 1992; Rembold and Murphy, 1993; Somlyo et al., 1989). However, large changes in the phosphatase rate leads to rather modest changes in the predicted relation between LC20 phosphorylation and force (Fig. 3). Pharmacological interventions that alter phosphatase activity support the fact the relation between LC20 phosphorylation and force is relatively unaffected (Siegman et al., 1989). The principal effect of a regulated MLCP would be expected to be on the Ca^{2+} sensitivity of LC20 phosphorylation rather than on the dependence of force or velocity on LC20 phosphorylation (Murphy, 1994; Gong et al., 1992; see Kamm and Grange, Chapter 27, this volume).

3. The Four-State Cooperative Model

A model involving cooperativity has been proposed as an alternative to the Hai and Murphy formulation (Vyas et al., 1992, 1994). The models are similar in that in both models cross-bridges exist in the same four basic states. The novelty of this model is a finite attachment rate of unphosphorylated cross-bridges (k_8 in Fig. 2B), which is dependent on the number of attached, phosphorylated cross-bridges (AMp). In effect, some slowly cycling "latch-bridges" are postulated to form cooperatively as a result of the attach-

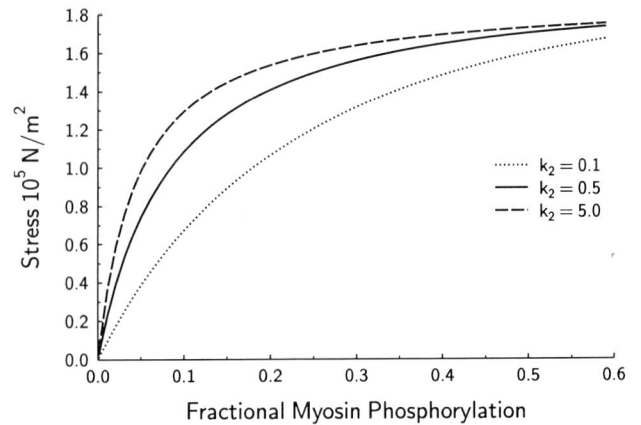

FIGURE 3 Effect of changes in phosphatase rate on the steady-state relation between force and phosphorylation in swine carotid artery according to the Hai and Murphy four-state model. The rate constants for phosphatase activity (k_2 and k_5) were varied over a 50-fold range from 0.1 s^{-1} (dotted line) to 5 s^{-1} (dashed line). The rate constants for kinase activity (k_1 and k_6) were varied to produce the desired range of phosphorylation. All other constants (k_3 and k_4, and k_7) were consistent with the original Hai and Murphy formulation.

ment of some phosphorylated cross-bridges. The molecular mechanism by which unphosphorylated cross-bridges are allowed to attach is not specified. Other concepts of cooperative attachment of dephosphorylated cross-bridges have been proposed (Himpens et al., 1988), and the consequences of cooperative regulation have been reviewed (Rembold and Murphy, 1993; Siegman et al., 1991; A. V. Somlyo et al., 1988). This model, like the Hai and Murphy model, was generated to be consistent with known physiology and to explain the phenomena in biochemical mechanistic terms. This cooperative model is dependent on the existence of a sensor for the number of attached cross-bridges that can discriminate between phosphorylation states, and possibly additional mechanisms to limit the positive feedback potential for cross-bridge attachment (Rembold and Murphy, 1993).

4. Four-State Models with Uniform Detachment Rates

Four-state models incorporating a constitutively active MLCP imply that a significant fraction of the increase in ATP consumption associated with contraction would be used for cross-bridge phosphorylation/-dephosphorylation (Walker et al., 1994; Butler et al., 1994). This is counterintuitive, given the extraordinary economy of force maintenance by smooth muscle. Paul (1990) suggested that a very slow cross-bridge detachment rate (where $k_4 = k_7$) would provide for the necessary economy observed in SM while reducing the ATP cost for phosphorylation. This hypothesis would also produce a time delay in force generation, accounting for the dissociation between phosphorylation transients and force generation. This model attempts to explain the latch phenomenon without invoking a special population of "latch-bridges" with slower detachment rates. A necessary result of this theory would be that once a cross-bridge was phosphorylated and attached, it would be committed to finishing the contraction cycle with a single detachment rate for a given load regardless of whether or not it was dephosphorylated. However, Paul's model does not predict the observed shortening velocities (Hai *et al.*, 1991), and little biochemical evidence suggests that such a great disparity exists between phosphorylation–dephosphorylation rates and actomyosin ATPase rates (Butler *et al.*, 1994). Studies of oxygen consumption suggest that the ATP cost of covalent regulation is very low in absolute terms, even if it may be a significant proportion of the suprabasal ATP consumption (Wingard *et al.*, 1994).

5. Regulation of Cross-bridge Detachment by [Mg-ADP]

Smooth muscle myosin has a high affinity for bound Mg^{2+}-ADP (Nishiye *et al.*, 1993; Fuglsang *et al.*, 1993), so the lifetime of attached cross-bridges, whether phosphorylated or not, would be prolonged by increases in ADP. Thus, slowed detachment rates could contribute to latch if ATP resynthesis did not match consumption (Nishiye *et al.*, 1993; Fuglsang *et al.*, 1993; Somlyo and Somlyo, 1994). This proposed mechanism for continuously variable cycling rates seems inconsistent with the high cellular ADP levels measured in smooth muscle and the fact that ATP consumption rates are lowest under conditions associated with latch phenomena (Murphy, 1994).

6. Other Proposed Thick Filament Regulatory Mechanisms

Ebashi (1990) hypothesized that a gene product related to MLCK acts to directly activate cross-bridges without phosphorylation of the LC20. This proposal is consistent with reports that a protein with a sequence identical to the carboxy terminal of MLCK stabilizes filament formation of SM myosin (Parker, 1993), and that polyclonal antibodies against MLCK completely inhibited force despite elevated phosphorylation (de Lanerolle *et al.*, 1991). No regulatory theory has been proposed to account for these findings, which imply that perhaps another protein similar to MLCK but without kinase activity may play a role in modifying SM contraction.

B. Thin Filament Regulation

A variety of actin-binding proteins are candidates for thin filament-linked regulation. This is somewhat misleading in that some hypotheses involve a degree of thick filament participation (e.g., the "cross-linking" proposals). Since the thin filament protein, troponin, is primarily responsible for the regulation of muscle contraction in skeletal muscle, it was initially assumed that an analogous system would operate in SM. However, the troponin subunits are absent in SM (Sobieszek and Small, 1976; Driska and Hartshorne, 1975). Perhaps the most compelling evidence for thin filament regulation are data showing that certain reconstitutions of SM thin filaments are Ca^{2+} sensitive in activating skeletal myosin ATPase (Marston, 1986, 1989; Seidel *et al.*, 1986; Gusev *et al.*, 1994). The investigation of potential thin filament regulation has resulted in a host of candidates. Major proposals include CP-based regulation, CD-based regulation, and PKC-based regulation. We also discuss regulation by PKC in this section, although this need not necessarily involve thin filaments. None of these proposals is currently regarded as alternatives to cross-bridge phosphorylation, but they are postulated as additional mechanisms or possible factors in instances of phosphorylation-independent contraction. All are reviewed in detail in Chapters 6, 7, 12, 13, and 24, this volume.

1. Caldesmon

CD has a high affinity for actin (Drabrowska and Galazkiewicz, 1986), which is influenced by TM and also binds to TM (Horiuchi and Chacko, 1989). The affinity for actin is Ca^{2+} sensitive and this sensitivity was thought to be conferred via the ubiquitous Ca^{2+}-binding protein, calmodulin, (CaM). CD also interacts with other proteins including myosin (Ikebe and Reardon, 1988). Original theories suggested that Ca^{2+}/calmodulin binding and/or phosphorylation of CD disassociated CD from thin filaments to remove a tonic inhibition, analogous to the older steric-blocking mechanism for the action of TN–TM in skeletal muscle. This Ca^{2+}-sensitive steric mechanism for CD action has been modified (Marston, 1986; Ngai and Walsh, 1987; Horiuchi and Chacko, 1989) and may involve other ancillary thin filament proteins such as gelsolin or filamin (Gusev *et al.*, 1994). The highly variable CD: actin stoichiometry in different smooth muscles is unexplained (Haeberle *et al.*, 1992), but appears inconsistent with a basic cross-bridge regulatory mechanism. CD interferes in binding of myosin to actin (Nowak *et al.*, 1989). Brenner *et al.* (1991) have shown that this protein specifically inhibits formation of the "weakly-attached" state (Fig. 2A) with little effect on the subsequent load-bearing, strongly attached state in studies of the mechanics of permeabilized SM. However, biochemical studies of the effects of caldesmon on the actomyosin ATPase kinetics led to the opposite conclusions (Marston *et al.*, 1994). Although CD and CD fragments can inhibit contraction in permeabilized preparations where no effect on myosin phosphorylation was observed (Pfitzer *et al.*, 1993; Katsuyama *et al.*, 1992), its specific role is not yet clarified. The difficulty with these proposed regulatory systems is a lack of *in vivo* or cell-based data to suggest predictable correlations between CD and LC20 phosphorylation and contraction. In addition, the affinity of CD for Ca^{2+}-CaM is low, such that little Ca^{2+}-CaM would be bound under physiological conditions (Walsh, 1991).

2. Calponin and Caltropin

Calponin is an approximately 30-kDa protein isolated from SM. CP is an actin- and calmodulin-binding protein capable of inhibiting actomyosin ATPase activity *in vitro*. The inhibition is relieved by reducing the affinity of CP for actin through phosphorylation of CP by either PKC or, interestingly, the Ca^{2+}-stimulated kinase that also phosphorylates CD (Winder and Walsh, 1989). As significant *in vivo* increases in CP phosphorylation have not been reliably measured (Bárány and Bárány, 1993), CP phosphorylation is unlikely to participate in whatever physiological role this protein may have. Other studies suggest that another Ca^{2+}-binding protein, caltropin, is more likely to be involved in mediating the effects of Ca^{2+} on CP interaction with actin (Wills *et al.*, 1994; Mani and Kay, 1993), although the data in support of this are biochemical.

Haeberle (1994) described a potential role for CP based on *in vitro* motility assays. CP increased the "force" exerted by myosin on the actin filaments while decreasing ATPase activity. It was suggested that the protein may act to stabilize the A·MD state of the cross-bridge cycle, although it is not clear how it may do this in a regulated manner. This stabilizing influence would have to be inhibited *in vivo* under conditions that promote high values of LC20 phosphorylation, since high phosphorylation is associated with rapid cross-bridge turnover.

The distribution of CP along actin filaments has been reported to be similar to that of troponin T in skeletal muscle (Takahashi *et al.*, 1988). This localization is intriguing because the TN–TM system confers cooperativity in cross-bridge attachment in striated tissue. However, the function of CP in SM remains unknown and there is no direct evidence that this role affects cross-bridge cycling. CP is also found in cytoskeletal

elements (North *et al.*, 1994). The consequences of this distribution are discussed by Gimona and Small (Chapter 7, this volume).

3. Protein Kinase C

The evidence that PKC has a regulatory role is largely based on force development without measurable increases in LC20 phosphorylation in response to activators of this enzyme (Singer *et al.*, 1989; Rasmussen *et al.*, 1987; Khalil and Morgan, 1992). However, such responses are actually very slow, weak "contractures" occurring after long delays. The substrate for PKC that might be involved in the response is unknown, but is not the myosin LC20 (Lee and Severson, 1994). We are aware of no evidence supporting a direct role for PKC in cross-bridge regulation during normal contraction and relaxation of SM. PKC activity may alter cytoplasmic Ca^{2+} either directly or indirectly through changes in the electrical properties of SM (Rasmussen *et al.*, 1990; Lee and Severson, 1994). PKC may also participate by altering the Ca^{2+} sensitivity of the MLCK/MLCP system through phosphatase regulation (Matsuo *et al.*, 1989).

VI. <u>CONCLUSIONS AND FUTURE DIRECTIONS</u>

The data show a complex picture for cross-bridge regulation in SM. It is unarguable that regulation of SM cross-bridge interactions differs from that of skeletal muscle despite having similar "motors." This involves regulation of cross-bridge cycling rates and thus any parameters affected by this, including shortening velocities, power output, and ATP consumption. There is little doubt that as more is learned about the biochemistry of smooth muscle, the potential regulatory elements will become even more complex. However, any proposed mechanism must be subjected to the criteria in Table I. When discussing a particular proposal we must ask: (1) Are the components identifiable and measurable *in vivo*? (2) Are changes of appropriate magnitude observable? (3) Does the system work in the right time frame? (4) Can reversibility be demonstrated? (5) Are correlations unique and, if not, can variation be explained? (6) Are the biochemical conditions under which a mechanism operates physiologically relevant? These criteria may be summed up with a single question. When *everything* is considered, from the reaction of the intact animal, through the biochemical and energetic analysis, to the most pure biochemistry, does it still make sense? For the few mechanisms briefly considered, there is no clear consensus, although most people in the field would acknowledge a central role for Ca^{2+}-dependent cross-bridge regulation and the necessity for both phosphorylated and dephosphorylated cross-bridges to contribute to force generation.

Native smooth muscle thin filaments and various reconstituted systems exhibit behavior *in vitro* much like the allosteric troponin Ca^{2+} switch of striated muscle (Marston *et al.*, 1994). However, contraction in cells is not Ca^{2+} dependent if the cross-bridges are phosphorylated (Itoh *et al.*, 1989; A. V. Somlyo *et al.*, 1988). We are aware of no convincing evidence for changes in any property of thin filaments associated with some aspect of cross-bridge cycling in intact cells. A plausible explanation is that although thin filaments have the potential to act as an allosteric switch, nevertheless they are constitutively "on" in smooth muscle unless the cellular environment is altered experimentally. Alternatively, Trybus (1991) has noted that the inhibitory properties of CD and CP are consistent with a role in maintenance of a relaxed state, functionally inhibiting recruitment of cross-bridges. Thin filament calcium switches are insufficient to explain regulation of the variable cross-bridge cycle rate and have not been directly implicated in apparently phosphorylation-independent contractions.

The predictive capacity of the simplest scheme (Fig. 2) implies that multiple mechanisms need not be necessary under most conditions. The basic proposition that LC20 phosphorylation–dephosphorylation, per se, is responsible for cross-bridge regulation in SM does not preclude a regulated phosphatase or cooperativity in cross-bridge attachments. It is also consistent with higher-level regulation at the level of signal transduction (Ca^{2+} sensitivity of LC20 phosphorylation).

The current research priorities include elucidation of the *in vivo* function of proteins such as CD and CP that need not involve cross-bridge cycling, and resolving the physiological relevance of the experimental situations in which LC20 phosphorylation does not explain contraction. Another potentially fertile topic is identification of mechanisms that modulate MLCK and MLCP activities and, of equal importance, establishing their *in vivo* significance.

Acknowledgments

We gratefully acknowledge Christine Palazzolo and Amy K. Browne for their aid in the production of this chapter and John S. Walker for his helpful discussions and preparation of several of the figures. Research by the authors was supported by the NIH (P01 HL 19242) and the American Heart Association (Virginia Affiliate).

References

Adam, L. P., Haeberle, J. R., and Hathaway, D. R. (1989). *J. Biol. Chem.* **264,** 7698–7703.

Adelstein, R. S., and Eisenberg, E. (1980). *Annu. Rev. Biochem.* **49,** 921–956.

Aksoy, M. O., Mras, S., Kamm, K. E., and Murphy, R. A. (1983). *Am. J. Physiol.* **245,** C255–C270.

Bárány, M., and Bárány, K. (1993). *Biochim. Biophys. Acta* **1179**(2), 229–233.

Brandt, P. W., Roemer, D., and Schachat, F. H. (1990). *J. Mol. Biol.* **212,** 473–480.

Brenner, B. (1987). *Annu. Rev. Physiol.* **49,** 655–672.

Brenner, B., Yu, L. C., and Chalovich, J. M. (1991). *Proc. Natl. Acad. Sci. U.S.A.* **88,** 5739–5743.

Brozovich, F. V., and Morgan, K. G. (1989). *Am. J. Physiol.* **257,** H1573–H1580.

Butler, T. M., Siegman, M. J., and Mooers, S. U. (1986). *Am. J. Physiol.* **251,** C945–C950.

Butler, T. M., Siegman, M. J., Mooers, S. U., and Narayan, S. R. (1990). *Am. J. Physiol.* **258,** C1092–C1099.

Butler, T. M., Narayan, S. R., Mooers, S. U., and Siegman, M. J. (1994). *Am. J. Physiol.* **267,** C1160–C1166.

Cassidy, P. S., Hoar, P. E., and Kerrick, W. G. L. (1979). *J. Biol. Chem.* **245,** 11148–11152.

Crist, J. R., He, X. D., and Goyal, R. K. (1992). *J. Physiol. (London)* **447,** 119–131.

Dabrowska, R., and Galazkiewicz, G. (1986). *Biomed. Biochim. Acta* **45,** 9993–10000.

de Lanerolle, P., Nishikawa, M., Yost, D. A., and Adelstein, R. S. (1984). *Science* **223,** 1415–1417.

de Lanerolle, P., Strauss, J. D., Felsen, R., Doerman, G. E., and Paul, R. J. (1991). *Circ. Res.* **68,** 457–465.

Driska, S. P., and Hartshorne, D. J. (1975). *Arch. Biochem. Biophys.* **167,** 203–212.

Driska, S. P. (1987). *Prog. Clin. Biol. Res.* **245,** 387–398.

Ebashi, S. (1990). *Prog. Clin. Biol. Res.* **327,** 159–166.

Filo, R. S., Bohr, D. F., and Rüegg, J. C. (1965). *Science* **147,** 1581–1583.

Fischer, W., and Pfitzer, G. (1989). *FEBS Lett.* **258,** 59–62.

Fuglsang, A., Khromov, A., Török, K., Somlyo, A. V., and Somlyo, A. P. (1993). *J. Muscle Res. Cell Motil.* **14,** 666–677.

Gerthoffer, W. T. (1987). *J. Pharmacol. Exp. Ther.* **240,** 8–15.

Gerthoffer, W. T., Murphey, K. A., and Gunst, S. J. (1989). *Am. J. Physiol.* **257,** C1062–C1068.

Gong, M. C., Fuglsang, A., Alessi, D., Kobayashi, S., Cohen, P., Somlyo, A. V., and Somlyo, A. P. (1992). *J. Biol. Chem.* **267,** 21492–21498.

Gusev, N. B., Pritchard, K., Hodgkinson, J. L., and Marston, S. B. (1994). *J. Muscle Res. Cell Motil.* **15,** 672–681.

Haeberle, J. R. (1994). *J. Biol. Chem.* **269,** 12424–12431.

Haeberle, J. R., Hathaway, D. R., and Smith, C. L. (1992). *J. Muscle Res. Cell Motil.* **13,** 81–89.

Hai, C.-M., and Murphy, R. A. (1988). *Am. J. Physiol.* **254,** C99–C106.

Hai, C.-M., and Murphy, R. A. (1989a). *Prog. Clin. Biol. Res.* **315,** 253–263.

Hai, C.-M., and Murphy, R. A. (1989b). *Annu. Rev. Physiol.* **51,** 285–298.

Hai, C.-M., Rembold, C. M., and Murphy, R. A. (1991). *Adv. Exp. Med. Biol.* **304,** 159–170.

Hartshorne, D. J. (1987). *In* "Physiology of the Gastrointestinal Tract" (L. R. Johnson, ed.), pp. 423–482. Raven Press, New York.

Hartshorne, D. J., and Gorecka, A. (1980). *In* "Handbook of Physiology" (D. F. Bohr, A. P. Somlyo, and H. V. Sparks, eds.), Sect. II, pp. 93–120. Am. Physiol. Soc., Bethesda, MD.

Hellstrand, P., and Arner, A. (1985). *Pflügers Arch.* **405,** 323–328.

Hewett, T. E., Martin, A. F., and Paul, R. J. (1993). *J. Physiol. (London)* **460,** 351–364.

Himpens, B., Matthijs, G., Somlyo, A. V., Butler, T. M., and Somlyo, A. P. (1988). *J. Gen. Physiol.* **92**, 713–729.

Holmes, K. C., Popp, D., Gebhard, W., and Kabsch, W. (1990). *Nature (London)* **347**, 44–49.

Horiuchi, K. Y., and Chacko, S. (1989). *Biochemistry* **28**, 9111–9116.

Huxley, A. F. (1957). *Prog. Biophys. Biophys. Chem.* **7**, 255–318.

Ikebe, M. (1989). *Biochemistry* **28**, 8750–8755.

Ikebe, M., and Reardon, S. (1988). *J. Biol. Chem.* **263**, 3055–3058.

Ikebe, M., Hartshorne, D. J., and Elzinga, M. (1986). *J. Biol. Chem.* **261**, 36–39.

Ikebe, M., Inagaki, M., Naka, M., and Hidaka, H. (1988). *J. Biol. Chem.* **263**, 10698–10704.

Ito, M., and Hartshorne, D. J. (1990). *Prog. Clin. Biol. Res.* **327**, 57–72.

Itoh, T., Ikebe, M., Kargacin, G. J., Hartshorne, D. J., Kemp, B. E., and Fay, F. S. (1989). *Nature (London)* **338**, 164–167.

Kamm, K. E., Hsu, L.-C., Kubota, Y., and Stull, J. T. (1989). *J. Biol. Chem.* **264**, 21223–21229.

Karaki, H., Mitsui, M., Nagase, H., Ozaki, H., Shibata, S., and Uemura, D. (1989). *Br. J. Pharmacol.* **98**, 590–596.

Katsuyama, H., Wang, C.-L.A., and Morgan, K. G. (1992). *J. Biol. Chem.* **267**, 14555–14558.

Kenney, R. E., Hoar, P. E., and Kerrick, W. G. L. (1990). *J. Biol. Chem.* **265**, 8642–8649.

Khalil, R. A., and Morgan, K. G. (1992). *News Physiol. Sci.* **7**, 10–15.

Krebs, E. G., and Beavo, J. A. (1979). *Annu. Rev. Biochem.* **48**, 923–959.

Krisanda, J. M., and Paul, R. J. (1988). *Am. J. Physiol.* **255**, C393–C400.

Kwon, S.-C., Rembold, C. M., and Murphy, R. A. (1992). *Biophys. J.* **61**, A163.

Larsson, L., and Moss, R. L. (1993). *J. Physiol. (London)* **472**, 595–614.

Lee, M. W., and Severson, D. L. (1994). *Am. J. Physiol.* **267**, C659–C678.

Malmqvist, U., and Arner, A. (1991). *Pflügers Arch.* **418**, 523–530.

Mani, R. S., and Kay, C. M. (1993). *Biochemistry* **3**, 11217–11223.

Marston, S. B. (1986). *Biochem. J.* **237**, 605–607.

Marston, S. B. (1989). *J. Muscle Res. Cell Motil.* **10**, 97–100.

Marston, S. B., Fraser, I. D. C., and Huber, P. A. J. (1994). *J. Biol. Chem.* **269**, 32104–32109.

Matsuo, K., Gokita, T., Karibe, H., and Uchida, M. K. (1989). *Biochem. Biophys. Res. Commun.* **165**, 722–727.

McDaniel, N. L., Chen, X.-L., Singer, H. A., Murphy, R. A., and Rembold, C. M. (1992). *Am. J. Physiol.* **263**, C461–C467.

Merkel, L., Gerthoffer, W. T., and Torphy, T. J. (1990). *Am. J. Physiol.* **258**, C524–C532.

Monical, P. L., Owens, G. K., and Murphy, R. A. (1993). *Am. J. Physiol.* **264**, C1466–C1472.

Morano, I., and Rüegg, J. C. (1991). *Pflügers Arch.* **418**, 333–337.

Moreland, R. S., Moreland, S., and Murphy, R. A. (1988). *Am. J. Physiol.* **255**, C473–C478.

Murphy, R. A. (1989). *Annu. Rev. Physiol.* **51**, 275–283.

Murphy, R. A. (1994). *FASEB J.* **8**, 311–318.

Ngai, P. K., and Walsh, M. P. (1987). *Biochem. J.* **244**, 417–425.

Nishiye, E., Somlyo, A. V., Török, K., and Somlyo, A. P. (1993). *J. Physiol. (London)* **460**, 247–271.

North, A. J., Gimona, M., Cross, R. A., and Small, J. V. (1994). *J. Cell Sci.* **107**, 437–444.

Nowak, E., Borovikov, Y. S., and Dabrowska, R. (1989). *Biochim. Biophys. Acta* **999**, 289–292.

Obara, K., and de Lanerolle, P. (1989). *J. Appl. Physiol.* **66**, 2017–2022.

Oishi, K., Takano-Ohmuro, H., Minakawa-Matsuo, N., Suga, O., Karibe, H., Kohama, K., and Uchida, M. K. (1991). *Biochem. Biophys. Res. Commun.* **176**, 122–128.

Parker, J. C. (1993). *Am. J. Physiol.* **265**, C1191–C1200.

Paul, R. J. (1989). *Annu. Rev. Physiol.* **51**, 331–349.

Paul, R. J. (1990). *Am. J. Physiol.* **258**, C369–C375.

Pfitzer, G., Fischer, W., and Chalovich, J. M. (1993). *Adv. Exp. Med. Biol.* **332**, 195–203.

Rasmussen, H., Takuwa, Y., and Park, S. (1987). *FASEB J.* **1**, 177–185.

Rasmussen, H., Haller, H., Takuwa, Y., Kelley, G., and Park, S. (1990). *Prog. Clin. Biol. Res.* **327**, 89–106.

Rayment, I., Rypniewski, W. R., Schmidt-Bäse, K., Smith, R., Tomchick, D. R., Benning, M. M., Winkelmann, D. A., Wesenberg, G., and Holden, H. M. (1993). *Science* **261**, 50–58.

Rembold, C. M., and Murphy, R. A. (1993). *J. Muscle Res. Cell Motil.* **14**, 325–333.

Seidel, J. C., Nath, N., and Nag., S. (1986). *Biochim. Biophys. Acta* **871**, 93–100.

Sellers, J. R., Spudich, J. A., and Sheetz, M. P. (1985). *J. Cell Biol.* **101**, 1897–1902.

Shirazi, A., Iizuka, K., Fadden, P., Mosse, C., Somlyo, A. P., Somlyo, A. V., and Haystead, T. A. J. (1994). *J. Biol. Chem.* **269**, 31598–31606.

Siegman, M. J., Butler, T. M., and Mooers, S. U. (1989). *Biochem. Biophys. Res. Commun.* **161**, 838–842.

Siegman, M. J., Butler, T. M., Vyas, T., Mooers, S. U., and Narayan, S. (1991). *Adv. Exp. Med. Biol.* **304**, 77–84.

Singer, H. A., Kamm, K. E., and Murphy, R. A. (1986). *Am. J. Physiol.* **251**, C465–C473.

Singer, H. A., Oren, J. W., and Benscoter, H. A. (1989). *J. Biol. Chem.* **264**, 21215–21222.

Sobieszek, A., and Small, J. V. (1976). *J. Mol. Biol.* **102**, 75–92.

Sobieszek, A., and Small, J. V. (1977). *J. Mol. Biol.* **112**, 559–576.

Somlyo, A. P., and Somlyo, A. V. (1990). *Annu. Rev. Physiol.* **52**, 857–874.

Somlyo, A. P., and Somlyo, A. V. (1994). *Nature (London)* **372**, 231–236.

Somlyo, A. P., Kitazawa, T., Himpens, B., Matthijs, G., Keisuke, H., Kobayashi, S., Goldman, Y. E., and Somlyo, A. V. (1989). *Adv. Protein Phosphatases* **5**, 181–195.

Somlyo, A. V., Goldman, Y. E., Fujimori, T., Bond, M., Trentham, D. R., and Somlyo, A. P. (1988). *J. Gen. Physiol.* **91**, 165–192.

Sweeney, H. L., and Stull, J. T. (1990). *Proc. Natl. Acad. Sci. U.S.A.* **87**, 414–418.

Takahashi, K., Hiwada, K., and Kokubu, T. (1988). *Hypertension* **11**, 620–626.

Tanner, J. A., Haeberle, J. R., and Meiss, R. A. (1988). *Am. J. Physiol.* **255**, C34–C42.

Taylor, E. W. (1987). *In* "Regulation and Contraction of Smooth Muscle, Progress in Clinical and Biological Research" (M. J. Siegman, A. P. Somlyo, and N. L. Stephens, eds.), pp. 59–66. Alan R. Liss, New York.

Trybus, K. M. (1989). *J. Cell Biol.* **109**, 2887–2894.

Trybus, K. M. (1991). *Cell Motil. Cytoskeleton* **18**, 81–85.

Umemoto, S., and Sellers, J. R. (1990). *J. Biol. Chem.* **265**, 14864–14869.

Vyas, T. B., Mooers, S. U., Narayan, S. R., Witherell, J. C., Siegman, M. J., and Butler, T. M. (1992). *Am. J. Physiol.* **263**, C210–C219.

Vyas, T. B., Mooers, S. U., Narayan, S. R., Siegman, M. J., and Butler, T. M. (1994). *J. Biol. Chem.* **269**, 7316–7322.

Walker, J. S., Wingard, C. J., and Murphy, R. A. (1994). *Hypertension* **23**, 1106–1112.

Walsh, M. P. (1991). *Biochem. Cell Biol.* **69**, 771–800.

Walsh, M. P., Bridenbaugh, R., Hartshorne, D. J., and Kerrick, W. G. L. (1982). *J. Biol. Chem.* **257**, 5987–5990.

Warshaw, D. M., Yamakawa, M., and Harris, D. (1989). *Prog. Clin. Biol. Res.* **315**, 329–345.

Warshaw, D. M., Desrosiers, J. M., Work, S. S., and Trybus, K. M. (1990). *J. Cell Biol.* **111,** 453–463.

Wills, F. L., McCubbin, W. D., and Kay, C. M. (1994). *Biochemistry* **33,** 5562–5569.

Winder, S. J., and Walsh, M. P. (1989). *Biochem. Soc. Trans.* **17,** 786–787.

Wingard, C. J., Paul, R. J., and Murphy, R. A. (1994). *J. Physiol. (London)* **481,** 111–117.

27

Calcium Sensitivity of Contraction

KRISTINE E. KAMM and ROBERT W. GRANGE

Department of Physiology
The University of Texas Southwestern Medical Center at Dallas
Dallas, Texas

I. Ca²⁺ DEPENDENCE OF CONTRACTION IN SMOOTH MUSCLE

The Ca^{2+} dependence of smooth muscle contraction was first definitively confirmed by the observation that both vascular smooth muscle strips and skeletal muscle fibers made permeable to Ca^{2+}-EGTA buffers by glycerination had a similar Ca^{2+} requirement for contraction (Filo *et al.*, 1965). This relation between the concentration of Ca^{2+} surrounding the myofilaments and contractile force in the steady state has since been documented in both intact and permeabilized smooth muscle preparations. The general features of this relation suggest that muscle is relaxed at an intracellular Ca^{2+} concentration ($[Ca^{2+}]_i$) of about 100 nM. As $[Ca^{2+}]_i$ increases above 100 nM, force increases steadily until, at a maximally effective $[Ca^{2+}]_i$, little augmentation in force occurs. The precise Ca^{2+} dependence of force, however, varies within a given smooth muscle depending on the mode of activation, suggesting that the sensitivity of the contractile apparatus to Ca^{2+} is not fixed but may be differentially modulated by second-messenger pathways. Although it is difficult to classify all agents according to the Ca^{2+} sensitivity they elicit because of variability among tissue types and preparations, as well as specific protocols under which agents are tested, a few generalizations can be drawn (Karaki, 1989; Kamm and Stull, 1989; Khalil and Morgan, 1992; Murphy, 1994; Somlyo and Somlyo, 1994). For example, agonists that stimulate Ca^{2+} release from intracellular stores result in greater force development at a given $[Ca^{2+}]_i$ than agents resulting in contractions primarily

due to an influx of Ca^{2+}. Conversely, activators of protein kinase C (PKC) are widely reported to elicit high force with only small or no changes in $[Ca^{2+}]_i$. The purpose of this chapter is to describe our current understanding of mechanisms by which the Ca^{2+} sensitivity of smooth muscle contraction may be modulated. Selected experimental examples are presented to illustrate important concepts, and relevant reviews are cited for further details.

A. Myosin Phosphorylation and Other Potential Regulatory Systems

Smooth muscle contraction arises from the MgATP-dependent cyclic interaction of myosin in thick filaments with actin in thin filaments. The ability of actin to activate myosin MgATPase is inhibited by the nonphosphorylated 20-kDa regulatory light chain (RLC) subunit of myosin. When the RLC is phosphorylated by a specific Ca^{2+}/calmodulin (CaM)-dependent enzyme, myosin light chain kinase (MLCK), inhibition is removed, allowing a large increase in actin-activated MgATPase activity that produces force and/or shortening in the muscle (Kamm and Stull, 1985; Hartshorne, 1987; Sellers and Adelstein, 1987; Murphy, 1994). Muscle relaxation occurs when the myosin RLC is dephosphorylated by a type 1 protein phosphatase (MLCP) that is bound to myosin by a targeting subunit (Alessi *et al.*, 1992; Shirazi *et al.*, 1994; Shimizu *et al.*, 1994). The net extent of RLC phosphorylation is determined by the relative activities of MLCK and MLCP, and insofar as force is dependent on RLC phosphorylation, these activities set the magnitude of smooth muscle contraction (Fig. 1).

In addition to the activities of MLCK and MLCP that confer thick filament regulation, smooth muscle contains two actin-binding proteins, caldesmon (CD) and calponin (CP), with biochemical properties consistent with a role in thin filament regulation. *In vitro* both CD and CP inhibit actin-activated myosin MgATPase; inhibition is reversed by Ca^{2+}/CaM binding or phosphorylation. Thus, CD and CP have been implicated in serving collateral regulatory roles in contraction (Marston and Redwood, 1991; Sobue and Sellers, 1991; Winder and Walsh, 1993). Although contractile regulation by myosin phosphorylation is well documented and generally accepted, the necessity and relative contribution of thin filament regulation in contracting smooth muscle are still being defined.

As illustrated in Fig. 1, the regulatory pathways modulating the Ca^{2+} dependence of force may exert their effects by altering the dependence of myosin RLC phosphorylation on $[Ca^{2+}]_i$ or the dependence of force on RLC phosphorylation. In both intact and permeabilized smooth muscle preparations, either of these relations can be modulated by the activities of specific second-messenger pathways.

B. Measurement of Alterations in the Ca^{2+} Sensitivity of Contraction

Investigations of the Ca^{2+} dependence of cellular processes are typically conducted in intact or permeabilized preparations where $[Ca^{2+}]_i$ is either measured or controlled, respectively. Following are comments on the relative merits of each approach.

1. Intact Muscle: Measurement of $[Ca^{2+}]_i$

Intact muscle preparations offer the advantage of unimpaired excitation–contraction coupling processes, and retained regulatory targets for second-messenger pathways. Although the Ca^{2+} concentration bathing myofibrils cannot be precisely controlled in intact cells, the advent of luminescent and fluorescent indicators to estimate $[Ca^{2+}]_i$ has greatly expanded our understanding of the Ca^{2+} dependence of smooth muscle contraction *in vivo*.

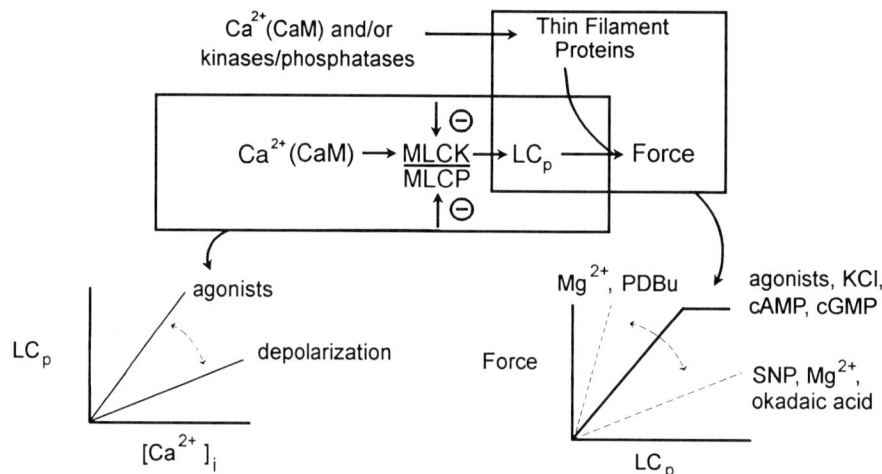

FIGURE 1   General scheme illustrating factors influencing the dependence of force on $[Ca^{2+}]_i$ for smooth muscle. Regulatory pathways that modulate the Ca^{2+} dependence of force may exert their effects by altering the dependence of myosin RLC phosphorylation (LC_p) on $[Ca^{2+}]$ (lower left) or the dependence of force on LC_p (lower right). The primary action of Ca^{2+} in regulation is to activate myosin MgATPase via Ca^{2+}/calmodulin (CaM)-dependent phosphorylation of RLC by myosin light chain kinase (MLCK). Steady-state values of LC_p result from the relative activities of MLCK and myosin light chain phosphatase (MLCP) on myosin. Alterations in the activities of MLCK or MLCP at fixed $[Ca^{2+}]_i$ will affect the Ca^{2+} sensitivity of LC_p. Studies indicate that both MLCK and MLCP may be inhibited ($\ominus$) by second-messenger pathways. As illustrated at the lower left, agonists generally result in greater LC_p at a given $[Ca^{2+}]_i$ than depolarization, thus agonists appear to sensitize force to Ca^{2+}. Although the majority of stimulus conditions result in a unique relation between LC_p and force in the steady state (lower right, solid line), a number of agents (PDBu, phorbol dibutyrate; SNP, sodium nitroprusside) have been shown to either increase or decrease the force/LC_p ratio, suggesting a role for collateral regulation via thin filaments in modulating contraction. The activity of thin filament regulatory proteins (CD and CP) may also be regulated by Ca^{2+}/CaM and/or phosphorylation. Adapted with permission from R. A. Word and J. T. Stull *in* "Heart Failure: Basic Science and Clinical Aspects," (J. K. Gwathmey, G. M. Briggs, and P. D. Allen, eds.) Marcel Dekker, Inc., New York, 1993.

The earliest simultaneous measurements of $[Ca^{2+}]_i$ and force in smooth muscle were made in strips loaded by microinjection or reversible permeabilization with the bioluminescent photoprotein aequorin (Neering and Morgan, 1980; Morgan and Morgan, 1982). Aequorin binds Ca^{2+} with high affinity in the range of $10^{-6} M Ca^{2+}$ (the Ca^{2+} affinity is also sensitive to $[Mg^{2+}]$) and is consumed during light emission (Blinks et al., 1982). Because the bioluminescent signal from aequorin is nonlinear with respect to $[Ca^{2+}]$ and rather insensitive to low $[Ca^{2+}]$, direct aequorin luminescence signals tend to exaggerate the magnitude of Ca^{2+} transients; therefore, results should be interpreted with respect to calculated $[Ca^{2+}]_i$. Currently, fura-2, a ratiometric fluorescent dye that binds Ca^{2+} in one-to-one stoichiometry, is widely used for measurement of $[Ca^{2+}]_i$ in smooth muscle. The Ca^{2+} affinity of fura-2 is about 10-fold greater than that of aequorin (Karaki, 1989; Somlyo and Himpens, 1989). Fura-2 is usually introduced as its lipophilic acetoxymethyl ester derivative fura-2 AM, which is subsequently hydrolyzed by intracellular esterases and trapped as fura-2. The loading efficiency is greater than that obtained with procedures for photoproteins. Fura-2 also has the advantages that measurements are made against a low background and multiple photon absorption/emission cycles are obtained; thus, it provides a stronger signal than luminescent indicators. Fura-2 fluorescence is suited to imaging Ca^{2+} in single cells (Moore et al., 1990). With prolonged loading, fura-2 may enter intracellular compartments, whereas aequorin appears to be retained in the cytoplasm (DeFeo et al., 1987); however, estimates of $[Ca^{2+}]_i$ are generally comparable between aequorin and fura-2 when appropriate loading procedures and calibration constants are used (DeFeo et al., 1987; Karaki, 1989; Gilbert et al., 1991).

Ca^{2+}-dependent enzymatic activities in smooth muscle tissues appear to increase more rapidly than $[Ca^{2+}]_i$ (Miller-Hance et al., 1988; Word et al., 1994). $[Ca^{2+}]_i$ started to increase immediately following electrical stimulation of either human myometrium or bovine trachealis, the half-times for maximal $[Ca^{2+}]_i$ being 2.7 and 2.0 sec, respectively, whereas approximate half-times to maximal RLC phosphorylation were less than 1.0 sec each (Word et al., 1994). Moreover, maximal fractional activation of Ca^{2+}/CaM-dependent phosphodiesterase occurred by 500 msec in electrically stimulated bovine trachealis (Miller-Hance et al., 1988). Although these results may indicate that Ca^{2+}-dependent processes are sensitive to small changes in $[Ca^{2+}]_i$ and become desensitized to Ca^{2+} as it increases to high values, it is also possible that inhomogeneities in $[Ca^{2+}]_i$ result in discrepancies

between measured bulk $[Ca^{2+}]_i$ and the effective $[Ca^{2+}]$ required for activation of CaM-dependent enzymes. Microfluorometry and digital imaging of individual aortic smooth muscle cells in culture have demonstrated discrete oscillations and inhomogeneous distributions of $[Ca^{2+}]_i$ in response to agonists (Johnson et al., 1991). Imaging showed discrete areas of elevated $[Ca^{2+}]_i$ that reached levels from 650 to 900 nM in response to angiotensin, whereas microfluorometric measurements indicated average increases to 245 nM at the same time after stimulation. Localized differences in $[Ca^{2+}]_i$ of isolated smooth muscle cells have been observed by others (Goldman et al., 1990; Etter et al., 1994). These results suggest that the Ca^{2+} dependence of cellular responses in intact tissue determined from an averaged indicator light emission should be interpreted with some caution.

2. Permeabilized Fibers: Control of $[Ca^{2+}]_i$

Fiber preparations in which cellular membranes are permeabilized (or skinned) to facilitate control of the intracellular milieu represent a hybrid between in vitro biochemistry and the intact muscle preparation. Two advantages of this approach are that the fibers generate force since the cytoskeleton remains highly organized, and the Ca^{2+} concentrations surrounding the myofibrils can be controlled with Ca^{2+}-EGTA buffer solutions. A disadvantage is that potential regulatory elements may be lost during permeabilization.

The extent of cell membrane removal can be controlled so that the pores through which bathing solutions diffuse can be varied from a molecular mass cutoff of 1 kDa to greater than 130 kDa. Extensive skinning with Triton X-100 and glycerol results in the dissolution of all cellular membranes, avoiding complications due to sequestration and release of Ca^{2+} from Ca^{2+} storage sites. However, such extensive skinning can result in loss of regulatory elements; CaM concentrations are reduced to less than half those in intact muscle following Triton X-100 treatment (Gardner et al., 1989; Tansey et al., 1994). Despite such losses, these preparations are suitable for studies in which exogenous proteins are exchanged into the contractile elements to evaluate function. Restricted permeabilization is obtained with saponin at low concentrations for short durations, which dissolves only plasma membrane. Such preparations are used to study sarcoplasmic reticulum (SR) function. Still milder detergents, for example, β-escin or the use of bacterial-derived proteins that form pores such as staphylococcal α-toxin or streptolysin O, have been used to perforate cell membranes, leaving receptor coupling to second-messenger pathways intact (Nishimura et al., 1988; Kitazawa et al., 1991). Although these preparations limit the size of agents that can enter cells (≤ 1 kDa), they have proven useful in studying the receptor-linked regulation of force in smooth muscle fibers at fixed $[Ca^{2+}]_i$.

Alterations in Ca^{2+} sensitivity of smooth muscle contraction have been demonstrated in each of these types of permeabilized fiber preparation. Addition of regulatory proteins to fibers permeabilized with Triton X-100 can sensitize or desensitize contractile force to $[Ca^{2+}]_i$, as seen, for example, with CaM (Sparrow et al., 1981; Tansey et al., 1994) or CD (Szpacenko et al., 1985; Pfitzer et al., 1993), respectively. Addition of the catalytic subunit of cAMP-dependent protein kinase (cAK) results in a desensitization of force to $[Ca^{2+}]_i$ (Meisheri and Rüegg, 1983; Rüegg and Paul, 1982). In receptor-coupled, permeabilized fibers, agonists have been shown to sensitize the Ca^{2+} dependence of force (Kitazawa et al., 1989; Nishimura et al., 1988). Although experiments in permeabilized preparations reveal regulatory functions of tested components, it is important to further test proposed pathways in intact cells and tissues to discriminate their possible roles in regulation under more physiological conditions.

II. MECHANISMS OF Ca^{2+} SENSITIZATION AND DESENSITIZATION

The mechanisms by which the Ca^{2+} sensitivity of force is modulated are best evaluated by establishing two distinct relations: first, the Ca^{2+} dependence of RLC phosphorylation and, second, the RLC phosphorylation dependence of force, as illustrated in Fig. 1. An effect on the former suggests that second-messenger pathways act to modify MLCK or MLCP activities at a given $[Ca^{2+}]_i$, whereas an effect on the latter suggests that regulatory elements in addition to RLC phosphorylation are recruited.

A. Ca^{2+} Dependence of Myosin Regulatory Light Chain Phosphorylation

Biochemical and physiological data have shown that both smooth muscle MLCK and MLCP activities are subject to modulation. MLCK can be phosphorylated at a regulatory site A near the CaM-binding domain, which densitizes its activation by Ca^{2+}/CaM (Conti and Adelstein, 1981). MLCK site A phosphorylation in smooth muscle cells desensitizes RLC phosphorylation to Ca^{2+} (Tansey et al., 1994). MLCP activity is diminished following dissociation of the catalytic subunit from its targeting subunit in vitro (Alessi et al., 1992), and although the specifics of its regulation in cells are not yet clearly established, there is ample evidence that agonists activate second-messenger

pathways that inhibit MLCP activity and thereby sensitize the Ca^{2+} dependence of RLC phosphorylation (Somlyo and Somlyo, 1994). Modulation of MLCK or MLCP activities provides distinct cellular mechanisms for affecting Ca^{2+} sensitivity of RLC phosphorylation.

1. Desensitization by Inhibition of Myosin Light Chain Kinase Activity

The effect of site A phosphorylation in MLCK is to decrease the sensitivity of the enzyme to activation by Ca^{2+}/CaM. Though MLCK can be phosphorylated at site A by cAK, protein kinase C (PKC), and the Ca^{2+}/CaM-dependent protein kinase II (CaMK II), the relative importance of these protein kinases in modulating smooth muscle contraction has only recently been defined.

a. Phosphorylation of MLCK Increases the $[Ca^{2+}/ CaM]$ Required for Activation Mammalian smooth and nonmuscle cells express a common MLCK with predicted mass of 126 kDa (Gallagher et al., 1991). It is a Ca^{2+}/CaM-regulated phosphotransferase with high specificity for myosin RLC, and contains substrate and CaM-binding domains. The catalytic core is highly homologous to other protein kinases, containing the structural determinants for ATP and RLC binding. C-terminal to the catalytic core is a shorter segment that binds CaM in a Ca^{2+}-dependent manner and with high affinity. CaM is a dumbbell-shaped molecule containing a pair of EF-type Ca^{2+}-binding sites on each lobe. Binding of Ca^{2+} to CaM activates MLCK (Blumenthal and Stull, 1980). Ca^{2+}/CaM binding results in enzyme activation by removing autoinhibition that is conferred by a region linking the catalytic core and the CaM-binding domain (Stull et al., 1995, and Chapter 9, this volume). The concentration of Ca^{2+}/ CaM required for half-maximal activation (K_{CaM}) of MLCK is 1 nM. The apparent Ca^{2+} sensitivity of RLC phosphorylation in cells and tissues, in the 100–400 nM range, results in large part from a combination of the Ca^{2+} affinity of CaM and the Ca^{2+}/CaM affinity of MLCK. Diminished Ca^{2+} sensitivity of RLC phosphorylation is brought about by increased K_{CaM} (Tansey et al., 1994).

In vitro, MLCK is phosphorylated at two sites (sites A and B) in the C-terminal region adjacent to the CaM-binding domain leading to a 10-fold increase in K_{CaM}. These sites are phosphorylated by cAK, PKC, and CaMK II (Conti and Adelstein, 1981; Nishikawa et al., 1984; Hashimoto and Soderling, 1990; Ikebe and Reardon, 1990). MLCK is desensitized to activation by Ca^{2+}/CaM following phosphorylation of sites A and B; however, phosphorylation at site B alone is insufficient to alter the Ca^{2+}/CaM activation properties of the enzyme. Site-directed mutagenesis of site A to a negative charge (serine to aspartate) is sufficient to increase K_{CaM}, whereas charge substitution at site B has no effect (Kamm et al., 1995). It is concluded that phosphorylation of site A is necessary and sufficient to account for the desensitization of MLCK to Ca^{2+}/CaM upon phosphorylation. Purified MLCK is dephosphorylated by different protein phosphatases, including types 1 and 2A (Pato and Kerc, 1990; Nomura et al., 1992).

b. MLCK Phosphorylation by CaM Kinase II in Intact Smooth Muscle
With the discovery that MLCK could be desensitized by phosphorylation, it was proposed that β-adrenergic receptor stimulation and attendant increases in cAMP formation would result in phosphorylation of MLCK and relaxation of smooth muscle (Conti and Adelstein, 1981). However, agents that elevate cAMP (or activate PKC) in smooth muscle tissues have very little effect on phosphorylation of site A in MLCK (Stull et al., 1990; reviewed in Stull et al., 1993; Van Riper et al., 1995). These results and others indicate that agents that relax smooth muscle in a cAMP-dependent manner do not exert their effects by MLCK phosphorylation-dependent desensitization to Ca^{2+}. Both cAMP and cGMP cause relaxation primarily by effecting lower $[Ca^{2+}]_i$ (Kotlikoff and Kamm, 1996). Desensitization of force to $[Ca^{2+}]_i$ in response to cAMP (Nishimura and van Breemen, 1989) or the catalytic subunit of cAK (Meisheri and Rüegg, 1983; Rüegg and Paul, 1982) has been reported in permeabilized smooth muscle fibers. Desensitization of RLC phosphorylation to $[Ca^{2+}]_i$ in response to isoproterenol in KCl-depolarized muscle has also been observed, but was not accounted for by MLCK phosphorylation (Tang et al., 1992; Van Riper et al., 1995). These findings suggest that under certain conditions cAK can act to desensitize contractile elements, perhaps by activating MLCP or affecting thin filament function, although no biochemical mechanisms in this regard have been identified.

In contrast to the effects of activators of cAK or PKC, elevation of $[Ca^{2+}]_i$ results in stoichiometric phosphorylation of MLCK at site A in smooth muscle that is associated with diminished enzymatic activation by Ca^{2+}/CaM (Stull et al., 1990; Van Riper et al., 1995). MLCK phosphorylation, as assessed by alterations in K_{CaM}, is Ca^{2+} dependent in both intact and permeabilized tracheal smooth muscle strips and cells (Tang et al., 1992; Tansey et al., 1992; 1994). The Ca^{2+} concentration required for half-maximal phosphorylation is 500 nM for MLCK and 250 nM for RLC. Treatment of cells with inhibitors of CaMK II activity abolishes MLCK phosphorylation in response to elevated $[Ca^{2+}]_i$ and potentiates RLC phosphorylation, reducing the half-maximal $[Ca^{2+}]_i$ for RLC phosphorylation to 170 nM (Tansey et al., 1992, 1994; Word et al., 1994). These results are consistent with a scheme in which MLCK is phosphorylated by CaMK II when $[Ca^{2+}]_i$ is elevated to high values, thus down-regulating activation by the Ca^{2+} signal. MLCK is phosphorylated and dephosphorylated during cycles of contraction and relaxation of both tracheal and myometrial smooth muscles, and MLCK phosphorylation coincides with alterations in the dependence of RLC phosphorylation on $[Ca^{2+}]_i$ (Word et al., 1994). Ca^{2+}-dependent phosphorylation of MLCK acts to inhibit high extents of RLC phosphorylation, thus limiting the ATP cost of rapidly cycling cross-bridges (see Section II.B.1).

Ca^{2+}-dependent MLCK phosphorylation, like RLC phosphorylation (discussed in the following), has greater sensitivity to $[Ca^{2+}]_i$ with agonists than with depolarization (Tang et al., 1992). Moreover, both GTPγS and carbachol increase the Ca^{2+} sensitivity of MLCK phosphorylation, as well as RLC phosphorylation in permeabilized strips, suggesting that both myosin RLC and MLCK are dephosphorylated by the same protein phosphatase (Tang et al., 1993).

2. Sensitization by Inhibition of Myosin Light Chain Phosphatase Activity

Regulation of MLCK has been extensively studied, but relatively little is known about regulatory mechanisms involving MLCP. However, recent studies have documented that inhibition of MLCP, most likely through a guanine nucleotide-binding protein (G-protein)-dependent pathway, leads to Ca^{2+} sensitization in smooth muscle: an increase in force for the same or smaller $[Ca^{2+}]_i$.

a. Myosin Phosphatases
There are four general classes of serine/threonine phosphatases. These are types 1, 2A, 2B, and 2C, which are also designated PP1, PP2, PP3, and MP1, respectively (Mumby and Walter, 1993). The type 1 phosphatases preferentially dephosphorylate the β subunit of phosphorylase kinase and are sensitive to both inhibitors 1 and 2, whereas the type 2 phosphatases preferentially dephosphorylate the α subunit of phosphorylase kinase and are insensitive to inhibitor 2. Although smooth muscle contains multiple forms of protein phosphatases with broad and overlapping substrate specificities (Pato, 1985), it appears that it is a type 1 phosphatase that is responsible for dephosphorylating the RLC of myosin in both skeletal and smooth muscles (Hubbard and Cohen, 1993, and Chapter 10, this volume).

The PP1 class of phosphatases are multimeric structures composed of a catalytic subunit complexed with one or more accessory subunits that target the catalytic subunit to a specific substrate (Hubbard and Cohen, 1993, and Chapter 10, this volume). The most well-studied PP1 is the glycogen-associated PP-1G, which is a heterodimer composed of a 37-kDa catalytic sub-unit (C_{sub}) complexed to a 124-kDa glycogen-targeting subunit (G_{sub}). The binding of C_{sub} to G_{sub} is regulated such that phosphorylation of G_{sub} at specific sites either enhances or diminishes binding and thereby activity of C_{sub} in a reversible manner. The phosphatase that dephosphorylates the myosin RLC in smooth muscle was purified from avian gizzard and has been designated smooth muscle-PP-1M (myosin-associated form) to differentiate it from the skeletal muscle-PP-1M (Alessi et al., 1992). A similar enzyme complex was isolated from pig bladder (Shirazi et al., 1994). PP-1M is a heterotrimer consisting of a 37-kDa catalytic subunit and two subunits of 130 and 20 kDa that specifically target the complex to myosin (Alessi et al., 1992; Shirazi et al., 1994; Shimizu et al., 1994). Association of C_{sub} with the targeting subunits enhances its activity toward the RLC in myosin (Alessi et al., 1992).

b. Agonist-Mediated Sensitization of Regulatory Light Chain Phosphorylation

In intact smooth muscle, the $[Ca^{2+}]_i$ required for half-maximal RLC phosphorylation is less when tissues are stimulated by agonists than by depolarization (Rembold and Murphy, 1988; Karaki et al., 1988; Gerthoffer et al., 1989; Suematsu et al., 1991; Tang et al., 1992). The Ca^{2+}-sensitizing effect of agonists on force is also observed in receptor-coupled permeabilized preparations and is mimicked by GTPγS (Nishimura et al., 1988; Kitazawa et al., 1989). GTP-dependent potentiation of force is associated with elevated RLC phosphorylation in permeabilized fibers (Fujiwara et al., 1989; Kitazawa et al., 1991; Kubota et al., 1992), and the effect appears to act through inhibition of MLCP rather than stimulation of MLCK. Both agonist and GTPγS decreased rates of relaxation and RLC dephosphorylation without affecting rates of RLC phosphorylation in α-toxin-permeabilized rabbit portal vein (Kitazawa et al., 1991). GTPγS was ineffective in potentiating contractions in permeabilized trachealis when MLCP was inhibited by okadaic acid, and in tracheal homogenates dephosphorylation of ^{32}P-labeled heavy meromyosin was inhibited by GTPγS, whereas MLCK activity was not affected (Kubota et al., 1992).

The agonist- and GTP-activated pathway for inhibition of MLCP is believed to rely on a guanine nucleotide-binding protein(s). Ligand occupancy of G-protein-linked agonist receptors initiates coupling to intracellular targets through the action of heterotrimeric G-proteins that are distinct from the smaller mono-meric GTP-binding proteins. A trimeric G-protein appears to be involved in the MLCP pathway, as fluoroaluminate, which activates only trimeric G-proteins, mimics the GTP effect (Kawase and van Breemen, 1992); however, the sensitivities of mechanical responses to fluoroaluminate and GTPγS are not identical (Hai and Ma, 1993). In rabbit portal vein permeabilized with α-toxin, receptor activation of the MLCP inhibitor pathway appears to diverge from the Ca^{2+} release pathway at a step prior to phospholipase C (Kobayashi et al., 1991). Divergence may arise by agonist receptors coupling to phospholipase C and other targets by the same or different G-proteins. In addition to the trimeric G-proteins, there is also evidence that monomeric GTP-binding proteins, rhoA p21 or ras p21, enhance Ca^{2+} sensitivity of smooth muscle contraction in permeabilized fibers, although it is not known if they act to inhibit MLCP (Hirata et al., 1992; Satoh et al., 1993).

Following on the strong premise that the MLCP inhibitory pathway is initiated by receptor activation of a trimeric G-protein, investigators have sought to identify second messengers that may effect Ca^{2+} sensitization (Somlyo and Somlyo, 1994). Diacylglycerol (DAG) via activation of PKC and arachidonic acid are two potential messengers arising from receptor–G-protein activation of phospholipase C and phospholipase A_2, respectively (Lee and Severson, 1994). PKC is considered a reasonable candidate because of the known Ca^{2+}-sensitizing effects of phorbol esters (Khalil and Morgan, 1992), which are direct activators of PKC. Phorbol esters can potentiate force development in skinned fibers in a Ca^{2+}-independent manner but RLC phosphorylation-dependent manner (Itoh et al., 1993). A Ca^{2+}-sensitizing effect of arachidonic acid (300 μM) was observed in permeabilized rabbit femoral artery (Gong et al., 1992). This effect was independent of the G-protein since GDPβS, which inhibits G-protein activation, did not abolish potentiation of force, indicating that arachidonic acid could act as a downstream messenger. In addition, arachidonic acid was shown to dissociate oligomeric MLCP into subunits and inhibit activity toward myosin but not phosphorylase (Gong et al., 1992). It has become clear that MLCP is a target of regulation in smooth muscle under physiological conditions. Detailed schemes of the second-messenger pathways affecting MLCP will be the object of future research.

B. Dependence of Force on Regulatory Light Chain Phosphorylation

The dependence of steady-state isometric force on myosin RLC phosphorylation has been documented in many smooth muscles and supports the model that Ca^{2+}-dependent myosin phosphorylation is the primary effector of contractile regulation. However, two related aspects of the force–RLC phosphorylation relation, as described next, have spawned numerous investigations to identify additional or collateral effectors of regulation.

1. Regulation of Nonphosphorylated Cross-bridges

Maximal force may be achieved with steady-state values of RLC phosphorylation as low as 0.2 mol phosphate/mol RLC (Di Blasi et al., 1992). Assuming maximal force results from the activity of the majority of cross-bridges, this result indicates that nonphosphorylated cross-bridges contribute to force maintenance. How are these nonphosphorylated cross-bridges regulated? Figure 2 summarizes aspects of the cross-bridge cycle in smooth muscle. The earliest and most formalized model is that of Murphy and coworkers (Murphy, 1994) in which dephosphorylated attached cross-bridges (latch bridges) detach slowly and thus contribute to force maintenance along with cycling phosphorylated cross-bridges. The latch bridge hypothesis requires only one regulatory mechanism, phosphorylation of myosin RLC. Nonphosphorylated attached cross-bridges are proposed to arise only from

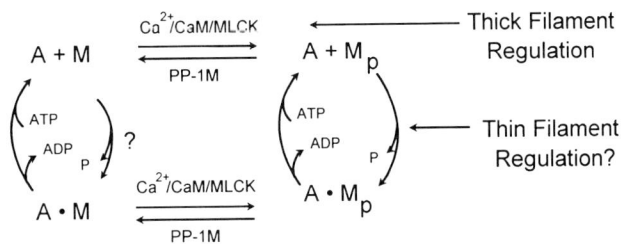

FIGURE 2 Scheme illustrating theories for regulation of the cross-bridge cycle in smooth muscle. The basic model for regulation consists of four states: nonphosphorylated and phosphorylated myosin, and nonphosphorylated and phosphorylated actin-bound myosin. Myosin cross-bridges (M) interact with actin (A) in the thin filament in a MgATP-dependent, cyclic fashion. The cycle generates force, which results from the fraction of attached cross-bridges (A·M or A·M_p). Thick filament regulation results from Ca^{2+}/CaM/MLCK-dependent phosphorylation of myosin RLC (M_p) that is reversed by dephosphorylation by MLCP (PP-1M). Myosin phosphorylation initiates MgATP-dependent cross-bridge cycling (cycle on right). Thin filament regulatory proteins can modify the fraction of cross-bridges in the attached state; however, the significance of thin filament regulation in vivo is not clear. The key element of the latch model is that dephosphorylation of attached cross-bridges (A·M_p) leads to a cross-bridge state that detaches slowly and supports force (A·M). Thus, a latch bridge can only be formed via RLC phosphorylation. Others have proposed that nonphosphorylated cross-bridges (M) can be activated to attach (A·M) by cooperative activation by force-generating cross-bridges and/or by the actions of thin filament proteins (?). Adapted with permission from Murphy et al., Fed. Proc. **42**, 51–56, 1983.

previously phosphorylated attached bridges, and not from direct attachment. A second hypothesis, derived primarily from results with permeabilized fibers, proposes that active, nonphosphorylated cross-bridges are regulated by a cooperative mechanism in which a few phosphorylated cross-bridges lead to attachment and cycling of many nonphosphorylated cross-bridges (Somlyo et al., 1988; Kenney et al., 1990; Vyas et al., 1992, 1994). When myosin RLC in permeabilized fibers is thiophosphorylated and thus resistant to the action of MLCP, the full range of forces and ATP consumption are achieved over a small range of RLC phosphorylation. Under these conditions, phosphorylated cross-bridges are prevented from becoming latch bridges, arguing that nonphosphorylated bridges must attach and cycle (Kenney et al., 1990; Vyas et al., 1992). Cooperative activation by phosphorylated cross-bridges, in analogy with cooperative effects of cross-bridge activation in striated muscle, has been proposed (Somlyo et al., 1988), and biochemical studies argue that strong myosin complexes (Myosin ADP) cooperatively activate smooth muscle actin–tropomyosin (Horiuchi and Chacko, 1989; Marston et al., 1994). Rembold and Murphy (1993) have mathematically modeled combinations of these regulatory schemes and conclude that isometric force produced by nonphosphorylated cross-bridges could be explained by the latch mechanism or cooperative regulation in which only phosphorylated cross-bridges effect activation. Thus Ca^{2+}-dependent RLC phosphorylation would remain the primary regulatory mechanism. Several questions arise in this regard. Do thin filament regulatory proteins differentially affect phosphorylated and nonphosphorylated cross-bridges? Can nonphosphorylated cross-bridges be activated in the absence of RLC phosphorylation? What are the relative contributions of the phosphorylated, nonphosphorylated, and latch bridge cycles to force maintenance in vivo?

2. Collateral Regulation: Role for Thin Filament Proteins

The second aspect of the force–RLC phosphorylation relation that suggests additional regulation is illustrated by alterations in the slope of the relation (Fig. 1, lower right). Alterations in isometric force at fixed values of RLC phosphorylation are probably the strongest evidence for the in vivo operation of thin filament regulation. Smooth muscle contains two thin filament proteins, caldesmon and calponin, that inhibit actin-activated MgATPase activity of phosphorylated myosin. This inhibitory activity is reversed by the binding of Ca^{2+}/CaM or by phosphorylation, and thus CD and CP may modulate the RLC phosphorylation dependence of force. CD is a long, flexible, 87-kDa protein containing binding sites for myosin as well as actin, tropomyosin, and Ca^{2+}/CaM. CD is phosphorylated *in vitro* by PKC, CaMK II, mitogen-activated protein kinase (MAPK), and others (Marston and Redwood, 1991; Sobue and Sellers, 1991). CP is a 34-kDa actin-binding protein found essentially only in smooth muscle and, like CD, binds tropomyosin, actin, and Ca^{2+}/CaM. Phosphorylation of purified CP by PKC or CaMK II reverses the inhibition of actin-activated myosin MgATPase (Winder and Walsh, 1993). CD and CP compete for binding to actin and appear to be localized on different populations of thin filaments in smooth muscle cells (Lehman, 1991; Makuch *et al.*, 1991; North *et al.*, 1994). There is evidence that both CD and CP undergo significant conformational changes during contraction, particularly when compared to other contractile and cytoskeletal proteins (Bárány *et al.*, 1992). The potential for simultaneous effects on MLCK, CD, and CP by Ca^{2+}/CaM has resulted in difficulty defining the relative contributions of not only thick and thin filament regulation, but also the respective roles of CD and CP. Because the affinities of CD and CP for Ca^{2+}/CaM appear to be two or three orders of magnitude lower than the affinity of Ca^{2+}/CaM for MLCK (bringing into doubt the importance of CaM in thin filament regulation) (Allen and Walsh, 1994), it has been proposed that other Ca^{2+}-binding proteins may effect Ca^{2+} sensitization of smooth muscle thin filaments (Pritchard and Marston, 1993).

CD is effective in inhibiting isometric force in Triton-permeabilized gizzard fibers (Szpacenko *et al.*, 1985; Pfitzer *et al.*, 1993). The actin-binding fragment of CD desensitizes the force–RLC phosphorylation relation, and the inhibitory effect is reversed by high concentrations of Ca^{2+}/CaM (Pfitzer *et al.*, 1993). In addition, CD peptides that act to displace CD from thin filaments are effective in enhancing force at fixed Ca^{2+} in isolated permeabilized smooth muscle cells (Katsuyama *et al.*, 1992). These results demonstrate that CD can modulate the interaction of phosphorylated cross-bridges with thin filaments in smooth muscle strips.

Depression of the force–RLC phosphorylation relation in intact smooth muscle was first described by Gerthoffer (1987), who demonstrated that elevating RLC phosphorylation by readdition of external Ca^{2+} in the presence of agonist was not accompanied by force development. Subsequently, others have shown depression of the force–RLC phosphorylation relation upon relaxation of activated smooth muscle with low concentrations of the phosphatase inhibitor okadaic acid or high external $[Mg^{2+}]$ (Tansey *et al.*, 1990; D'Angelo *et al.*, 1992, respectively, and Chapter 25, this volume). In these studies, relaxation was accompanied by

reductions in $[Ca^{2+}]_i$ consistent with Ca^{2+}-dependent thin filament regulation; conversely, sodium nitroprusside has been shown to relax activated vascular smooth muscle in the presence of elevated RLC phosphorylation without lowering $[Ca^{2+}]_i$ (McDaniel *et al.*, 1992). Redistribution of phosphate in the RLC into sites that do not result in actin activation of MgATPase might account for apparent discrepancies in the force–RLC phosphorylation relation; however, peptide mapping confirmed in each of these cases that RLC was phosphorylated only on Ser 19, the activating MLCK site (Tansey *et al.*, 1990; D'Angelo *et al.*, 1992; McDaniel *et al.*, 1992).

Sensitization of force to RLC phosphorylation in response to activators of PKC has proven to be an intriguing topic, since agonist stimulation results in the simultaneous formation of two second messengers, inositol 1,4,5-trisphosphate that releases Ca^{2+} from SR and DAG that activates PKC. It has been speculated that PKC exerts a significant influence on contractile regulation in the latter or sustained phase of agonist-induced contraction (Rasmussen *et al.*, 1987). Force development in the absence of RLC phosphorylation can also be promoted with very high $[Mg^{2+}]$ (20 mM) in skinned fibers, however, this occurs by a Mg^{2+}-dependent conformational change in myosin that mimics phosphorylation and is believed to have no physiological significance (Hartshorne, 1987). The tumor-promoting phorbol esters, though also not physiological activators, have similar effects to DAG in activating PKC and have been widely used to examine the role of PKC in biological functions (Nishizuka, 1984). Addition of phorbol esters to resting, intact smooth muscle preparations often, but not always (Kamm *et al.*, 1989), results in slowly developing, sustained, and large contractions (Khalil and Morgan, 1992). In some preparations the contraction is accompanied by small increases in $[Ca^{2+}]_i$ and/or RLC phosphorylation (Rembold and Murphy, 1988; Ozaki *et al.*, 1990; Singer, 1990; Itoh *et al.*, 1993), however, in others force develops with no increase in $[Ca^{2+}]_i$ (Jiang and Morgan, 1989; Sato *et al.*, 1992) or RLC phosphorylation (Singer and Baker, 1987; Adam *et al.*, 1989; Jiang and Morgan, 1989; Sato *et al.*, 1992; Fulginiti *et al.*, 1993). The bases for these different responses are not yet defined, but tissues exhibiting force in the absence of increased RLC phosphorylation have provided models for examining the PKC effects in more detail.

How might such Ca^{2+}- and RLC phosphorylation-independent contractions come about? It is widely assumed that disinhibition of thin filaments plays a role in this process, and that nonphosphorylated cross-bridges are recruited by one of the mechanisms discussed here. Alternatively, bridging of thin and thick filaments by CD or cross-linking of actin filaments in the "cytoskeletal" domain have been proposed (Rasmussen *et al.*, 1987); however, the mechanical properties of tissues in RLC phosphorylation-independent contractions suggest that force is maintained by cycling cross-bridges (Sato *et al.*, 1992; Fulginiti *et al.*, 1993). Morgan and colleagues have investigated the PKC pathway in the ferret aorta preparation, which is exquisitely sensitive to phorbol esters and shows similar Ca^{2+}- and RLC phosphorylation-independent responses to α-agonist (Khalil and Morgan, 1992). With a combination of immunolocalization and inhibitors it was shown that Ca^{2+}-independent isoforms of PKC differentially translocate with α-agonist stimulation, with the ε-PKC associating with the sarcolemma (Khalil *et al.*, 1992). The translocation of ε-PKC was followed by a transient association of MAPK with the sarcolemma, after which MAPK became diffusely distributed in the cytoplasm. It was proposed that Ca^{2+}-independent activation involves a kinase cascade initiated by the agonist-dependent activation of ε-PKC that leads to activation and translocation of MAPK. MAPK would phosphorylate CD, resulting in disinhibition of thin filaments and recruitment of nonphosphorylated cross-bridges (Khalil and Morgan, 1993). Intermediates in the kinase cascade activated by PKC and Ras are present in amounts consistent with a potential role in regulation of smooth muscle contraction via phosphorylation of CD (Adam *et al.*, 1995). Moreover, CD has been reported to become phosphorylated during smooth muscle contraction (Park and Rasmussen, 1986; Adam *et al.*, 1989), and this occurs at MAPK sites in arterial smooth muscle (Adam and Hathaway, 1993). MAPK activation requires phosphorylation of both threonine and tyrosine residues that is carried out by a dual-specificity kinase, MAPK kinase. Tyrosine kinase inhibitors have been shown to suppress agonist-induced contractions in gut and vascular smooth muscles; whether these effects are exerted on the MAPK pathway remains to be determined (Di Salvo *et al.*, 1993).

The biochemical evidence that phosphorylation of CP results in disinhibition of thin filaments is clear (Winder and Walsh, 1990; Nakamura *et al.*, 1993); however, measurements of CP phosphorylation *in vivo* are inconsistent. No phosphorylation was observed in relaxed or stimulated swine carotid artery or chicken gizzard muscle (Bárány *et al.*, 1991; Gimona *et al.*, 1992), whereas carbachol-induced phosphorylation was observed in strips of toad stomach (Winder *et al.*, 1993). The stoichiometry of CP phosphorylation was not measured in the latter studies. CP decreases actin filament translocation velocity and enhances force in *in vitro* motility assays, suggesting that it may function to reduce the detachment rate of cross-bridges, thus contributing to the latch state of smooth muscle (Haeberle, 1994). At present, the relative contributions of CD and CP to the physiological regulation of smooth muscle contraction remain undefined.

III. SUMMARY AND PERSPECTIVES

Of what import are changes in the Ca^{2+} sensitivity of RLC phosphorylation to the function of smooth muscle? The simultaneous activation of MLCK and inhibition of MLCP upon activation by agonists (e.g., neurotransmitters) promotes rapid rates of RLC phosphorylation allowing rapid initial force development at the cost of high MgATP consumption by phosphorylated cross-bridges. In the tonic phase of contraction, maximum force is supported by few phosphorylated cross-bridges and many nonphosphorylated cross-bridges; desensitization of MLCK may promote the economy of contraction by preventing the recruitment of rapidly cycling cross-bridges that consume MgATP but do not enhance force.

The sufficiency of myosin phosphorylation for contraction in smooth muscle is indisputably demonstrated in experiments where the addition of proteolyzed CaM-independent MLCK to permeabilized fibers brings about contraction at very low $[Ca^{2+}]$ (Walsh *et al.*, 1982). Furthermore, injection of CaM-independent MLCK into single smooth muscle cells results in contraction under conditions where $[Ca^{2+}]_i$ remains at resting values (Itoh *et al.*, 1989). These results tend to rule out a model in which thin filaments are fully inhibited in resting muscle, and require Ca^{2+}-dependent disinhibition to support contraction. However, these experiments do not rule out the possibility that thin filament-binding proteins modulate the contractile response. Alterations in the RLC phosphorylation–force relation in intact muscle indicate that collateral regulation can occur. Moreover, experiments with permeabilized fibers demonstrate that thin filament-binding proteins can inhibit force independent of RLC phosphorylation (Pfitzer *et al.*, 1993). Additional physiological investigations will be required to fully establish a role for thin filament proteins in regulating actin–myosin interactions in smooth muscle. Moreover, it should be recognized that the relative contribution of thin filament regulation to the Ca^{2+} sensitivity of force may vary greatly among different smooth muscles.

Acknowledgment

Dr. Grange is the recipient of a Research Training Fellowship Award from the American Lung Association.

References

Adam, L. P., and Hathaway, D. R. (1993). *FEBS Lett.* **322,** 56–60.

Adam, L. P., Franklin, M. T., Raff, G. J., and Hathaway, D. R. (1995). *Circ. Res.* **76,** 183–190.

Adam, L. P., Haeberle, J. R., and Hathaway, D. R. (1989). *J. Biol. Chem.* **264,** 7698–7703.

Alessi, D., Macdougall, L. K., Sola, M. M., Ikebe, M., and Cohen, P. (1992). *Eur. J. Biochem.* **210,** 1023–1035.

Allen, B. G., and Walsh, M. P. (1994). *Trends Biochem. Sci.* **19,** 362–368.

Bárány, M., Rokolya, A., and Bárány, K. (1991). *FEBS Lett.* **279,** 65–68.

Bárány, K., Polyák, E., and Bárány, M. (1992). *Biochem. Biophys. Res. Commun.* **187,** 847–852.

Blinks, J. R., Wier, W. G., Hess, P., and Prendergast, F. G. (1982). *Prog. Biophys. Mol. Biol.* **40,** 1–114.

Blumenthal, D. K., and Stull, J. T. (1980). *Biochemistry* **19,** 5608–5614.

Conti, M. A., and Adelstein, R. S. (1981). *J. Biol. Chem.* **256,** 3178–3181.

D'Angelo, D. K. G., Singer, H. A., and Rembold, C. M. (1992). *J. Clin. Invest.* **89,** 1988–1994.

DeFeo, T. T., Briggs, G. M., and Morgan, K. G. (1987). *Am. J. Physiol.* **253,** H1456–H1461.

Di Blasi, P., Van Riper, D., Kaiser, R., Rembold, C. M., and Murphy, R. A. (1992). *Am. J. Physiol.* **262,** C1388–C1391.

Di Salvo, J., Steusloff, A., Semenchuk, L., Satoh, S., Kolquist, K., and Pfitzer, G. (1993). *Biochem. Biophys. Res. Commun.* **190,** 968–974.

Etter, E. F., Kuhn, M. A., and Fay, F. S. (1994). *J. Biol. Chem.* **269,** 10141–10149.

Filo, R. S., Bohr, D. F., and Rüegg, J. C. (1965). *Science* **147,** 1581–1583.

Fujiwara, T., Itoh, T., Kubota, Y., and Kuriyama, H. (1989). *J. Physiol. (London)* **408,** 535–547.

Fulginiti, J., III, Singer, H. A., and Moreland, R. S. (1993). *J. Vasc. Res.* **30,** 315–322.

Gallagher, P. J., Herring, B. P., Griffin, S. A., and Stull, J. T. (1991). *J. Biol. Chem.* **266,** 23936–23944.

Gardner, J. P., Stout, M. A., and Harris, S. R. (1989). *Pflügers Arch.* **414,** 484–491.

Gerthoffer, W. T. (1987). *J. Pharmacol. Exp. Ther.* **240,** 8–15.

Gerthoffer, W. T., Murphey, K. A., and Gunst, S. J. (1989). *Am. J. Physiol.* **257,** C1062–C1068.

Gilbert, E. K., Weaver, B. A., and Rembold, C. M. (1991). *FASEB J.* **5,** 2593–2599.

Gimona, M., Sparrow, M. P., Strasser, P., Herzog, M., and Small, J. V. (1992). *Eur. J. Biochem.* **205,** 1067–1075.

Goldman, W. F., Bova, S., and Blaustein, M. P. (1990). *Cell Calcium* **11,** 221–231.

Gong, M. C., Fuglsang, A., Alessi, D., Kobayashi, S., Cohen, P., Somlyo, A. V., and Somlyo, A. P. (1992). *J. Biol. Chem.* **267,** 21492–21498.

Haeberle, J. R. (1994). *J. Biol. Chem.* **269,** 12424–12431.

Hai, C.-M., and Ma, C. B. B. (1993). *Am. J. Physiol.* **265,** L73–L79.

Hartshorne, D. J. (1987). *In* "Physiology of the Gastrointestinal Tract" (L. R. Johnson, ed.), pp. 423–482. Raven Press, New York.

Hashimoto, Y., and Soderling, T. R. (1990). *Arch. Biochem. Biophys.* **278,** 41–45.

Hirata, K.-I., Kikuchi, A., Sasaki, T., Kuroda, S., Kaibuchi, K., Matsuura, Y., Seki, H., Saida, K., and Takai, Y. (1992). *J. Biol. Chem.* **267,** 8719–8722.

Horiuchi, K. Y., and Chacko, S. (1989). *Biochemistry* **28,** 9111–9116.

Hubbard, M. J., and Cohen, P. (1993). *Trends Biochem. Sci.* **18,** 172–177.

Ikebe, M., and Reardon, S. (1990). *J. Biol. Chem.* **265,** 8975–8978.

Itoh, H., Shimomura, A., Okubo, S., Ichikawa, K., Ito, M., Konishi, T., and Nakano, T. (1993). *Am. J. Physiol.* **265,** C1319–C1324.

Itoh, T., Ikebe, M., Kargacin, G. J., Hartshorne, D. J., Kemp, B. E., and Fay, F. S. (1989). *Nature (London)* **338,** 164–167.

Jiang, M. J., and Morgan, K. G. (1989). *Pflügers Arch.* **413,** 637–643.

Johnson, E. M., Thelers, J.-M., Capponi, A. M., and Vallotton, M. B. (1991). *J. Biol. Chem.* **266,** 12618–12626.

Kamm, K. E., and Stull, J. T. (1985). *Annu. Rev. Pharmacol. Toxicol.* **25,** 593–620.

Kamm, K. E., and Stull, J. T. (1989). *Annu. Rev. Physiol.* **51,** 299–313.

Kamm, K. E., Hsu, L.-C., Kubota, Y., and Stull, J. T. (1989). *J. Biol. Chem.* **264,** 21223–21229.

Kamm, K. E., Luby-Phelps, K., Tansey, M. G., Gallagher, P. J., and Stull, J. T. (1995). *In* "Regulation of the Contractile Cycle in Smooth Muscle." (M. Ito, ed.) pp. 139–158. Springer-Verlag, Tokyo.

Karaki, H. (1989). *Trends Pharmacol. Sci.* **10,** 320–325.

Karaki, H., Sato, K., and Ozaki, H. (1988). *Eur. J. Pharmacol.* **151,** 325–328.

Katsuyama, H., Wang, C.-L.A., and Morgan, K. G. (1992). *J. Biol. Chem.* **267,** 14555–14558.

Kawase, T., and van Breemen, C. (1992). *Eur. J. Pharmacol.* **214,** 39–44.

Kenney, R. E., Hoar, P. E., and Kerrick, W. G. L. (1990). *J. Biol. Chem.* **265,** 8642–8649.

Khalil, R. A., and Morgan, K. G. (1992). *News Physiol. Sci.* **7,** 10–15.

Khalil, R. A., and Morgan, K. G. (1993). *Am. J. Physiol.* **265,** C406–C411.

Khalil, R. A., Lajoie, C., Resnick, M. S., and Morgan, K. G. (1992). *Am. J. Physiol.* **263,** C714–C719.

Kitazawa, T., Kobayashi, S., Horiuti, K., Somlyo, A. V., and Somlyo, A. P. (1989). *J. Biol. Chem.* **264,** 5339–5342.

Kitazawa, T., Gaylinn, B. D., Denney, G. H., and Somlyo, A. P. (1991). *J. Biol. Chem.* **266,** 1708–1715.

Kobayashi, S., Gong, M. C., Somlyo, A. V., and Somlyo, A. P. (1991). *Am. J. Physiol.* **260,** C364–C370.

Kotlikoff, M. I., and Kamm, K. E. (1996). *Annu. Rev. Physiol.* **58** (in press).

Kubota, Y., Nomura, M., Kamm, K. E., Mumby, M. C., and Stull, J. T. (1992). *Am. J. Physiol.* **262,** C405–C410.

Lee, M. W., and Severson, D. L. (1994). *Am. J. Physiol.* **267,** C659–C678.

Lehman, W. (1991). *J. Muscle Res. Cell Motil.* **12,** 221–224.

Makuch, R., Birukov, K., Shirinsky, V., and Dabrowska, R. (1991). *Biochem. J.* **280,** 33–38.

Marston, S. B., and Redwood, C. S. (1991). *Biochem. J.* **279,** 1–16.

Marston, S. B., Fraser, I. D. C., Huber, P. A. J., Pritchard, K., Gusev, N. B., and Torok, K. (1994). *J. Biol. Chem.* **269,** 8134–8139.

McDaniel, N. L., Chen, X.-L., Singer, H. A., Murphy, R. A., and Rembold, C. M. (1992). *Am. J. Physiol.* **263,** C461–C467.

Meisheri, K. D., and Rüegg, J. C. (1983). *Pflügers Arch.* **399,** 315–320.

Miller-Hance, W. C., Miller, J. R., Wells, J. N., Stull, J. T., and Kamm, K. E. (1988). *J. Biol. Chem.* **263,** 13979–13982.

Moore, E. D. W., Becker, P. L., Fogarty, K. E., Williams, D. A., and Fay, F. S. (1990). *Cell Calcium* **11,** 157–179.

Morgan, J. P., and Morgan, K. G. (1982). *Pflügers Arch.* **395,** 75–77.

Mumby, M. C., and Walter, G. (1993). *Physiol. Rev.* **73,** 673–699.

Murphy, R. A. (1994). *FASEB J.* **8,** 311–318.

Murphy, R. A., Aksoy, M. O., Dillon, P. F., Gerthoffer, W. T., and Kamm, K. E. (1983). *Fed. Proc.* **42,** 51–56.

Nakamura, F., Mino, T., Yamamoto, J., Naka, M., and Tanaka, T. C. (1993). *J. Biol. Chem.* **268,** 6194–6201.

Neering, I. R., and Morgan, K. G. (1980). *Nature (London)* **288,** 585–587.

Nishikawa, M., de Lanerolle, P., Lincoln, T. M., and Adelstein, R. S. (1984). *J. Biol. Chem.* **259,** 8429–8436.

Nishimura, J., and van Breemen, C. (1989). *Biochem. Biophys. Res. Commun.* **163,** 929–935.

Nishimura, J., Kolber, M., and van Breemen, C. (1988). *Biochem. Biophys. Res. Commun.* **157,** 677–683.

Nishizuka, Y. (1984). *Nature (London)* **308,** 693–698.

Nomura, M., Stull, J. T., Kamm, K. E., and Mumby, M. C. (1992). *Biochemistry* **31,** 11915–11920.

North, A. J., Gimona, M., Cross, R. A., and Small, J. V. (1994). *J. Cell Sci.* **107,** 437–444.

Ozaki, H., Ohyama, T., Sato, K., and Karaki, H. (1990). *Jpn. J. Pharmacol.* **52,** 509–512.

Park, S., and Rasmussen, H. (1986). *J. Biol. Chem.* **261,** 15734–15739.

Pato, M. D. (1985). *Adv. Protein Phosphatases* **1,** 367–382.

Pato, M. D., and Kerc, E. (1990). *Arch. Biochem. Biophys.* **276,** 116–124.

Pfitzer, G., Zeugner, C., Troschka, M., and Chalovich, J. M. (1993). *Proc. Natl. Acad. Sci. U.S.A.* **90,** 5904–5908.

Pritchard, K., and Marston, S. B. (1993). *Biochem. Biophys. Res. Commun.* **190,** 668–673.

Rasmussen, H., Takuwa, Y., and Park, S. (1987). *FASEB J.* **1,** 177–185.

Rembold, C. M., and Murphy, R. A. (1988). *Circ. Res.* **63,** 593–603.

Rembold, C. M., and Murphy, R. A. (1993). *J. Muscle Res. Cell Motil.* **14,** 325–333.

Rüegg, J. C., and Paul, R. J. (1982). *Circ. Res.* **50,** 394–399.

Sato, K., Hori, M., Ozaki, H., Takano-Ohmuro, H., Tsuchiya, T., Sugi, H., and Karaki, H. (1992). *J. Pharmacol. Exp. Ther.* **261,** 497–505.

Satoh, S., Rensland, H., and Pfitzer, G. (1993). *FEBS Lett.* **324,** 211–215.

Sellers, J. R., and Adelstein, R. S. (1987). *In* "The Enzymes" (P. D. Boyer and E. G. Krebs, eds.), pp. 381–418. Academic Press, Orlando, FL.

Shimizu, H., Ito, M., Miyahara, M., Ichikawa, K., Okubo, S., Konishi, T., Naka, M., Tanaka, T., Hirano, K., Hartshorne, D. J., and Nakano, T. (1994). *J. Biol. Chem.* **269,** 30407–30411.

Shirazi, A., Iizuka, K., Fadden, P., Mosse, C., Somlyo, A. P., Somlyo, A. V., and Haystead, T. A. J. (1994). *J. Biol. Chem.* **269,** 31598–31606.

Singer, H. A. (1990). *J. Pharmacol. Exp. Ther.* **1074,** 1068–1074.

Singer, H. A., and Baker, K. M. (1987). *J. Pharmacol. Exp. Ther.* **243,** 814–821.

Sobue, K., and Sellers, J. R. (1991). *J. Biol. Chem.* **266,** 12115–12118.

Somlyo, A. P., and Himpens, B. (1989). *FASEB J.* **3,** 2266–2276.

Somlyo, A. P., and Somlyo, A. V. (1994). *Nature (London)* **372,** 231–236.

Somlyo, A. V., Goldman, Y. E., Fujimori, T., Bond, M., Trentham, D. R., and Somlyo, A. P. (1988). *J. Gen. Physiol.* **91,** 165–192.

Sparrow, M. P., Mrwa, U., Hofmann, F., and Rüegg, J. C. (1981). *FEBS Lett.* **125,** 141–145.

Stull, J. T., Hsu, L.-C., Tansey, M. G., and Kamm, K. E. (1990). *J. Biol. Chem.* **265,** 16683–16690.

Stull, J. T., Tansey, M. G., Tang, D.-C., Word, R. A., and Kamm, K. E. (1993). *Mol. Cell. Biochem.* **127/128;** 229–237.

Stull, J. T., Krueger, J. K., Zhi, G., and Gao, G.-H. (1995). *In* "Calcium as Cell Signal" (K. Kohama, ed.), pp. 175–184. Igaku-Shoin Ltd., Tokyo.

Stull, J. T., Krueger, J. K., Zhi, G., and Gao, G.-H. (1995). *In* "International Symposium on Regulation of the Contractile Cycle in Smooth Muscle." Springer-Verlag, Tokyo (in press).

Suematsu, E., Resnick, M., and Morgan, K. G. (1991). *Am. J. Physiol.* **261,** C253–C258.

Szpacenko, A., Wagner, J., Dabrowska, R., and Rüegg, J. C. (1985). *FEBS Lett.* **192,** 9–12.

Tang, D.-C., Stull, J. T., Kubota, Y., and Kamm, K. E. (1992). *J. Biol. Chem.* **267,** 11839–11845.

Tang, D.-C., Kubota, Y., Kamm, K. E., and Stull, J. T. (1993). *FEBS Lett.* **331,** 272–275.

Tansey, M. G., Hori, M., Karaki, H., Kamm, K. E., and Stull, J. T. (1990). *FEBS Lett.* **270,** 219–221.

Tansey, M. G., Word, R. A., Hidaka, H., Singer, H. A., Schworer, C. M., Kamm, K. E., and Stull, J. T. (1992). *J. Biol. Chem.* **267,** 12511–12516.

Tansey, M. G., Luby-Phelps, K., Kamm, K. E., and Stull, J. T. (1994). *J. Biol. Chem.* **269,** 9912–9920.

Van Riper, D. A., Weaver, B. A., Stull, J. T., and Rembold, C. M. (1995). *Am. J. Physiol.* **268,** H2466–H2475.

Vyas, T. B., Mooers, S. U., Narayan, S. R., Witherell, J. C., Siegman, M. J., and Butler, T. M. (1992). *Am. J. Physiol.* **263,** C210–C219.

Vyas, T. B., Mooers, S. U., Narayan, S. R., Siegman, M. J., and Butler, T. M. (1994). *J. Biol. Chem.* **269,** 7316–7322.

Walsh, M. P., Bridenbaugh, R., Hartshorne, D. J., and Kerrick, W. G. L. (1982). *J. Biol. Chem.* **257,** 5987–5990.

Winder, S. J., and Walsh, M. P. (1990). *J. Biol. Chem.* **265,** 10148–10155.

Winder, S. J., and Walsh, M. P. (1993). *Cell. Signal.* **5,** 677–686.

Winder, S. J., Allen, B. G., Fraser, E. D., Kang, H.-M., Kargacin, G. J., and Walsh, M. P. (1993). *Biochem. J.* **296,** 827–836.

Word, R. A., and Stull, J. T. (1993). *In* "Heart Failure: Basic Science and Clinical Aspects" (J. K. Gwathmey, C. M. Briggs, and P. D. Allen, eds.), pp. 145–165. Marcel Dekker, Inc., New York.

Word, R. A., Tang, D.-C., and Kamm, K. E. (1994). *J. Biol. Chem.* **269,** 21596–21602.

28

Pharmacological Regulation of Smooth Muscle by Ion Channels, Kinases, and Cyclic Nucleotides

PAUL J. SILVER and DOUGLAS S. KRAFTE

Department of Cardiovascular Pharmacology
Sterling Winthrop, Inc.
Collegeville, Pennsylvania

I. INTRODUCTION

The purpose of this chapter is to acquaint the reader with the various types of smooth muscle relaxant agents that are used, or under development, as therapy (pharmaceutical drugs), or those that are useful for scientific experiments (tools). The focus will be on agents that act on some of the mechanisms of smooth muscle regulation described in Chapters 18, and 25–27 in this volume (see also Somlyo and Somlyo, 1994). We will not review well-known, receptor-mediated mechanisms (such as α-adrenergic antagonists or β-adrenergic agonists); the reader is referred to any standard textbook on pharmacology for discussion of these types of relaxants.

The outline of this chapter will follow a "system's approach" to regulation of smooth muscle relaxation and will be divided into sections dealing with ion flux regulation, key protein kinases, and regulation of cyclic nucleotides. For detail on the normal biochemical/physiological function of these systems, the reader is referred to Chapters 9, 11, 12, 16, 20, and 22 in this volume. Where appropriate, differences between vascular, airway, gastrointestinal, and reproductive smooth muscle will be noted.

II. ION FLUX REGULATION: GENERAL CONSIDERATIONS

Ionic gradients in smooth muscle cells are similar to those found in most other cell types with high levels of potassium inside the cell relative to the outside. Gradients are reversed for sodium and calcium levels, with lower levels on the intracellular side of the membrane. Smooth muscle cells are excitable cells, but they generally do not utilize the transmembrane sodium gradient for activity as do neurons, skeletal muscle cells, and cardiac cells. Consequently, voltage-dependent sodium channels are not important regulators of smooth muscle cell function and modulators of sodium channel function do not play a role in smooth muscle pharmacology. The sodium gradient is, however, utilized in co- and countertransport processes to maintain ionic homeostasis within these cells. Important ion fluxes that do regulate cell function tend to involve the flow of either potassium and/or calcium across the cell membrane and modulators of these processes are of pharmacological importance to those interested in studying smooth muscle cell function.

Opening of potassium channels drives the membrane potential to more negative values consistent with those of resting, quiescent cells. In normal excitation–contraction coupling schemes, potassium channel agonists would be considered smooth muscle relaxants, whereas potassium channel blockers would promote contraction. A variety of potassium channels appear to play a role in regulation of smooth muscle contraction, including Ca^{2+}-activated and ATP-gated channels. The situation is reversed when one considers calcium channels. Calcium influx is necessary for maintenance of contractile responses and calcium channel antagonists are, therefore, relaxants whereas agonists promote contraction. Calcium entry in smooth muscle can occur through either L-type voltage-dependent calcium channels or receptor-operated channels.

From a pharmacological perspective, therapeutically useful agents have been developed that interact with L-type calcium channels and ATP-gated potassium channels (i.e., pharmacological drugs). Each of these channel types will be discussed in more detail. Other types of channels, such as Ca^{2+}-activated potassium channels and receptor-operated calcium channels, do not yet have therapeutically useful ligands, but do represent molecular targets of therapeutic interest. Such channels and the available pharmacological tools will also be discussed briefly.

A. ATP-Gated Potassium Channels

ATP-gated channels are found in a variety of cell types including vascular smooth muscle (for reviews see Gopalakrishnan et al., 1993; Evans and Taylor, 1994). ATP closes this class of channels and agonists appear to operate by shifting the concentration–response relationship for ATP inhibition to the right, thereby removing inhibition at physiological ATP concentrations. The resultant effect of an agonist is hyperpolarization and smooth muscle relaxation. Therapeutically, agonists in this class have been used for the treatment of hypertension, angina, and asthma. Applications in other areas where smooth muscle relaxation would be beneficial, such as irritable bladder syndrome, have also been discussed.

Three major classes of compounds are known to activate ATP-gated channels. These are the benzopyrans, cyanoquanidines, and nitronicatinamide analogs, with representative examples in each class being cromakalim, pinacidil, and nicorandil (Gopalakrishnan et al., 1993); their structure is shown in Fig. 1. In general, the effective concentration range of this class of drugs on smooth muscle can be summarized with representative agents as: cromakalim (10–100 nM) > pinacidil (200–500 nM) > nicorandil (5–7 μM). Evans and Taylor (1994) have also reviewed this area and subdivided further to include a series of agents typified by aprikalim as well as novel structures that do not fit into any of these classes.

Data from studies in humans are most readily available for pinicidil since it has been clinically utilized for the treatment of hypertension (Ahnfeldt-Ronne, 1988). Pinicidil is effective in mild to moderate hypertension, but exhibits a number of side effects. Reflex tachycardia in response to decreases in blood pressure is one problem that can be managed with β-adrenergic blockers. Headache is another common side effect of this class of agents. Pinacidil inhibits bronchoconstriction in response to exogenous stimuli (Longman and Hamilton, 1992) and inhibits contraction of human bladder (Fovaeus et al., 1989). One of the issues regard-

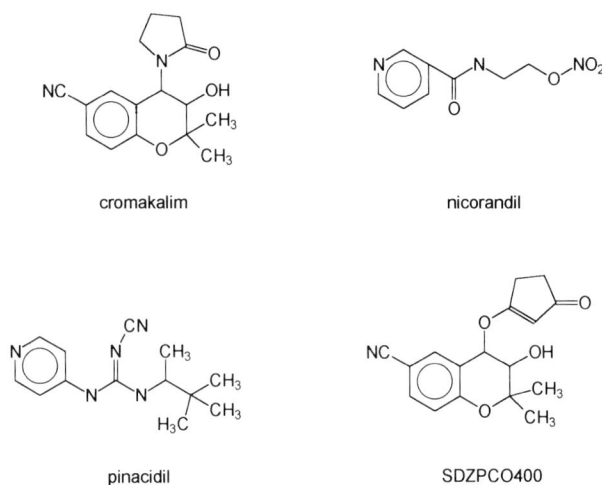

FIGURE 1   Representative ATP-gated potassium channel openers.

ing the use of pinicidil as well as other agonists of ATP-gated potassium channels is selectivity among the smooth muscle of different tissues. Pinacidil is more potent in inhibiting vascular smooth muscle contraction than that of the bladder (Edwards et al., 1991), and the resultant systemic hypotension precludes targeting of nonvascular smooth muscle for therapeutic purposes, at least for this particular molecule.

Among the benzopyrans, cromakalim has been the most widely studied and has been demonstrated to have antihypertensive effects as a consequence of vascular smooth muscle relaxation. This compound has been reported to be effective in the management of nocturnal asthma in humans (Williams et al., 1990), although development has been discontinued. Systemic hypotension was not, however, a problem in this study, suggesting it may be possible to target this ATP-gated channel agonist to the lung. A closely related molecule, SDZPCO400 (Fig. 1), has been reviewed (Morley, 1994) and reported to suppress airway hyperreactivity, a condition where bronchial vascular smooth muscle is more susceptible to spasmogens. The compound was more effective in hyperreactive airways than those of normal animals and the duration of action was longer in the hyperreactive case as well. This interesting observation may involve indirect regulation of bronchial smooth muscle through vagal nerves since the compound was not effective following vagal sectioning (Chapman et al., 1992). Effects on neuronal regulation of airway smooth muscle may be an interesting approach to managing hyperreactivity for this class of ion channel modulators.

Of the nitronicatinamide analogs, the representative compound noted here, nicorandil, has been used in humans to treat angina pectoris by causing coronary vasodilation. The mechanism of action of nicorandil, however, probably involves actions in addition to opening of ATP-gated potassium channels. This compound also stimulates soluble guanylate cyclase (see the following), which will have relaxant properties. Combined with relaxation of smooth muscle on both the arterial and venous side of the circulation, nicroandil can reduce the load to the heart (Kinoshita and Sakai, 1990) and also promote coronary blood flow, leading to a reduction in angina.

The future for agonists of ATP-gated potassium channels is likely to lie in the development of agents that can relax smooth muscle in nonvascular beds while not affecting blood pressure. Interestingly, there have been reports that this class of molecules can act at other types of potassium channels in smooth muscle (Wickenden et al., 1991), which will also achieve relaxation. If such observations are generally true, it may be possible to achieve tissue selectivity by targeting potassium channels other than ATP-gated channels to achieve smooth muscle relaxation.

B. Voltage-Gated Calcium Channels

Calcium channel antagonists relax smooth muscle by inhibiting calcium entry and, therefore, have similar functional effects and therapeutic applications to the ATP-gated potassium channel agonists just discussed. Calcium channel antagonists have been widely used therapeutically to treat cardiovascular disorders and were available prior to the discovery of compounds such as cromakalim.

Although multiple types of calcium channels have been identified, L-type channels are the important subtype in regulating smooth muscle tone and are the target of clinically utilized calcium channel antagonists. The binding sites for the different antagonists that relax smooth muscle appear to reside on the α_1 subunit (Catterall and Striessnig, 1992). Even though the pharmacology of calcium channels antagonists preceded the molecular cloning of calcium channel genes, it is interesting to note that the gene expressed in smooth muscle is also expressed in heart, whereas different genes are expressed in skeletal muscle and neuronal and endocrine tissues (Hullin et al., 1993). This parallels the therapeutic targets of available calcium channel blockers and the tissue selectivity of these agents.

There are three major classes of calcium channel antagonists. Dihydropyridines (e.g., nifedipine) have significant effects on vascular smooth muscle and

have been utilized in the treatment of hypertension and angina. Benzothiazepines (e.g., diltiazem), which are also useful in the same settings as dihydropyridines, and phenylalkylamines (e.g., verapamil) utilized more in the treatment of cardiac rhythm disturbances. Relaxation of vascular smooth muscle and regulation of blood pressure and coronary resistance by these agents have been extensively studied and will not be discussed further here. However, other applications (Fisher and Grotta, 1993) of the dihydropyridine class of antagonists involve smooth muscle relaxation, particularly in the cerebrovascular and renal systems.

Nimodipine is a member of the dihydropyridine class of antagonists that improves the outcome of patients following subarachnoid hemorrhage (Wadsworth and McTavish, 1992). This agent relaxes cerebrovascular smooth muscle and reduces the incidence of cerebral vasospasm. Nicardipine has also been demonstrated to have similar efficacy (Haley et al., 1991) although at higher concentrations (Flamm et al., 1988).

In the kidney, dihydropyridines have been reported to reduce vasoconstriction on the afferent arterioles without affecting efferent resistance (Loutzenhiser and Epstein, 1985). This profile appears to be unique to the dihydropyridine class since members of the other classes did not have preferential effects on the afferent side. The resulting increase in renal perfusion and glomerular filtration rates leads to improved kidney function. In humans with induced renal insufficiency, nifedipine was shown to improve renal performance (Neumayer et al., 1989; Russo et al., 1990).

C. Other Smooth Muscle Ion Channels

Receptor-operated calcium channels represent yet another channel type present in smooth muscle that play an important physiological role. Receptor ligands such as vasopressin, angiotensin II, and ATP all activate calcium entry that is insensitive to the classic voltage-dependent calcium channel antagonists discussed earlier. Potent and selective pharmacological agents are not yet available for these channels and investigors often resort to inorganic cations such as Cd^{2+} for inhibitors. Rüegg et al. (1989) have reported that 2-nitro-4-carboxyphenyl-N,N-diphenyl carbamate (NCDC) will inhibit receptor-operated calcium entry in rat aortic smooth muscle cells and mesenteric resistance vessels. The IC_{50} value for inhibition, however, was $> 10\,\mu M$ in most cases. Another tool has been reported by Krautwurst et al. (1993) to inhibit receptor-operated calcium entry. This compound, LOE 908, has an IC_{50} value <1 μM and will potentially be useful in further exploring this class of ion channels.

Calcium-activated potassium channels are important regulators of smooth muscle tone, but fewer pharmacological tools are available compared to ATP-gated potassium channels, particularly with respect to channel agonists. Some ATP-gated channel agonists, however, may actually also open calcium-activated channels, which would contribute to their efficacy in relaxing smooth muscle (Gelband and McCullough, 1993). McManus et al. (1993) have reported a series of agents derived from natural products that are agonists of calcium-activated potassium channels. Such agents are particularly useful in defining the functional role of particular potassium channels in the regulation of smooth muscle tone. These studies are also aided by the availability of antagonists of both ATP-gated (e.g., glibenclamide) and Ca-activated (e.g., iberiotoxin) channels.

III. MODIFICATION OF KEY KINASES AND PHOSPHORYLATION

The major kinases that play a key role in regulating contraction/relaxation of smooth muscles are discussed in Chapters 9, and 11–13 of this volume. In this review, we focus on inhibitors of myosin light chain kinase (MLCK), protein kinase C (PKC), and Ca^{2+}-calmodulin kinase II. In addition, since inhibitors of smooth muscle cell migration and proliferation represent an area of current pharmaceutical interest for prevention of restenosis following angioplasty, or for treating atherosclerosis, a discussion of recently identified selective inhibitors of platelet-derived growth factor (PDGF) and other tyrosine kinases is included.

A. Modulation of Myosin Light Chain Phosphorylation

There are no current therapeutic agents in clinical use that function solely via inhibition of smooth muscle contractile protein interactions. Thus, although this area remains a potential site of action for the discovery and development of new therapeutic agents, the question of specificity looms large for potential discoveries. Direct pharmacological regulation of smooth muscle contractile protein interactions can theoretically occur at several sites. Among these are the sites of regulation on the thin filament (caldesmon) or on the thick filament (phosphorylation of myosin light chain). Most efforts at regulation have focused on modulating myosin light chain (MLC) phosphorylation.

Alteration of MLC phosphorylation can occur at four possible sites: inhibition of Ca^{2+} binding to calmodulin, inhibition of calmodulin activation of MLCK, direct inhibition of MLCK catalytic activity, or activation of myosin light chain phosphatase (MLCP). Although inhibitors of MLCP, such as okadaic acid, are well characterized, there are no known specific activators of MLCP.

Direct inhibitors of MLCK act via one of two modes. A series of small peptide substrate (MLC) and/or calmodulin inhibitors have been described by two groups (Moreland and Hunt, 1987; Foster and Gaeta, 1988). Since these are peptides, they are not active as inhibitors in intact cell or tissue studies, but can be used as tools in skinned smooth muscle preparations, as well as with purified contractile proteins.

Other direct MLCK inhibitors act via competition with ATP. Among the first small molecules described that work via this mechanism are the isoquinoline sulfonamides (Hidaka et al., 1984). These agents also inhibit other kinases such as PKC, and the cyclic nucleotide-dependent kinases, although some selectivity is apparent with certain structural modifications. They do not inhibit other ATP-dependent enzymes. Among the most selective of these agents for MLCK are ML-7 and ML-9, with IC_{50} values in the 0.3–0.4 μM range (Saitoh et al., 1987). Staurosporine, K252A, and related analogs are also potent inhibitors of MLCK and other protein kinases, including PKC (see Section III.B and Chapter 24, this volume). Mechanistically, these agents are also ATP-competitive inhibitors. Interestingly, an analog of K252A, the propylether derivative KT 5926, is a selective inhibitor of MLCK (Nakanishi et al., 1990). Mechanistically, it appears to be an ATP-competitive (K_i = 18 nM) and MLC noncompetitive (K_i = 12 nM) inhibitor. Inhibition of PKC, cAMP-dependent protein kinase (cAK) and cGMP-dependent protein kinase (cGK) occurs in the submicromolar to low micromolar range. Selective inhibition of MLC phosphorylation in platelets could also be observed, suggesting that this compound might be useful in delineating MLCK from PKC-dependent phosphorylation events in smooth muscle. Calmodulin antagonists remain the most common way to inhibit MLCK activity. The largest group of calmodulin antagonists are those that compete with the regulated enzyme for the Ca^{2+}–calmodulin complex. One agent, HT-74, reportedly inhibits Ca^{2+} binding to calmodulin (Tanaka et al., 1986). Numerous chemical classes of calmodulin antagonists have been reported, and include (nonexclusively) the phenothiazine and diphenylbutylpiperidine antipsychotics, tricyclic and nontricyclic antidepressants, certain peptide venoms (mastosparan, melittin), naphthalene sulfonamides such as W-7, the antifungal analog calmidazolium, amiodarone and other lipophilic class III antiarrhythmic agents, and certain Ca^{2+} entry blockers (Van Belle, 1981; Prozialeck, 1983; Hidaka and Tanaka, 1983; Mannold, 1984; Silver et al., 1985, 1988a, 1989). RO 22-4839, a papaverine analog with cerebral vasodilator activity, has been shown to preferentially inhibit calmodulin activation of MLCK ($IC_{50} = 3$ μM), but did not inhibit calmodulin activation of Ca^{2+}/Mg^{2+}-ATPase or adenylate cyclase up to concentrations of 300 μM, suggesting a degree of selectivity for MLCK (Nakajima and Katoh, 1987).

B. Protein Kinase C

Since this enzyme is an important regulator of smooth muscle tone, it represents a logical target for novel drug discovery. In fact, there are multiple potential diseases in which PKC may be involved (see Bradshaw et al., 1993, for review). The ubiquitous nature of PKC suggests that selective development of agents for chronic diseases will not be possible, due to lack of specificity and the resultant high degree of side effects. However, acute use in critical situations, such as cancer, may be possible.

There are several sites for pharmacological development of modulators, most notably substrate inhibitors of the catalytic domain (either ATP or the phosphoprotein substrate) or inhibitors of the regulatory domain. ATP-competitive substrate inhibitors include the aforementioned isoquinoline sulfonamides, although no real specificity for kinase inhibition is realized with most agents of this chemical class. More potent inhibitors (nanomolar to submicromolar) obtained from fermentation broths include staurosporine, K252A, and analogs (such as RK-286C; Osada et al., 1990), which are also ATP-competitive inhibitors with roughly equipotency for inhibiting PKC and cAK (Kase et al., 1986; Silver et al., 1989). These agents have been reported to possess smooth muscle relaxant activity and depressor activity in animals (Hachisu et al., 1987; Buchholz et al., 1991; Dundore et al., 1992).

Selective PKC-directed inhibitor analogs of staurosporine have been synthesized. CGP 41251 is roughly 10 times less potent than staurosporine as a PKC inhibitor ($IC_{50} = 50$ nM), but is approximately 40 times more selective for PKC relative to cAK (Meyer et al., 1989). CGP 41251 retains the ability to inhibit PKC-mediated events in intact cellular systems. Other analogs of staurosporine, such as 7-O-methyl UCN-01, also display selectivity for PKC (Takahashi et al., 1990). A fuller discussion of the structure/activity relationships for

PKC inhibition and selectivity of staurosporine analogs is provided by Murray and Coates (1994).

An inhibitor of PKC-mediated histone s2 substrate phosphorylation of PKC, chelerythrine, has been described (Herbert, 1990). This commercially available compound is a tertacyclic alkaloid. Potency for PKC inhibition is in the submicromolar range, and efficacy in intact cells or platelets versus PKC-mediated activation is evident. Activity versus other substrates is not known.

As diacylglycerol (DAG) is the most critical regulator of PKC activity, a logical target would be to discover DAG binding antagonists. This offers the potential to obtain more selective inhibitors than the aforementioned ATP-competitive inhibitors. Moreover, there are now at least eight isozymes of PKC described in the literature (see Nishizuka, 1988, for review). Several of these isozymes differ by virtue of a different regulatory domain and, thus, to activation by phorbol esters (Cabot and Jaken, 1984), suggesting that selective DAG antagonists of these different isozymes can be designed. Some potent naturally occurring DAG displacers have been identified, including the bryostatins, which are derived from the marine bryozoan *Bugula neritina* (Kraft et al., 1988).

Calphostin C (also known as UCN-1028C), produced by *Cladosporium cladosporioides*, is a novel, potent (IC_{50} - 50 nM) inhibitor of PKC that acts via inhibition of the regulatory subunit (Kobayashi et al., 1989). Mechanistically, inhibition of phorbol ester binding is evident, and anti-PKC activity in intact cells occurs. Interestingly, inhibition of phorbol ester binding may be light dependent with coincubation of the enzyme and inhibitor needed (Bruns, 1991). This may limit the usefulness of this inhibitor in intact tissue or *in vivo* settings. Other inhibitors of the regulatory site of PKC have been described, including trifluoperazine and other calmodulin antagonists (Mori et al., 1980; Wise et al., 1982), sphingosine (Hannun et al., 1986), tamoxifen (Su et al., 1985), and some peptides found in bee and wasp venoms (Raynor et al., 1991). However, these agents are nonselective, have other mechanisms, and most likely work via hydrophobic interactions.

A synthetic inhibitor of the regulatory domain of PKC, NPC 15437, has also been reported (Sullivan et al., 1992). Although lacking in potency ($IC_{50} = 20$ μM), this peptide compound does not inhibit other kinases up to concentrations of 300 μM. Mechanistically, this agent is a competitive inhibitor of phorbol ester activation ($K_i = 5$ μM) and phosphatidylserine activation ($IC_{50} = 12$ μM), and a mixed inhibitor to activation by calcium. Inhibition of PKC-mediated phosphorylation in platelets by NPC 15437 also occurs.

C. Calcium/Calmodulin-Dependent Protein Kinase II

An inhibitor of Ca^{2+} calmodulin (CaM) kinase II, KN-62, has been described (Okazaki *et al.*, 1994). This isoquinoline derivative inhibits this kinase by an ATP-competitive mechanism, with a K_i value of approximately 1 μM. Although this agent has been used in several intact models to investigate the role of Ca^{2+}/ CaM kinase II, it does not possess a high degree of selectivity versus other CaM-regulated kinases (K_i = 3–10 μM range). Thus, a more potent and selective inhibitor of Ca^{2+}/CaM kinase II is needed.

In sum, inhibitors of key serine/threonine kinases in smooth muscle, namely, MLCK and PKC, now exist. Most have demonstrated antikinase activity in platelets, so it should be possible to use these as tools in various smooth muscle preparations. However, though useful as tools, particularly when used in the right concentration range, it is highly unlikely that any of these agents will be developed as therapeutic agents for smooth muscle-related diseases. It is possible that some of these agents, or analogs, may be developed for more serious diseases, such as AIDS or cancer, where cytotoxicity or cytostasis and a link to kinase inhibition are desired.

D. Inhibition of Tyrosine Kinases

It is now accepted that several growth regulatory factors influence vascular smooth muscle cell proliferation, migration, and constriction. Foremost among these factors are PDGF, β-fibroblast growth factor (β-FGF), and TGF-β, which have been clearly implicated in lesions of atherosclerosis, as well as in restenosis of vessels following angioplasty (see Ross, 1991; Libby *et al.*, 1988, for some reviews). Thus, selective inhibitors of some, or all, of these tyrosine kinases represent a potential therapeutic avenue for new agents for the prevention of vascular smooth muscle cell proliferation and migration, and so may be useful for atherosclerosis or restenosis.

Initial inhibitors were the tyrphostins, which are analogs of erbstatin, that nonselectively inhibit most tyrosine kinases by an ATP-competitive mechanism (for some reviews, see Levitzki, 1990; Levitzki and Gilon, 1991; Fry, 1994). Although some selectivity was claimed, these agents rarely possessed potency below 1 μM. They were useful as initial tools to show that certain tyrosine kinases were involved in certain proliferative diseases. In this regard, certain tyrphostins have been used to link inhibition of PDGF kinase with inhibition of PDGF-mediated autophosphorylation, intracellular phosphorylation, DNA synthesis, and

cell growth in vascular smooth muscle cells (Bilder *et al.*, 1991).

These and other positive results have spurred several research groups into looking for more selective and more potent PDGF tyrosine kinase inhibitors. Two groups, one at Rhone Poulenc Rorer and one at Sterling, have reported on obtaining such compounds. The Rorer group disclosed a series of substituted quinoline derivatives, with potency as low as 20 nM for some analogs (Maguire *et al.*, 1994). Most of these compounds were inactive versus cAK and did not inhibit EGF tyrosine kinase. However, no inhibitory activity in intact vascular smooth muscle cells versus a PDGF-mediated event was reported.

At about the same time, the Sterling group reported on a series of pyrindinyl-substituted quinoline compounds with similar potency and selectivity versus human vascular smooth muscle cell PDGF kinase (Dolle *et al.*, 1994). Interestingly, these compounds are ATP-competitive inhibitors (K_i = 14 nM) but do not inhibit serine/threonine kinases (PKC or cAK). Moreover, in a broader panel of tyrosine kinases assayed in the same format, there was little or no inhibition of EGF, erbB2, or p561ck. In addition, inhibition of PDGF-dependent thymidine incorporation in smooth muscle cells was evident for PDGF-inhibitory compounds. Structural analogs, which did not inhibit PDGF kinase, also did not inhibit thymidine incorporation, suggesting a positive correlation for inhibition of kinase activity and inhibition of proliferation. Similar results are evident with a second series of inhibitors identified at Sterling (D. Sawutz, C. Bode, M. Briggs, and P. Silver, personal observations).

In toto, these results suggest that it is possible to synthesize selective inhibitors of tyrosine kinases. If these agents can be shown to be safe and effective in relevant animal models, inhibitors of growth factors, such as PDGF or erbB2, may eventually be useful as therapeutic agents.

IV. CYCLIC NUCLEOTIDES AND RELAXATION

Cyclic nucleotide activity is regulated by the actions of three different enzymes. These are the cyclases, which catalyze the formation of cyclic nucleotides from triphosphate precursors, the low K_m cyclic nucleotide phosphodiesterases (PDEs), which catalyze the degradation of the cyclic nucleotides to their 5' analogs, and the protein kinases, which are the effectors of these second messengers. In smooth muscle, increases in either cAMP or cGMP produce relaxation. Thus any agent that activates or potentiates either cyclase or protein kinase activity, or inhibits PDE activity, will promote relaxation.

Adenylate cyclase is membrane bound, and catalytic activity is regulated by the GTP-regulatory proteins. Among agents that activate adenylate cyclase, increase cAMP content, and promote relaxation in smooth muscle are β-adrenergic agonists, prostacyclin, prostaglandin E2, and forskolin and related analogs. Guanylate cyclase is both cytosolic and particulate in smooth muscle. Among the activators of guanylate cyclase are the nitrovasodilators and nitric oxide (cytosolic), and the atriopeptins and mimetics (particulate).

There are no well-characterized activators of the cyclic nucleotide protein kinases, except for several series of cAMP or cGMP mimetics. These agents often are poorly permeable in intact cell systems, and thus are of limited usefulness.

In the past ten years, efforts have been directed at characterizing and designing inhibitors of the cyclic nucleotide PDEs. The remainder of this section will focus on three families of isozymes that play important roles in modulating smooth muscle tone.

There are now as many as seven distinct families of PDE isozymes (Beavo, 1988; Beavo and Reifsnyder, 1990; Michaeli et al., 1993). These isozymes are distinct based on protein sequences and catalytic activity for the different cyclic nucleotides. In smooth muscle, cAMP hydrolysis is regulated by either PDE III, PDE IV, or both enzymes. Cyclic GMP hydrolysis is primarily regulated by PDE V, although PDE I (which is calmodulin regulated) may also play a role in relaxation.

PDE III is present in most smooth muscle types (Silver and Harris, 1988; Harris et al., 1989b; Christensen and Torphy, 1994; Nicholson and Shahid, 1994). This enzyme has a low K_m for cAMP (around 0.3 μM), and activity can be modulated (inhibited) by submicromolar concentrations of cGMP. In fact, cGMP-dependent relaxants may exert some of their relaxant effects via inhibition of cAMP PDE III and subsequent elevation of cAMP levels. Numerous inhibitors of this PDE have been developed, primarily for the acute therapy of heart failure. Among the agents marketed for this use are amrinone, milrinone, and enoximone. Among the numerous other agents are medorinone, imazodan, CI-930, trequinsin, and several more potent (K_i = 10–50 nM) inhibitors, such as WIN 62005 and WIN 65282 (Dundore et al., 1995; Pagani et al., 1995). Inhibition of PDE III is accompanied by increases in cAMP content and activation of cAMP protein kinase, in a manner analogous to what occurs with other cAMP-dependent relaxants (Silver and Harris, 1988). Selective inhibition of this same isozyme in cardiac muscle leads to increases in the rate and force of contraction in the heart. Thus afterload reduction (peripheral vaso-

dilation) and positive inotropy occur with the same PDE III inhibitor. These cardiovascular effects limit the use of PDE III inhibitors in other smooth muscle-related diseases, such as asthma.

PDE IV is another family of isozymes present in most smooth muscles. This PDE also hydrolyzes cAMP, but there is no regulation by cGMP. Selective inhibitors of this isozyme include rolipram and RO 20-1724.

PDE IV inhibitors, as well as some combined PDE III/IV inhibitors, are currently under development for the treatment of asthma (Christensen and Torphy, 1994; Nicholson and Shahid, 1994). The basis for this development includes bronchodilatory activity directly related to inhibition of PDE IV, and/or PDE III, in airway smooth muscle, as well as the anti-inflammatory activity of these compounds. PDE IV is the major cAMP PDE isozyme present in inflammatory cell types and inhibition of PDE IV has been linked to elevation of cAMP and inhibition of histamine or leukotriene release from mast cells, inhibition of oxygen free radical release from eosinophils or neutrophils, inhibition of adhesion, migration, or activation of eosinophils, and inhibition of TNF-α release from human monocytes (see Nicholson and Shahid, 1994, or Christensen and Torphy, 1994, for some reviews). There are now known to be at least four subtypes of PDE IV that are encoded by different cDNAs (Bolger et al., 1993; Davis et al., 1989; Livi et al., 1990). Although sequence homology of 75–90% is evident among subtypes, key differences, as well as cellular distribution, are apparent. Currently there are no selective inhibitors of the PDE IV subtypes.

PDE IV is inducible, and agents that increase cAMP levels increase the synthesis of PDE IV in some cell types, most likely via stimulation of transcription (Torphy et al., 1992a). PDE IV is apparently inducible in uterine smooth muscle and appears during pregnancy in human myometrium (Leroy et al., 1987).

One of the selective inhibitors of PDE IV, rolipram (Fig. 2; IC$_{50}$ value of 100–500 nM), also binds in a stereoselective and saturable manner to a high-affinity binding site (K_d = 1–2 nM) initially identified in the brain (Schneider et al., 1986). Rank order of displacement of binding versus PDE IV inhibition for a series of PDE IV inhibitors is not positively correlated (Silver et al., 1990; Harris et al., 1989b). Moreover, bronchorelaxation and in vivo bronchodilation in guinea pigs (Harris et al., 1989b; Silver et al., 1990) and inhibition of gastrointestinal motility in rats (Silver et al., 1990) are correlated with displacement of binding, but not with PDE IV inhibition. Some cell types do not contain this high-affinity binding site, but do contain PDE IV. Alternatively, this high-affinity binding site is coexpressed with human monocytic recombinant PDE IV

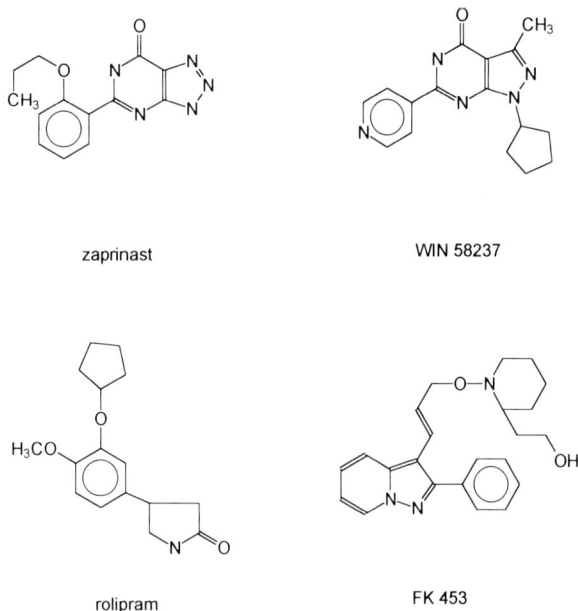

FIGURE 2   Representative modulators of cyclic nucleotides.

(Torphy *et al.*, 1992b), and the binding site is distinct from the catalytic site on this enzyme.

Among the combined PDE III/IV inhibitors are zardaverine. The basis for these compounds includes better bronchodilation (relative to PDE IV inhibitors) and lower cardiovascular effects (relative to PDE III inhibitors).

Two forms of cGMP PDE are present in smooth muscle (Silver and Harris, 1988; Lugnier *et al.*, 1986). Both have a K_{ms} for cGMP in the 0.2–1 μM range. One isozyme, called PDE I, which is present in several other tissues (such as brain and heart), also hydrolyzes cAMP in some cases and is activated by calmodulin. Calmodulin increases V_{max} but does not alter the K_m for cGMP. Vinpocetine, a cerebral vasodilator, is reported as a selective inhibitor of PDE I, but this is questionable given the other known mechanisms of this compound (Souness *et al.*, 1989).

A second isozyme, PDE V, is also found in most smooth muscles. This enzyme is related to the cGMP PDE present in rods and cones and is not activated by calmodulin (Beavo, 1988). Zaprinast (Fig. 2; IC_{50} = 0.5–1 μM) is an inhibitor of this PDE, but selectivity versus PDE I (IC_{50} = 5–10 μM) is not great.

Zaprinast, as well as other cGMP PDE inhibitors, potentiates the vasorelaxant and depressor actions of guanylate cyclase activators, including nitric oxide, nitrovasodilators, and atrial natriuretic factor (ANF) (Martin *et al.*, 1986; Harris *et al.*, 1989a). In addition, zaprinast can lower blood pressure and promote natriuresis following intravenous administration to

rats or dogs (McMahon *et al.*, 1989; Dundore *et al.*, 1993; Wilkins *et al.*, 1990; Pagani *et al.*, 1992). A structurally distinct series of PDE V inhibitors, the pyrazolpyrimidines, has also been described (Silver *et al.*, 1994; Bacon *et al.*, 1994). These compounds, typified by WIN 58237 (Fig. 2; K_i = 170 nM) and several analogs with low nanomolar potency, are competitive inhibitors with respect to the substrate, cGMP. Interestingly, some analogs are also inhibitors of PDE IV.

WIN 58237 is more potent than zaprinast in intact cell and *in vivo* models (Silver *et al.*, 1994), suggesting greater ability to penetrate cellular membranes. *In vivo*, WIN 58237 lowers blood pressure in spontaneously hypertensive rats following intravenous, or oral, dosing. Increases in cGMP content, *in vivo*, in abdominal aortic segments and in plasma are observed with WIN 58237. Some of the hypotensive response, and the increase in vascular cGMP, can be blocked with the nitric oxide inhibitor N-nitrol-arginine, implicating nitric oxide in this depressor effect.

Other PDE V inhibitors are described in the literature, including FK 453 (Fig. 2) (Satake *et al.*, 1992) and 4-[2-*n*-butyl-5-chloro-1-(2-chlorobenzyl)] imidazoylmethyl acetate (Booth *et al.*, 1990). Both are less potent (IC_{50} = 5–10 μM), but do possess vasorelaxant and depressor activities.

Therapeutically, PDE V inhibitors are being developed as potentiators of nitric oxide, or in conjunction with nitrovasodilators. In several *in vitro* and *in vivo* models, it has been demonstrated that zaprinast, WIN 58237, and other PDE V inhibitors can reinstate vasorelaxant responsiveness after tolerance to nitrovasodilators has occurred by preventing cGMP breakdown (Silver *et al.*, 1991, 1994; Pagani *et al.*, 1993). Clinically, this may represent a way to maintain patients on continuous nitroglycerin therapy. In pulmonary medicine, Zapol and colleagues (Rossaint *et al.*, 1993) have shown that inhaled nitric oxide can increase ventilation (bronchodilation) and perfusion (vasodialation of pulmonary vessels) in several preclinical models of airway constriction. Inhaled nitric oxide is also used in several clinical diseases involving pulmonary constriction. Zaprinast can potentiate the effects of nitric oxide in preclinical models. Another area in which nitric oxide plays a key role is in penile erection (Rajfer *et al.*, 1992). Impotence may be related to decreased levels of nitric oxide. Nitric oxide donors, and PDE V inhibitors, can reverse impotence by cGMP-related vasodilation.

References

Ahenfeldt-Ronne, I. (1988). *Drugs* **36**, 4–9.
Bacon, E. R., Singh, B., and Lesher, G. Y. (1994). U.S. Pat. 5,294,612.

Beavo, J. A. (1988). *Adv. Second Messenger Phosphoprotein Res.* **22,** 1–38.

Beavo, J. A. and Reifsnyder, D. K. (1990). *Trends Pharmacol. Sci.* **11,** 150–155.

Bilder, G. .E, Krawiec, J. A., McVety, K., Gazit, A., Gilon, C., Lyall, R., Zilberstein, A., Levitski, A., Perrone, M. H., and Schreiber, A. B. (1991). *Am. J. Physiol.* **260,** C721–C730.

Bolger, G., Michaeli, T., Martins, T., St. John, T., Steiner, B., Rodgers, L., Riggs, M., Wigler, M., and Ferguson, K. (1993). *Mol. Cell. Biol.* **13,** 6558–6571.

Booth, R. F. G., Lunt, D. O., Lad, N., Buckham, S. P., Oswald, S., Clough, D. P., Floyd, C. D., and Dickens, J. (1990). *Biochem. Pharmacol.* **40,** 2315–2321.

Bradshaw, D., Hill, C. H., Nixon, J. S., and Wilkinson, S. E. (1993). *Agents Actions* **38,** 135–147.

Bruns, R. F. (1991). *Biochem. Biophys. Res. Commun.* **176,** 288–293.

Bucholz, R. A., Dundore, R. L., and Silver, P. J. (1991). *Adv. Exp. Med. Biol.* **308,** 199–204.

Cabot, M. C., and Jaken, S. (1984). *Biochem. Biophys. Res. Commun.* **125,** 163–169.

Catterall, W. A., and Striessnig, J. (1992). *Trends Phamracol. Sci.* **13,** 256–262.

Chapman, I. D., Kristersson, A., Schaeublin, G., Mazzoni, L., Boubekeur, K., Murphy, N., and Morley, J. (1992). *Br. J. Pharmacol.* **106,** 423–429.

Christensen, S. B., and Torphy, T. J. (1994). *Annu. Rep. Med. Chem.* **29,** 185–194.

Davis, R. L., Takaysasu, H., Eberwine, M., and Myers, J. (1989). *Proc. Natl. Acad. Sci. U.S.A.* **86,** 3604–3608.

Dolle, R. E., Dunn, J. A., Bobko, M., Singh, B., Kuster, J. E., Baizman, E., Harris, A. L., Sawutz, D. G., Miller, D., Wang, S., Flatynek, C. R., Xie, W., Sarup, J., Boder, D. C., Pagani, E. D., and Silver, P. J. (1994). *J. Med. Chem.* **37,** 2627–2729.

Dundore, R. E., Brousseau, A. C., Habeeb, P. G., Pratt, P F., Becker, L. T., Clas, D. M., Silver, P. J., and Buchholz, R. A. (1992). *J. Cardiovasc. Pharmacol.* **20,** 525–532.

Dundore, R. E., Clas, D. M., Wheeler, L. T., Habeeb, P. G., Bode, D. C., Buchholz, R. A., Silver, P. J., and Pagani, E. D. (1993). *Eur. J. Pharmacol.* **249,** 293–297.

Dundore, R. L., Pagani, E. D., Bode, D. C., Bacon, E. R., Singh, B., Lesher, G. Y., Buchholz, R. A., and Silver, P. J. (1995). *J. Cardiovasc. Pharmacol.* (in press).

Edwards, G., Henshaw, M., Miller, M., and Weston, A. H. (1991). *Br. J. Pharmacol.* **102,** 679–686.

Evans, J. M., and Taylor, S. G. (1994). *Prog. Med. Chem.* **31,** 411–446.

Fisher, M., and Grotta, J. (1993). *Drugs* **46,** 961–975.

Flamm, E. S., Adams, H. P., Beck, D. W., Pinto, R. S., Marler, J. R., Walker, M. D., Godersky, J. C., Loftus, C. M., Biller, J., Boarini, D. J., O'Dell, C., Banwart, K., and Kongable, G. (1988). *J. Neurosurg.* **68,** 393–400.

Foster, C. J., and Gaeta, FCA (1988). *Biophys. J.* **53,** 182a.

Fovaeus, M., Andersson, K.-E., and Hedlund, H. (1989). *J. Urol.* **141,** 637–640.

Fry, D. W. (1994). *Exp. Opin. Invest. Drugs* **3,** 577–595.

Gelband, C. H., and McCullough, J. R. (1993). *Am. J. Physiol.* **263,** C1119–C1127.

Gopalakrishnan, M., Janis, R. A., and Triggle, D. J. (1993). *Drug Dev. Res.* **28,** 95–127.

Hachisu, M., Hiranuma, T., Sagawa, S., Koyama, M., and Nishio, M. (1987). *Jpn. J. Pharmacol.* **43**(Suppl.), 295P.

Haley, E. C., Kassel, N. F., Torner, J. C., and Kongable, G. (1991). *Neurology* **41**(Suppl. 1), 346.

Hannun, Y. A., Loomis, C. R., Merrill, A. H., and Bell, R. M. (1986). *J. Biol. Chem.* **261,** 12604–12609.

Harris, A. L., Lemp, B. M., Bentley, R. G., Perrone, M. H., Hamel,

L. T., and Silver, P. J. (1989a). *J. Pharmacol. Exp. Ther.* **249,** 394–400.

Harris, A. L., Connell, M. J., Ferguson, E. W., Wallace, A. M., Gordon, R. J., Pagani, E. D., and Silver, P. J. (1989b). *J. Pharmacol. Exp. Ther.* **251,** 199–206.

Herbert, J. M. (1990). *Biochem. Biophys. Res. Commun.* **172,** 993–999.

Hidaka, H., and Tanaka, T. (1983). *In* "Methods in Enzymology" (A. R. Means and B. W. O'Malley, eds.), Vol. 102, pp. 185–194. Academic Press, New York.

Hidaka, H., Inagaki, M., Kawamoto, S., and Sasaki, Y. (1984). *Biochemistry* **23,** 5036–5041.

Hullin, R., Biel, M., Flockerzi, V., and Hofmann, F. (1993). *Trends Cardiovasc. Med.* **3,** 48–53.

Kase, H., Iwahashi, K., and Matsuda, Y. (1986). *J. Antiobiot.* **39,** 1059–1065.

Kinoshita, M., and Sakai, K. (1990). *Cardiovasc. Drugs Ther.* **4,** 1075–1088.

Kobayashi, E., Nakano, H., Morimoto, M., and Tamaoki, T. (1989). *Biochem. Biophys. Res. Commun.* **159,** 548–553.

Kraft, A. S., Reeves, J. A., and Ashendel, C. L. (1988). *J. Biol. Chem.* **263,** 8437–8442.

Krautwurst, D., Hescheler, J., Arndts, D., Losel, W., Hammer, R., and Schultz, G. (1993). *Mol. Pharmacol.* **43,** 655–659.

Leroy, M. J., Blot, P., and Ferre, F. (1987). *Gynecol. Obstet. Invest.* **24,** 190–199.

Levitzki, A. (1990). *Biochem. Pharmacol.* **40,** 913–918.

Levitzki, A., and Gilon, C. (1991). *Trends Pharmacol. Sci.* **12,** 171–174.

Libby, P., Warner, S. J. C., Salomon, R. N., and Birinyi, L. K. (1988). *N. Engl. J. Med.* **318,** 1493–1498.

Livi, G. P., Kmetz, P., McHale, M. M., Cieslinski, L. B., Sathe, G. M., Taylor, D. P., Davis, R., Torphy, T. J., and Balearek, J. M. (1990). *Mol. Cell. Biol.* **10,** 2678–2686.

Longman, S. D., and Hamilton, T. C. (1992). *Med. Res. Rev.* **12,** 73–148.

Loutzenhiser, R., and Epstein, M. (1985). *Am. J. Physiol.* **249,** F619–F629.

Lugnier, C., Schoeffter, P., LeBec, A., Strouthou, E., and Stoclet, J. C. (1986). *Biochem. Pharmacol.* **35,** 1743–1751.

Maguire, M. P., Sheets, K. R., McVety, K., Spada, A. P., and Zilberstein, A. (1994). *J. Med. Chem.* **37,** 2129–2137.

Mannold, R. (1984). *Drugs Future* **9,** 677–690.

Martin, W., Furchgott, R. F., Villani, G. M., and Jothianandan, D. (1986). *J. Pharmacol. Exp. Ther.* **237,** 539–547.

McMahon, E. G., Palomo, M. A., Mehta, P., and Olins, G. M. (1989). *J. Pharmacol. Exp. Ther.* **251,** 1000–1005.

McManus, O. B., Harris, G. H., Giangiacomo, K. M., Feigenbaum, P., Reuben, J. P., Addy, M. E., Burka, J. F., Kaczorowski, G. J., and Garcia, M. L. (1993). *Biochemistry* **32,** 6128–6133.

Meyer, T., Regenass, V., Fabbro, D., Alteri, E., Rosel, J., Muller, M., Caravatti, G., and Matter, A. (1989). *Int. J. Cancer* **43,** 851–856.

Michaeli, T., Bloom, T. J., Martins, T., Loughney, K., Ferguson, K., Rigges, M., Rodgers, L., Beavo, J. A., and Wigler, M. (1993). *J. Biol. Chem.* **268,** 12925–12932.

Moreland, S., and Hunt, J. T. (1987). *FASEB J.* **46,** 1098.

Mori, T., Takai, Y., Minakuchi, R., Yu, B., and Nishizuka, Y. (1980). *J. Biol. Chem.* **255,** 8378–8380.

Morley, J. (1994). *Trends Pharmacol. Sci.* **15,** 463–468.

Murray, K. J., and Coates, W. J. (1994). *Annu. Rev. Med. Chem.* **29,** 255–264.

Nakajima, T., and Katoh, A. (1987). *Mol. Pharmacol.* **32,** 140–146.

Nakanishi, S., Yanada, K., Iwahashi, K., Kuroda, K., and Kase, H. (1990). *Mol. Pharmacol.* **37,** 482–488.

Neumayer, H. H., Junge, W., Kufner, A., and Wenning, A. (1989). *Nephrol., Dial., Transplant,* **4,** 1030–1036.

Nicholson, C. D., and Shahid, M. (1994). *Pulmonary Pharmacol.* **7**, 1–17.

Nishizuka, Y. (1988). *Nature (London)* **34**, 661–665.

Okazaki, K., Ishikawa, T., Inui, M., Tada, M., Goshima, K., Okamoto, T., and Hidaka, H. (1994). *J. Pharmacol. Exp. Ther.* **270**, 1319–1324.

Osada, H., Takahashi, H., Tsunoda, K., Kusakabe, H., and Isono, K. (1990). *J. Antibiot.* **43**, 163–167.

Pagani, E. D., Buchholz, R. A., and Silver, P. J. (1992). *In* "Cellular and Molecular Alterations in the Failing Human Heart" (G. Hasenfuss, C. Holubarsh, H. Just, and N. Alpert, eds.), pp. 73–86. Steinkopff Verlag, Darmstadt.

Pagani, E. D., Van Aller, G. S., O'Connor, B., and Silver, P. J. (1993). *Eur. J. Pharmacol.* **243**, 141–147.

Pagani, E. D., Bode, D. C., Dundore, R. L., Bacon, E. R., Buchholz, R. A., and Silver, P. J. (1995). *J. Cardiovasc. Pharmacol.* (in press).

Prozialeck, W. C. (1983). *Annu. Rep. Med. Chem.* **18**, 203–212.

Rafjer, J., Aronson, W. J., Bush, P. A., Dorey, F. J., and Ignarro, L. J. (1992). *N. Engl. J. Med.* **326**, 90–94.

Raynor, R. L., Zheng, B., and Kuo, J. F. (1991). *J. Biol. Chem.* **266**, 2753–2758.

Ross, R. (1991). *Trends Cardiovasc. Med.* **1**, 277–282.

Rossaint, R., Falke, K. J., Lopez, F., Slama, K., Pison, U., and Zapol, W. M. (1993). *N. Engl. J. Med.* **328**, 399–405.

Rüegg, U. T., Wallnofer, A., Weir, S., and Cauvin, C. (1989). *J. Cardiovasc. Pharmacol.* **14**(Suppl. 6), S49–S58.

Russo, D., Testa, A., Della Volpe, L., and Sansone, G. (1990). *Nephron* **55**, 254–257.

Saitoh, M., Ishikawa, T., Matsushima, S., Naka, M., and Hidaka, H. (1987). *J. Biol. Chem.* **262**, 7796–7803.

Satake, N., Zhou, Q., Sato, N., Matsuo, M., Sawada, T., and Shibata, S. (1992). *Pharmacology* **44**, 206–214.

Schneider, H. H., Schmiecher, R., Brezinski, M., and Seidler, J. (1986). *Eur. J. Pharmacol.* **127**, 105–115.

Silver, P. J., and Harris, A. L. (1988). *In* "Resistance Arteries" (W. Halpern, ed.), pp. 284–291. Perinatology Press, New York.

Silver, P. J., Dachiw, J., Ambrose, J. M., and Pinto, P. B. (1985). *J. Pharmacol. Exp. Ther.* **234**, 629–635.

Silver, P. J., Fenichel, R., and Wendt, R. L. (1988a). *J. Cardiovasc. Pharmacol.* **11**, 299–307.

Silver, P. J., Lepore, R. E., O'Connor, B., Lemp, B. M., Hamel, L. T., Bentley, R. G., and Harris, A. L. (1988b). *J. Pharmacol. Exp. Ther.* **247**, 34–42.

Silver, P. J., Connell, M. J., Dillon, K. M., Cumiskey, W. R., Volberg, W. A., and Ezrin, A. M. (1989). *Cardiovasc. Drug Ther.* **3**, 675–682.

Silver, P. J., Harris, A. L., Buchholz, R. A., Miller, M., Gordon, R., Dundore, R., and Pagani, E. D. (1990). *In* "Purines in Cellular Signaling. Targets for New Drugs" (K. A. Jacobson, ed.), pp. 358–364. Springer-Verlag, New York.

Silver, P. J., Pagani, E. D., Degaravilla, L., Van Aller, G. S., Volberg, M. L., Pratt, P. F., and Buchholz, R. A. (1991). *Eur. J. Pharmacol.* **199**, 141–142.

Silver, P. J., Dundore, R. L., Bode, D. C., DeGaravilla, L., Buchholz, R. A., Van Aller, G. S., Hamel, L. T., Bacon, E. R., Singh, B., Lesher, G. Y., Hlasta, D., and Pagani, E. D. (1994). *J. Pharmacol. Exp. Ther.* **271**, 1143–1149.

Somlyo, A. P., and Somlyo, A. V. (1994). *Nature (London)* **312**, 231–236.

Souness, J. F., Brazdil, R., Diocce, B., and Jordan, R. (1989). *Br. J. Pharmacol.* **98**, 725–734.

Sullivan, J. P., Connor, J. R., Shearer, B. G., and Burch, R. M. (1992). *Mol. Pharmacol.* **41**, 38–44.

Su, H. D., Mazzei, G. J., Vogler, W. R., and Kuo, J. F. (1985). *Biochem. Pharmacol.* **34**, 3649–3653.

Takahashi, I., Kobayashi, E., Nakano, H., Murakata, C., Saitoh, H., Suzuki, K., and Tamaoki, T. (1990). *J. Pharmacol. Exp. Ther.* **255**, 1218–1221.

Tanaka, T., Umekuwa, H., and Saitoh, M. (1986). *Mol. Pharmacol.* **29**, 264–269.

Torphy, T. J., Zhou, H. L., and Cieslinki, L. B. (1992a). *J. Pharmacol. Exp. Ther.* **263**, 1195–1205.

Torphy, T. J., Stadel, J. M., Burman, M., Cieslinski, L. B., McLaughlin, M. M., White, J. R., and Livi, G. P. (1992b). *J. Biol. Chem.* **267**, 1798–1804.

Van Belle, H. (1981). *Cell Calcium* **2**, 483–494.

Wadsworth, A. N., and McTavish, D. (1992). *Drugs Aging* **2**, 262–286.

Wickenden, A. D., Grimwood, S., Grant, T. L., and Todd, M. H. (1991). *Br. J. Pharmacol.* **103**, 1148–1152.

Wilkins, M. R., Settle, S. L., and Needleman, P. (1990). *J. Clin. Invest.* **85**, 1274–1279.

Williams, A. J., Lee, T. H., and Cochrane, G. M. (1990). *Lancet* **2**, 334–336.

Wise, B. C., Glass, D. B., Choo, C. H., Raynor, R. L., Katoh, R. C., Schatzman, R. S., Turner, R. S., Kibler, R. F., and Kou, J. F. (1982). *J. Biol. Chem.* **257**, 8489–8495.

ENERGETICS

29

Energetics of Smooth Muscle Contraction

PER HELLSTRAND

Department of Physiological Sciences, University of Lund, Sweden

I. INTRODUCTION

Smooth muscle has developed to satisfy functional requirements set by the various internal organs of which it is an integral part. Organ-specific subdivisions of smooth muscle, such as "vascular" or "intestinal," cover a wide range of quite varying properties. Even though the basic principle of operation seems to be similar in all kinds of muscle, the characteristics of their energy conversion are predictably quite variable. This chapter aims at a general description of the energetic properties of smooth muscle, including metabolic supply, effects of metabolic inhibition, and energy turnover by activation mechanisms and the cross-bridge cycle. Aspects of smooth muscle energetics are also covered by Dillon (Chapter 30) and by Strauss and Murphy (Chapter 26) in this volume. Comprehensive reviews of the field have been published some years ago (Butler and Davies, 1980; Paul, 1980, 1987, 1989; Hellstrand and Paul, 1982; Butler and Siegman, 1985; Hartshorne, 1987). Therefore the emphasis here is on recent developments, although these are put into perspective by brief overviews of earlier work.

II. CELLULAR ENERGY STORES AND METABOLIC TURNOVER RATE

A. Oxidative and Glycolytic Metabolism

Although skeletal muscle fibers, particularly of the "fast glycolytic" type, are able to utilize their phosphocreatine (PCr) and glycogen stores for sizable bursts of energy turnover at rates well exceeding that of oxidative recovery processes, these stores are small in smooth muscle, and considering the slow time course of smooth muscle contraction, increased rate of energy utilization is tightly associated with an increased rate of metabolic recovery. Smooth muscle has a substantial rate of lactate production (J_{lac}) even under fully oxygenated conditions, and there is now substantial evidence that glycolysis preferentially supports membrane ion pumping, whereas oxygen consumption (J_{O_2}) correlates with contraction (Paul, 1989). However, this compartmentation seems to be relative, since oxidative metabolism is able to support membrane transport under glucose-free conditions or in the presence of metabolic substrates that cannot fuel glycolysis (Hellstrand *et al.*, 1984; Takai and Tomita, 1986; Lövgren and Hellstrand, 1987).

Lactate produced by smooth muscle of swine carotid artery has been shown to be quantitatively derived from extracellular glucose, whereas glycogenolysis provides carbohydrates for oxidative metabolism (Lynch and Paul, 1983, 1987). Interestingly, the lactate dehydrogenase (LDH) isoenzyme pattern varies between slow (aorta) and fast (portal vein and bladder) smooth muscle in rat, such that aorta has a higher proportion of the H subunit, which has a higher affinity for lactate and is product-inhibited at lower concentrations of lactate (Malmqvist *et al.*, 1991). Thus the slow aortic muscle is more adapted for oxidative metabolism than the faster bladder and portal vein muscles.

Most studies of muscle energetics *in vitro* have utilized glucose as substrate. Under *in vivo* conditions the utilization of metabolic substrate is presumably more varied. Chace and Odessey (1981) showed that metabolism in rabbit aorta can be sustained by fatty acids, ketone bodies, and amino acids, as well as carbohydrate energy substrates, and that the presence of other substrates in addition to glucose reduces the oxidation of glucose, while not influencing the rate of lactate production. In the absence of exogenous substrate, fatty acids are the preferred endogenous energy source (Odessey and Chace, 1982). Barron and co-workers have investigated the role of fatty acids in the control of glycolysis and glycogen metabolism in porcine carotid arteries. Fatty acids added to the normal glucose-containing medium under aerobic conditions were found to decrease glycolysis in resting arteries, but did not inhibit the increase in glycolysis caused by KCl-induced contraction. However, they prevented the decrease in glycogen content seen during contraction in the presence of glucose alone (Barron *et al.*, 1991). The basis of the effect of fatty acids on glycogenolysis was investigated by determination of phosphorylase *b* kinase and phosphorylase *a* phosphatase activities during KCl-induced contraction (Barron and Kopp, 1991). Whereas phosphorylase kinase increased in a biphasic manner during contraction, reaching a peak at 45 s and then declining, phosphatase activity increased at a similar rate but remained elevated for the duration of the contraction (30 min). Incubation with 0.5 mM palmitate increased resting kinase and phosphatase activities, but neither enzyme then increased its activity during contraction. These studies provide further evidence that the utilization of carbohydrate is partitioned between separate glycolytic pathways fueled by glucose and glycogen.

B. Phosphagens in Smooth Muscle

The immediate energy supply for contraction and active transport processes, as well as for phosphorylation reactions participating in cellular regulation, is provided by breakdown of nucleoside phosphates, mainly ATP. ATP is almost fully complexed as MgATP and its concentration is tightly regulated, so that under normal conditions of energy flow through intermediary metabolism it is kept constant during rest and contraction. This is evidenced both by chemical analysis of tissue extracts and by *in vivo* measurements using ^{31}P-NMR (Hellstrand and Paul, 1983; Krisanda and Paul, 1983; Vogel *et al.*, 1983; Hellstrand and Vogel, 1985; Kushmerick *et al.*, 1986; Nakayama *et al.*, 1988).

The tissue concentration of ATP in smooth muscle is in the range of 0.5–2 μmol/g wet weight (Butler and Davies, 1980), whereas the ATP turnover (J_{ATP}) in stimulated muscles ranges between 1 and 8 μmol/g (Paul, 1987). Thus, backup by recovery processes is essential. However, the immediate pool of high-energy phosphate bonds, PCr, is only slightly larger than the ATP pool itself, and not able to sustain ATP

utilization for long. In the spontaneously contracting rat portal vein, each phasic contraction, lasting 6–8 s, is associated with a net breakdown of about 25% of the PCr concentration, while ATP is unchanged (Hellstrand and Paul, 1983). In this respect, smooth muscles with phasic contractile characteristics resemble the pattern of skeletal muscle, where contraction is associated with net PCr breakdown. In contrast, the slowly contracting swine carotid artery was found to increase its J_{O_2} during contraction with no net decrease in PCr and ATP concentrations (Krisanda and Paul, 1983). Thus smooth muscle encompasses a range of tissues with quite variable patterns of energetic support, although as a group they are in most aspects "slower" than striated muscle (Hellstrand and Paul, 1982).

C. Role of Phosphocreatine

The role of PCr in muscle contraction is traditionally thought to be that of an energy store that will buffer MgATP by rephosphorylating ADP through the creatine kinase reaction. However, a role for PCr in the transport of high-energy phosphate bonds from mitochondria to sites of utilization in the cytoplasm and membranes, as well as for free creatine in the reverse transport of phosphate acceptor, has been advanced (for a review, see Wallimann *et al.*, 1992). By thus increasing the effective concentration of high-energy phosphate donors and acceptors, the PCr system will facilitate the diffusion of energy throughout the cell.

Mitochondrial creatine kinase (CK) is localized in the contact area between the mitochondrial inner and outer membranes, whereas cytosolic CK is also in contact with the glycolytic system. There is evidence that ATP-consuming systems such as contractile proteins and membrane pumps have structurally associated CK that will create a microenvironment where levels of ATP and ADP are regulated in the vicinity of the ATPase sites. It has been demonstrated that functional CK activity, able to produce ATP for contraction from PCr and ADP, remains bound in guinea pig taenia coli and myometrium after permeabilization by Triton X-100, a method that destroys mitochondria (Hellstrand and Arner, 1986; Clark *et al.*, 1992, 1993). In intact guinea pig smooth muscle, however, the specific mitochondrial isoform of CK has been demonstrated (Ishida *et al.*, 1991). This suggests that the PCr/CK system plays a role in the transport of high-energy phosphates from the mitochondrial compartment to the sites of energy utilization, correlating with the close association of PCr levels with the state of oxidative metabolism in mammalian smooth muscle (Lövgren and Hellstrand, 1985; Ishida and Paul, 1990). However, there may be a less tight functional coupling of mitochondrial CK and oxidative phosphorylation in smooth than in striated muscle (Clark *et al.*, 1994).

The significance of PCr as an intermediate energy store even in phasic smooth muscle with a comparatively high rate of energy turnover is open to question. Ekmehag and Hellstrand (1988) studied rat portal veins where PCr contents had been lowered to 14% of control by feeding the animals β-guanidinopropionic acid (BGPA), which inhibits creatine uptake. Essentially no effects were found on spontaneous activity, isometric force, ATP contents, J_{O2}, J_{lac}, or maximal shortening velocity (V_{max}). Neither was there any difference in the effect of respiratory inhibition by cyanide on these parameters. Scott and Coburn (1989) examined rabbit aortic rings that had either been depleted of PCr or loaded by exposure to excess creatine. They did not find any differences in the effects of metabolic inhibition on contractile responses in these different preparations, whereas changes in muscle force did correlate with the ratio of free creatine/PCr. Thus the PCr/CK system in the smooth muscle with its relatively slow metabolism may be principally a mechanism for facilitating the flow of high-energy phosphates and for controlling the free ADP/ATP ratio at sites of energy utilization.

III. RESPONSE TO HYPOXIA AND METABOLIC INHIBITION

The adaptation of tissue blood flow to local metabolic requirements is a dramatic example of the control of smooth muscle contraction by metabolic factors. The exact mechanisms involved are still not known with certainty, and in particular it is not clear how much is due to a direct response of the vascular smooth muscle to its chemical environment and how much is due to "vasodilator metabolites" released from the tissue parenchyma (Sparks, 1980). The possibility that structures other than smooth muscle cells, notably the endothelium, act as oxygen sensors is a further area for investigation. The emphasis here will be on studies relating the metabolic supply of smooth muscle and its contractile activity.

A. Relationship between Oxygen Supply and Contractile Force

Lowered oxygen tension (P_{O_2}) or inhibition of cell respiration has been found to decrease contractile force in a variety of smooth muscles during spontaneous myogenic activity, or stimulation by neuroeffectors or high-K$^+$ solution (Hellstrand *et al.*, 1977; Ishida and Paul, 1990; Coburn *et al.*, 1992). The sensitivity of

contractile force to metabolic inhibition depends on the way contraction is initiated. Spontaneous or agonist-induced contractions are more sensitive than are high-K$^+$ contractures, which depend on a sustained membrane depolarization (Hellstrand *et al.*, 1977; Coburn *et al.*, 1992). Thus the signaling pathway leading to activation of contraction is a target for metabolic inhibition, as well as the actual ATP turnover associated with cross-bridge activity. However, the immediate step regulating contraction, myosin LC20 phosphorylation, shows a variable dependence on metabolic supply. In rat portal vein, LC20 phosphorylation during high-K$^+$ contraction was decreased by 0.2 m*M* cyanide (Ekmehag and Hellstrand, 1989), whereas hypoxia did not decrease phosphorylation during high-K$^+$ contracture in rabbit aorta (Coburn *et al.*, 1992) or bovine tracheal muscle (Hai *et al.*, 1993).

B. Membrane and Receptor Mechanisms in the Metabolic Response

Much evidence suggests that membrane activation is influenced by the energetic status of the smooth muscle cell. Thus one or, more probably, several links in the membrane-associated activation sequence may be metabolically sensitive. Evidence has been reported for decreased agonist-stimulated formation of inositol 1,4,5-trisphosphate (IP$_3$) and diacylglycerol and for decreased levels of guanine nucleotides in contracting rabbit aorta during hypoxia (Coburn *et al.*, 1988, 1993). However, responses to receptor activation could still be elicited during metabolic depletion, indicating that G-protein-coupled activation mechanisms were operative.

Inhibited electrical and mechanical activity during respiratory inhibition has been shown to be associated with an increase in the basal and, to some extent, high-K$^+$-stimulated intracellular calcium concentration ($[Ca^{2+}]_i$) in rat portal vein and guinea pig gastric antrum (Swärd *et al.*, 1993; Huang *et al.*, 1993). Possibly this may in itself inhibit inflow of Ca^{2+} during action potentials, by Ca^{2+}-dependent inhibition of Ca^{2+} channels or activation of Ca^{2+}-dependent K$^+$ channels. These findings suggest that force development is partially uncoupled from $[Ca^{2+}]_i$ during respiratory inhibition.

New interest has focused on the possibility that ATP-regulated K$^+$ (K$_{ATP}$) channels might be involved in the membrane response to metabolic inhibition (Standen *et al.*, 1989; Daut *et al.*, 1990). These channels are blocked by ATP and activated by nucleoside diphosphates, and thus would be expected to increase their conductance as [ADP]/[ATP] increases, leading to hyperpolarization. This suggests an attractive hy-

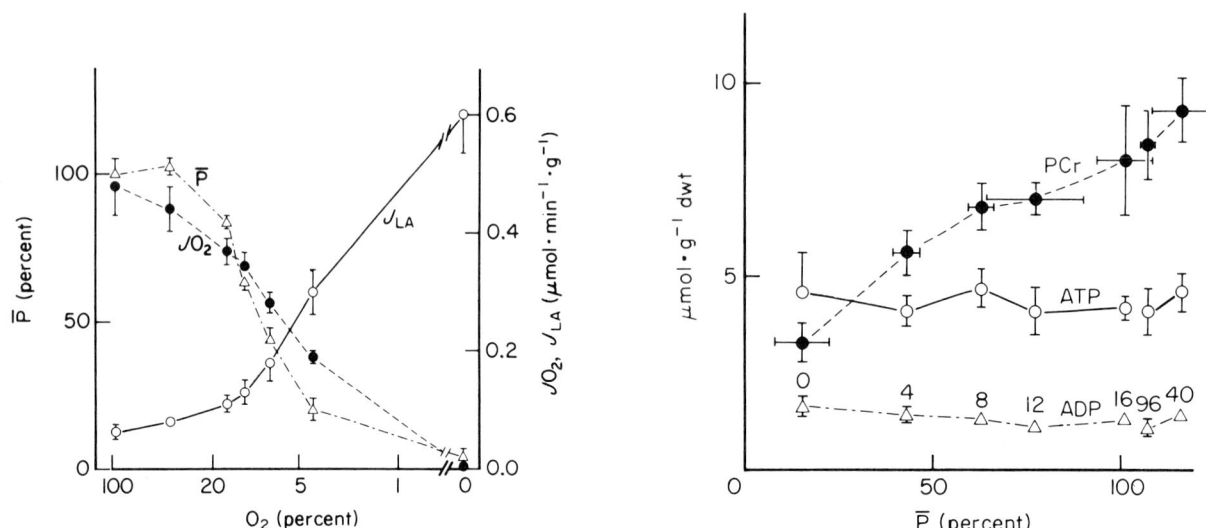

FIGURE 1   Left: spontaneous activity ($\bar{P}$,%), oxygen consumption (J_{O2}), and lactate production (J_{LA}) during exposure of rat portal veins ($n = 6$) to different O_2 concentrations (%) in medium. Right: tissue contents of PCr (●), ATP (○), and extractable ADP (△) during incubation of portal veins ($n = 6$) plotted against spontaneous activity ($\bar{P}$,%). Inset numbers show O_2 concentrations (%) in medium. From Lövgren and Hellstrand (1985).

pothesis to account for the coupling of metabolism and excitability. Work on vascular K_{ATP} channels has been reviewed (Nichols and Lederer, 1991; Quayle and Standen, 1994; Knot et al., Chapter 16, this volume). Most studies of the functional role of K_{ATP} channels rely on sulfonylurea compounds, such as glibenclamide, as channel blockers. However, the pharmacology is complex and cautious interpretation is warranted. K_{ATP} channels can also be affected by several vasoactive substances in addition to nucleoside phosphates.

From an energetic point of view, a crucial aspect is whether ATP, ADP, and other nucleoside phosphates vary in the vicinity of the ion channels in such a way that the channel activity is influenced. Elucidation of this question requires knowledge of metabolic and electrophysiological alterations under given conditions in the same tissue. Such information is to some extent available for the rat portal vein. Noack et al. (1992) described a slowly developing noninactivating and glibenclamide-sensitive K^+ current in portal vein cells when intracellular ATP was reduced by removal of glucose and substrates for oxidative phosphorylation. This current was concluded to be carried by the same small-conductance channels (10–20 pS) as those activated by the K_{ATP} channel-activating compound, levcromakalim. However, both metabolic depletion and levcromakalim also inhibited the delayed rectifier K^+ current, an effect possibly related to decreased channel phosphorylation.

Lövgren and Hellstrand (1985) determined force, [ATP], and [PCr] during progressive hypoxia or respiratory inhibition in the rat portal vein (Fig. 1). When spontaneous activity was almost abolished using either intervention, ATP contents were unchanged, whereas PCr had declined to about 40% of the control value. Based on a [PCr]/[total creatine] ratio of 0.75 in relaxed normoxic portal vein (Hellstrand and Paul, 1983) and assuming equilibrium of the CK reaction, this would suggest a seven-fold increase in the free-ADP concentration in hypoxia. At this level of metabolic inhibition the cell membrane depolarizes slightly, which would not be expected from opening of K^+ channels. However, as shown by Lydrup et al. (1994), glibenclamide (10^{-7} M) is able to partially restore spontaneous activity while reducing the amount of depolarization by a few mV (Fig. 2). Depletion of high-energy phosphate stores by combined blockade of oxidative and glycolytic metabolism caused a clearcut hyperpolarization, which was reversed to a depolarization by glibenclamide (Fig. 3). The concentration of glibenclamide used in these experiments was low (10^{-7} M) in an attempt to avoid effects on activity in normal medium, whereas higher concentrations of glibenclamide may increase spontaneous activity (Longmore et al., 1990). The mechanisms behind the membrane effects of inhibited oxidative metabolism are obviously complex, and it seems clear that activation of K_{ATP} channels, although contributing, does not provide the whole answer in this tissue at least.

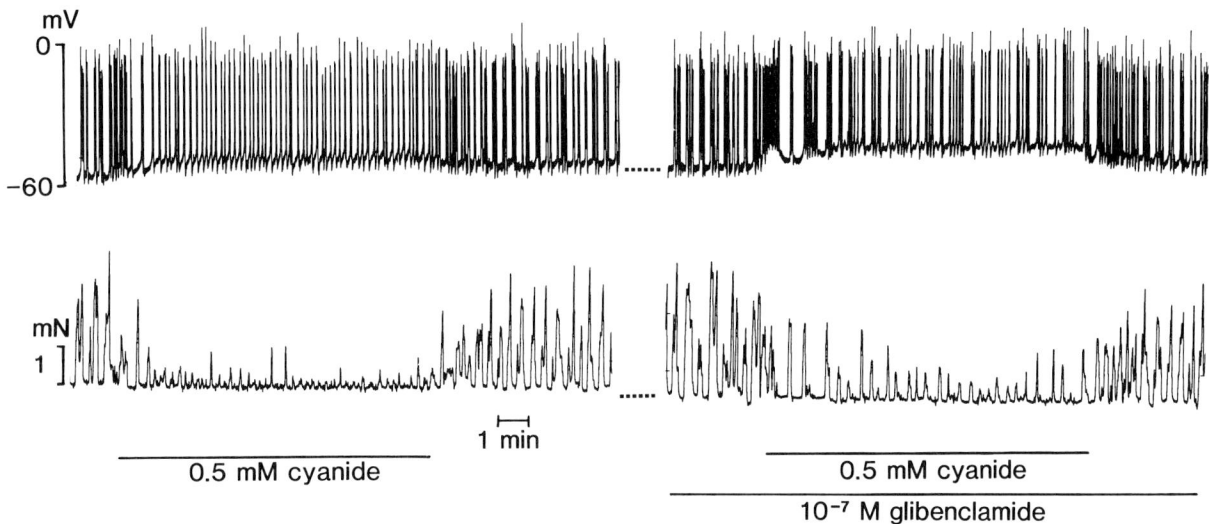

FIGURE 2 Electrical (upper) and mechanical (lower) recordings from a rat portal vein during exposure to cyanide ± glibenclamide. Segments are from one continuous microelectrode recording. From Lydrup et al. (1994).

IV. ENERGETICS OF THE CONTRACTILE SYSTEM

A. Energy Turnover in Resting Muscle

The energy turnover associated with contraction can be divided into several components, more or less interrelated. The most obvious division is in "resting" and "activated" energy turnover, where typically the activated J_{ATP} is two to three times the resting value in living smooth muscle. Subtracting the resting from the

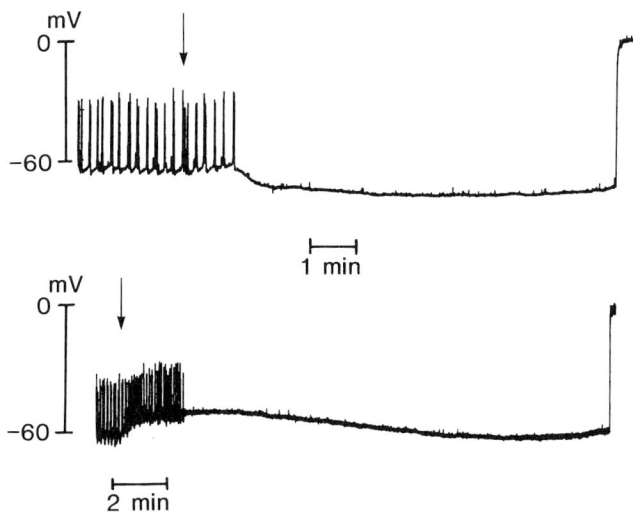

FIGURE 3 Microelectrode recordings from two portal veins incubated with β-hydroxybutyrate but no carbohydrate substrate. Upper: control; lower: with $10^{-7} M$ glibenclamide. Arrows mark addition of 2mM cyanide. From Lydrup et al. (1994).

activated J_{ATP} yields the "tension-related" J_{ATP}, which usually increases with contractile force. Another basis for partitioning is in terms of ATP turnover related to the cross-bridge interaction itself, that associated with LC20 phosphorylation, and that associated with Ca^{2+} handling, and so on. A problem in determining the background against which tension-related J_{ATP} should be measured is that it is not known to what extent "resting" J_{ATP} includes processes that might be altered by activation, for example, Ca^{2+} stimulation. For this and other reasons it is desirable to use permeabilized preparations for experiments on contractile energetics, as the extent of non-tension-related processes could be reduced by the use of inhibitors of metabolism and transport ATPases. However, even with extensive permeabilization using Triton X-100 there is a resting J_{ATP} that is typically around 30% of the total activated J_{ATP} as measured by several different techniques (Arner and Hellstrand, 1983, 1985; Lönnbro and Hellstrand, 1991; Trinkle-Mulcahy et al., 1994).

The question of basal J_{ATP} becomes even more complex in "conservatively" permeabilized preparations, such as those obtained using staphylococcus α-toxin, where the resting J_{ATP} is large enough to seriously impede the supply of high-energy phosphates, even in the presence of a high concentration of PCr in the solution. This was shown by Trinkle-Mulcahy et al. (1994) to be due to a high Ca^{2+}-insensitive ectonucleotidase activity in the preparation, which is also present in intact cells (Juul et al., 1991; Sjöblom-Widfeldt et al., 1993), and may actually be a diphosphohydrolase splitting both ATP and ADP (Yagi et al.,

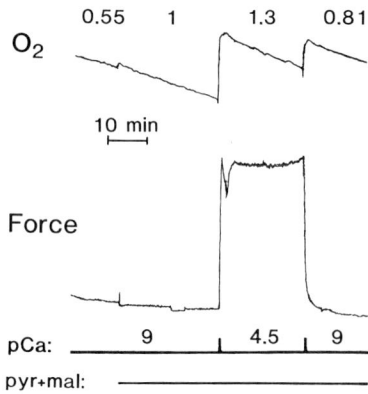

FIGURE 4 O_2 (upper) and force (lower) recordings from an α-toxin-permeabilized guinea pig ileum strip incubated in a closed chamber with a polarographic electrode (cf. Hellstrand and Paul, 1983). Solutions contain PCr (12 mM) and MgATP (3.2 mM). Pyruvate (10 mM) + malate (5 mM) added as indicated. Arbitrary units. Numbers above O_2 record show O_2 consumption (background subtracted) relative to first period after addition of pyruvate + malate.

1991; Picher *et al.*, 1994). This ATPase could be substantially (~70% in rabbit portal vein) reduced by pretreatment of the preparation by the ectonucleotidase inhibitor 4,4'-diisothiocyanatostilbene-2,2'-disulfonic acid (DIDS), followed by inclusion of sodium azide in solutions used for permeabilized preparations. This treatment dramatically reduced the ADP/ATP ratio measured in permeabilized muscle and increased contractile force. This method may greatly improve the usefulness of α-toxin-permeabilized preparations for energetics studies.

Being a nearly "physiological" preparation, mildly permeabilized muscle may have many of its metabolic processes intact. Thus, saponin-permeabilized cardiac and skeletal muscle has functioning mitochondria (Veksler *et al.*, 1987; Kunz *et al.*, 1993). Using α-toxin-permeabilized guinea pig ileum pretreated with DIDS but devoid of mitochondrial inhibitors, we have shown O_2 uptake in the presence of ATP, PCr, and CK (I. Nordström and P. Hellstrand, unpublished observations). This uptake is increased on addition of mitochondrial substrates, and is slightly stimulated when contraction is initiated by the addition of Ca^{2+} (Fig. 4). This approach may allow investigation of cellular metabolic regulation, as well as contribute to the energetic supply when diffusion from the extracellular medium is limiting.

B. Time Course of Energy Turnover during Contraction

A greater rate of energy turnover during the development than during the maintenance of tension has been a universal finding in different preparations of intact smooth muscle (for references, see Paul, 1980, 1989; Butler and Siegman, 1985). During maintained contraction J_{ATP} falls to varying extent, and during relaxation it decreases more rapidly than force. This pattern has been found in Triton-skinned smooth muscle as well (Butler *et al.*, 1990; Kühn *et al.*, 1990; Lönnbro and Hellstrand, 1991; Zhang and Moreland, 1994), although the initial transients tend to be larger in intact than in permeabilized muscle. These time-dependent changes may reflect energy turnover associated with the initial activation (Ca^{2+} translocation and LC20 phosphorylation), but in addition there may be a true decrease with time in the rate of energy turnover associated with cross-bridge cycling.

Dillon *et al.* (1981) suggested that myosin LC20 dephosphorylation during maintained contraction results in the appearance of slowly cyling ("latch") bridges, which would impede shortening. This would account for observations of decreased shortening velocity during the course of maintained contraction (summarized by Hellstrand and Paul, 1982). By implication, energy turnover during isometric contraction would be expected to decrease as well. Much interest in smooth muscle energetics over the last 10–15 years has been devoted to tests of this concept.

C. Rate of Phosphate Turnover on Myosin LC20 during Contraction

Two ATPase processes are operating in parallel during contraction: turnover of phosphate due to myosin LC20 phosphorylation and to actin–myosin cross-bridge interaction. The rate of myosin LC20 dephosphorylation at 37°C has been determined during relaxation from phasic contractions in intact bovine trachealis muscle by Kamm and Stull (1985) and in swine carotid artery by Driska *et al.* (1989). The values found, 0.2–0.3 s^{-1}, are likely to be underestimates, as noted by the authors, since immediate arrest of light chain kinase activity during relaxation is difficult to ascertain. The duration of the cross-bridge cycle in swine carotid artery, measured from the isometric tension cost and the myosin content, is about 0.75 s (Hellstrand and Paul, 1982). Thus there would be minimally one LC20 phosphorylation–dephosphorylation cycle per every five cross-bridge cycles (Driska *et al.*, 1989).

A high phosphatase rate was taken as the basis of a new form of the latch model by Driska (1987) and Hai and Murphy (1988). In this model, LC20 phosphorylation is considered to be the only switch for cross-bridge turnover, and the nonlinear relation between force (J_{ATP}) and maximal shortening velocity (V_{max}) arises because of a significant rate of LC20 dephosphorylation during active cross-bridge cycles, producing dephosphorylated, force-generating cross-bridges with a low tendency to dissociate. The model predicts a phosphatase rate at least of the same magnitude as that of the cross-bridge ATPase (Hai and Murphy, 1989; Walker *et al.*, 1994). As noted earlier, data from dephosphorylation experiments on intact smooth muscle do not quite indicate such a high value (Paul, 1990). However, the experimental testing of this prediction continues.

Butler *et al.* (1994) measured the steady-state rate of phosphatase activity during Ca^{2+}-induced activation of rabbit portal vein permeabilized by a combination of freeze-glycerination and Triton X-100. The muscle was equilibrated with ^{32}P-labeled ATP, and when contraction had reached its plateau a small amount of ^{33}P-labeled ATP was released by flash photolysis, whereafter the muscle was frozen following varying time intervals. After separation of phosphorylated light chains, the ^{33}P label incorporated was determined as a function of time. Since the system is in steady state, this equals the rate of turnover of phosphate on LC20. The rate constant found was $0.37 s^{-1}$ at 20°C. From the myosin concentration (55 μM), the degree of LC20 phosphorylation (47%), and the finding that only 76% of the LC20 phosphate exchanged as a monoexponential process, the calculated ATP usage by LC20 phosphorylation is 0.44 mM min^{-1}. This can be compared with the total suprabasal ATPase rate of 1.24 mM min^{-1}, indicating that 35% of this usage was due to LC20 phosphate turnover. This estimate is thus in approximate agreement with the conclusion from studies on intact muscle mentioned earlier. There is evidence that suprabasal J_{ATP} includes a significant amount of other Ca^{2+}-activated ATPases. In thiophosphorylated portal veins, Vyas *et al.* (1992, 1994) found a suprabasal ATPase rate of 0.26 mM min^{-1}. Taking this as an estimate of the cross-bridge turnover rate, there would be more than one exchange of LC20 phosphate per cross-bridge cycle, as required in the models by Driska (1987) and Hai and Murphy (1988).

Other results suggest a lower rate of phosphate turnover on LC20. Kitazawa *et al.* (1991) measured phosphatase activity in α-toxin-permeabilized portal vein from the rate of dephosphorylation in rigor solution with inhibited myosin light chain kinase (MLCK) activity. Their value was $0.017 s^{-1}$ at 15°C, which, corrected for temperature (Mitsui *et al.*, 1994), is about nine times smaller than the value of Butler *et al.* Reconciliation of these findings will require additional studies. However, the conditions of the experiments were quite different and it should also be pointed out that in the experiments of Butler *et al.* (1994), only 76% of the light chain phosphate turned over at the fast rate, al-

though over a longer period complete replacement of LC20 phosphate during contraction has been found (Bárány *et al.*, 1991).

As pointed out by Butler *et al.* (1994), the α-toxin-permeabilized preparation used by Kitazawa *et al.* might contain endogenous phosphatase inhibitors that have been lost in the Triton-skinned preparation. This could at least partially explain the observed difference in phosphatase rate, and is an intriguing possibility relating to the demonstration that receptor agonists are able to increase Ca^{2+} sensitivity in neuroeffector-coupled permeabilized preparations by a mechanism involving inhibited myosin light chain phosphatase (MLCP) activity (Kitazawa *et al.*, 1991; Kubota *et al.*, 1992). A further reason to speculate along these lines is the Ca^{2+}-sensitizing effect of polyamines (spermine and spermidine) found in β-escin-permeabilized ileum muscle and demonstrated to be associated with MLCP inhibition (Swärd *et al.*, 1994, 1995). Polyamines are endogenous intracellular substances present in significant quantity, although accurate estimates in intact cells are needed before the functional implications of this effect can be evaluated. Spermine only slowly penetrate α-toxin-permeabilized preparations (K. Swärd and P. Hellstrand, unpublished observations), suggesting that endogenous polyamine levels may be largely retained in this type of preparation in acute experiments.

D. Energy Turnover in Active Cross-bridge Cycling

With all modes of activation tested, J_{ATP} increases with contractile force. However, the slope of the force–J_{ATP} relationship ("the energetic tension cost") is variable. Generally the tension cost varies in parallel with other indicators of cross-bridge turnover rate, notably the maximal shortening velocity (Bárány, 1967; Rüegg, 1971; Paul, 1980; Hellstrand and Paul, 1982). Smooth muscle, particularly of the "tonic" type represented primarily by large muscular arteries, is at the low end of the scale, having both a low energetic tension cost and a low V_{max}. This is clearly advantageous for its physiological function of maintaining force (blood pressure) with little demand on speed of contraction. "Phasic" muscles have a higher tension cost associated with greater V_{max}. This group includes most muscles of the gastrointestinal and reproductive organs, and also airway smooth muscle and some blood vessels distal to the muscular arteries.

The shape of the relation between force and J_{ATP} is of interest for evaluating theories of cross-bridge regulation (Paul, 1989, 1990; Walker *et al.*, 1994; Strauss and Murphy, Chapter 26, this volume). A linear relationship is easily understood mechanistically if more force is caused by recruitment of a proportionate amount of additional cross-bridges, all having the same rate of ATP turnover. Linear relationships have been observed in intact muscle during activation at physiological levels of extracellular [Ca^{2+}] (Paul, 1980). However, nonlinear relations, showing greater slope of the force–J_{ATP} relationship with activation level, have been found in some studies when force is varied by increasing [Ca^{2+}] over a wide interval in both intact (Siegman et al., 1984; Krisanda and Paul, 1988; Walker et al., 1994), and permeabilized (Arner and Hellstrand, 1983, 1985; Lönnbro and Hellstrand, 1991; Zhang and Moreland, 1994) smooth muscle. These findings might imply either that the cross-bridge cycle is more rapid at higher level of activation, or that additional ATP-consuming reactions are recruited. The latter possibility could include the LC20 phosphorylation–dephosphorylation cycle or Ca^{2+}-stimulated ATPases unrelated to the cross-bridge interaction. However, in the permeabilized taenia coli and carotid arteries, activated J_{ATP} was only slightly (maximally 12%) smaller in thiophosphorylated than in Ca^{2+}-activated preparations, suggesting that turnover of LC20 phosphate does not contribute substantially to Ca^{2+}-activated J_{ATP} (Hellstrand and Arner, 1985; Zhang and Moreland, 1994) in these experiments.

Independent evidence for an affect of activation on cross-bridge cycling rate comes from measurements showing increased unloaded shortening velocity with increasing activation by Ca^{2+} in Triton-skinned smooth muscle (rat portal vein: Arner, 1983; swine carotid artery: Paul et al., 1983; guinea pig taenia coli: Paul et al., 1983; Arner and Hellstrand, 1985). This evidence from skinned muscle can be directly compared with the J_{ATP} measurements performed on the same preparations, also suggesting modulation of the cross-bridge cycling rate. In contrast to data from intact preparations, which suggest that V_{max} can be altered as a function of time, [Ca^{2+}]$_i$, or LC20 phosphorylation, the data from the skinned muscle should exclude the time factor and associated possible changes in the metabolic state of the tissue.

One mechanism that may account for the observed relationships of force, J_{ATP}, and V_{max} is cooperative attachment of nonphosphorylated cross-bridges as a result of the presence of other attached cross-bridges. Such a phenomenon is suggested by the experiments of Arner et al. (1987a) and Somlyo et al. (1988), showing that the relaxation from rigor after photolytic release of ATP from caged ATP is biphasic and not compatible with a rapid irreversible dissociation of rigor bonds. Vyas et al. (1992, 1994) have measured the rate of turnover of bound ADP on myosin in permeabilized rat

portal vein at varying degrees of thiophosphorylation of the myosin. It was found that a small degree (less than 20%) of myosin LC20 thiophosphorylation cooperatively activates the maximum number of myosin molecules, whereas the ATPase rate is approximately linearly related to the degree of thiophosphorylation. In chicken gizzard, Kenney et al. (1990) showed a highly nonlinear relation between LC20 phosphorylation and force development, although J_{ATP} was linearly related to LC20 phosphorylation in both Ca^{2+}-activated and thiophosphorylated preparations. Thus force may be a property relating to cooperative attachment of cross-bridges, whereas only phosphorylated cross-bridges undergo active cycles involving ATP turnover.

V. KINETICS OF THE CROSS-BRIDGE INTERACTION

A. Normalization of Mechanical Data

Maximal shortening velocities (V_{max}) measured in mammalian smooth muscle usually fall in the range 0.1–0.7 L_0 s^{-1} at 37°C, 50–100 times slower than that found in skeletal muscle. Several methods can be used to obtain V_{max}, and generally there is a decrease in shortening velocity with the time and/or extent of shortening (Arner and Hellstrand, 1985; Harris and Warshaw, 1990; Hellstrand and Nordström, 1993). This will influence velocity, for example, during shortening following a "quick release," which thus needs to be measured in a consistent manner. Therefore, values from different laboratories should be compared with methodological differences in mind. Normalization of mechanical data in smooth muscle has to be done in terms of an "optimal" muscle length (L_0), since "sarcomere" dimensions are unknown. In a frog skeletal muscle the sarcomere length at optimal filament overlap is about 2.2 μm, and that of a half-sarcomere, containing one layer of cross-bridge connections between its end points, is thus 1.1 μm. Therefore, the conventional unit of cross-bridge dimensions, 1 nm/half-sarcomere, corresponds to roughly 0.1% L_0. This is often taken as a basis for comparisons with smooth muscle, although filament lengths in this tissue may be somewhat longer (Ashton et al., 1975; Cooke et al., 1989; Small et al., 1990), and the cytoarchitecture is less well defined. There is thus considerable uncertainty in relating cell and tissue data to cross-bridge dimensions.

Other evidence suggests that the thick filaments in smooth muscle may be side-polar, that is, myosin molecules oriented in either direction are found along the whole length of the filament, with no central bare zone

(Cooke *et al.*, 1989). This may not have major implications for the "contractile unit" length, since a unit consisting of two oppositely oriented thin filaments joined by a thick filament will minimally have the length of the thick filament (provided the thin filaments do not limit contraction) and will include two sets of cross-bridge connections in series irrespective of which model is used. The maximal length above which active tension cannot be generated will for both models be the summed lengths of two thin and one thick filament. Long actin filaments (Small *et al.*, 1990) would thus permit a large capacity of shortening in smooth muscle cells.

B. Myosin Isoforms

Progress has been made in defining molecular factors responsible for the variability in shortening velocity and ATPase rates between smooth muscles. Three types of isoform variations have been found in smooth muscle myosin (for reviews, see Somlyo, 1993, and Chapters 1 and 2 in this volume): (1) two variants in the tail region of the myosin heavy chain, one heavier (~204 kDa; SM-1 and one lighter (~200 kDa; SM−2); (2) a seven-amino acid insert in the globular head region; and (3) two variants of 17-kDa alkali light chains, one acidic (LC17a) and one basic (LC17b). Particularly the two latter forms of variability are associated with changes in cross-bridge turnover rate, although it is reported that estrogen treatment of ovariectomized rats leads to higher V_{max} and relative SM-1 contents in myometrial smooth muscle (Hewett *et al.*, 1993).

The seven-amino acid insert in the myosin head region has been identified in visceral smooth muscle but is lacking in (tonic) vascular muscle, and the presence of the insert correlates with an increased ATPase activity and greater motility *in vitro* (Kelley *et al.*, 1993). However, also the LC17 isoforms correlate with differences in ATPase and shortening properties. Helper *et al.* (1988) showed that vascular muscle from large arteries contains a mixture of LC17a and LC17b, whereas gastrointestinal (stomach and jejunum) smooth muscle contains only LC17a. These differences correlate with greater actin-activated ATPase activity in the latter. Malmqvist and Arner (1991) found a strong correlation between LC17 isoform distribution and V_{max} in a range of different smooth muscles (Fig. 5). Hasegawa and Morita (1992) showed that reconstitution of porcine aorta with varying proportions of the LC17 isoforms modifies its actin-activated ATPase activity according to the same pattern. Myosin was found to bind more strongly to F-actin with increasing proportion of LC17b. Thus at least two principles are operating in determining the ATPase properties of myosin:

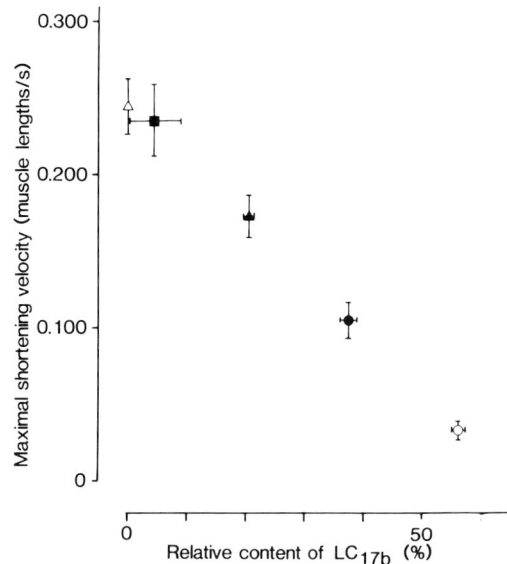

FIGURE 5 V_{max} at 22°C in skinned smooth muscle fibers plotted against relative contents, of myosin LC_{17b}. △, Rabbit rectococcygeus; ▲, guinea pig tenia coli; ●, rat uterus; ○, rabbit thoracic aorta; ■, rabbit trachea. From *Pflügers Arch.*, Energetics of smooth muscle contraction, U. Malmqvist and A. Arner, **418**, 523–530 (1991) copyright Springer-Verlag GMbH & Co. KG.

peptide inserts in the globular head region and LC17 isoform variability. The relative roles and interrelations of these two mechanisms remain to be established, and rapid progress in this area can be expected.

C. Cross-bridge Mechanics

The number of attached cross-bridges determines the force generated by a muscle. However, attached cross-bridges will not under all conditions contribute the same amount of force. For instance, active cross-bridge turnover is arrested in rigor, and slippage of cross-bridges or compliance in associated structures will reduce overall force. Huxley and Simmons (1971) proposed that cross-bridges are associated with intrinsic elasticity. A small extension or release of the muscle fiber (ΔL) will then give rise to a force deflection (ΔF), which is the sum of all individual cross-bridge responses. It was shown that when the length change is made more rapid, the stiffness ($\Delta F/\Delta L$) increases and approaches linear characteristics. Therefore, it was assumed that the individual cross-bridge compliance is linear and that deviations from linearity reflect recoil occurring in the bridges during the length change itself, even if very rapid. Using length changes complete in about 0.2 ms, Ford *et al.* (1977) concluded that an individual (linear) cross-bridge elasticity would be unloaded for a shortening of about 0.4 nm per half-sarcomere (roughly 0.4% L_0). In the model of Huxley and Simmons (1971) it is assumed that extensibility of the thick and thin filaments does not contribute significantly to the overall sarcomere compliance. However, X-ray diffraction evidence (Huxley *et al.*, 1994; Wakabayashi *et al.*, 1994) indicates that filament extensibility could account for as much as about half of the measured sarcomere compliance. These findings may necessitate reconsideration of current models of cross-bridge mechanics (Goldman and Huxley, 1994).

In experiments on chemically skinned fiber bundles from guinea pig taenia coli muscle, Arheden and Hellstrand (1991) showed that stiffness is steeply dependent on rate of length change (Fig. 6). The fastest length steps employed, complete in about 0.3 ms, gave a value for the linearized total fiber compliance of less than 0.8% L_0. This is smaller by about 1.5 times than the compliance of single toad stomach cells reported by Warshaw *et al.* (1988). Owing to the strong rate dependence of stiffness, the difference is likely to arise at least in part from the fact that slower length ramps

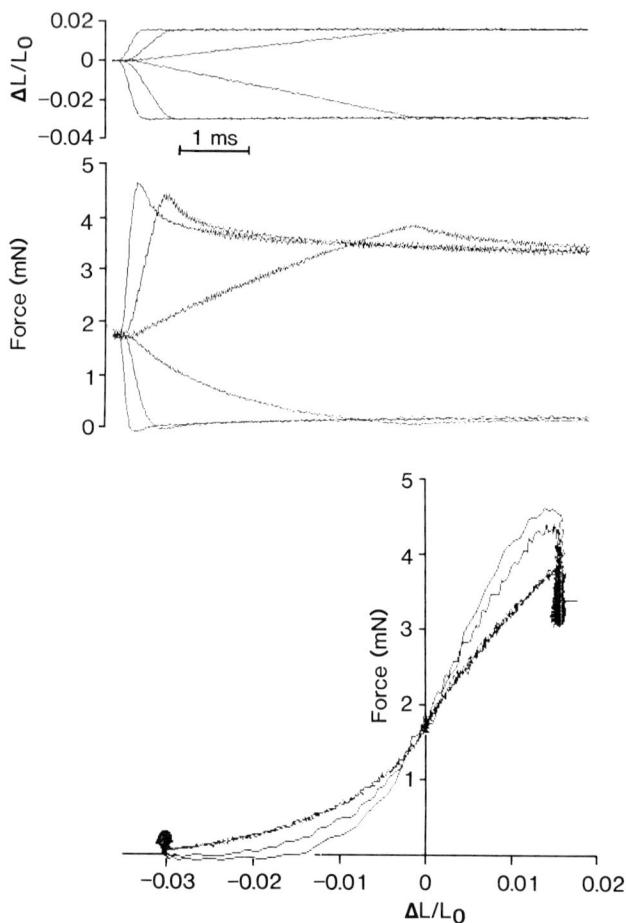

FIGURE 6 Force responses (middle) to length ramps (upper) in a skinned guinea pig taenia coli fiber. Lower panel shows plots of force versus length during ramps. Stiffness increases with rate of length change. From Arheden and Hellstrand (1991).

(2.5 ms) were used on the single cells. There is probably compliance in structures outside the cross-bridges, which contributes to the total cell or tissue compliance. Even though quantitative estimates are highly uncertain, the available evidence indicates that the molecular distances involved in the cross-bridge interaction may be essentially the same in smooth as in striated muscle.

D. Attached Cross-bridges in Isometric Contraction and during Shortening

The adjustment of the cross-bridge system to shortening has been studied by stiffness measurements using a stretch of the muscle after varying periods of unloaded shortening, initiated by a rapid release to slack length (Arheden and Hellstrand, 1991; Hellstrand and Nordström, 1993). As shown in Fig. 7, stiffness can be measured as a tangent to the force–length relation during a shortening ramp or stretch, and then again during a restretch following varying time periods of unloaded shortening. The first value will represent stiffness during isometric contraction and might thus be called, somewhat unlogically, "isometric stiffness." The second value should represent stiffness during unloaded contraction. This approach is particularly useful in the analysis of smooth muscle contraction, since stiffness can be compared at identical force levels at both points in time. Thus any contribution from series-coupled compliance should be equal. Release–stretch experiments of the kind illustrated in Fig. 7 indicate that stiffness drops to about 40% of the isometric value a few milliseconds after the release. Interestingly, this is about the same drop in stiffness as occurs when a frog skeletal muscle fiber is released to shorten at low load (Ford *et al.*, 1985). Comparison of responses in skinned preparations in active contraction and in rigor suggests that a release induces detachment of cross-bridges as they are pushed into a state where they would exert negative tension, in accordance with the Huxley (1957) cross-bridge model.

The mechanical basis of the latch phenomenon was investigated in intact taenia coli muscle (Hellstrand and Nordström, 1993). Muscle strips were activated by high-K^+ solution to produce sustained contractions at constant force. Data obtained at 20 s and at 5 min after initiation of contraction were compared. At 5 min, $[Ca^{2+}]_i$, LC20 phosphorylation, and V_{max} were significantly lower than at 20 s, indicating that a latch state had developed. There was no difference in isometric stiffness between the two points in time, suggesting that the average force per cross-bridge is unaffected by latch. When the muscle was released to slack length, stiffness dropped at a rate of 150–200 s^{-1}, but stabi-

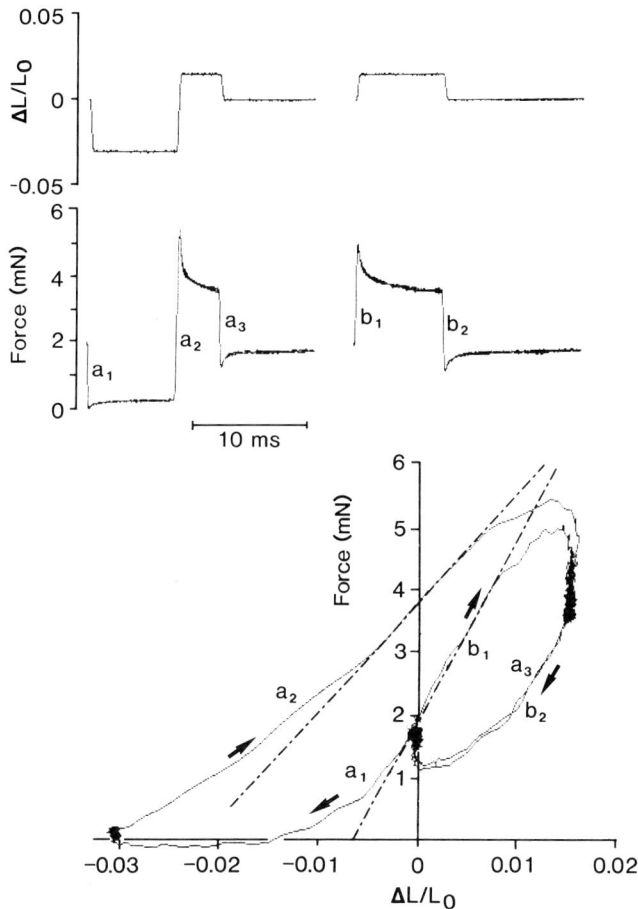

FIGURE 7   Release (a_1) and restretch after 8 ms (a_2) in a contracted skinned taenia coli fiber. The fiber was stretched above the initial length and released 2 ms later (a_3). A stretch (b_1) from the initial length and return release (b_2) are also shown. Panels as in Fig. 6. Dashed lines in lower panel indicate slope of length–force relation (stiffness) in initial release (a_1) or stretch (b_1), and in restretch after unloaded shortening (a_2). Note lower slope (=stiffness) after shortening. From Arheden and Hellstrand (1991).

FIGURE 8   Stiffness relative to isometric after release to slack length in one intact taenia coli fiber after activation for 20 s (○) or 5 min (●). Method as illustrated in Fig. 7. From Hellstrand and Nordström (1993).

sites on the thin filament before releasing their bound ADP (Cooke *et al.*, 1994). If the rate of ADP release is affected by latch (see Section V.E.), one would expect to see a greater number of attached cross-bridges during "equilibrated" shortening, whereas the rapid net detachment induced by the length step would not involve this rate-limiting process.

E. Influence of ADP and Phosphate on Cross-bridge Kinetics

The rate of relaxation from rigor by photolytic release of ATP from caged ATP is reduced by the presence of MgADP (Arner *et al.*, 1987b). This effect was found by Nishiye *et al.* (1993) to involve decreased amplitude of an initial rapid phase thought to represent ATP-induced dissociation of the rigor cross-bridges, and slowing of a later phase that might reflect cooperative cross-bridge reattachment and isomerization of ADP-bound (force-generating) states. MgADP decreased the amplitude of the rapid phase with high affinity ($K_D = 1.3$ μM in guinea pig portal vein). Arheden and Arner (1992) showed that MgADP potently inhibits the relaxation from rigor induced by Mg^{2+}-pyrophosphate (K_i ~2 μM). Taken together, these results indicate tight binding of MgADP to cross-bridges in smooth muscle. The binding affinity is higher in "tonic" than in "phasic" smooth muscle ($K_D = 1.1$ μ M in rabbit femoral artery and 4.9 μM in rabbit bladder; Fuglsang *et al.*, 1993), correlating with differences in myosin LC17 isoform contents and presence of a seven-amino acid insert in the myosin head region (see Section V.B). The binding affinity is higher than reported in cardiac and skeletal muscle, and a pattern

lized at a somewhat higher value during unloaded shortening in the 5-min than in the 20-s contraction (Fig. 8). Thus cross-bridges detach rapidly on transition from isometric to unloaded contraction, but this process is arrested as load has been discharged, and during the following active shortening stiffness only slowly declines further, suggesting a progressive decrease in the equilibrium number of attached cross-bridges. Stiffness was greater and unloaded shortening speed smaller in the 5-min than in the 20-s contractions. It has been proposed that an isomerization step limits the rate of ADP release from the cross-bridge (Siemankowski *et al.*, 1985), and that during rapid shortening cross-bridges may detach owing to mechanical strain and interact with multiple binding

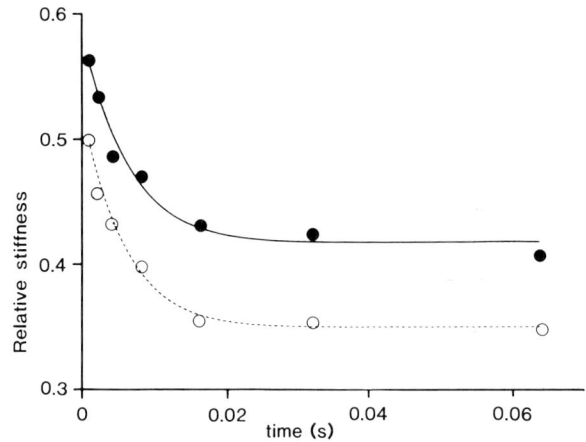

is apparent where lower cross-bridge cycling rate is associated with tighter ADP binding (Fuglsang et al., 1993).

Inorganic phosphate (P_i) decreases the maximal force in skinned smooth muscle and increases the rate of relaxation (Schneider et al., 1981; Itoh et al., 1986; Gagelmann and Güth, 1987). The rate of relaxation from rigor induced by photolytic release of ATP is also increased by P_i (Arner et al., 1987a). P_i does not influence the amplitude of the initial rapid relaxation but increases the rate of relaxation of the later phase, opposite to the effect of MgADP (Nishiye et al., 1993). The maximal shortening velocity is unaffected by P_i in maximally activated skinned taenia coli, and increased at submaximal activation (Österman and Arner, 1995). It was suggested that P_i dissociates cross-bridge states that impose a mechanical resistance to shortening at low levels of activation. Basically, the effects of P_i can be understood from the concept that P_i release marks the entry into the strongly bound cross-bridge states, able to carry force (Eisenberg and Greene, 1980). Thus any effect of added P_i involves a reversal of this reaction, tending to decrease the number of strongly bound cross-bridges. This might also apply to unphosphorylated cross-bridges formed either by dephosphorylation while attached or by cooperative attachment.

On the basis of the effects of ADP on cross-bridge kinetics, the tight ADP binding in smooth muscles was proposed to explain the latch phenomenon, as metabolic alterations during contraction might involve an increase in MgADP concentration (Nishiye et al., 1993). A problem with this concept is that calculated changes in [MgADP] during contraction (Hellstrand and Paul, 1983) and the effect of MgADP on V_{max} (Arner et al., 1987b) and on in vitro filament sliding velocity (Warshaw et al., 1991) occur in a concentration range well above the estimated K_D values of MgADP binding to myosin. Comparison of MgADP effects on the slow relaxation phase following photolytic release of ATP in tonic and phasic muscles might reconcile these findings. Whereas the rate of the slow relaxation in bladder muscle was essentially unaffected by MgADP, that in the femoral artery was slowed by increasing [MgADP] in the range 0–200 μM (Fuglsang et al., 1993). This process, which is limited by the rate of ADP release, might be a better correlate to shortening velocity than MgADP binding to cross-bridges. It is quite interesting in this perspective that the rabbit bladder, in contrast to most other smooth muscles studied, does not decrease its rate of energy expenditure during the course of maintained contraction (Wendt and Gibbs, 1987).

VI. PERSPECTIVES

The mechanisms regulating energy turnover in smooth muscle are becoming known in much greater detail owing to developments of the last few years. The molecular basis of the latch mechanism may eventually be elucidated, and energetic experiments define constraints on models to explain this phenomenon. The influence of metabolic factors on membrane and excitation–contraction processes touch on signaling pathways regulating a multitude of cellular functions. Diversity is a hallmark of smooth muscle, and it is expected that increasing knowledge of processes regulating growth and differentiation will eventually lead to an understanding of how functional demands on the tissue are reflected in its phenotypic differentiation. Defining the energetic consequences of functional challenge is a key element in this development.

Acknowledgments

Original research by the author was supported by the Swedish Medical Research Council (Project 14x-28), the Medical Faculty, University of Lund, and Astra-Hässle AB, Mölndal, Sweden.

References

Arheden, H., and Arner, A. (1992). Biophys. J. 61, 1480–1494.

Arheden, H., and Hellstrand, P. (1991). J. Physiol. (London) 442, 601–630.

Arner, A. (1983). Pflügers Arch. 397, 6–12.

Arner, A., and Hellstrand, P. (1983). Circ. Res. 53, 695–702.

Arner, A., and Hellstrand, P. (1985). J. Physiol. (London) 360, 347–365.

Arner, A., Goody, R. S., Rapp, G., and Rüegg, J. C. (1987a). J. Muscle Res. Cell Motil. 8, 377–385.

Arner, A., Hellstrand, P., and Rüegg, J. C. (1987b). Prog. Clin. Biol. Res. 245, 43–57.

Ashton, F. T., Somlyo, A. V., and Somlyo, A. P. (1975). J. Mol. Biol. 98, 17–29.

Bárány, M. (1967). J. Gen. Physiol. 50, 197–218.

Bárány, M., Rokolya, A., and Bárány, K. (1991). Arch. Biochem. Biophys. 287, 199–203.

Barron, J. T., and Kopp, S. J. (1991). Biochim. Biophys. Acta 1073, 550–554.

Barron, J. T., Kopp, S. J., Tow, J. P., and Parrillo, J. E. (1991). Biochim. Biophys. Acta 1093, 125–134.

Butler, T. M., and Davies, R. E. (1980). In "Handbook of Physiology" (D. F. Bohr, A. P. Somlyo, and H. V. Sparks, eds.), Sect. 2, Vol. II, pp. 237–252. Am. Physiol. Soc., Bethesda, MD.

Butler, T. M., and Siegman, M. J. (1985). Annu. Rev. Physiol. 47, 629–643.

Butler, T. M., Siegman, M. J., Mooers, S. U., and Narayan, S. R. (1990). Am. J. Physiol. 258, C1092–C1099.

Butler, T. M., Narayan, S. R., Mooers, S. U., and Siegman, M. J. (1994). Am. J. Physiol. 267, C1160–C1166.

Chace, K. V., and Odessey, R. (1981). *Circ. Res.* **48**, 850–858.

Clark, J. F., Khuchua, Z., and Ventura-Clapier, R. (1992). *Biochim. Biophys. Acta* **1100**, 137–145.

Clark, J. F., Khuchua, Z., Kuznetsov, A., Saks, V. A., and Ventura-Clapier, R. (1993). *J. Physiol. (London)* **466**, 553–572.

Clark, J. F., Kuznetsov, A. V., Khuchua, Z., Veksler, V., Ventura-Clapier, R., and Saks, V. (1994). *FEBS Lett.* **347**, 147–151.

Coburn, R. F., Baron, C. B., and Papadopoulos, M. T. (1988). *Am. J. Physiol.* **255**, H1476–H1483.

Coburn, R. F., Moreland, S., Moreland, R. S., and Baron, C. B. (1992). *J. Physiol. (London)* **448**, 473–492.

Coburn, R. F., Azim, S., Fillers, W. .S, and Baron, C. B. (1993). *Am. J. Physiol.* **264**, L1–L6.

Cooke, P. H., Fay, F. S., and Craig, R. (1989). *J. Muscle Res. Cell Motil.* **10**, 206–220.

Cooke, R., White, H., and Pate, E. (1994). *Biophys. J.* **66**, 778–788.

Daut, J., Maier-Rudolph, W., von-Beckerath, N., Mehrke, G., Gunther, K., and Goedel-Meinen, L. (1990). *Science* **247**, 1341–1344.

Dillon, D. F., Aksoy, M. O., Driska, S. P., and Murphy, R. A. (1981). *Science* **211**, 495–497.

Driska, S. P. (1987). *Prog. Clin. Biol. Res.* **245**, 387–398.

Driska, S. P., Stein, P. G., and Porter, R. (1989). *Am. J. Physiol.* **256**, C315–C321.

Eisenberg, E., and Greene, L. E. (1980). *Annu. Rev. Physiol.* **42**, 293–309.

Ekmehag, B. L., and Hellstrand, P. (1988). *Acta Physiol Scand.* **133**, 525–533.

Ekmehag, B. L., and Hellstrand, P. (1989). *Acta Physiol. Scand.* **136**, 367–376.

Ford, L. E., Huxley, A. F., and Simmons, R. M. (1977). *J. Physiol. (London)* **269**, 441–515.

Ford, L. E., Huxley, A. F., and Simmons, R. M. (1985). *J. Physiol. (London)* **361**, 131–150.

Fuglsang, A., Khromov, A., Török, K., Somlyo, A. V., and Somlyo, A. P. (1993). *J. Muscle Res. Cell Motil.* **14**, 666–677.

Gagelmann, M., and Güth, K. (1987). *Biophys. J.* **51**, 457–463.

Goldman, Y. E., and Huxley, A. F. (1994). *Biophys. J.* **67**, 2131–2136.

Hai, C.-M., and Murphy, R. A. (1988). *Am. J. Physiol.* **255**, C86–C94.

Hai, C.-M., and Murphy, R. A. (1989). *Annu. Rev. Physiol.* **51**, 285–298.

Hai, C.-M., Watson, C., Wallach, S. J., Reyes, V., Kim, E., and Xu, J. (1993). *Am. J. Physiol.* **264**, L553–L559.

Harris, D. E., and Warshaw, D. M. (1990). *J. Gen. Physiol.* **96**, 581–601.

Hartshorne, D. J. (1987). *In* "Physiology of the Gastrointestinal Tract" (L. R. Johnson, ed.), 2nd ed., pp. 423–482. Raven Press, New York.

Hasegawa, Y., and Morita, F. (1992). *J. Biochem. (Tokyo)* **111**, 804–809.

Hellstrand, P., and Arner, A. (1985). *Pflügers Arch.* **405**, 323–328.

Hellstrand, P., and Arner, A. (1986). *J. Muscle Res. Cell Motil.* **7**, 378–379.

Hellstrand, P., and Nordström, I. (1993). *Am. J. Physiol.* **265**, C695–C703.

Hellstrand, P., and Paul, R. J. (1982). *In* "Vascular Smooth Muscle: Relations between Energy Metabolism and Mechanics" (M. F. Crass, III and C. D. Barnes, eds.), pp. 1–35. Academic Press, New York.

Hellstrand, P., and Paul, R. J. (1983). *Am. J. Physiol.* **244**, C250–C258.

Hellstrand, P., and Vogel, H. J. (1985). *Am. J. Physiol.* **248**, C320–C329.

Hellstrand, P., Johansson, B., and Norberg, K. (1977). *Acta Physiol. Scand.* **100**, 69–83.

Hellstrand, P., Jorup, C., and Lydrup, M.-L. (1984). *Pflügers Arch.* **401**, 119–124.

Helper, D. J., Lash, J. A., and Hathaway, D. R. (1988). *J. Biol. Chem.* **263**, 15748–15753.

Hewett, T. E., Martin, A. F., and Paul, R. J. (1993). *J. Physiol. (London)* **460**, 351–364.

Huang, S. M., Chowdhury, J. U., Kobayashi, K., and Tomita, T. (1993). *Jpn. J. Physiol.* **43**, 229–238.

Huxley, A. F. (1957). *Prog. Biophys. Biophys. Chem.* **7**, 255–318.

Huxley, A. F., and Simmons, R. M. (1971). *Nature (London)* **233**, 533–538.

Huxley, H. E., Stewart, A., Sosa, H., and Irving, T. (1994). *Biophys. J.* **67**, 2411–2421.

Ishida, Y., and Paul, R. J. (1990). *J. Physiol. (London)* **424**, 41–56.

Ishida, Y., Wyss, M., Hemmer, W., and Wallimann, T. (1991). *FEBS Lett.* **283**, 37–43.

Itoh, T., Kanmura, Y., and Kuriyama, H. (1986). *J. Physiol. (London)* **376**, 231–252.

Juul, B., Luscher, M. E., Aalkjaer, C., and Plesner, L. (1991). *Biochim. Biophys. Acta* **1067**, 201–207.

Kamm, K. E., and Stull, J. T. (1985). *Am. J. Physiol.* **249**, C238–C247.

Kelley, C. A., Takahashi, M., Yu, J., and Adelstein, R. S. (1993). *J. Biol. Chem.* **268**, 12848–12854.

Kenney, R. E., Hoar, P. E., and Kerrick, W. G. L. (1990). *J. Biol. Chem.* **265**, 8642–8649.

Kitazawa, T., Masuo, M., and Somlyo, A. P. (1991). *Proc. Natl. Acad. Sci. U.S.A.* **88**, 9307–9310.

Krisanda, J. M., and Paul, R. J. (1983). *Am. J. Physiol.* **244**, C385–C390.

Krisanda, J. M., and Paul, R. J. (1988). *Am. J. Physiol.* **255**, C393–C400.

Kubota, Y., Nomura, M., Kamm, K. E., Mumby, M. C., and Stull, J. T. (1992). *Am. J. Physiol.* **262**, C405–C410.

Kühn, H., Tewes, A., Gagelmann, M., Güth, K., Arner, A., and Rüegg, J. C. (1990). *Pflügers Arch.* **416**, 512–518.

Kunz, W. S., Kuznetsov, A. V., Schulze, W., Eichhorn, K., Schild, L., Striggow, F., Bohnensack, R., Neuhof, S., Grasshoff, H. Neumann, H. W., and Gellerich, F. N. (1993). *Biochim. Biophys. Acta* **1144**, 46–53.

Kushmerick, M. J., Dillon, P. F., Meyer, R. A., Brown, T. R., Krisanda, J. M., and Sweeney, H. L. (1986). *J. Biol. Chem.* **261**, 14420–14429.

Longmore, J., Newgreen, D. T., and Weston, A. H. (1990). *Eur. J. Pharmacol.* **190**, 75–84.

Lönnbro, P., and Hellstrand, P. (1991). *J. Physiol. (London)* **440**, 385–402.

Lövgren, B., and Hellstrand, P. (1985). *Acta Physiol. Scand.* **123**, 485–495.

Lövgren, B., and Hellstrand, P. (1987). *Acta Physiol. Scand.* **129**, 211–219.

Lydrup, M.-L., Swärd, K., and Hellstrand, P. (1994). *J. Vasc. Res.* **31**, 82–91.

Lynch, R. M., and Paul, R. J. (1983). *Science* **222**, 1344–1346.

Lynch, R. M., and Paul, R. J. (1987). *Am. J. Physiol.* **252**, C328–C334.

Malmqvist, U., and Arner, A. (1991). *Pflügers Arch.* **418**, 523–530.

Malmqvist, U., Arner, A., and Uvelius, B. (1991). *Pflügers Arch.* **419**, 230–234.

Mitsui, T., Kitazawa, T. T., and Ikebe, M. (1994). *J. Biol. Chem.* **269**, 5842–5848.

Nakayama, S., Seo, Y., Takai, A., Tomita, T., and Watari, H. (1988). *J. Physiol. (London)* **402**, 565–578.

Nichols, C. G., and Lederer, W. J. (1991). *Am. J. Physiol.* **261**, H1675–H1686.

Nishiye, E., Somlyo, A. V., Török, K., and Somlyo, A. P. (1993). *J. Physiol. (London)* **460**, 247–271.

Noack, T., Edwards, G., Deitmer, P., and Weston, A. H. (1992). *Br. J. Pharmacol.* **107**, 945–955.

Odessey, R., and Chace, K. V. (1982). *Am. J. Physiol.* **243**, H128–H132.

Österman, Å., and Arner, A. (1995). *J. Physiol. (London)* **484**, 369–383.

Paul, R., J. (1980). *In* "Handbook of Physiology" (D. F. Bohr, A. P. Somlyo, and H. V. Sparks, eds.), Sect. 2, Vol. II, pp. 201–235. Am. Physiol. Soc. Bethesda, MD.

Paul, R. J. (1987). *In* "Physiology of the Gastrointestinal Tract" (L. R. Johnson, ed.), 2nd ed. Vol. 1, pp. 483–506. Raven Press, New York.

Paul, R. J. (1989). *Annu. Rev. Physiol.* **51**, 331–349.

Paul, R. J. (1990). *Am. J. Physiol.* **258**, C369–C375.

Paul, R. J., Doerman, G., Zeugner, C., and Rüegg, J. C. (1983). *Circ. Res.* **53**, 342–351.

Picher, M. B., Liveau, R., Potier, M., Savaria, D., Rousseau, E., and Beaudoin, A. R. (1994). *Biochim. Biophys. Acta* **1200**, 167–174.

Quayle, J. M., and Standen, N. B. (1994). *Cardiovasc. Res.* **28**, 797–804.

Rüegg, J. C. (1971). *Physiol. Rev.* **51**, 201–248.

Schneider, M., Sparrow, M., and Rüegg, J. C. (1981). *Experientia* **37**, 980–982.

Scott, D. P., and Coburn, R. F. (1989). *Am. J. Physiol.* **257**, H597–H602.

Siegman, M. J., Butler, T. M., Mooers, S. U., and Michalek, A. (1984). *Pflügers Arch.* **401**, 385–390.

Siemankowski, R. F., Wiseman, M. O., and White, H. D. (1985). *Proc. Natl. Acad. Sci. U.S.A.* **82**, 658–662.

Sjöblom-Widfeldt, N., Arner, A., and Nilsson, H. (1993). *J. Vasc. Res.* **30**, 38–42.

Small, J. V., Herzog, M., Barth, M., and Draeger, A. (1990). *J. Cell Biol.* **111**, 2451–2461.

Somlyo, A. P. (1993). *J. Muscle Res. Cell Motil.* **14**, 557–563.

Somlyo, A. P., Goldman, Y. E., Fujimori, T., Bond, M., Trentham, D. R., and Somlyo, A. P. (1988). *J. Gen. Physiol.* **91**, 165–192.

Sparks, H. V. (1980). *In* "Handbook of Physiology" (D. F. Bohr, A. P. Somlyo, and H. V. Sparks, eds.), Sect. 2, Vol. II, pp. 475–513. Am. Physiol. Soc., Bethesda, MD.

Standen, N. B., Quayle, J. M., Davies, N. W., Brayden, J. E., Huang, Y., and Nelson, M. T. (1989). *Science* **245**, 177–180.

Swärd, K., Josefsson, M., Lydrup, M.-L., and Hellstrand, P. (1993). *Acta Physiol. Scand.* **148**, 265–272.

Swärd, K., Nilsson, B.-O., and Hellstrand, P. (1994). *Am. J. Physiol.* **266**, C1754–C1763.

Swärd, K., Pato, M. D., Nilsson, B.-O., Nordström, I., and Hellstrand, P. (1995). *Am. J. Physiol.* **269**, C563–C571.

Takai, A., and Tomita, T. (1986). *J. Physiol. (London)* **381**, 65–75.

Trinkle-Mulcahy, L., Siegman, M. J., and Butler, T. M. (1994). *Am. J. Physiol.* **266**, C1673–C1683.

Veksler, V. I., Kuznetsov, A. V., Sharov, V. G., Kapelko, V. I., and Saks, V. A. (1987). *Biochim. Biophys. Acta* **892**, 191–196.

Vogel, H. J., Lilja, H., and Hellstrand, P. (1983). *Biosci. Rep.* **3**, 863–870.

Vyas, T. B., Mooers, S. U., Narayan, S. R., Witherell, J. C., Siegman, M. J., and Butler, T. M. (1992). *Am. J. Physiol.* **263**, C210–C219.

Vyas, T. B., Mooers, S. U., Narayan, S. R., Siegman, M. J., and Butler, T. M. (1994). *J. Biol. Chem.* **269**, 7316–7322.

Wakabayashi, K., Sugimoto, Y., Tanaka, H., Ueño, Y., Takezawa, Y., and Amemiya, Y. (1994). *Biophys. J.* **67**, 2422–2435.

Walker, J. S., Wingard, C. J., and Murphy, R. A. (1994). *Hypertension* **23**, 1106–1112.

Walliman, T., Wyss, M., Brdiczka, D., Nicolay, K., and Eppenberger, H. M. (1992). *Biochem. J.* **281**, 21–40.

Warshaw, D. M., Rees, D. D., and Fay, F. S. (1988). *J. Gen. Physiol.* **91**, 761–779.

Warshaw, D. M., Desrosiers, J. M., Work, S. S., and Trybus, K. M. (1991). *J. Biol. Chem.* **266**, 24339–24343.

Wendt, I. R., and Gibbs, C. L. (1987). *Am. J. Physiol.* **252**, C88–C96.

Yagi, K., Shinbo, M., Hashizume, M., Shimba, L. S., Kurimura, S., and Miura, Y. (1991). *Biochem. Biophys. Res. Commun.* **180**, 1200–1206.

Zhang, Y., and Moreland, R. S. (1994). *Am. J. Physiol.* **267**, H1032–H1039.

30

^{31}P Nuclear Magnetic Resonance Spectroscopy

PATRICK F. DILLON

Department of Physiology
Michigan State University
East Lansing, Michigan

I. INTRODUCTION

Nuclear magnetic resonance (NMR) allows the measurement of smooth muscle energetics without damage to the tissue. Although the concentrations of the metabolites involved in supplying the energy for contraction and ion pumping are relatively low in smooth muscle when compared with striated muscles, the low metabolic rate of smooth muscle allows a longer time for data collection without tissue compromise.

This review will focus on several points relevant to the uses and results of ^{31}P-NMR spectroscopy of smooth muscle. These will include the technical requirements necessary to produce high-quality spectra; the identification of the different peaks in the spectrum; the ionic information available from the shifting of certain peaks under different physiological and pharmacological conditions; the dynamic and equilibrium relationships between the phosphorus compounds involved in energy exchange; the relation between the different metabolites of smooth muscle and its ability to generate force; and how foreign compounds can be tested for toxicity and also provide additional information not available from the naturally occurring compounds. The nondestructive nature of NMR allows many of these areas to be addressed simultaneously, a considerable temporal and cost-effective aspect of NMR.

II. TECHNICAL REQUIREMENTS OF SMOOTH MUSCLE SPECTROSCOPY

Spectral quality is determined by a number of factors: magnetic field strength, relaxation rates of the nuclei of interest, tissue mass, concentration of metabolites, sample homogeneity, and time. The relative influence and the degree of experimenter control vary with each factor. For those elements that can be directly controlled, the purpose of the experiment will determine the quality of the spectra required for proper interpretation. Not all the information available from the phosphorus spectrum requires the same degree of experimental precision.

The field strength is determined by the size of the magnet. Most high-field magnets in current use for biological experiments are between 300 and 500 MHz (7.1–11.75 tesla) with 5- to 10-cm bore sizes. There are some medium-strength magnets with bore sizes up to 40 cm, but in general these are less useful for smooth muscle experiments. Signal strength varies with the square of the field strength. It is advantageous to use the largest magnet available to produce the best signals in the shortest time. Given the high cost of NMR systems, most are multiuser with both time and cost sharing.

The relaxation rates of different nuclei are a function of the chemical nature of the compound and to a lesser extent of the field strength. The exponential time constant for relaxation, known as the T_1, determines how often the spin of the nucleus can be perturbed to generate the spectral signal. In most biological applications, T_1 varies between 1 and 3 seconds. Collection of a fully relaxed spectrum requires about five T_1's. Signal-to-noise values can be improved by collecting more often, but the relative size of the peaks in such partially saturated spectra will not reflect the relative concentrations of the different peaks. In these cases, relative concentrations can be determined if the T_1 values of the different peaks are known.

Signal strength will increase linearly with the mass of the tissue. Smooth muscles vary considerably in size, from very small blood vessels to quite large bladder, intestinal, and uterine tissues. Using tissue that is as uniform as possible will enhance the reliability of the experiment. In the case of very small tissues, it is usually necessary to combine several samples to produce an acceptable signal within a reasonable time.

Signal strength will also vary linearly with the concentration of the metabolite of interest. In smooth muscle, ATP and phosphocreatine (PCr) have concentrations that vary between 0.5 and 2 μmoles/g tissue wet weight (approximately 0.5–2 mM), with visceral smooth muscles having higher concentrations than vascular smooth muscles. For tissue masses near 1 g, a reasonable spectrum of these metabolites can be obtained in 15 minutes. Similar times are needed for some monoesters, such as phosphoethanolamine, which also have millimolar concentrations. Detection of lower-concentration metabolites, such as ADP, requires longer collection time. Under limited oxygen and energy supply, ADP peaks of about 100 μM can be seen, but collection durations of 30 minutes or more are required. Since signal-to-noise strength only increases with the square root of the collection time, there are temporal limits on data collection. It is unreasonable to use NMR to detect phosphorus compounds below about 50 μM in 1 g of tissue, since the collection time required would be too great.

Whereas an experimenter may have little or no control over the available magnet, the T_1's, and the tissue concentration, there can be considerable control over the shimming of the tissue sample to maximize the sample homogeneity within the magnetic field. The more homogeneous the sample, the sharper will be the lines of the spectrum. There are limits on the homogeneity of the experiment due to tissue variations and the physics of the sample coil, but for experiments that require very narrow line widths, the time taken to adjust or shim the receiver coil for optimal signal reception is worthwhile. Differentiating intracellular and extracellular pH (Section III.A) or detecting the β peak of ADP (Section V.A) requires very narrow lines. The stability of smooth muscle preparations provides a particular advantage over other tissues in this regard. The extra time needed for precise shimming of the sample will not compromise the integrity of the preparation in most cases.

Information on the concentration of ions bound to NMR visible peaks can be obtained from the position or chemical shift of the peak. In particular, the tissue pH can be obtained from the chemical shift of the P_i peak, and the free Mg^{2+} concentration can be measured from the position of the β-ATP resonance (Section IV). Since an NMR spectrum is a collection of digital points, the distance between the points in units of frequency will determine the accuracy of the chemical shift measurement. In general, the ion concentration difference determined by the distance between two adjacent points should be less than a physiologically significant change in the ion concentration.

III. SPECTRAL PEAK IDENTIFICATION

The peak positions in an NMR spectrum are determined by the total magnetic field to which the nuclei are exposed. The magnetic field of the magnet is the primary factor distinguishing the resonant frequency of the nuclei of different elements, that is, hydrogen, phosphorus, and so on. There are wide frequency differences between different elements. For any given element, such as phosphorus, different chemical compounds have resonances at different frequencies because of the distinctive electron shell of each atom. Electrons are charged and moving, and will therefore generate a magnetic field. If the electron cloud is altered chemically, the frequency will change and the peak will appear in a different place. This is the basis of the chemical shift discussed in Section II.

A. Frequency Dependence

Figure 1 shows a phosphorus NMR spectrum of porcine carotid artery (Fisher and Dillon, 1988). The naturally occurring peaks are numbered 1–10. The two peaks on the left side of the spectrum were added to the experimental preparation. There is no absolute frequency standard within an NMR spectrum. Peak positions are defined relative to a particular peak that has a defined frequency. This peak position should be impervious to tissue metabolic changes. Phosphocreatine (peak 6) is most commonly used as an internal reference peak in those tissues that contain it. PCr has a pK below 6.0, and will therefore not shift its position when there are small changes in tissue pH. This spectrum also demonstrates an alternative method of presenting a frequency standard. A capillary tube containing a compound with a resonant peak away from the natural peaks generates a distinct resonance. In this case, acidified phenylphosphonate (PPA) is used as a frequency and signal-to-noise reference. PPA can also be added to a perfusion solution as an extracellular pH marker (Fisher and Dillon, 1987a; see Section VII.A).

The number of scans (NS) generating this spectrum is 480. RD indicates that the relaxation delay between each scan was 15 seconds. With this delay, there was

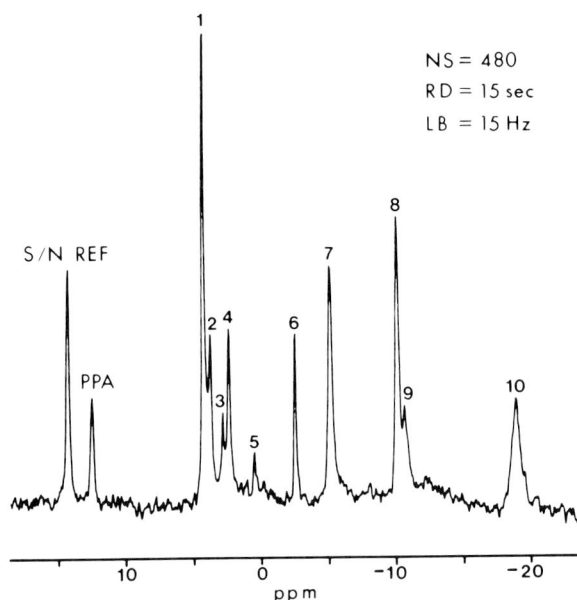

NS = 480
RD = 15 sec
LB = 15 Hz

FIGURE 1 Phosphorus NMR spectrum of porcine carotid artery. S/N REF. refers to the resonance of a sealed capillary tube of phenylphopshonate (PPA), pH 4.0. The individual naturally occurring peaks are identified by number in the text. From Fisher and Dillon (1988, Fig. 1, p. 123).

sufficient time for virtually complete recovery of all T_1's between scans. Therefore, this spectrum is fully relaxed and the relative peak heights (and areas) correspond to the concentrations of the different compounds in the tissue. The line broadening (LB) of 15 Hz is used to smooth the spectrum and improve the signal-to-noise. If the line broadening is greater than the natural width of the peaks, the spectrum will be overprocessed and look too smooth, losing small peaks.

The natural peaks in the spectrum shown in Fig. 1 are phosphoethanolamine (1), phosphocholine (2), extracellular phosphate (3), intracellular phosphate (4), glycerolphosphorylcholine (5), PCr (6), γ-ATP (7), α-ATP (8), NAD and NADH (9), and β-ATP (10). Peaks 6–10 were easily identified in the first spectra of smooth muscle, because these peaks are commonly present in other tissues. Distinguishing the intracellular and extracellular phosphate peaks requires a well-shimmed sample. There is about a 0.3 pH unit difference between the intracellular pH (7.1) and the extracellular pH (7.4). It has become commonplace to reduce the phosphate content of perfusion solutions to minimize the influence of the extracellular phosphate on the mean position of a P_i peak, and thereby use it as an intracellular pH marker.

Peaks 1, 2, and 5 were not as readily identifiable in initial smooth muscle spectra. Although they ap-

peared to have well-defined frequency positions, there are a number of compounds that could have similar positions at a pH of 7.1. The monoester region (peaks 1–2) could contain other compounds such as phosphorylated sugars. It was important to distinguish exactly which compounds were present in these relatively high concentrations in smooth muscle.

B. Extract Spectra

The presence of metal compounds such as calcium and magnesium, coupled with structural inhomogeneities within tissues, combines to broaden peaks within a biological NMR sample. When tissues are ground into a uniform mixture, the metals can be removed by adding chelating agents that are then separated by centrifugation. Concentration of the extract can then produce a spectrum such as that from rabbit urinary bladder in Fig. 2 (Kushmerick et al., 1986). The lines are much sharper than in Fig. 1, and the splitting of peaks by adjacent nuclei can be seen in the doublets of α- and γ-ATP and the triplet of β-ATP. In the monoester region at 5 ppm, many small peaks now appear that would never be seen in a tissue spectrum. The large peak in this region is phosphoethanolamine. Tritration of the extract solution causes the position of the peak to shift as its electron cloud is altered, and a pK for this peak is determined. Final confirmation requires high-performance liquid chromatography (HPLC), enzymatic analysis, proton NMR spectroscopy, or some other method of confirming the chemical nature of the compound.

Close examination of the β-ATP region revealed another small resonance within this area. This peak was subsequently identified as UTP (Kopp et al., 1990; Kushmerick et al., 1986). Chemical analysis revealed that in both bladder and uterine tissues (which are predominately smooth muscle), 20–30% of the nucleotide triphosphates are not ATP, but are about evenly split between GTP and UTP. Tissue NMR would not be able to separate these different compounds. Changes in the "ATP" concentration then do not represent solely ATP changes. It is assumed that all the triphosphates are in equilibrium with one another, and that all change together. No work is presently available testing this assumption. The chemical concentrations of perfused bladder and uteri from Kushmerick et al. (1986) are included in Table I. There is good agreement between these measurements and the NMR spectra, indicating that most metabolites are not chemically bound in the tissues (making them NMR invisible). An exception to this is the ADP measurement, which is discussed in Section V. A. The major peak in the monoester region is phospho-

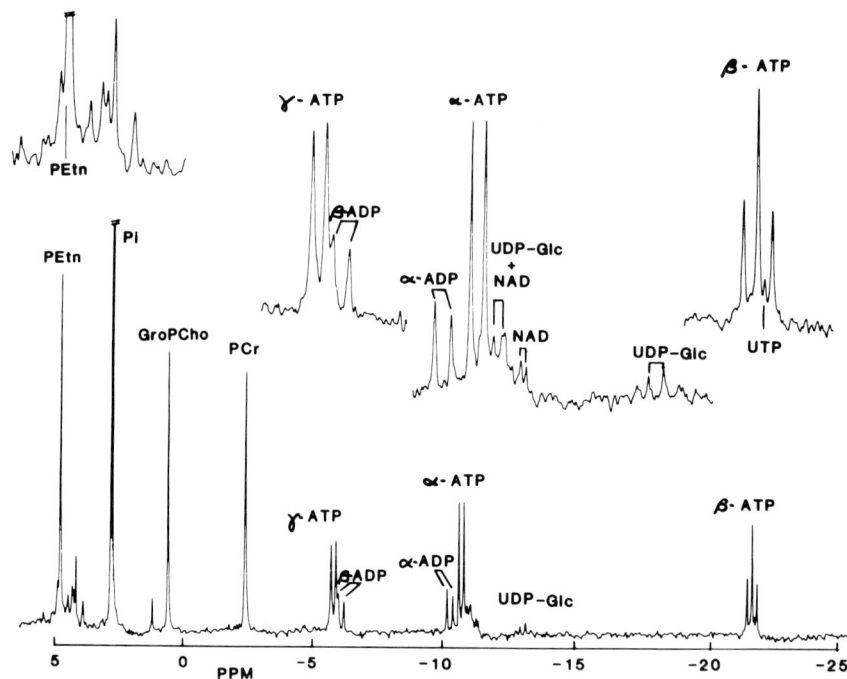

FIGURE 2   Phosphorus NMR spectrum of a tissue extract of rabbit urinary bladder. Spectra at the top are enlargements of the main spectrum. From Kushmerick *et al.* (1986, Fig. 2, p. 14424).

ethanolamine, with relatively minor contributions from sugar phosphates. Changes in this resonance associated with pregnancy (Dawson and Wray, 1985) are discussed in the following section.

TABLE I   Metabolite Contents of Perfused Bladders and Uteri[a]

Compound	Bladder[b]	Uterus[b]
ATP	0.97 ± 0.17	1.19 ± 0.18
ADP	0.23 ± 0.05	0.28 ± 0.06
PCr	1.64 ± 0.08	1.29 ± 0.15
Free creatine	1.33 ± 0.06	1.27 ± 0.22
PCr/total creatine	0.55 ± 0.02	0.51 ± 0.02
GTP	0.12 ± 0.03	0.18 ± 0.04
GDP	0.04 ± 0.02	0.04 ± 0.01
UTP	0.15 ± 0.03	0.25 ± 0.05
UDP-glucose	0.26 ± 0.06	0.22 ± 0.05
NAD	0.29 ± 0.05	0.33 ± 0.09
Glucose-6-P + fructose-6-P + Glucose-1-P	0.07 ± 0.02	0.05 ± 0.02
P_i	2.02 ± 0.48	2.82 ± 0.43
Phosphoethanolamine	1.09 ± 0.07	Not assayed

[a]From Kushmerick *et al.* (1986, Table V, p. 14423).
[b]μmoles/g (w/w) ± S.E.M., $N = 4$.

C. Tissue Variations

Although the general features of most smooth muscle tissue spectra are consistent, there are differences in the spectra from different tissues that reflect the different chemical content of the tissues. The PCr peak is usually smaller than the ATP peaks in vascular smooth muscle (Adams and Dillon, 1989; Clark and Dillon, 1989; Fisher and Dillon, 1988; Hardin *et al.*, 1992a) but greater than the ATP peak in visceral smooth muscle tissues (Kushmerick *et al.*, 1986; Hellstrand and Vogel, 1985; Dawson and Wray, 1985; Fisher and Dillon, 1987a, 1989; Vermue and Nicolay, 1983; Nakayama and Tomita, 1990, 1991; Nakayama *et al.*, 1994). There are some reports of lower PCr peaks in uterine tissues (Wray, 1990; Harrison *et al.*, 1994). There is generally good agreement between NMR peak areas and chemical analyses of tissues, although some differences in PCr and P_i have been reported (Dawson and Wray, 1985). Because PCr concentrations will be susceptible to changes in the metabolic state of the tissue, conclusions drawn on the health of a preparation based on the size of the PCr peak must take into account the normal variations that occur in tissues.

Dawson and Wray (1985) reported metabolite differences in pregnant and nonpregnant uteri. They reported increases in PCr, P_i, and a large phosphate monoester peak during pregnancy, but no change in pH. They concluded that the changes in the peaks were associated with hormonal changes in the animals, not mechanical work differences. Changes in phosphoethanolamine have been reported in proliferating and secretory tissues (Awapara *et al.*, 1950; Tallan *et al.*, 1954). The changes that occur in the uterus during pregnancy may reflect similar changes in phosphoethanolamine. These pregnancy experiments demonstrate that long-term changes in the biochemistry of smooth muscle can also be followed using NMR. They also reiterate that conclusions on the metabolic state of the tissue based on peak areas alone must be drawn with care, especially if the pH remains above 7.0.

Changes in tissue composition assessed using NMR often take advantage of the nondestructive nature of NMR. It is also possible to use extracts of smooth muscle to examine the changes that occur in many metabolites simultaneously. Studies on carbohydrate metabolism (Barron *et. al.*, 1989, 1991) used phosphorus NMR spectra of tissue extracts. The tissues were collected under different conditions to examine the structural partitioning of carbohydrate metabolism and to show that different fatty acids alter glycogen synthesis. The authors concluded from the extract spectra that ATP production had specific carbohydrate preferences, in agreement with the work of Lynch and Paul (1983). Another unique work using phosphorus NMR on smooth muscle demonstrated the mono- and bisphosphorylation of chicken gizzard myosin serine 19 (Levine *et al.*, 1988), the site believed to regulate myosin ATPase activity. ^{31}P-NMR used as a tool of chemical analysis should continue to make significant contributions to areas of smooth muscle research not directly related to tissue metabolism.

IV. ION-DEPENDENT CHEMICAL SHIFT

The peak position or chemical shift of an NMR peak within the phosphorus spectrum is determined by the electron cloud around the nucleus. When the electron cloud is altered, the peak will shift position. Binding of an ion to an NMR visible compound changes the electron cloud and produces a position proportional to the concentration of the ion. Different ion concentrations are used to generate a titration curve from which a particular position can be converted into an ion concentration. To be an effective method in tissue spectroscopy, the physiological concentration of the ion must be within a factor of 10 of the pK of the binding of the ion and must be the predominant ion binding to the compound. This means that P_i, with a pK of 6.77 (Kushmerick *et al.*, 1986), is useful in monitoring tissue

pH. Similarly, Mg^{2+} binding to ATP and ADP will also produce chemical shifts useful in measuring the free Mg^{2+} concentration.

A. pH Dependence of Inorganic Phosphate

Changes in tissue pH are generally used as an index of the viability of the preparation. Phosphorus NMR provides a noninvasive method for measuring tissue pH, without adding dyes or requiring electrode impalement. The pH measurement generated by NMR cannot account for intracellular pH gradients or different pH values in different cells, but only the average value for the entire preparation. Measurement of intracellular pH using the chemical shift of the P_i peak requires low extracellular phosphate amounts relative to the amount of phosphate in the cells, or sufficient separation of the two phosphate peaks (at different pH's) to allow computer separation of the peaks.

During the development of the different smooth muscle NMR preparations, every paper noted that a shift of the P_i peak accompanied acidosis of the tissue under a variety of conditions (Vermue and Nicolay, 1983; Hellstrand and Vogel, 1985; Kushmerick *et al.*, 1986; Spurway and Wray, 1987). Vermue and Nicolay (1983) found a large decline in pH from 7.1 to 6.6 during potassium-induced contractions of guinea pig taenia cecum, but no changes in pH were found during contractions of porcine carotid arteries (Adams and Dillon, 1989; Fisher and Dillon, 1988) or strips of rabbit taenia coli and bladder (Hellstrand and Vogel, 1985). A small change in pH (7.1 to 6.9) was generated during maximal carbachol stimulations of vascularly perfused rabbit bladder (Kushmerick *et al.*, 1986). Acidification of the tissue indicates an inability of the muscle to remove protons generated by tissue metabolism. This is due to excessive H^+ production, low oxygen or substrate availability, or a combination of these. The pH changes that occur during ischemia are on a time scale approximating the changes in PCr, and are much faster than the changes in ATP (Vermue and Nicolay, 1983; Hellstrand and Vogel, 1985).

Changes in tissue pH can also occur owing to changes in extracellular pH. Changes of 1 pH unit (7.4 to 6.4) extracellularly produced a decline in rat uterine intracellular pH of 0.29 units (Wray, 1988). There were smaller changes in intracellular pH corresponding to smaller alterations in extracellular pH. These changes were not accompanied by changes in ATP or phosphocreatine. This evidence is in agreement with $Na^+–H^+$ exchange being a mechanism for removal of hydrogen ions from smooth muscle. Application of isosmotic 30 mM NH_4Cl to rat vascular smooth muscle produced an increase in tissue pH from 7.19 to over 7.6 in the presence of small amounts of cyanide and fluoride (Spurway and Wray, 1987). Outside agents are capable of altering tissue pH, but physiological changes in pH seem to require a degree of metabolic impairment, usually coupled with mechanical stimulation, to produce large changes in pH. It appears that smooth muscle can recover easily from pH values as low as 6.6.

B. Free Mg^{2+} Dependence of ATP and ADP

One of the major advances produced by smooth muscle NMR spectroscopy is the determination of the free Mg^{2+} concentration. Smooth muscle has a total magnesium content of about 6–7 μmoles/g (Kushmerick et al., 1986), corresponding to about 6 mM Mg^{2+} if it were all free in solution. It was known that magnesium could complex with ATP and produce a change in the ATP spectrum (Cohn and Hughes, 1962). NMR spectra from smooth muscle showed a shifted β-ATP peak relative to that of striated muscle, indicating less than complete magnesium saturation of ATP. When control solutions with different free Mg^{2+} concentrations were used to calibrate the ATP shift, the free concentration of magnesium in intact smooth muscle was shown to be in the 0.5 mM range (Hellstrand and Vogel, 1985; Kushmerick et al., 1986). This range was confirmed when spectra of smooth muscle were found to produce free ADP peaks during metabolic inhibition (Kushmerick et al., 1986). Figure 3 shows the titration curves for both ATP and ADP. ATP binds Mg^{2+} an order of magnitude tighter than ADP because of its longer chain of phosphates. In spectra in which both ATP and ADP could be seen, their respective chemical shifts indicated the same free Mg^{2+} (Fig. 3), making errors in this estimate very unlikely.

Initial attempts to alter the intracellular free Mg^{2+} concentration using high concentrations of extracellular Mg^{2+} were unsuccessful (Hellstrand and Vogel, 1985; Dillon, 1985). Figure 4 shows that smooth muscle is able to maintain a low free Mg^{2+} concentration even when faced with a large increase in intracellular magnesium. Hydrolysis of ATP during metabolic impairment of rabbit bladder smooth muscle produced a large fall in ATP. As noted earlier, ADP does not bind magnesium as well as ATP, and ATP hydrolysis liberates magnesium. Figure 4 shows that the amount of free Mg^{2+} appearing during ATP hydrolysis is much lower than the amount being liberated. The additional magnesium must be either complexed or pumped from the cell. An intracellular concentration of 0.5 mM is lower than extracellular concentrations of 1–2 mM. With smooth muscle membrane potentials of about -60 mV, both the chemical and electrical gradients for magnesium are inward. This means that to maintain

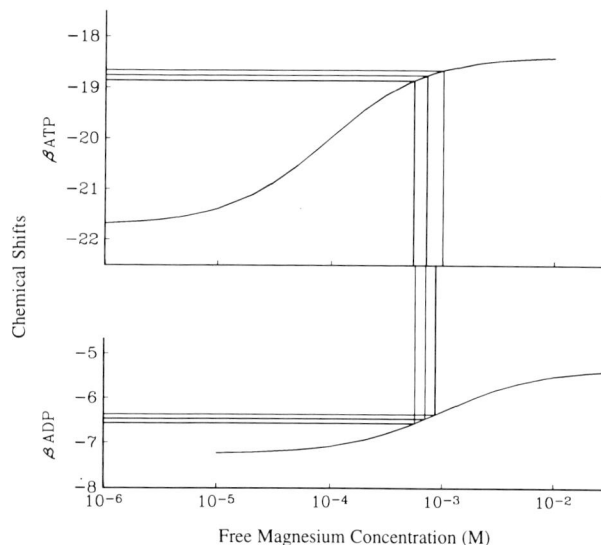

FIGURE 3 Titration of the chemical shift of β-ATP and β-ADP by magnesium. The horizontal lines indicate the mean ± SD of the chemical shift of the β-ATP and β-ADP peaks in spectra of rabbit bladder. Both nucleotides indicate the same free Mg^{2+} (vertical lines) in this tissue. The chemical shifts are expressed in terms of ppm.

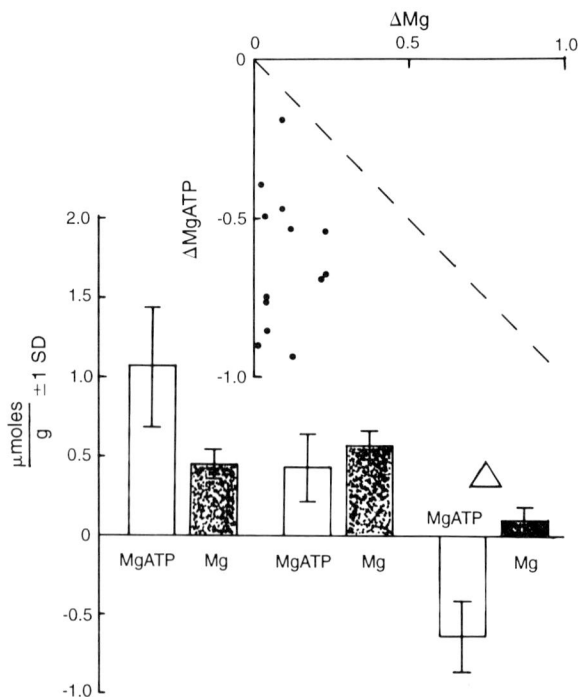

FIGURE 4 Changes in free Mg^{2+} (ΔMg) during a reduction in MgATP ($\Delta MgATP$) during metabolic impairment of bladder smooth muscle. The dashed line indicates the line of unity expected if all the liberated Mg^{2+} appeared as free Mg^{2+}.

the free Mg^{2+} concentration in the submillimolar range, smooth muscle must use energy, either in the form of a direct Mg^{2+} pump or by linked counter-transport with sodium, or both.

In a series of papers examining free Mg^{2+} regulation in smooth muscle, Nakayama and Tomita (1990, 1991) and Nakayama et al. (1994) demonstrated that the free Mg^{2+} depended on both calcium and sodium. Although removal of extracellular magnesium did not decrease intracellular magnesium, there was a large decline in intracellular magnesium if both extracellular magnesium and calcium were removed. Further studies showed that very prolonged exposure to 40 mM extracellular magnesium could produce a modest increase in intracellular free Mg^{2+}. Replacement of sodium externally with N-methyl-D-glucamine produced a rise in intracellular free Mg^{2+} to 2.1 mM, near the limit of NMR magnesium measurement. Further experiments indicated that Na^+–Mg^{2+} exchange is the primary method of maintaining low physiological free Mg^{2+}.

Lynch and Paul (1983) demonstrated the preference of the Na^+–K^+-ATPase for glycolytically generated ATP. With the linkage of magnesium regulation to sodium, it is not surprising that the free Mg^{2+} concentration also shows substrate dependence. As shown in Table II, when oxygenated porcine carotid arteries are exposed to different substrates, there is a shift in the free Mg^{2+}. The lower values seen with pyruvate indicate a more efficient use of this fuel in supplying the energy that must be used to maintain the large magnesium electrochemical gradient. Interestingly, in the same tissues the pH falls in the presence of pyruvate relative to that seen with glucose (Table II). These interactions between metabolism and ion concentration will remain a fertile area for smooth muscle NMR work.

There is another intriguing finding in the regulation of smooth muscle free Mg^{2+}. Kopp et al. (1990) found a large increase in intracellular free magnesium when segments of porcine carotid artery were inflated. Values for the free Mg^{2+} concentration in flaccid arteries were similar to those seen in other flaccid arteries (Fisher and Dillon, 1988). Similar increases in free magnesium were not found in other inflated arteries (Adams and Dillon, 1989), nor in inflated rabbit uteri (Kushmerick et al., 1986) nor in stretched rabbit bladder (Kushmerick et al., 1986; Fisher and Dillon, 1987b, 1989). Since free magnesium plays an important role in the regulation of contraction of smooth muscle (Chacko and Rosenfeld, 1982), variation in its concentration as a functional of length and/or activation could be important in the regulation of contraction and deserves further study.

V. CREATINE KINASE EQUILIBRIUM AND KINETICS

Creatine kinase catalyzes the conversion of PCr and ADP into ATP and creatine. When muscles use high amounts of ATP, this reaction buffers the ATP concentration. This reaction is also used to estimate the intracellular free ADP concentration. In addition, it is one of the few reactions whose pseudo-first-order rate constants can be measured using phosphorus NMR. Measurement of the creatine kinase rate constants has been done in many tissues, including smooth muscle.

A. Quantification of Free ADP and Creatine Kinase Equilibrium Constant

An ADP peak in smooth muscle was used initially as a measure (along with ATP) of the free magnesium in smooth muscle (Kushmerick et al., 1986). It was noted that the amount of ADP appearing, or rather not appearing, in NMR spectra did not agree with the chemical analysis of smooth muscle (see Table I) (Kushmerick et al., 1986; Hellstrand and Vogel, 1985). This difference was attributed to the large amount of ADP bound to actin filaments (Seraydarian et al., 1962). ADP released from actin during tissue extraction would have been bound, and therefore NMR invisible, under in vivo conditions. Bound ADP does not participate in the creatine kinase reaction. The appearance of the small β-ADP resonance next to the γ-ATP peak allowed the opportunity to quantitate the free ADP value in smooth muscle. This in turn would allow the calculation of the tissue creatine kinase equilibrium constant.

The appearance of a free ADP peak requires a low free magnesium concentration. If the concentration of free Mg^{2+} were above 1 mM, as it is in striated muscles, the β-ADP resonance would be too close to the γ-ATP resonance to successfully separate them. When a preparation has sufficiently narrow line

TABLE II Substrate Dependence of Porcine Carotid Artery pH and Free Mg^{2+}

Substrate	pH[a]	Free Mg^{2+} [a]
15 mM glucose	7.10 ± 0.03	0.48 ± 0.02
10 mM pyruvate	7.02 ± 0.02	0.40 ± 0.02[b]
30 mM pyruvate	6.92 ± 0.04[b]	0.36 ± 0.01[b]

[a]Value ± S.E.M., N = 5.
[b]Significantly different from 15 mM glucose, $P < 0.05$.

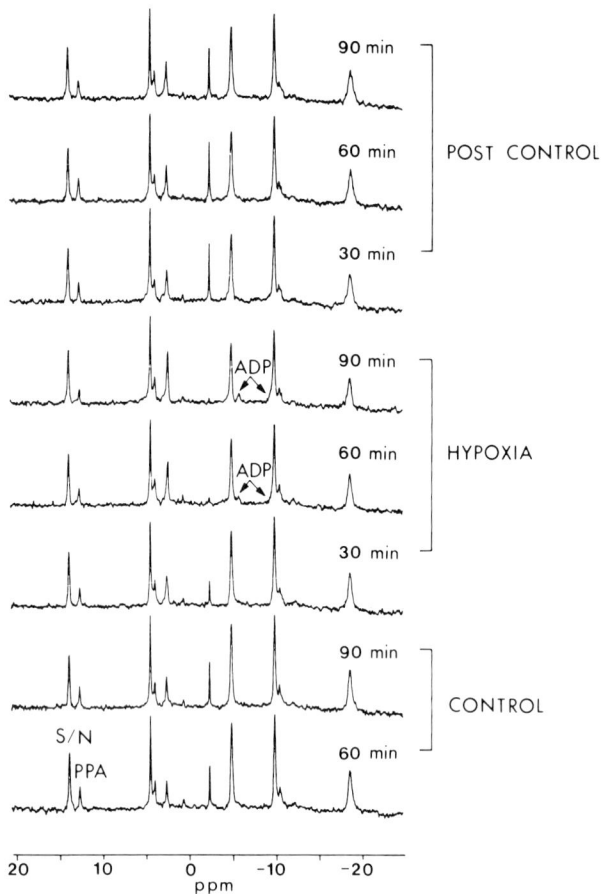

FIGURE 5 The appearance of ADP resonances in spectra of porcine carotid arteries. From Fisher and Dillon (1988, Fig. 2, p. 124).

mined from chemical analysis, and decreases in PCr are presumed to indicate increases in creatine. Calculation of the equilibrium constant then involves only accurate measurement of ADP. The phosphorus spectrum from hypoxic smooth muscle produces this information. A K_{eqCK} value of $7.6 \times 10^8 \ M^{-1}$ was determined for porcine carotid smooth muscle (Fisher and Dillon, 1988), within the range of K_{eqCK} values determined *in vitro* (Lawson and Veech, 1979).

The tissue K_{eqCK} was used to compare calculated ADP concentrations [by rearranging Eq. (1)] to those ascertained from NMR spectra, showing good agreement. This value of K_{eqCK} has subsequently been used to estimate ADP concentrations in smooth muscle tissues in which the concentration of ADP is too small to be seen (Hardin *et al.*, 1992b). There is close agreement between the two studies of smooth muscle ADP showing that the free ADP concentration is about 25 μM under control conditions (Fisher and Dillon, 1988; Hardin *et al.*, 1992b), much lower than the chemically measured ADP concentration (Table I). It has been proposed from *in vitro* studies that MgADP may play a role in controlling smooth muscle contractions (Nishiye *et al.*, 1993) when its concentration is below 10 μM. Calculation of the MgADP concentration in smooth muscle places this value above 10 μM under all conditions, making MgADP regulation of contraction unlikely. Theoretical calculations of ADP and myosin ATPase activity confirm this conclusion (Clark *et al.*, 1995b).

B. Creatine Kinase Kinetics

Kinetic measurements using saturation transfer require that one specific peak be continuously irradiated. When there follows a general irradiation of all the peaks, every peak will generate its characteristic spectral signal except the continuously irradiated one. The T_1 requirement for recovery between scans (Section II) is not met during continuous saturation, and no signal is generated by the irradiated peak. The chemical compound is still present, but no peak appears on the NMR spectrum. When that compound exchanges its saturated phosphate with another compound, the saturation of the phosphate goes along (as long as the rate of exchange is faster than the T_1 recovery rate). The receiving compound now has a combination of saturated and unsaturated phosphate, and its average peak size will be reduced. If there is very rapid exchange between two compounds during the continuous saturation period, most of the receiving compound will also have saturated phosphate, and its peak size will be very small. If the exchange is slow, there will be little diminution of the receiving peak.

widths, these two resonances can be distinguished (Fig. 5) (Fisher and Dillon, 1988). The α-ADP resonance cannot be separated from the α-ATP resonance. Magnesium has little effect on the chemical shift of either α resonance. Normally, the concentration of ADP is low compared with ATP, PCr, and creatine. Hypoxia decreases the concentration of PCr and increases the concentration of creatine. This results in an increase in the concentration of ADP, an increase that moves it into the NMR visible range. Fisher and Dillon (1988) used this visible ADP peak to quantitate the ADP concentration, and then used that value to estimate the tissue creatine kinase equilibrium constant, K_{eqCK}.

The calculation of the tissue K_{eqCK} is defined by

$$K_{eqCK} = [ATP][creatine]/[ADP][PCr][H^+] \qquad (1)$$

where the brackets denote metabolite concentrations. The relative ATP and PCr concentrations are readily available from the NMR spectrum. The chemical shift of the P_i peak is used to define the pH, which is readily converted to a H^+ concentration. Creatine is deter-

The decrease in size of the receiving peak is, therefore, a function of the reaction rate (Brown, 1980).

In examining the exchange of phosphate between creatine and ADP catalyzed by creatine kinase, Yoshizaki et al. (1987) found the flux rates for this reaction in bullfrog stomach smooth muscle to be approximately 100 times larger than the ATP turnover rate. They concluded that the enzyme appeared to be at equilibrium. A discrepancy in the forward (PCr to ADP phosphate transfer) and backward (ATP to creatine phosphate transfer) fluxes could have been accounted for on the basis of either compartmentation of nucleotides or the additional reactions that ATP takes part in. They were unable to differentiate between the two models. Since ATP can participate in multiple reactions, a single saturation is not sufficient to account for both the creatine kinase reaction, with the ATP transferring phosphate to creatine, and ATPase activity, with ATP splitting into ADP and P_i. Multisite saturation of smooth muscle is feasible (Clark et al., 1991), showing that the PCr, ATP, and P_i saturations can be used to accurately calculate the creatine kinase forward and backward fluxes independently of the ATPase reaction. Measurements of porcine carotid artery smooth muscle creatine kinase kinetics using multisite saturation show no statistical discrepancy between the forward and backward fluxes, indicating that the creatine kinase reaction is at equilibrium (Clark et al., 1995a). Both the stomach and artery studies were done on unstimulated muscles, having about 50% of the ATPase activity of stimulated smooth muscle (Paul and Peterson, 1975). No creatine kinase kinetic studies of stimulated smooth muscles have been reported. The finding that ATP can be directly transferred between pyruvate kinase and creatine kinase (Dillon and Clark, 1990; Dillon et al., 1995) provides additional complications in the analysis of creatine kinase kinetics. Even in the absence of a PEP visible resonance, saturation of the PEP peak position could alter the PCr peak, and provide NMR evidence of the direct coupling of enzymes in tissues.

VI. METABOLIC CONTROL
OF FORCE GENERATION

One of the great advantages of phosphorus NMR is the simultaneous measurement of many metabolites. These include ATP, PCr, P_i, ADP, and pH. In addition, all of the information is available to calculate the free energy of ATP in smooth muscle. Since energy is required to drive muscle contraction, an obvious question to ask is which of these metabolites plays the most important role in controlling muscle contractions. The primary control of smooth muscle contraction is, of course, not metabolic. The calcium activation of myosin light chain kinase (Gorecka et al., 1976) and the attachment of phosphorylated and dephosphorylated (latch) cross-bridges (Dillon et al., 1981) form the primary regulatory system of smooth muscle activation. After several minutes of activation, however, these systems reach a steady state (Dillon et al., 1981) and the effects of different metabolic agents can be assessed. Since typical phosphorus NMR data collection periods require tens of minutes for smooth muscle preparations, the requirement for steady-state activation is easily met.

A. Metabolic Changes during Stimulation

There are conflicting reports as to whether there are NMR observable changes in tissue metabolites during contraction. There are reports of decreases in PCr (Vermue and Nicolay, 1983; Hellstrand and Vogel, 1985; Kushmerick et al., 1986) in some smooth muscles during contractions. In some experiments, however, changes in PCr during contractions are not seen (Adams and Dillon, 1989; Hardin et al., 1992b). During stimulation alone, changes in ATP are not seen. The changes in pH that can occur were documented in Section IV.A. There are many reports of changes in metabolites in the presence of ischemia or withdrawal of substrates (Hellstrand and Vogel, 1985; Kushmerick et al., 1986; Hardin et al., 1992a,b; Harrison et al., 1994; Spurway and Wray, 1987). PCr and pH are the most sensitive indicators of tissue activation, but it appears that impairment of oxygen metabolism or substrate limitation is required to radically alter the metabolite concentrations.

Figure 6 shows, however, that relatively subtle differences can be seen when different substrates are used to supply smooth muscle. In these control spectra of K^+-stimulated porcine carotids, there is a change in the ratio of PCr/P_i. PCr (at -2.52 ppm) is virtually constant, with the changes occurring in P_i. (Intracellular P_i is the right side of the doublet at 2.6 ppm.) In these experiments from the same set showing the substrate differences in pH and free Mg^{2+} concentrations (Section IV.B), the changes in phosphate indicate the close relationship between substrate supply and ion concentrations.

B. Free Energy Calculation

The calculation of the free energy of ATP hydrolysis (in kcal) is

$$\Delta G = -RT(\ln K_{eqATP}) \tag{2}$$

NS = 1800
RD = 1 sec

PCr/Pi = 2.4   K⁺ STIMULATED
                    30mM PYRUVATE

PCr/Pi = 2.1   K⁺ STIMULATED
                    10mM PYRUVATE

PCr/Pi = 1.1   K⁺ STIMULATED
                    15mM GLUCOSE

20.0   10.0   0.0   −10.0   −20.0
                 ppm

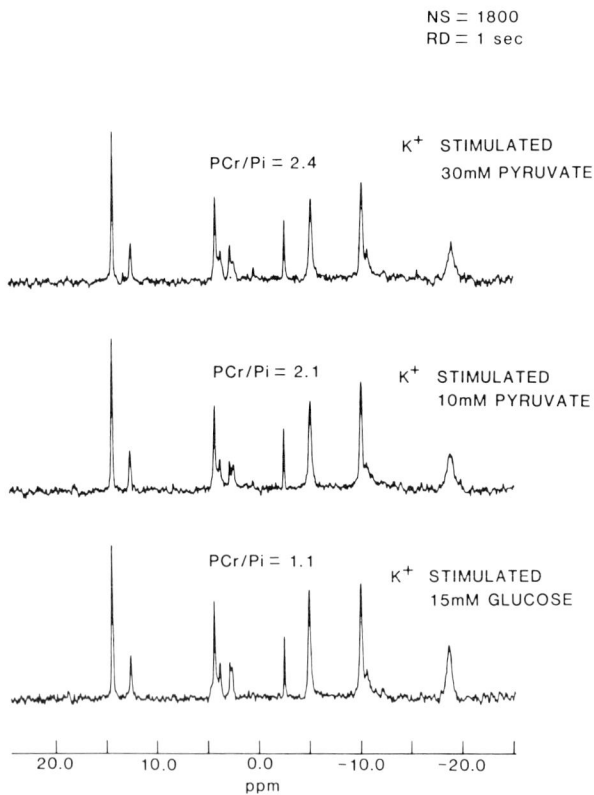

FIGURE 6   Porcine carotid artery spectra of oxygenated, K⁺-stimulated porcine carotid arteries showing substrate-induced differences.

and requires that

$$K_{\mathrm{eqATP}} = [\mathrm{P_i}][\mathrm{ADP}]/[\mathrm{ATP}] \qquad (3)$$

be known. The estimations of intracellular P_i (Section III.A) and ADP (Section V.A) have been discussed. Calculation of the free energy will give a global estimate of the energetic state of the tissue. Although in the broadest sense the free energy will influence the contraction of smooth muscle, other functions will also be affected by the free energy. It is reasonable to suspect that particular metabolites may be more specifically tied to muscle contraction (or ion pumping, etc.) and thus may play a more instrumental role in the control of a particular function than free energy itself. A decline in free energy, however, will lead to a reduction in contraction strength along with a general decline in all cellular functions.

C. Hierarchy of Metabolic Effectors of Force Generation

Smooth muscles are excellent at maintaining contractions for long periods of time. The stability of the contractions may be tied to special mechanisms, such

as latch (Dillon *et al.*, 1981), that reduce the need for energy use. It is possible to alter the contractions of smooth muscle using individual agents that alter energetics, proteins, ions, or membranes. This volume discusses many such examples. Because of the stability of the metabolites during activation discussed in Section VI.A, using NMR to measure the effect on force generation requires metabolic impairment.

When porcine carotid arteries were K⁺-stimulated and made hypoxic, they went through a range of changes in which ATP, PCr, P_i, pH, ADP, and free energy changed (Fisher and Dillon, 1987b). Force measurements under the same conditions also produced a series of changes, allowing comparison of the simultaneously measured metabolites. In this case correlation coefficients between the different metabolites and force indicated the following order of importance: P_i > free energy = pH > PCr > ATP > ADP. Global indicators, free energy and pH, were in the middle as expected. There have been associations of P_i with actomyosin ATPase regulation in skeletal muscle (Nosek *et al.*, 1987; Webb *et al.*, 1986), and it appears P_i may play a similar role in smooth muscle. Although ADP is also a product of myosin ATPase, it does not appear to play a relatively strong role in the metabolic control of smooth muscle contraction.

The use of NMR for ion measurement can be used to test circumstances in which there are different amounts of force generation, as in studies on aortic smooth muscle from control and hypertensive rats on magnesium, calcium, and sodium concentrations (Jelicks and Gupta, 1990). NMR spectra from phosphorus (for Mg^{2+}), sodium (for Ca^{2+} and Na^+), and deuterium (a control for sodium spectra) were used. The authors were able to conclude that in primary hypertension there were alterations in cell membrane ion transport. In studies demonstrating the glucose dependence of norepinephrine-induced contractions of porcine carotid arteries, phosphorus NMR was used to show that there was no change in the metabolic state of the tissue, in either metabolites, pH, or free Mg^{2+} (Adams and Dillon, 1989). Thus, in addition to those conditions under which smooth muscle alters its function because of a change in its metabolic state, NMR can be used to eliminate such a change as the cause of the altered function.

VII. APPLICATION OF FOREIGN SUBSTANCES TO SMOOTH MUSCLE

It is often useful to add NMR-sensitive compounds to a tissue preparation. Such materials may provide information not otherwise available, such as the size of an extracellular space compartment or its pH. Smooth muscle tissues can also be used to test the toxicity of materials. Changes in ion concentrations or metabolite levels can be useful in ascertaining how much of a compound can be used without compromising the tissue.

A. Testing the Toxicity of Foreign Substances

Measurements of force development are much easier and more cost-effective than NMR experiments. As such, though it is common to use smooth muscle strips to test different agents, the need to use NMR for toxicity testing is more limited. Agents that are hypothesized to specifically affect an NMR visible species are more likely to be tested, but only in conjunction with force measurements.

An example of such testing is the examination of the mechanical and metabolic toxicity of 3-(trimethylsilyl)propanesulfonic acid (TMSPS) (Clark and Dillon, 1989). TMSPS is a water-soluble NMR frequency marker used for both proton and carbon spectroscopy. Mechanical experiments using strips of porcine carotid artery showed a statistically significant fall in force to about 94% of control in the presence of 10 mM TMSPS, but no effect at concentrations of 3 mM or below. In phosphorus NMR spectra, there was a small but significant decline in PCr using 6 mM TMSPS on unstimulated tissues and a larger decline with K$^+$ stimulation. These results set the upper end of the range over which TMSPS can be used without compromising the tissue.

Phosphorus NMR is also used for vascularly perfused preparations, such as heart, skeletal muscle, bladder, or uterus. In these cases, the contributions to the spectra by the perfusate may be important. Phenylphosphonate or similar derivatives are often used for such experiments. PPA has a direct C–P bond, unlike phosphates that have O–P bonds. As a result, PPA resonates downfield (to the left) of the naturally occurring phosphorus compounds. In mechanical and NMR experiments, the effects of PPA on the isolated, perfused bladder were measured (Fisher and Dillon, 1987a). PPA in concentrations up to 20 mM did not produce a significant reduction in force generation. Its NMR peak position was shown to be pH sensitive with a pK of 7.09, making it ideal for measurements of extracellular pH. It did not produce any alteration in the natural phosphorus spectrum, and it was able to be washed into and out of the perfused tissue without measurable residue, indicating that it did not cross cell membranes. In addition to being a pH indicator, it is therefore also useful as a marker for extracellular space.

B. Effect of Insulin on Extracellular Space

It had been proposed that Mg^{2+} was a second messenger for the effects of insulin on the bladder (Lostroh and Krahl, 1973). The advent of phosphorus NMR allowed the measurement of free Mg^{2+} in smooth muscle (Section IV.B). The addition of insulin in physiological concentrations to the isolated, perfused bladder did not produce any change in the free magnesium concentration, negating the hypothesis (Fisher and Dillon, 1989). The value of the wealth of information provided by phosphorus NMR experiments was borne out by changes observed in the extracellular space in these experiments. Application of insulin caused a large increase in the size of the PPA peak (Fig. 7). This increase was accompanied by an increase in the size of the PCr peak and an increase in the tissue pH. The PPA peak was washed from both the control and insulin tissues with the same time constant, leaving no PPA residue in either. This indicated that the PPA stayed in the extracellular space. The addition of

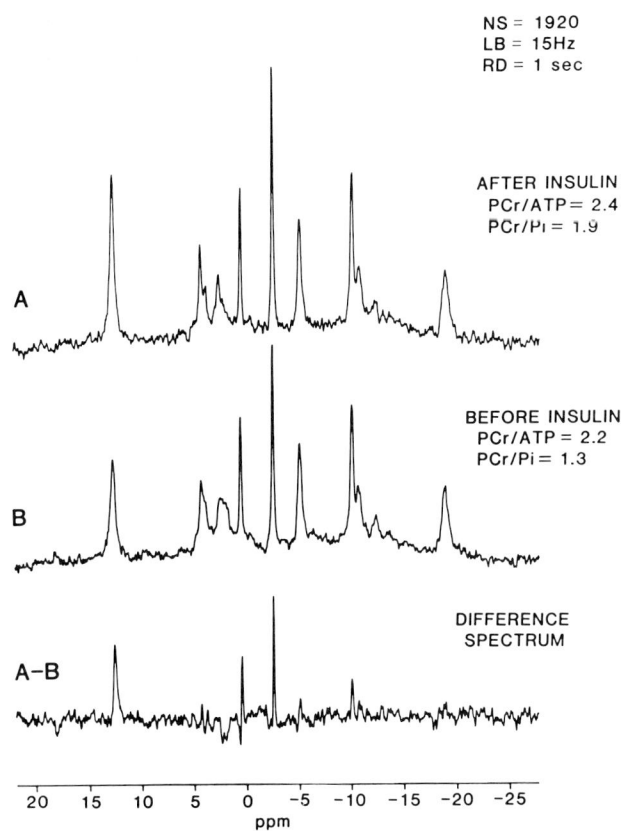

FIGURE 7 Phosphorus NMR spectra of isolated, perfused rabbit bladder before (B) and after (A) the addition of insulin. A–B is the difference spectrum showing large changes in PPA at 13 ppm and PCr at −2.5 pp. Other peak size changes were not significant over multiple tissue experiments. From Fisher and Dillon (1989, Fig. 2, p. 59).

insulin increased the perfusion space in the perfused bladder. There was a rise in the calculated ADP in the insulin-treated bladder, with the increase in PCr more than offset by the fall in H$^+$ (see Section V.A for the ADP calculation). There was no change in free energy, indicating that perhaps the free energy is regulated at the expense of ADP, PCr, and pH. Functionally, the changes induced by insulin may indicate a bladder in which the energy state is not greater, but more well buffered. A reduction in blood flow to an insulin-treated bladder may preserve the tissue better than a non-insulin-treated bladder. This hypothesis remains to be tested.

VIII. CONCLUSIONS

The use of phosphorus NMR to measure the decline in PCr and ATP during ischemia was not surprising to anyone. Such measurements were confirmatory rather than groundbreaking. However, finer elements of tissue energetics, such as measurements of free energy and ADP, are significant advances produced because of the existence of phosphorus NMR. The ability to address questions of tissue ion concentrations noninvasively promises to continue to be an important area, as well as more refined experiments relating tissue mechanics to metabolism. The use of surface coils to study smooth muscles *in situ* has not been developed strongly as yet, although there are indications that this work is moving forward (Harrison *et al.*, 1994). Improvements in spectroscopy systems promise to reduce the collection times and/or tissue mass required for experimentation. More rapid collection time may allow the kind of temporal resolution necessary to measure changes that occur during activation transitions, rather than the current required steady state. The acquisition of NMR spectroscopy systems by many institutions promises to make opportunities for the type of experiments included herein more widespread. These experiments will greatly improve our understanding of the metabolic and ionic milieu of smooth muscle.

References

Adams, G. R., and Dillon, P. F. (1989). *Blood Vessels* **26**, 77–83.
Awapara, J., Landua, A. J., and Fuerst, R. (1950). *J. Biol. Chem.* **183**, 545–548.
Barron, J. T., Kopp, S. J., Tow, J. P., and Messer, J. V. (1989). *Biochim. Biophys. Acta* **976**, 42–52.
Barron, J. T., Kopp, S. J., Tow, J. P., and Parrillo, J. E. (1991). *Biochim. Biophys. Acta* **1093**, 125–134.
Brown, T. R. (1980). *Phos. Trans. R. Soc. London. Ser. B* **289**, 441–444.
Chacko, S., and Rosenfeld, A. (1982). *Proc. Natl. Acad. Sci. U.S.A.* **79**, 292–296.
Clark, J. F., and Dillon, P. F. (1989). *Biochim. Biophys. Acta* **1014**, 235–238.
Clark, J. F., Harris, G. I., and Dillon, P. F. (1991). *Magn. Reson. Med.* **17**, 274–278.
Clark, J. F., and Dillon, P. F. (1995a). *J. Vasc. Res.* **32**, 24–30.
Clark, J. F., Kemp, G. J., and Radda, G. K. (1995b). *J. Theor. Biol.* **173**, 207–211.
Cohn, M., and Hughes, T. R., Jr. (1962). *J. Biol. Chem.* **237**, 176–181.
Dawson, M. J., and Wray, S. (1985). *J. Physiol. (London)* **368**, 19–31.
Dillon, P. F. (1985). *J. Physiol (London)* **368**, 204P.
Dillon, P. F., and Clark, J. F. (1990). *J. Theor. Biol.* **143**, 275–284.
Dillon, P. F., Aksoy, M. O., Driska, S. P., and Murphy, R. A. (1981). *Science* **211**, 495–497.
Dillon, P. F., Weberling, M. K., LeTarte, S. M., Clark, J. F., Sears, P. R., and Root-Bernstein, R. S. (1995). *J. Biol. Phys.* (in press).
Fisher, M. J., and Dillon, P. F. (1987a). *Circ. Res.* **60**, 472–477.
Fisher, M. J., and Dillon, P. F. (1987b). *Biophys. J.* **51**, 336a.
Fisher, M. J., and Dillon, P. F. (1988). *NMR Biomed.* **1**, 121–126.
Fisher, M. J., and Dillon, P. F. (1989). *Magn. Reson. Med.* **9**, 53–65.
Gorecka, A., Aksoy, M. O., and Hartshorne, D. J. (1976). *Biochem. Biophys. Res. Commun.* **71**, 325–331.
Hardin, C. D., Wiseman, R. W., and Kushmerick, M. J. (1992a). *J. Physiol. (London)* **458**, 139–150.
Hardin, C. D., Wiseman, R. W., and Kushmerick, M. J. (1992b). *Biochim. Biophys. Acta* **1133**, 133–141.
Harrison, N., Larcombe-McDougall, J. B., Earley, L., and Wray, S. (1994). *J. Physiol. (London)* **476**, 349–354.
Hellstrand, P., and Vogel, H. J. (1985). *Am. J. Physiol.* **248**, C320–C329.
Jelicks, L. A., and Gupta, R. K. (1990). *J. Biol. Chem.* **265**, 1394–1400.
Kopp, S. J., Barron, J. T., and Tow, J. P. (1990). *Biochim. Biophys. Acta* **1055**, 27–35.
Kushmerick, M. J., Dillon, P. F., Meyer, R. A., Brown, T. R., Krisanda, J. M., and Sweeney, H. L. (1986). *J. Biol. Chem.* **261**, 14420–14429.
Lawson, J. W. R., and Veech, R. L. (1979). *J. Biol. Chem.* **254**, 6528–6537.
Levine, B. A., Griffiths, H. S., Patchell, V. B., and Perry, S. V. (1988). *Biochem. J.* **254**, 277–286.
Lostroh, A. J., and Krahl, M. E. (1973). *Biochim. Biophys. Acta* **291**, 260–263.
Lynch, R. M., and Paul, R. J. (1983). *Science* **222**, 1344–1345.
Nakayama, S., and Tomita, T. (1990). *J. Physiol. (London)* **421**, 363–378.
Nakayama, S., and Tomita, T. (1991). *J. Physiol. (London)* **435**, 559–572.
Nakayama, S., Nomura, H., and Tomita, T. (1994). *J. Gen. Physiol.* **103**, 833–851.
Nishiye, E., Somlyo, A. V., Török, K., and Somlyo, A. P. (1993). *J. Physiol. (London* **460**, 247–271.
Nosek, T. M., Fender, K. Y., and Godt, R. E. (1987). *Science* **236**, 191–193.
Paul, R. J., and Peterson, J. W. (1975). *Am. J. Physiol.* **228**, 915–922.
Seraydarian, K., Mommaerts, W. F. H. M., and Wallner, A. (1962). *Biochim. Biophys. Acta* **65**, 443–460.
Spurway, N. C., and Wray, S. (1987). *J. Physiol. (London)* **393**, 57–71.
Tallan, H. H., Moore, S., and Stein, W. H. (1954). *J. Biol. Chem.* **211**, 927–939.
Vermue, N. A., and Nicolay, K. (1983). *FEBS Lett.* **156**, 293–297.
Webb, M. R., Hibberd, M. G., Goldman, Y. E., and Trentham, D. R. (1986). *J. Biol. Chem.* **261**, 15557–15564.
Wray, S. (1988). *Biochim. Biophys. Acta* **972**, 299–301.
Wray, S. (1990). *J. Physiol. (London)* **423**, 411–423.
Yoshizaki, K., Radda, G. K., Inubushi, T., and Chance, B. (1987). *Biochim. Biophys. Acta* **928**, 36–44.

Subject Index

Abbreviations

1-ptase	1-Phosphatase
1D	One dimensional
2D	Two dimensional
3D	Three dimensional
aPKC	Atypical PKC isoforms
A	Actin
Acrylodan	6-Acryloyl-2(dimethyl-amino)naphthalene
AA	Amino acids
ACh	Acetylcholine
AKAP	A kinase anchor protein
AM	Actomyosin
ANP	Atrial natriuretic peptide
4-AP	4-Aminopyridine
AVP	[Arg^8]-vasopressin
BGPA	β-Guanidinopropionic acid
BK	Ca^{2+}-activated "big" K$^+$ channel
cADPR	Cyclic adenosine diphosphate ribose
cAK	Cyclic AMP-dependent protein kinase
cDNA	Complementary deoxyribonucleic acid
cGK	Cyclic GMP-dependent protein kinase
cPKC	Coventional PKC isoforms
[Ca^{2+}]$_i$	Intracellular free calcium concentration
Ca^{2+}-ATPase	Calcium transport ATPase
CaBP	Calcium binding protein
CaM	Calmodulin
CaM-kinase II	Calcium/calmodulin-dependent protein kinase II
CaT	Caltropin
CBFβ	β subunit of core binding factor
CCK	Cholecystokinin
CD	Caldesmon
CDh	Caldesmon, human
CDl	Caldesmon, isoforms
CGRP	Calcitonin gene-related peptide
CICR	Ca^{2+}-induced Ca^{2+} release
CK	Creatine kinase
CP	Calponin
CPA	Cyclopiazonic acid
CREB	cAMP response elements
DiC8	Dioctanoglycerol
DAG	Diacylglycerol
DG1 and DG2	Type 1 and type 2 cGMP-dependent protein kinases from *Drosophila*
DHP	Dihydropyridine
DIDS	4,4'-Diisothiocyanatostil-bene-2,2'-disulfonic acid
DS	Desmin
DTT	Dithiothreitol
E_{Cl}	Chloride equilibrium potential
E_K	Potassium equilibrium potential
Et	Endothelin-1
EDC	1-Ethyl-3-[3-(dimethyl-amino)propyl]carbodiimide

EDHF	Endothelium-derived hyperpolarizing factor	IP_2	Inositol bisphosphate
EDRF	Endothelium-derived relaxing factor	IP_3	Inositol 1,4,5-trisphosphate
		IP_3R	Inositol 1,4,5-trisphosphate receptor
EDTA	Ethylenediaminetetraacetic acid	IP_4	Inositol tetrakisphosphate
EGF	Epidermal growth factor	ISIT	Intensified silicon intensifier target
EGTA	Ethylene glycol bis(β-aminoethyl ether) N,N'-tetraacetic acid	J_{ATP}	Rate of ATP turnover
		J_{lac}	Rate of lactate production
ELC	Essential light chain	J_{O2}	Rate of oxygen consumption
ER	Endoplasmic reticulum	JNK	c-jun N-terminal kinase
ERK	Extracellular signal-regulated kinase	K_{ATP}	ATP-regulated K^+ channel
ΔF	Change in force	K_{Ca}	Calcium-activated K^+ channel
F-actin	Filamentous actin	K_{CaM}	Concentration of Ca^{2+}/calmodulin required for half-maximal activation
FGF	Fibroblast growth factor		
FPLC	Fast-protein liquid chromatography	K_{IR}	Inward rectifier K^+ channel
FSK	Forskolin	K_V	Voltage-dependent K^+ channel
FTP	Formycin triphosphate	ΔL	Change in length
g	Small GTP binding protein	L_0	Optimal length for force development
G	Heterotrimeric GTP binding protein		
GAP	GTPase-activating protein	L_S	Muscle slack length
GPC	Glycerol 3-phosphoryl-choline	LC	Myosin light chain
		LC17	17-kDa smooth muscle myosin light chain, essential
GPE	Glycerol 3-phosphoryl-ethanolamine		
GPI	Glycerophosphorylinositol	LC17a and LC17b	17-kDa smooth muscle myosin light chain, acidic form and basic form, respectively
HC	Myosin heavy chain		
HEPES	4-(2-Hydroxyethyl)-piperazine-1-ethane-sulfonic acid		
		LC20	20-kDa smooth muscle myosin light chain, phosphorylatable
HMM	Heavy meromyosin		
HPLC	High-performance liquid chromatography	LDH	Lactate dehydrogenase
		LMM	Light meromyosin
I1 and I2	Protein phosphatase inhibitor 1 and 2	mRNA	Messenger ribonucleic acid
		MAP2	Microtubule-associated protein 2
IgG	Immunoglobulin		
IB	Immunoblotting	MAPK	Mitogen-activated protein kinase
IC_{50}	Concentration required to achieve 50% inhibition	MARCKS protein	Myristoylated alanine-rich C kinase substrate
IF	Immunofluorescence		
IGF-I	Insulinlike growth factor-I	MBP	Myelin basic protein
IP	Inositol monophosphate	MEK	MAP kinase kinase

MEKK	MEK kinase	PCA	Perchloric acid
ML-9	1-5-Chloronaphthalene-sulfonyl-1H-hexahydo-1,4-diazepine	PCr	Phosphocreatine
		PCR	Polymerase chain reaction
		PDBu	Phorbol 12,13-dibutyrate
MLC	Myosin light chain, phosphorylatable	PDE	Cyclic nucleotide phosphodiesterase
MLCK	Myosin light chain kinase	PDGF	Platelet-derived growth factor
MLCP	Myosin light chain phosphatase	PDPK	Proline-directed protein kinase
MOPS	3-(N-Morpholino)propane-sulfonic acid	PG	Prostaglandin
		PI	Phosphatidylinositol
MPE	Monophosphate esters	PIP	Phosphatidylinositol 4-phosphate
nPKC	New PCK isoforms		
NAD kinase	ATP:NAD 2'-phosphotransferase	PIP_2	Phosphatidylinositol 4,5-bisphosphate
NCDC	2-Nitro-4-carboxyphenyl-N,N-diphenylcarbamate	PI-PLC	Phosphoinositide-specific phospholipase C
NE	Norepinephrine	PKC	Protein kinase C
NGD^+	Nicotinamide guanine dinucleotide	PKI	Protein kinase inhibitor
NGF	Nerve growth factor	PKM	Constitutively active fragment of PKC
NMR	Nuclear magnetic resonance	PLB	Phospholamban
NO	Nitric oxide	PM	Plasma membrane
NOS	Nitric oxide synthase	PMA	Phorbol 12-myristate 13-acetate
NS	Number of scans		
NTCB	2-Nitro-5-thiocyanobenzoic acid	PMCA	Plasma membrane calcium transporting ATPase
OAG	1-Oleoyl-2-acetylglycerol	PMSF	Phenylmethanesulfonyl fluoride
OMOR	Optical memory disk recorders	PP	Protein phosphatase
p42^{MAPK} and p44^{MAPK}	42-kDa and 44-kDa isoforms of MAPK	PP1 and PP2	Type-I and type-II protein phosphatase
p90^{rsk}	90-kDa ribosomal S6 kinase	PP1c	Catalytic subunit of type-I protein phosphatase
^{31}P	The NMR visible form of phosphorus	PP2A, PP2B, PP2C	Subtypes of type-II protein phosphatase
^{31}P-NMR	^{31}P-nuclear magnetic resonance	PPA	Phenylphosphonate
P_i	Inorganic phosphate	PPI	Polyphosphatidylinositol
P_o	Open-state probability	PS	Phosphatidylserine
P_{O_2}	Oxygen tension	PSS	Physiological salt solution
PA	Phosphatidic acid	rasGAP	rasGTPase activating protein
PAA	Phenylalkylamine		
PAGE	Polyacrylamide gel electrophoresis	RyR	Ryanodine receptor
		RACK	Receptor for activated protein kinase C
PC	Phosphatidylcholine		

RD	Relaxation delay	SMC	Smooth muscle cells
RIIβ	Regulatory subunit IIβ of cAMP-dependent protein kinase	SNAP	S-Nitroso-N-acetylpenacillamine
		SNP	Sodium nitroprusside
RLC	Regulatory light chain	SOD	Superoxide dismutase
RT–PCR	Reverse transcriptase–polymerase chain reaction	SR	Sarcoplasmic reticulum
		tBu-BHQ	2,5-Di(tert-butyl)-1,4-benzohydroquinone
sGC	Soluble guanylate cyclase		
S-1 and S-2	Myosin subfragment-1 and -2, respectively	TCA	Trichloroacetic acid
		TM	Tropomyosin
SAPK	Stress-activated protein kinase	TMSPS	3-(Trimethylsilyl)propane-sulfonic acid
SDS–PAGE	Sodium dodecyl sulfate–polyacrylamide gel electrophoresis	TN	Troponin
		TnC, TnI, and TnT	Troponin C, I, and T, respectively
SERCA and SERCA2	Sarco(endo)plasmic reticulum calcium transporting ATPase and its type 2	TNS	2-p-Toluidinylnaphtha-lene-6-sulfonate
		TPM1–TPM4	Tropomyosin genes
SH_2 and SH_3	src homology 2 and 3	V_m	Membrane potential
SHR	Spontaneously hypertensive rat	V_{max}	Maximal shortening velocity
SIT	Silicon intensifier target	VASP	Vasodilator-stimulated phosphoprotein
SK	Ca^{2+}-activated "small" K^+ channel		
		VI	Vimentin
SL	Sarcolemma	VIP	Vasointestinal peptide
SM	Smooth muscle	VSM	Vascular smooth muscle
SM-1 and SM-2	Smooth muscle myosin heavy chain isoform-1 and isoform-2	VSMC	Vascular smooth muscle cells

DATE DUE

AUG 1 6 1996	
MAR 0 3 1997	
MAY 1 5 1997	
AUG 1	
OCT 2 1 1997	
DEC 3 1 2011	

DEMCO, INC. 38-2931